Petroleum Geology of the Black Sea

Geological Society books refereeing procedures

The Society makes every effort to ensure that the scientific and production quality of its books matches that of its journals. Since 1997, all book proposals have been refereed by specialist reviewers as well as by the Society's Books Editorial Committee. If the referees identify weaknesses in the proposal, these must be addressed before the proposal is accepted.

Once the book is accepted, the Society Book Editors ensure that the volume editors follow strict guidelines on refereeing and quality control. We insist that individual papers can only be accepted after satisfactory review by two independent referees. The questions on the review forms are similar to those for *Journal of the Geological Society*. The referees' forms and comments must be available to the Society's Book Editors on request.

Although many of the books result from meetings, the editors are expected to commission papers that were not presented at the meeting to ensure that the book provides a balanced coverage of the subject. Being accepted for presentation at the meeting does not guarantee inclusion in the book.

More information about submitting a proposal and producing a book for the Society can be found on its website: www.geolsoc.org.uk.

It is recommended that reference to all or part of this book should be made in one of the following ways:

SIMMONS, M.D., TARI, G.C. & OKAY, A.I. (eds) 2018. *Petroleum Geology of the Black Sea*. Geological Society, London, Special Publications, **464**.

REES, E.V.L., SIMMONS, M.D. & WILSON, J.W.P. 2018. Deep-water plays in the western Black Sea: insights into sediment supply within the Maykop depositional system. *In:* SIMMONS, M.D., TARI, G.C. & OKAY, A.I. (eds) *Petroleum Geology of the Black Sea*. Geological Society, London, Special Publications, **464**, 247–265. First published online September 15, 2017, https://doi.org/10.1144/SP464.13

GEOLOGICAL SOCIETY SPECIAL PUBLICATION NO. 464

Petroleum Geology of the Black Sea

EDITED BY

M. D. SIMMONS
Halliburton, UK

G. C. TARI
OMV, Austria

and

A. I. OKAY
Eurasia Institute of Earth Sciences, Turkey

2018
Published by
The Geological Society
London

THE GEOLOGICAL SOCIETY

The Geological Society of London (GSL) was founded in 1807. It is the oldest national geological society in the world and the largest in Europe. It was incorporated under Royal Charter in 1825 and is Registered Charity 210161.

The Society is the UK national learned and professional society for geology with a worldwide Fellowship (FGS) of over 10 000. The Society has the power to confer Chartered status on suitably qualified Fellows, and about 2000 of the Fellowship carry the title (CGeol). Chartered Geologists may also obtain the equivalent European title, European Geologist (EurGeol). One fifth of the Society's fellowship resides outside the UK. To find out more about the Society, log on to www.geolsoc.org.uk.

The Geological Society Publishing House (Bath, UK) produces the Society's international journals and books, and acts as European distributor for selected publications of the American Association of Petroleum Geologists (AAPG), the Indonesian Petroleum Association (IPA), the Geological Society of America (GSA), the Society for Sedimentary Geology (SEPM) and the Geologists' Association (GA). Joint marketing agreements ensure that GSL Fellows may purchase these societies' publications at a discount. The Society's online bookshop (accessible from www.geolsoc. org.uk) offers secure book purchasing with your credit or debit card.

To find out about joining the Society and benefiting from substantial discounts on publications of GSL and other societies worldwide, consult www.geolsoc.org.uk, or contact the Fellowship Department at: The Geological Society, Burlington House, Piccadilly, London W1J 0BG: Tel. +44 (0)20 7434 9944; Fax +44 (0)20 7439 8975; E-mail: enquiries@geolsoc.org.uk.

For information about the Society's meetings, consult *Events* on www.geolsoc.org.uk. To find out more about the Society's Corporate Affiliates Scheme, write to enquiries@geolsoc.org.uk.

Published by The Geological Society from:
The Geological Society Publishing House, Unit 7, Brassmill Enterprise Centre, Brassmill Lane, Bath BA1 3JN, UK

The Lyell Collection: www.lyellcollection.org
Online bookshop: www.geolsoc.org.uk/bookshop
Orders: Tel. +44 (0)1225 445046, Fax +44 (0)1225 442836

The publishers make no representation, express or implied, with regard to the accuracy of the information contained in this book and cannot accept any legal responsibility for any errors or omissions that may be made.

British Library Cataloguing in Publication Data

A catalogue record for this book is available from the British Library.
ISBN 978-1-78620-358-8
ISSN 0305-8719

Distributors
For details of international agents and distributors see:
www.geolsoc.org.uk/agentsdistributors

Typeset by Nova Techset Private Limited, Bengaluru & Chennai, India
Printed and bound by CPI Group (UK) Ltd, Croydon CR0 4YY

Contents

Acknowledgements

The editors of this volume wish to thank all those who have contributed articles and those scientists who kindly gave up their time to peer review articles. Gratitude is expressed to staff at the Geological Society Publishing House for the high standard of professional care that has been applied to the production of this volume.

The Geological Society thanks the following companies for their generous sponsorship of the Petroleum Group and the conference that led to this volume. The Petroleum Group is thanked for its contribution towards the colour printing in this volume.

Petroleum Group corporate sponsors at the time of the conference:

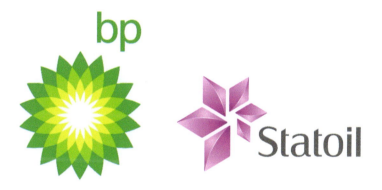

Sponsors of the Petroleum Geology of the Black Sea conference:

Petroleum geology of the Black Sea: introduction

M. D. SIMMONS[1]*, G. C. TARI[2] & A. I. OKAY[3]

[1]*Halliburton, 97 Jubilee Avenue, Milton Park, Abingdon OX14 4RW, UK*

[2]*OMV Exploration & Production GmbH, Trabrennstrasse 6–8, A-1020 Vienna, Austria*

[3]*Eurasia Institute of Earth Sciences, İstanbul Technical University, 34469 Maslak, İstanbul, Turkey*

*Correspondence: mike.simmons@halliburton.com

Abstract: The exploration for petroleum in the Black Sea is still in its infancy. Notwithstanding the technical challenges in drilling in its deep-water regions, several geological risks require better understanding. These challenges include reservoir presence and quality (partly related to sediment provenance), and the timing and migration of hydrocarbons from source rocks relative to trap formation. In turn, these risks can only be better understood by an appreciation of the geological history of the Black Sea basins and the surrounding orogens. This history is not without ongoing controversy. The timing of basin formation, uplift of the margins and facies distribution remain issues for robust debate. This Special Publication presents the results of 15 studies that relate to the tectonostratigraphy and petroleum geology of the Black Sea. The methodologies of these studies encompass crustal structure, geodynamic evolution, stratigraphy and its regional correlation, petroleum systems, source to sink, hydrocarbon habitat and play concepts, and reviews of past exploration. They provide insight into the many ongoing controversies regarding the geological history of the Black Sea region and provide a better understanding of the geological risks that must be considered for future hydrocarbon exploration. The Black Sea remains one of the largest underexplored rift basins in the world. Although significant biogenic gas discoveries have been made within the last decade, thermogenic petroleum systems must be proven through the systematic exploration of a wide variety of play concepts.

The Black Sea, located between Russia, Georgia, Turkey, Bulgaria, Romania and Ukraine, covers an area of approximately 423 000 km^2 with a maximum water depth of 2245 m. Sedimentary thickness can exceed 14 km. The Black Sea holds an abiding fascination for petroleum geologists and is a true frontier basin with very few wells drilled in its deep-water sector. Abundant seepage, outcrops of potential source rocks around its margins, large potential traps imaged on seismic data, and a variety of potential reservoir and play concepts point towards considerable potential to reward the successful explorer. This volume brings together several geoscience studies (Fig. 1) that provide additional information about the origins of the Western and Eastern Black Sea basins, their tectonostratigraphic history, sedimentary fill and petroleum potential.

The Black Sea and its surrounding regions have a long history of geological research: for example, it has long been regarded as the type example of an euxinic basin in which bottom water anoxia and free hydrogen sulphide (H$_2$S) result in an absence of benthonic life and the preservation of organic matter (Wignall 1994). Somewhat ironically, the term 'euxinic' is derived from an ancient Greek name for the sea, *Pontus Euxinus*, meaning the welcoming or hospitable sea. Another ancient Greek name (probably derived from an Iranian name), *Pontos Axeinos* (the dark or somber sea: King 2004), may better reflect the widespread anoxia in its deep waters. Nonetheless, it is indeed a welcoming region for the geologist wishing to unravel its geological history and petroleum potential, although its secrets are not given up lightly.

The Black Sea comprises two distinct depositional basins: the Western Black Sea and the Eastern Black Sea separated by the Mid Black Sea High (the Andrusov Ridge and the Tetyaev and Archangelsky highs) (Fig. 2) (Finetti *et al.* 1988). The Eastern Black Sea contains the Tuapse Trough, the foreland basin to the Caucasus fold and thrust belt. The Tuapse Trough is separated from the main part of the Eastern Black Sea by the Shatsky Ridge.

From the Oligocene onwards, the Black Sea and its constituent basins formed part of Paratethys, a remnant of the closure of Tethys. It lies at the southern margin of Laurasia, which formed the northern margin of Tethys. The basins of the Black Sea are

From: SIMMONS, M. D., TARI, G. C. & OKAY, A. I. (eds) 2018. *Petroleum Geology of the Black Sea.*
Geological Society, London, Special Publications, **464**, 1–18.
First published online May 4, 2018, https://doi.org/10.1144/SP464.15
© 2018 The Author(s). Published by The Geological Society of London.
Publishing disclaimer: www.geolsoc.org.uk/pub_ethics

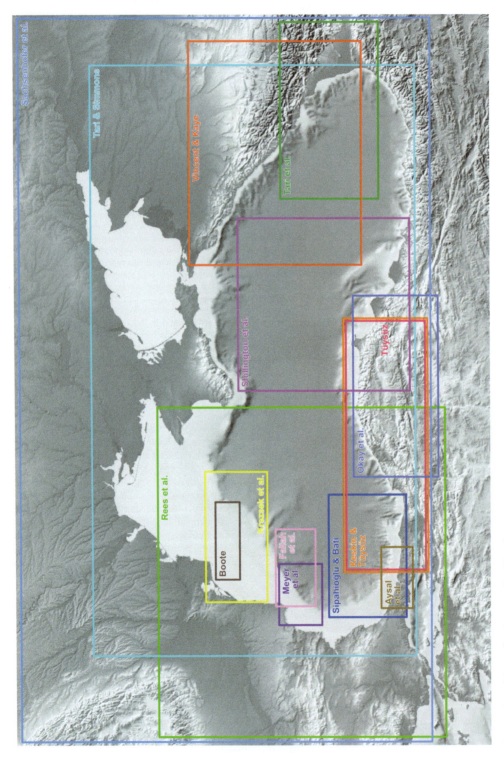

Fig. 1. Location of the studies that comprise this volume.

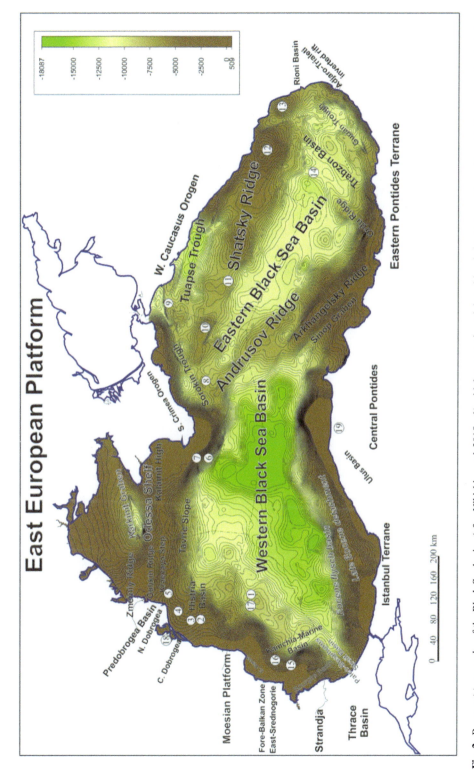

Fig. 2. Basement topography of the Black Sea basins (after Nikishin *et al.* 2015*a*, *b*) with key tectonic and depositional elements as mentioned in the text. 1, Polshkov Ridge; 2, Tindala-Midia Ridge; 3, Tomis Ridge; 4, Lebada Ridge; 5, St George Ridge; 6, Sevastopol Swell; 7, Lomonosov Massif; 8, Tetyaev Ridge; 9, Anapa Swell; 10, North Black Sea High; 11, South-Doobskaya High; 12, Gudauta High; 13, Ochamchira High; 14, Ordu-Pitsunda Flexure; 15, Rezovo-Limankoy Folds; 16, Kamchia Basin; 17, East Moesian Trough; 18, Babadag Basin; 19, Küre Basin.

extensional in origin (Zonenshain & Le Pichon 1986; Okay *et al.* 1994), relating at least in part to the northwards subduction of strands of Tethys beneath Laurasia, but are surrounded by compressive belts. Crimea and the Greater Caucasus, formed by the inversion of the Mesozoic Caucasus Basin in the Cenozoic, border the Black Sea to the north and east. A small basin, the Rioni, lies to the east in Georgia. The Balkanides and Pontides orogenic zones, formed from an accretion of terranes and island arcs, lie to the south and SE (Fig. 2).

Many oil and gas fields lie around the margins of the Black Sea (Robinson *et al.* 1996; Benton 1997), both in shallow-marine areas and onshore. More recently, the first efforts to explore in the deep-water offshore have met with mixed success, despite the presence of a variety of play types (Tari *et al.* 2011; **Tari & Simmons 2018**). In 2017, respected analysts Wood Mackenzie reported an estimate of 1.35 BBOE (billion barrels of oil equivalent) of yet-to-find reserves for the Black Sea (Wood Mackenzie 2017). This may be a modest estimate, given the presence of widespread source rocks, seepage and large potential traps. By contrast, in 2000, the USGS World Petroleum Assessment estimated in excess of 7 BBOE (https://pubs.usgs.gov/dds/dds-060/index.html#TOP). However, the current contribution of the Black Sea to global petroleum production is minor, especially when compared to the neighbouring Caspian Sea region. Interestingly, similarities exist between the two regions in terms of key petroleum geology elements. Partial isolation from the world's oceans from the Eocene onwards led to deposition of the Kuma and Maykop suites and their equivalents. These are important source rocks in both the South Caspian Sea and the Black Sea. Both the Black Sea and the South Caspian have been influenced by the Cenozoic influx of sediment from palaeo-river systems that occupy very different drainage pathways than the current equivalent systems. These sediments can form key proven and potential reservoir targets.

To date, the limited exploration in the deep water of the Black Sea has mostly resulted in discoveries of biogenic gas as at Domino in the Romanian offshore (a play first described by Bega & Ionescu 2009). Thermogenic petroleum systems, although proven, are yet to yield major finds, although at the time of writing several play concepts are being tested. On the shelf of the Black Sea, additions to the discoveries listed by Benton (1997) continue to be made. These include, for example, Subbotina, discovered in 2005 offshore Crimea. A thrust anticline with reservoirs in Maykop Suite sands, the field is reported to have recoverable reserves of 100 Mt (million tons) of oil (*c.* 680 MMBO (million barrels of oil)) and 3.5 TCF (trillion cubic ft) of gas (Stovba *et al.* 2009). Similar (and larger) untested anticlinal

structures extend southwards into the deep water (Tari *et al.* 2011).

Geological history: ongoing uncertainties

The geological history of the Black Sea region is related to the history of the amalgamation of the tectonic terranes that have accreted around it (Figs 2–4). Anatolia to the south of the Black Sea is a notable collage of different continental and oceanic fragments (Okay & Tüysüz 1999; Okay 2008; Hippolyte *et al.* 2015). Uncertainty exists with regard to the timing of the amalgamation events and their consequences (Sosson *et al.* 2010, 2017). Nonetheless, the subduction of branches of Neotethys to the south of the present-day Black Sea led to phases of arc volcanism and extension (although the two are not always clearly related) on the southern margin of Laurasia within the Mesozoic (Fig. 3) (Nikishin *et al.* 2003; Dinu *et al.* 2005; Okay & Nikishin 2015). This was followed by uplift and compressional deformation as strands of Neotethys progressively closed (Allen & Armstrong 2008; Vincent *et al.* 2016).

The relative timing of the opening of the two basins remains controversial. Evidence of rifting exists in the Western Black Sea during the Early Cretaceous (Barremian–Aptian) (Görür 1988, 1997; Nikishin *et al.* 2003; Hippolyte *et al.* 2010). Deep seismic studies indicate that oceanic crust is present in the Western Black Sea (Belousov *et al.* 1988; Görür 1988; Okay *et al.* 1994; Graham *et al.* 2013; Tari *et al.* 2015*b*; Schleder *et al.* 2015; Nikishin *et al.* 2015*a*, *b*). Interpolation between the onshore stratigraphy and seismic data suggests that the ocean crust in the Western Black Sea is Santonian in age (Okay *et al.* 2013; Nikishin *et al.* 2015*a*, *b*). In much of the Central Pontides, a large hiatus of *c.* 20 myr exists between the deposition of the Barremian–Aptian sequence and the start of arc volcanism in the Santonian (Fig. 3). In the southern margin of the Pontides near the Tethyan subduction zone, this stratigraphic gap is represented by accretion of oceanic edifices involving deformation and metamorphism of the Barremian–Aptian depositional sequence (Okay *et al.* 2013). An uncertainty in the geology of the Black Sea is whether the Barremian–Aptian rift succession is related to the opening of the Western Black Sea, or represents an unrelated earlier event.

The main phase of the opening of the Eastern Black Sea has been variously interpreted as coeval with the Western Black Sea (Okay *et al.* 1994; Nikishin *et al.* 2003, 2015*a*, *b*; Stephenson & Schellart 2010); as late Campanian–Danian (Vincent *et al.* 2016), as Paleocene–Early Eocene (Robinson *et al.* 1995, 1996; Spadini *et al.* 1996; Shillington *et al.* 2008) or as Eocene (Kazmin

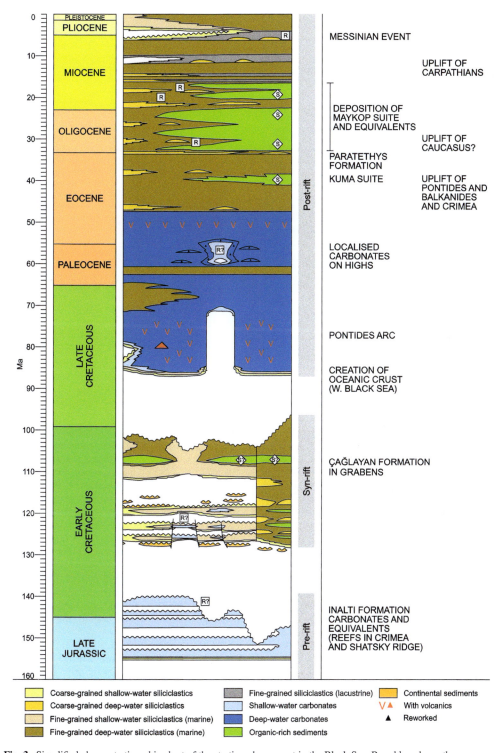

Fig. 3. Simplified chronostratigraphic chart of the stratigraphy present in the Black Sea. Based largely on the Western/Central Pontides and an interpretation of the Western Black Sea but may be largely applicable to the Eastern Black Sea depending on the timing of the opening of that basin.

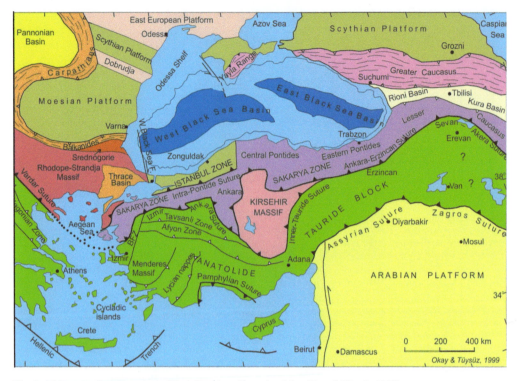

Fig. 4. Major tectonic terranes in the Black Sea/Anatolia region (after Okay & Tüysüz 1999).

et al. 2000). It is notable that the rift-related successions of the Western and Central Pontides (e.g. Çağlayan Formation) have no equivalents in the Eastern Pontides (Okay & Şahintürk 1997; Hippolyte *et al.* 2015).

Given that the Western Greater Caucasus Basin to the north of the Shatsky Ridge opened in the Early Jurassic as a result of Neotethyan subduction, there may have also been a proto-Eastern Black Sea south of the Shatsky Ridge during the Jurassic, or at least an initial phase of rifting (Vincent *et al.* 2016).

A commonly cited model for the Cretaceous opening of the Western Black Sea was described by Okay *et al.* (1994) that involved the Istanbul Terrane (modern-day Western Pontides) splitting away from Moesia along two major faults – the West Black Sea Fault and the West Crimea Fault – as a consequence of subduction of the Neo-Tethyan Ocean to the south. Notwithstanding evidence for Albian-aged volcanism south of the Karkinit Trough (offshore Ukraine), in Crimea and on the Shatsky Ridge (Kazmin *et al.* 2000; Nikishin *et al.* 2013, 2015*a*, 2017), it is now argued that initial rifting may not be back-arc related (Tari 2015; **Okay *et al.* 2017**) and, moreover, that the nature of the key faults cited by Okay *et al.* (1994), especially the West Crimea Fault, can be reappraised (Tari

et al. 2015*b*) in the light of excellent regional 2D seismic data that have been gathered across the Black Sea for both petroleum exploration and academic purposes (Graham *et al.* 2013; Nikishin *et al.* 2015*a*, *b*). For example, the West Crimea Fault can be shown to be a major transform fault running NW–SE on the margin of the Andrusov High (Fig. 2). This enables the Western Black Sea to open by means of a 15–20° counterclockwise rotation of the Istanbul Zone around a Euler pole located to the SW of the Black Sea (Stephenson & Schellart 2010; Schleder *et al.* 2015; Tari *et al.* 2015*b*). This discussion highlights the ongoing uncertainties associated with the Black Sea tectonics and geodynamic history, and the role that petroleum industry data can play in resolving these issues.

The formation of the semi-isolated Paratethys at the end of the Eocene through a combination of eustatic sea-level fall (Miller *et al.* 2005) and orogenesis (Schulz *et al.* 2005) is an important element in the petroleum geology of the Black Sea region. Isolation led to deposition of the Oligocene–Early Miocene Maykop Suite and equivalents, which includes important source rocks deposited in a restricted basin and potential reservoir sands derived from the surrounding orogens (e.g. from the Balkanides, Pontides and Caucasus).

The timing of the uplift of one of these orogens, the Greater Caucasus, is a matter of dispute. Vincent *et al.* (2007, 2016) favour a base Oligocene age, based on the onset of massive microfossil reworking and palynological data, indicating palaeoaltitudes of the Greater Caucasus of *c.* 2 km at this time. Palaeocurrent vectors and heavy mineral compositions also support this notion (Vincent *et al.* 2007), as does some fission-track data (Vincent *et al.* 2013*a*, *b*). Several other authors (e.g. Lozar & Polino 1997; Adamia *et al.* 2011) draw similar conclusions, and abundant evidence exists that Crimea experienced Eocene, then Oligocene, uplift (Panek *et al.* 2009; Nikishin *et al.* 2017). Conversely, Avdeev & Niemi (2011), Forte *et al.* (2014) and Cowgill *et al.* (2016) have suggested that the major phase of uplift occurred no earlier than *c.* 5 myr ago, whereas Rolland (2017) prefers a Miocene (*c.* 15 Ma) age for the onset of Caucasus uplift, based on geodynamic context and fission-track constraints on 'hard' Arabia–Eurasia collision (e.g. Okay *et al.* 2010). The resolution of this controversy has important consequences for the petroleum prospectivity of the Eastern Black Sea plays within the Maykop Suite or younger, because the timing of inversion would also affect the timing of structural uplift within the Black Sea (e.g. of the Shatsky Ridge). A young age for uplift would limit the amount of clastic (potential reservoir) sediment entering the basin from its margins during the deposition of the Maykop Suite, for example. Seismic data (Afanasenkov *et al.* 2007; Nikishin *et al.* 2010; Mityukov *et al.* 2011) demonstrate thickening of the Maykop Suite into the Tuapse and Indolo-Kuban foreland basins to the south and north of the Greater Caucasus, respectively, supporting the notion of Vincent *et al.* (2016) that the Caucasus were forming during the Oligocene.

Contents of this Special Publication

In addition to this introduction, 15 studies (Fig. 1) are included that relate to the tectonostratigraphy and petroleum geology of the Black Sea region. The methodologies used include examination of crustal structure from seismic data, geodynamic evolution, stratigraphy and its regional correlation, petroleum systems, source to sink, hydrocarbon habitat and play concepts, and reviews of past exploration. Their intent is to provide insight into the many ongoing controversies regarding the geological history of the Black Sea region and to provide a better understanding of the geological risks that must be considered during future hydrocarbon exploration.

The papers are grouped thematically. The first section corresponds to crustal structure and tectonostratigraphy, including geodynamic aspects of the evolution of the Black Sea region; it includes papers about the evolution of the Pontides that, as well as sedimentary geology, encompass Late Cretaceous volcanism and magmatic intrusions. The second section includes papers that examine hydrocarbon plays in the Western Black Sea, including sediment routing systems. The third section features a set of papers that focuses on regional petroleum systems and source rocks. The Messinian and Holocene stratigraphy of the Black Sea is addressed in two papers, whereas the petroleum potential of the Eastern Black Sea is discussed in a review of the Rioni Basin, with a closing paper reviewing the exploration history of the deep-water Black Sea to date.

Crustal structure and tectonostratigraphy

Shillington et al. (2017) review the crustal structure of the Mid Black Sea High, formed of two en echelon basement ridges: the Archangelsky and Andrusov ridges (Fig. 2). This high separates the Western Black Sea Basin from the Eastern Black Sea Basin. Using a densely sampled wide-angle seismic profile, Shillington *et al.* demonstrate that the basement ridges are covered by approximately 1–2 km of pre-rift sedimentary rocks and 20 km of thinned continental crust that are suspected of being related to the Pontides geology. Thinning factors of 1.5–2.0 are implied by thickness variations between the Mid Black Sea High and the adjacent crust in the Turkish Pontides. The velocity structure suggests little magmatic addition during rifting. That the Western and Eastern Black Sea basins are separated by continental crust lends support to the notion that the two basins have different times and mechanisms of opening (Okay *et al.* 1994).

Okay et al. (2017) evaluate the broad geological evolution of the Central Pontides, whereas **Tüysüz (2017)** focuses on the Cretaceous evolution of this region. An understanding of the geology of the Pontides is important for understanding the evolution of the Black Sea because it represents the active northern margin of Tethys; consequently, its stratigraphy records many events that reflect the geodynamic evolution of the region. Using new field observations, sedimentology, detrital zircon geochronology and biostratigraphy, the framework of tectonostratigraphic evolution is established in these two papers, although there are contrasts in interpretation.

The Central Pontides represents two terranes (the Istanbul Zone in the west and the Sakarya Zone in the east) that, in the view of **Okay et al. (2017)**, were amalgamated before the Late Jurassic, given the uniformity of deposition on both terranes after this time. Shallow-marine carbonate deposition (İnalti Formation) was widespread and is confined to Kimmeridgian–Berriasian time from new biostratigraphic analyses (Fig. 3). Shallow-marine carbonate deposition was widespread around the Black Sea at this time (including the formation of reefs in Crimea and the Caucasus, but not in the Pontides) (Muratov 1969; Krajewski & Olszewska 2007; Guo et al. 2011), and it is likely that the Black Sea region was the site of a widespread carbonate platform on the northern margin of Tethys with only relatively minor bathymetric differences.

Uplift and erosion occurred in the Valanginian and Hauterivian, and was followed in the Barremian–Aptian by synrift deposition of a more than 2 km-thick succession of turbidites (the Çağlayan Formation) in basinal depocentres (Okay et al. 2013); whereas on the present-day Black Sea coast, shallow-marine and continental deposition occurred (Yılmaz & Altiner 2007; Masse et al. 2009) (Fig. 3). Sediment provenance of the turbidites from the East European Platform (Akdoğan et al. 2017) demonstrates that while rifting was occurring, the Black Sea was yet to open. Albian-aged accretion of Tethyan oceanic crust is demonstrated by deformation and metamorphism in the present-day southern Central Pontides (Okay et al. 2013; Hippolyte et al. 2015).

In contrast, **Tüysüz (2017)** favours a later amalgamation of the Sakarya and Istanbul terranes. In his view, an Intra-Pontide Ocean (Şengör & Yılmaz 1981; Yılmaz et al. 1997) between the two terranes existed until Cenomanian times, arguing that openmarine deposition of Berriasian age in the southern part of the Zonguldak-Ulus Basin on the Istanbul Terrane suggests an older formation of this basin than interpreted by **Okay et al. (2017)** (and others, e.g. Hippolyte et al. 2015, 2016) and that it faced into the Intra-Pontide Ocean. Mid-ocean ridge basalts and volcanic-arc-related magmatic rocks interbedded with radiolarian cherts (Kervansaray Formation) of late Tithonian–Berriasian age offer evidence for oceanic separation of the Istanbul and Sakarya terranes at this time. Both **Okay et al. (2017)** and **Tüysüz (2017)** agree that the Çağlayan Basin on the Sakarya Terrane formed later in the Early Cretaceous; the contrast being that **Okay et al. (2017)** regard the Çağlayan and Zonguldak-Ulus basins as being one post-terrane amalgamation rift-related basin, whilst **Tüysüz (2017)** suggests that they have separate geological histories. The presence and age of closure of the Intra-Pontide Ocean remains a contentious topic in Turkish geology.

Keskin & Tüysüz (2017) suggest another possibility that the Istanbul and Sakarya Terranes did not collide until the very latest Cretaceous; this occurred with the southwards translation of the Istanbul Terrane, driven by rollback of the subducting slab of the Tethys Ocean beneath Laurasia.

A new depositional cycle occurred in the Late Cretaceous, as demonstrated by the deposition of Turonian–Santonian pelagic limestones and arc-related volcanic rocks that are no younger than mid-Campanian (Fig. 3). This magmatic arc can be traced for approximately 2000 km from the Georgian Lesser Caucasus to the Balkans (Okay & Şahintürk 1997; Okay 2008; Okay & Nikishin 2015) and is the result of the northwards subduction of the Tethys Ocean beneath Laurasia (e.g. Şengör & Yılmaz 1981). Submarine volcanoes related to this arc have been identified on seismic data by Nikishin et al. (2015a). The Western Black Sea Basin opened behind and to the north of this arc (e.g. Görür 1988, 1997) from the Santonian onwards.

Okay et al. (2017) argue that Barremian–Aptian extension and Late Cretaceous Black Sea opening are unrelated because contractional deformation and metamorphism occurred in the Albian. The driving mechanism for rifting in the Early Cretaceous remains enigmatic (Tari 2015), but a wide-style of rifting is increasingly envisaged for the Early Cretaceous. This is, perhaps, related to a flat subduction of the Intra-Pontide Ocean (**Tüysüz 2017**), whereas the Late Cretaceous extension is clearly narrow and back-arc related (**Keskin & Tüysüz 2017**). Nonetheless, an unconformity typically occurs between the Lower and Upper Cretaceous sequences in the Central Pontides, notwithstanding a few local exceptions. Cenomanian and Turonian age sediments are often missing (Hippolyte et al. 2015), which is in contrast to, for example, successions in Bulgaria and Crimea where relatively continuous and thick deposition occurred. **Tüysüz (2017)** suggests that the amalgamation of the Istanbul and Sakarya terranes at this time may have caused uplift and erosion; although in the view of **Okay et al. (2017)**, these terranes had already amalgamated before the Late Jurassic.

Keskin & Tüysüz (2017) consider in detail the evolution of magmatic activity relating to the development of the Late Cretaceous volcanic arc based on outcrop studies and geochemical data. They recognize a first phase of magmatism during the middle Turonian–early Santonian (Dereköy Formation), relating to subduction of the Tethys Ocean beneath the southern margin of Laurasia. During the Late Santonian, volcanism briefly ceased and pelagic limestones were widely deposited. Interpreted as reflecting increased subsidence and intensified extension, this may in turn be related to the beginning of oceanic spreading in the Western Black Sea Basin as a consequence of a southwards rollback of the subducting slab. Further magmatism occurred in the Campanian (Cambu Formation). This includes magmas derived from an enriched asthenospheric mantle source similar to ocean island basalts. Upwelling of the asthenospheric mantle may occur during the mature stages of rifting. Support for this model of magmatism is provided by the detailed geochemical and isotopic analysis of Late Cretaceous and Early Cenozoic intrusions (dyke complexes) reported by **Aysal** *et al.* **(2017)**. Older calc-alkaline dykes were probably derived from a shallow mantle source, such as a metasomatized lithospheric mantle wedge, that contained a subduction signal during the initial stages of rifting. Younger alkaline and lamprophyre dykes were derived from a possibly deeper and, consequently, presumably asthenospheric mantle source. This may, in turn, reflect an initial thinning of the lithosphere during back-arc rifting and subsequent upwelling of the asthenospheric mantle that created ocean island basalt type magmas.

Subsequent Late Cretaceous and Early Cenozoic deposition is dominated by siliciclastic and calcareous turbidite deposition, although shallow-marine carbonates are present on depositional highs (Fig. 3). Arc-related volcanism may have ended during the Campanian because of a subduction jump from the north to the south of the Anatolide–Tauride–South Armenian microplate (Rolland *et al.* 2012; Hippolyte *et al.* 2015). Uplift related to the accretion of the Kırşehir Massif ended the majority of marine deposition in the Central Pontides in the Middle Eocene (**Okay** *et al.* **2017**). In the Eastern Pontides, thick successions of Eocene arc-related volcanics were deposited as a result of the impinging Anatolide–Tauride block (Hippolyte *et al.* 2015). Extension related to this event provides further evidence of a young age for the Eastern Black Sea as compared to the Western Black Sea.

Hydrocarbon plays of the Western Black Sea

Only a limited number of significant hydrocarbon discoveries have been made in the Eastern Black Sea Basin. In contrast, the Western Black Sea Basin has had more successful discoveries that are both historical and recent (Ionescu *et al.* 2002; Georgiev 2012). Many of these discoveries have been in the Romanian offshore (Moroşanu 2012); **Boote (2017)** and **Krezsek** *et al.* **(2017)** provide insights into the geological factors that have led to some of this success.

The Histria/Istria 'Depression' or basin contains a Late Mesozoic–Cenozoic succession that represents a polyphase history of sedimentation and subsidence, divisible into second-/third-order sequences bounded by major erosional unconformities, visible on regional 2D seismic data and calibrated from wells (**Boote 2017**). Notable events include those related to Aptian–Albian rifting, major incisions caused by relative sea-level fall at or around the Eocene–Oligocene boundary, and subsequent similar events in the Middle and Late Miocene (Dinu *et al.* 2005; **Boote 2017**).

Rifting in the Early Cretaceous led to uplift and erosion of the Late Jurassic carbonate platform in a similar manner to that observed in the Pontides.

Early Cretaceous shallow-marine carbonates and siliciclastics were deposited on the rift shoulders, which in turn were cut by a major west–east-trending incised valley cut in the Late Aptian. A spectacular Eocene–Oligocene boundary age incision follows a similar trend and represents a base-level fall of the order of 2000 m or more. The presence of incised valley systems provides insights into the sediment conduits (a theme explored by **Rees** *et al.* **2017**) and the potential distribution of reservoir sediments, and provides locations for subcrop plays beneath the valley bases and onlap plays within the subsequent sedimentary fill.

Insights into the Cretaceous rift history of the Western Black Sea can be gained from an analysis of the successions in onshore and offshore Romania (Munteanu *et al.* 2011; **Krezsek** *et al.* **2017**). Initial rifting occurred during the Aptian with the deposition of fluvial and lacustrine clastic successions and local marine carbonates. A second rifting phase occurred during the Cenomanian, marked by shallow-marine transgression. Continental break-up occurred during the mid-Turonian associated with regional uplift and erosion; this was followed by a Late Cretaceous succession of deep-water chalks and marls. Rifting thus was approximately 30 myr in duration, which is long for a single synrift episode in a basin (Tari 2015). Instead, several episodes of rifting may have occurred, even separated by a shortening event. This would reflect a transition from wide to narrow style rifting (*sensu* Buck 1991; Hopper & Buck 1996) and in turn a change from relatively flat subduction with no volcanic arc, to higher-angle subduction and the creation of a volcanic arc.

Maykop Suite is the name given to distinctive, often organic carbon-rich sediments, deposited during the Oligocene–Lower Miocene within a region that spans the Black Sea and its margins, the Greater Caucasus and the South Caspian Sea (Bazhenova *et al.* 2003; Sachsenhofer *et al.* 2017a). Deposition relates to the initial isolation of Paratethys at the end of the Eocene and beginning of the Oligocene. The Oligocene–Miocene time period encompasses several eustatic and regional changes in sea level (Popov *et al.* 2010), which are recorded within the Maykop Suite by the cyclic deposition of fine-grained organic-rich sediments and sandstone packages (Nikishin *et al.* 2015a) (Fig. 3). These sandstones have long been considered to be an exploration target in the deep-water Western Black Sea, confirmed by recent success within the Han Asparuh Block, offshore Bulgaria (**Tari & Simmons 2018**).

To assess the prospectivity of plays within Maykop Suite sandstones, **Rees** *et al.* **(2017)** assess sediment provenance and sediment conduits into the Western Black Sea during the time of Maykop Suite deposition. By taking into account the geodynamic history, the palaeotopography of the hinterland surrounding the Western Black Sea can be constrained and sediment provenance areas identified. Subsurface and outcrop data can then be used to further recognize sediment pathways within the basin. The drainage pattern present today on the western margin of the Black Sea (dominated by the Danube) bears little relationship to the drainage planform that would have been present during Maykop Suite deposition, with the Danube only reaching the Black Sea in the relatively recent geological past (de Leeuw *et al.* 2017). Instead, during the time of Maykop Suite deposition, a major river (the palaeo-Kamchia) can be envisaged running axially through the foredeep to the north of the Balkanides and transporting sediment derived from granitic, gneissic and older sandstone source areas. Known shelf-edge canyons (**Mayer** *et al.* **2017**) in offshore Bulgaria facilitated this sediment reaching the deep-water offshore (Tari *et al.* 2009) where a sedimentary fan with a length in excess of 150 km is likely to have developed.

Farther to the south, the presence of widespread Late Cretaceous volcaniclastics related to arc magmatism would provide poor-quality sediment in limited volumes. Nonetheless, areas such as the NE Moesian Platform, the Strandja Massif and parts of the Balkanides contain crystalline basement (gneiss) and Variscan and Late Cretaceous granitic plutons that would yield high-quality quartz-rich sediment when eroded.

Regional petroleum systems and source rocks

A key aspect of the petroleum geology of the Black Sea is the assumed widespread presence of potential source rocks. The Oligocene–Early Miocene Maykop Suite (Fig. 3) is important among these but the importance of older source rocks is increasingly being stressed, including the Eocene Kuma Suite (Fig. 3) and a variety of potential Mesozoic stratigraphic units (Mayer *et al.* in press). The source-rock potential of the Maykop Suite and its equivalents are reviewed at a regional scale by **Sachsenhofer** *et al.* (**2017***b*), and on more local scales by **Mayer** *et al.* (**2017**) and **Vincent & Kaye** (**2017**). **Vincent & Kaye** (**2017**) also consider the potential of the Kuma Suite.

Sachsenhofer *et al.* (**2017***b*) note significant regional differences in source-rock potential across Central and Eastern Paratethys. The initial isolation of Paratethys at the beginning of the Oligocene created excellent source rocks in Central Paratethys, whereas coeval sediments in Eastern Paratethys are less organically rich (see also **Sachsenhofer** *et al.* **2017***a*). Upper Oligocene and Lower Miocene sediments are generally less rich, although localized upwelling created important diatomaceous source rocks in the Western Black Sea. The potentially poor yields of hydrocarbons from the Maykop Suite (often less than 2 t HC m^{-2}) in Eastern Paratethys suggests that other source rocks may be contributing to charge at some locations. By contrast, potential yields from Maykop Suite equivalent source rocks in the Carpathian Basin (e.g. Menilite Formation) are up to 10 t HC m^{-2}.

Both **Mayer** *et al.* (**2017**) and **Sachsenhofer** *et al.* (**2017***b*) emphasize the importance of local depositional conditions in the creation of good-quality potential source rocks. For example, an erosional unconformity at the base of the Oligocene in offshore Bulgaria creates a spectacular shelf-edge canyon (the Kaliakra Canyon). Source-rock quality appears to be better within the canyon, especially during the Early Miocene when diatomaceous shales with a total organic carbon (TOC) content of 2.5% and hydrogen index (HI) values of up to 530 mg HC g^{-1} TOC were deposited. This may relate to localized upwelling. Conversely, in the Early Oligocene, oxygen-depleted, brackish environments were best developed outside of the canyon and are associated with blooms of calcareous nannoplankton. These source-rock horizons are immature on the shelf, but are within the oil and gas window in the deeper parts of the basin. Long-distance migration within Maykopian sandstones from this highly productive kitchen is proven by published biomarker data with large quantities of hydrocarbons expelled from the Miocene onwards (Robinson *et al.* 1996; Olaru-Florea *et al.* 2014).

Vincent & Kaye (**2017**) examined potential Eocene–Early Miocene source rocks from outcrops in the western Greater Caucasus in Russia and the margins of the Rioni Basin in west Georgia. These outcrops are important because they are more directly relevant to the Eastern Black Sea than the classic outcrops south of the town of Maykop on the northern side of the Caucasus (e.g. along the Belaya River) (Sachsenhofer *et al.* 2017*a*) that were probably deposited in a separate sub-basin. A significant number of their samples have good to excellent organic richness and source potential with potential yields of 0.7–2.5 t HC m^{-2}, especially the base of the Maykop Suite and from within the Kuma Suite. In western Georgia, the basal Maykop Suite that is organically rich is between 60 and 200 m thick. The thickness of the better source-rock quality of the Kuma Suite is unconstrained, but its regional potential has been highlighted in previous studies (e.g. Beniamovski *et al.* 2003; Distanova 2007; Peshkov *et al.* 2016), and with TOC values of up to 10.3%, and S$_1$ and S$_2$ values of up to 30 kg t^{-1}, it merits strong consideration in modelling potential charge, especially in the Eastern Black Sea (Mayer *et al.* in press). It is worth noting that equivalent sediments in the Pontides can have good source-rock potential (Aydemir *et al.* 2009; Menlikli *et al.* 2009), suggesting that anoxia was widespread during Kuma Suite deposition. However, the effects of bathymetric highs, such as the Shatsky Ridge, on the formation of sub-basins that may have each provided a distinctive depositional character must be considered.

Messinian: recent stratigraphy

The major relative sea-level fall within the Messinian that was first described from the Mediterranean (Hsu *et al.* 1973) also has a significant expression across Paratethys (van Baak *et al.* 2017) and notably in the Black Sea (Tari *et al.* 2015a, 2016). However, the sea-level fall did not lead to desiccation of the Black Sea and was not of the magnitude once envisaged (cf. 1600 m (Hsu & Giovanoli 1979; Robinson *et al.* 1995) with 500–600 m (Krezsek *et al.* 2016)). This has petroleum significance in that a relatively modest fall in sea level translates to a lower risk of trap failure because of hydraulic seal fracture (Tari *et al.* 2016). Sediments cored in DSDP Leg 380 and previously interpreted as indications of shallow-water deposition can be interpreted as deep-water mass-transport deposits. Using 3D and 2D seismic data, **Sipahioğlu & Batı (2017)** describe the effect of this event in the Turkish sector of the Western Black Sea. They recognize a series of canyons, including the prominent SW–NE-trending Karaburun Canyon, which incises the shelf and acts as a major conduit of sediment to the abyssal floor of the basin. Blind canyons infilled with mass-transport complexes are also recognized and are typically confined to the continental rise where the steep shelf-slope morphology is governed by the presence of the underlying Late Cretaceous volcanic arc.

The modern Black Sea is often considered as an important analogue for source-rock deposition because of a high level of nutrients, high organic productivity, and widespread water stratification and anoxia (Wignall 1994; Arthur & Sageman 2004). Accordingly, it is interesting to investigate controls that enhance or reduce the quality of this analogue. **Fallah *et al.* (2017)** investigated 40 Holocene drop-core samples from offshore Bulgaria with regard to their composition and geochemical parameters, and integrated these data with high-resolution bathymetric surveys. Source-rock quality is relatively poor and is attributed to fine-grained sedimentary input from the Danube River. The sediment input from the Danube has served to limit organic productivity by reducing the thickness of the photic zone and by diluting organic-rich sediment with low TOC sediments. This highlights that anoxia alone is not sufficient to generate potential source rocks. Distance from sedimentary input points and basin geometry are also key factors to consider.

Petroleum potential of the Eastern Black Sea and deep-water exploration review

The Rioni Basin is an underexplored foreland basin located at the Georgian margin of the Black Sea and flanked by the fold belts of the Greater Caucasus and the Achara-Trialet Belt. In a review of this basin, **Tari *et al.* (2018)** argue that the proven plays are not fully understood nor systematically explored using modern technology. The northern Rioni Basin has stratigraphic similarities with the offshore Shatsky Ridge (at the time of writing, a major unexplored structure in the Eastern Black Sea), and the southern Rioni Basin is both stratigraphically and structurally akin to the offshore Gurian fold belt in the Eastern Black Sea.

The existing oil fields in the onshore Rioni Basin are generally small (2–4 MMbbl (million barrels) recoverable), but they demonstrate working petroleum systems, as do abundant seeps in the offshore (Dembicki 2014). The discovery of Supsa dates back to the 1880s where stacked Miocene (Sarmatian) clastic horizons occur in an anticlinal trap formed by the north-vergent leading edge of the Achara-Trialet thrust-fold belt. Charge is interpreted as being derived from both the Oligocene Maykop and the Eocene Kuma suites (Mayer *et al.* in press). The Shromisubani Field is a subthrust accumulation with reservoirs in Miocene (Maeotian) clastics. In the northern Rioni Basin, the Chaladidi Field is formed by two accumulations (i.e. two adjacent compressional ramp anticlines) with reservoirs in fractured Late Cretaceous and Paleocene chalky carbonates. The Okumi discovery is unusual in containing a light oil with an unknown, but suspected, Jurassic source, reservoired in Late Jurassic sands sealed beneath Kimmeridgian–Tithonian evaporites. This discovery opens up the possibility of a new petroleum system that may be operating in the offshore Shatsky Ridge. Other, more speculative, plays may include Middle Jurassic sandstones in synrift fault blocks (as encountered in the subcommercial discovery at Ochamchira-1).

Tari & Simmons (2018) review the history of deep-water exploration in the Black Sea, which is still in its infancy, with approximately 20 wells at the time of writing (end of 2017) having targeted a large variety of plays. Success has been mostly associated with biogenic gas in Miocene–Pliocene reservoirs associated with the palaeo-Dnieper/Dniester or in Oligocene deep-water clastics associated with similar-aged thermogenic source rocks. Synrift and early post-rift targets (mostly exploring for shallow-marine carbonate reservoirs) have met with little success because of the lack of predicted reservoir.

The first deep-water wells drilled in the Black Sea were Limanköy-1 and Limanköy-2 drilled in 1999 in Turkish waters. Encouraged by the presence of amplitude variation with offset (AVO) anomalies, Pliocene sandstone reservoirs were targeted in three-way fault-bound closures. These reservoirs were found to be non-permeable diatomites and diatomaceous shales (Menlikli *et al.* 2009). Secondary Early Miocene sandstones contained small amounts of gas from both thermogenic and biogenic sources. Although this was non-commercial, it usefully demonstrated the presence of a thermogenic petroleum system in the deep-water sector of the Black Sea.

The next deep-water well was HPX (Hopa)-1 drilled in Turkish waters near the maritime border with Georgia. Drilled on a four-way closure on the offshore continuation of the Achara-Trialet (Gurian) fold belt of the Lesser Caucasus, the well targeted Upper Miocene deep-water sand units presumed to be charged from the Oligocene Maykop Suite associated with seeps in the vicinity of the well. The well was unsuccessful, with reservoir quality suspected as the major issue with sediments derived from the Lesser Caucasus likely to be lithic-rich. Targeting older sandstones within the Gurian fold belt may prove to be more successful because these may be derived from more quartz-rich provenance areas. The understanding of the sediment-dispersal patterns in relation to favourable sediment provenance areas at key times of potential reservoir sedimentation is a key issue in reducing risk in Black Sea exploration (e.g. see Maynard *et al.* 2012; Vincent *et al.* 2013*a*, *b*; **Rees *et al.* 2017**).

The failure of HPX-1 resulted in a 5 year hiatus in deep-water exploration. In 2010, attention shifted to plays on structural highs on the Andrusov Ridge separating the Western and Eastern Black seas. The potential of structures on this high had been highlighted earlier by TPAO/BP Eastern Black Sea Project Study Group (1997). Sinop-1 targeted Late Cretaceous–Paleocene carbonate reservoirs on such a high, with charge expected to be from the laterally adjacent Maykop Suite. Although such carbonates are present at outcrop in the Pontides, and can be thick and porous (Menlikli *et al.* 2009; Aydemir & Demirer 2013), their presence was found to be unproven in the basin centre. Furthermore, a thick succession of Cretaceous volcanics and volcaniclastics was found to be present, precluding the exploration of deeper objectives. Yassihöyük-1, drilled directly after Sinop-1, encountered similar issues, with Late Cretaceous–Paleocene reservoir quality carbonates present only in very limited thicknesses (Aydemir & Demirer 2013). As noted by Posamentier *et al.* (2014), volcanoes and volcanic-rich successions can easily be mistaken for carbonate build-ups, even with good-quality seismic data.

Sürmene-1, drilled in the Turkish Eastern Black Sea in 2011, tested a new play concept: a four-way closure located above a Cretaceous palaeo-volcano.

Reservoirs were Miocene sheet sands thought to be derived from the Greater, rather than Lesser, Caucasus. Despite multiple oil shows, the well was not successful in opening up a new play fairway in the Eastern Black Sea, perhaps, once again, because of reservoir quality issues. Sile-1, drilled in the Turkish Western Black Sea in 2015, tested a similar play concept: a four-way closure above a large Cretaceous palaeo-volcano. The results of this well remain unpublished at the time of writing.

Kastamonu-1 tested one of the many large and elongated shale-cored anticlines present in the central Western Black Sea between the Central Pontides and Crimea. Drilled offshore Turkey in 2011, it tested thermogenic gas from Pliocene and Miocene reservoirs, arguably the first technical success of the deep-water exploration campaign. Low gas saturations may have been the result of late crestal faulting breaching the trap. Given that Kastamonu-1 was drilled on 2D seismic data alone, analysis of similar prospects with 3D seismic data may identify those without the influence of neotectonics.

In 2012, the first deep-water well was drilled in offshore Romania: Domino-1. This well tested a Late Neogene inversion structure above a basement high with reservoirs occurring in the Miocene–Pliocene palaeo-Dnieper/Dniester depositional system and associated with biogenic gas (Bega & Ionescu 2009; Moroşanu 2012). The well proved to be an economic discovery, and satellite discoveries have been made in the Neptun Deep exploration block in which Domino is located. The full extent of the play remains to be defined.

In 2016, exploration drilling began in the deep-water offshore Bulgaria with the Polshkov-1 well drilled on compactional anticline formed over the prominent Polshkov High (Tari *et al.* 2009), a structural high formed during the rifting of the Western Black Sea. The main reservoir target was channelized deep-water sands within the Maykop Suite associated with the palaeo-Kamchia depositional system described by **Rees *et al.* (2017)**. Deeper plays in the syn- and pre-rift also exist (Robinson *et al.* 1996; Tari *et al.* 2009), and the exploration campaign continues in this block at the time of writing following the successful encountering of hydrocarbons in Polshkov-1.

Looking ahead, the first deep-water well in the Russian sector of the Black Sea is being drilled. Maria-1 is targeting a well-known apparent Late Jurassic carbonate build-up on the Shatsky Ridge (e.g. see Afanasenkov *et al.* 2005, 2007) with charge assumed from the juxtaposed Maykop and Kuma suites (seepage is described by Andreev 2005). If successful, this well has the potential to be a spectacular play opener because many other similar carbonate structures have been mapped in the Eastern Black Sea (Afanasenkov *et al.* 2005). Reservoir quality is a significant risk in carbonate plays, but karstification as noted in age-equivalent outcrops on Crimea (Nikishin *et al.* 2017) may offer some encouragement.

Other plays to be explored in the Russian sector include Maykop Suite deep-water clastics in the Tuapse Trough (Glumov & Viginskiy 1999; Afanasenkov *et al.* 2007; Meisner *et al.* 2009), although the timing of the Caucasus uplift is a key component in determining the volume of sediment flux from potentially granitic and gneissic sediment sources (see Vincent *et al.* 2016 for a discussion of this controversy). Maykop sediments entering into the Eastern Black Sea are likely to be locally derived if the Greater Caucasus were emergent during deposition; these mountains also prevented sediment derived from the Russian Shield (e.g. via the palaeo-Don) from reaching large parts of the basin and being ponded, instead, in the Indolo-Kuban Basin. Spectacular Maykop Suite turbidite fans derived from the Greater Caucasus have been imaged on 3D seismic data from within the Tuapse Trough (Mityukov *et al.* 2011). Vincent *et al.* (2013*a*, *b*) and Khlebnikova *et al.* (2014) have demonstrated that Maykop sandstones in the Russian western Caucasus are significantly more quartz-rich than those located farther SE in western Georgia. This finding suggests that the reservoir quality of Maykop plays within the Tuapse Trough (sediment derived in part from Jurassic granitoids) carries lower risk than offshore Georgia (sediment derived in part from Eocene volcanics). Seepage in the Tuapse Trough has been described by Andreev (2005).

Offshore Crimea, the Tetyaev High has compaction closure above it, leading to the potential within Maykop sandstones analogous to the Polshkov play, offshore Bulgaria. Subbotina, discovered on the Crimea shelf (Stovba *et al.* 2009), demonstrates that the Maykop Suite can offer working plays in the region. Structural analogues to Subbotina extend southwards towards the Shatsky Ridge.

Offshore Turkey, potential also exists with Maykop sandstones (Menlikli *et al.* 2009; Sipahioğlu *et al.* 2013). These plays may function as stratigraphic traps in deep-water fan systems or by onlap onto highs, such as the Kozlu Ridge; however, sediment provenance must be considered (Maynard *et al.* 2012; **Rees *et al.* 2017**) as a guide to likely reservoir quality.

More widely, can synrift plays work in the deep-water Black Sea? The Early Cretaceous synrift sequences exposed in the Pontides have both potential source rocks and reservoirs associated with them (Görür & Tüysüz 1997; Şen 2013), and their potential offshore (in the region of the Kozlu Ridge) was discussed by Menlikli *et al.* (2009), who suggested a lateral charge may be possible from the Maykop Suite. Synrift sediments form reservoirs on the Romanian shelf (e.g. Lebada, Midia) where traps

are created by Eocene inversion (Munteanu *et al.* 2011) and, consequently, juxtaposition with Maykop Suite equivalent source rocks (Robinson *et al.* 1996; Cranganu & Saramet 2011). Synrift sandstones may be present in faulted culminations on the Shatsky and Andrusov ridges, and provide an alternative objective to the pre-rift carbonates currently targeted at Maria. The Early Cretaceous synrift succession so well exposed on Crimea (Nikishin *et al.* 2017) may have a seismic expression on the Shatsky Ridge. By comparison with Crimea, nummulitic banks may be present on highs within the Shatsky Ridge (e.g. the Gudauta High).

The deep-water Black Sea is at a similar stage of exploration as the Eastern Mediterranean. After the discovery of large biogenic gas accumulations, the thermogenic petroleum systems must be proven by systematic exploration of the deep-water and multiple play concepts that exist.

Summary

Controversy and uncertainty continue to be key features of Black Sea geoscience. Several deep-water exploration wells have failed because of an inability to predict correctly reservoir presence and reservoir quality. This is true for both carbonate plays and siliciclastic plays. The correct prediction of carbonate presence requires an understanding of the uplift and subsidence history of the highs they might be deposited upon. Does the apparent lack of Late Cretaceous carbonate on the Andrusov Ridge (i.e. at Sinop-1) suggest that that both the Western Black Sea and Eastern Black Sea were rapidly subsiding at this time? This possibility appeals to the fundamental question regarding the relative timing of the opening of the Western and Eastern Black seas. Was this synchronous (e.g. Nikishin *et al.* 2003, 2015*a*, *b*; Stephenson & Schellart 2010), or do they relate to completely separate phases of tectonic evolution (Robinson *et al.* 1995, 1996; Spadini *et al.* 1996; Kazmin *et al.* 2000; Shillington *et al.* 2008; Vincent *et al.* 2016)?

Siliciclastic reservoir presence and quality have also proven to be difficult to predict (e.g. at HPX-1 and at Sürmene-1). This prediction requires an understanding of the source to sink relationships in the basins, specifically sediment conduits and the nature of the rocks being eroded to create potential reservoirs. In turn, this requires an understanding of the uplift history of the orogens from which the sediment is suggested as being sourced: for example, compare the timing of uplift of the Caucasus (cf. Cowgill *et al.* 2016 and Vincent *et al.* 2016).

Petroleum charge can be an issue that limits exploration success. **Mayer *et al.* (2017), Sachsenhofer *et al.* (2017*b*)** and **Vincent & Kaye (2017)** all demonstrate that the Maykop Suite and its equivalents may not be the high-quality source rock in the Black Sea basins that is often assumed. How important are older source rocks, such as the Kuma Suite, and, given the likely depth of burial of that unit, is gas a more likely hydrocarbon phase than oil?

Notwithstanding some recent success with the exploration for post-rift Cenozoic plays, can plays within the synrift and pre-rift Mesozoic stratigraphy be successful? These and many other questions will be answered in part by seismic and drilling campaigns in the years to come, by detailed outcrop studies, and by the application of the latest tools to model geodynamic history and the 3D imaging of the subsurface. The Black Sea will remain a focus for geological research as it continues to yield its secrets.

The editors of this volume wish to thank all those who have contributed articles and those scientists who kindly gave up their time to peer review articles. Financial support from the Geological Society Petroleum Group and from Neftex Petroleum Consultants Ltd (now part of Halliburton) and OMV allowed for colour figures to be reproduced throughout. Gratitude is expressed to staff at the Geological Society Publishing House for the high standard of professional care that has been applied to the production of this volume. This paper is published with the permission of Halliburton.

References

ADAMIA, S., ALANIA, V., CHABUKIANI, A., KUTELIA, Z. & SADRADZE, N. 2011. Great Caucasus (Cavcasioni): a long-lived north-Tethyan back-arc basin. *Turkish Journal of Earth Sciences*, **20**, 611–628.

AFANASENKOV, A.P., NIKISHIN, A.M. & OBUKHOV, A.N. 2005. The system of Late Jurassic carbonate buildups in the northern Shatsky swell (Black Sea). *Doklady Earth Sciences*, **403**, 696–699.

AFANASENKOV, A.P., NIKISHIN, A.M. & OBUKHOV, A.N. 2007. *Geology of the Eastern Black Sea*. Scientific World, Moscow [in Russian with an English summary].

AKDOĞAN, R., OKAY, A.I., SUNAL, G., TARI, G., MENHOLD, D. & KYLANDER-CLARK, A.R.C. 2017. Provenance of a large Lower Cretaceous turbidite submarine fan complex on the active Laurasian margin: central Pontides, northern Turkey. *Journal of Asian Earth Sciences*, **134**, 309–329.

ALLEN, M.B. & ARMSTRONG, H.A. 2008. Arabia–Eurasia collision and the forcing of mid-Cenozoic global cooling. *Palaeogeography, Palaeoclimatology, Palaeoecology*, **265**, 52–58.

ANDREEV, V.M. 2005. Mud volcanoes and oil shows in the Tuapse Trough and Shatsky Ridge. *Doklady Earth Science*, **402**, 516–519.

ARTHUR, M.A. & SAGEMAN, B.B. 2004. Sea-level control on source rock development: perspectives from the Holocene Black Sea, the mid-Cretaceous western interior basin of North America and the Late Devonian Appalachian Basin. *In*: HARRIS, N.B. (ed.) *The Deposition of Organic Carbon-Rich Sediments: Models, Mechanisms*

and Consequences. SEPM, Special Publications, **82**, 35–59.

AVDEEV, B. & NIEMI, N.A. 2011. Rapid Pliocene exhumation of the central Greater Caucasus constrained by low-temperature thermochronometry. *Tectonics*, **30**, TC2009, https://doi.org/10.1029/2010TC002808

AYDEMIR, V. & DEMIRER, A. 2013. Upper Cretaceous and Paleocene shallow water carbonates along the Pontide Belt. *In*: *Proceedings of the 19th International Petroleum and Natural Gas Congress and Exhibition of Turkey*, 284–290.

AYDEMIR, V., IZTAN, Y.H., SHIKHLINSKY, S., BATI, Z. & GÜRGEY, A. 2009. Middle Eocene aged source rock evidence from Black Sea, Turkey; BSP14 biozone. *In*: DALKILIÇ, M. (ed.) *2nd International Symposium on the Geology of the Black Sea Region, Abstracts*. General Directorate of Mineral Research and Exploration (MTA), Ankara, 28–29.

AYSAL, N., KESKIN, M., PEYTCHEVA, I. & DURU, O. 2017. Geochronology, geochemistry and isotope systematics of a mafic–intermediate dyke complex in the İstanbul Zone. New constraints on the evolution of the Black Sea in NW Turkey. *In*: SIMMONS, M.D., TARI, G.C. & OKAY, A.I. (eds) *Petroleum Geology of the Black Sea*. Geological Society, London, Special Publications, **464**. First published online September 25, 2017, https://doi.org/10.1144/SP464.4

BAZHENOVA, O.K., FADEEVA, N.P., SAINT-GERMES, M.L. & TIKHOMIROVA, E.E. 2003. Sedimentation conditions in the eastern Paratethys ocean in the Oligocene–Early Miocene. *Moscow University Geology Bulletin*, **58**, 11–21.

BEGA, Z. & IONESCU, G. 2009. Neogene structural styles of the NW Black Sea region, offshore Romania. *The Leading Edge*, **28**, 1082–1089.

BELOUSOV, V.V., VOLVOVSKY, B.S. *ET AL*. 1988. Structure and evolution of the Earth's crust and upper mantle of the Black Sea. *Bollettino di Geofisica Teorica ed Applicata*, **30**, 109–196.

BENIAMOVSKI, V.N., ALEKSEEV, A.S., OVECHKINA, M.N. & OBERHÄNSLI, H. 2003. Middle to Upper Eocene dysoxic–anoxic Kuma Formation (northeast Peri-Tethys): biostratigraphy and paleoenvironments. *In*: WING, S.L., GINGERICH, P.D., SCHMITZ, B. & THOMAS, E. (eds) *Causes and Consequences of Globally Warm Climate in the Early Paleogene*. Geological Society of America, Special Papers, **369**, 95–112.

BENTON, J. 1997. Exploration history of the Black Sea province. *In*: ROBINSON, A.G. (ed.) *Regional and Petroleum Geology of the Black Sea and Surrounding Region*. AAPG Memoirs, **68**, 7–18.

BOOTE, D. 2017. The geological history of the Istria 'Depression', Romanian Black Sea shelf: tectonic controls on second-/third-order sequence architecture. *In*: SIMMONS, M.D., TARI, G.C. & OKAY, A.I. (eds) *Petroleum Geology of the Black Sea*. Geological Society, London, Special Publications, **464**. First published online October 9, 2017, https://doi.org/10.1144/SP464.8

BUCK, W.R. 1991. Modes of continental lithospheric extension. *Journal of Geophysical Research: Solid Earth*, **96**, 20 161–20 178.

COWGILL, E., FORTE, A.M. *ET AL*. 2016. Relict basin closure and crustal shortening budgets during continental collision: An example from Caucasus sediment provenance. *Tectonics*, **35**, 2918–2947.

CRANGANU, C. & SARAMET, M. 2011. Hydrocarbon generation and accumulation in the Histria Basin of the Western Black Sea. *In*: RYANN, A.L. & PERKINS, N.J. (eds) *The Black Sea: Dynamics, Ecology and Conservation*. Nova Science, Hauppauge, NY, 243–264.

DE LEEUW, A., MORTON, A., VAN BAAK, C.G. & VINCENT, S.J. 2017. Timing of arrival of the Danube to the Black Sea: provenance of sediments from DSDP Site 380/380A. *Terra Nova*, **30**, 114–124, https://doi.org/10.1111/ter.12314

DEMBICKI, H. 2014. Confirming the presence of a working petroleum system in the eastern Black Sea basin, offshore Georgia using SAR imaging, sea surface slick sampling, and geophysical seafloor characterization. AAPG Search and Discovery Article 10610, AAPG Annual Convention and Exhibition, April 6–9, 2014, Houston, Texas, USA.

DINU, C., WONG, H.K., TAMBREA, D. & MATENCO, L. 2005. Stratigraphic and structural characteristics of the Romanian Black Sea shelf. *Tectonophysics*, **410**, 417–435.

DISTANOVA, L.R. 2007. Conditions of oil source potential formation in the Eocene deposits in the basins of the Crimea and Caucasus regions. *Moscow University Geology Bulletin*, **62**, 59–64.

FALLAH, M., MAYER, J., TARI, G. & BAUR, J. 2017. Holocene source rock deposition in the Black Sea, insights from a dropcore study offshore Bulgaria. *In*: SIMMONS, M.D., TARI, G.C. & OKAY, A.I. (eds) *Petroleum Geology of the Black Sea*. Geological Society, London, Special Publications, **464**. First published online September 15, 2017, https://doi.org/10.1144/SP464.11

FINETTI, I., BRICCHI, G., DEL BEN, A. & PIPAN, M. 1988. Geophysical study of the Black Sea area. *Bollettino di Geofisica Teorica ed Applicata*, **30**, 197–324.

FORTE, A.M., COWGILL, E. & WHIPPLE, K.X. 2014. Transition from a singly- to doubly-vergent wedge in a young orogen: the Greater Caucasus. *Tectonics*, **33**, 2077–2101.

GEORGIEV, G. 2012. Geology and hydrocarbon systems in the Western Black Sea. *Turkish Journal of Earth Sciences*, **21**, 723–754.

GLUMOV, I.F. & VIGINSKIY, V.A. 1999. Oil-gas prospects of Tuapse downwarp, Trans-Caucasus offshore (Black Sea). *Razdevka i Okhrana Nedr*, **1**, 17–20 [in Russian].

GRAHAM, R., KAYMAKCI, N. & HORN, B. 2013. The Black Sea: something different? *GeoExpro*, **10**, 57–62.

GUO, L., VINCENT, S.J. & LAVRISHCHEV, V. 2011. Upper Jurassic reefs from the Russian western Caucasus, implications for the eastern Black Sea. *Turkish Journal of Earth Sciences*, **20**, 629–653.

GÖRÜR, N. 1988. Timing of opening of the Black Sea Basin. *Tectonophysics*, **147**, 247–262.

GÖRÜR, N. 1997. Cretaceous syn- to postrift sedimentation on the southern continental margin of the western Black Sea basin. *In*: ROBINSON, A.G. (ed.) *Regional and Petroleum Geology of the Black Sea and Surrounding Region*. AAPG Memoirs, **68**, 227–240.

GÖRÜR, N. & TÜYSÜZ, O. 1997. Petroleum geology of the southern continental margin of the Black Sea. *In*: ROBINSON, A.G. (ed.) *Regional and Petroleum Geology of the Black Sea and Surrounding Region*. AAPG Memoirs, **68**, 241–254.

HIPPOLYTE, J.C., MÜLLER, C., KAYMAKCI, V. & SANGU, E. 2010. Dating of the Black Sea Basin: new nannoplankton ages

from its inverted margin in the Central Pontides (Turkey). *In*: SOSSON, M., KAYMAKCI, N., STEPHENSON, R.A., BERGERAT, F. & STAROSTENKO, V. (eds) *Sedimentary Basin Tectonics from the Black Sea and Caucasus to the Arabian Platform*. Geological Society, London, Special Publications, **340**, 113–136, https://doi.org/10.1144/SP340.7

HIPPOLYTE, J.C., MÜLLER, C., SANGU, E. & KAYMAKCI, N. 2015. Stratigraphic comparisons along the Pontides (Turkey) based on new nannoplankton age determinations in the Eastern Pontides: geodynamic implications. *In*: SOSSON, M., STEPHENSON, R.A. & ADAMIA, S.A. (eds) *Tectonic Evolution of the Eastern Black Sea and Caucasus*. Geological Society, London, Special Publications, **428**, 323–358, https://doi.org/10.1144/SP428.9

HIPPOLYTE, J.C., ESPURT, N., KAYMACKCI, N., SANGU, E. & MÜLLER, C. 2016. Cross-section anatomy and geodynamic evolution of the Central Pontide orogenic belt (northern Turkey). *International Journal of Earth Sciences*, **105**, 81–106.

HOPPER, J.R. & BUCK, W.R. 1996. The effect of lower crustal flow on continental extension and passive margin formation. *Journal of Geophysical Research: Solid Earth*, **101**, 20 175–20 194.

HSU, K.J. & GIOVANOLI, F. 1979. Messinian event in the Black Sea. *Palaeogeography, Palaeoclimatology, Palaeoecology*, **29**, 75–93.

HSU, K.J., RYAN, W.B.F. & CITA, M.B. 1973. Late Miocene desiccation of the Mediterranean. *Nature*, **242**, 240–244.

IONESCU, G., SISMAN, M. & CATARAIANI, R. 2002. Source and reservoir rocks and trapping mechanism on the Romanian Black Sea shelf. *In*: DINU, C. & MOCANU, V. (eds) *Geology and Tectonics of the Romanian Black Sea Shelf and its Hydrocarbon Potential*. Bucharest Geoscience Forum (BGF), Special Volume, **2**, 67–83.

KAZMIN, V.G., SCHREIDER, A.A. & BULYCHEV, A.A. 2000. Early stages of evolution of the Black Sea. *In*: BOZKURT, E., WINCHESTER, J.A. & PIPER, J.D.A. (eds) *Tectonics and Magmatism in Turkey and the Surrounding Area*. Geological Society, London, Special Publications, **173**, 235–249, https://doi.org/10.1144/GSL.SP.2000.173.01.12

KESKIN, M. & TÜYSÜZ, O. 2017. Stratigraphy, petrogenesis and geodynamic setting of Late Cretaceous volcanism on the SW margin of the Black Sea, Turkey. *In*: SIMMONS, M.D., TARI, G.C. & OKAY, A.I. (eds) *Petroleum Geology of the Black Sea*. Geological Society, London, Special Publications, **464**. First published online September 28, 2017, https://doi.org/10.1144/SP464.5

KHLEBNIKOVA, O.A., NIKISHIN, A.M., MITYUKOV, A.V., RUBTSOVA, E.V., FOKIN, P.A., KOPAEVICH, L.F. & ZAPOROZHETS, N.I. 2014. The composition of the sandstone from the Oligocene turbidite of the Tuapse marginal trough. *Moscow University Geology Bulletin*, **69**, 399–409.

KING, C. 2004. *The Black Sea*. Oxford University Press, Oxford.

KRAJEWSKI, M. & OLSZEWSKA, B. 2007. Foraminifera from the Late Jurassic and Early Cretaceous carbonate platform facies of the southern part of the Crimea Mountains, Southern Ukraine. *Annales Societatis Geologorum Poloniae*, **77**, 291–311.

KREZSEK, C., SCHLEDER, Z., BEGA, Z., IONESCU, G. & TARI, G. 2016. The Messinian sea-level fall in the western Black

Sea: small or large? Insights from offshore Romania. *Petroleum Geoscience*, **22**, 392–399, https://doi.org/10.1144/petgeo2015-093

KREZSEK, C., BERCEA, R.-I., TARI, G. & IONESCU, G. 2017. Cretaceous sedimentation along the Romanian margin of the Black Sea: inferences from onshore to offshore correlations. *In*: SIMMONS, M.D., TARI, G.C. & OKAY, A.I. (eds) *Petroleum Geology of the Black Sea*. Geological Society, London, Special Publications, **464**. First published online September 7, 2017, https://doi.org/10.1144/SP464.10

LOZAR, F. & POLINO, R. 1997. Early Cenozoic uprising of the Great Caucasus revealed by reworked calcareous nannofossils. *Terra Nova*, **9**, 17–30.

MASSE, J.-P., TÜYSÜZ, O., FENERCI-MASSE, M., ÖZER, S. & SARI, B. 2009. Stratigraphic organisation, spatial distribution, palaeoenvironmental reconstruction, and demise of Lower Cretaceous (Barremian–Lower Aptian) carbonate platforms of the Western Pontides (Black Sea region, Turkey). *Cretaceous Research*, **30**, 1170–1180.

MAYER, J., RUPPRECHT, B.J. *ET AL.* 2017. Source potential and depositional environment of Oligocene and Miocene rocks offshore Bulgaria. *In*: SIMMONS, M.D., TARI, G.C. & OKAY, A.I. (eds) *Petroleum Geology of the Black Sea*. Geological Society, London, Special Publications, **464**. First published online September 25, 2017, https://doi.org/10.1144/SP464.2

MAYER, J., SACHSENHOFER, R., UNGUREANU, C. & TARI, G. In press. Charge and migration in the Black Sea area and surroundings – insights from oil and source rock geochemistry. *Journal of Petroleum Geology*.

MAYNARD, J.R., ARDIC, C. & MCALLISTER, N. 2012. Source to sink assessment of Oligocene to Pleistocene sediment supply in the Black Sea. *Gulf Coast Section Society of Economic Paleontologists and Mineralogists Conference Transactions*, **32**, 664–700.

MEISNER, A., KRYLOV, O. & NEMČOK, M. 2009. Development and structural architecture of the Eastern Black Sea. *The Leading Edge*, **28**, 1046–1055.

MENLIKLI, C., DEMIRER, A., SIPAHIOĞLU, Ö., KÖRPE, L. & AYDEMIR, V. 2009. Exploration plays in the Turkish Black Sea. *The Leading Edge*, **28**, 1066–1075.

MILLER, K.G., KOMINZ, M.A. *ET AL.* 2005. The Phanerozoic record of global sea-level change. *Science*, **310**, 1293–1298.

MITYUKOV, A., AL'MENDINGER, O., MYASOEDOV, N., NIKISHIN, A. & GAIDUK, V. 2011. The sedimentation model of the Tuapse Trough (Black Sea). *Doklady Earth Sciences*, **440**, 1245–1248.

MOROŞANU, I. 2012. The hydrocarbon potential of the Romanian Black Sea continental plateau. *Romanian Journal of Earth Sciences*, **86**, 91–109.

MUNTEANU, I., MATENCO, L., DINU, C. & CLOETINGH, S. 2011. Kinematics of back-arc inversion of the western Black Sea basin. *Tectonics*, **30**, TC5004.

MURATOV, M.V. 1969. *Geology of the USSR. Volume VIII, Crimea*. Nedra, Moscow [in Russian].

NIKISHIN, A.M., KOROTAEV, M.V., ERSHOV, A.V. & BRUNET, M.F. 2003. The Black Sea basin: tectonic history and Neogene–Quaternary rapid subsidence modelling. *Sedimentary Geology*, **156**, 149–168.

NIKISHIN, A.M., ERSHOV, A.V. & NIKISHIN, V.A. 2010. Geological history of the western Caucasus and adjacent foredeeps based on analysis of the regional balanced section. *Doklady Earth Sciences*, **430**, 155–157.

NIKISHIN, A.M., KHOTYLEV, A.O., BYCHKOV, A.Y., KOPAEVICH, L.F., PETROV, E.I. & YAPASKURT, V.O. 2013. Cretaceous volcanic belts and the Black Sea Basin history. *Moscow University Geology Bulletin*, **68**, 141–154.

NIKISHIN, A.M., OKAY, A.I., TÜYSÜZ, O., DEMIRER, A., AMELIN, N. & PETROV, E. 2015a. The Black Sea basins structure and history: New model based on new deep penetration regional seismic data. Part 1: basin structure and fill. *Marine and Petroleum Geology*, **59**, 638–655.

NIKISHIN, A.M., OKAY, A.I., TÜYSÜZ, O., DEMIRER, A., AMELIN, N. & PETROV, E. 2015b. The Black Sea basins structure and history: New model based on new deep penetration regional seismic data. Part 2: Tectonic history and paleogeography. *Marine and Petroleum Geology*, **59**, 656–670.

NIKISHIN, A.M., WANNIER, M. *ET AL.* 2017. Mesozoic to recent geological history of southern Crimea and the Eastern Black Sea. *In*: SOSSON, M., STEPHENSON, R.A. & ADAMIA, S.A. (eds) *Tectonic Evolution of the Eastern Black Sea and Caucasus*. Geological Society, London, Special Publications, **428**, 241–268, https://doi.org/10.1144/SP428.1

OKAY, A.I. 2008. Geology of Turkey: a synopsis. *Anschnitt*, **21**, 19–42.

OKAY, A.I. & NIKISHIN, A.M. 2015. Tectonic evolution of the southern margin of Laurasia in the Black Sea region. *International Geology Review*, **57**, 1051–1076.

OKAY, A.I. & ŞAHINTÜRK, Ö. 1997. Geology of the Eastern Pontides. *In*: ROBINSON, A.G. (ed.) *Regional and Petroleum Geology of the Black Sea and Surrounding Region*. AAPG Memoirs, **68**, 291–312.

OKAY, A.I. & TÜYSÜZ, O. 1999. Tethyan sutures of northern Turkey. *In*: DURAND, B., JOLIVET, L., HORVÁTH, F. & SÉRANNE, M. (eds) *The Mediterranean Basins: Tertiary Extension with the Alpine Orogen*. Geological Society, London, Special Publications, **156**, 475–515, https://doi.org/10.1144/GSL.SP.1999.156.01.22

OKAY, A.I., ŞENGÖR, A.C. & GÖRÜR, N. 1994. Kinematic history of the opening of the Black Sea and its effect on the surrounding regions. *Geology*, **22**, 267–270.

OKAY, A.I., ZATTIN, M. & CAVAZZA, W. 2010. Apatite fission-track data for the Miocene Arabia–Eurasia collision. *Geology*, **38**, 35–38.

OKAY, A.I., SUNAL, G., SHERLOCK, S., ALTINER, D., TÜYSÜZ, O., KYLANDER-CLARK, A.R.C. & AYGÜL, M. 2013. Early Cretaceous sedimentation and orogeny on the active margin of Eurasia: southern Central Pontides, Turkey. *Tectonics*, **32**, 1–25.

OKAY, A.I., ALTINER, D., SUNAL, G., AYGÜL, M., AKDOĞAN, R., ALTINER, S. & SIMMONS, M. 2017. Geological evolution of the Central Pontides. *In*: SIMMONS, M.D., TARI, G.C. & OKAY, A.I. (eds) *Petroleum Geology of the Black Sea*. Geological Society, London, Special Publications, **464**. First published online September 15, 2017, https://doi.org/10.1144/SP464.3

OLARU-FLOREA, R., UNGUREANU, C. *ET AL.* 2014. Understanding of the petroleum system(s) of the Western Black Sea: insights from 3-D basin modeling. AAPG Search and Discovery Article 10686, AAPG International Conference & Exhibition, 14–17 September 2014, Istanbul, Turkey.

PANEK, T., DANIŠÍK, M., HRADECKÝ, J. & FRISCH, W. 2009. Morpho-tectonic evolution of the Crimean Mountains (Ukraine) as constrained by apatite fission track data. *Terra Nova*, **21**, 271–278.

PESHKOV, G.A., BARABANOV, N.N., BOLSHAKOVA, M.A., BORDUNOV, S.I., KOPAEVICH, L.F. & NIKISHIN, A.M. 2016. The oil and gas potential of the Kuma rocks of the Bakhchisarai region of Crimea. *Moscow University Geology Bulletin*, **71**, 262–268.

POPOV, S.V., ANTIPOV, M.P., ZASTROZHNOV, A.S., KURINA, E.E. & PINCHUK, T.N. 2010. Sea-level fluctuations on the north shelf of the Eastern Paratethys in the Oligocene–Neogene. *Stratigraphy and Geological Correlation*, **18**, 200–224.

POSAMENTIER, H., AYDEMIR, V. *ET AL.* 2014. Volcanic deposits in the Black Sea – seismic recognition criteria for differentiating volcanics from carbonates. AAPG Search and Discovery Article 90189, AAPG 2014 Annual Conference and Exhibition, 6–9 April 2014, Houston, Texas, USA.

REES, E.V.L., SIMMONS, M.D. & WILSON, J.W.P. 2017. Deep-water plays in the western Black Sea: insights into sediment supply within the Maykop depositional system. *In*: SIMMONS, M.D., TARI, G.C. & OKAY, A.I. (eds) *Petroleum Geology of the Black Sea*. Geological Society, London, Special Publications, **464**. First published online September 15, 2017, https://doi.org/10.1144/SP464.13

ROBINSON, A., SPADINI, G., CLOETINGH, S., RUDAT, J., CLOETINGH, S., DURAND, B. & PUIGDEFABREGAS, C. 1995. Stratigraphic evolution of the Black Sea; inferences from basin modelling. *Marine and Petroleum Geology*, **12**, 821–835.

ROBINSON, A.G., RUDAT, J.H., BANKS, C.J. & WILES, R.L.F. 1996. Petroleum geology of the Black Sea. *Marine and Petroleum Geology*, **13**, 195–223.

ROLLAND, Y. 2017. Caucasus collisional history: review of data from East Anatolia to West Iran. *Gondwana Research*, **49**, 130–146.

ROLLAND, Y., PERINCEKB, D., KAYMAKCIC, N., SOSSON, M., BARRIER, E. & AVAGYANE, A. 2012. Evidence for *c.* 80–75 Ma subduction jump during Anatolide–Tauride–Armenian block accretion and *c.* 48 Ma Arabia–Eurasia collision in Lesser Caucasus–East Anatolia. *Journal of Geodynamics*, **56–57**, 76–85.

SACHSENHOFER, R.F., POPOV, S.V. *ET AL.* 2017a. The type section of the Maikop Group (Oligocene–lower Miocene) at the Belaya River (North Caucasus): depositional environment and hydrocarbon potential. *AAPG Bulletin*, **101**, 289–319.

SACHSENHOFER, R.F., POPOV, S.V. *ET AL.* 2017b. Oligocene and Lower Miocene source rocks in the Paratethys: palaeogeographical and stratigraphic controls. *In*: SIMMONS, M.D., TARI, G.C. & OKAY, A.I. (eds) *Petroleum Geology of the Black Sea*. Geological Society, London, Special Publications, **464**. First published online September 7, 2017, https://doi.org/10.1144/SP464.1

SCHLEDER, Z., KREZSEK, C., TURI, V., TARI, G., KOSI, W. & FALLAH, M. 2015. Regional structure of the western Black Sea Basin: constraints from cross-section balancing. *In: Transactions of the GCSEPM Foundation Perkins–Rosen 34th Annual Research Conference on Petroleum Systems in Rift Basins*, 396–411.

SCHULZ, H.M., BECHTEL, A. & SACHSENHOFER, R.F. 2005. The birth of the Paratethys during the Early Oligocene: from Tethys to an ancient Black Sea analogue? *Global and Planetary Change*, **49**, 163–176.

ŞEN, Ş. 2013. New evidences for the formation of and for petroleum exploration in the fold-thrust zones of the central Black Sea Basin of Turkey. *AAPG Bulletin*, **97**, 465–485.

ŞENGÖR, A.M.C. & YILMAZ, Y. 1981. Tethyan evolution of Turkey: a plate tectonic approach. *Tectonophysics*, **75**, 181–241.

SHILLINGTON, D.J., WHITE, N., MINSHULL, T.A., EDWARDS, G.R.H., JONES, S., EDWARDS, R.A. & SCOTT, C.L. 2008. Cenozoic evolution of the eastern Black Sea: a test of depth-dependent stretching models. *Earth and Planetary Science Letters*, **265**, 360–378.

SHILLINGTON, D.J., MINSHULL, T.A., EDWARDS, R.A. & WHITE, N. 2017. Crustal structure of the Mid Black Sea High from wide-angle seismic data. *In*: SIMMONS, M.D., TARI, G.C. & OKAY, A.I. (eds) *Petroleum Geology of the Black Sea*. Geological Society, London, Special Publications, **464**. First published online October 4, 2017, https://doi.org/10.1144/SP464.6

SIPAHIOĞLU, N.Ö. & BATI, Z. 2017. Messinian canyons in the Turkish western Black Sea. *In*: SIMMONS, M.D., TARI, G.C. & OKAY, A.I. (eds) *Petroleum Geology of the Black Sea*. Geological Society, London, Special Publications, **464**. First published online September 25, 2017, https://doi.org/10.1144/SP464.12

SIPAHIOĞLU, N.O., KORUCU, Ö., AKTEPE, S. & BENGÜ, E. 2013. Westerly-sourced Late Oligocene–Middle Miocene axial sediment dispersal system in Turkish Western Black Sea: myth or reality? Paper presented at the 19th International Petroleum and Natural Gas Congress and Exhibition of Turkey, 15–17 May 2013, Ankara, Turkey.

SOSSON, M., KAYMAKCI, N., STEPHENSON, R., BERGERAT, F. & STAROSTENKO, V. 2010. Sedimentary basin tectonics from the Black Sea and Caucasus to the Arabian Platform: introduction. *In*: SOSSON, M., KAYMAKCI, N., STEPHENSON, R., BERGERAT, F. & STAROSTENKO, V. (eds) *Sedimentary Basin Tectonics from the Black Sea and Caucasus to the Arabian Platform*. Geological Society, London, Special Publications, **340**, 1–10, https://doi.org/10.1144/SP340.1

SOSSON, M., STEPHENSON, R. & ADAMIA, S. 2017. Tectonic evolution of the Eastern Black Sea and Caucasus: an introduction. *In*: SOSSON, M., STEPHENSON, R.A. & ADAMIA, S.A. (eds) *Tectonic Evolution of the Eastern Black Sea and Caucasus*. Geological Society, London, Special Publications, **428**, 1–9, https://doi.org/10.1144/SP428.16

SPADINI, G., ROBINSON, A. & CLOETINGH, S. 1996. Western v. Eastern Black Sea tectonic evolution: pre-rift lithospheric controls on basin formation. *Tectonophysics*, **266**, 139–154.

STEPHENSON, R. & SCHELLART, W.P. 2010. The Black Sea back-arc basin: insights to its origin from geodynamic models of modern analogues. *In*: SOSSON, M., KAYMAKCI, N., STEPHENSON, R., BERGERAT, F. & STAROSTENKO, V. (eds) *Sedimentary Basin Tectonics from the Black Sea and Caucasus to the Arabian Platform*. Geological Society, London, Special Publications, **340**, 11–21, https://doi.org/10.1144/SP340.2

STOVBA, S., KHRIACHITCHEVSKAIA, O. & POPADYUK, I. 2009. Hydrocarbon-bearing areas in the eastern part of the Ukrainian Black Sea. *The Leading Edge*, **28**, 1042–1045.

TARI, G. 2015. Is the Black Sea really a back-arc basin? *In*: *Transactions of the GCSEPM Foundation Perkins– Rosen 34th Annual Research Conference on Petroleum Systems in Rift Basins*, 510–520.

TARI, G., DAVIES, J., DELLMOUR, R., LARRATT, E., NOVOTNY, B. & KOZHUHAROV, E. 2009. Play types and hydrocarbon potential of the deepwater Black Sea, NE Bulgaria. *The Leading Edge*, **28**, 1076–1081.

TARI, G., MENLIKLI, C. & DERMAN, S. 2011. Deepwater play types of the Black Sea: a brief overview. AAPG Search and Discovery Article 10310, *AAPG European Region Annual Conference*, 17–19 October 2010, Kiev, Ukraine.

TARI, G., FALLAH, M., KOSI, W., FLOODPAGE, J., BAUR, J., BATI, Z. & SIPAHIOGLU, N.O. 2015*a*. Is the impact of the Messinian Salinity Crisis in the Black Sea comparable to that of the Mediterranean. *Marine and Petroleum Geology*, **66**, 135–148.

TARI, G., FALLAH, M., KOSI, W., SCHLEDER, Z., TURI, V. & KREZSEK, C. 2015*b*. Regional rift structure of the Western Black Sea Basin: map-view kinematics. *In*: *Transactions of the GCSEPM Foundation Perkins–Rosen 34th Annual Research Conference on Petroleum Systems in Rift Basins*, 372–395.

TARI, G., FALLAH, M. *ET AL.* 2016. Why are there no Messinian evaporites in the Black Sea? *Petroleum Geoscience*, **22**, 381–391, https://doi.org/10.1144/ petgeo2016-003

TARI, G.C. & SIMMONS, M.D. 2018. History of deepwater exploration in the Black Sea and an overview of deepwater petroleum play types. *In*: SIMMONS, M.D., TARI, G.C. & OKAY, A.I. (eds) *Petroleum Geology of the Black Sea*. Geological Society, London, Special Publications, **464**. First published online May 4, 2018, https://doi.org/10.1144/SP464.16

TARI, G.C., VAKHANIA, D. *ET AL.* 2018. Stratigraphy, structure and petroleum exploration play types of the Rioni Basin, Georgia. *In*: SIMMONS, M.D., TARI, G.C. & OKAY, A.I. (eds) *Petroleum Geology of the Black Sea*. Geological Society, London, Special Publications, **464**. First published online May 4, 2018, https://doi. org/10.1144/SP464.14

TPAO/BP EASTERN BLACK SEA PROJECT STUDY GROUP 1997. A promising area in the Eastern Black Sea. *The Leading Edge*, **16**, 911–916.

TÜYSÜZ, O. 2017. Cretaceous geological evolution of the Pontides. *In*: SIMMONS, M.D., TARI, G.C. & OKAY, A.I. (eds) *Petroleum Geology of the Black Sea*. Geological Society, London, Special Publications, **464**. First published online September 8, 2017, https://doi.org/10. 1144/SP464.9

VAN BAAK, C.G., KRIJGSMAN, W. *ET AL.* 2017. Paratethys response to the Messinian salinity crisis. *Earth-Science Reviews*, **172**, 193–223.

VINCENT, S.J. & KAYE, M.N.D. 2017. Source rock evaluation of Middle Eocene–Early Miocene mudstones from the NE margin of the Black Sea. *In*: SIMMONS, M.D., TARI, G.C. & OKAY, A.I. (eds) *Petroleum Geology of the Black Sea*. Geological Society, London, Special Publications, **464**. First published online September 25, 2017, https://doi.org/10.1144/SP464.7

VINCENT, S.J., MORTON, A.C., CARTER, A., GIBBS, S. & BARABADZE, T.G. 2007. Oligocene uplift of the western Greater Caucasus; an effect of initial Arabia–Eurasia collision. *Terra Nova*, **19**, 160–166.

VINCENT, S.J., HYDEN, F. & BRAHAM, W. 2013*a*. Along-strike variations in the composition of sandstones derived from the uplifting western Greater Caucasus – Causes and implications for reservoir quality prediction in the eastern Black Sea. *In*: SCOTT, R.A., SMYTH, H.R., MORTON, A.C. & RICHARDSON, N. (eds) *Sediment Provenance Studies in Hydrocarbon Exploration and Production*. Geological Society, London, Special Publications, **386**, 111–127, https://doi.org/10.1144/ SP386.15

VINCENT, S.J., MORTON, A.C., HYDEN, F. & FANNING, M. 2013*b*. The composition, provenance and reservoir potential of Cenozoic siliciclastic depositional systems supplying the northern margin of the eastern Black Sea. *Marine and Petroleum Geology*, **45**, 331–348.

VINCENT, S.J., BRAHAM, W., LAVRISCHEV, V.A., MAYNARD, J.R. & HARLAND, M. 2016. The formation and inversion of the western Greater Caucasus Basin and the uplift of the western Greater Caucasus: implications for the wider Black Sea region. *Tectonics*, **35**, 2948–2962, https://doi.org/10.1002/2016TC004204

WIGNALL, P.B. 1994. *Black Shales*. Clarendon Press, Oxford.

WOOD MACKENZIE 2017. Black Sea: unlocking its full potential. Wood Mackenzie, Edinburgh, https://www. woodmac.com/news/editorial/black-sea-potential/ [last accessed 8 January 2018].

YILMAZ, İ.Ö. & ALTINER, D. 2007. Cyclostratigraphy and sequence boundaries of inner platform mixed carbonate–siliciclastic successions (Barremian–Aptian) (Zonguldak, NW Turkey). *Journal of Asian Earth Sciences*, **30**, 253–270.

YILMAZ, Y., TUYSUZ, O., YIGITBAS, E., GENC, S.C. & ŞENGÖR, A.M.C. 1997. Geology and tectonic evolution of the Pontides. *In*: ROBINSON, A.G. (ed.) *Regional and Petroleum Geology of the Black Sea and Surrounding Region*. AAPG Memoirs, **68**, 183–226.

ZONENSHAIN, L.P. & LE PICHON, X. 1986. Deep basins of the Black Sea and Caspian Sea as remnants of Mesozoic back-arc basins. *Tectonophysics*, **123**, 181–212.

Crustal structure of the Mid Black Sea High from wide-angle seismic data

D. J. SHILLINGTON[1]*, T. A. MINSHULL[2], R. A. EDWARDS[3] & N. WHITE[4]

[1]*Lamont-Doherty Earth Observatory of Columbia University, 61 Route 9W, Palisades, NY 10964, USA*

[2]*Ocean and Earth Science, National Oceanography Centre Southampton, University of Southampton, European Way, Southampton SO14 3ZH, UK*

[3]*National Oceanography Centre, European Way, Southampton SO14 3ZH, UK*

[4]*Bullard Laboratories, University of Cambridge, Madingley Rise, Madingley Road, Cambridge CB3 0EZ, UK*

*Correspondence: djs@ldeo.columbia.edu

Abstract: The Mid Black Sea High comprises two en echelon basement ridges, the Archangelsky and Andrusov ridges, that separate the western and eastern Black Sea basins. The sediment cover above these ridges has been characterized by extensive seismic reflection data, but the crustal structure beneath is poorly known. We present results from a densely sampled wide-angle seismic profile, coincident with a pre-existing seismic reflection profile, which elucidates the crustal structure. We show that the basement ridges are covered by approximately 1–2 km of pre-rift sedimentary rocks. The Archangelsky Ridge has higher pre-rift sedimentary velocities and higher velocities at the top of basement (*c.* 6 km s^{-1}). The Andrusov Ridge has lower pre-rift sedimentary velocities and velocities less than 5 km s^{-1} at the top of the basement. Both ridges are underlain by approximately 20-km-thick crust with velocities reaching around 7.2 km s^{-1} at their base, interpreted as thinned continental crust. These high velocities are consistent with the geology of the Pontides, which is formed of accreted island arcs, oceanic plateaux and accretionary complexes. The crustal thickness implies crustal thinning factors of approximately 1.5–2. The differences between the ridges reflect different sedimentary and tectonic histories.

Several episodes of extension and shortening have shaped the Black Sea region since Permian times (e.g. Yilmaz *et al.* 1997; Nikishin *et al.* 2003; Robertson *et al.* 2004), which led to the addition of a series of volcanic arcs, oceanic plateaux and accretionary complexes to the Eurasian margin (e.g. Okay *et al.* 2013). The basin is thought to have formed in a back-arc extensional environment because of its close spatial association with the subduction of both the Palaeo- and Neo-Tethys oceans (e.g. Letouzey *et al.* 1977), but the timing and style of this opening history remain controversial, partly because the thick sediment cover means that the oldest sedimentary fill has not been drilled (Zonenshain & Le Pichon 1986; Okay *et al.* 1994, 2017; Banks *et al.* 1997; Nikishin *et al.* 2015*a*). The Black Sea is commonly subdivided into eastern and western basins; these sub-basins are separated by the Mid Black Sea High (MBSH), a system of buried basement ridges that runs SW–NE (Fig. 1) (e.g. Okay *et al.* 1994; Nikishin *et al.* 2015*a*).

The opening of the western basin may be estimated from the ages of arc volcanic rocks in the Western Pontides and from associated plate reconstructions; this evidence suggests a Middle–Upper Cretaceous age (Görür 1988; Okay *et al.* 1994, 2017). Based on seismic refraction and gravity data, the crust in the centre of the basin is 7–8 km thick and has velocities consistent with the presence of oceanic crust, suggesting that rifting culminated in seafloor spreading (Letouzey *et al.* 1977; Belousov *et al.* 1988; Starostenko *et al.* 2004).

The age and nature of the eastern basin are more controversial. The basin is thought to have formed by rotation of the Shatsky Ridge relative to the MBSH (Figs 1 & 2) (Okay *et al.* 1994; Nikishin *et al.* 2003). The main phase of opening has been interpreted as Jurassic, Cretaceous (Zonenshain & Le Pichon 1986; Okay *et al.* 1994; Nikishin *et al.* 2003, 2015*b*), Early Eocene–Paleocene (Robinson *et al.* 1995; Banks *et al.* 1997; Shillington *et al.* 2008) or Eocene (Kazmin *et al.* 2000; Vincent *et al.* 2005). Based on gravity and early seismic data, the crust in the centre of this basin was inferred to have a thickness of approximately 10–11 km and

From: SIMMONS, M. D., TARI, G. C. & OKAY, A. I. (eds) 2018. *Petroleum Geology of the Black Sea.*
Geological Society, London, Special Publications, **464**, 19–32.
First published online October 4, 2017, https://doi.org/10.1144/SP464.6

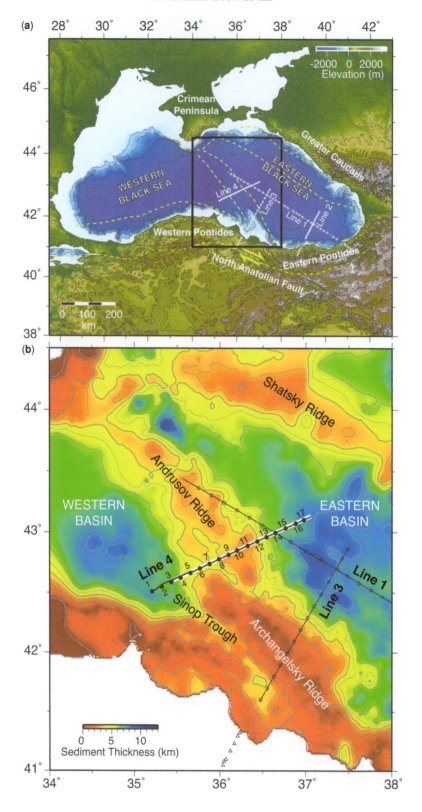

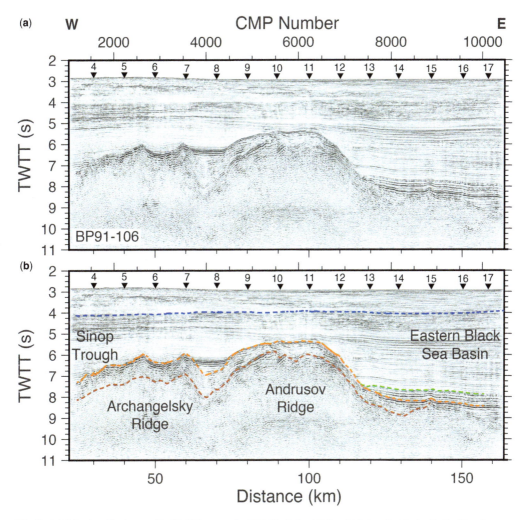

Fig. 2. (**a**) Seismic reflection profile 91–106 across the Mid-Black Sea High, which is coincident with the Line 4 OBS profile (courtesy of BP and TPAO) (see Fig. 1 for the location). (**b**) Seismic reflection profile with interfaces used in seismic inversion. The blue, green and orange dotted lines show interpreted horizons used to invert for post- and syn-rift sedimentary structure by Scott *et al.* (2009). The red dotted line shows the interpreted pre-rift sedimentary horizon used in the inversions presented here.

lower seismic velocities than those of typical oceanic crust, suggesting the presence of thinned continental crust (Belousov *et al.* 1988; Starostenko *et al.* 2004). However, results from a wide-angle seismic experiment in 2005 suggest that the crustal structure varies along the basin, with the western

Fig. 1. (**a**) Elevation/bathymetry of the Black Sea region from GEBCO showing the location of the 2005 onshore/offshore seismic refraction experiment. Shot lines are indicated with white lines, OBS are shown with white circles and seismometers deployed onshore are shown with white triangles. OBS from Line 4, which are used in this study, are indicated with solid circles. Major tectonic elements indicated with dashed yellow lines (Zonenshain & Le Pichon 1986). The black box indicates the area shown in (b). (**b**) Close-up of the Mid Black Sea High showing sediment thickness (Shillington *et al.* 2008) and OBS locations and shot lines from the 2005 experiment in black. Note that the Mid Black Sea High separates the western and eastern basins of the Black Sea and comprises two ridges: the Archangelsky Ridge and the Andrusov Ridge. Seismic reflection profile 91–106 (Fig. 2) is shown with thick white line. It is coincident with Profile 4 but shorter; it extends SW to between OBS 3 and 4.

part floored by thinned continental crust (7–9 km thick), and thicker, higher velocity crust below the eastern part that is attributed to magmatically robust early seafloor spreading resulting in early oceanic crust that is thicker and has higher velocities than average oceanic crust (Shillington *et al.* 2009).

The MBSH itself is divided into the en echelon Archangelsky and Andrusov ridges, which have different sediment thicknesses and are inferred to have different structure and origin (Robinson *et al.* 1996; Nikishin *et al.* 2015a) (Fig. 1b). These ridges are poorly explored compared to the basins on either side. The Andrusov Ridge is inferred to have formed during early opening of the eastern basin (Okay *et al.* 1994; Robinson *et al.* 1996; Nikishin *et al.* 2015b). This rifting event is inferred to have been amagmatic in this part of the basin (Shillington *et al.* 2009). Alternatively, the Andrusov Ridge is interpreted as a marginal ridge associated with the opening of the western basin along the West Crimean Transform Fault (Tari *et al.* 2015). The Archangelsky Ridge was formed by the opening of the Sinop Trough, which is linked to the western basin and is interpreted to have opened in Cretaceous–Paleocene times (Robinson *et al.* 1996; Espurt *et al.* 2014), with ongoing extension into the Miocene (Rangin *et al.* 2002; Espurt *et al.* 2014). An Upper Cretaceous sedimentary sequence and lower Cretaceous platform carbonate rocks have been dredged where the pre-rift sequences crop out on the flank of Archangelsky Ridge, providing an upper limit on its age of formation (Rudat *et al.* 1993; Robinson *et al.* 1996).

After their formation, both ridges have also experienced compressional deformation (Rangin *et al.* 2002; Espurt *et al.* 2014). This region has probably experienced multiple episodes of compression, continuing to the present; apatite fission-track data and palaeostress measurements onshore show that inversion of rifting structure onshore occurred as early as 55 Ma (Saintot & Angelier 2002; Espurt *et al.* 2014), following extension leading to the opening of eastern Black Sea. Active compression continues around margins of the easternmost Black Sea today based on seismicity and onshore geology, particularly in the Caucus (Saintot & Angelier 2002; Gobarenko *et al.* 2016).

Published constraints on crustal structure beneath the ridges are sparse. Seismic refraction data acquired in the 1960s were recently re-analysed using modern ray-tracing techniques (Yegorova & Gobarenko 2010). This analysis suggests a crustal thickness of approximately 20 km beneath both ridges and crustal velocities in the range 6.0–7.0 km s^{-1}, interpreted as representing thinned continental crust. A profile crossing the southern part of Archangelsky Ridge acquired in 2005 suggests that here, crustal thickness reaches about 25 km (Shillington *et al.* 2009). In this paper we present results from a modern, densely sampled wide-angle seismic profile acquired in 2005 that crosses the Andrusov Ridge close to its southern tip and the Archangelsky Ridge at its northern tip (Fig. 1).

Wide-angle seismic data

An onshore-offshore wide-angle seismic dataset was collected in 2005 using the R/V *Iskatel* to determine the deep structure of the eastern basin and Mid Black Sea High (Minshull *et al.* 2005). Seventeen four-component short-period ocean-bottom seismometers (OBS) from GeoPro GmbH were deployed on Profile 4 across the Andrusov Ridge (Fig. 1; Table 1), and they recorded seismic shots generated from an airgun array with a total volume of 3140 in^3 that was triggered every 90 s (shot spacing *c.* 150 m). Profile 4 was co-located with existing industry seismic reflection data: reflection profile 91–106 (Figs 1 & 2).

Data analysis

Data processing

Water-wave arrivals were used to relocate OBS positions on the seafloor, using a seafloor depth determined by echosounder at the position of each deployment and a water velocity of 1.47 km s^{-1}. Relocated positions were typically less than 75 m from deployment positions, but three OBS have relocated positions that differ by 200–300 m from deployment positions. We applied a minimum phase band-phase filter with corners at 3, 5, 15 and 20 Hz to suppress noise, and applied offset-dependent gains and a reduction velocity of 8 km s^{-1}.

Table 1. *Relocated OBS positions*

OBS	Latitude (°N)	Longitude (°E)
1	42.511005	35.212699
2	42.549179	35.322692
3	42.589855	35.432331
4	42.625923	35.543609
5	42.663829	35.654865
6	42.701	35.766201
7	42.738536	35.876911
8	42.777019	35.987457
9	42.813636	36.09951
10	42.851139	36.21101
11	42.887451	36.323356
12	42.925537	36.434604
13	42.960388	36.540924
14	42.995098	36.645301
15	43.035087	36.765087
16	43.073787	36.882984
17	43.10337	36.974617

Phase identification

We identified refractions and wide-angle reflections from the pre-rift sedimentary section, the crust and the upper mantle that could be consistently identified on a majority of the receiver gathers. Phase interpretations and velocity models of the overlying syn- and post-rift sedimentary section have been presented elsewhere (Scott *et al.* 2009). Travel-time picks were made manually of the following phases: reflections off the base of an interpreted pre-rift sedimentary layer (PprP); crustal refractions (Pg); reflections from the base of the crust (PmP); and upper-mantle refractions (Pn) (Fig. 3; Table 2). Reflections from the base of the interpreted pre-rift sedimentary section are observed from near-vertical incidence to offsets up to approximately 30 km and have picking uncertainties of 30–50 ms. Crustal refractions are observed as first arrivals at offsets from approximately 12 to 100 km and have picking uncertainties of 30–75 ms. Reflections from the base of the crust are observed at offsets between approximately 35 and 100 km; the offsets where PmP reflections are observed vary significantly over the line, indicating variations in crustal thickness. Likewise, the amplitude and character of PmP reflections is also highly variable and thus picks of this phase have relatively high uncertainties of 125 ms. We observed limited and relatively low-amplitude refractions interpreted to arise from the upper mantle in some receiver gathers; these refractions are weak and variable, and have a picking uncertainty of 125 ms. Figure 3 shows examples of OBS data, phase identifications and associated ray paths.

Wide-angle reflections interpreted to originate from the base of the interpreted pre-rift sedimentary layer can be linked to a coincident industry seismic reflection profile (BP91–106, Fig. 2). Picks of this interface were thus also made on the reflection profile (Fig. 2, red dotted line) and included in the inversion. We assigned an uncertainty of 100 ms to these picks to account for uncertainties in associating multichannel seismic reflection (MCS) and wide-angle reflections, and for small-scale variations in interface geometry that cannot be recovered by inversion.

Velocity modelling

The travel-time picks described above were used to invert for velocities of the pre-rift sedimentary section, crust and upper mantle. We used JIVE3D, a regularized tomographic inversion code (Hobro *et al.* 2003), which solves for a minimum structure layer-interface model that fits the data within its uncertainties. Velocities within each layer and interface depths are defined by splines and vary smoothly; interfaces represent velocity discontinuities. The forward problem involves tracing a fan of

rays from each OBS position through specified layers in the model to generate predicted travel times (i.e. ray shooting); the ray that arrives within a distance tolerance of the target with the minimum travel time is used. Inversion involves a sequence of linear steps to reduce the difference between observed and predicted travel times (e.g. Figs 4d & 5d), and satisfy other smoothing criteria. In each step, smoothing is reduced and structure is allowed to develop to improve data fit. Smoothing is implemented during inversion by minimizing a function of data misfit and model roughness.

We employed a layer-stripping approach for this line. The previously determined velocity structure of the post- and syn-rift sediment from Scott *et al.* (2009) was held fixed. We first inverted for the interpreted pre-rift sediment layer using picks of wide-angle reflections from OBS data and vertically incident reflections from the coincident seismic reflection profile (Fig. 2). This layer was then held fixed during the inversion for crustal and upper-mantle structure. The inversion converged more quickly and stably for both the pre-rift sedimentary section and for the crustal–mantle sections when we inverted for them separately. However, inverting for all layers simultaneously yielded the same overall velocity structure. We also performed two different inversions for crust–mantle structure. The first inversion used only first-arriving refractions from the crust and mantle. The second inversion included interpreted wide-angle reflections from the base of the crust (PmP) in addition to the first arrivals. The purpose of performing two inversions for the crust and upper mantle structure was to assess which features in the model arise from the inclusion of wide-angle reflections from the base of the crust; identifying PmP is associated with more uncertainty and subjectivity than first arrivals. We are most confident of features that are present in both the first-arrival and reflection/refraction tomographic inversions, and more cautious of features that are primarily constrained by the PmP reflections.

We used a grid spacing of 1 × 0.5 km in the pre-rift interval, and 1 × 1 km in the crust and upper mantle. For both inversions, we applied twice as much horizontal smoothing than vertical smoothing and allowed more interface roughness than velocity roughness. A simple 1D velocity model and constant interfaces were used for the starting models in both inversions.

The inversion for the pre-rift layer used 825 picks from the OBS data and 129 picks from the MCS data. The final model has a chi-squared misfit of 1.29 and RMS residual of 72 ms if only the OBS picks are included. Larger misfits are associated with the MCS picks since they include smaller-scale variations in interface geometry than can be recovered by the inversion. If these are included, the overall

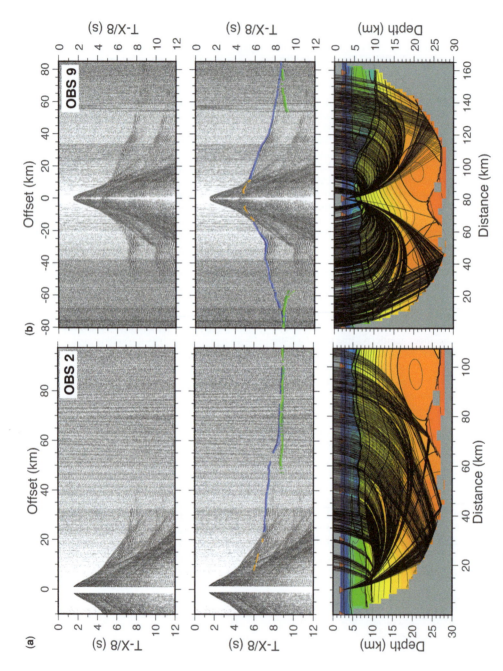

Fig. 3. Receiver gather without picks (top panel). Data with observed picks and picking errors (closed circles and bars) and predicted picks (solid, lighter coloured circles) (middle panel). Orange, PprP; blue, Pg; green, PmP; red, Pn. Ray paths through the final model from the reflection/refraction tomography model (lower panel). (a) OBS 2 and (b) OBS9.

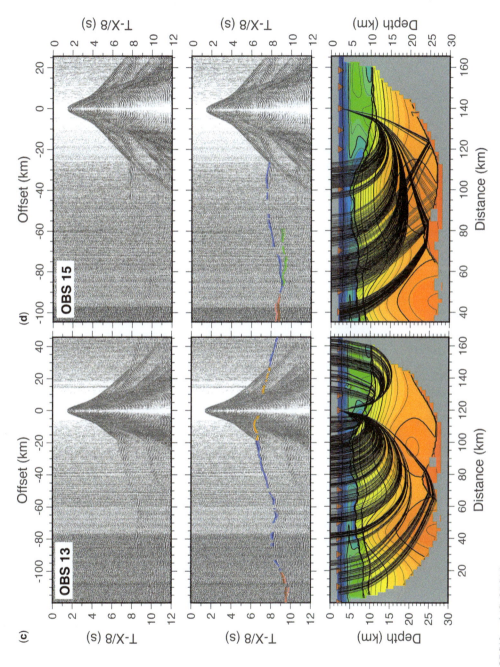

Fig. 3. (c) OBS13 and (d) OBS15.

Table 2. *Misfits by phase*

Phase	Number picks	Chi squared	RMS misfit (s)
PprP	866	3.442881645	0.129567735
Pg	5334	2.038537344	0.106868266
PmP	1502	2.182268919	0.182210481
Pn	249	2.604642144	0.200007259

chi-squared misfit is 1.65, and the RMS residual is 90 ms.

The first-arrival inversion for the crust and upper mantle structure used 5732 picks. The final model has a chi-squared value of 0.96 and an RMS residual of 76 ms. The reflection/refraction inversion used 7085 picks. The final model has a chi-squared value of 2.23 and an RMS residual of 127 ms.

Based on ray coverage, data fit and testing of different inversion parameterizations, we discuss the confidence that should be placed in different features of our final models. The upper-crustal structure is very well sampled by ray coverage associated with our travel-time picks, and refractions from this part of the model have relatively low misfits (Figs 3–5). Similar features are apparent in both the reflection/refraction tomography and the first-arrival tomography. Thus, we consider the variations in upper-crustal velocity structure between the Andrusov and Archangelsky ridges to be a robust result (Figs 4b & 5b). The lowermost crustal sections beneath the Andrusov and Archangelsky ridges are only constrained by sparse turning wave coverage and relatively sparse reflections from the base of the crust (Figs 3 & 5). Because the uppermost part of the lower crust is sampled by reversed refracted arrivals, we are confident that high velocities are required. However, we cannot constrain the velocity gradient of the lowermost crust or absolute velocity at the very base of the lower crust, and there are thus trade-offs between velocities in the lowermost crust and depth to the base of the crust. Both wide-angle reflections and vertically incident reflections constrain the interpreted pre-rift sedimentary layer on top of the MBSH. We find relatively high data misfits for phases defining this layer (Table 2), which we attribute to substantial lateral variability that cannot be accounted for in the analysis of OBS spaced at approximately 15 km. However, we think that the large-scale patterns of thickness and velocity are well constrained.

Although we obtained an excellent misfit for the first-arrival tomography model (chi-squared value of 0.96), our favoured model from reflection/refraction tomography has a higher chi-squared value of 2.23. We relaxed the data misfit criteria to obtain a relatively smooth model; models with better data fit

were substantially rougher. We feel this choice is justified by the likely three-dimensionality of velocity structure beneath these complex ridges, and the complexity of sedimentary, crustal and upper-mantle phases observed on OBS.

Results and discussion

The final velocity models across the Mid Black Sea High provide constraints on the deep sedimentary and crustal structure of this composite ridge.

Sedimentary rocks overlying the Mid Black Sea High

The flat-lying post-rift sedimentary rocks exhibit a low-velocity zone in the Miocene Maikop Formation (Figs 5 & 6) that extends across the eastern basin, and also appears to be present above parts of the MBSH and in the Sinop Trough (Fig. 6) (Scott *et al.* 2009). The low-velocity zone is attributed to fluid overpressure, and fluid pressures close to lithostatic have been inferred (Scott *et al.* 2009), although the application of a more sophisticated approach in the eastern basin (Marin-Moreno *et al.* 2013a, b) suggests that fluid pressures are lower than those derived from the empirical approaches of Scott *et al.* (2009).

Wide-angle reflections in the OBS data (Fig. 3) and reflections in the reflection profile (Fig. 6) define a distinct layer with a thickness of 1–2 km and velocities of 3.0–4.75 km s^{-1} on top of the Andrusov and Archangelsky ridges (Fig. 5). Based on the character of this layer in the reflection profile, dredging on the Archangelsky Ridge and drilling of the Andrusov Ridge, we interpret this layer to represent a sequence of pre-rift Upper Cretaceous sedimentary rocks (Rudat *et al.* 1993; Aydemir & Demirer 2013). This layer is characterized by brightly reflective layering in the reflection profile, which is consistent with a sedimentary origin (Figs 2 & 6). Drilling on Andrusov Ridge at Sinop-1 recovered a relatively thin layer of Upper Cretaceous carbonate rocks (Aydemir & Demirer 2013). Aydemir & Demirer (2013) suggested that the thickness of this interval would be strongly controlled by basement topography at the time of deposition and thus be highly variable, which may explain why we appear to observe a thicker Upper Cretaceous layer on Profile 4. A similar sequence overlies the Shatsky Ridge to the north (Fig. 1) (Robinson *et al.* 1996; Nikishin *et al.* 2015a).

The base of this layer is marked by a bright, continuous reflection in the reflection profile (Fig. 6), which has been interpreted to mark the top of Lower Cretaceous platform carbonate rocks (Rudat *et al.* 1993; Robinson *et al.* 1996). Based on dredging

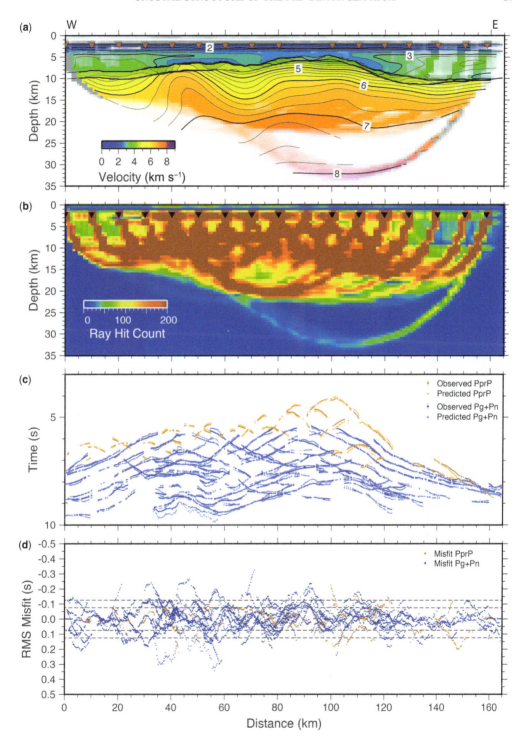

Fig. 4. (**a**) Result of inversion for pre-rift sedimentary reflections (PprP) and first-arriving refractions from crust and upper mantle (Pg and Pn). Velocities contoured at 0.25 km s^{-1}. The velocity model is masked by the density of ray coverage. (**b**) Density of ray coverage over the velocity model in (a). (**c**) Observed and predicted travel-time picks. The uncertainty of observed picks is indicated with bars. (**d**) Travel-time residuals for picks.

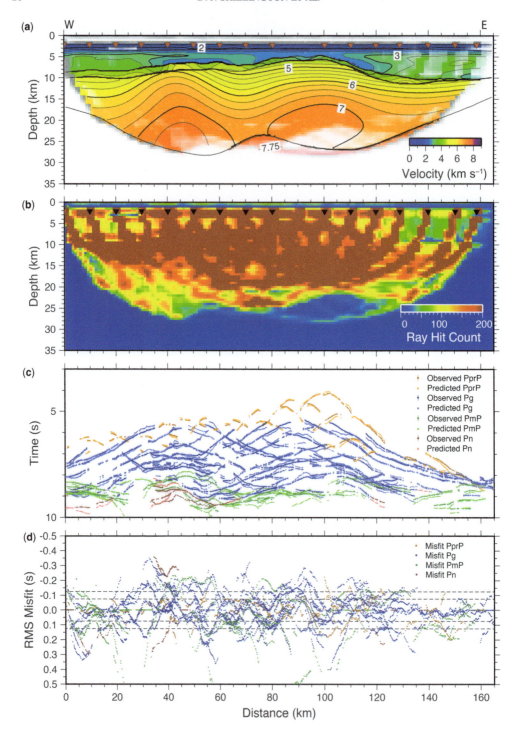

Fig. 5. (**a**) Result of inversion for pre-rift sedimentary reflections (PprP), first-arriving refractions from crust and upper mantle (Pg and Pn), and reflections from the base of the crust (PmP). Velocities are contoured at 0.25 km s^{-1}. The velocity model is masked by the density of ray coverage. (**b**) Density of ray coverage over the velocity model in (a). (**c**) Observed and predicted travel-time picks. The uncertainty of observed picks is indicated with bars. (**d**) Travel-time residuals for picks.

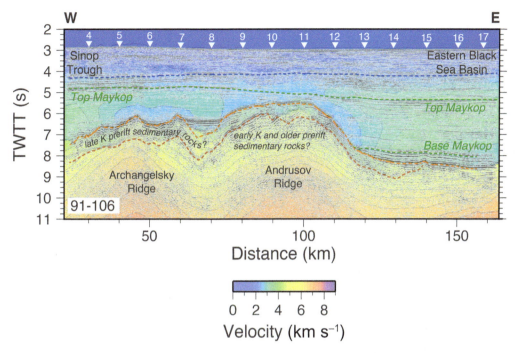

Fig. 6. Overlay of reflection profile 91–106 on the final velocity model from reflection/refraction tomography (Fig. 5), which was converted to two-way travel time.

results on the shallow part of the Archangelsky Ridge, we interpret the uppermost basement beneath this reflection as being composed of Lower Cretaceous platform carbonate rocks and other older pre-rift sedimentary rocks. Platform carbonate rocks are expected to have similar P-wave velocities to upper crystalline crust (Christensen & Mooney 1995), so it is not possible for us to definitively identify carbonate rocks or quantify their thickness, but the nearby dredging results suggest that pre-rift sedimentary rocks are likely to be present in the uppermost basement here. The uppermost basement beneath the prominent reflection described above reaches 6–6.25 km s^{-1} beneath the top of the Archangelsky Ridge, and drops to approximately 4.5 km s^{-1} beneath the Andrusov Ridge. The overlying layer interpreted to represent Upper Cretaceous prerift sedimentary rocks also has significantly higher velocities beneath Archangelsky Ridge than beneath Andrusov Ridge. These differences may be attributed to several factors. First, although Archangelsky Ridge is generally a shallower feature (Fig. 1), at the location of Profile 4 it is more deeply buried, so the pre-rift sedimentary rocks may have undergone greater compaction and diagenesis. Secondly, seismic reflection data suggest that the Andrusov Ridge is disrupted by more faults than the Archangelsky Ridge (Robinson et al. 1996), and fracturing

associated with these faults may reduce the velocity by creating zones of higher porosity and/or causing an elongation of pores, which have a bigger impact on elastic properties (Töksöz et al. 1976). Thirdly, other differences in lithology may contribute to observed variations in velocity. Finally, the low-velocity layer in the post-rift directly abuts the Andrusov Ridge, but is separated from Archangelsky Ridge by a layer of higher-velocity material. Therefore, it is possible that fluid overpressure is transmitted into pre-rift sedimentary rocks on the Andrusov Ridge but not on the Archangelsky Ridge.

Crustal structure and implications for tectonic evolution

The Andrusov and Archangelsky ridges exhibit distinctly different crustal velocity structures. As described in the previous subsection, the Archangelsky Ridge has higher velocities in the uppermost basement (6–6.25 km s^{-1}) and a relatively low velocity gradient (c. 0.075 km s^{-1} km^{-1}). In contrast, the Andrusov Ridge has velocities in the shallow basement as low as 4.5 km s^{-1} and a high velocity gradient in the upper 10 km of 0.25 km s^{-1} km^{-1}. These differences might be associated with different degrees of fracturing of platform carbonate rocks (see the previous subsection) or of crystalline rocks, or might arise because the pre-rift sedimentary sequence within the basement is thicker beneath Andrusov Ridge, as perhaps suggested by seismic reflection data (Fig. 2).

Beneath both ridges, the velocity gradient is smaller in the lower crust, and velocities reach a maximum of 7.2–7.3 km s^{-1} at the base of the crust (Fig. 5). These velocities are somewhat higher than those observed beneath Archangelsky Ridge on Profile 3 (*c.* 6.75–7 km s^{-1}) (Shillington *et al.* 2009) (Fig. 1), and may indicate the presence of a more mafic pre-rift crust (e.g. Christensen & Mooney 1995). Rifting to form the eastern Black Sea occurred in a series of terranes accreted to the Euroasian margin, which include volcanic arcs and oceanic plateaux, both of which are typified by high-velocity lower crust in modern analogues (Shillington *et al.* 2004; Kodaira *et al.* 2007; Calvert 2011).

These velocities are also only slightly lower than lower-crustal velocities observed in crust within the centre of the eastern part of the eastern Basin (Shillington *et al.* 2009), which were interpreted as evidence for new magmatic crust formed during magma-rich rifting and early spreading. However, the relationship between lower-crustal velocity and crustal thickness suggests that syn-rift magmatism is not responsible for the high lower-crustal velocities beneath the MBSH. In the eastern part of the Eastern Basin (Shillington *et al.* 2009) and at other volcanic rifted margins worldwide (e.g. Holbrook & Kelemen 1993; White *et al.* 2008), high-velocity lower crust (*c.* 7.4–7.5 km s^{-1}) interpreted to represent mafic synrift intrusions is most prominent in the area of crustal thinning. In contrast, the highest velocities observed beneath the MBSH occur in the thickest crust and do not increase towards the thinned margins of the ridge. Consequently, we propose that high lower-crustal velocities beneath the MBSH represent high velocities associated with accreted volcanic arcs and oceanic plateaux in the pre-rift crust. Hence, our observations from Profile 4 is consistent with the view that extension in the western part of the Eastern Black Sea Basin was largely amagmatic (Shillington *et al.* 2009).

The crustal layer, which may include platform carbonate rocks and possibly other pre-rift sedimentary rocks, thickens beneath both ridges to reach a maximum of 20–23 km (Fig. 5). Between the two ridges, it decreases to approximately 16 km, providing evidence that the modest increase in sediment thickness between the two ridges (Fig. 1) is associated with crustal-scale extension. Although the Archangelsky Ridge is deeply buried at the location of Profile 4 (Fig. 1), it clearly remains a major crustal feature at this location. Uppermost mantle velocities are a little below 8 km s^{-1}. Based on teleseismic receiver functions, gravity data and limited wide-angle seismic constraints, the crustal thickness onshore Turkey in the vicinity of Archangelsky Ridge is approximately 35 km (Özacar *et al.* 2010; Yegorova *et al.* 2013), with thicker crust farther east where it is affected more by compressional deformation. Therefore, the crust along Profile 4 has been thinned by a factor of 1.5–2. The degree of thinning is somewhat lower than inferred by Shillington *et al.* (2008) based on the relationship between sediment thickness and thinning factor on a well-constrained profile; this relationship gives a thinning factor of 2–2.5 along most of Profile 4 (Fig. 7). One possible explanation for this difference is that the 'crust' of the MBSH may include sections of pre-rift sedimentary rocks that are not a part of the unthinned crustal section onshore (Okay *et al.* 2017).

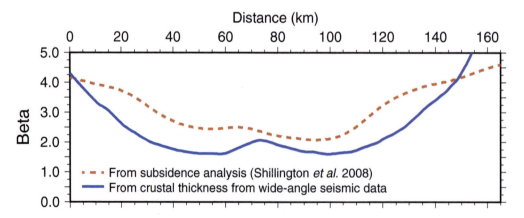

Fig. 7. Comparison of the crustal thinning factor (beta = initial thickness/rifted thickness) along Line 4 from subsidence analysis based on sediment thickness (Shillington *et al.* 2008) and from this study assuming an initial crustal thickness of 35 km.

Conclusions

From our analysis of data from a wide-angle seismic profile across the Mid Black Sea High, comprising the en echelon Archangelsky and Andrusov ridges, we conclude that:

- The basement highs are covered by at least a 1- to 2-km-thick layer of pre-rift sedimentary rocks overlying a higher-velocity basement that may include pre-rift sedimentary rocks, including platform carbonates that cannot be readily distinguished from the underlying crystalline crust.
- The pre-rift sedimentary rocks and upper basement have higher velocities on the Archangelsky Ridge and lower velocities on the Andrusov Ridge. These differences could be explained by different amounts of faulting or changes in the abundance and/or composition of pre-rift sedimentary rocks.
- The lower crust has a low velocity gradient, and velocities exceed 7.0 km s^{-1} at its base; the velocity structure is consistent with the presence of a mafic pre-rift crust with little magmatic addition during rifting.
- The crust is 20–23 km thick beneath the ridges and approximately 16 km thick between them, representing thinning factors of 1.5–2.0 compared to adjacent crust in NE Turkey.

We thank T. Besevli, G. Coskun, A. Demirer, M. Erduran, S. Jones, R. O'Connor, B. Peterson, A. Price, K. Raven and M. Shaw-Champion, and the officers, crew and technical team aboard R/V Iskatel for their support during the acquisition and analysis of this dataset. This work was supported by the Natural Environment Research Council (UK) (NER/T/S/2003/00114 and NER/T/S/2003/00885), BP and the Turkish Petroleum Company (TPAO). BP and TPAO generously provided access to the seismic reflection data. We also thank N. Hodgson and an anonymous reviewer for constructive comments that greatly improved the manuscript.

References

AYDEMIR, V. & DEMIRER, A. 2013. Upper Cretaceous and Paleocene shallow water carbonates along the Pontide Belt. In: 19th International Petroleum and Natural Gas Congress and Exhibition of Turkey, 15–17 May 2013, Turkish Association of Petroleum Geologists, Ankara, 284–290.

BANKS, C.J., ROBINSON, A.G. & WILLIAMS, M.P. 1997. Structure and Regional Tectonics of the Achara-Trialet Fold Belt and the Adjacent Rioni and Kartli Foreland Basins. American Association of Petroleum Geologists, Tulsa, OK.

BELOUSOV, V.V., VOLVOVSKY, B.S. ET AL. 1988. Structure and evolution of the earth's crust and upper mantle of the Black Sea. Bollettino Di Geofisica Teorica ed Applicata, 30, 109–196.

CALVERT, A.J. 2011. The seismic structure of island arc crust. In: BROWN, D. & RYAN, P.D. (eds) Arc–Continent Collision. Springer, Berlin, 87–119.

CHRISTENSEN, N.I. & MOONEY, W.D. 1995. Seismic velocity structure and composition of the continental crust – a global view. Journal of Geophysical Research: Solid Earth, 100, 9761–9788.

ESPURT, N., HIPPOLYTE, J.C., KAYMAKCI, N. & SANGU, E. 2014. Lithospheric structural control on inversion of the southern margin of the Black Sea Basin, Central Pontides, Turkey. Lithosphere, 6, 26–34.

GOBARENKO, V.S., MUROVSKAYA, A.V., YEGOROVA, T.P. & SHEREMET, E.E. 2016. Collision processes at the northern margin of the Black Sea. Geotectonics, 50, 407–424.

GÖRÜR, N. 1988. Timing of opening of the Black Sea basin. Tectonophysics, 147, 247–262.

HOBRO, J.W.D., SINGH, S.C. & MINSHULL, T.A. 2003. Three-dimensional tomographic inversion of combined reflection and refraction seismic traveltime data. Geophysical Journal International 152, 79–93.

HOLBROOK, W.S. & KELEMEN, P.B. 1993. Large igneous province on the US Atlantic margin and implications for magmatism during continental breakup. Nature, 364, 433–436.

KAZMIN, V.G., SCHREIDER, A.A. & BULYCHEV, A.A. 2000. Early stages of evolution of the Black Sea. In: BOZKURT, E., WINCHESTER, J.A. & PIPER, J.D.A. (eds) Tectonics and Magmatism in Turkey and the Surrounding Area. Geological Society, London, Special Publications, 173, 235–249, https://doi.org/10.1144/GSL.SP. 2000.173.01.12

KODAIRA, S., SATO, T., TAKAHASHI, N., ITO, A., TAMURA, Y., TATSUMI, Y. & KANEDA, Y. 2007. Seismological evidence for variable growth of crust along the Izu intra-oceanic arc. Journal of Geophysical Research, 112, B05104.

LETOUZEY, J., BIJU-DUVAL, B., DORKEL, A., GONNARD, R., KRISTCHEV, K., MONTADERT, L. & SUNGURLU, O. 1977. The Black Sea: A marginal basin, geophysical and geological data. In: BIJU-DUVAL, B. & MONTADERT, L. (eds) International Symposium of the Mediterranean Basins. Editions Technip, Paris, 363–376.

MARIN-MORENO, H., MINSHULL, T.A. & EDWARDS, R.A. 2013a. Inverse modelling and seismic data constraints on overpressure generation by disequilibrium compaction and aquathermal pressuring: application to the Eastern Black Sea Basin. Geophysical Journal International, 194, 814–833.

MARIN-MORENO, H., MINSHULL, T.A. & EDWARDS, R.A. 2013b. A disequilibrium compaction model constrained by seismic data and application to overpressure generation in the Eastern Black Sea Basin. Basin Research, 25, 331–347.

MINSHULL, T.A., WHITE, N.J. ET AL. 2005. Seismic data reveal eastern Black Sea structure. Eos, 86, 416–417.

NIKISHIN, A.M., KOROTAEV, M.V., ERSHOV, A.V. & BRUNET, M.-F. 2003. The Black Sea basin: tectonic history and Neogene–Quaternary rapid subsidence modelling. Sedimentary Geology, 156, 149–168.

NIKISHIN, A.M., OKAY, A.I., TUYSUZ, O., DEMIRER, A., AMELIN, N. & PETROV, E. 2015a. The Black Sea basins structure and history: new model based on new deep penetration regional seismic data. Part 1: basins structure and fill. *Marine and Petroleum Geology*, **59**, 638–655.

NIKISHIN, A.M., OKAY, A., TUYSUZ, O., DEMIRER, A., WANNIER, M., AMELIN, N. & PETROV, E. 2015*b*. The Black Sea basins structure and history: new model based on new deep penetration regional seismic data. Part 2: tectonic history and paleogeography. *Marine and Petroleum Geology*, **59**, 656–670.

OKAY, A.I., SENGOR, A.M.C. & GÖRÜR, N. 1994. Kinematic history of the opening of the Black Sea and its effect on the surrounding regions. *Geology*, **22**, 267–270.

OKAY, A.I., SUNAL, G., SHERLOCK, S., ALTINER, D., TÜYSÜZ, O., KYLANDER-CLARK, A.R.C. & AYGÜL, M. 2013. Early Cretaceous sedimentation and orogeny on the active margin of Eurasia: Southern Central Pontides, Turkey. *Tectonics*, **32**, 1247–1271.

OKAY, A.I., ALTINER, D., SUNAL, G., AYGÜL, M., AKDOĞAN, R., ALTINER, S. & SIMMONS, M. 2017. Geological evolution of the Central Pontides. *In*: SIMMONS, M.D., TARI, G.C. & OKAY, A.I. (eds) *Petroleum Geology of the Black Sea*. Geological Society, London, Special Publications, **464**, https://doi.org/10.1144/SP464.3

ÖZACAR, A.A., ZANDT, G., GILBERT, H. & BECK, S.L. 2010. Seismic images of crustal variations beneath the East Anatolian Plateau (Turkey) from teleseismic receiver functions. *In*: SOSSON, M., KAYMAKCI, N., STEPHENSON, R.A., BERGERAT, F. & STAROSTENKO, V. (eds) *Sedimentary Basin Tectonics from the Black Sea and Caucasus to the Arabian Platform*. Geological Society, London, Special Publications, **340**, 485–496, https://doi.org/10.1144/SP340.21

RANGIN, C., BADER, A.G., PASCAL, G., ECEVITOGLU, B. & GÖRÜR, N. 2002. Deep structure of the Mid Black Sea High (offshore Turkey) imaged by multi-channel seismic survey (BLACKSIS cruise). *Marine Geology*, **182**, 265–278.

ROBERTSON, A.H.F., USTAÖMER, T., PICKETT, E.A., COLLINS, A.S., ANDREW, T. & DIXON, J.E. 2004. Testing models of Late Palaeozoic–Early Mesozoic orogeny in Western Turkey: support for an evolving open-Tethys model. *Journal of the Geological Society, London*, **161**, 501–511, https://doi.org/10.1144/0016-764903-080

ROBINSON, A.G., BANKS, C.J., RUTHERFORD, M.M. & HIRST, J.P.P. 1995. Stratigraphic and structural development of the Eastern Pontides, Turkey. *Journal of the Geological Society, London*, **152**, 861–872, https://doi.org/10.1144/gsjgs.152.5.0861

ROBINSON, A.G., RUDAT, J.H., BANKS, C.J. & WILES, R.L.F. 1996. Petroleum geology of the Black Sea. *Marine and Petroleum Geology*, **13**, 195–223.

RUDAT, J.H., MACGREGOR, D.S. & IGNATOV, A.M. 1993. Unconventional exploration techniques in a high cost deepwater basin: a case study from the Black Sea. *SEG Technical Program Expanded Abstracts*, **1993**, 1351.

SAINTOT, A. & ANGELIER, J. 2002. Tectonic paleostress fields and structural evolution of the NW-Caucasus fold-and-thrust belt from Late Cretaceous to Quaternary. *Tectonophysics*, **357**, 1–31.

SCOTT, C.L., SHILLINGTON, D.J., MINSHULL, T.A., EDWARDS, R.A., BROWN, P.J. & WHITE, N.J. 2009. Wide-angle seismic data reveal extensive overpressures in Eastern

Black Sea. *Geophysical Journal International*, **178**, 1145–1163, https://doi.org/10.1111/j.1365-246X.2009.04215.x

SHILLINGTON, D.J., VAN AVENDONK, H.J.A., HOLBROOK, W.S., KELEMEN, P.B. & HORNBACH, M.J. 2004. Composition and structure of the central Aleutian island arc from arc-parallel wide-angle seismic data. *Geochemistry, Geophysics, Geosystems*, **5**, Q10006, https://doi.org/10.1029/2004GC000715

SHILLINGTON, D.J., WHITE, N., MINSHULL, T.A., EDWARDS, G.R.H., JONES, S., EDWARDS, R.A. & SCOTT, C.L. 2008. Cenozoic evolution of the eastern Black Sea: a test of depth-dependent stretching models. *Earth and Planetary Science Letters*, **265**, 360–378.

SHILLINGTON, D.J., SCOTT, C.L., MINSHULL, T.A., EDWARDS, R.A., BROWN, P.J. & WHITE, N. 2009. Abrupt transition from magma-starved to magma-rich rifting in the eastern Black Sea. *Geology*, **37**, 7–10, https://doi.org/10.1130/G25302A.1

STAROSTENKO, V., BURYANOV, V. *ET AL.* 2004. Topography of the crust-mantle boundary beneath the Black Sea Basin. *Tectonophysics*, **381**, 211–233.

TARI, G., SCHLEDER, ZS., FALLAH, M., TURI, V., KOSI, W. & KREZSEK, Cs. 2015. Regional rift structure of the Western Black Sea Basin: map-view kinematics. *In*: *Transactions of the GCSSEPM Foundation Perkins-Rosen 34th Annual Research Conference 'Petroleum Systems in Rift Basins'*, Houston, Texas, 372–396.

TÖKSÖZ, M.N., CHENG, C.H. & TIMUR, A. 1976. Velocities of seismic waves in porous rocks. *Geophysics*, **41**, 621–645.

VINCENT, S.J., ALLEN, M.B., ISMAIL-ZADEH, A.D., FLECKER, R., FOLAND, K.A. & SIMMONS, M.D. 2005. Insights from the Talysh of Azerbaijan into the Paleogene evolution of the South Caspian region. *GSA Bulletin*, **117**, 1513–1533.

WHITE, R.S., SMITH, L.K., ROBERTS, A.W., CHRISTIE, P.A.F., KUSZNIR, N.J. & iSIMM TEAM 2008. Lower-crustal intrusion on the North Atlantic continental margin. *Nature*, **452**, 460–465.

YEGOROVA, T. & GOBARENKO, V. 2010. Structure of the Earth's crust and upper mantle of the West- and East-Black Sea Basins revealed from geophysical data and its tectonic implications. *In*: SOSSON, M., KAYMAKCI, N., STEPHENSON, R.A., BERGERAT, F. & STAROSTENKO, V. (eds) *Sedimentary Basin Tectonics from the Black Sea and Caucasus to the Arabian Platform*. Geological Society, London, Special Publications, **340**, 23–42, https://doi.org/10.1144/SP340.3

YEGOROVA, T., GOBARENKO, V. & YANOVSKAYA, T. 2013. Lithosphere structure of the Black Sea from 3-D gravity analysis and seismic tomography. *Geophysical Journal International*, **193**, 287–303.

YILMAZ, Y., TÜYSÜZ, O., YIGITBAS, E., CAN GENÇ, S. & SENGÖR, A.M.C. 1997. Geology and tectonic evolution of the Pontides. *In*: ROBINSON, A.G. (ed.) *Regional and Petroleum Geology of the Black Sea and Surrounding Region*. American Association of Petroleum Geologists Memoirs, **68**, 183–226.

ZONENSHAIN, L.P. & LE PICHON, X. 1986. Deep basins of the Black Sea and Caspian Sea as remnants of Mesozoic back-arc basins. *Tectonophysics*, **123**, 181–211.

Geological evolution of the Central Pontides

ARAL I. OKAY[1,2]*, DEMIR ALTINER[3], GÜRSEL SUNAL[2], MESUT AYGÜL[1],
REMZIYE AKDOĞAN[2], SEVINÇ ALTINER[3] & MIKE SIMMONS[4]

[1]*Eurasia Institute of Earth Sciences, Istanbul Technical University,
34469 Maslak, Istanbul, Turkey*

[2]*Department of Geology, Istanbul Technical University, 34469 Maslak, Istanbul, Turkey*

[3]*Department of Geological Engineering, Middle East Technical
University, 06800 Ankara, Turkey*

[4]*Halliburton, 97 Jubilee Avenue, Milton Park, Abingdon OX14 4RW, UK*

Correspondence: okay@itu.edu.tr

Abstract: Before the Late Cretaceous opening of the Black Sea, the Central Pontides constituted part of the southern margin of Laurasia. Two features that distinguish the Central Pontides from the neighbouring Pontide regions are the presence of an extensive Lower Cretaceous submarine turbidite fan (the Çağlayan Formation) in the north, and a huge area of Jurassic–Cretaceous subduction–accretion complexes in the south.

The Central Pontides comprise two terranes, the Istanbul Zone in the west and the Sakarya Zone in the east, which were amalgamated before the Late Jurassic (Kimmeridgian), most probably during the Triassic. The basement in the western Central Pontides (the Istanbul Zone) is made up of a Palaeozoic sedimentary sequence, which ends with Carboniferous coal measures and Permo-Triassic red beds. In the eastern Central Pontides, the basement consists of Permo-Carboniferous granites and an Upper Triassic forearc sequence of siliciclastic turbidites with tectonic slivers of pre-Jurassic ophiolite (the Küre Complex). The Küre Complex is intruded by Middle Jurassic granites and porphyries, which constitute the western termination of a major magmatic arc.

Upper Jurassic–Lower Cretaceous shallow-marine limestones (the İnaltı Formation) lie unconformably over both the Istanbul and Sakarya sequences in the Central Pontides. Two new measured stratigraphic sections from the İnaltı Formation constrain the age of the İnaltı Formation as Kimmeridgian–Berriasian. After a period of uplift and erosion during the Valanginian and Hauterivian, the İnaltı Formation is unconformably overlain by an over 2 km-thick sequence of Barremian–Aptian turbidites. Palaeocurrent measurements and detrital zircons indicate that the major part of the turbidites was derived from the East European Platform, implying that the Black Sea was not open before the Aptian. The Çağlayan turbidites pass northwards to a coeval carbonate–clastic shelf exposed along the present Black Sea coast. In the southern part of the Central Pontides, the Lower Cretaceous turbidites were deformed and metamorphosed in the Albian. Albian times also witnessed accretion of Tethyan oceanic crustal and mantle sequences to the southern margin of Laurasia, represented by Albian eclogites and blueschists in the Central Pontides.

A new depositional cycle started in the Late Cretaceous with Coniacian–Santonian red pelagic limestones, which lie unconformably over the older units. The limestones pass up into thick sequences of Santonian–Campanian arc volcanic rocks. The volcanism ceased in the middle Campanian, and the interval between late Campanian and middle Eocene is represented by a thick sequence of siliciclastic and calciclastic turbidites in the northern part of the Central Pontides. Coeval sequences in the south are shallow marine and are separated by unconformities. The marine deposition in the Central Pontides ended in the Middle Eocene as a consequence of collision of the Pontides with the Kırşehir Massif.

Supplementary material: The palaeontological data (foraminifera, nannofossil and pollen) are available at: https://doi.org/10.6084/m9.figshare.c.3842359

The Pontides and the Anatolide–Tauride Block are two major tectonic units of Anatolia, representing the former active and passive margins of the Tethyan Ocean, respectively. The Pontides consists of three terranes, which were amalgamated during the Mesozoic; these are the Strandja Massif in the west, and the Istanbul and the Sakarya zones in the east (Fig. 1) (Okay & Tüysüz 1999). The Central Pontides constitute the northwards arched central segment of the Pontides, and includes parts of the

From: SIMMONS, M. D., TARI, G. C. & OKAY, A. I. (eds) 2018. *Petroleum Geology of the Black Sea.*
Geological Society, London, Special Publications, **464**, 33–67.
First published online September 15, 2017, https://doi.org/10.1144/SP464.3

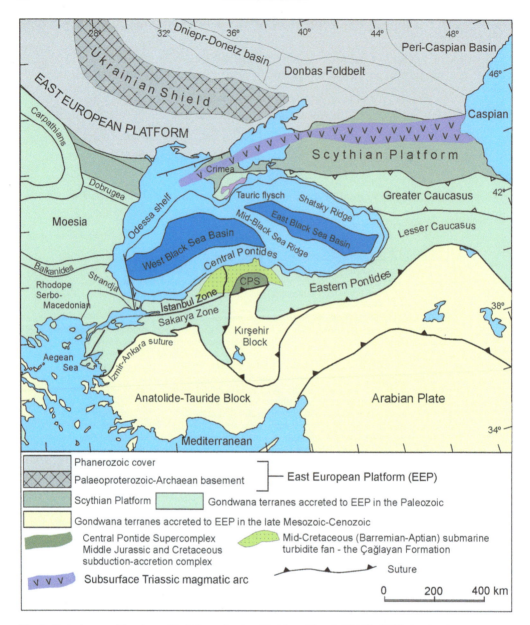

Fig. 1. Tectonic map of the circum-Black Sea region (modified from Okay & Nikishin 2015) showing the distribution of the mid-Cretaceous turbidites and Jurassic–Cretaceous subduction–accretion complexes in the Central Pontides. CPS, Central Pontide Supercomplex.

Istanbul and Sakarya zones. It is distinguished from the Western and Eastern Pontides by the presence of a very large mid-Cretaceous turbidite fan, which is bordered to the south by an extensive Jurassic and Cretaceous subduction–accretion complex, the Central Pontide Supercomplex (Fig. 1).

The Central Pontides and their offshore extension have been a target for petroleum exploration for many years with dry wells drilled both onshore and offshore (Robinson et al. 1996; Görür & Tüysüz 1997; Şen 2013). The accepted geological framework and understanding of the Central Pontides during this phase of exploration changed drastically as a result of new data. The large metamorphic and ophiolitic area in the southern Central Pontides, which was long considered as pre-Jurassic basement, was shown to be made up of Jurassic and Lower Cretaceous subduction–accretion complexes and accreted oceanic island arcs (Fig. 1) (Okay *et al.* 2006, 2013; Marroni *et al.* 2014; Aygül *et al.* 2015*a*, *b*, 2016; Çelik *et al.* 2016). New studies have also shown the presence of Permo-Carboniferous granitic magmatism, and Jurassic high-temperature–low-pressure (HT–LP) metamorphism and associated magmatism in the Central Pontides (Nzegge *et al.* 2006; Okay *et al.* 2015; Gücer *et al.* 2016). New precise palaeontological data have led to the establishment of a more robust biostratigraphy (Hippolyte *et al.* 2010, 2015, 2016; Gand *et al.* 2011; Okay *et al.* 2015; Stolle 2016; Tüysüz *et al.* 2016). Detrital zircon U–Pb dates provided a new perspective to the Mesozoic palaeogeography (Karslıoğlu *et al.* 2012; Okay *et al.* 2013; Akdoğan *et al.* 2017). The present paper synthesizes these recent data into an internally consistent Late Palaeozoic–Cenozoic evolution of the Central Pontides. We also provide new biostratigraphic data on the Mesozoic and Cenozoic sequences of the Central Pontides, including several measured stratigraphic sections with the biostratigraphy controlled by the palaeontological study of over 300 thin sections.

Stratigraphic framework

The Central Pontides can be divided into crescent-shaped northern and southern sectors. The northern sector consists of sequences deposited on continental crust, whereas most of the southern Central Pontides are made up of Mesozoic oceanic subduction–accretion complexes, collectively called as the Central Pontide Supercomplex (Fig. 1).

The northern Central Pontides comprises two types of pre-Jurassic basement; in the western part of the Central Pontides (Istanbul Zone), the basement consists of late Neoproterozoic crystalline rocks and an overlying Palaeozoic sedimentary succession (Fig. 2). Permian and Triassic sequences are represented by fluviatile red beds and lacustrine deposits with acidic magmatic rocks, reminiscent of the Rotliegende and Keuper series of Central Europe (e.g. Stolle 2016). In the eastern Central Pontides, which is part of the Sakarya Zone, the pre-Jurassic basement consists of low-grade metamorphic rocks intruded by Upper Carboniferous and Lower Permian granites (Okay *et al.* 2015). The bulk of the Triassic is represented by a thick sequence of distal forearc turbidites (the Akgöl Formation), which were deposited partly on oceanic crust and partly on Lower–Middle Triassic pelagic carbonates (Fig. 2). During the Middle Jurassic, the eastern Central Pontides were part of a magmatic arc; several shallow-level intrusions were emplaced into the Triassic turbidites. Heat flow associated

with the arc resulted in high-temperature metamorphism of the middle crust, which crops out in two large massifs (Fig. 3) (Okay *et al.* 2014; Gücer *et al.* 2016).

The Upper Jurassic–Lower Cretaceous shallow-marine carbonates (the İnaltı Formation) are the first major unit in the Central Pontides, and were deposited over both the Istanbul and Sakarya zones; they provide a time constraint for the amalgamation of these two tectonic terranes (Figs 2 & 3). The carbonates are unconformably overlain by a thick sequence of Lower Cretaceous (upper Barremian–Aptian) siliciclastic turbidites (the Çağlayan Formation). Turbidite deposition was followed by a major phase of shortening, uplift and erosion in the Albian, especially evident in the south. This event is linked to accretion of subduction complexes to the southern margin of Laurasia, represented by Albian eclogites and blueschists in the Central Pontide Supercomplex.

A new depositional cycle started with the Upper Cretaceous (Turonian–Santonian) deep-marine red limestones (the Kapanboğazı Formation); these generally lie unconformably above the Lower Cretaceous turbidites, as well as on the older formations. The Turonian–Santonian transgression also marks the inception of Late Cretaceous arc magmatism; the volcanic rocks are intercalated and interfinger with Santonian–Campanian pelagic limestones. The arc volcanism waned in the Late Campanian–Maastrichtian, and was succeeded in the northern parts of the Central Pontides by the deposition of Maastrichtian–Paleocene turbidites (the Gürsökü, Akveren and Atbaşı formations). The coeval sequences in the southern parts of the Central Pontides are shallow-marine sandstones and limestones. The marine deposition in the Central Pontides ended with a thick sequence of Lower–Middle Eocene turbidites in the north, and coeval carbonates in the south.

The southern part of the Central Pontides consists mainly of metamorphic and ophiolitic rocks representing Jurassic and Cretaceous oceanic subduction–accretion complexes (Fig. 1) (Okay *et al.* 2006, 2013; Marroni *et al.* 2014; Aygül *et al.* 2015*a*, *b*, 2016; Çelik *et al.* 2016). The metamorphism is in eclogite, blueschist and greenschist facies, with two age peaks in the Middle–early Late Jurassic and Early Cretaceous.

The pre-Late Jurassic stratigraphy of the western Central Pontides: the Istanbul Zone

Neoproterozoic crystalline basement and the Palaeozoic sequence

In the western Central Pontides, the crystalline basement is composed of late Neoproterozoic granites

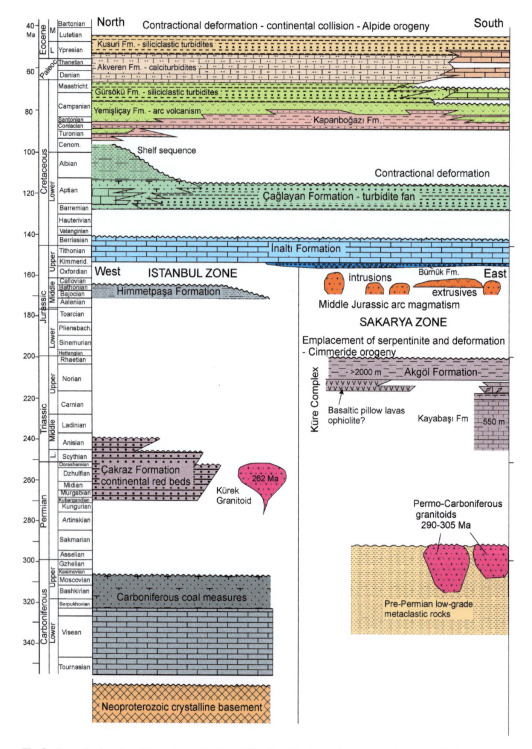

Fig. 2. Generalized stratigraphic section of the Central Pontides. The timescale is from Cohen *et al.* (2013).

and gneisses, which crop out in a small area in the Karadere region north of Araç (Fig. 3) (Chen et al. 2002). The granites have yielded latest Neoproterozoic zircon U–Pb (590–560 Ma) and Rb–Sr biotite ages (548–545 Ma), the latter indicate that the sequence has not been reheated since the Neoproterozoic. Larger outcrops of the late Neoproterozoic Pan-African basement exists in the Bolu Massif further west (Fig. 3) (Ustaömer et al. 2005).

The late Neoproterozoic metagranites are stratigraphically overlain by a Palaeozoic sedimentary sequence ranging in age from Ordovician to Carboniferous (Fig. 4). The lower part of the sequence, which is Ordovician–Silurian in age, is well exposed in the Karadere region, and consists mainly of dark shale and siltstone with graptolite faunas (Boztuğ 1992; Dean et al. 2000). The upper part of the Palaeozoic sequence, Devonian–Late Carboniferous in age, crops out mostly along the Black Sea coast in the Zonguldak and Amasra regions (Fig. 3). It consists mostly of shallow-marine Devonian–Lower Carboniferous limestones (Dil 1976; Dil et al. 1976), similar to those of NW Europe. The transition from carbonate to clastic deposition occurs at the end of the Visean, and the Namurian–Westphalian (middle Serpukhovian–Moscovian) is represented by coal series several kilometres in thickness (Fig. 4) (Kerey 1985), which crop out along the Black Sea coastal area.

Permo-Triassic red beds and Middle Jurassic shallow-marine clastic rocks

The Palaeozoic sedimentary rocks are unconformably overlain by a thick Permo-Triassic continental sequence called the Çakraz Formation (Figs 2 & 3). The lower parts of the Çakraz Formation consist of fluviatile red sandstone, siltstone and conglomerate with middle Permian vertebrate imprints and spores (Gand et al. 2011; Stolle 2016). The upper part is made up of lacustrine siltstone and marl possibly of Triassic age (Alişan & Derman 1995). The Permian red beds (New Red Sandstone) are widespread in the Balkans and in Central Europe, and represent continental deposition following the Variscan Orogeny. In the western part of the Istanbul Zone, close to the city of Istanbul, the Palaeozoic sedimentary sequence is intruded by Upper Permian (c. 254 Ma) granite (Yılmaz 1975). A granite of similar age (c. 262 Ma), albeit in an allochthonous position, crops out in the southern part of the western Central Pontides (Fig. 3) (Okay et al. 2013). There are also possibly Permian rhyolitic domes within the Permian red beds SW of Cide (Akyol et al. 1974).

The Permo-Triassic red beds of the Çakraz Formation are unconformably overlain by continental to shallow-marine sandstone, siltstone, shale and conglomerate, about 375 m in thickness. This Himmetpaşa Formation contains ammonites and palynomorphs of Middle Jurassic age (Akyol et al. 1974; Derman et al. 1995).

The pre-Late Jurassic stratigraphy of the eastern Central Pontides: the Sakarya Zone

Palaeozoic subduction–accretion complex and Permo-Carboniferous granites

The oldest sequence, which crops out in the eastern part of the Central Pontides, is a pre-Permian succession of black to brown slate and phyllite interbedded with metasiltstone and fine-grained meta-sandstone, which crops out close to the Black Sea coast (Fig. 3) (Boztuğ & Yılmaz 1983; Okay et al. 2015). There are also tectonic lenses of serpentinite within this low-grade metasedimentary sequence. The metasedimentary rocks and the serpentinite lens are intruded by Upper Carboniferous granitoids, providing an upper age for the deposition, deformation and metamorphism (Okay et al. 2015). The metasedimentary sequence with serpentinite lenses probably represents a Palaeozoic subduction–accretion complex.

Two Permo-Carboniferous granites with U–Pb zircon ages of 303–291 Ma crop out close to the Black Sea coast (Fig. 3) (Nzegge et al. 2006; Nzegge 2008; Okay et al. 2015). They range from hornblende–biotite granodiorite to two-mica granite, and are peraluminous, calc-alkaline and high-K in composition (Nzegge et al. 2006). Their $\varepsilon Nd_{(t)}$, $\delta^{18}O$ values and $Sr_{(i)}$ ratios suggest derivation by dehydration melting of metapelitic and mafic crust (Nzegge et al. 2006), which is in accordance with the presence of primary muscovite and enclaves of high-temperature metamorphic rocks in one of the granites. On the tectonic discrimination diagrams, most samples plot in the field of volcanic arc granitoids (Nzegge et al. 2006). Their peraluminous nature, high K and Sr contents, high Rb/Sr values, and initial Sr ratios point to crustal melting and suggest an episode of crustal thickening.

Permo-Carboniferous granites also crop out in other parts of the Sakarya Zone (Topuz et al. 2010; Ustaömer et al. 2012; Ustaömer et al. 2013) and in the Greater Caucasus (Hanel et al. 1992; Somin 2011). Carboniferous and Permian detrital zircons are generally the dominant population in the upper Palaeozoic and Mesozoic sedimentary rocks of the Pontides and of the Crimea (Okay et al. 2013; Nikishin et al. 2015c; Akdoğan et al. 2017), suggesting that Permo-Carboniferous granites are widely present under the Mesozoic cover in the Central Pontides and possibly in the Scythian Platform, and extend westwards into the Balkans; their generation

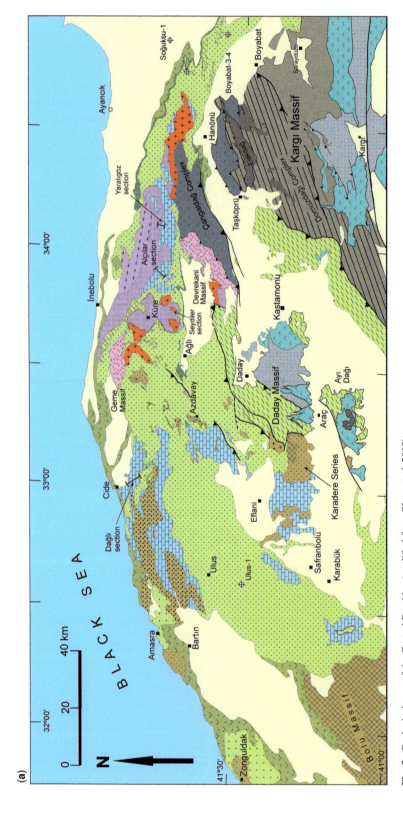

Fig. 3. Geological map of the Central Pontides (modified from Okay *et al.* 2013).

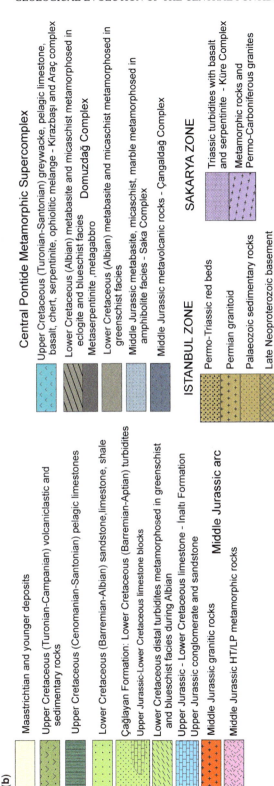

(b)

Central Pontide Metamorphic Supercomplex

Upper Cretaceous (Turonian-Santonian) greywacke, pelagic limestone, basalt, chert, serpentinite, ophiolitic melange - Kirazbaşı and Araç complex

Lower Cretaceous (Albian) metabasite and micaschist metamorphosed in eclogite and blueschist facies   Domuzdağ Complex
Metaserpentinite, metagabbro

Lower Cretaceous (Albian) metabasite and micaschist metamorphosed in greenschist facies

Middle Jurassic metabasite, micaschist, marble metamorphosed in amphibolite facies - Saka Complex

Middle Jurassic metavolcanic rocks - Çangaldağ Complex

SAKARYA ZONE

Triassic turbidites with basalt and serpentinite - Küre Complex

Metamorphic rocks and Permo-Carboniferous granites

ISTANBUL ZONE

Permo-Triassic red beds

Permian granitoid

Palaeozoic sedimentary rocks

Late Neoproterozoic basement

Maastrichtian and younger deposits

Upper Cretaceous (Turonian-Campanian) volcaniclastic and sedimentary rocks

Upper Cretaceous (Cenomanian-Santonian) pelagic limestones

Lower Cretaceous (Barremian-Albian) sandstone, limestone, shale

Çağlayan Formation: Lower Cretaceous (Barremian-Aptian) turbidites

Upper Jurassic-Lower Cretaceous limestone blocks

Lower Cretaceous distal turbidites metamorphosed in greenschist and blueschist facies during Albian

Upper Jurassic - Lower Cretaceous limestone - İnaltı Formation
Upper Jurassic conglomerate and sandstone

Middle Jurassic granitic rocks   Middle Jurassic arc

Middle Jurassic HT/LP metamorphic rocks

Fig. 3. *Continued.*

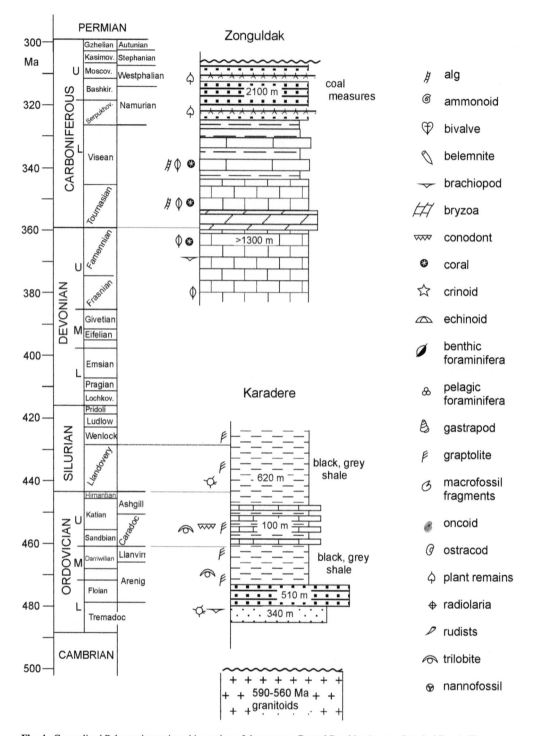

Fig. 4. Generalized Palaeozoic stratigraphic section of the western Central Pontides (eastern Istanbul Zone). The stratigraphic and palaeontological data are from Dil *et al.* (1976) and Dean *et al.* (2000).

is related to the Variscan Orogeny (e.g. Okay & Topuz 2017).

Triassic: Cimmeride Orogeny

The latest Triassic Cimmeride Orogeny in the Pontides, specifically in the Sakarya Zone, was a short-lived accretionary rather than collisional event (Okay 2000; Robertson & Ustaömer 2011; Topuz et al. 2013). It was caused by the accretion of a major oceanic plateau or a number of oceanic islands along with trench deposits to the southern active margin of Laurasia (Fig. 5b). In NW Turkey, the Triassic subduction–accretion units are known as the Karakaya Complex, and include Triassic eclogites and blueschists (Fig. 1) (Okay & Göncüoğlu 2004). Similar sequences crop out in the northern part of the Central Pontides, where there are known as the Küre Complex (Fig. 3) (Ustaömer & Robertson 1994). The bulk of the Küre Complex consists of dark siliciclastic distal turbidites of Late Triassic (Norian) age, more than 2000 m in thickness (Kozur et al. 2000; Okay et al. 2015). The geochemistry of the shale and sandstone suggests deposition in an active continental margin (Ustaömer & Robertson 1994). This Akgöl Formation has a composite stratigraphic basement (Fig. 2); in some regions, the distal turbidites are stratigraphically underlain by a pelagic Middle–Upper Triassic (Anisian–Carnian) limestone sequence with conodonts and ammonites, yet in some other areas they are underlain by pillow basalts (Okay et al. 2015). The basalts are tholeiitic and mostly mid-ocean ridge (MORB) type (Ustaömer & Robertson 1994). Economic chalcopyrite and pyrite mineralization, exploited in the ancient Küre mine, has developed along the contact between basalt and black shale. There are also several kilometres of large tectonic serpentinite slivers within the turbidites. The serpentinite and the Upper Triassic turbidites are intruded by a Middle Jurassic (162 ± 4 Ma) granite indicating a pre-Jurassic, most probably Triassic, age for the ultramafic rocks and associated gabbro and diabase (Okay et al. 2015). Serpentinite, gabbro and basalt constitute a dismembered ophiolite on which the Upper Triassic turbidites of the Akgöl Formation were deposited (Ustaömer & Robertson 1994). The Küre Complex was deformed by folding and thrusting during the latest Triassic–?earliest Jurassic before the intrusion of the Middle Jurassic granites and deposition of the Upper Jurassic conglomerates and limestones.

Detrital zircon ages from the turbiditic sandstones of the Akgöl Formation are predominantly Triassic (Karslıoğlu et al. 2012). A similar dominance of Triassic detrital zircons is recorded in the Upper Triassic Tauric flysch of the Crimea (Nikishin et al. 2015c; Nikishin pers. comm.), which shares a similar lithology and stratigraphic position with the Akgöl Formation. Ustaömer et al. (2016) reported an extensive zircon dataset from the sandstones of the Karakaya Complex; the ages are also predominantly Triassic. The abundance of Triassic detrital zircons and the presence of Triassic subduction–accretion complexes in the Pontides suggest that a now-hidden Triassic magmatic arc is present in the Scythian Platform north of the Black Sea (Fig. 5b), as implied by some well data (Tikhomirov et al. 2004). The arc is now buried beneath Cenozoic sediments. The Triassic arc extends from the northern Crimea to Central Asia (Fig. 1) (Natal'in & Şengör 2005; Nikishin et al. 2012). The Akgöl Formation and its lateral equivalent, the Tauric flysch in the Crimea, were deposited in a forearc between the subduction–accretion units of the Karakaya Complex and the Triassic magmatic arc (Fig. 5b).

Middle Jurassic arc magmatism

During the Middle Jurassic, a major Andean type magmatic arc was established along the southern margin of Laurasia; the arc extended for 2800 km from Makran in Iran to the Pontides in Turkey (Fig. 5c) (Şengör et al. 1991; Şen 2007; Dokuz et al. 2010; Genç & Tüysüz 2010; McCann et al. 2010; Adamia et al. 2011). Jurassic arc magmatism is observed in the eastern Central Pontides and in Crimea, where they are represented by shallow-level intrusions, dacite and andesite porphyries, with lesser amounts of volcanic rocks and granites (Fig. 3) (Yılmaz & Boztuğ 1986; Meijers et al. 2010b; Okay et al. 2014). The geochemistry of the Jurassic porphyries and volcanic rocks has a distinct arc signature with a crustal melt component (Boztuğ et al. 1984, 1995; Okay et al. 2014). A crustal melt component is also suggested by cordierite and garnet in the magmatic assemblage and the abundance of inherited zircons in the porphyries. The age of magmatism, based on zircon U–Pb data, is predominantly Middle–early Late Jurassic (175–155 Ma: Okay et al. 2014). Radiolaria in the mudstones intercalated with basaltic flows also give Callovian–Oxfordian ages (Bragin et al. 2002). The magmatic arc developed largely on the continental crust; the possible exception is the Çangaldağ Complex, a thick pile of basic and intermediate volcanic rocks of Middle Jurassic age, which is considered as a magmatic arc built on oceanic crust (Fig. 3) (Ustaömer & Robertson 1999; Okay et al. 2014). In contrast to the other Jurassic magmatic rocks, the Çangaldağ Complex underwent a low-grade greenschist- to blueschist-facies regional metamorphism in the Early Cretaceous (Okay et al. 2013; Çimen et al. 2016).

The high heat flow, which produced the arc magmatism, also led to the metamorphism of the

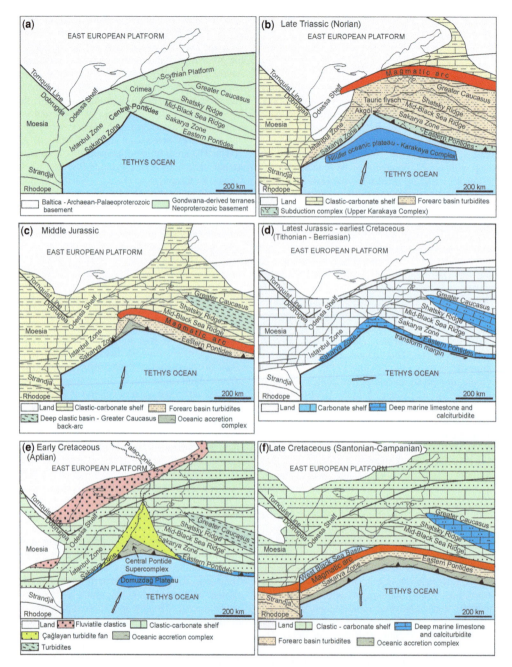

Fig. 5. Palaeogeographical maps for the Central Pontides and the surrounding regions for the Mesozoic. (**a**) A probable configuration of the Black Sea terranes in the early Mesozoic. (**b**) Late Triassic. The major event is the collision of the Nilüfer oceanic plateau with the Laurasian margin deforming the active margin and causing the Cimmeride Orogeny. (**c**) Middle Jurassic. Subduction, arc magmatism and accretion continue but do not extend to the western part of the Black Sea region. (**d**) Latest Jurassic–earliest Cretaceous. This interval is characterized by the cessation of subduction and widespread carbonate deposition in the Black Sea region. (**e**) Early Cretaceous. Shallow subduction leads to the uplift of the East European Platform, which is eroded into the Çağlayan turbidite fan, at the same time there is a major oceanic accretion in the south. (**f**) Late Cretaceous. Subduction, arc magmatism and extension led to the opening of the western Black Sea as a back-arc basin. The palaeogeographical maps are based on Barrier & Vrielynck (2008), Okay *et al.* (2013) and Nikishin *et al.* (2015*b*).

basement. In the Central Pontides, there are two large outcrops of Middle Jurassic (*c.* 172 Ma) HT–LP (4 kbar and 720°C) metamorphic rocks in the Geme and Devrekani complexes (Fig. 3) (Okay *et al.* 2014; Gücer *et al.* 2016). They consist predominantly of gneiss and migmatite with cordierite and sillimanite. The metamorphic rocks are intruded by the Middle Jurassic granitic veins, stocks and larger magmatic bodies. The detrital zircons in the gneisses indicate that the metamorphic rocks represent Hercynian basement, which was remobilized under the Middle Jurassic magmatic arc.

Middle Jurassic subduction–accretion: high-pressure amphibolite-facies metamorphism

Part of the Central Pontide Supercomplex is made up of micaschists and amphibolites with Middle–early Late Jurassic (170–159 Ma) Ar/Ar mica ages (Okay *et al.* 2013; Marroni *et al.* 2014; Aygül *et al.* 2016; Çelik *et al.* 2016). The metamorphism is in high-pressure amphibolite facies (10 kbar and 620°C), and the metamorphic rocks are tectonically intercalated with Cretaceous subduction–accretion complexes and lie tectonically beneath the metamorphosed flysch of the Lower Cretaceous Çağlayan Formation (Fig. 3). This Saka Complex has formed during the Middle Jurassic subduction, which also gave rise to the arc magmatism. Similar Jurassic subduction–accretion complexes are also described from other parts of the Pontides (Çelik *et al.* 2011; Topuz *et al.* 2013).

Upper Jurassic–Lower Cretaceous platform limestones

Pre-Jurassic sequences in the northern Central Pontides are unconformably overlain by Upper Jurassic–Lower Cretaceous limestones, the İnaltı Formation (Fig. 2). The İnaltı Formation is the first lithostratigraphic unit that indisputably extends both over the Istanbul and Sakarya zones, and thereby provides an upper age (Kimmeridgian) for the juxtaposition of these two terranes in the Central Pontides. Upper Jurassic–Lower Cretaceous limestones are ubiquitous in the Balkans and in the Pontides, and also extend eastwards into the Caucasus (Fig. 5d) (Altıner *et al.* 1991; Taslı 1993; Tari *et al.* 1997; Rojay & Altıner 1998; Koch *et al.* 2008; Guo *et al.* 2011; Kaya & Altıner 2015; Nikishin *et al.* 2015c). The Late Jurassic–earliest Cretaceous (Kimmeridgian–Berriasian) was a tectonically quite period in the northern Tethys, with little tectonic activity fostering carbonate accumulation.

In the western Central Pontides, Upper Jurassic–Lower Cretaceous limestones lie unconformably above the Permo-Triassic red beds of the Çakraz Formation or above the Himmetpaşa Formation, whereas in the eastern Central Pontides they lie unconformably, through a basal continental clastic unit, above the Upper Triassic Akgöl Formation (Fig. 2). The basal clastic unit (Bürnük Formation) is made up of fluviatile red sandstone, siltstone and conglomerate, with a thickness varying from a few metres up to 500 m; the clasts in the conglomerate are predominantly subvolcanic rocks derived from the Middle Jurassic intrusives.

The İnaltı Formation in the Central Pontides is reported to be Late Jurassic (Aydın *et al.* 1995), or Late Jurassic–Early Cretaceous (Oxfordian–Berriasian/Valanginian) in age (Derman & Sayılı 1995; Tüysüz 1999). There are no published detailed measured stratigraphic sections from the İnaltı Formation. To determine the age of the İnaltı Formation in the Central Pontides, four stratigraphic sections were measured (Fig. 3): one in the western part and three in the eastern part. Two of these sections are described below:

- The Alçılar section – The Alçılar section is located east of Küre in the eastern Central Pontides (Figs 3 & 6a). Here, limestones of the İnaltı Formation overlie the fluviatile conglomerates and sandstones of the Bürnük Formation, and are overlain unconformably by the conglomerates and breccias of the Çağlayan Formation (Fig. 7). The Bürnük Formation itself lies unconformably above the Upper Triassic turbidites of the Akgöl Formation. The İnaltı Formation in the Alçılar section is 600 m thick and is made up of Kimmeridgian–lower Berriasian shallow-marine limestones (Fig. 7). Altogether, 107 limestone samples from the Alçılar section were examined petrographically and palaeontologically. The basal part of the section is made up of slightly nodular, thin- to medium-bedded micritic limestones (Fig. 6a) of Kimmeridgian age with gastropods and other macrofossil fragments. The bulk of the section consists of medium-bedded, grey Tithonian limestone with macrofossil fragments, oncoids and locally corals. Limestones in the top part of the section are massive and rich in oncoids (Fig. 6a); this part of the section is middle Tithonian–early Berriasian in age. The limestones are rich in benthic foraminifera and partly in algae, including *Freixialina planispiralis*, *Charentia* spp., *Pseudocyclammina sphaeroidalis*, *P. lituus*, *Rectocyclammina* sp., *Everticyclammina virguliana*, *E. praekelleri*, *Alveosepta* sp., *Anchispirocyclammina lusitanica*, *Kastamonina abanica*, *Coscinoconus alpinus*, *C. elongatus*, *C. delphinesis*, *C.* spp., *Frentzenella* ? *odukpaniensis*,

(a)

İnaltı Formation
Upper Jurassic (Kimmeridgean) -
Lower Cretaceous (Berriasian)
limestones
Alçılar section

(b)

angular unconformity

Kapanboğazı Formation
Santonian pelagic micritic limestone

Tasmaca Formation
Upper Albian siltstone and shale

Fig. 6. Field photographs illustrating sequences and unconformities in the northern Central Pontides. (**a**) Upper Jurassic–Lower Cretaceous limestones of the Inaltı Formation north of Kastamonu; the Alçılar section (Figs 3a & 7) was measured on this cliff face. (**b**) Upper Albian siltstone and shales (the Tasmaca Formation) overlain with an angular unconformity by the Santonian red pelagic limestones (the Kapanboğazı Formation), Amasra. Note the yellowish basal sandstones below the red limestones. Two shale samples from this locality (AM5 and AM9: see the Supplementary material) have yielded late Albian nannofossils, younger than the Aptian ages reported by Hippolyte *et al.* (2010) from this locality.

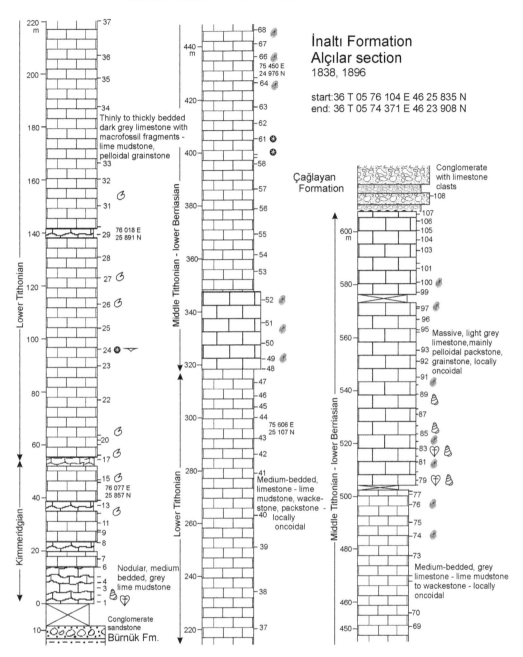

Fig. 7. The Alçılar measured stratigraphic section in the Upper Jurassic–Lower Cretaceous İnaltı Formation. For the location, see Figure 3; for the symbols, see Figure 4.

Mohlerina basiliensis, *Protopeneroplis ultragra-nulata*, '*Quinqueloculina*' *robusta*, *Crescentiella morronensis*, *Campbelliella striata* and *Salpingo-porella annulata*, which allowed the section to be divided into three zones (Figs 8 & 9). The lower zone of Kimmeridgian age is characterized by the coexistence of the benthic foraminifera *Freix-ialina planispiralis* and *Kastamonina abanica*, and the absence of *Campbelliella striata* (samples 1–16). The second zone is of early Tithonian age (samples 17–48), and is distinguished by *Campbelliella striata* and *Kastamonina abanica*.

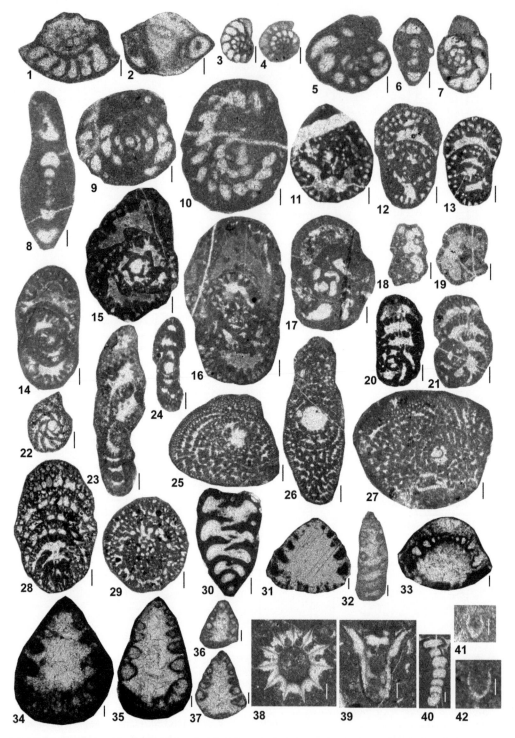

Fig. 8. Foraminifera, calpionellids and dascylad algae from the İnaltı Formation from the Alçılar and Dağlı measured sections. (**1**) & (**2**) *Protopeneroplis ultragranulata* (Gorbachik): (1) sample AA 107, A3 Zone, middle Tithonian–lower Berriasian; and (2) sample AA 96, A3-a Subzone, middle Tithonian–lowermost Berriasian. (**3**) & (**4**) *Freixialina planispiralis* Ramalho: (3) sample AA 9/2, A1 Zone, Kimmeridgian; and (4) sample AA 38, A2 Zone,

The third and highest zone is middle Tithonian–early Berriasian in age (samples 59–107) and is characterized by *Protopeneroplis ultragranulata*. The top of the third zone is calibrated by the B or C zone of calpionellids. A subzone within zone 3 is distinguished by the presence of *Anchispirocyclina lusitanica* and *Protopeneroplis ultragranulata* (middle Tithonian–lowermost Berriasian, samples 49–97). The presence of *Calpionella alpina* in the topmost part of the section indicates deepening of the platform in the Berriasian.

- The Dağlı section – The Dağlı section is located in the western Central Pontides in the Istanbul Zone along the main Cide–Azdavay highway (Fig. 3). The base of the section is a thrust fault: however, in nearby localities, limestones of the İnaltı Formation lie on the Triassic continental red beds (the Çakraz Formation: Fig. 3). The Dağlı section is 1400 m thick, and consists of shallow-marine limestones and dolomites (Fig. 10). The basal part of the section consists of medium-bedded micritic limestones, which are overlain by 500 - m-thick massive dolomites; the dolomites are in turn overlain by thickly bedded to massive limestone, 800 m thick. The limestones are mainly pelloidal grainstones intercalated with subordinate lime mudstone beds. In the Dağlı section, the İnaltı Formation is overlain by the Lower Cretaceous Çağlayan Formation along a fault contact.

Seventy-one samples from the Dağlı section were petrographically and palaeontologically examined. The foraminifera fauna and algal flora include *Belorussiella* sp., *Charentia* spp., *Redmondoides lugeoni*, *Pseudocyclammina lituus*, *Rectocyclammina* sp., *Alveosepta jaccardi*, *Kastamonina abanica*, *Coscinoconus alpinus*, *C. elongatus*, *C. cherchiae*, *C. delphinensis*, *C.* spp., *Mohlerina basieliensis*, *Protopeneroplis ultragranulata*, '*Quinqueloculina*' *robusta*, *Hectina praeantiqua*, *Clypeina sulcata*, *Campbelliella striata* and *Salpingoporella annulata*, and indicate a late Kimmeridgian–early Berriasian age (Fig. 8). Based on the foraminifera, it is possible to differentiate three zones (Fig. 9). The basal zone is of late Kimmeridgian age, and is characterized by *Alveosepta jaccardi* and *Clypeina sulcata* (samples 1–16); the second zone of early Tithonian age is distinguished by the coexistence of *Campbelliella striata* and *Clypeina sulcata*, and the absence of *Protopeneroplis ultragranulata* (samples 17–25). The third zone (middle Tithonian–lower Berriasian) is characterized by *Protopeneroplis ultragranulata* (samples 26–71); it is subdivided into two subzones: subzone D3-a is characterized by *Campbelliella striata* and *Protopeneroplis ultragranulata* (middle–upper Tithonian, samples 26–67), and subzone D3-b by *Clypeina sulcata* and *Protopeneroplis ultragranulata* (lower Berriasian, samples DG 67–71).

Fig. 8. (*Continued*) lower Tithonian. (**5**) *Mayncina* ? sp. Sample DG 38, D3-a Subzone, middle–upper Tithonian. (**6**) & (**7**) *Charentia* sp.: (6) sample AA 4, A1 Zone, Kimmeridgian; and (7) sample AA 8, A1 Zone, Kimmeridgian. (**8**) & (**9**) *Charentia* sp.: (8) sample AA 74, A3-a Subzone, middle Tithonian–lowermost Berriasian; and (9) sample AA 95, A3-a Subzone, middle Tithonian–lowermost Berriasian. (**10**) *Charentia* sp. Sample DG 43, D3-a Subzone, middle–upper Tithonian. (**11**)–(**13**) *Pseudocyclammina sphaeroidalis* Hottinger: (11) sample AA 20/3, A2 Zone, lower Tithonian; (12) sample AA 10, A1 Zone, Kimmeridgian; and (13) AA 37, A2 Zone, lower Tithonian. (**14**)–(**16**) *Pseudocyclammina lituus* (Yokohama): (14) sample AA 73, A3-a Subzone, middle Tithonian–lowermost Berriasian; (15) sample AA 38, A2 Zone, lower Tithonian; and (16) sample AA 42, A2 Zone, lower Tithonian. (**17**) *Everticyclammina virguliana* (Koechlin). Sample AA 93, A3-a Subzone, middle Tithonian–lowermost Berriasian. (**18**) & (**19**) *Everticyclammina praekelleri* Banner and Highton. Sample AA 33, A3-a Subzone, lower Tithonian. (**20**) *Bramkampella* or *Pseudocyclammina* sp. Sample AA 34, A2 Zone, lower Tithonian. (**21**) *Pseudocyclammina* sp. Sample AA 42, A2 Zone, lower Tithonian. (**22**)–(**24**) *Alveosepta jaccardi* (Schrodt): (22) sample DG 2/A, D1 Zone, upper Kimmeridgian; (23) sample DG 2/B, D1 Zone, upper Kimmeridgian; and (24) sample DG 1/B, D1 Zone, upper Kimmeridgian. (**25**)–(**27**) *Anchispirocyclina lusitanica* (Egger): (25) sample AA 90, A3-a Subzone, middle Tithonian–lowermost Berrriasian; (26) sample 96, A3-a Subzone, middle Tithonian–lowermost Berrriasian; and (27) sample 25, A3-a Subzone, middle Tithonian–lowermost Berrriasian. (**28**) & (**29**) *Kastamonina abanica* Sirel. 28: sample AA 33, A2 Zone, lower Tithonian; and (29) sample AA 9/2, A1 Zone, Kimmeridgian. (**30**) *Redmondoides lugeoni* (Septfontaine). Sample DG 64, D3-a Subzone, middle–upper Tithonian. (**31**) *Coscinoconus alpinus* Leupold. Sample AA 38, A2 Zone, lower Tithonian. (**32**) *Coscinoconus elongatus* Leupold. Sample DG 66, D3-a Subzone, middle–upper Tithonian. (**33**) *Frentzenella* ? *odukpaniensis* (Dessauvagie). Sample AA 97, A3-a Subzone, middle Tithonian–lowermost Berriasian. (**34**) *Coscinoconus campanellus* (Arnaud-Vanneau, Boisseau and Darsac). Sample DG 58, D3-a Subzone, middle–upper Tithonian. (**35**) *Coscinoconus cherchiae* (Arnaud-Vanneau, Boisseau and Darsac). Sample DG 71, D3-b Subzone, lower Berriasian. (**36**) *Coscinoconus sagittarius* (Arnaud-Vanneau, Boisseau and Darsac) ? Sample DG 71, D3-b Subzone, lower Berriasian. (**37**) *Coscinoconus delphinensis* (Arnaud-Vanneau, Boisseau and Darsac). Sample DG 54, D3-a Subzone, middle–upper Tithonian. (**38**) *Clypeina sulcata* (Alth). Sample DG 26, D3-a Subzone, middle–upper Tithonian. (**39**) *Campbelliella striata* (Carozzi). Sample DG 27, D3-a Subzone, middle–upper Tithonian. (**40**) *Troglotella incrustans* Wernli and Fookes. Sample AA 77, A3-a Subzone, middle Tithonian–lowermost Berriasian. (**41**) *Calpionella alpina* Lorenz. Sample AA 105, uppermost part of the A3 Zone, lower Berriasian. (**42**) *Tintinopsella* ? sp. Sample 108, reworked calpionellid in a pebble derived from the İnaltı Formation, probably Berriasian. The scale bar is 100 μm for (1)–(10), (31)–(37) and (40)–(42); and 250 μm for (11)–(30), (38) and (39).

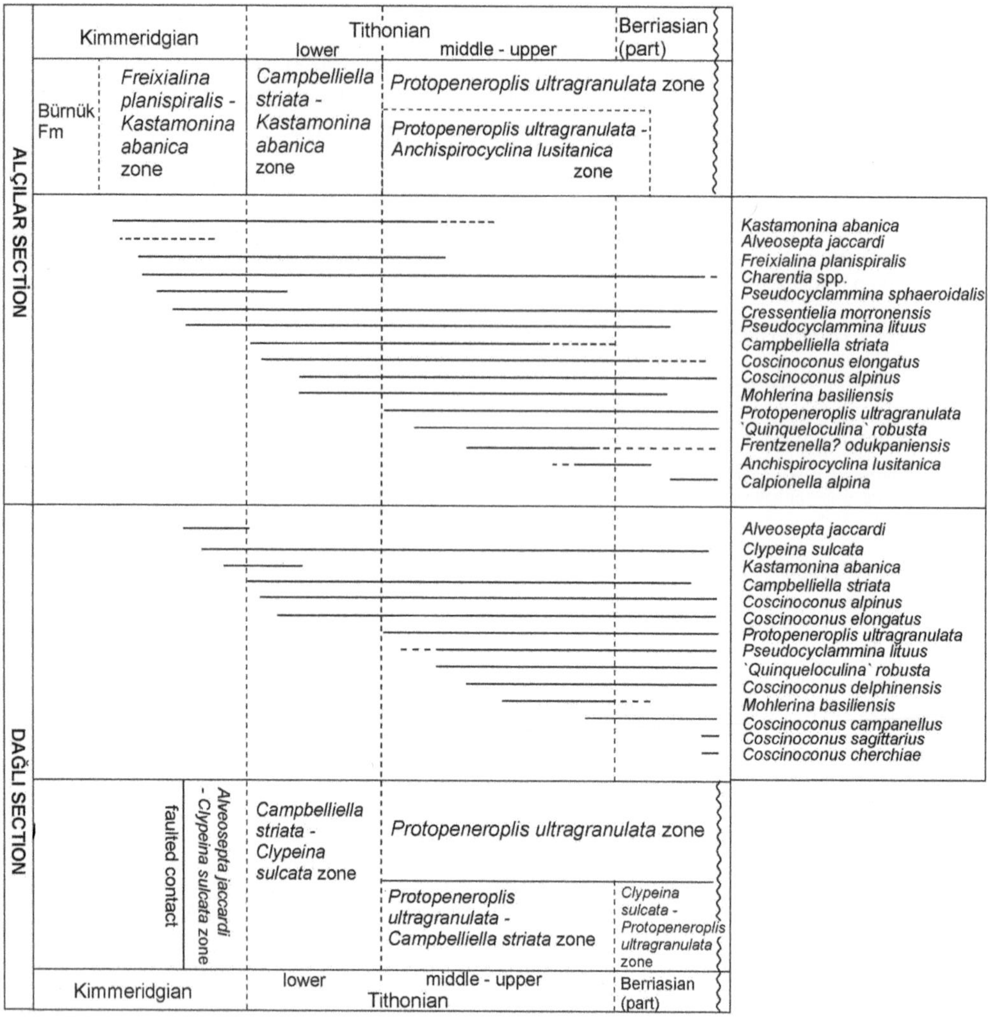

Fig. 9. Stratigraphic distribution of some diagnostic foraminifera in the Alçılar and Dağlı sections.

In the other measured sections (Yaralıgöz and Seydiler) the age of the İnaltı Formation is also Kimmeridgian–Berriasian, which is compatible with the Middle Jurassic age of the underlying magmatic rocks. The Upper Jurassic–Lower Cretaceous carbonates on both sides of the Black Sea, in the Central Pontides and in Crimea, are shallow marine and have a similar Kimmeridgian–Berriasian age range (Krajewski & Olszewska 2007). There is no evidence for the existence of a deep-marine Late Jurassic–Early Cretaceous sea between the Central Pontides and Crimea in the present position of the Black Sea, as suggested by some models (e.g. Barrier & Vrielynck 2008; Nikishin *et al.* 2012). In the Pontides, the Upper Jurassic–Lower Cretaceous shallow-marine carbonates must have passed

southwards to a continental margin. Such a margin is missing in the Central Pontides but exists in the north-vergent thrust sheets in the Eastern Pontides (Okay & Şahintürk 1997).

Early Cretaceous evolution of the Central Pontides: development of a submarine fan and subduction–accretion complexes

After the Jurassic, the Central Pontides became a single tectonic unit, and the differentiation of the Istanbul and Sakarya zones ceased to exist. The deposition of the Upper Jurassic–lowermost Cretaceous shallow-marine carbonates (the İnaltı Formation) was followed by uplift and erosion. Valanginian

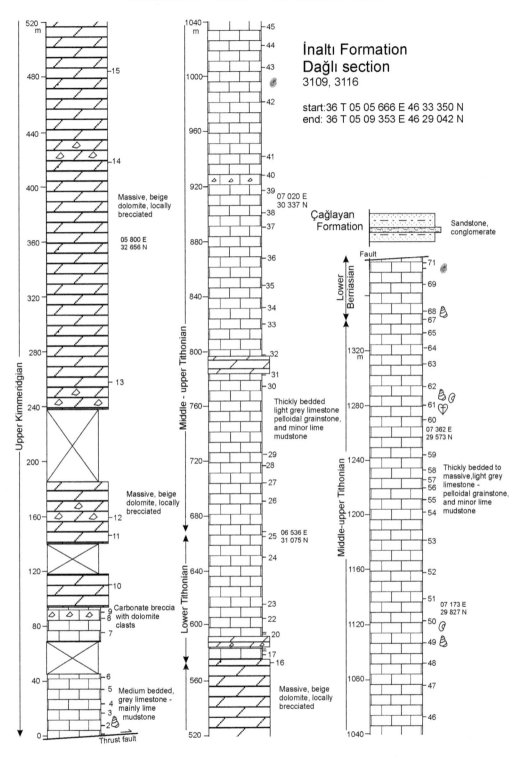

Fig. 10. The Dağlı measured stratigraphic section in the Upper Jurassic–Lower Cretaceous İnaltı Formation. For the location, see Figure 3; for the symbols, see Figure 4.

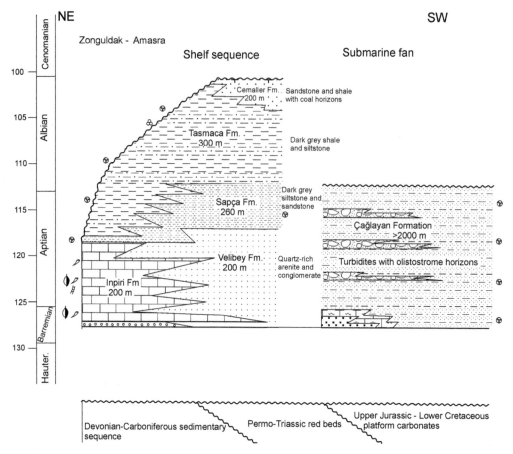

Fig. 11. Generalized stratigraphic section of the Lower Cretaceous shelf and slope sequences of the Central Pontides. For the symbols, see Figure 4.

and Hauterivian rocks have not been recognized in the Pontides. The uplift and erosion was followed by a major phase of clastic deposition during the Early Cretaceous (Barremian–Aptian), when there was a differentiation into a clastic–carbonate shelf in the north along the present Black Sea coast and a major turbidite fan in the south extending to the Tethyan Ocean (Fig. 5e) (Okay *et al.* 2013).

Early Cretaceous shelf

The Lower Cretaceous shelf sequence crops out along the Black Sea coast between Zonguldak and Amasra (Fig. 3). It lies unconformably above the Upper Jurassic–Lower Cretaceous shallow-marine carbonates or above the Palaeozoic and Permo-Triassic sedimentary series of the Istanbul Zone. The sequence starts with quartz-arenites and conglomerates, which interfinger with upper Barremian–Aptian shallow-marine carbonates (Fig. 11) (Yılmaz & Altıner 2007; Masse *et al.* 2009). Detrital zircons

from the quartz-arenites indicate a source in Permo-Carboniferous and late Neoproterozoic granitoids (Akdoğan *et al.* 2017). The quartz-arenites and shallow-marine carbonates pass up into glauconite-bearing dark siltstone and sandstone of Aptian age, which are in turn overlain by dark shale and siltstones of the Tasmaca Formation (Fig. 6b). Several shale samples from this Tasmaca Formation collected from Zonguldak and Amasra regions have yielded late Albian nannofossils and foraminifera (for more detail, see the Supplementary material). The Early Cretaceous transgression is followed by a regression during the topmost Albian–Cenomanian, when shallow-marine to continental sandstone and shale with coal horizons were laid down (Fig. 11).

Early Cretaceous submarine turbidite fan: the Çağlayan Formation

The Lower Cretaceous shelf sequence of the Central Pontides passed southwards to a Barremian–Aptian submarine turbidite fan succession; the fan extended from the Laurasian margin southwards to the Tethyan Ocean (Fig. 5e). The turbidite fan sequence, the Çağlayan Formation, has a thickness of more than 2 km and its outcrops stretch out over an area of 400 × 60 km (Fig. 3) (Gedik & Korkmaz 1984; Tüysüz 1999; Hippolyte et al. 2010; Okay et al. 2013; Akdoğan et al. 2017). The western part of the turbidite fan is ascribed to the Ulus Formation (e.g. Tüysüz 1999): however, it has the same age range and facies as the Çağlayan Formation and there is no palaeo-high separating the two; here they are considered as part of the same basin.

The Çağlayan Formation consists predominantly of sandstone and shale but also includes significant amounts of debris flows and olistostromes. The blocks in these mass flows are mainly Upper Jurassic–Lower Cretaceous limestones but there are also large clasts of Permo-Triassic red sandstone, Carboniferous sandstone, Middle Jurassic dacite-porphyry and metamorphic rock (Okay et al. 2013). The age of the Çağlayan Formation, based on nanno-fossils, is Barremian–late Aptian (Hippolyte et al. 2010). Two shale samples collected during this study from north of Boyabat (Fig. 3) yielded upper-most Barremian and middle Aptian nannofossils and pollen (see the Supplementary material), con-firming these ages.

The large metamorphic area south of the Lower Cretaceous turbidites in the Central Pontides, the Central Pontide Supercomplex, was regarded as a pre-Jurassic basement, and hence a potential source area. However, recent isotopic dating has shown that the Central Pontide Supercomplex consists predominantly of Early Cretaceous subduction–accretion complexes (Okay et al. 2006, 2013; Aygül et al. 2016). Palaeocurrent measurements in the turbiditic sandstones of the Çağlayan Formation indicate that the material was coming predominantly from the north to NW (Akdoğan et al. 2017). Detrital zircons from the turbiditic sandstones also indicate a major source area from the Palaeoproterozoic–Archaean granitoids of the Ukrainian Shield north of the Black Sea (Okay et al. 2013; Akdoğan et al. 2017). This shows that the Black Sea did not exist, or at least did not form a barrier between the Pontides and the Ukrainian Shield during the Early Cretaceous. During this time, a large river was flowing from the Ukrainian Shield south into the Tethys Ocean (Fig. 5e). Most of the sediment brought down by the river was trapped on the conti-nental slope but some reached the trench and was subducted, resulting in an Albian accretionary com-plex made up of metamorphosed distal turbidites (see below).

The uplift preceding the deposition of the Çağl-ayan Formation is commonly ascribed to rifting lead-ing to the opening of the Black Sea (e.g. Görür

1997). However, a period of contractional deforma-tion and metamorphism followed the deposition of the Çağlayan Formation during the Albian (see below); thus, the extension related to the deposition of the Çağlayan Formation and the later opening of the Black Sea as a back arc-basin are two separate events (Okay et al. 2013).

Albian: subduction and HP–LT metamorphism

A major phase of subduction and accretion took place during the Albian in the southern part of the Central Pontides (Fig. 5e); Albian subduction–accretion complexes make up the bulk of the Central Pontide Supercomplex. Most of the accreted rocks consist of oceanic crustal sequences metamorphosed in eclogite, blueschist and greenschist facies (Okay et al. 2006, 2013; Aygül et al. 2015a, b, 2016). Metabasites are the dominant lithology, followed by micaschist and meta-ultramafic rocks with minor amounts of metachert and metagabbro. The distal parts of the mid-Cretaceous turbidite fan, the Çağl-ayan Formation, were also entrained in the trench, and were metamorphosed in blueschist and greens-chist facies (Okay et al. 2013; Aygül et al. 2015b, 2016). They consist of phyllite, slate, marble, meta-sandstone and serpentinite lenses; detrital zircons from the meta-sandstones are as young as Middle Jurassic (171 Ma: Okay et al. 2013). The Ar/Ar phengite ages from the metamorphic rocks are mainly Albian (113–102 Ma: Okay et al. 2006, 2013; Aygül et al. 2016). The metamorphic rocks are unconformably overlain by lower Turonian pelagic limestones (Fig. 12) (Okay et al. 2006).

Albian metamorphism is not observed north of the Central Pontide Supercomplex: however, the intense deformation of the Çağlayan Formation most probably also took place in the Albian. Albian depositional ages are only recorded close to the Black Sea coast in the Lower Cretaceous shelf sequences (Figs 6b & 12); further south, Albian was a time of uplift and erosion.

Although there is clear indication of Albian sub-duction in the Central Pontides, in terms of oceanic eclogites and blueschists, there is no evidence for an Albian magmatic arc. The Çağlayan Formation, deposited in a basin above the subduction zone, does not contain any Early Cretaceous zircons (Okay et al. 2013; Akdoğan et al. 2017). The absence of an Albian magmatic arc may be related to flat subduc-tion, which also induced uplift and exhumation of the East European Platform north of the present Black Sea, thus supplying detritus to the Çağlayan Basin (Akdoğan et al. 2017). Most of the East European Platform was an erosional area in the Hauterivian–Albian interval (Baraboshkin et al. 2003; Nikishin et al. 2012).

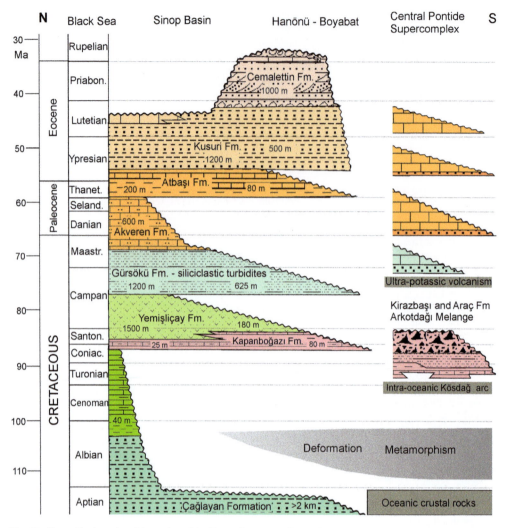

Fig. 12. Generalized stratigraphic section of the Upper Cretaceous–Cenozoic sequences of the Central Pontides. For the symbols, see Figure 4.

Late Cretaceous – a new depositional cycle: sedimentation, arc volcanism and the opening of the Black Sea

In the Central Pontides, a new sedimentary–volcanic cycle started in the Late Cretaceous, which was characterized by thicker and more continuous sequences in the north, reflecting the presence of the Western Black Sea Basin (Figs 5e & 12) (Hippolyte *et al.* 2016). In the early part of the Late Cretaceous, in the Turonian–early Campanian, carbonate deposition and arc volcanism were dominant, which switched to turbidite sedimentation from the late Campanian onwards, reflecting the uplift of the Central Pontide Supercomplex. In the Central Pontides, magmatism

started in the Santonian; the volcanic rocks form part of the magmatic arc that can be followed for more than 2000 km from the Lesser Caucasus in Georgia to the Balkans in Romania (e.g. Okay & Şahintürk 1997; Adamia *et al.* 2011; Gallhofer *et al.* 2015).

In the northern parts of the Central Pontides, Upper Cretaceous–Eocene sequences are more continuous and show increasingly deeper marine conditions towards the Black Sea, attesting to its influence (Fig. 12). The coeval sequences in the south are shallower marine and are marked by unconformities. A marine volcanosedimentary sequence containing ultrapotassic lavas and dykes was deposited in the Late Cretaceous on the accretionary complex in the south (Gülmez *et al.* 2016).

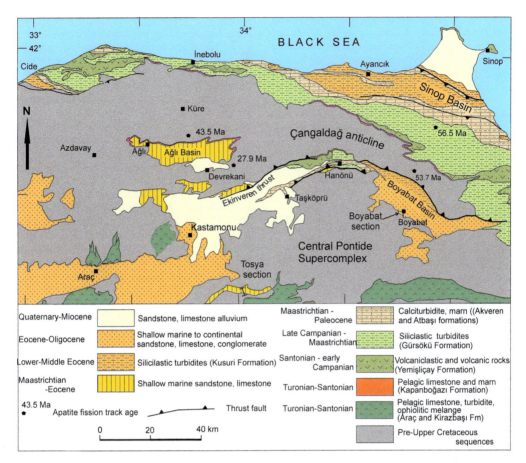

Fig. 13. Geological map of the Central Pontides showing the distribution of the Upper Cretaceous and younger sequences (based on Uğuz *et al.* 2002; Okay *et al.* 2013). The apatite fission-track ages are from Cavazza *et al.* (2012) and Espurt *et al.* (2014).

The post-Eocene uplift and erosion in the Central Pontides have isolated the Upper Cretaceous–Eocene sequences into several domains (Fig. 13). The Sinop Basin along the Black Sea coast shows the most complete record, other domains in the south (the Hanönü, Ağlı and Boyabat regions) provide examples of Upper Cretaceous–Eocene sedimentation along the southern margin of the continental Central Pontides.

Turonian–Campanian: carbonate deposition and arc volcanism

The Turonian–Santonian sequences along the Black Sea coastal area consist mainly of pelagic limestone, marl and shale, which are overlain or intercalated with volcaniclastic rocks. Locally, the transition from Lower to Upper Cretaceous sequences is conformable (Tüysüz *et al.* 2016), whereas in most

areas, such as on the Boyabat–Sinop road, Amasra or Hanönü regions (Fig. 13), Cenomanian–Turonian sequences are missing and Coniacian–Santonian red pelagic limestones lie unconformably above the deformed Çağlayan Formation (Figs 6b & 12) (Okay *et al.* 2006, 2013; Hippolyte *et al.* 2010). The pink Coniacian–Santonian pelagic limestones, typically a few tens of metres thick, are overlain and interfinger with volcaniclastic and volcanic rocks. Within the Cenomanian–Santonian sequence, Tüysüz *et al.* (2012, 2016) differentiated two distinct pelagic limestone horizons, of Cenomanian–Turonian and Santonian ages, respectively, separated by a Turonian–Santonian volcanic interval: however, interfingering of deep-marine carbonates, volcaniclastic and volcanic rocks is more likely.

The earliest precisely dated Upper Cretaceous volcanic rocks in the Central Pontides are Santonian in age; Middle Turonian basalts intercalated with deep-marine limestones are described from the Eastern Pontides (Taner & Zaninetti 1978). The bulk of the volcanism in the Pontides extends from the Santonian to the early Campanian (Hippolyte *et al.* 2016). In the Central Pontides, the arc volcanism consists predominantly of submarine pyroclastic rocks of andesitic and basaltic origin; lava flows are rare. The thickness of the volcanic sequence ranges from less than 100 m in the south to more than 1000 m along the Black Sea coast. Seismic reflection profiles in the Black Sea show that the volcanic centres are located offshore (Nikishin *et al.* 2015a), which suggests the opening of the Western Black Sea Basin through the splitting of the magmatic arc.

Turonian–Santonian debris flows and mélanges in the Central Pontide Supercomplex

During the early part of the Late Cretaceous, the Central Pontide Supercomplex formed a wide submarine accretionary complex on the northern margin of the Tethys Ocean. The accretionary complex consisted of oceanic crustal rocks and of intra-oceanic arcs, such as the Cenomanian–Turonian Kösdağ arc (Fig. 12) (Aygül *et al.* 2015a). Olistostrome-bearing, deep-marine Turonian–Santonian sequences were deposited on this accretionary complex. These sequences, known as the Kirazbaşı or Araç formations or Arkotdağı Mélange (Okay *et al.* 2006, 2013; Göncüoğlu *et al.* 2014), occur as tectonized slices between metamorphic thrust sheets (Fig. 13). They are often described as ophiolitic mélange (Göncüoğlu *et al.* 2014): however, at several localities they lie stratigraphically above the Albian metabasites and phyllites (Yiğitbaş *et al.* 1990; Okay *et al.* 2006). The sequence starts with Turonian pelagic limestones, less than 10 m thick, which are overlain by greywacke and shale followed by debris flows with blocks of radiolarian chert and Albian pelagic limestone (Fig. 12). The debris flows are overlain by ophiolitic mélange of basalt, radiolarian chert, serpentinite and limestone. The chert blocks yielded Middle Jurassic–Early Cretaceous radiolaria ages (Tüysüz & Tekin 2007; Göncüoğlu *et al.* 2012).

Maastrichtian–Eocene: turbidite deposition

Thick sequences of turbidites were deposited in the northern parts of the Central Pontides during the Maastrichtian–Eocene interval, probably sourced from the Central Pontide Supercomplex. These are best preserved in the Sinop Basin (Gedik & Korkmaz 1984; Hippolyte *et al.* 2010), which was separated from the West Black Sea Basin by the Sinop horst, where Santonian volcanic rocks crop out (Fig. 13)

(Leren *et al.* 2007; Espurt *et al.* 2014). The Sinop horst formed during the Late Cretaceous opening of the Black Sea by footwall uplift, and generated the Sinop Basin on its lee side. More discontinuous and shallower marine sequences were formed in the southern parts of the Central Pontides and on parts of the Central Pontide Supercomplex, which are best preserved in the Hanönü region.

The Sinop Basin. In the Sinop Basin, the Coniacian–middle Campanian volcaniclastic rocks are succeeded by middle Campanian–lower (?) Maastrichtian siliciclastic turbidites, the Gürsökü Formation, 700–1200 m in thickness (Gedik & Korkmaz 1984; Hippolyte *et al.* 2010). The bulk of the Gürsökü Formation in the Sinop Basin consists of sheet-like, Bouma-type turbidites (Leren *et al.* 2007). The sandstones are litharenites with volcanic and minor calcareous clasts; the amount of limestone clasts increases upwards in the section. The palaeocurrents indicate predominantly eastwards axial flow (Leren *et al.* 2007). A mudstone sample from the Gürsökü Formation collected during this study from the north of Boyabat contains abundant nannofossils of early–middle Campanian age (Zone UC15, for more details, see the Supplementary material).

The Gürsökü Formation passes upwards with a gradual increase in the carbonate clasts into the calcareous turbidites of the Maastrichtian–Paleocene Akveren Formation (Fig. 12). The Akveren Formation consists of calcarenite beds alternating with marl and calcareous mudstone (Leren *et al.* 2007). A mudstone sample from the Akveren Formation north of Boyabat contains Danian (NP2) nannofossils. The calciturbidites of the Akveren Formation also show eastwards palaeocurrents, and are overlain by Upper Paleocene (Thanetian) reefal limestones. A renewed transgression in the Early Eocene leads to the deposition of pink calcareous mudstones and subordinate thin calc-arenite beds of the Atbaşı Formation.

The pink mudstones of the Atbaşı Formation gradually pass upwards into Eocene siliciclastic turbidites and associated mudstones of the Kusuri Formation (Gedik & Korkmaz 1984; Görür & Tüysüz 1997). The sequence, which is 1200 m thick, is Early and early Middle Eocene (Ypresian–Lutetian) in age (Janbu *et al.* 2007). The age is confirmed from a shale sample close to Ayancık, which contains Middle Eocene (NP15) nannofossils. The Kusuri Formation starts with mudstones intercalated with thin carbonate-rich turbidite sandstone beds. These are overlain by sandstone-rich turbidites showing west- to NW-directed palaeocurrents (Janbu *et al.* 2007). The sequence becomes calcareous towards the top, and ends with shallow-marine Lutetian limestones, which crop out on the Black Sea coastal area. The Eocene turbidite sequence shows evidence for syndepositional contractional deformation (for more details, see the Supplementary material).

Hanönü and Boyabat regions. The Hanönü region on the southern limb of the Çangaldağ anticline preserves a record of Late Cretaceous–Eocene sedimentation on the southern margin of the continental Central Pontides and on the Central Pontide Supercomplex (Figs 13 & 14) (Okay *et al.* 2006). The region lies within the Eocene fold and thrust belt, and shows a comparable Upper Cretaceous–Eocene stratigraphy with the Sinop Basin but marked by several unconformities (Fig. 12).

The Çağlayan Formation with very large blocks of metamorphic rocks is unconformably overlain by red pelagic limestone, shale and debris flows (the Kapanboğazı Formation). Two measured stratigraphic sections in the Kapanboğazı Formation in the Hanönü area (Vakıf and Kaşharman) indicate a Santonian age; Cenomanian–Coniacian stages are missing (Fig. 15). The Kapanboğazı Formation is overlain by volcaniclastic and volcanic rocks (the Yemişliçay Formation) with lenses of Santonian pelagic limestone. Lying unconformably above the volcaniclastic sequence are Maastrichtian proximal turbidites of the Gürsökü Formation. The base of the Gürsökü Formation steps down to the Çağlayan Formation, indicating a period of uplift and erosion in the Campanian (Fig. 14). The turbidites consist of thick-bedded coarse sandstone and conglomerate with minor shale, and contain transported large benthic foraminifera. A sample from the basal parts of the Gürsökü Formation contains benthic foraminifera such as *Siderolites* cf. *calcitrapoides*, *Cideina* cf. *soezerii* and *Lepiorbitoides* sp., indicating a Maastrichtian age (Fig. 15, Ballıkaya section) (see also Özkan-Altıner & Özcan 1999).

The Gürsökü Formation is unconformably overlain by the Upper Paleocene–Lower Eocene white marl and pelagic limestone with calciturbidite beds, the equivalent of the Akveren Formation of the Sinop Basin (Fig. 12). Measured stratigraphic sections of the Akveren Formation in the Hanönü area indicate a Thanetian–Early Eocene age (Figs 14 & 15). The base of the Akveren Formation in the Hanönü area is a major angular unconformity and steps down to the Santonian volcaniclastic rocks, indicating tilting, uplift and erosion in the Early Paleocene (Fig. 16a).

Lower–Middle Eocene (Lutetian) turbidites (the Kusuri Formation) crop out in the Boyabat Basin east of the Hanönü region (Fig. 13). The turbidites pass up into a thick fluviatile to deltaic sequence of sandstone, conglomerate and mudstone of Oligocene age (the Cemalettin Formation). Within this continental sequence, Sanders *et al.* (2014) described a shallow-marine mudstone horizon of latest Eocene–earliest Oligocene age, which if confirmed would constitute the youngest marine bed in the southern Central Pontides.

The Upper Cretaceous–Eocene sequence in the Hanönü region is bounded in the south by the Gökırmak Fault, a major Late Cretaceous–Paleocene normal fault reactivated as an Eocene thrust (Figs 13 & 14). South of the Gökırmak Fault, the Çangaldağ Complex, consisting of Middle Jurassic metabasite, meta-andesites, phyllite and marble, is unconformably overlain by shallow-marine Maastrichtian, Paleocene and Eocene sandstone–limestone sequences (Fig. 12). The Maastrichtian–Eocene sequence is not continuous but different parts lie isolated on the metamorphic basement (Figs 13 & 17), indicating periodic uplift and erosion, as also deduced by Aydemir & Demirer (2013). Özcan *et al.* (2007) described two Eocene shallow-marine limestone sequences, of Ypresian and Lutetian ages, respectively, lying unconformably above the metamorphic rocks in the Kastamonu region. Similar shallow-marine Paleocene sandstone–limestone sequences, a few hundred metres thick, crop out south of Boyabat and south of Kastamonu lying unconformably on the metamorphic rocks of the Central Pontide Supercomplex (Figs 16b & 17; Özgen-Erdem *et al.* 2005). Such Maastrichtian and Paleocene carbonate built-ups must have been the source of the calciturbidites of the Akveren Formation.

Collision, uplift and erosion

Marine sedimentation in the Pontides largely ended at the end of the Middle Eocene (Lutetian) as a result of collision of the Central Pontides with the Kırşehir Block. Continent–continent collision is a prolonged process, as exemplified by the India–Asia collision, which started in the Eocene and is still continuing (e.g. Najman *et al.* 2010). It starts when the distal parts of the passive continental margin enter the subduction zone, which leads to the uplift of the active margin. This process leads to the development of retro forearc basins filled with turbidites. A critical stage in the collision is reached when the whole of orogen is lifted above sea level, this corresponds to the end of the Lutetian (Middle Eocene) in the Central Pontides.

The first signs of impending collision in the Central Pontides are the deposition of the Upper Campanian–Maastrichtian turbidites in retro fore-arc basins. The turbidites were probably sourced from the Central Pontide Supercomplex, which was undergoing uplift due to underthrusting by the Kırşehir Block. Palaeomagnetic studies have also shown that the northwards concave shape of the Central Pontides started to form in the Maastrichtian as a result of collision with the Kırşehir Block (Channel *et al.* 1996; Meijers *et al.* 2010a). The

Fig. 14. *Continued.*

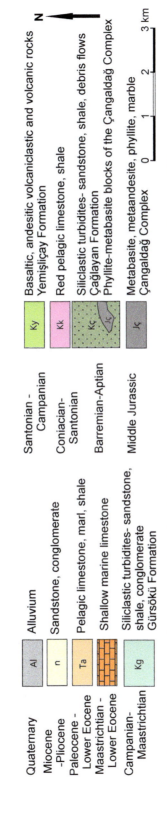

Fig. 14. Geological map of the Hanönü region in the southern Central Pontides. For the location, see Figure 13. Modified from Okay *et al.* (2006).

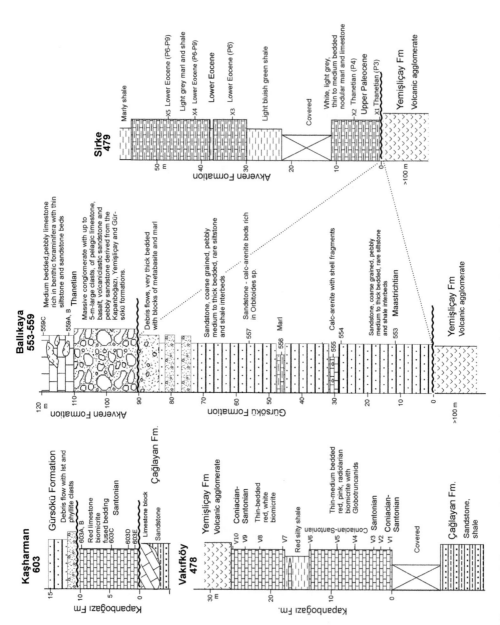

Fig. 15. Measured Cretaceous–Paleocene stratigraphic sections in the Hanönü area of the southern Central Pontides. For the location, see Figure 13; for a list of fossils, see the Supplementary material.

Fig. 16. Field photographs illustrating sequences and unconformities in the southern Central Pontides. (**a**) Upper Paleocene–Lower Eocene marl and shale lying with an angular unconformity over the tilted Santonian volcaniclastic rocks of the Yemişliçay Formation, Sirke, Hanönü. This section is logged in Figure 15. (**b**) Lower–Middle Paleocene shallow-marine sandstones and limestones lying unconformably over the metabasites of the Domuzdağ Complex, Pervanekaya Hill, Boyabat. This section is logged in Figure 17.

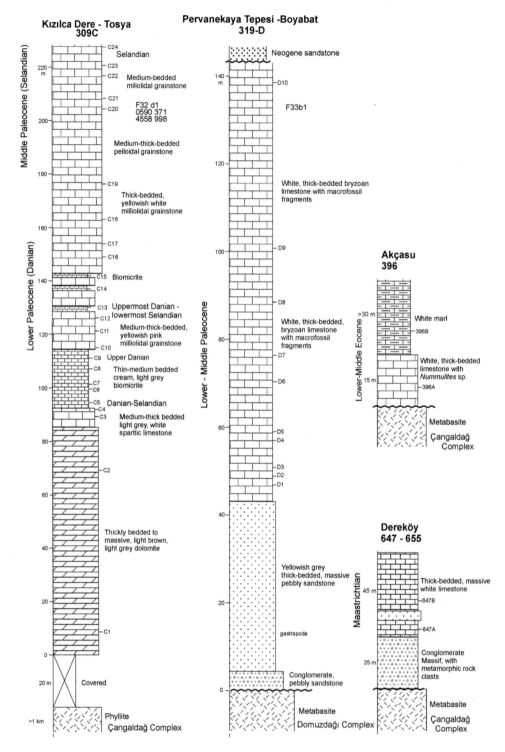

Fig. 17. Measured stratigraphic sections of the Upper Cretaceous–Eocene sequences lying on the metamorphic rocks of the Central Pontide Supercomplex. For the locations of the Tosya and Boyabat sections, see Figure 3; for the others, see Figure 13. For a list of fossils, see the Supplementary material.

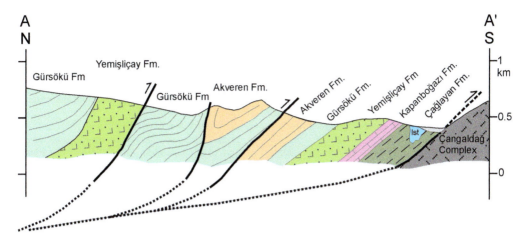

Fig. 18. Geological cross-section from the Hanönü region in the southern Central Pontides showing the south-vergent Eocene fold and thrust belt. For the location of the section, see Figure 13.

convergence rate between the Pontides and the Anatolide–Tauride and Kırşehir blocks probably decreased in the Paleocene, when there was deposition of carbonates and shales. An acceleration of convergence rate during the Early–Middle Eocene led to the deposition of widespread siliciclastic turbidites due to renewed uplift in the source region.

Thrust faulting in the Central Pontides started in the Late Eocene and continued episodically since then, propagating northwards to the Black Sea and southwards to the Çankırı Basin (Kaymakçı et al. 2009; Espurt et al. 2014; Hippolyte et al. 2016). The M 6.6 Bartın earthquake on 3 September 1968, with the epicentre located offshore in the Black Sea 10 km north of Amasra, occurred on a north-vergent thrust fault (Alptekin et al. 1986; Yıldırım et al. 2011), indicating active ongoing shortening. The shortening direction (σ_1) in the Central Pontides shows a change from NW–SE to NNE–SSW, going from clockwise apparently with little variation since the Late Eocene (Sunal & Tüysüz 2002; Hippolyte et al. 2016).

In the southern Central Pontides, the thrusting was south-vergent. A major structure was the Ekinveren Thrust, which can be followed for more than 100 km along strike (Fig. 13) (Yıldırım et al. 2011), and overthrust sequences as young as Miocene in the Kastamonu Basin. Another major structure is the Gökırmak Fault, located along the northern boundary of the Central Pontide Supercomplex (Figs 12 & 13). It was a major normal fault, controlling the sedimentation during the Cretaceous and Paleocene, which was reactivated as a thrust fault in the Eocene. In the Hanönü region, the Gökırmak Fault consists of several splays, which form an imbricate thrust stack (Fig. 18). The Oligocene fluvio-deltaic clastic rocks of the Boyabat Basin

are folded and thrust-faulted, indicating continuing shortening in the Oligocene (e.g. Espurt et al. 2014). In the northern Central Pontides, the Eocene and younger thrusting was north-vergent and led to the uplift of the Sinop Basin (Fig. 13) (Sunal & Tüysüz 2002). The present structure of the Central Pontides is a doubly-vergent orogen, which formed from the Eocene onwards (Aydın et al. 1995; Şengör 1995; Espurt et al. 2014). Based on seismic reflection sections and surface geology, Espurt et al. (2014) estimated the north–south crustal shortening in the Central Pontides as 33%. The few apatite fission-track ages from the Central Pontides are Late Paleocene–Oligocene, with a general younging towards the north (Fig. 13) (Cavazza et al. 2012; Espurt et al. 2014).

Conclusions

• Before the Late Cretaceous opening of the Black Sea, the Central Pontides constituted part of the southern margin of Laurasia. During most of the Mesozoic, the margin was active due to Tethyan Ocean subducting northwards below the Laurasia. In the Central Pontides, there were pronounced phases of oceanic subduction–accretion during the Late Triassic, Middle–early Late Jurassic, Early Cretaceous (Albian) and Late Cretaceous (Santonian–Campanian). The evidence includes subduction–accretion complexes, magmatic arcs and forearc sequences of these ages. The Late Jurassic–earliest Cretaceous (Tithonian–Berriasian) was a tectonically quiescent period.

• The Central Pontides comprises two tectonic terranes: the Istanbul Zone in the west and the Sakarya Zone in the east. They were amalgamated before the Late Jurassic (Kimmeridgian), most probably during the Triassic. The Istanbul Zone has a thick well-developed Palaeozoic sedimentary basement, whereas the basement of the Sakarya Zone includes Permo-Carboniferous granites and Triassic and older subduction–accretion complexes.

- Two features that distinguish the Central Pontides from the western and eastern Pontides are: the presence of a vast region of Jurassic–Cretaceous subduction–accretion complexes in the south; and a large mid-Cretaceous submarine turbidite fan, the Çağlayan Formation, in the north. These unique features are probably related to its location at a cusp between the subduction segments (Okay *et al.* 2013; Nikishin *et al.* 2015*b*)
- The first stratigraphic unit that clearly extends both over the İstanbul and Sakarya zones in the Central Pontides is the İnaltı Formation, comprising Upper Jurassic–lowermost Cretaceous shallow-marine limestones,. Two new detailed measured stratigraphic sections (Figs 7 & 10) indicate a Kimmeridgian–Berriasian age range for the Inaltı Formation showing that the Istanbul and Sakarya zones were amalgamated before the Kimmeridgian.
- Carbonate deposition during the earliest Cretaceous (Berriasian) was followed by widespread uplift and erosion during the Valanginian and Hauterivian.
- In the Central Pontides, clastic deposition took place during the Lower Cretaceous (Barremian–Aptian), when the region was differentiated into a clastic–carbonate shelf in the north and a large submarine turbidite fan in the south (the Çağlayan Formation). Turbidites of the Çağlayan Formation crop out over an area of 400 × 60 km. Detrital zircons from the sandstone turbidites indicate a dominant source area in the north in the East European Platform, which implies that the Black Sea was not yet open in the Aptian.
- Large segments of subducted oceanic crust and mantle were accreted to the southern margin of the Central Pontides in the Albian, represented by eclogites and blueschist with oceanic protoliths with metamorphic ages of 114–105 Ma. Distal parts of the Çağlayan turbidite fan were entrapped in the Albian subduction zone and were metamorphosed. During the Albian subduction–accretion event, the Central Pontides was uplifted; the southern parts close to the subduction zone were deformed.
- A new sedimentary cycle, controlled by the opening of the Western Black Sea Basin, started in the Late Cretaceous with the deposition of Turonian–Santonian pelagic limestones, which interfinger and are overlain by arc volcanic rocks. In the middle–late Campanian, volcanism was replaced by the deposition of siliciclastic turbidites in response to uplift of the Central Pontide Supercomplex. In the northern parts of the Central Pontides, sedimentation was continuous from the Campanian to the Middle Eocene, whereas it was marked by several angular unconformities in the south reflecting the growth of the orogen.
- In the Central Pontides, the marine sedimentation ended in the Middle Eocene (Lutetian) as a result of the collision with the Kırşehir Massif and/or the Anatolide–Tauride Block. The Eocene and younger contraction created a doubly vergent orogen.

The work in the Central Pontides has been supported over the last 15 years by the TÜBİTAK projects 101Y032 and 109Y049 and partly by TÜBA. We thank E. Sirel and Mike Bidgood for additional foraminifera determinations, B. Riding for the characterization of the palynomorphs, and Paul Bown and Steve Starkie for nannofossil determinations. Michael Wagreich and Anatoly Nikishin provided constructive reviews.

References

ADAMIA, S., ZAKARIADZE, G., CHKHOTUA, T., SADRADZE, N., TSERETELI, N., CHABUKIANI, A. & GVENTSADZE, A. 2011. Geology of the caucasus: a review. *Turkish Journal of Earth Sciences*, **20**, 489–544.

AKDOĞAN, R., OKAY, A.I., SUNAL, G., TARI, G., MEINHOLD, G. & KYLANDER-CLARK, A.R.C. 2017. Provenance of a large Lower Cretaceous turbidite submarine fan complex on the active Laurasian margin: Central Pontides, northern Turkey. *Journal of Asian Earth Sciences*, **134**, 309–329.

AKYOL, Z., ARPAT, E. *ET AL.* 1974. *Geological Map of the Cide-Kurucaşile Region, scale 1: 50 000.* Maden Tetkik ve Arama Enstitüsü, Ankara.

ALIŞAN, C. & DERMAN, A.S. 1995. The first palynological age, sedimentological and stratigraphic data for Çakraz Group (Triassic), Western Black Sea. *In:* ERLER, A., ERCAN, T., BINGÖL, E. & ÖRÇEN, S. (eds) *Geology of the Black Sea Region.* Maden Tetkik ve Arama Enstitüsü, Ankara, 93–98.

ALPTEKIN, Ö., NÁBĚLEK, J.L. & TOKSÖZ, M.N. 1986. Source mechanism of the Bartin earthquake of September 3, 1968 in northwestern Turkey: evidence for active thrust faulting at the southern Black Sea margin. *Tectonophysics*, **122**, 73–88.

ALTINER, D., KOÇYIĞIT, A., FARINACCI, A., NICOSIA, U. & CONTI, M.A. 1991. Jurassic-Lower Cretaceous stratigraphy and paleogeographic evolution of the southern part of north-western Anatolia. *Geologica Romana*, **28**, 13–80.

AYDEMIR,V. & DEMIRER, A. 2013. Upper Cretaceous and Paleocene shallow water carbonates along the Pontide belt. *In: Proceedings of the 19th International Petroleum and Natural Gas Congress of Turkey, Ankara,* Turkish Association of Petroleum Geologists, Ankara, 284–290.

AYDIN, M., DEMIR, O., ÖZÇELIK, Y., TERZIOĞLU, N. & SATIR, M. 1995. A geological revision of İnebolu, Devrekani,

Ağlı and Küre areas: new observations in PaleoTethys – NeoTethys sedimentary successions. *In*: ERLER, A., ERCAN, T., BINGÖL, E. & ÖRÇEN, S. (eds) *Geology of the Black Sea Region*. Maden Tetkik ve Arama Enstitüsü, Ankara, 33–38.

AYGÜL, M., OKAY, A.I., OBERHÄNSLI, R., SCHMIDT, A. & SUDO, M. 2015*a*. Late Cretaceous infant intra-oceanic arc volcanism, the Central Pontides, Turkey: Petrogenetic and tectonic implications. *Journal of Asian Earth Sciences*, **111**, 312–327.

AYGÜL, M., OKAY, A.I., OBERHAENSLI, R. & ZIEMANN, M.A. 2015*b*. Thermal structure of low-grade accreted Lower Cretaceous distal turbidites, the Central Pontides, Turkey: insights for tectonic thickening of an accretionary wedge. *Turkish Journal of Earth Sciences*, **24**, 461–474.

AYGÜL, M., OKAY, A.I., OBERHÄNSLI, R. & SUDO, M. 2016. Pre-collisional accretionary growth of the southern Laurasian margin, Central Pontides, Turkey. *Tectonophysics*, **671**, 218–234.

BARABOSHKIN, E.Yu., ALEKSEEV, A.S. & KOPAEVICH, L.F. 2003. Cretaceous palaeogeography of the North-Eastern Peri-Tethys. *Palaeogeography, Palaeoclimatology, Palaeoecology*, **196**, 177–208.

BARRIER, E. & VRIELYNCK, B. 2008. *MEBE Atlas of the Paleotectonic maps of the Middle East*. Commission for the Geological Map of the World, Paris.

BOZTUĞ, D. 1992. Lithostratigraphic units and tectonics of the southwestern part of Daday-Devrekani massif, western Pontides, Turkey. *Bulletin of the Mineral Research and Exploration, Turkey*, **114**, 1–22.

BOZTUĞ, D. & YILMAZ, O. 1983. Mineralogical, petrological and geochemical investigations of the Büyükçay-Elmalıçay granitoid (Kastamonu) and its country rocks. *Yerbilimleri*, **10**, 71–88.

BOZTUĞ, D., DEBON, F., LE FORT, P. & YILMAZ, O. 1984. Geochemical characteristics of some plutons from the Kastamonu granitoid belt (Northern Anatolia, Turkey). *Schweizerische Mineralogische und Petrographische Mitteilungen*, **64**, 389–404.

BOZTUĞ, D., DEBON, F., LE FORT, P. & YILMAZ, O. 1995. High compositional diversity of the Middle Jurassic Kastamonu Plutonic Belt, northern Anatolia, Turkey. *Turkish Journal of Earth Sciences*, **4**, 67–86.

BRAGIN, N.Y., TEKIN, U.K. & ÖZÇELIK, Y. 2002. Middle Jurassic radiolarians from the Akgöl Formation, central Pontids, northern Turkey. *Neues Jahrbuch für Geologie und Paläontologie, Monatshefte*, **2002**, 609–628.

CAVAZZA, C., FEDERICI, I., OKAY, A.I. & ZATTIN, M. 2012. Apatite fission-track thermochronology of the Western Pontides (NW Turkey). *Geological Magazine*, **149**, 133–140.

ÇELIK, Ö.F., MARZOLI, A., MARSCHIK, R., CHIARADIA, M., NEUBAUER, F. & ÖZ, İ. 2011. Early–Middle Jurassic intra-oceanic subduction in the İzmir–Ankara–Erzincan Ocean, Northern Turkey. *Tectonophysics*, **539**, 120–134.

ÇELIK, Ö.F., CHIARADIA, M., MARZOLI, A., ÖZKAN, M., BILLOR, Z. & TOPUZ, G. 2016. Jurassic metabasic rocks in the Kızılırmak accretionary complex (Kargı region, Central Pontides, Northern Turkey). *Tectonophysics*, **672–673**, 34–49.

CHANNEL, J.E.T., TÜYSÜZ, O., BEKTAS, O. & SENGÖR, A.M.C. 1996. Jurassic–Cretaceous paleomagnetism and paleogeography of the Pontides (Turkey). *Tectonics*, **15**, 201–212.

CHEN, F., SIEBEL, W., SATIR, M., TERZIOĞLU, N. & SAKA, K. 2002. Geochronology of the Karadere basement (NW Turkey) and implications for the geological evolution of the Istanbul Zone. *International Journal of Earth Sciences*, **91**, 469–481.

ÇIMEN, O., GÖNCÜOĞLU, M.C. & SAYIT, K. 2016. Geochemistry of the metavolcanic rocks from the Çangaldağ Complex in the Central Pontides: implications for the Middle Jurassic arc-back-arc system in the Neotethyan Intra-Pontide Ocean. *Turkish Journal of Earth Science*, **25**, 419–512, https://doi.org/10.3906/yer-1603-11

COHEN, K.M., FINNEY, S.C., GIBBARD, P.L. & FAN, J.-X. 2013. The ICS international chronostratigraphic chart. *Episodes*, **36**, 199–204.

DEAN, W.T., MONOD, O., RICKARDS, R.B., DEMIR, O. & BULTYNCK, P. 2000. Lower Palaeozoic stratigraphy and palaeontology, Karadere-Zirze area, Pontus Mountains, northern Turkey. *Geological Magazine*, **137**, 555–582.

DERMAN, A.S. & SAYILI, A. 1995. İnaltı formation: a key unit for regional geology. *In*: ERLER, A., ERCAN, T., BINGÖL, E. & ÖRÇEN, S. (eds) *Geology of the Black Sea Region*. Maden Tetkik ve Arama Enstitüsü, Ankara, 104–108.

DERMAN, A.S., ALIŞAN, C. & ÖZÇELIK, Y. 1995. Himmetpaşa formation: a new palynological age and stratigraphic significance. *In*: ERLER, A., ERCAN, T., BINGÖL, E. & ÖRÇEN, S. (eds) *Geology of the Black Sea Region*. Maden Tetkik ve Arama Enstitüsü, Ankara, 99–103.

DIL, N. 1976. Assemblages caractéristiques de foraminifères du Dévonien Supérieur et du Dinantien de Turquie (Bassin Carbonifère de Zonguldak. *Annales de la Société Géologique de Belgique*, **99**, 373–400.

DIL, N., TERMIER, G., TERMIER, H. & VACHARD, D. 1976. A l'étude stratigraphique et paléontologique du Viséen supérieur et du Namurien inférieur du bassin houiller de Zonguldak (nord-ouest de la Turquie). *Annales de la Société géologique de Belgique*, **99**, 401–449.

DOKUZ, A., KARSLI, O., CHEN, B. & UYSAL, I. 2010. Sources and petrogenesis of Jurassic granitoids in the Yusufeli area, Northeastern Turkey: implications for pre- and post-collisional lithospheric thinning of the eastern Pontides. *Tectonophysics*, **480**, 258–279.

ESPURT, N., HIPPOLYTE, J.-C., KAYMAKCI, N. & SANGU, E. 2014. Lithospheric structural control on inversion of the southern margin of the Black Sea Basin, Central Pontides, Turkey. *Lithosphere*, **6**, 26–34.

GALLHOFER, D., VON QUADT, A., PEYTCHEVA, I., SCHMID, S.M. & HEINRICH, C.A. 2015. Tectonic, magmatic, and metallogenic evolution of the Late Cretaceous arc in the Carpathian-Balkan orogen. *Tectonics*, **34**, 1813–1836, https://doi.org/10.1002/2015TC003834

GAND, G., TÜYSÜZ, O. *ET AL.* 2011. New Permian tetrapod footprints and macroflora from Turkey (Çakraz Formation, northwestern Anatolia): biostratigraphic and palaeoenvironmental implications. *Comptes Rendus Palevol*, **10**, 617–625.

GEDIK, A. & KORKMAZ, S. 1984. Geology and petroleum potential of the Sinop basin. *Jeoloji Mühendisliği*, **19**, 53–79 [in Turkish].

GENÇ, Ş.C. & TÜYSÜZ, O. 2010. Tectonic setting of the Jurassic bimodal magmatism in the Sakarya Zone (Central and Western Pontides), Northern Turkey: a geochemical and isotopic approach. *Lithos*, **118**, 95–111.

GÖNCÜOĞLU, M.C., MARRONI, M., SAYIT, K., TEKIN, U.K., OTTRIA, G., PANDOLFI, L. & ELLERO, A. 2012. The Ayli Dağ ophiolite sequence (central-northern Turkey): a fragment of Middle Jurassic oceanic lithosphere within the Intra-Pontide suture zone. *Ofioliti*, **37**, 77–92.

GÖNCÜOĞLU, M.C., MARRONI, M. *ET AL.* 2014. The Arkot Dağ Mélange in Araç area, central Turkey: evidence of its origin within the geodynamic evolution of the Intra-Pontide suture zone. *Journal of Asian Earth Sciences*, **85**, 117–139.

GÖRÜR, N. 1997. Cretaceous syn- to postrift sedimentation on the southern continental margin of the western Black Sea Basin. *In*: ROBINSON, A.G. (ed.) *Regional and Petroleum Geology of the Black Sea and Surrounding Region*. American Association of Petroleum Geologists, Memoirs, **68**, 227–240.

GÖRÜR, N. & TÜYSÜZ, O. 1997. Petroleum geology of the southern continental margin of the Black Sea. *In*: ROBINSON, A.G. (ed.) *Regional and Petroleum Geology of the Black Sea and Surrounding Region*. American Association of Petroleum Geologists, Memoirs, **68**, 241–254.

GÜCER, M.A., ARSLAN, M., SHERLOCK, S. & HEAMAN, L.M. 2016. Permo-Carboniferous granitoids with Jurassic high temperature metamorphism in Central Pontides, Northern Turkey. *Mineralogy and Petrology*, **110**, 943–964, https://doi.org/10.1007/s00710-016-0443-5

GÜLMEZ, F., GENÇ, Ş.C., PRELEVIĆ, D., TÜYSÜZ, O., KARACIK, Z., RODEN, M.F. & BILLOR, Z. 2016. Ultrapotassic volcanism from the waning stage of the Neo-Tethyan subduction: a key study from the İzmir–Ankara–Erzincan Suture Belt, Central Northern Turkey. *Journal of Petrology*, **57**, 561–593.

GUO, L., VINCENT, S.J. & LAVRISHCHEV, V. 2011. Upper Jurassic reefs from the Russian western Caucasus: implications for the eastern Black Sea. *Turkish Journal of Earth Sciences*, **20**, 629–653.

HANEL, M., GURBANOV, A.G. & LIPPOLT, H.J. 1992. Age and genesis of granitoids from the Main-Range and Bechasyn zones of the western Great Caucasus. *Neues Jahrbuch für Mineralogie Monatshefte*, **12**, 529–544.

HIPPOLYTE, J.-C., MÜLLER, C., KAYMAKÇI, N. & SANGU, E. 2010. Dating of the Black Sea Basin: new nannoplankton ages from its inverted margin in the Central Pontides (Turkey). *In*: STEPHENSON, R.A., KAYMAKCI, N., SOSSON, M., STAROSTENKO, V. & BERGERAT, F. (eds) *Sedimentary Basin Tectonics from the Black Sea and Caucasus to the Arabian Platform*. Geological Society, London, Special Publications, **340**, 113–136, https://doi.org/10.1144/SP340.7

HIPPOLYTE, J.-C., MÜLLER, C., SANGU, E. & KAYMAKCI, N. 2015. Stratigraphic comparisons along the Pontides (Turkey) based on new nannoplankton age determinations in the Eastern Pontides: geodynamic implications. *In*: SOSSON, M., STEPHENSON, R.A. & ADAMIA, S.A. (eds) *Tectonic Evolution of the Eastern Black Sea and Caucasus*. Geological Society, London, Special Publications, **428**. First published online 27 October 2015, https://doi.org/10.1144/SP428.9

HIPPOLYTE, J.-C., ESPURT, N., KAYMAKÇI, N., SANGU, E. & MÜLLER, C. 2016. Cross-sectional anatomy and geodynamic evolution of the Central Pontide orogenic belt

(northern Turkey). *International Journal of Earth Sciences*, **105**, 81–106.

JANBU, N.E., NEMEC, W., KIRMAN, E. & ÖZAKSOY, V. 2007. Facies anatomy of a sand-rich channelized turbiditic system: the Eocene Kusuri Formation in the Sinop Basin, north-central Turkey. *In*: NICHOLS, G., WILLIAMS, E. & PAOLA, C. (eds) *Sedimentary Processes, Environments and Basins*. International Association of Sedimentologists, Special Publications, **38**, 457–517.

KARSLIOĞLU, Ö., USTAÖMER, T., ROEBRTSON, A.H.F. & PEYTCHEVA, I. 2012. Age and provenance of detrital zircons from a sandstone turbidite of the Triassic–Early Jurassic Küre Complex, Central Pontides. *In*: *Abstracts Book of International Earth Sciences Colloquium on theAegean Region, IAESCA-2012, Izmir*, 57.

KAYA, M.Y. & ALTINER, D. 2015. Microencrusters from the Upper Jurassic–Lower Cretaceous İnaltı Formation (Central Pontides, Turkey): remarks on the development of reefal/perireefal facies. *Facies*, **61**, 1–25.

KAYMAKÇI, N., ÖZÇELIK, Y., WHITE, S.H. & VAN DIJK, P.M. 2009. Tectono-stratigraphy of the Çankırı Basin: Late Cretaceous to early Miocene evolution of the Neotethyan Suture Zone in Turkey. *In*: VAN HINSBERGEN, D.J.J., EDWARDS, M.A. & GOVERS, R. (eds) *Collision and Collapse at the Africa–Arabia–Eurasia Subduction Zone*. Geological Society, London, Special Publications, **311**, 67–106, https://doi.org/10.1144/SP311.3

KEREY, I.E. 1985. Facies and tectonic setting of the Upper Carboniferous rocks of northwestern Turkey. *In*: DIXON, J.E. & ROBERTSON, A.H.F. (eds) *The Geological Evolution of the Eastern Mediterranean*. Geological Society, London, Special Publications, **17**, 123–128, https://doi.org/10.1144/GSL.SP.1984.017.01.06

KOCH, R., BUCUR, I.I., KIRMACI, M.Z., EREN, M. & TASLI, K. 2008. Upper Jurassic and Lower Cretaceous carbonate rocks of the Berdiga Limestone – Sedimentation on an onbound platform with volcanic and episodic siliciclastic influx. Biostratigraphy, Facies and Diagenesis (Kircaova, Kale-Gümüshane area; NE-Turkey). *Neues Jahrbuch für Geologie und Paläontologie*, **247**, 23–61.

KOZUR, H., AYDIN, M., DEMIR, O., YAKAR, H., GÖNCÜOĞLU, M.C. & KURU, F. 2000. New stratigraphic and palaeogeographic results from the Palaeozoic and early Mesozoic of the Middle Pontides (northern Turkey) in the Azdavay, Devrekani, Küre and İnebolu areas. Implications for the Carboniferous-Early Cretaceous geodynamic evolution and some related remarks to the Karakaya oceanic rift basin. *Geologica Croatica*, **53**, 209–268.

KRAJEWSKI, M. & OLSZEWSKA, B. 2007. Foraminifera from the Late Jurassic and Early Cretaceous carbonate platform facies of the southern part of the Crimea mountains, southern Ukraine. *Annales Societatis Geologorum Poloniae*, **77**, 291–311.

LEREN, B.L.S., JANBU, N.E., NEMEC, W., KIRMAN, E. & ILGAR, A. 2007. Late Cretaceous to early Eocene sedimentation in the Sinop-Boyabat Basin, north-central Turkey: a deep-water turbiditic system evolving into littoral carbonate platform. *In*: NICHOLS, G., WILLIAMS, E. & PAOLA, C. (eds) *Sedimentary Processes, Environments and Basins*. International Association of Sedimentologists, Special Publications, **38**, 401–457.

MARRONI, M., FRASSI, C. *ET AL.* 2014. Late Jurassic amphibolite facies metamorphism in the Intra-Pontide Suture

Zone (Turkey): an eastward extension of the Vardar Ocean from the Balkans into Anatolia? *Journal of the Geological Society, London*, **171**, 605–608, https://doi.org/10.1144/jgs2013-104

MASSE, J-P., TÜYSÜZ, O., FENERCI-MASSE, M., ÖZER, S. & SARI, B. 2009. Stratigraphic organisation, spatial distribution, palaeoenvironmental reconstruction, and demise of Lower Cretaceous (Barremian–lower Aptian) carbonate platforms of the Western Pontides (Black Sea region, Turkey). *Cretaceous Research*, **30**, 1170–1180.

MCCANN, T., CHALOT-PRAT, F. & SAINTOT, A. 2010. The Early Mesozoic evolution of the Western Greater Caucasus (Russia): Triassic–Jurassic sedimentary and magmatic history. *In*: SOSSON, M., KAYMAKCI, N., STEPHENSON, R.A., BERGERAT, F. & STAROSTENKO, V. (eds) *Sedimentary Basin Tectonics from the Black Sea and Caucasus to the Arabian Platform*. Geological Society, London, Special Publications, **340**, 181–238, https://doi.org/10.1144/SP340.10

MEIJERS, M.J.M., KAYMAKCI, N., VAN HINSBERGEN, D.J.J., LANGEREIS, C.G., STEPHENSON, R.A. & HIPPOLYTE, J.-C. 2010a. Late Cretaceous to Paleocene oroclinal bending in the central Pontides (Turkey). *Tectonics*, **29**, TC4016, https://doi.org/10.1029/2009TC002620

MEIJERS, M.J.M., VROUWE, B. *ET AL.* 2010b. Jurassic arc volcanism on Crimea (Ukraine): implications for the paleo-subduction zone configuration of the Black Sea region. *Lithos*, **119**, 412–426.

NAJMAN, Y., APPEL, E. *ET AL.* 2010. Timing of India–Asia collision: Geological, biostratigraphic, and palaeomagnetic constraints. *Journal of Geophysical Research*, **115**, B12416, https://doi.org/10.1029/2010jb007673

NATAL'IN, B.A. & ŞENGÖR, A.M.C. 2005. Late Palaeozoic to Triassic evolution of the Turan and Scythian platforms: the pre-history of the Palaeo-Tethyan closure. *Tectonophysics*, **404**, 175–202.

NIKISHIN, A., ZIEGLER, P., BOLOTOV, S. & FOKIN, P. 2012. Late Palaeozoic to Cenozoic Evolution of the Black Sea–Southern Eastern Europe Region: a view from the Russian Platform. *Turkish Journal of Earth Sciences*, **21**, 571–634.

NIKISHIN, A.M., OKAY, A.I., TÜYSÜZ, O., DEMIRER, A., AMELIN, N. & PETROV, E. 2015a. The Black Sea basins structure and history: new model based on new deep penetration regional seismic data. Part 1: Basins structure and fill. *Marine and Petroleum Geology*, **59**, 638–655.

NIKISHIN, A.M., OKAY, A.I., TÜYSÜZ, O., DEMIRER, A., WANNIER, M., AMELIN, N. & PETROV, E. 2015b. The Black Sea basins structure and history: new model based on new deep penetration regional seismic data. Part 2: Tectonic history and paleogeography. *Marine and Petroleum Geology*, **59**, 656–670.

NIKISHIN, A.M., WANNIER, M. *ET AL.* 2015c. Mesozoic to recent geological history of southern Crimea and the Eastern Black Sea region. *In*: SOSSON, M., STEPHENSON, R.A. & ADAMIA, S.A. (eds) *Tectonic Evolution of the Eastern Black Sea and Caucasus*. Geological Society, London, Special Publications, **428**. First published online 27 October 2015, https://doi.org/10.1144/SP428.1

NZEGGE, O.M. 2008. *Petrogenesis and geochronology of the Deliklitaş, Sivrikaya and Devrekani granitoids and basement, Kastamonu belt – Central Pontides (NW Turkey): evidence for Late Palaeozoic-Mesozoic plutonism and geodynamic interpretation*. PhD thesis, University of Tübingen, Tübingen, Germany.

NZEGGE, O.M., SATIR, M., SIEBEL, W. & TAUBALD, H. 2006. Geochemical and isotopic constraints on the genesis of the Late Palaeozoic Deliktaş and Sivrikaya granites from the Kastamonu granitoid belt (Central Pontides, Turkey). *Neues Jahrbuch für Mineralogie, Abhandlungen*, **183**, 27–40.

OKAY, A.I. 2000. Was the Late Triassic orogeny in Turkey caused by the collision of an oceanic plateau? *In*: BOZKURT, E., WINCHESTER, J.A. & PIPER, J.A.D. (eds) *Tectonics and Magmatism in Turkey and Surrounding Area*. Geological Society, London, Special Publications, **173**, 25–41, https://doi.org/10.1144/GSL.SP.2000.173.01.02

OKAY, A.I. & GÖNCÜOĞLU, M.C. 2004. Karakaya Complex: a review of data and concepts. *Turkish Journal of Earth Sciences*, **13**, 77–95.

OKAY, A.I. & NIKISHIN, A.M. 2015. Tectonic evolution of the southern margin of Laurasia in the Black Sea region. *International Geology Review*, **57**, 1051–1076.

OKAY, A.I. & ŞAHINTÜRK, Ö. 1997. Geology of the Eastern Pontides. *In*: ROBINSON, A.G. (ed.) *Regional and Petroleum Geology of the Black Sea and Surrounding Region*. American Association of Petroleum Geologists, Memoirs, **68**, 291–311.

OKAY, A.I. & TOPUZ, G. 2017. Variscan orogeny in the Black Sea region. *International Journal of Earth Sciences*, **106**, 569–592.

OKAY, A.I. & TÜYSÜZ, O. 1999. Tethyan sutures of northern Turkey. *In*: DURAND, B., JOLIVET, L., HORVÁTH, F. & SÉRANNE, M. (eds) *The Mediterranean Basins: Tertiary Extension within the Alpine Orogen*. Geological Society, London, Special Publications, **156**, 475–515, https://doi.org/10.1144/GSL.SP.1999.156.01.22

OKAY, A.I., TÜYSÜZ, O., SATIR, M., ÖZKAN-ALTINER, S., ALTINER, D., SHERLOCK, S. & EREN, R.H. 2006. Cretaceous and Triassic subduction–accretion, HP/LT metamorphism and continental growth in the Central Pontides, Turkey. *Geological Society of America Bulletin*, **118**, 1247–1269.

OKAY, A.I., SUNAL, G., SHERLOCK, S., ALTINER, D., TÜYSÜZ, O., KYLANDER-CLARK, A.R.C. & AYGÜL, M. 2013. Early Cretaceous sedimentation and orogeny on the southern active margin of Eurasia: Central Pontides, Turkey. *Tectonics*, **32**, 1247–1271.

OKAY, A.I., SUNAL, G., TÜYSÜZ, O., SHERLOCK, S., KESKIN, M. & KYLANDER-CLARK, A.R.C. 2014. Low-pressure–high temperature metamorphism during extension in a Jurassic magmatic arc, Central Pontides, Turkey. *Journal of Metamorphic Geology*, **32**, 49–69.

OKAY, A.I., ALTINER, D. & KILIÇ, A.M. 2015. Triassic limestone, turbidites and serpentinite–the Cimmeride orogeny in the Central Pontides. *Geological Magazine*, **152**, 460–479.

ÖZCAN, E., LESS, G. & KERTESZ, B. 2007. Late Ypresian to middle Lutetian Orthophragminid record from central and northern Turkey: taxonomy and remarks on zonal scheme. *Turkish Journal of Earth Sciences*, **16**, 281–318.

ÖZGEN-ERDEM, N., İNAN, N., AKYAZI, M. & TUNOĞLU, C. 2005. Benthonic foraminiferal assemblages and microfacies analysis of Paleocene–Eocene carbonate rocks in the Kastamonu region, Northern Turkey. *Journal of Asian Earth Sciences*, **25**, 403–417.

ÖZKAN-ALTINER, & ÖZCAN, E. 1999. Upper Cretaceous planktonic foraminiferal biostratigraphy from NW Turkey: calibration of the stratigraphic ranges of larger benthonic foraminifera. *Geological Journal*, **34**, 287–301.

ROBERTSON, A.H.F. & USTAÖMER, T. 2011. Testing alternative tectono-stratigraphic interpretations of the Late Palaeozoic–Early Mesozoic Karakaya Complex in NW Turkey: support for an accretionary origin related to northward subduction of Palaeotethys. *Turkish Journal of Earth Sciences*, **21**, 961–1007.

ROBINSON, A.G., RUDAT, J.H., BANKS, C.J. & WILES, R.L.F. 1996. Petroleum geology of the Black Sea. *Marine and Petroleum Geology*, **13**, 195–223.

ROJAY, B. & ALTINER, D. 1998. Middle Jurassic–Lower Cretaceous biostratigraphy in the Central Pontides (Turkey): remarks on paleogeography and tectonic evolution. *Rivista Italiana di Paleontologia e Stratigrafia*, **104**, 167–180.

SANDERS, W.J., NEMEC, W., ALDINUCCI, M., JANBU, N.E. & GHINASSI, M. 2014. Latest evidence of Palaeoamasia (Mammalia, Embrithopoda) in Turkish Anatolia. *Journal of Vertebrate Paleontology*, **34**, 1155–1164.

ŞEN, C. 2007. Jurassic volcanism in the Eastern Pontides: is it rift related or subduction related? *Turkish Journal of Earth Sciences*, **16**, 523–539.

ŞEN, Ş. 2013. New evidences for the formation of and for petroleum exploration in the fold-thrust zones of the central Black Sea Basin of Turkey. *AAPG Bulletin*, **97**, 465–485.

ŞENGÖR, A.M.C. 1995. The larger tectonic framework of the Zonguldak Coal Basin in Northern Turkey: an outsider's view. *In*: YALÇIN, M.N. & GÜRDAL, G. (eds) *Zonguldak Basin Research Wells – I Kozlu-K20/G*. Special Publication of Tübitak, MAM, 1–26.

ŞENGÖR, A.M.C., CIN, A., ROWLEY, D.B. & NIE, S.Y. 1991. Magmatic evolution of the Tethysides: a guide to reconstruction of collage history. *Palaeogeography, Palaeoclimatology, Palaeoecology*, **87**, 411–440.

SOMIN, M. 2011. Pre-Jurassic basement of the Greater Caucasus: brief overview. *Turkish Journal of Earth Sciences*, **20**, 545–610.

STOLLE, E. 2016. Çakraz Formation, Çamdağ area, NW Turkey: early/mid-Permian age, Rotliegend (Germany) and Southern Alps (Italy) equivalent – a stratigraphic re-assessment via palynological long-distance correlation. *Geological Journal*, **51**, 223–235.

SUNAL, G. & TÜYSÜZ, O. 2002. Palaeostress analysis of Tertiary postcollisional structures in the western Pontides, northern Turkey. *Geological Magazine*, **139**, 343–359.

TANER, M.F. & ZANINETTI, L. 1978. Etude paléontologique dans le Crétace volcano-sédimentaire de Güneyce (Pontides orientales, Turquie). *Rivista Italiana di Paleontologia e Stratigrafia*, **84**, 187–198.

TARI, G., DICEA, O., FAULKERSON, J., GEORGIEV, G., POPOV, S., STEFANESCU, M. & WEIR, G. 1997. Cimmerian and Alpine stratigraphy and structural evolution of the Moesian Platform (Romania/Bulgaria). *In*: ROBINSON, A.G. (ed.) *Regional and Petroleum Geology of the Black Sea and Surrounding Region*. American Association of Petroleum Geologists, Memoirs, **68**, 63–90.

TASLI, K. 1993. Micropaléontologie, stratigraphie et environnement de dépôt des séries Jurassiques a faciès de plat-forme de la région de Kale-Gümüşhane (Pontides Orientales, Turquie). *Revue de Micropaléontologie*, **36**, 45–65.

TIKHOMIROV, P.L., CHALOT-PRAT, F. & NAZAREVICH, B.P. 2004. Triassic volcanism in the Eastern Fore-Caucasus: Evolution and geodynamic interpretation. *Tectonophysics*, **381**, 119–142.

TOPUZ, G., ALTHERR, R. *ET AL.* 2010. Carboniferous high-potassium I-type granitoid magmatism in the Eastern Pontides: the Gümüşhane pluton (NE Turkey). *Lithos*, **116**, 92–110.

TOPUZ, G., GÖÇMENGIL, G., ROLLAND, Y., ÇELIK, Ö.F., ZACK, T. & SCHMITT, A.K. 2013. Jurassic accretionary complex and ophiolite from northeast Turkey: no evidence for the Cimmerian continental ribbon. *Geology*, **41**, 255–258.

TÜYSÜZ, O. 1999. Geology of the Cretaceous sedimentary basins of the Western Pontides. *Geological Journal*, **34**, 75–93.

TÜYSÜZ, O. & TEKIN, U.K. 2007. Timing of imbrication of an active continental margin facing the northern branch of Neotethys, Kargı Massif, northern Turkey. *Cretaceous Research*, **28**, 754–764.

TÜYSÜZ, O., YILMAZ, İ.Ö., ŠVÁBENICKÁ, L. & KIRICI, S. 2012. The unaz formation: a key unit in the Western Black Sea Region, N Turkey. *Turkish Journal of Earth Sciences*, **21**, 1009–1028.

TÜYSÜZ, O., MELINTE-DOBRINESCU, M.C., YILMAZ, İ.Ö., KIRICI, S., ŠVÁBENICKÁ, L. & SKUPIEN, P. 2016. The Kapanboğazı formation: a key unit for understanding Late Cretaceous evolution of the Pontides, N Turkey. *Paleogeography, Paleoclimatology, Paleoecology*, **441**, 565–581.

UĞUR, M.F., SEVIN, M. & DURU, M. 2002. *Geological Map Series of Turkey, Sinop Sheet, 1: 500 000*. Maden Tetkik ee Arama Genel Müdürlüğü, Ankara.

USTAÖMER, P.A., MUNDIL, R. & RENNE, P.R. 2005. U/Pb and Pb/Pb zircon ages for arc-related intrusions of the Bolu Massif (W Pontides, NW Turkey): evidence for Late Precambrian (Cadomian) age. *Terra Nova*, **17**, 215–223.

USTAÖMER, P.A., USTAÖMER, T. & ROBERTSON, A.H.F. 2012. Ion probe U–Pb dating of the Central Sakarya Basement: a peri-Gondwana Terrane intruded by late Lower Carboniferoussubduction/collision-related granitic rocks. *Turkish Journal of Earth Sciences*, **21**, 905–932.

USTAÖMER, T. & ROBERTSON, A.H.F. 1994. Late Paleozoic marginal basin and subduction–accretion: the Paleotethyan Küre Complex, Central Pontides, northern Turkey. *Journal of the Geological Society, London*, **151**, 291–305, https://doi.org/10.1144/gsjgs.151.2.0291

USTAÖMER, T. & ROBERTSON, A.H.F. 1999. Geochemical evidence used to test alternative plate tectonic models for the pre-Upper Jurassic (Palaeotethyan) units in the Central Pontides, N Turkey. *Geological Journal*, **34**, 25–53.

USTAÖMER, T., ROBERTSON, A.H.F., USTAÖMER, P.A., GERDES, A. & PEYTCHEVA, I. 2013. Constraints on Variscan and Cimmerian magmatism and metamorphism in the Pontides (Yusufeli–Artvin area), NE Turkey from

U–Pb dating and granite geochemistry. *In*: ROBERTSON, A.H.F., PARLAK, O. & ÜNLÜGENÇ, U.C. (eds) *Geological Development of Anatolia and the Easternmost Mediterranean Region*. Geological Society, London, Special Publications, **372**, 49–74, https://doi.org/10.1144/SP372.13

USTAÖMER, T., USTAÖMER, P.A., ROBERTSON, A.H.F. & GERDES, A. 2016. Implications of U–Pb and Lu–Hf isotopic analysis of detrital zircons for the depositional age, provenance and tectonic setting of the Permian–Triassic Palaeotethyan Karakaya Complex, NW Turkey. *International Journal of Earth Sciences*, **105**, 7–38.

YIĞITBAŞ, E., TÜYSÜZ, O. & SERDAR, H.S. 1990. Geological features of the Late Cretaceous active continental margin of the Central Pontides. *In*: *Proceedings of the 8th Petroleum Congress of Turkey*, Turkish Association of Petroleum Geologists, Ankara, 141–151.

YILDIRIM, C., SCHILDGEN, T.F., ECHTLER, H., MELNICK, D. & STRECKER, M.R. 2011. Late Neogene and active orogenic uplift in the Central Pontides associated with the North Anatolian Fault: implications for the northern margin of the Central Anatolian Plateau, Turkey. *Tectonics*, **30**, TC5005, https://doi.org/10.1029/2010TC002756

YILMAZ, İ. 1975. Determination de l'age d'un granite de Sancaktepe, dans la presquile de Kocaeli (Nord de la Mer de Marmara). *Comptes Rendus de l'Académie des Sciences D*, **281**, 1563–1565.

YILMAZ, İ.Ö. & ALTINER, D. 2007. Cyclostratigraphy and sequence boundaries of inner platform mixed carbonate-siliciclastic successions (Barremian–Aptian) (Zonguldak, NW Turkey). *Journal of Asian Earth Sciences*, **30**, 253–270.

YILMAZ, O. & BOZTUĞ, D. 1986. Kastamonu granitoid belt of northern Turkey: first arc plutonism product related to the subduction of the Paleo-Tethys. *Geology*, **14**, 179–183.

Cretaceous geological evolution of the Pontides

OKAN TÜYSÜZ

Eurasia Institute of Earth Sciences and Faculty of Mines, İstanbul Technical University, 34469 Maslak, İstanbul, Turkey
tuysuz@itu.edu.tr

Abstract: The Pontides forming the southern continental margin of the Black Sea consist of the Strandja, İstanbul and Sakarya zones. The Zonguldak-Ulus Basin, located in the NE part of the İstanbul Zone, has traditionally been viewed as opening during the Barremian and deepening until the Albian under the control of normal faults. New outcrop data indicate that the southern and eastern parts of this basin facing towards the Intra-Pontide Ocean in the south were already open during the Berriasian or earlier. Uplift and erosion of the Zonguldak-Ulus Basin during the Cenomanian is attributed to collision of the İstanbul and the Sakarya zones along the Intra-Pontide Suture. The Sinop Basin in the Sakarya Zone opened during Hauterivian–Barremian time. Sedimentation in this basin continued in a deepening environment until the development of the Pontide Magmatic Belt during the Turonian. The contact between the İstanbul and the Sakarya zones is represented by a shear zone that consists of siliciclastic distal turbidites, debris-flow deposits and radiolarian cherts imbricated with Middle Jurassic–Lower Cretaceous magmatic arc fragments. This shear zone is interpreted as being the eastern continuation of the Intra-Pontide Suture, separating the İstanbul and the Sakarya zones.

The Western Black Sea Basin to the north of the Pontides possibly opened in two stages. In the first stage, coeval with the opening of the Zonguldak-Ulus Basin, the rifting was a wide-rift style and caused thinning of the continental crust. During the Turonian–Santonian, the Pontide Magmatic Belt started to develop as an extensional arc, and caused break-up of the already thinned crust and the start of oceanic spreading in the Western Black Sea Basin.

The Pontides in northern Turkey (Ketin 1966) form a prominent segment of the Alpine mountain belt. This tectonic unit, forming the southern continental margin of the Black Sea, extends between the Balkans in the west and the Caucasus in the east (Fig. 1a) (Okay & Tüysüz 1999). According to Ketin's original description, the Pontides are delimited by the Black Sea (*Pontus Euxinus*) to the north and the İzmir-Ankara-Erzincan ophiolitic belt (later renamed as the İzmir–Ankara–Erzincan Suture by Şengör & Yılmaz 1981) to the south. The Pontides have a vital importance in understanding the evolutionary history and petroleum potential of the Black Sea, as offshore data are limited or inaccessible for the academic scientific community.

Different subdivisions and different evolutionary models have been proposed for the Pontides and the Black Sea to the north of it. Şengör & Yılmaz (1981) subdivided the Pontides (*sensu* Ketin 1966) into Rhodope-Pontide and the Sakarya continental fragments separated by the Intra-Pontide Suture, the remnant of a branch of the Tethys Ocean. Şengör *et al.* (1985) described the Kırklareli, Strandja, İstanbul, Küre and Bayburt nappes, from west to east, constituting the pre-Middle Jurassic basement of the Rhodope-Pontide Fragment and interpreted these nappes as amalgamated due to the closing of the Palaeotethys and its back-arc basin, the Karakaya

Ocean, during the Cimmeride Orogeny. Okay (1989) subdivided the Pontides into the Strandja (consisting of the Kırklareli and the Strandja nappes), İstanbul (corresponding to İstanbul nappe) and Sakarya (corresponding to the Küre and Bayburt nappes, and the Sakarya Continent) zones (Fig. 1a).

The Intra-Pontide Suture (Şengör & Yılmaz 1981), separating the Sakarya and the İstanbul zones, was later modified or destroyed by the North Anatolian Fault Zone, an active dextral shear zone (see Şengör *et al.* 2005). The age and even the occurrence of such a suture is one of the most contentious subjects within the geology of Turkey. Bozkurt *et al.* (2013*a*, *b*) suggested that the Intra-Pontide Suture developed during the Permo-Triassic because of the closure of the Palaeotethys Ocean. Yılmaz *et al.* (1997) regarded the Intra Pontide Ocean as a remnant of Palaeotethys that remained open until the late Cretaceous. Based on the late Jurassic–early Cretaceous (158–111 Ma) regional metamorphism in the western part of the Intra-Pontide Suture and unconformably overlying Campanian sediments Akbayram *et al.* (2013) proposed that the Intra-Pontide Ocean opened possibly during the early Triassic and closed during the early Cretaceous. Marroni *et al.* (2014) also described Oxfordian metamorphic ages from the eastern part of this suture, around Araç (Fig. 1b). In a very recent paper, Akbayram *et al.* (2016)

From: Simmons, M. D., Tari, G. C. & Okay, A. I. (eds) 2018. *Petroleum Geology of the Black Sea*.
Geological Society, London, Special Publications, **464**, 69–94.
First published online September 8, 2017, https://doi.org/10.1144/SP464.9

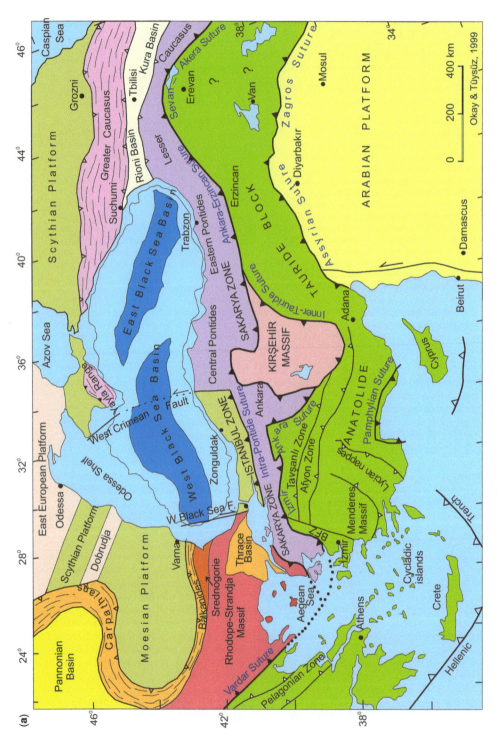

Fig. 1. (a) Tectonic units of Turkey and surroundings (after Okay & Tüysüz 1999).

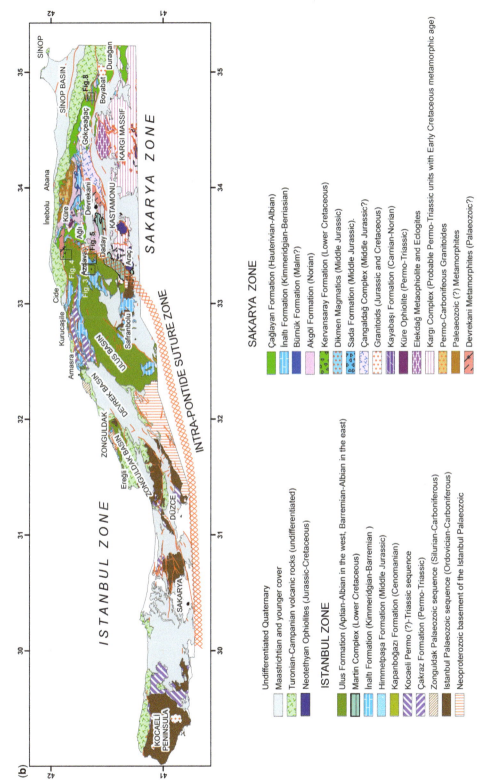

Fig. 1. (*Continued*) (**b**) Simplified geological map of the Western and Central Pontides. Red lines indicate faults; red hachured line represents the Intra-Pontide Suture (modified after the 1:500 000 scale geology map of Turkey, General Directorate of Mineral Research and Exploration, Ankara).

concluded that this ocean closed during the early Eocene. According to Göncüoğlu *et al.* (2014), the Intra-Pontide Ocean opened during the early Jurassic and was consumed by a north-dipping subduction during the middle Jurassic–early Cretaceous, which also caused then opening of a suprasubduction oceanic basin. The convergence in this suprasubduction basin started at the early–late Cretaceous boundary by an obduction process and its final closure was during the late Paleocene. This model also accounts a back-arc origin for the Western Black Sea Basin (WBSB).

Okay *et al.* (1994) proposed that the İstanbul Zone rifted off from its original location around the present Odessa Shelf during the early Cretaceous, and started to move southwards along the West Black Sea and West Crimean transform faults. According to this model, the WBSB opened as a back-arc basin behind the southwards-moving Istanbul Zone and collided with the Sakarya Zone along the Intra-Pontide Suture during the Eocene. This model also suggests that the Western and Eastern Black Sea basins opened via different mechanisms. Tüysüz (1999) and Tüysüz *et al.* (2016) advocated that the İstanbul Zone emplaced into its present position during Cenomanian time, and that there was no magmatic arc on the İstanbul Zone until the Turonian, so the back-arc nature of the Black Sea Basin should be younger than the early Cretaceous (see also Tari 2015).

Most authors agree that the rifting in the Black Sea began during the early Cretaceous (Letouzey *et al.* 1977; Kazmin *et al.* 1986; Zonenshain & Le Pichon 1986; Görür 1988; Banks & Robinson 1997 and references therein) and reached an oceanic stage during the late Cretaceous (Görür *et al.* 1993; Tüysüz 1999; Tüysüz *et al.* 2012, 2016). Okay *et al.* (2013) and Akdoğan *et al.* (2017) proposed that there was no thoroughgoing Black Sea Basin during the early Cretaceous, when a large submarine turbidite fan fed by the Ukranian Shield was depositing in a deep-marine environment on the previously amalgamated İstanbul and the Sakarya zones, and the Black Sea opened after the deposition of these Lower Cretaceous sediments.

Although recent studies provided new and important data from the onshore, such as new geological maps (Okay *et al.* 2006, 2013, 2015) together with palaeontological (Hippolyte *et al.* 2010, 2015), radiometric (Okay *et al.* 2006, 2013; Akdoğan *et al.* 2017), geochemical (Çimen *et al.* 2016; Keskin & Tüysüz 2017) and palaeomagnetic data (Cengiz Cinku *et al.* 2015), and seismic reflection data from the offshore (Nikishin *et al.* 2015a, b), it seems that we are still struggling to reach a consensus on a widely accepted evolutionary model for the Black Sea and Pontides; and there are still many open-ended questions, as summarized by Tari (2015).

The goal of this paper is to contribute to the understanding of stratigraphic and structural history of the Pontides during the critical Cretaceous period in the light of field mapping and field observations on the Western and Central Pontides (the Istanbul and the Sakarya zones, respectively), and to propose an evolutionary model that fits with these observations.

Pre-rift succession

Istanbul Zone

The youngest rocks underlying the Lower Cretaceous rift succession are the Upper Jurassic–Lower Cretaceous shallow-marine carbonates of the İnaltı Formation (Tüysüz *et al.* 2004) (Fig. 2). These limestones sit directly on older units with a basal thinly bedded micritic horizon, or, in places, with a thin basal conglomerate consisting of rounded and unsorted coarse grains. At the base of the carbonates there are, in places, reefal complexes, but most of the succession is represented by homogenous neritic carbonates reaching up to 400–500 m in thickness, represented by light-coloured, medium to thickly bedded intraclastic or peloidal micrites, sparites and some dolomites. Deposition of these shallow platform-type carbonates around Cide and Zonguldak started during the Kimmeridgian and lasted until the early Valanginian (Masse *et al.* 2009 and references therein). In the lower parts of these platform carbonates, Kimmeridgian–Valanginian algae and foraminifera such as *Arabidocium bicazensis* and *Conicospirillina basiliensis* indicate inner-shelf deposition, while in the upper parts of the succession the presence of the calpionellids *Calpionellites darderi* (early Valanginian), *Calpionellopsis oblanga* and *Tintinopsella carpathica* (late Berriasian) infer relatively deeper marine conditions (unpublished data provided by BP 1992).

Sakarya Zone

The Sakarya Zone displays very different stratigraphic characteristics compared to the İstanbul Zone. The main differences between these two zones are as follows:

- The İstanbul and Zonguldak Palaeozoic sequences (Görür *et al.* 1997; Lom *et al.* 2016 and references therein), the Permo-Triassic red beds (Alişan & Derman 1995; Tüysüz *et al.* 2004; Gand *et al.* 2011; Stolle 2016), the Kocaeli Triassic sequence (Erguvanlı 1947; Yurttaş-Özdemir 1971; Assereto 1972; Gedik 1975), the Middle Jurassic siliciclastics (Akyol *et al.* 1974; Derman *et al.* 1995) and the Upper Barremian–Aptian

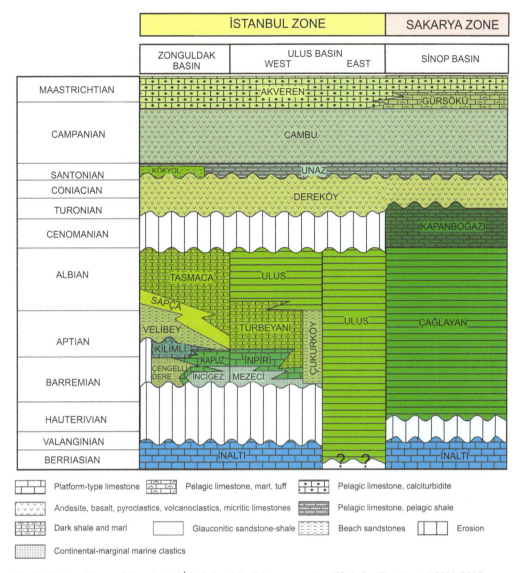

Fig. 2. Stratigraphic correlation chart of İstanbul and the Sakarya zones (modified after Tüysüz *et al.* 2004, 2016).

platform carbonates (Masse *et al.* 2009) of the İstanbul Zone are absent in the Sakarya Zone.

- Nearly all pre-Upper Jurassic units of the Sakarya Zone are metamorphosed and penetratively deformed, in contrast to the coeval units of the İstanbul Zone (Tüysüz 1990; Okay *et al.* 2006, 2014 and references therein).
- All pre-Upper Jurassic units in the İstanbul Zone were deposited on a continental basement and represent passive margin sediments (Lom *et al.* 2016), while most of the coeval units of the Sakarya Zone are oceanic or related with active margin events (Okay *et al.* 2006, 2014, 2015).

Intensely deformed and mostly metamorphosed pre-Upper Jurassic units of the Sakarya Zone in the Central Pontides crop out in the Küre-Devrekani area (Fig. 1b), and consist of different units having different ages and lithological characteristics, and are commonly intruded by Middle Jurassic magmatic rocks (Yılmaz & Boztuğ 1986; Tüysüz 1990, 1993; Okay *et al.* 2006, 2014, 2015 and references therein). Structural and stratigraphic relationships between some of these different units are not clear due to covering sediments, dense vegetation and late Palaeogene compressional deformation (Sunal & Tüysüz 2002).

In the Sakarya Zone, the Upper Jurassic continental clastics and overlying platform carbonates form a uniform cover for the complex basement described above. In contrast to the İstanbul Zone, a continental clastic unit (the Bürnük Formation) is present, reaching up to a few hundred metres in thickness beneath carbonates (the İnaltı Formation). The Bürnük Formation is represented by coarse- to fine-grained red to purple conglomerates, arkosic sandstones and mudstones deposited in an alluvial-fan/floodplain environment (Tüysüz et al. 1990; Tüysüz 1993). In some places, a slight unconformity separates these red beds from the overlying limestones.

At the base of the limestones, which show frequent vertical and lateral facies changes, large gastropods, bivalves and in situ corals are common. Back-reef, reef core, fore-reef and lagoon facies can be recognized in different locations. In contrast to the lower parts, the upper parts of the formation are rather homogenous and very similar to the limestones described in the İstanbul Zone. The İnaltı Formation is Kimmeridgian–Berriasian in age (Kaya & Altıner 2014, 2015).

Early Cretaceous syn-rift of the İstanbul Zone

Late Barremian transgression, development and demise of a carbonate platform on the Zonguldak High

In the northern parts of the Zonguldak and Ulus basins, between Ereğli and Cide (Fig. 1b), an Urgonian-type limestone sequence (Charles & Flandrin 1929; Altınlı 1951) resting unconformably on the Upper Jurassic platform carbonates and older units crops out along a 130 km-long strip parallel to the Black Sea coast. Facies properties of these carbonates indicate deposition in restricted lagoon, open lagoon-shelf and carbonate shoal environments (Varol & Akman 1988).

At the base of these limestones are red sandstones, mudstones and conglomerates deposited in alluvial-fan/marginal-marine environments (Fig. 2) (the İncigez Formation and lower parts of the Çengellidere Formation in the Zonguldak Basin, and the Mezeci Clastic Member in the Western Ulus Basin: see Tüysüz et al. 2004). Recently, Masse et al. (2009) described details of the stratigraphy and facies distribution of the Urgonian limestones, which are known as the Kapuz and İnpiri formations in the Zonguldak and the Western Ulus basins, respectively (Fig. 2). They determined two types of succession with dissimilarities regarding their thickness, facies and terrigenous content. In the east, between Amasra and Cide (Fig. 1b), there are relatively thin, terrigenous-rich, rudist-free or rudist-poor

successions representing a mixed carbonate–siliciclastic ramp; whereas in the west, between Amasra and Zonguldak, thick (>100 m) rudist-bearing carbonate successions crop out. Masse et al. (2009) attributed these differences to a fluviatile system in the east that controlled the balance between carbonate and terrigenous sedimentation, together with tectonic effects. A relatively narrow ramp-type morphology with abundant terrigenous input was present in the east, and a wide and nearly flat carbonate platform developed in the west.

The Upper Barremian–Lower Aptian carbonates are only seen in the İstanbul Zone and they are absent in the Sakarya Zone. Based on similarity of facies and fauna, Masse et al. (2002, 2004) concluded that the Pontides were a continuation of the Balkan subprovince during the deposition of these carbonates.

The Urgonian-type limestones at the base of the Zonguldak syn-rift succession are conformably overlain by Aptian organic-rich grey-black shales and marls of the Kilimli Formation (Fig. 2) (Tokay 1952, 1954/1955; Akman 1992; Hippolyte et al. 2010) in the northern part of the Zonguldak Basin. In the south, the Velibey Formation, consisting of yellowish, medium to thickly bedded quartz arenites with conglomerate and local limestone lenses, unconformably overlies the Carboniferous and Devonian units. The limestone lenses are of mainly bioclastic type, rich in rudist and Orbitolina fossils (Tokay 1952; Ketin 1955; Derman 1990a). The quartz-rich (>85%), well-sorted, rounded and grain-supported nature of the sandstones, together with symmetrical ripple marks and rudist-rich carbonate lenses, all indicate that the Velibey Formation deposited in a beach/shallow-shelf environment consisting of some patch reefs.

The Velibey Formation grades upwards into the Sapça Formation (Fig. 2), the 'Green Sandstone' of Altınlı (1951), comprising glauconitic sheet sands with parallel lamination and bioturbation at the base and overlying proximal to medial turbiditic sandstones alternating with marls, sandy limestones and blue to black shales. Upsection are distal turbidites consisting dominantly of shales with turbidite channels, boulder beds, slumps and debris flows fed by the underlying rocks (Derman 1990a). As first indicated by Altınlı (1951), there are abundant Lower Aptian limestone blocks (the İnpiri Formation) within this formation as mass flows, in addition to Upper Jurassic and Devonian limestones. The best outcrops of these blocks can be seen along the Zonguldak–Ereğli main road. An Albian age has been assigned to the Sapça Formation by Tokay (1952), Aydın et al. (1987) and by Canca (1994), but recent nannoplankton data by Hippolyte et al. (2010) indicate an early Aptian–early Albian age for this formation, and at least partly coeval deposition with the

Velibey sandstones, suggesting that formation boundaries may be diachronous (Fig. 2).

The overlying Tasmaca Formation (Fig. 2) comprises blue to black organic-rich shales alternating with thinly to medium-bedded, clayey, micritic and partly sandy limestones. Arni (1931) assigned a late Albian age to these 'micaeous blue marls' (Altınlı 1951) based on ammonite and *Inoceramus* fossils collected by Ralli (1895). Recently, a late Aptian–early Albian age is determined based on nannoplankton by Hippolyte *et al.* (2010). The upper parts of the Tasmaca Formation (the Cemaller Formation of Hippolyte *et al.* 2010; see Tüysüz *et al.* 2004 for details), consisting of abundant exotic blocks, is late Albian in age (Hippolyte *et al.* 2010).

The Zonguldak stratigraphic section described above (Fig. 2) implies that a new transgression started on the İstanbul Zone after a relatively short (*c.* 10 myr) erosional period between the early Valanginian and late Barremian. This transgression covered an area in the east of the İstanbul Zone between Ereğli and Cide (Fig. 1b), and resulted in deposition of shallow-marine carbonates and marginal-marine sediments during the late Barremian.

According to Tüysüz (1999), an Upper Barremian–Lower Aptian carbonate platform developed on an east- or ENE-trending, extensional fault-controlled archipelago, the Zonguldak Archipelago, probably a horst separating the Zonguldak-Ulus Basin in the south and the WBSB in the north. In the southern flank of this archipelago, the platform carbonates are first replaced by fluvio-deltaic siliciclastics (the Velibey Formation and upper parts of the Çengellidere Formation: Fig. 2), indicating a shallowing event during the early Aptian (late Bedoulian: Masse *et al.* 2009), then passing into deepening-upwards turbiditic sediments. In the north, the carbonate platform is covered by ammonite-bearing marls (the Kilimli and Türbeyanı formations: Fig. 2), indicating an abrupt deepening or drowning event and demise of the carbonate platform (Masse *et al.* 2009). Temporal distribution of these differential vertical movements indicate a northwards downwarping, possibly due to northwards tilting of the archipelago that carbonate platform was deposited on. This tilting event is also evidenced by the nature of the clasts within the fluvio-deltaic units; they are mainly derived from Upper Jurassic platform carbonates in lower horizons and from the Palaeozoic sediments in upper horizons, indicating erosion removing first Upper Jurassic platform carbonates and then eroding the Palaeozoic substratum. The pre-late Barremian units (i.e. pre-rift units of the İstanbul Zone) were not intensely folded but generally tilted due to normal faulting. All these data imply that the area presently covered by the Zonguldak-Ulus Basin was possibly affected by an extensional tectonic regime

and the basin started to subside during the early Aptian. Most previous authors also accepted that the initial opening of the Black Sea, at least for the WBSB, was coeval with this basin subsidence (Görür 1988, 1997 and references therein).

Aptian and Albian times witnessed a deepening of the basin towards the south and deposition of deepening-upwards siliciclastic turbidites. The occurrence of debris flows and exotic blocks within the turbiditic sediments of the Sapça and the Tasmaca formations imply a tectonically active environment, possibly due to continuing extension and normal faulting. All of the debris-flow horizons and olistoliths are derived from Upper Jurassic–Lower Cretaceous platform carbonates and underlying basement units, together with Urgonian-type limestones. This indicates that the Upper Barremian carbonate platform on the Zonguldak Archipelago was also disrupted by ongoing faulting. Sedimentation in the Zonguldak Basin lasted until the late Albian, and was ended by regional Cenomanian uplift.

The Uppermost Cretaceous–Eocene sediments of the Devrek Basin are superimposed on both the Zonguldak and Ulus basins (Fig. 1b). The Ulus Basin consists mainly of basinal turbidites similar to southern parts of the Zonguldak Basin. It is generally accepted that the Zonguldak and Ulus basins were parts of a single, SE-deepening Zonguldak-Ulus Basin before the development of the Devrek Basin (see Tüysüz 1999).

Ulus Basin

The NE–SW-trending Ulus Basin, the largest sedimentary basin on the İstanbul Zone, can be subdivided into two parts with respect to the stratigraphic characteristics of its sedimentary fill and their contact relationships with the underlying units. In the Western Ulus Basin, west of Azdavay (Fig. 1b), Lower Cretaceous clastics sit unconformably on the Upper Jurassic platform carbonates and the underlying older units of the Istanbul Zone. In the Eastern Ulus Basin, east of Azdavay, basinal sediments also rest unconformably on the Upper Jurassic platform carbonates in the westernmost part of the basin (Fig. 3). Towards the eastern part, there are sediments older than the western part, which were deposited in a deeper marine environment and intensely imbricated with Upper Jurassic magmatic units. In contrast to the Western Ulus Basin, the Eastern Ulus Basin has a very complex structure due to syn- and post-depositional tectonics.

Western Ulus Basin. In the NE margin of the Ulus Basin, south of Amasra and Cide (Fig. 1b), basin fill starts with the Çukurköy Formation, consisting of alluvial-fan deposits with complexly channelled

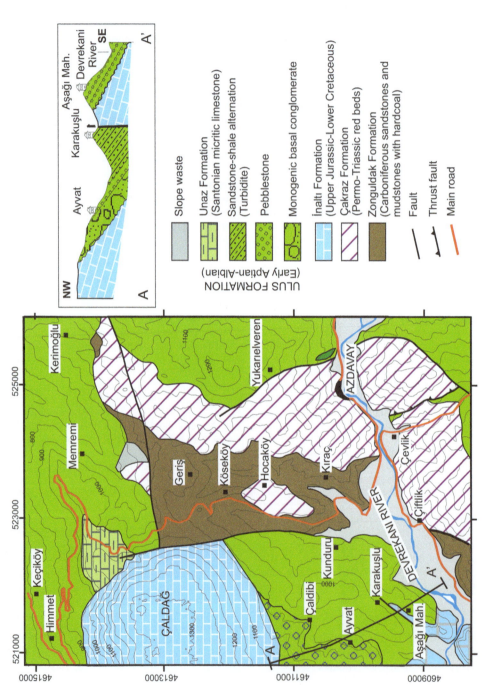

Fig. 3. Geology map of the Azdavay region (for the location of this map see Fig. 1b).

coarse red clastics passing upward into beach sandstones and conglomerates (Mezeci clastics). Upsection are limestone and dolomitic limestone interbeds and lenses (İnpiri limestones), and then blueish marls (Türbeyanı marls) (Fig. 2, Tüysüz 1999; Tüysüz et al. 2004). This typical marginal-marine clastic and carbonate sequence is late Barremian–Aptian in age according to fossils from its basal part (Tüysüz 1999) and corresponds to the mixed platform deposits of Masse et al. (2009). The Türbeyanı marls contain, in places (UTM Zone 36, 4617400–453200), abundant ammonite fossils, such as *Cheloniceras (C.) rauffi* and *Deshayesites* sp., indicating an early Aptian age. There are no fossil data from the upper parts of the Çukurköy Formation that possibly include Albian-aged sediments. Towards the southern part of this basin, this 100–150 m-thick basal part, which can be correlated with the Zonguldak region, grades quickly into proximal turbiditic sandstone–shale alternations, the Ulus Formation (Fig. 2) (Akyol et al. 1974). Upsection, and towards the centre of the basin in the south, are distal turbidites with radiolarian red-grey pelagic micritic carbonates and carbonaceous shale interbeds consisting of blocks and debris-flow horizons. The blocks, reaching up to hundreds of metres in size, were mainly derived from the Upper Barremian–Aptian and Upper Jurassic carbonates and partly from underlying Permo-Triassic and Palaeozoic units. Some of the blocks are surrounded by a pebbly envelope, implying synsedimentary gravity sliding mechanisms from the uplifted basin margins reaching to the deeper parts of the basin. Palaeocurrent measurements from the northern part of the Ulus Basin also indicate a source to the north. Similar to the Zonguldak region, sedimentation in the Western Ulus Basin possibly lasted until late Albian time (Hippolyte et al. 2010).

In the SW, the Sünnice Massif (Fig. 1b), comprising the Neoproterozoic crystalline basement of the İstanbul Zone (Ustaömer & Kipman 1998; Yiğitbaş et al. 2008), is unconformably overlain by Upper Jurassic platform carbonates. Basin fill here rests unconformably on both of these units, and starts with coarse conglomerates and boulder beds (Ahmetusta conglomerates: Derman 1990a) grading upwards rapidly into flysch-type sediments with abundant blocks and debris flows. Some of the pebbles and blocks of the conglomerates were derived from Palaeozoic units, but some others consist only of Upper Jurassic platform carbonates. The latter were deposited as grain flows, also known as 'limestone megabreccia' (Derman 1990a, b). These coarse-grained, poorly sorted monogenic breccias are possibly deposited in front of uplifted blocks as talus breccias (Akyol et al. 1974; Derman 1990a; Tüysüz 1993; Timur & Aksay 2002).

The stratigraphic thickness of the Ulus Formation is not known due to its intensely folded structure, but in 1988 Turkish Petroleum Co. drilled 3200 m within this formation.

The data presented above imply that the Western Ulus Basin started to open during the Barremian together with the Zonguldak Basin. The occurrence of fault-controlled deposits, such as megabreccias, blocks and thick debris-flow horizons, and the upwards-deepening character of the basin fill indicate fault control on the basin development. Similar to the Zonguldak Basin, marginal facies (the Çukurköy Formation) is only seen along the northern margin, between Amasra and Cide (Fig. 1b), while turbidites with debris flows and boulder beds (the Ulus Formation) are more extensive in the south indicating a south- or SE-deepening asymmetrical structure for the basin. The Zonguldak Basin consists mainly of marginal sediments, whilst basinal sediments are dominant in the Western Ulus Basin.

Eastern Ulus Basin. The Eastern Ulus Basin shows some important differences compared to the western basin in the lithology and age of the basin fill and structural deformation. In the western part of this sector, around Azdavay (Figs 1b & 3), the basin fill starts above the Upper Jurassic platform carbonates with very badly sorted monogenic conglomerates or breccias consisting of pebbles and blocks of these carbonates, and quickly grades upwards into a chaotic siliciclastic turbidite succession consisting of abundant debris flows and olistoliths. Debris-flow horizons reach up to tens or, sometimes, hundreds of metres in thickness. They consist of angular and very badly sorted blocks from Carboniferous, Permo-Triassic and Upper Jurassic units, with or without very small amounts of matrix. The angular shape of the blocks and pebbles indicates a close source region. There are also individual blocks within the turbidites, reaching up to 1 km in size (Figs 3 & 4). Diameters of the blocks decrease towards the east, possibly indicating a source in the west. To the north of Azdavay (Fig. 3) and around Ömerköy village (Fig. 4), some of the blocks consist of both Carboniferous and unconformably overlying Permo-Triassic red beds. Coal production from these blocks has taken place since the 1970s in both locations. Some exploration wells have passed through the Carboniferous and Permo-Triassic sequences, and have entered Lower Cretaceous turbidites at depths of 100–500 m (Canca 1994). This possibly indicates the occurrence of huge gravity slidings during the deposition of the turbidites from the İstanbul Zone. Bayrak (1991) studied the characteristics of the coal within these blocks and concluded that there is a good correlation with the autochtonous Carboniferous coals of the Amasra-Kurucaşile region in the west.

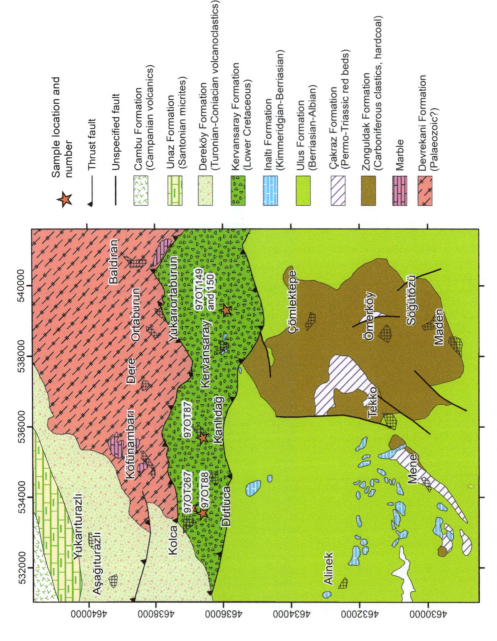

Fig. 4. Geology map of the Kervansaray region (for the location of this map see Fig. 1b).

Eastwards, radiolarian shales and cherts occur as synsedimentary interlayers within the distal turbidites. Some of the chert/red pelagic shale horizons reaching up to hundreds of metres in thickness are seen as tectonic slices delimited by shear zones. In addition to these red pelagic units, some pillow lava slices are also seen. These greenish basaltic lava flows include red pelagic shale and chert intercalations. The contact relationships between the chert–shale–basalt slices and the sheared distal turbidites of the Ulus Formation are tectonic in nature, wherever they are seen. These contacts are represented mainly by shear zones ranging from a few metres to a few tens of metres, characterized by intense folding and penetrative cleavage with mainly dextral transpressional kinematic indicators. Some late-stage sinistral structures were also observed. Due to intense deformation, a lack of detailed age data and dense vegetation, it is not clear for these tectonic slices whether they were originally deposited and imbricated together with the distal turbidites, or tectonically transported from their original position and inserted into the distal turbidites. In spite of superimposed deformation phases, some preserved synsedimentary structures indicating transpressional deformations were also determined in some outcrops. The intensity of penetrative deformation increases towards the east, where different lithologies tectonically imbricated with each other, and a slight metamorphism that caused the development of tiny muscovite and sericite minerals along the foliation planes appears to have affected them. Based on field observations, these deformations started during the deposition of the turbidites and continued after the deposition as progressive phases. The basin fill, especially its eastern part, was deformed by four or five main deformation phases.

In contrast, none of these deformational structures, radiolarian cherts and lava slices, such large blocks, such thick and continuous debris-flow and lava horizons have been observed in the Western Ulus Basin.

The Ulus Formation in the Eastern Ulus Basin is unconformably overlain by the Middle Turonian Dereköy and Upper Santonian Unaz formations (Figs 2 & 3), which were not penetratively deformed.

Age data from the Eastern Ulus Formation itself are very rare; they were obtained only from black micritic limestones alternating with distal turbidites. In sample 97OT119 (UTM Zone 36, 4631800–527925), we determined *Calpionella elliptica* and Radiolaria, together with some pelagic lamellibranches, indicating a Berriasian age. A few tens of metres below this sample, a possible Berriasian–Valanginian age was determined in sample 97OT116, based on the presence of the foraminifera *Protopeneroplis trochanqulata*. In sample 99OT113 (UTM Zone 36 4616620–520235), *Nannoconus kampneri*,

Nannoconus bucheri and *Conusphaera mexicana* within the shales of the Ulus Formation enveloping the coal-bearing Carboniferous sandstone blocks indicate a late Tithonian–Berriasian age. Turkish Petroleum Co. geologists also list nannofossil and radiolarian data (unpublished data) close to this area, indicating a similar age. These data show that the Ulus Formation in the eastern part of the basin is older than the western part and was deposited in a deeper environment. This has important implications for the presence of an open basin facing into the Intra-Pontide Ocean during earliest Cretaceous times and is discussed further below.

One of the interesting and conspicuous properties of this chaotic sedimentary unit is the absence of any clasts or blocks from the Central Pontide (Sakarya Zone) basement, such as serpentinites, basalts, granites or Permian limestones that are abundant in the coeval turbidites of the Central Pontides (the Çağlayan Formation of the Sinop Basin: Fig. 2). All of the blocks and pebbles within the turbidites of the Ulus Formation are derived from the easily recognizable rocks of the İstanbul Zone. Conversely, neither any clast nor block from the Palaeozoic or Permo-Triassic units of the İstanbul Zone has been observed within the Çağlayan Formation, even in very close areas. This observation leads to the interpretation that during early Cretaceous time, turbidites in the İstanbul and Sakarya were depositing in front of different source regions. Zircon data from the Lower Cretaceous turbidites (Akdoğan *et al.* 2017) of the Istanbul and the Sakarya zones also indicate different sources.

Magmatic units along the eastern margin of the Eastern Ulus Basin

The eastern part of the Ulus Basin is a tectonically key region. This area forms the contact between the İstanbul and the Sakarya zones, and is a shear zone as shown by the narrowly localized complex deformations along it. The eastern part of this shear zone is delimited by the Kervansaray, Sada and Dikmen magmatic units.

Kervansaray Formation

The Kervansaray Formation consists of pillow lava and pyroclastics alternating with red, black, olive and grey shales, and cherts. There are also some neritic limestone blocks within these lithologies. This formation is seen as a south-vergent tectonic slice between the Ulus Formation and the Devrekani metamorphics, Palaeozoic (?) basement of the Sakarya Zone (Geme Complex of Okay *et al.* 2014) in the NE part of the Eastern Ulus Basin (Fig. 4).

The lowest visible part of the Kervansaray Formation is dominated by deep-marine sedimentary rocks such as pelagic shales, radiolarian cherts and fine-grained sandstones (distal turbidites). The first volcanic horizons are expressed as randomly distributed volcanic pebbles within the shale. They are 5–20 cm in size and subrounded in shape. Upsection, they increase in frequency and form horizons a few metres thick comprising of greenish spilitic basalt fragments embedded within black-greenish tuffaceous matrix. In the uppermost parts of the formation there are thick (more than 100 m) pillow lava flows consisting of red-grey siliceous shale interbeds, 3–4 cm-thick greenish chert bands and palagonitic tuffs filling spaces between the pillows.

The Kervansaray Formation was intensely sheared, folded and foliated. S–C structures within the shales indicate dextral transpressional deformation. Intense chloritization and the development of some epidote and tiny actinolite crystals in the basaltic lava are attributed to a hydrothermal metamorphism

There are some exotic limestone blocks ranging in size from a few centimetres to a few tens of metres within the shales and pyroclastics of the Kervansaray Formation, close to Kervansaray village (Fig. 4). Two samples from these blocks (97OT149 and 97OT150, UTM Zone 36, 4635950–539300) yielded the following fossils: *Cladocoropsis mirabilis*, *Koskinobulina socialis*, *Bacinella irregularis*, *Tubiphytes morronensis* and *Pseudocyclammina* sp., which are characteristic taxa encountered within the Upper Kimmeridgian–Lower Valanginian shallow-marine platform carbonates of the İnaltı Formation (Altıner 1991). The Kervansaray Formation is unconformably overlain by the middle Turonian Dereköy Formation in the north (Fig. 4). In contrast to the Ulus and the Kervansaray formations, this Turonian unit is not affected by penetrative deformations.

The K–Ar radiometric dating by Terzioğlu *et al.* (2000) indicates the age of the magmatic rocks of the Kervansaray Formation as 146 ± 3 Ma (late Tithonian–Berriasian). These age data indicate that the Kervansaray Formation is coeval with the eastern Ulus Formation. Although these two formations are separated from each other by a tectonic contact, both are in the same age range, consist of the same blocks and both comprise distal turbidites. The only difference between them is the dominant occurrence of volcanics within the Kervansaray Formation. Based on the occurrence of smaller tectonic slices within the eastern Ulus Formation that have similar lithology, such as radiolarian cherts and basaltic lava, both formations were deposited in the same deep-marine environment and were tectonically imbricated and interpenetrated into each other via an accretionary mechanism. Soft-sediment

deformations within both formations are possibly related with the early stages of this accretionary mechanism. Synsedimentary blocks of the Upper Kimmeridgian–Lower Valanginian neritic limestone blocks within both formations, together with the similar age range, may indicate more or less co-occurrence of deep- and shallow-marine environments in and around the Eastern Ulus Basin.

The geochemical properties of three samples from the Kervansaray basalts, the locations of which are given on Figure 4, show that these lavas range lithologically from basalt to trachybasalt and andesite, with mid-ocean ridge basalt (MORB) (97OT87 and 97OT88) and volcanic arc (97OT267) affinities. Based on trace element geochemistry, Kibaroğlu & Satır (2000) also suggested that the occurrence of MORB, island-arc tholeiite and ocean-island alkali basalts developed in a subduction-related tectonic setting.

Based on the limited geochemical features, on the occurrence of radiolarian cherts alternating with volcanics and distal turbidites, on the age data described above, and on the nature of the deformation, the Kervansaray Formation most probably represents a tectonic slice originally belonging to an oceanic realm that existed between the İstanbul and Sakarya zones during the late Jurassic–early Cretaceous. In addition, the easternmost part of the Ulus Formation was possibly deposited on an oceanic substratum, close to an accretionary subduction zone (similar to the Martin Complex of Okay *et al.* 2013).

Sada Formation

The Sada Formation crops out as a narrow strip bounded by shear zones along the SE of the Eastern Ulus Basin (Fig. 5). Similar to the Kervansaray Formation, this unit is represented by a distal turbidite succession consisting of mafic to acidic lava and pyroclastic intercalations. There are also some neritic or reefal limestone blocks and pebbles within this unit.

Turbidites of the Sada Formation are similar to those in the eastern Ulus and Kervansaray formations, but their shale/sandstone ratio is higher. Shales are black, dark grey red or greenish in colour and partly silicified. Sandstones are white, red, grey and rich in quartz. The volcanics and volcanoclastics alternating with turbidites consist of dacites, andesites and andesitic basalts with pillow structures. In spite of the highly sheared structure of the Sada Formation, field observations clearly show that lava and pyroclastic flows are coeval with the deposition of the shales. Some spherical lava blocks, 3–5 m in diameter, rolled down the depositional slope during sedimentation, as indicated by a 10–15 cm-thick crust consisting of baked mud fragments enveloping the pillow lava inside the block. In other places, lava

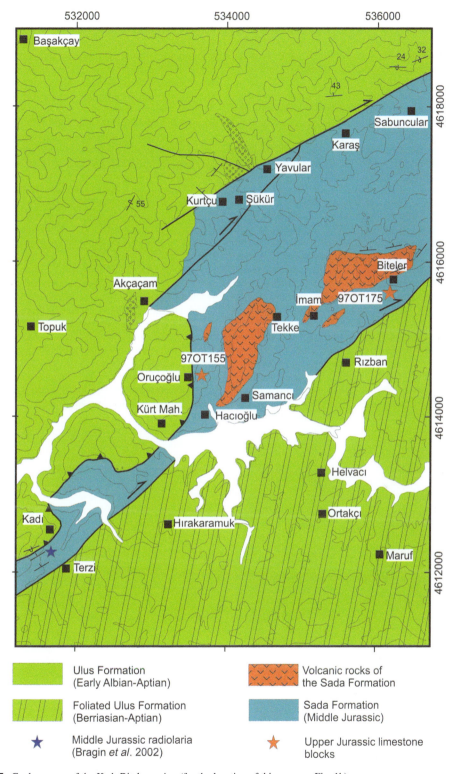

Fig. 5. Geology map of the Kadı-Biteler region (for the location of this map see Fig. 1b)

and pyroclastics alternate with shales with thickness of interbeds or lenses ranging from a few centimetres to a few tens of metres. Geochemical properties of the Sada Formation (Okay *et al.* 2014) imply that it was also produced by a magmatic arc, similar to the Kervansaray Formation.

In some locations, there are limestone blocks and pebbles within the shales of the Sada Formation (Figs 5 & 6). Most of these light-coloured limestones are rich in coral, algae and foraminifera. *Conicospirillina basiliensis* (97OT162, UTM Zone 36, 4611100–529100), *Cladocoropsis mirabilis* (97OT155, UTM Zone 36, 4614500–533600), *Conicospirinella basiliensis* and *Thaumotoporella parvovesiculifera* (98OT136, UTM Zone 36, 4620875–541800), and *Pseudocyclammina* cf. *lituus*, *Clypenia jurassica, Bacinella irregularis* and *Cladocoropsis mirablis* (98OT138 and 98OT139, UTM Zone 36, 4620870–541820) from these limestone blocks and pebbles indicate an age range of Kimmeridgian–Valanginian that can be compared with the Upper Jurassic–Lower Cretaceous platform carbonates (İnaltı Formation) described above. Based on the age of the limestone blocks, the Sada Formation should be younger than or coeval with the Valanginian. However, Bragin *et al.* (2002) found a rich radiolarian fauna from the silicified shales alternating with basaltic pillow lava, indicating a late Bathonian–early Callovian age. Although they correlated this formation with the Akgöl Formation of the Sakarya Zone, this formation is Triassic in age (Okay *et al.*

2015). Due to intense imbrication and multiphase deformations that affected the whole formation, it was difficult to follow the stratigraphy of the Sada Formation in the field from base to top. It is also clear from the field observations that the formation is an association of tectonically imbricated slices, consisting of distal turbidites, lava and pyroclastics (Fig. 7). These contradicting age data can be explained in two ways: (1) the Sada Formation was originally a product of an abyssal magmatic arc that was active between the middle Jurassic and early Cretaceous – in this case, the Kervansaray Formation represents a younger part of this arc; and (2) the Sada Formation is a tectonic mixture of two separate units developed during the middle Jurassic and early Cretaceous. By regarding the structural properties of the formation and due to a lack of age data from all different horizons or tectonic slices within the formation, the first interpretation is favoured herein.

In most places, a shear zone, represented by intensely sheared and foliated shales and mylonites reaching up to tens of metres, separates the coarser-grained turbidites of the eastern Ulus Formation and the Sada Formation. The width of the shear zone ranges between 0.5 and 1.5 km. Tectonic planes within the shear zone are steeply dipping and, according to S–C structures, are dominantly dextral and transpressional in nature. Some sinistral structures are also seen within the unit. Non-deformed Santonian sediments unconformably overlie the Sada Formation.

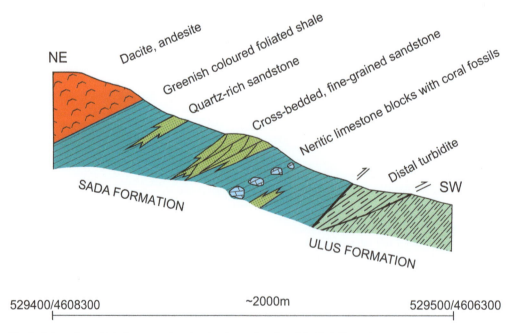

Fig. 6. Schematic geological cross-section showing the stratigraphy of the Sada Formation.

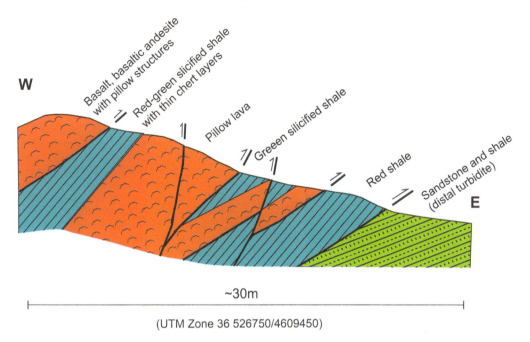

Fig. 7. Schematic geological cross-section showing the internal structure of the Sada Formation.

Dikmen magmatites

The Dikmen Magmatic Belt (Fig. 1b) is represented by light-coloured, porphyritic, acidic to intermediate volcanic rocks, such as andesite, trachyte and latite, and some small granodiorite and monzogranite intrusions. As this unit crops out in a highly vege- tated area and it was strongly weathered, it is diffi- cult to follow its internal structure in the field. However, it is clear from the field observations that sparse mafic lithologies of the unit, resembling those of the Kervansaray Formation, were cut by felsic ones.

The Dikmen magmatites crop out along the east- ern margin of the Ulus Formation as a north-trending narrow belt. The contact between the Dikmen mag- matites and the Sada Formation and the eastern Ulus Formation in the east is a shear zone. Any pri- mary stratigraphic contact between these two units has not been observed in the field. The Dikmen mag- matites are unconformably overlain by the Lower Cretaceous Çağlayan Formation to the east.

Recently, Okay *et al.* (2015) analysed some bio- tite grains from a dacite sample using the Ar–Ar laser probe technique and dated it as 163 Ma (Callovian). The geochemical properties of the Dikmen magma- tites, consisting of lavas and granites (Okay *et al.* 2015), imply that the Dikmen magmatites are of magmatic arc in origin, but some lavas display a tho- leiitic character. Granitoides of this formation were also produced in a volcanic arc setting.

Although the Dikmen magmatites are older than the Kervansaray magmatics, the structural posi- tion of this magmatic belt shows that it also attached to the eastern Ulus Formation as a narrow tectonic slice.

The Cenomanian stratigraphic gap in the İstanbul Zone

After the late Albian, the whole Zonguldak High was uplifted and strata were eroded until Turonian times (Fig. 2). Altınlı (1951), who first noticed this uncon- formity, concluded that, in spite of absence of the formations between the Albian and Turonian, there is an apparent conformity between these sedimentary sequences separated by this 'stratigraphic gap'. This shows that the event that created this stratigraphic gap is either not linked with a severe tectonic event or this event was far from the Zonguldak region and caused a regional uplift rather than any deforma- tional phase.

The Dereköy Formation, the lowest Upper Cre- taceous unit (Fig. 2), rests disconformably on the Upper Albian older deposits of the Zonguldak Basin. At the base of this unit is a conglomerate, in places reaching up to 200 m in thickness, grading upwards into volcanic/volcanogenic sandstone–shale and pyroclastic alternations. Within this volcanic/volca- noclastic unit, there are also some red pelagic lime- stone interlayers. The lowest pelagic limestone horizon within this unit contains a rich middle Turonian pelagic fauna (Tüysüz *et al.* 2012).

Marine sedimentation continued between the late Barremian and the late Albian in the eastern part of the İstanbul Zone, except the area between İstanbul and Ereğli, which possibly remained emergent all along the late Jurassic–late Santonian time (see Tüysüz *et al.* 2012). Stratigraphic data and the distribution of lithological units show that the whole İstanbul Zone, consisting of the Zonguldak-Ulus Basin, was emergent during the Cenomanian without any noteworthy deformational phase. The cause of this emergence will be discussed further below.

Early Cretaceous syn-rift of the Sakarya Zone

Çağlayan Formation

The Çağlayan Formation, which is dominated by siliciclastic turbidites with abundant exotic blocks and debris-flow horizons, comprises the main basin fill of the Sinop Basin on the Sakarya Zone in the Central Pontides (Fig. 1b). Some authors, such as Görür (1997), Okay *et al.* (2013) and Akdoğan *et al.* (2017), regarded this formation as the eastern equivalent of the Ulus Formation due to their similar lithology and age. Based on the dissimilarity of the underlying formations, as well as some lithological differences, Tüysüz (1999) advocated that these two formations were deposited in different environments.

The base of the Çağlayan Formation rests unconformably on the Upper Jurassic and older units of the Sakarya Zone. In one of the typical outcrops to the west of Bürnük village (Figs 8 & 9), the Çağlayan Formation rests unconformably on Upper Jurassic–Lower Cretaceous platform carbonates, the İnaltı Formation. At the base of the formation are quartz-rich, well-sorted, rounded and grain-supported sandstones. This medium to thickly bedded, light-coloured basal part, which is a few tens of metres in thickness, consists of some well-rounded and badly sorted limestone pebbles and blocks. Upsection, there are different debris-flow horizons alternating with quartz-rich sandstones that were possibly deposited in a beach environment as indicated by thick-shelled large gastropod and bivalve fossils, symmetrical ripple marks, and highly rounded nature of the clasts. Spore-rich assemblages from the lower part of the Çağlayan Formation comprising *Trilobosporites* spp., *Callialasporites trilobatus, C. dampieri*, and *Pilosisporites trichopaillosus and P. verus* indicate an early Aptian age. Freshwater or brackish-water algae, in addition to dinocycsts and spore taxa, indicate a deltaic depositional environment supplied with rich miospores and woody kerogen from nearby

terrestrial sources, and freshwater input into the depositional environment (unpublished data provided by BP 1992). This basal part of the Çağlayan Formation, cropping out close to the southern parts of the Sinop Basin, grades upwards into shelf sandstones with debris-flow horizons and then into proximal to medial turbidites.

In the lower parts of the Çağlayan Formation there are some thick horizons consisting totally of reworked clasts from the underlying Bürnük and İnaltı formations. These reworking events, which are common in the lower parts of the Çağlayan Formation, possibly imply repeated uplifting and erosion of the basement in some regions (horsts) which provided clasts to the neighbouring depositional areas (graben). Tüysüz *et al.* (1990) concluded that such phenomena took place not only during the beginning of the deposition but continued all through the deposition of the Çağlayan Formation, except in its uppermost part which is represented by distal turbidites (Tüysüz *et al.* 2016). In addition to these fast erosion and deposition events, rapid changes of the lithology and thickness of the Çağlayan Formation were attributed to an extensional tectonic regime and there is a general agreement that the Çağlayan Formation is a syn-rift deposit (Görür & Tüysüz 1997; Hippolyte *et al.* 2010, 2015).

In the lower parts of the Çağlayan Formation there are some massive sandstone horizons that were regarded as potential reservoir rocks for the exploration wells drilled by Turkish Petroleum Co. during the 1970s and 1980s. The thickness of these quartz-rich sandstones reaches, in places, up to 150–200 m. Within the Çağlayan Formation there are frequent debris-flow horizons consisting of pebbles and blocks of ophiolites and metamorphites, in addition to abundant Upper Jurassic limestones. In places, there are limestone breccias a few tens or hundreds of metres thick consisting of angular and poorly sorted limestone pebbles with carbonate cement. One of the interesting block types are black coloured, *Schwagerina* and *Mizzia* fossil-rich bituminous Permian limestones, occurring together with Upper Jurassic limestones, reaching up to a few hundreds of metres in size. In the SE of Durağan (Fig. 1b) these blocks form 60–70% of the areal distribution of the formation (Tüysüz & Yiğitbaş 1994). The ratio and size of the blocks increase towards the south, implying that the formation was fed by a source to the south. There is no record on the occurrence of these limestones in the İstanbul Zone, but large outcrops were described from the Kargı Massif in the south of the Sinop Basin (Tüysüz & Yiğitbaş 1994), Amasya region (Tüysüz 1996) and in many other parts of the Sakarya Zone (Akdeniz 1988; Altıner *et al.* 2000; Okuyucu & Göncüoğlu 2010; Sevin *et al.* 2015). Based on these data, Tüysüz (1999) concluded that the source for the Çağlayan

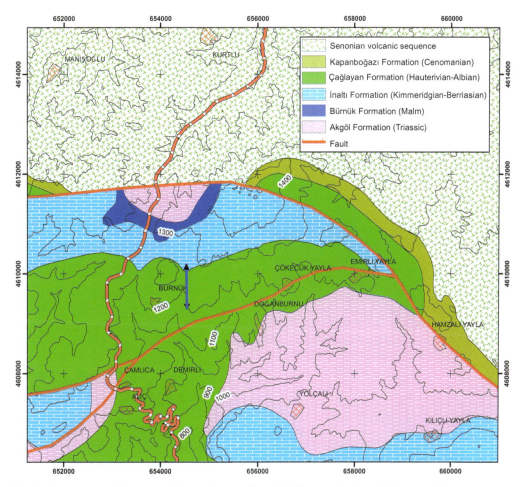

Fig. 8. Geology map of the Bürnük village and surroundings. The blue arrow indicates the location of the stratigraphic section in Figure 9.

Formation was the Tokat and Kargı massifs in the south. In contrast to this view, based on the Archaean–Palaeoproterozoic and Permian detrital zircon ages, Okay *et al.* (2013) and Akdoğan *et al.* (2017) interpreted the Çağlayan Formation as having been sourced from the East European Platform to the north.

Although there are no data on the stratigraphic thickness of the Çağlayan Formation due to its chaotic internal structure and intense post-Cretaceous compressional deformation that created north- and south-vergent thrusts and duplex structures, at least 3000 m of stratigraphic thickness can be estimated.

Age data from the Çağlayan Formation are rare. Recently, Hippolyte *et al.* (2010) indicated a Hauterivian age from the Çağlayan Formation near Ağlı (Fig. 1b). In the Yemişliçay River valley, south of Abana (UTM Zone 36, 4619200–643600), the co-occurrence of dinocyst taxa *Muderongia*

tetracantha and *Pseudoceratium pelliferum* are indicative of the early Barremian, while in the upper parts of the section *Muderongia simplex* is indicative of a Barremian–Aptian age. The spores *Aequitriradites verrucosus* and *A. spinulosus* confirm an Aptian age (unpublished data provided by BP). Tüysüz *et al.* (2016) described Albian fossils from the uppermost part of the formation. Based on these data, the Çağlayan Formation is the Hauterivian–Albian in age.

Kapanboğazı Formation: continuous Albian–Cenomanian sedimentation in the Sakarya Zone

The Kapanboğazı Formation consists of red pelagic limestones, shales and cherts. This name was erroneously used in subsequent studies for all Upper

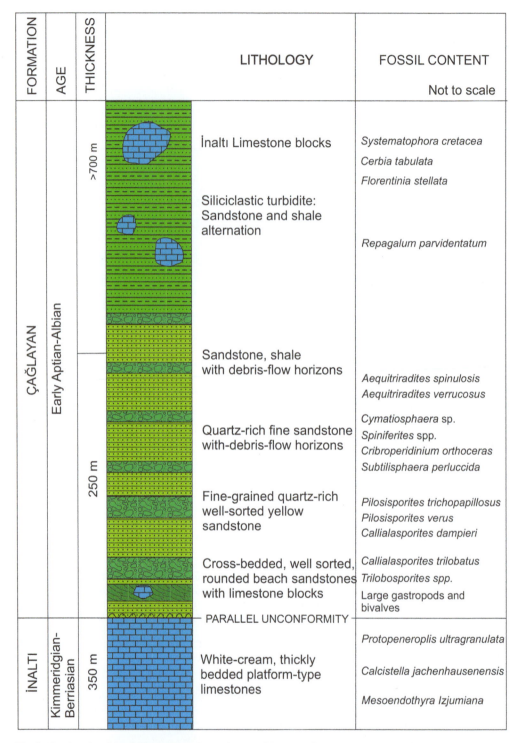

Fig. 9. Stratigraphy of the Çağlayan Formation to the east of Bürnük village. For the location of the stratigraphic section, see Figure 8. Fossils of the İnaltı Formation are taken from Kaya & Altıner (2015). Some of the fossils of the Çağlayan Formation are from unpublished data provided by BP.

Cretaceous red pelagic sediments, which are repeatedly deposited in different horizons within the Upper Cretaceous sequence, without reference to their age and stratigraphic position. Recently, Tüysüz et al. (2016) showed that the Kapanboğazı Formation resting conformably on the distal turbidites of the Çağlayan Formation is older than the second main pelagic red-bed interval, the Unaz Formation (Tüysüz et al. 2012). Radiolarian, foraminifera, nannofossil and dinoflagellate data from the Kapanboğazı Formation indicate a Cenomanian age.

Tüysüz et al. (2016) showed that, in contrast to the İstanbul Zone, deep-marine conditions prevailed on the Sakarya Zone during the whole Albian–Cenomanian–Turonian interval. Deposition of the siliciclastic turbidites of the Çağlayan Formation started during the Hauterivian (Hippolyte et al. 2010) and continued until the end of the Albian. The Albian black shales and distal turbidites are replaced during the Cenomanian by red shales and radiolarian cherts deposited below the calcite compensation depth (CCD), and then by the volcanic and volcanoclastics alternating with pelagic carbonates during the middle Turonian. This uppermost part of the Kapanboğazı Formation grades upwards into the Pontide magmatic sequence (see Keskin & Tüysüz 2017), which started during the middle Turonian (Tüysüz et al. 2012) and lasted until end of the Campanian. There is a general agreement that this magmatic belt developed on the Strandja, İstanbul and the Sakarya zones of the Pontides in response to the northwards subduction of the Neotethys Ocean along the south of the Pontides (Tüysüz et al. 2012, 2016; Keskin & Tüysüz 2017).

Interpretation of tectonic evolution and discussion

There is general agreement that completely different pre-late Jurassic geology of the İstanbul and Sakarya zones indicate their distinct evolutionary trend through quite different geological settings until the late Jurassic (Şengör & Yılmaz 1981; Okay & Tüysüz 1999; Tüysüz 1999 and references therein). The presence of Upper Jurassic–Lower Cretaceous limestones document the development of a neritic carbonate platform since the Kimmeridgian on both zones. Deposition of these carbonates lasted until the Berriasian in the Sakarya Zone (Kaya & Altıner 2014, 2015) and until the early Valanginian in the İstanbul Zone.

There are different views for the late Jurassic and early Cretaceous tectonic evolution of the Pontides, which is not only critical for the Pontides, but also has a vital importance for understanding the evolutionary history of the Black Sea. Two models for the timing of the assembling of the Pontides are most favoured: (1) the İstanbul and the Sakarya zones collided with each other at the end of the Liassic as a result of the closing of the Palaeotethys; the Upper Jurassic platform carbonates are thus deposited uniformly on both zones (Şengör et al. 1980); and (2) these zones juxtaposed during the beginning (Tüysüz 1999) or end of the late Cretaceous (Okay et al. 1994) as a result of closing of the Intra Pontide Ocean.

Data presented in this paper show that the Eastern Ulus Basin between the Istanbul and Sakarya zones was covered by a deep-marine environment during the late Tithonian–Berriasian time, when neritic platform carbonates (the İnaltı Formation) were depositing on both zones. This implies that there was no single Kimmeridgian–Valanginian carbonate platform overall the Pontides, but a deep-marine environment, possibly an eastern continuation of the Intra-Pontide Ocean (Tüysüz 1999; Tüysüz et al. 2016), was separating the Istanbul and Sakarya zones. The Urgonian-type carbonates imply that the Istanbul Zone was a continuation of the Balkan subprovince during the late Barremian (Masse et al. 2009), but the absence of these carbonates in the Sakarya Zone may indicate that these two zones were possibly in a different palaeogeographical position at that time.

Okay et al. (2013) described a much deeper and slightly metamorphosed southern equivalent of the eastern Ulus Formation (the Martin Complex) consisting of metabasite and chert interlayers, probably deposited on an oceanic substratum. Although they did not find any direct evidence, based on (a) youngest zircons (171 Ma); (b) the occurrence of Kimmeridgian–Berriasian limestone blocks within the 'interpretatively' equivalent formations; and (c) Ar/Ar data indicating an Albian (107 ± 4 Ma) metamorphic age, they assigned a Valanginian–Aptian depositional age for the Martin Complex.

The occurrence of large blocks/slices exceeding kilometres in size and frequent debris-flow horizons reaching tens to hundreds of metres in thickness within the eastern Ulus Formation indicate fast uplifting of the Upper Jurassic and older basement of the basin. In addition to the highly chaotic nature of these deposits, in light of: (a) the geographical position of the basin delimiting the Istanbul Zone in the east; (b) the nature of the syn- and post-sedimentary structures indicating progressive transpressional deformations; (c) the fining- and deepening-eastwards nature of the chaotic deposits; (d) the occurrence of deep-marine sediments such as radiolarian cherts and distal turbidites; and (e) the lack of any clast or block from the Sakarya Zone, it is interpreted that the eastern part of the Ulus Basin was under the control of a north-trending transform(?) fault delimiting the İstanbul Zone in the east. These very chaotic deposits have not been reported within the Martin Complex, which was possibly depositing on the abyssal plain to the south.

In the eastern part of the Eastern Ulus Basin, radiolarian cherts and distal turbidites were intensely imbricated together with coeval (the Kervansaray Formation) and older (the Sada and Dikmen formations) oceanic sediments and magmatic arc fragments. The nature of the basin fill, together with style of deformation, the intensity of which increases towards the east and SE (note also increasing degree of metamorphism of the Martin Complex towards the SE: Okay et al. 2013), can be attributed to an accretionary complex developed within the Intra-Pontide Ocean along an intra-oceanic subduction zone.

Sediments of the Eastern Ulus Basin rest unconformably on the Kimmeridgian–Berriasian neritic carbonates in the west (Fig. 3) and grade into turbiditic sediments towards the east, where Upper Tithonian–Berriasian deep-marine turbidites crop out. In the field, it was impossible to recognize any obvious contact separating these two units having very similar lithology. However, the occurrence of abundant Upper Jurassic limestone blocks and fragments within the eastern deep-marine sequence show that deposition was continuous, and a deep-marine environment that existed in the east during the Berriasian transgressed over the western margins of the basin during the late Barremian.

The Upper Barremian platform carbonates and marginal sequences at the base of the sedimentary fill of the Zonguldak and Western Ulus basins in the north demonstrate rifting during the late Barremian–early Aptian. These basins have started to subside since the beginning of the Aptian, as indicated by continuously deepening upwards and southwards sedimentation that lasted until the end of the Albian. Although there is no direct evidence from the onshore WBSB in the north, the nature and palaeogeographical distribution of the Upper Barremian–Aptian deposits imply that the opening of the WBSB was also coeval (see Masse et al. 2009 for details). If this interpretation is valid, buried Upper Barremian platform deposits developed in a similar way can be expected to be encountered along the southern margin of the WBSB.

Sedimentary fill of the Zonguldak and Ulus basins, and their deeper equivalents (the Eastern Ulus and Martin units), are interpreted as a passive margin sedimentary prism possibly facing into the Intra-Pontide Ocean (Fig. 10). Sedimentary fill of the Zonguldak and the western Ulus basins to the north represent marginal and shallower parts (Fig. 10), while the eastern Ulus Formation and the Martin Complex of Okay et al. (2013) represent deeper parts, possibly reaching to an abyssal plain in the south.

Okay et al. (2013) showed that the Martin Complex in the south was affected by a regional metamorphism during the Albian (107 ± 4 Ma), and then imbricated with non-metamorphosed Upper Cretaceous active margin sediments and ophiolites. Similar imbrication was also described from the southern part of the Kargı Massif (Yiğitbaş et al. 1990; Okay et al. 2006; Tüysüz & Tekin 2007). This imbrication and Albian metamorphism along the southern margin of the İstanbul Zone are interpreted to have been related to the northwards subduction of the Intra-Pontide Ocean and the collision of the İstanbul and Sakarya zones to the south, which also fits the models proposed by Akbayram et al. (2013) and Marroni et al. (2014). This collision caused uplifting of the southern parts of the İstanbul Zone and development of a metamorphic belt along its southern periphery (Okay et al. 2006). The Zonguldak-Ulus Basin in the north of the Istanbul Zone was regionally uplifted at the end of the Albian, and remained emerged during the Cenomanian. No pronounced deformation associated with this uplift has been determined.

To the east of this collision zone, sedimentation was continuous between the Hauterivian and Cenomanian in the Sinop Basin (Tüysüz et al. 2016). Deep-marine sediments of the Kapanboğazı Formation were deposited on the distal turbidites of the Çağlayan Formation, indicating subsidence of the Sinop Basin possibly below the CCD, while the Istanbul Zone was emergent. The collision between the İstanbul and Sakarya zones along the Intra-Pontide Suture in the east is interpreted as a soft collision due to the lack of metamorphism, in contrast to the southern parts of the suture. As the medial Turonian volcanic/volcanogenic deposits of the Pontide magmatic arc cover both units (Tüysüz et al. 2012; Keskin & Tüysüz 2017), emplacement of the İstanbul Zone should have been completed before this time.

Recently, Akdoğan et al. (2017) investigated the provenance of the Lower Cretaceous turbidites of the Pontides using palaeocurrent measurements, U–Pb zircon ages, REE abundances of dated zircons and the geochemistry of detrital rutile grains. They collected samples from the western, central and eastern parts of the Pontides, corresponding to the Zonguldak-Western Ulus, the Eastern Ulus and the Sinop basins, respectively. They show that the majority of the zircons in the Western Ulus and Zonguldak basins are Carboniferous and Neoproterozoic, implying local Palaeozoic and older continental basement of the Istanbul Zone as a source. The absence of these ages in the Sinop Basin support the model described above. Zircons in the Sinop Basin consist mainly of two groups, Archaean–Palaeoproterozoic and Permian, which are interpreted as sourced from the East European

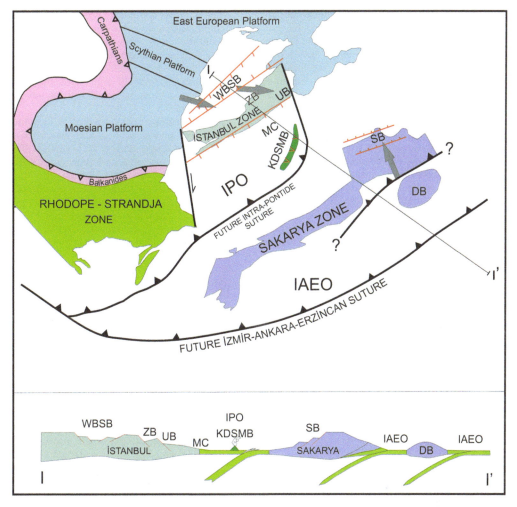

Fig. 10. Schematic palaeogeography map of the Pontides during the Early Cretaceous. IPO, Intra-Pontide Ocean (northern branch of the Neotethys); IAEO, İzmir–Ankara–Erzincan Ocean (main branch of the Neotethys); WBSB, Western Black Sea Basin; ZB, Zonguldak Basin; U, Ulus Basin; MC, depositional area of the Martin Complex; SB, Sinop Basin; DB, Domuzdağ Block; KDSMB, Kervansaray Dikmen, Sada Magmatic Belt.

Platform. Although their results clearly show different characteristics of the three basins in terms of palaeocurrent directions and detrital zircon ages, they concluded that all the Lower Cretaceous turbidites, consisting of the Çağlayan and Ulus formations, were deposited in a single basin, the Çağlayan Basin, which was fed by one or more rivers draining the East European Platform south to the Tethyan Ocean.

Data presented in this paper contradict this interpretation. Basal parts of the Çağlayan Formation in the Sinop Basin were deposited in very-shallow-marine conditions, as indicated by their sedimentary nature and fossil content. For a southwards-flowing river and for a south-facing large fan system, one would expect deep-marine sediments along its southern periphery and an underlying oceanic substratum, as is present in the Nile and Ganges fans. In fact, in the Sinop Basin, the Çağlayan Formation was deposited in a north-facing basin and its deeper parts were located in the north (see Tüysüz *et al.* 2016). Archaean–Palaeoproterozoic and the Permian zircons within these sediments were possibly sourced from the Sakarya Zone and the Domuzdağ Block in the south, which was also a possible source for the Permian limestone blocks within the Çağlayan Formation (see Tüysüz & Yiğitbaş 1994). Archaean and Palaeoproterozoic zircon populations in the Sakarya Zone (Aysal *et al.* 2012) support this idea.

A tentative geodynamic sketch showing the early Cretaceous palaeogeography of the region and evolutionary scenario for the assembling of the İstanbul and Sakarya zones, and the opening of the WBSB is given on Figure 10. The İstanbul Zone was located in the north of the present WBSB. The Intra-Pontide Ocean was subducting towards the north along an intra-oceanic subduction zone. This zone was delimited in the east by a possible transform fault. The Sada, Dikmen and Kervansaray magmatic belt(s) developed in response to this intra-oceanic subduction between the middle Jurassic and the early Cretaceous. These magmatic units were imbricated with Lower Cretaceous deep-marine turbidites via an accretionary mechanism. The WBSB and the Zonguldak-Ulus Basin were opening during the late Barremian–early Aptian in the north, while deep-marine sedimentation has been continuing in the south and SE since the late Tithonian–Berriasian or earlier (?). As indicated by the zircon data (Akdoğan et al. 2017), the Lower Cretaceous sediments filling the Zonguldak-Ulus Basin were being sourced off the northern areas. The Kimmeridgian–Berriasian shallow-marine platform carbonates were depositing on both the İstanbul and Sakarya zones separately.

The Sinop Basin started to open possibly during the Hauterivian on the Sakarya Zone. The northern part of this zone consists of a Middle Jurassic magmatic arc and its continental basement (Okay et al. 2014) covered by Middle Jurassic carbonates, while subduction and accretion along its southern margin has been continuous possibly since the early Jurassic or earlier (Tüysüz & Tekin 2007). The Domuzdağ continental block accreted to the Sakarya Zone possibly during the Albian (Okay et al. 2006).

Emplacement of the İstanbul Zone into its present position was possibly during the late Albian. Collision of the Sakarya and İstanbul zones along the Intra-Pontide Suture caused uplift of the İstanbul Zone during the Cenomanian, while the Sinop Basin to the east remained as a north-facing deep-marine environment. The Pontide Magmatic Belt developed after the assembling of the Istanbul and the Sakarya zones in response to northwards subduction of the Ankara-Erzincan Ocean.

Tari (2015) concluded that the opening of the WBSB can be explained by asymmetrical, wide-style phase of rifting at the southern margin of the European Plate without invoking back-arc extension, followed by a subsequent Turonian–Coniacian narrow-style phase of rifting that was driven by subduction roll-back in the south. Our data favour late Barremian–early Aptian rifting without any arc magmatism. This first period of wide-style rifting should cause rupturing of the İstanbul Zone from its original position in the north of present WBSB. A possible explanation of the reason for such extensional system could be a far and flat subduction within the Intra-Pontide Ocean. Yılmaz et al. (1997) regarded the Intra-Pontide as a remnant of the Palaeotethys Ocean. If this view is valid, subduction of an old oceanic basin in the south could have created an extensional system in its north without any arc magmatism.

The main back-arc opening of the WBSB following the first continental rifting during the early Cretaceous was possibly during the Turonian. This back-arc opening was followed by the breaking-up of continental crust during the Coniacian and oceanic spreading during the late Santonian. Stratigraphic and geochemical evidence of these stages are presented by Tüysüz et al. (2012) and Keskin & Tüysüz (2017), respectively. Nikishin et al. (2015a, b) supported this model based on offshore seismic data.

Conclusions

The Pontide tectonic belt forming the southern continental margin of the Black Sea consists of the İstanbul and the Sakarya zones. Many models proposed for the tectonic evolution of the Pontides accept that these continental fragments amalgamated before the deposition of the Upper Jurassic platform carbonates. Based on data presented in this paper, it is concluded that:

- There was an oceanic environment between these two zones during the late Jurassic–early Cretaceous time, possibly an eastern continuation of the Intra-Pontide Ocean.
- The eastern part of the Ulus Basin is an accretionary complex developed as a result of northwards subduction of the Intra-Pontide Ocean.
- There are Middle Jurassic–Lower Cretaceous magmatic arc fragments within the accretionary complex, possibly developed due to intra-oceanic subduction within the Intra-Pontide Ocean.
- The Intra-Pontide Ocean closed in the Central Pontides definitely before the Turonian, possibly during the late Albian.
- The WSBS started to open during the late Barremian, together with the Zonguldak-Ulus Basin in the south. Following the early Cretaceous wide-style rifting, a narrow-style back-arc rifting during the Turonian–Coniacian caused the break-up of the already thinned continental crust and oceanic spreading from the late Santonian. A narrow-style phase of rifting driven by subduction roll-back in the south.

I gratefully acknowledge the constructive criticism and suggestions for improvement by two anonymous reviewers and especially by volume editor Mike Simmons.

References

AKBAYRAM, K., OKAY, A.İ. & SATIR, M. 2013. Early Cretaceous closure of the Intra-Pontide Ocean in western Pontides (northwestern Turkey). *Journal of Geodynamics*, **65**, 38–55.

AKBAYRAM, K., ŞENGÖR, A.M.C. & ÖZCAN, E. 2016. The evolution of the Intra-Pontide suture: Implications of the discovery of late Cretaceous–early Tertiary mélanges. *In*: SORKHABI, R. (ed.) *Tectonic Evolution, Collision, and Seismicity of Southwest Asia: In Honor of Manuel Berberian's Forty-Five Years of Research Contributions.* Geological Society of America, Special Papers, **525**.

AKDENIZ, N. 1988. Permian and Carboniferous of Demirözü and their significance in the regional structure. *Geological Bulletin of Turkey*, **3**, 71–80.

AKDOĞAN, R., OKAY, A.I., SUNAL, G., TARI, G., MEINHOLD, G. & KYLANDER-CLARK, A.R.C. 2017. Provenance of a large Lower Cretaceous turbidite submarine fan complex on the active Laurasian margin: Central Pontides, northern Turkey. *Journal of Asian Earth Sciences*, **134**, 309–329.

AKMAN, A.Ü. 1992. The age and the opening mechanisms of the basins surrounding Black Sea in Nortwestern Turkey. *In*: *9th Petroleum Congress and Exhibition of Türkiye. Proceedings, UCTEA Chamber of Geophysical Engineers, Turkish Association of Petroleum Geologists, UCTEA Chamber of Petroleum Engineers*, 17–21 February 1992, Ankara, 319–332.

AKYOL, Z., ARPAT, E. ET AL. 1974. *Geological Map of Cide-Kurucaşile Area. 1:50 000 Scale Geology Maps of Turkey*. Institute of the Mineral Research and Exploration, Ankara.

ALIŞAN, C. & DERMAN, A.S. 1995. The first palynological age, sedimentological and stratigraphic data for Çakraz Group (Triassic), Western Black Sea. *In*: ERLER, A., ERCAN, T., BINGÖL, E. & ÖRÇEN, S. (eds) *Geology of the Black Sea Region*. Directorate of the Mineral Research and Exploration, Ankara, 93–98.

ALTINER, D. 1991. Microfossil biostratigraphy (mainly foraminifers) of the Jurassic–Cretaceous carbonate successions in northwestern Anatolia (Turkey). *Geologica Romana*, **27**, 167–213.

ALTINER, D., ÖZKAN-ALTINER, S. & KOÇYIĞIT, A. 2000. Late Permian foraminiferal biofacies belts in Turkey: paleogeography and tectonic implication. *In*: BOZKURT, E., WINCHESTER, J.A. & PIPER, J.A.D. (eds) *Tectonic and Magmatism in Turkey and Surrounding Area*. Geological Society, London, Special Publications, **173**, 83–96, https://doi.org/10.1144/GSL.SP.2000.173.01.04

ALTINLI, İ.E. 1951. The geology of the western portion of Filyos River. *Revue de la Faculté des Sciences de l'Université d'Istanbul Série B*, **XVI**, 154–188.

ARNI, V.P. 1931. Zur Stratigraphie und Tektonik der Kreidescichten östlich Eregli an der Schwarzmeerküste. *Eclogae Geologicae Helvetiae*, **24**, 305–345.

ASSERETO, R. 1972. Notes on the Anisian biostratigraphy of the Gebze area (Kocaeli Peninsula, Turkey). *Istituti di Geologia e Paleontologia dell'Universita degli Studi di Milano*, **113**, 436–444.

AYDIN, M., SERDAR, H.S., ŞAHINTÜRK, Ö., YAZMAN, M., ÇOKUĞRAŞ, R., DEMIR, O. & ÖZÇELIK, Y. 1987. Geology of Çamdağ (Sakarya)–Sünnicedağ (Bolu) region. *Geological Society of Turkey Bulletin*, **30**, 1–14.

AYSAL, N., ÖNGEN, S., PEYTCHEVA, I. & KESKIN, M. 2012. Origin and evolution of the Havran Unit, Western Sakarya basement (NW Turkey): new LA-ICP-MS U–Pb dating of the metasedimentary–metagranitic rocks and possible affiliation to Avalonian microcontinent. *Geodinamica Acta*, **25**, 226–247.

BANKS, C.J. & ROBINSON, A.G. 1997. Mesozoic strike-slip back-arc basins of the Western Black Sea region. *In*: ROBINSON, A.G. (ed.) *Regional and Petroleum Geology of the Black Sea and Surrounding Region*. American Association of Petroleum Geologists Memoirs, **68**, 53–62.

BAYRAK, A. 1991. Origin of the allochthonous coals within the Ulus Formation. *İstanbul Üniversitesi Yerbilimleri, Special Publications*, **8**, 89–94.

BOZKURT, E., WINCHESTER, J.A. & SAYIR, M. 2013b. The Çele mafic complex: evidence for Triassic collision between the Sakarya and İstanbul Zones, NW Turkey. *Tectonophysics*, **595–596**, 198–214.

BOZKURT, E., WINCHESTER, J.A., SATIR, M. & CROWLEY, Q.G. 2013a. The Almacık mafic-ultramafic complex: exhumed Sakarya subcrustal mantle adjacent to the İstanbul Zone, NW Turkey. *Geological Magazine*, **150**, 254–282.

BRAGIN, N.Y., TEKIN, U.K. & ÖZÇELIK, Y. 2002. Middle Jurassic radiolarians from the Akgöl Formation, central Pontids, northern Turkey. *Neues Jahrbuch für Geologie und Paläontologie, Monatshefte*, **10**, 609–628.

CANCA, N. 1994. *1:100 000 Scale Geological Maps: Western Black Sea Region Hard Coal Basin*. Directorate of the Mineral Research and Exploration, Ankara.

CENGIZ CINKU, M., HISARLI, Z.M. ET AL. 2015. Evidence of Late Cretaceous oroclinal bending in north-central Anatolia: palaeomagnetic results from Mesozoic and Cenozoic Rocks along the İzmir–Ankara–Erzincan Suture Zone. *In*: PUEYO, E.L., CIFELLI, F., SUSSMAN, A.J. & OLIVA-URCIA, B. (eds) *Palaeomagnetism in Fold and Thrust Belts: New Perspectives*. Geological Society, London, Special Publications, **425**, 1–25, https://doi.org/10.1144/SP425.2

CHARLES, F. & FLANDRIN, J. 1929. Contribution à l'étude des terrains crétacés de l'Anatolie du Nord (Asdie Mineure). *Annales de l'Université de Grenoble*, **6**, 289–375.

ÇIMEN, O., GÖNCÜOĞLU, M.C. & SAYIT, K. 2016. Geochemistry of the meta-volcanic rocks from the Çangaldağ Complex in Central Pontides: implications for the Middle Jurassic arc–back–arc system in the Neotethyan Intra–Pontide Ocean. *Turkish Journal of Earth Sciences*, **25**, 491–512.

DERMAN, A.S. 1990a. Late Jurassic and Early Cretaceous evolution of the Western Black Sea region. *8th Petroleum Congress of Turkey, Proceedings, Geology, Turkish Association of Petroleum Geologists/UCTEA Chamber of Petroleum Engineers*, Ankara, Turkey, 328–339.

DERMAN, A.S. 1990b. Sedimentation in a faulted basin margin: a sample from Cretaceous in Western Black Sea Region. *8th Petroleum Congress of Turkey, Proceedings, Geology, Turkish Association of Petroleum Geologists/UCTEA Chamber of Petroleum Engineers*, Ankara, Turkey, 314–321.

DERMAN, A.S., ALIŞAN, C. & ÖZÇELIK, Y. 1995. Himmetpaşa Formation: New palynological age data and significance. *In*: ERLER, A., ERCAN, T., BINGÖL, E. & ÖRÇEN, S. (eds) *Geology of the Black Sea Region*. Directorate of the Mineral Research and Exploration, Ankara, 99–104.

ERGUVANLI, K. 1947. New fossiliferous beds in Kocaeli Triassic formations. *Bulletin of the Geological Society of Turkey*, **1**, 158–163.

GAND, G., TÜYSÜZ, O. *ET AL.* 2011. New Permian tetrapod footprints and macroflora from Turkey (Çakraz Formation, northwestern Anatolia): biostratigraphic and palaeoenvironmental implications. *Comptes Rendus Palevol*, **10**, 617–625.

GEDIK, İ. 1975. Die Conodonten der Trias auf der Kocaeli-Halbinsel (Türkei). *Palaeontographica*, **150**, 99–160.

GÖNCÜOĞLU, M.C., MARRONI, M. *ET AL.* 2014. The Arkot Dağ Mélange in Araç area, central Turkey: evidence of its origin within the geodynamic evolution of the Intra-Pontide suture zone. *Journal of Asian Earth Sciences*, **85**, 117–139.

GÖRÜR, N. 1988. Timing of opening of the Black Sea Basin. *Tectonophysics*, **147**, 247–262.

GÖRÜR, N. 1997. Cretaceous syn- to post-rift sedimentation on the southern continental margin of the Western Black Sea Basin. *In*: ROBINSON, A.G. (ed.) *Regional and Petroleum Geology of the Black Sea and Surrounding Region*. American Association of Petroleum Geologists Memoirs, **68**, 227–240.

GÖRÜR, N. & TÜYSÜZ, O. 1997. Petroleum geology of the southern continental margin of the Black Sea. *In*: ROBINSON, A.G. (ed.) *Regional and Petroleum Geology of the Black Sea and Surrounding Region*. American Association of Petroleum Geologists Memoirs, **68**, 241–254.

GÖRÜR, N., TÜYSÜZ, O., AYKOL, A., SAKINÇ, M., YIĞITBAŞ, E. & AKKÖK, R. 1993. Cretaceous red pelagic carbonates of northern Turkey: their place in the opening history of the Black Sea. *Eclogae Geologicae Helvetiae*, **86**, 819–838.

GÖRÜR, N., MONOD, O. *ET AL.* 1997. Palaeogeographic and tectonic position of the Carboniferous rocks of the western Pontides (Turkey) in frame of the Variscan belt. *Bulletin Societe de Géologie France*, **168**, 197–205.

HIPPOLYTE, J.-C., MÜLLER, C., KAYMAKCI, N. & SANGU, E. 2010. Dating of the Black Sea Basin: New nannoplankton ages from its inverted margin in the Central Pontides (Turkey). *In*: STEPHENSON, R.A., KAYMAKCI, N., SOSSON, M., STAROSTENKO, V. & BERGERAT, F. (eds) *Sedimentary Basin Tectonics from the Black Sea and Caucasus to the Arabian Platform*. Geological Society, London, Special Publications, **340**, 113–136, https://doi.org/10.1144/SP340.7

HIPPOLYTE, J.C., ESPURT, N., KAYMAKÇI, N., SANGU, E. & MÜLLER, C. 2015. Cross-sectional anatomy and geodynamic evolution of the Central Pontide orogenic belt (northern Turkey). *International Journal of Earth Sciences (Geologische Rundschau)*, **105**, 81–106.

KAYA, M. & ALTINER, D. 2014. *Terebella lapilloides Münster, 1833 from the Upper Jurassic–Lower Cretaceous İnaltı carbonates, northern Turkey: its taxonomic position and paleoenvironmental– paleoecological significance. Turkish Journal of Earth Sciences*, **23**, 166–183.

KAYA, M. & ALTINER, D. 2015. Microencrusters from the Upper Jurassic–Lower Cretaceous İnaltı Formation (Central Pontides, Turkey): remarks on the

development of reefal/peri-reefal facies. *Facies*, **61**, 18, https://doi.org/10.1007/s10347-015-0445-5.

KAZMIN, V.G., SBORTSHIKOV, I.M., RICOU, L.E., ZONENSHAIN, L., BOULIN, J. & KNIPPER, A.L. 1986. Volcanic belts as markers of the Mesozoic–Cenozoic active margin of Eurasia. *In*: AUBOUIN, J., LE PICHON, X. & MONIN, A.S. (eds) *Evolution of the Tethys. Tectonophysics*, **123**, 123–152.

KESKIN, M. & TÜYSÜZ, O. 2017. Stratigraphy, petrogenesis and geodynamic setting of the Late Cretaceous volcanism on the SW margin of the Black Sea, Turkey. *In*: SIMMONS, M.D., TARI, G.C. & OKAY, A.I. (eds) *Petroleum Geology of the Black Sea*. Geological Society, London, Special Publications, **464**, https://doi.org/10.1144/SP464.5

KETIN, İ. 1955. Über die Geologie der Gegend von Ovacuma östlich Zonguldak. *Revue de la Sciences de l'Université d'İstanbul*, **20**, 147–154.

KETIN, İ. 1966. Tectonic Units of Asia Minor. *Bulletin of General Directorate of Mineral Research and Exploration*, **66**, 20–34.

KIBAROĞLU, M. & SATIR, M. 2000. Geochemistry of basaltic rocks of the Küre-Ophiolitic Complex and volcanics of Kervansaray, Central Pontides, Northern Turkey. *International Earth Science Colloquium on the Aegean Region, IESCA-2000*, Dokuz Eylül University Engineering Faculty, Department of Geology, İzmir, Turkey, Abstract Book, 109.

LETOUZEY, J., BIJU-DUVAL, B., DORKEL, A., GONNARD, R., KRISTCHEV, K., MONTADERT, L. & SUNGURLU, O. 1977. The Black Sea: a marginal basin; geophysical and geological data. *In*: BIJU-DUVAL, B. & MONTADERT, L. (eds) *International Symposium on the Structural History of the Mediterranean Basins*. Editions Technip, Paris, 363–376.

LOM, N., ÜLGEN, S.C., SAKINÇ, M. & ŞENGÖR, A.M.C. 2016. Geology and stratigraphy Istanbul region. *In*: ŞEN, Ş. (ed.) *Late Miocene Mammal Locality of Küçükçekmece, European Turkey. Geodiversitas*, **38**, 175–195.

MARRONI, M., FRASSI, C. *ET AL.* 2014. Late Jurassic amphibolites-facies metamorphism in the Intra-Pontide Suture Zone (Turkey): an eastward extension of the Vardar Ocean from the Balkans into Anatolia? *Journal of the Geological Society, London*, **171**, 605–608, https://doi.org/10.1144/jgs2013-104

MASSE, J.-P., FENERCI-MASSE, M. & ÖZER, S. 2002. Late Aptian Rudist faunas from the Zonguldak region, western Black Sea, Turkey (taxonomy, biostratigraphy, palaeoenvironment and palaeobiogeography). *Cretaceous Research*, **23**, 523–536.

MASSE, J.-P., ÖZER, S. & FENERCI, M. 2004. Upper Barremian–Lower Aptian Rudist Faunas from the Western Black Sea Region (Turkey). *Courier Forschungsinstitut Senckenberg*, **247**, 75–88.

MASSE, J.-P., TÜYSÜZ, O., FENERCI-MASSE, M., ÖZER, S. & SARI, B. 2009. Stratigraphic organisation, spatial distribution, palaeoenvironmental reconstruction, and demise of Lower Cretaceous (Barremian–lower Aptian) carbonate platforms of the Western Pontides (Black Sea region, Turkey). *Cretaceous Research*, **30**, 1170–1180.

NIKISHIN, A.M., OKAY, A.İ., TÜYSÜZ, O., DEMIRER, A., AMELIN, N. & PETROV, E. 2015*a*. The Black Sea basins structure and history: new model based on new deep penetration regional seismic data. Part 1: basins structure and fill. *Marine and Petroleum Geology*, **59**, 638–655.

NIKISHIN, A.M., OKAY, A.İ., TÜYSÜZ, O., DEMIRER, A., WANNIER, M., AMELIN, N. & PETROV, E. 2015*b*. The Black Sea basins structure and history: new model based on new deep penetration regional seismic data. Part 2: tectonic history and paleogeography. *Marine and Petroleum Geology*, **59**, 656–670.

OKAY, A.İ. 1989. Tectonic units and sutures in the Pontides, northern Turkey. *In*: ŞENGÖR, A.M.C. (ed.) *Tectonic Evolution of the Tethyan Region*. Kluwer Academic, Dordrecht, The Netherlands, 109–115.

OKAY, A.İ. & TÜYSÜZ, O. 1999. Tethyan sutures of northern Turkey. *In*: DURAND, B., JOLIVET, L., HORVÁTH, F. & SÉRANNE, M. (eds) *The Mediterranean Basins: Tertiary Extension Within the Alpine Orogen*. Geological Society, London, Special Publications, **156**, 475–515, https://doi.org/10.1144/GSL.SP.1999.156.01.22

OKAY, A.İ., ŞENGÖR, A.M.C. & GÖRÜR, N. 1994. Kinematic history of the opening of the Black Sea and its effect on the surrounding regions. *Geology*, **22**, 267–270.

OKAY, A.İ., TÜYSÜZ, O., SATIR, M., ÖZKAN-ALTINER, S., ALTINER, D., SHERLOCK, S. & EREN, R.H. 2006. Cretaceous and Triassic subduction–accretion, HP/LT metamorphism and continental growth in the Central Pontides, Turkey. *Geological Society of America Bulletin*, **118**, 1247–1269.

OKAY, A.İ., SUNAL, G., SHERLOCK, S., ALTINER, D., TÜYSÜZ, O., KYLANDER-CLARK, A.R.C. & AYGÜL, M. 2013. Early Cretaceous sedimentation and orogeny on the active margin of Eurasia: southern Central Pontides, Turkey. *Tectonics*, **32**, 1–25.

OKAY, A.İ., SUNAL, G., TÜYSÜZ, O., SHERLOCK, S., KESKIN, M. & KYLANDER, R.C. 2014. Low- pressure–high-temperature metamorphism during extension in a Jurassic magmatic arc, Central Pontides, Turkey. *Journal of Metamorphic Geology*, **32**, 49–69.

OKAY, A.İ., ALTINER, D. & KILIÇ, A.M. 2015. Triassic limestone, turbidites and serpentinite–the Cimmeride orogeny in the Central Pontides. *Geological Magazine*, **152**, 460–479.

OKUYUCU, C. & GÖNCÜOĞLU, M.C. 2010. Middle–late Asselian (Early Permian) fusulinid fauna from the post-Variscan cover in NW Anatolia (Turkey): biostratigraphy and geological implications. *Geobios*, **43**, 225–240.

RALLI, G. 1895. Le Basin houiller d'Héraclée. *Societe Geologique de Belgique*, **22**, 151–267.

STOLLE, E. 2016. Çakraz Formation, Çamdağ area, NW Turkey: early/mid-Permian age, Rotliegend (Germany) and Southern Alps (Italy) equivalent – a stratigraphic reassessment via palynological long-distance correlation. *Geological Journal*, **51**, 223–235.

SUNAL, G. & TÜYSÜZ, O. 2002. Palaeostress analysis of Tertiary post-collisional structures in the Western Pontides, Northern Turkey. *Geological Magazine*, **139**, 343–359.

ŞENGÖR, A.M.C. & YILMAZ, Y. 1981. Tethyan evolution of Turkey: a plate tectonic approach. *Tectonophysics*, **75**, 181–241.

ŞENGÖR, A.M.C., YILMAZ, Y. & KETIN, I. 1980. Remnants of a pre-Late Jurassic ocean in Northern Turkey: fragments of Permian–Triassic Palaeotethys. *Geological Society of America Bulletin*, **91**, 599–609.

ŞENGÖR, A.M.C., YILMAZ, Y. & SUNGURLU, O. 1985. Tectonics of the Mediterranean Cimmerides: nature and evolution of the western termination of Palaeotethys. *In*: DIXON, J.E. & ROBERTSON, A.H.F. (eds) *The Geological Evolution of the Eastern Mediterranean*. Geological Society, London, Special Publications, **17**, 77–112, https://doi.org/10.1144/GSL.SP.1984.017.01.04

ŞENGÖR, A.M.C., TÜYSÜZ, O. *ET AL.* 2005. The North Anatolian Fault: a new look. *Annual Review of Earth and Planetary Sciences*, **33**, 1–75.

SEVIN, M., DÖNMEZ, M., ATICI, G., ESATOĞLU, A.H., SARIFAKIOĞLU, E., ARIKAN, S. & SOYCAN, H. 2015. Late Permian unconformity around Ankara and new age data in the basement rocks, Ankara, Turkey. *Bulletin of General Directorate of Mineral Research and Exploration*, **151**, 133–151.

TARI, G. 2015. Is the Black Sea really a back-arc basin? *In*: POST, P.J., COLEMAN, JR. J.L. *ET AL.* (eds) *Petroleum Systems in 'Rift' Basins*, Transactions of the GCSSEPM Foundation Perkins-Rosen 34th Annual Research Conference, 13–16 December 2015, Houston, Texas, USA, 510–520.

TERZIOĞLU, M.N., SATIR, M. & SAKA, K. 2000. Geochemistry and geochronology of basaltic rocks of Küre basin, Central Pontides (N Turkey). *International Earth Science Colloquium on the Aegean Region, IESCA-2000, 25–29 September 2000*, Dokuz Eylül University Engineering Faculty, Department of Geology, İzmir, Turkey, Abstract Book, 219.

TIMUR, E. & AKSAY, A. 2002. *1:100 000 Scale Geology Maps of Turkey, Zonguldak F-29 Sheet, No.30*. Mineral Research and Exploration Institute of Turkey, Ankara.

TOKAY, M. 1952. Contribution a l'étude géeologique de la région comprise entre Ereğli, Alaplı, Kızıltepe ET ALacaağzı. *Bulletin of the Mineral Research and Exploration Institute of Turkey*, **42/43**, 35–78.

TOKAY, M. 1954/1955. Geologie de la region de Bartın (Zonguldak-Turqui du nord). *Bulletin of the Mineral Research and Exploration Institute of Turkey*, **46/47**, 46–63.

TÜYSÜZ, O. 1990. Tectonic evolution of a part of the Tethyside orogenic collage: the Kargı massif, Northern Turkey. *Tectonics*, **9**, 141–160.

TÜYSÜZ, O. 1993. A Geotraverse from the Black Sea to the Central Anatolia: tectonic evolution of Northern Neotethys (Karadeniz'den Orta Anadolu'ya bir Jeotravers: Kuzey Neotetisin tektonik evrimi). *Bulletin of the Turkish Petroleum Geologists Association*, **5**, 1–33.

TÜYSÜZ, O. 1996. Geology of Amasya and surroundings (Amasya ve çevresinin jeolojisi). *11th Petroleum Congress of Turkey, Proceedings, Geology, Turkish Association of Petroleum Geologists/UCTEA Chamber of Petroleum Engineers, Ankara, Turkey*, 32–48.

TÜYSÜZ, O. 1999. Geology of the Cretaceous sedimentary basins of the Western Pontides. *Geological Journal*, **34**, 75–93.

TÜYSÜZ, O. & TEKIN, U.K. 2007. Timing of imbrication of an active continental margin facing the northern branch of Neotethys, Kargı Massif, northern Turkey. *Cretaceous Research*, **28**, 754–764.

TÜYSÜZ, O. & YIĞITBAŞ, E. 1994. The Karakaya Basin: a Palaeo-Tethyan marginal basin and its age of opening. *Acta Geologica Hungarica*, **37**, 327–350.

TÜYSÜZ, O., YILMAZ, Y., YIĞITBAŞ, E. & SERDAR, H.S. 1990. Upper Jurassic–Lower Cretaceous stratigraphy of the Central Pontides and its tectonic meanings. *8th Petroleum Congress of Turkey, Proceedings, Geology, Turkish Association of Petroleum Geologists/UCTEA Chamber of Petroleum Engineers, Ankara, Turkey*, 340–351.

TÜYSÜZ, O., AKSAY, A. & YIĞITBAŞ, E. 2004. Batı Karadeniz Bölgesi Litostratigrafi Birimleri [Litostratigraphic units of the western Black Sea region]. *Maden Tetkik ve Arama Genel Müdürlüğü, Stratigrafi Komitesi Litostratigrafi Birimleri Serisi-1, Ankara [General Directorate of Mineral Research and Exploration, Stratigraphy Comittee, Lithostratigraphic Units Series-1, Ankara, Turkey]*, 92 [in Turkish].

TÜYSÜZ, O., YILMAZ, İ.Ö., ŠVABENICKÁ, L. & KIRICI, S. 2012. The Unaz Formation: a key unit in the Western Black Sea region, N Turkey. *Turkish Journal of Earth Sciences*, **21**, 1009–1028.

TÜYSÜZ, O., MELINTE-DOBRINESCU, M.C., YILMAZ, İ.Ö., KIRICI, S., ŠVABENICKÁ, L. & SKUPIEN, P. 2016. The Kapanboğazı Formation: a key unit for understanding Late Cretaceous evolution of the Pontides, N Turkey. *Palaeogeography, Palaeoclimatology, Palaeoecology*, **441**, 565–581.

USTAÖMER, P.A. & KIPMAN, E. 1998. An example for a Pre-Early Ordovician arc magmatism from North Turkey: geochemical study of the Çaşurtepe Formation (Bolu, W Pontides). *Bulletin of the Mineral Research and Exploration Institute of Turkey*, **120**, 37–53.

VAROL, B. & AKMAN, Ü. 1988. Facies properties of the Barremian -?Aptian carbonates and their characteristic Dacycladacean algae (East of Amasra, Zonguldak, Turkey). *METU Journal of Pure and Applied Sciences*, **21**, 307–319.

YIĞITBAŞ, E., TÜYSÜZ, O. & SERDAR, H.S. 1990. Geology of Late Cretaceous active continental margin in Central Pontides. *8th Petroleum Congress of Turkey. Proceedings, Geology, Turkish Association of Petroleum Geologists/UCTEA Chamber of Petroleum Engineers, Ankara, Turkey*, 141–151.

YIĞITBAŞ, E., WINCHESTER, J.A. & OTTLEY, C. 2008. The Geochemistry and Setting of the Demirci Paragneisses of the Sünnice Massif, NW Turkey. *Turkish Journal of Earth Sciences*, **17**, 421–431.

YILMAZ, O. & BOZTUĞ, D. 1986. Kastamonu granitoid belt of northern Turkey: first arc plutonism product related to subduction of Paleotethys. *Geology*, **14**, 179–183.

YILMAZ, Y., TÜYSÜZ, O., YIĞITBAŞ, E., GENÇ, Ş.C. & ŞENGÖR, A.M.C. 1997. Geology and tectonic evolution of the Pontides. *In*: ROBINSON, A.G. (ed.) *Regional and Petroleum Geology of the Black Sea and Surrounding Region*. American Association of Petroleum Geologists Memoirs, **68**, 183–226.

YURTTAŞ-ÖZDEMIR, Ü. 1971. Macrofauna and biostratigraphy of the Tepeköy Triassic on the Kocaeli peninsula. *Bulletin of the Mineral Research and Exploration, Ankara, Turkey*, **77**, 57–98.

ZONENSHAIN, L.P. & LE PICHON, X. 1986. Deep basins of the Black Sea and Caspian Sea as remnants of Mesozoic back-arc basins. *Tectonophysics*, **123**, 181–211.

Stratigraphy, petrogenesis and geodynamic setting of Late Cretaceous volcanism on the SW margin of the Black Sea, Turkey

MEHMET KESKİN[1]* & OKAN TÜYSÜZ[2]

[1]*Department of Geological Engineering, Faculty of Engineering, İstanbul University, 34320 Avcılar, İstanbul, Turkey*

[2]*Eurasia Institute of Earth Sciences and Faculty of Mines, İstanbul Technical University, 34469 Maslak, İstanbul, Turkey*

Correspondence: keskin@istanbul.edu.tr

Abstract: The Western Pontide Magmatic Belt consists of two different magmatic series corresponding to two distinct periods of intense volcanism, separated by a pelagic limestone marker horizon resting on a regional unconformity. The first stage of magmatism and associated extensional tectonic regime prevailed in the region between the Middle Turonian and Early Santonian. During the first stage, magmas were derived from a depleted mantle source containing a clear subduction signature. The extrusives intercalated with marine clastic sediments and pelagic carbonates associated with thick debris-flow horizons and olistoliths. Based on geochemistry and depositional features, the first stage is interpreted as an extensional ensialic arc setting developed in response to northwards subduction of the Tethys Ocean beneath the southern margin of Laurasia. During the Late Santonian, the volcanism stopped and the whole region suddenly subsided with the deposition of a thin, but laterally continuous, pelagic limestone horizon. This subsidence may imply the break-up of the Laurasian continental lithosphere and the beginning of oceanic spreading in the Western Black Sea Basin. The intensified extension is interpreted to be linked to the southwards rollback of the subducting slab.

During the second stage in the Campanian, magmas were derived from two contrasting mantle sources: (1) a depleted lithospheric mantle enriched by a subduction component; and (2) an enriched asthenospheric mantle which is similar to that of the ocean island basalts (OIB). The depleted lithospheric source may be linked to the subcontinental lithospheric mantle of Laurasia, which was metasomatized by the previous Tethyan subduction event rather than by an active arc magmatism. Lavas derived from the depleted source are abundant throughout the stratigraphic column, whereas those from the enriched source dominate the end of the second stage. The presence of the alkaline lavas may indicate thinning of the lithosphere and upwelling of the asthenospheric mantle in the matured stages of rifting. We argue that the main cause of both rifting and temporal change in magma generation was the steepening and rollback of the northwards subducting slab of the Tethys Ocean. The aforementioned rollback also caused the Istanbul Zone to be moved to the south, and colliding with the Sakarya Zone in the south during the Maastrichtian. Based on geochemical, stratigraphic, palaeontological and sedimentary data, we suggest that the oceanic Western Black Sea Basin opened as an intra-arc basin during Turonian–Santonian time.

Supplementary material: The full geochemical dataset in MS Excel workbook format is available at https://doi.org/10.6084/m9.figshare.c.3841255

The Pontide Magmatic Belt (PMB) connecting the Srednogoria Magmatic Belt of Bulgaria to the Caucasus along the southern coast of the Black Sea is one of the main tectonic elements of the Mediterranean Tethysides. There is a general consensus that this magmatic belt is a result of northwards subduction of the northern branch of Tethys Ocean (Boccaletti *et al.* 1974, 1978; Peccerillo & Taylor 1975; Manetti *et al.* 1979; Şengör & Yılmaz 1981; Yılmaz *et al.* 1997; Tüysüz 1999; Tüysüz *et al.* 2012, 2016), and the Black Sea opened as a back-arc and/or intra-arc basin to the north of the PMB (Letouzey *et al.* 1977; Zonenshain & Le Pichon 1986; Finetti *et al.* 1988; Görür 1988; Manetti *et al.* 1988; Tüysüz *et al.* 2012, 2016). There is also a general agreement that the western and eastern basins of the Black Sea are separated by the continental Andrusov and Archangelsky ridges (see the seismic reflection profiles of Finetti *et al.* 1988; Spadini *et al.* 1996) and hence have different time and mechanisms of opening (Okay *et al.* 1994, 2013; Robinson *et al.* 1996; Spadini *et al.* 1997; Shillington *et al.* 2008, 2009; Stephenson & Schellart 2010; Munteanu *et al.* 2011).

From: SIMMONS, M. D., TARI, G. C. & OKAY, A. I. (eds) 2018. *Petroleum Geology of the Black Sea.*
Geological Society, London, Special Publications, **464**, 95–130.
First published online September 28, 2017, https://doi.org/10.1144/SP464.5

Görür (1988) suggested that the Black Sea had opened during the Aptian, which predates the development of the arc (PMB) itself. Other researchers (Tüysüz 1999; Tüysüz *et al.* 2012; Nikishin *et al.* 2015*a, b*) argued that the Black Sea had opened coevally with the development of the PMB, during Cenomanian–Santonian time. Hippolyte *et al.* (2010) proposed an approximately 40 myr single rifting phase between the Barremian and Coniacian, while Tüysüz (1999) and Tari (2015) proposed two stages of opening: an Aptian–Albian non-volcanic wide rift style extensional system; and a superimposed narrow-rift style volcanic back-arc rifting, the latter corresponding to the subject of our study.

Due to very limited direct observations from the Black Sea itself, which are mainly limited to surficial sediments (Ross 1974, 1978; Ross & Degens 1974; Ross *et al.* 1978), and due to the limited available seismic data (Finetti *et al.* 1988; Nikishin *et al.* 2015*a, b*), most of the tectonic models on the evolution of the Black Sea rely on data from the onshore areas rather from the basins themselves. Thus, any reliable data from this region play an important role in our understanding of the geological evolution of the Black Sea and surrounding tectonic units.

Detailed data on the stratigraphy and age of the Western Part of the PMB (WPMB) were published by Tüysüz (1999), Hippolyte *et al.* (2010, 2015) and Tüysüz *et al.* (2012, 2016). In contrast to the Eastern Pontides, published geochemical data from the WPMB are limited. The only published geochemical data are on the Cretaceous volcanic rocks located north of Istanbul (Keskin *et al.* 2003, 2010), while more geochemical data were published on local intrusive bodies in this zone (Aykol & Tokel 1991; Karacık & Tüysüz 2010; Şahin *et al.* 2012). In this paper, we present, for the first time, geochemical data from volcanic rocks of the WPMB, between Istanbul and the town of İnebolu (Fig. 1), in order to reveal the timing and evolution of the WPMB and its role in the geological evolution of the Western Pontides and the Western Black Sea. We collected more than 250 lava samples from various stratigraphic levels for both geochemical and palaeontological purposes. As most of the lava horizons alternate with sedimentary layers, we have a good palaeontological (planktonic foraminifera) control on the ages of these levels. Fifty-eight representative lava samples were analysed for major oxides, trace and rare earth elements (REEs).

Our aims in this paper are: (i) to briefly overview the stratigraphy of the WPMB based on previously published data (Tüysüz *et al.* 2012, 2016); (ii) to present new geochemical data; (iii) to determine the geodynamic setting of these volcanosedimentary units on the basis of their stratigraphic, structural and geochemical properties; (iv) to constrain the nature of the source(s) from which magmas were derived; (v) to assess the nature and relative contributions of magmatic processes in the evolution of these lavas; (vi) to evaluate the genesis, evolution, spatial and temporal variations of the volcanism; and (vii) to propose a model for magma genesis and sedimentation, as well as for the timing and mechanism of the opening of the Western Black Sea Basin.

Regional tectonic setting

Anatolia, forming a part of the Alpide–Himalayan collision zone, is represented by an orogenic collage consisting of different continental fragments (tectonic units), Gondwanian (southern) and Laurasian (northern) in origin, separated by Tethyan suture zones (Şengör & Yılmaz 1981; Okay & Tüysüz 1999; Okay *et al.* 2001). The Pontides (*sensu* Ketin 1966), the northernmost tectonic unit of Turkey, consists of three different continental fragments (Okay 1989): the Strandja, the Istanbul and the Sakarya zones, from west to east.

Up to now, different models have been proposed on the nature and location of the contacts between these zones (see Okay & Tüysüz 1999; Tüysüz 1999; Tüysüz *et al.* 2016 for details). Some authors (Şengör & Yılmaz 1981; Okay *et al.* 2006, 2013) advocated that the Istanbul and the Sakarya zones juxtaposed before the Late Jurassic. However, based on the stratigraphic differences of these two zones, Tüysüz (1999) and Tüysüz *et al.* (2016) concluded that these two zones were separated by the Tethys Ocean (Intra-Pontide branch) until the Cenomanian. According to this model, the Istanbul Zone rifted off from the Odessa Shelf in the north and moved southwards during the late Barremian–Albian interval in response to the northwards subduction of the Tethys Ocean (Intra-Pontide branch) beneath the southern margin of Laurasia. This rifting gave way to the opening of the Western Black Sea Basin in the north, behind the southwards-moving Istanbul Zone (Okay *et al.* 1994; Schleder *et al.* 2015), and the opening of the Zonguldak-Ulus Basin on the Istanbul Zone (Tüysüz 1999; Masse *et al.* 2009). Details of the pre-Late Cretaceous evolution of the region are discussed in Tüysüz (2017). As our study area sits mainly on the Istanbul Zone, geographically corresponding to the Western Pontides, we will not deal with the pre-Late Cretaceous tectonics of the region, and relationships between the Istanbul and Sakarya zones, which have been discussed in the references given above.

Stratigraphy of the WPMB

The Upper Cretaceous rocks of the WPMB in the Istanbul Zone sit on a regional unconformity

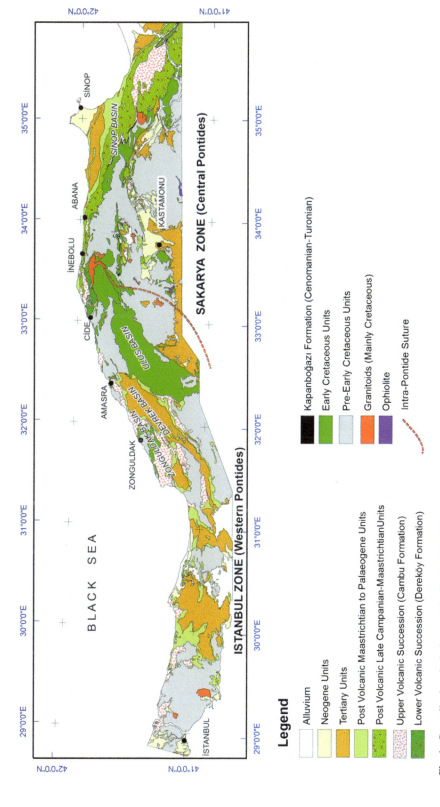

Fig. 1. Generalized geological map of the Western Pontides.

surface. In the western parts of the Istanbul Zone, west of Ereğli (Fig. 1), the Upper Cretaceous units cover Triassic and older units; while in the east (east of Ereğli), WPMB developed on Late Barremian–Albian sediments of the Ulus and the Zonguldak sedimentary basins. These two basins were possibly a single southwards-deepening basin before the development of the overlying latest Cretaceous Devrek Basin. The Zonguldak Basin consists mainly of marginal sediments, while the Ulus Basin was filled by basinal turbidites (see Tüysüz 1999; Hippolyte *et al.* 2010, 2015 for details). The stratigraphy of these basins reflect: (a) the opening of the basins since the Late Barremian; (b) establishment of a short-lived carbonate platform during the late Barremian–early Aptian (see Masse *et al.* 2009); (c) the demise of the platform and deepening of the basins until the Albian; and (d) uplift and erosion during the late Albian–Cenomanian.

In contrast to the Istanbul Zone, sedimentation was continuous during Albian and Cenomanian times (Tüysüz *et al.* 2016) in the Sinop Basin of the Sakarya Zone (Fig. 1). Here, Hauterivian–Albian turbidites grade upwards into Cenomanian–Turonian red pelagic carbonates alternating with radiolarian cherts (i.e. the Kapanboğazı Formation: Tüysüz *et al.* 2016) and then into the volcanic-volcanoclastic rocks of the WPMB. In places, a local unconformity can be seen below the volcanic rocks.

The WPMB is composed mainly of extrusive rocks and associated sediments. Apart from reconnaissance studies before the 1950s, one of the oldest and most detailed, but local, studies on the stratigraphy of these volcanosedimentary rocks was conducted by Akyol *et al.* (1974). Recently, Hippolyte *et al.* (2010, 2015) and Tüysüz *et al.* (2012, 2016) described the Upper Cretaceous stratigraphy of the WPMB based on nannofossil and planktonic foraminifera data, respectively. A detailed account of the volcanostratigraphy and petrology of the WPMB in the north of Istanbul was given by Keskin *et al.* (2003, 2010).

The WPMB itself consists of two different sequences separated by a regional unconformity (Fig. 2). The lower unit, the Dereköy Formation (Şahintürk & Özçelik 1983), is Turonian–early Santonian in age and rests on an unconformity surface covering older units. The first volcanic products of the WPMB are seen within this unit. The upper unit, Late Santonian–Campanian in age, rests unconformably on the Dereköy Formation and older rocks (Fig. 2). Although formal stratigraphic names for this region were published by Tüysüz *et al.* (2004), we use the formation names used in Tüysüz *et al.* (2012, 2016) for simplicity and to prevent confusion with recently published data.

Lower succession: the Dereköy Formation

The Dereköy Formation unconformably overlies Albian and older units to the east of Amasra town (Fig. 1), but is absent in the west of Amasra, possibly due to latest Albian–Cenomanian uplift and erosion (Tüysüz *et al.* 2012). It starts at the base with a thick basal conglomerate deposited in a very high-energy environment, as indicated by badly sorted and rounded pebbles and a grain-supported matrix. The thickness of this conglomerate reaches, in places, up to 100 m. The conglomerates are overlain by a cyclic alternation of fine-grained turbiditic sandstones, white to reddish marls, thin tuff layers and abundant debris-flow horizons (Fig. 3a, b). The lowermost red pelagic limestone horizon within this unit contains planktonic foraminifera that indicate the Middle Turonian (Tüysüz 1999; Tüysüz *et al.* 2012). No fossils have been collected from the lowermost conglomerates of the formation, implying that initiation of the deposition of this formation may be as old as Early Turonian or even Late Cenomanian. Akyol *et al.* (1974) described some Late Cenomanian fossils from this unit.

An important feature of the Dereköy Formation is the occurrence of debris-flow horizons and olistoliths (Figs 3b and 4). The best outcrops of this type of chaotic units are seen between Kurucaşile and Cide, where both the Upper Barremian–Aptian and the Kimmeridgian–Valanginian platform carbonate blocks are embedded within a pelagic matrix. The thick and abundant debris-flow horizons and olistholiths, and the angular shape of the badly sorted pebbles/blocks, embedded within a pelagic matrix, together with frequent thickness and facies changes and synsedimentary growth faults imply that deposition of the Dereköy Formation occurred in a tectonically active and deep-marine environment, possibly under the control of normal faults.

In the west of Amasra, the upper unit rests unconformably on the Albian and older units without the Dereköy Formation at the base. The upper unit is again, but locally, seen directly on the Upper and Middle Jurassic units to the east of Amasra, indicating that the Dereköy Formation was either not deposited or uplifted and eroded in such places before the deposition of the upper unit.

Upper parts of the Dereköy Formation are mainly represented by an alternation of calc-alkaline and acidic to intermediate porphyritic lavas and pyroclastics, red to whitish pelagic micritic limestones and turbiditic volcanoclastics cut by some debris-flow horizons and boulder blocks. The Dereköy Formation overall is a fining- and deepening-upwards succession. Planktonic foraminifera from different pelagic limestone horizons of this formation indicate an age between Middle Turonian and Early Santonian (Tüysüz 1999; Tüysüz *et al.* 2012).

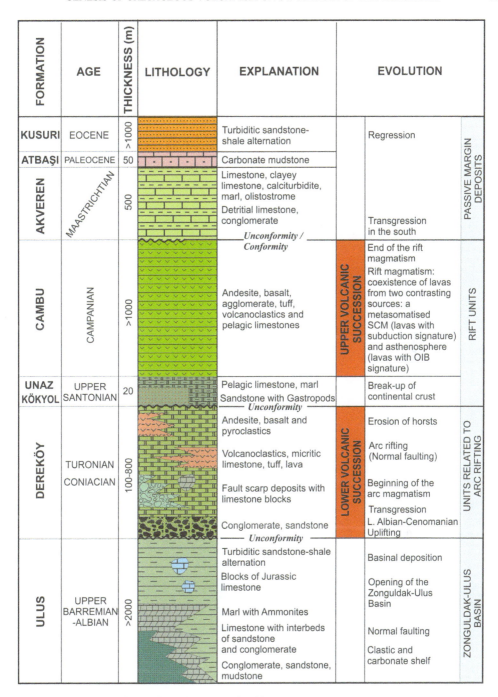

Fig. 2. Generalized stratigraphic section of the Western Pontides.

Based on Coniacian–Santonian nannofossil dates and by disregarding benthic foraminifera listed by Tüysüz (1999), Hippolyte *et al.* (2010) concluded that the age of the Dereköy Formation is possibly Santonian in age, as is the rest of the volcanic sequence. We believe that this a misinterpretation

of the stratigraphy possibly due to sampling strategy and/or misunderstanding of the complex structure of the region (e.g. Sunal & Tüysüz 2002). As indicated by Tüysüz et al. (2012), the contact between the lower and upper units is a pronounced regional unconformity that can be traced all along the Western Pontides, separating two distinct volcanic successions (Fig. 4). Possibly, Hippolyte et al. (2015) also misinterpreted stratigraphy of the Central Pontides and put a synthetic stratigraphic gap between the Albian and Coniacian. However, based on planktic foraminifera, calcareous nannofossils, radiolaria and non-calcareous dinoflagellates, Tüysüz et al. (2016) evidenced a continuous and deep-marine deposition in the Sinop Basin on the Central Pontides from the Albian to the Turonian.

Upper succession

The upper succession consists of three formations: (1) the Kökyol Formation; (2) the Unaz Formation; and (3) the Cambu Formation, from bottom to top.

The Kökyol Formation. This formation (Şahintürk & Özçelik 1983) locally crops out around Amasra (Fig. 1) and sits on a regional unconformity surface. Below the unconformity are mainly the Dereköy Formation to the east of Amasra, and Lower Cretaceous and older (Jurassic, Triassic and Palaeozoic) units to the west of Amasra.

The Kökyol Formation is represented by light-coloured thickly bedded, rounded to subrounded, well-sorted carbonaceous sandstones and conglomerates. In places, it also includes thickly bedded sandy limestones. The presence of big gastropod and bivalve fossils within the formation implies a shallow-marine depositional environment. The total thickness of the formation is about 5–20 m but it reaches up to 100 m in places. No characteristic fossils have been found in the Kökyol Formation to date, but a Middle–Upper Santonian age can be assigned based on its stratigraphic position. Hippolyte et al. (2010) described Coniacian–Santonian nannofossils from this formation. There is no tuff or lava horizon within this formation, indicating that the volcanism was not active during its deposition.

The Unaz Formation. This formation sits on the Kökyol Formation along a sharp, but conformable, contact. To the east of Amasra, it unconformably overlies the Dereköy Formation and older units without the Kökyol Formation (Fig. 4). To the east of Cide, the contact between the Dereköy and overlying Unaz formations is a conglomerate, a few tens of centimetres thick, with a pelagic limestone matrix (see Tüysüz et al. 2012 for details).

The Unaz Formation (Akyol et al. 1974) is represented by rather homogenous red to pinkish, sometimes whitish, thinly bedded, bioclastic micritic limestones. Shale interbeds, a few millimetres thick, and slump structures are common (Fig. 4). The thickness of the formation ranges between 5 and 40 m but locally reaches up to 150 m. These limestones can be traced all along the Pontides as a marker horizon. The Unaz Formation is rich in planktonic foraminifera, indicating a Late Santonian age. Tüysüz et al. (2012) provided detailed sections of the Unaz Formation and concluded that this limestone indicates a sudden deepening of the whole Pontides. There was neither volcanism nor an emergent area during its deposition, as indicated by the absence of siliciclastic and volcanic fragments within this formation.

The Cambu Formation. The Unaz Formation grades upwards into the Cambu Formation (Şahintürk & Özçelik 1983), consisting of pyroclastics, volcanoclastics, pelagic limestones, shales, and basaltic and andesitic lava horizons (Fig. 3c). It is clear from the field observations that the volcanism was more voluminous and more intense during the deposition of the Cambu Formation compared to the Dereköy.

Abundant planktonic foraminifera within the red pelagic limestone horizons that were repeatedly deposited within the formation indicate a Campanian age. The thickness of the Cambu Formation ranges between 500 and 1800 m.

The Cambu Formation is overlain by the Akveren (west of Cide) and the Gürsökü (east of Cide) formations (Figs 2 & 4), both of which consist of calciturbidites. Although there are some thin tuff layers within these formations that are Late Campanian–Maastrichtian in age, the volcanism mainly ceased at the end of the Campanian, except for some weak impulses of tuff eruptions during the Maastrichtian.

Fig. 3. Field photographs from the Western Pontides. (a) Fine-grained sand and mudstone intercalations in the Dereköy Formation, formed in a deep-marine environment: south of the town of İnebolu, north of Kastamonu. (b) Chert beds intercalated with chaotic beds of debris flows in the Dereköy Formation: Evrenye village, east of İnebolu. (c) Contact relationship between the Dereköy Formation and the overlying Unaz Formation. Pelagic limestone beds of the Unaz Formation overlie the Dereköy Formation with an angular unconformity along the roadcut seen on the photograph: Çayaltı village, SW of Cide. (d) Pillow lavas and lava tubes located close to the uppermost part of the Cambu Formation: Güble village, east of İnebolu. The pillows occasionally reach up to 3 m in diameter. (e) An almost 50 m-thick, columnar-jointed trachy-andesitic lava flow in the Cambu Formation: Güzelcehisar beach, NW of Bartın.

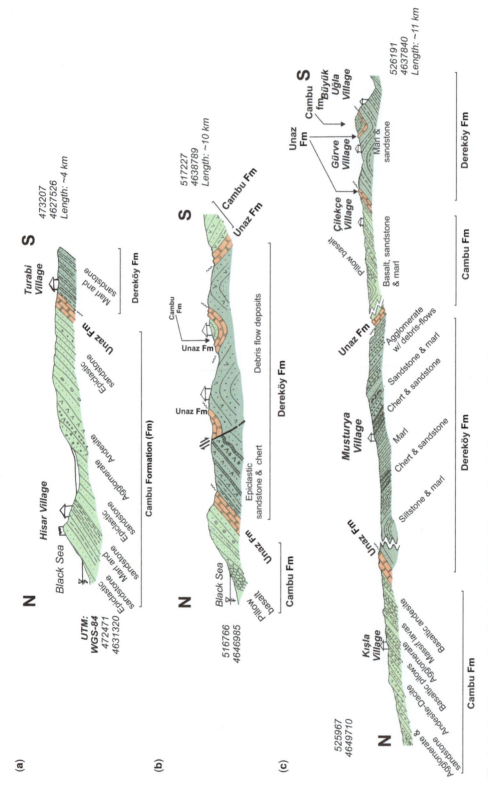

Fig. 4. Geological cross-sections displaying the lithological properties and relationships of the units exposed in the study area. They are arranged from west to east (from **a** to **e**). For the ages and relationships of the formations, see Figure 2. UTM coordinates of each cross-section are displayed on the sides of the section.

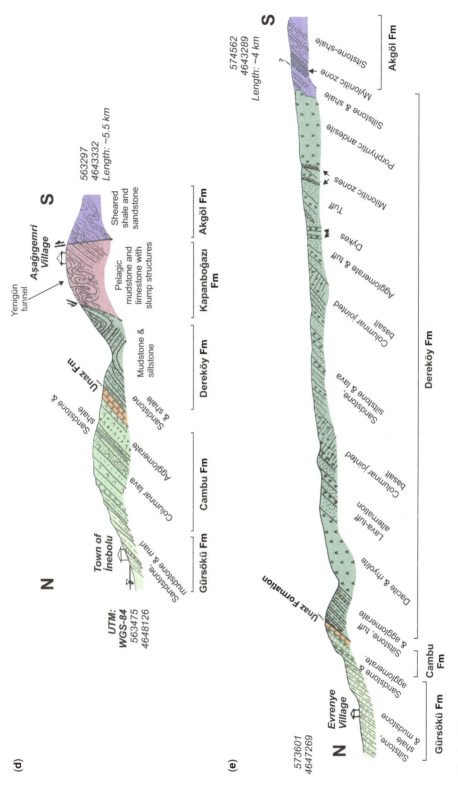

Fig. 4. *Continued.*

Petrology

The lavas and volcaniclastics of the first stage (i.e. the Dereköy Formation) have dominantly dacitic to rhyolitic compositions, except for a few andesitic lavas (Fig. 5a). They are characterized by an anhydrous phenocryst assemblage basically represented by plagioclases (Plg) and pyroxenes (Px) (i.e. Plg + Px-phyric lavas), but may contain quartz phenocrysts in evolved lavas (Table 1). Most of the pyroxenes are represented by clinopyroxene (i.e. augite) phenocrysts, although orthopyroxenes also exist. Plagioclase phenocrysts commonly display rhythmic zoning. In terms of their textural properties, lavas of the Dereköy Formation are porphyritic in texture with a variable microlite and partly devitrified glassy groundmass. Therefore, various types of microlitic textures (i.e. intersertal, hyaloplitic and pilotaxitic) are common. Crystal contents range from 10 to 35 vol%.

The second-stage lavas (i.e. the Cambu Formation) can be divided into three series on the basis of dominant phenocrysts assemblages, they contain: (1) Plg + Px-phyric lavas; (2) Plg + amphibole (Amp)-phyric lavas; and (3) sanidine + Plg + biotite (Bio)-phyric lavas. None of the second stage lavas contains olivine. Sanidine + Plg + Bio-phyric lavas, on the other hand, are quite scarce, exposed only around the south of the town of Amasra. Most of the lavas of the Cambu Formation are represented by Plg + Amp-phyric (hydrous assemblage) and Plg + Px-phyric lavas (anhydrous assemblage), the latter being the dominant one (Fig. 5b–f). The most common mineral in these rocks is Plg, followed in abundance by clinopyroxene (Cpx), orthopyroxene (Opx) and amphibole (Amp) (Fig. 5b–f). Amp and biotite (Bio) occur in relatively minor and variable proportions, while apatite and zircon are rarely present as accessories. The crystal content of these lavas range from about 5 to 45 vol%. They are generally porphyritic in texture, but glomeroporphyritic, seriate and various types of microlitic textures also exist. In some of these lavas, zoned Plg phenocrysts contain numerous patchy glass inclusions, sieve texture, zonal inclusions (Fig. 5d) of various microcrystals and corroded rims. These lavas also display petrographical evidence of an early Opx crystallization followed by resorption and then mantling by a Cpx rim. These textures are typical of the magmas that evolved under the influence of the magma-mixing process.

Geochemistry

Analytical methods

Major elements of 58 representative volcanic rock samples from the WPMB were analysed by inductively coupled plasma optical emission spectrometry (ICP-OES), while trace elements including REE for the same samples were analysed by inductively coupled plasma mass spectrometry (ICP-MS) at the ACME laboratories, Canada (Table 2). Seven international USGS standards (i.e. AGV-1, BCR-2, G-2, GSP-2, W-2, SO-2 and SO-15, and also an internal standard, CT-3) were analysed together with the samples. The standard error for each element was calculated on the basis of results from regression analyses of analysed geo-standards and listed at the end of the Table 2. Samples showing heavier alteration were rejected.

Geochemical classification

We applied the formal nomenclature schemes of Irvine & Baragar (1971), Peccerillo & Taylor (1976) and Le Bas et al. (1986) for the classification of our samples (Fig. 6). We also used the Zr/TiO_2 v. SiO_2 diagram of Winchester & Floyd (1977) to check the extent to which weathering affected the major and trace element geochemistry. On these diagrams, we divide the samples into a number of data series on the basis of petrographical classification and use major element concentrations calculated on loss on ignition (LOI)-free basis.

Both aphyric and Plg + px-phyric lavas of the Dereköy Formation span a compositional range from andesite to rhyolite on the total alkali v. silica (TAS) diagram of Le Bas et al. (1986), except for

Table 1. *The stages of volcanism, corresponding formations, magma series and mineral combinations in each magma series*

Formation	Stage	Abbreviation	Magma series	Mineral content	Type
Cambu	Late	C_A	Alkaline and transitional (High-K series)	Plagioclase + clinopyroxene	Blocks and lavas
		C_{CA}	Calc-alkaline (Medium-K series)	Plagioclase ± sanidine amphibole + pyroxene	
Dereköy	Early	D_{CA}	Calc-alkaline (Low- and High-K series)	Plagioclase + clinopyroxene	Blocks and lavas

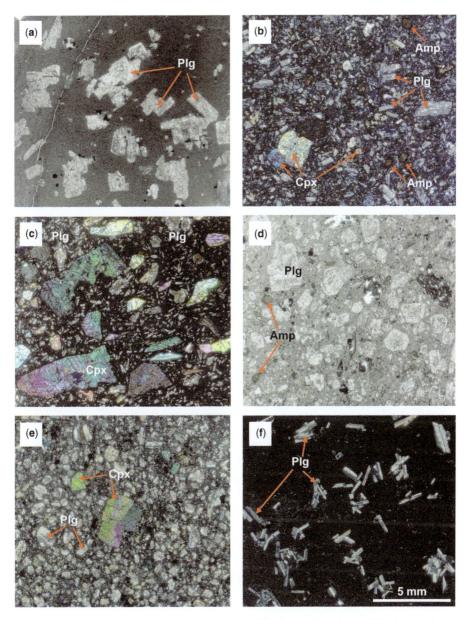

Fig. 5. Microphotographs of the thin sections of major volcanic units from the Late Cretaceous volcanosedimentary succession. (**a**) OT-255 (PPL). A calc-alkaline dacite sample from the Dereköy Formation containing large, partly weathered and euhedral plagioclase phenocrysts set in a glassy groundmass. Vitrophyric, porphyritic texture. (**b**) Gezi II-7 (CPL). An alkaline basalt sample from the Cambu Formation (High-K basalt). Large anhedral augite, subhedral plagioclase and fine brown hornblende phenocrysts are set in a glassy and partly microcrystalline groundmass. Porphyritic, seriate texture. Note that this particular sample is the most primitive lava from the alkaline lava series. (**c**) OT-248 (CPL). Calc-alkaline basaltic andesite from the Cambu Formation. Anhedral to subhedral clinopyroxene (mostly augite) megacrysts, and random plagioclase phenocrysts set in a hyaloplitic groundmass. Porphyritic. This particular sample is the most primitive sample of the calc-alkaline lava series. (**d**) OT-239 (PPL). Calc-alkaline dacite from the Cambu Formation. Porphyritic texture. Note the zonal glass inclusions in anhedral to subhedral plagioclase phenocrysts. (**e**) OT-226 (CPL). Calc-alkaline basaltic andesite from the Cambu Formation. Subhedral clinopyroxene magacrysts with twinning, and rounded and partly resorbed plagioclase phenorysts. Porphyritic. (**f**) OT-98 (CPL). Alkaline andesite (High-K) from the Cambu Formation. Vitrophyric, porphyritic. Abbreviations: Cpx, clinopyroxene; Plg, plagioclase; Amp, amphibole.

Table 2. *Major element, trace element and REE ICP-MS data of a subset of representative samples from the study area in Western Pontides*

Inebolu-Cide area

Dereköy Formation (first stage of volcanism)

Series	Low-K calc-alkaline			High-K calc-alkaline			Alkaline		
Sample No.	OT73	OT201	OT225	OT72	OT209	OT215	Gezi II-7	OT204	Gezi II-10
UTM coordinates									
X	538300	515950	579425	522350	569925	574675	380720	562500	447050
Y	4647850	4637800	4644475	4643800	4645925	4644225	4576570	4646525	4619025
Stratigraphic position	Middle	Middle	Middle	Uppermost	Uppermost	Middle			
Geochemical description	A	D	R	T	D	R	Hi-K BA	Hi-K A	T
Type	Lava	Block	Block	Lava	Lava	Dyke	Lava	Lava	Lava
Texture	Porph	Porph	Porph	Porph	Aphy	Aphy	Mic.Porp	Mic.Porp	Porph
Phenocrysts	Qtz Plg Px	Plg	Plg Px	Plg Px	–	–	Amp	Plg Px	Sanidine
SiO_2	62.63	73.42	79.04	66.67	68.26	72.55	54.08	59.84	62.83
TiO_2	0.496	0.736	0.403	0.553	0.776	0.501	1.090	0.832	0.394
Al_2O_3	18.07	14.15	11.28	17.40	14.62	13.45	16.34	18.08	18.34
Fe_2O_3	5.08	2.29	1.87	3.29	3.59	2.97	8.59	5.41	3.92
MnO	0.062	0.041	0.050	0.052	0.092	0.061	0.100	0.144	0.091
MgO	4.35	1.51	0.54	0.54	0.54	0.90	3.45	1.76	1.11
CaO	4.12	3.38	0.78	2.44	4.20	1.05	6.46	3.35	2.80
Na_2O	4.27	3.38	5.05	4.00	3.28	3.64	4.39	4.43	4.57
K_2O	0.82	0.96	0.90	4.95	4.18	4.69	2.16	5.75	5.76
P_2O_5	0.10	0.14	0.08	0.09	0.46	0.18	0.26	0.40	0.18
LOI	3.30	2.10	1.00	3.60	1.60	1.90	3.00	1.60	0.60
Sc	13	18	9	9	15	11	22	9	6
Cr	6	117	8	6	3	4	44	4	4
V	97	98	34	31	89	35	186	170	73
Ni	3	68	5	3			18		3
Co	11	19	4	3	13	6	24	11	7
Cu	31.0	32.1	18.2	15.0	24.6	26.7	29.4	64.1	29.9
Zn	60	52	35	52	81	49	71	89	68
Rb	17	16	21	189	94	70	79	120	148
Sr	513	378	151	324	315	141	361	642	397
Y	19	26	27	27	27	42	29	30	22
Zr	143	108	171	182	126	190	166	120	127
Nb	4	9	6	10	8	10	18	7	7
Ba	190	1633	77	983	767	545	414	1462	709
La	16.7	17.3	21.6	32.9	25.5	27.3	21.5	26.5	22.3
Ce	34	32	44	53	50	51	41	51	45
Pr	5	4	6	7	7	8	4	7	5
Nd	17.0	16.6	24.7	26.8	26.7	29.8	19.1	28.9	21.7
Sm	3.46	4.08	5.37	5.55	5.66	6.68	4.40	6.03	4.37
Eu	1.00	1.03	0.95	1.41	1.39	1.36	1.26	1.79	1.15
Gd	3.27	3.96	5.09	5.43	5.34	6.46	4.38	5.54	3.95
Tb	0.47	0.59	0.81	0.80	0.78	1.00	0.69	0.83	0.56
Dy	2.90	4.04	5.02	4.79	4.55	6.05	4.60	5.01	3.55
Ho	0.69	0.85	1.18	1.04	0.98	1.39	0.98	1.05	0.72
Er	2.05	2.33	3.26	2.78	2.65	3.94	2.84	2.96	2.17
Tm	0.39	0.40	0.61	0.53	0.45	0.77	0.45	0.55	0.35
Yb	2.11	2.10	3.06	2.49	2.37	3.58	3.03	2.75	2.57
Lu	0.32	0.27	0.44	0.38	0.31	0.57	0.33	0.42	0.29
Hf	3.87	2.99	4.71	5.00	3.59	6.42	4.61	3.30	3.15
Ta	0.27	0.42	0.48	0.55	0.35	0.73	0.74	0.34	0.43
Pb		14	9	21	21	24	6	38	20
Th	4.8	4.1	5.2	11.1	7.5	5.5	4.4	7.2	9.4
U	2.4	2.7	3.6	6.9	7.9	4.1	1.0	4.7	3.5

The full geochemical dataset can be found in the Supplementary material. UTM coordinates (i.e. UTM Zone 36 ED50) of the samples Hi-K A, high-K (potassium) andesite; D, dacite; R, rhyolite; T, trachyte.

Cambu Formation (second stage of volcanism)

	Mildly alkaline			Calc alkaline					
OT233	OT299	OT293	OT248	OT226	Gezi II-4	OT241	OT238		
537800	527550	544350	522800	550200	369300	544300	544900		
4648500	4648950	4649400	4648850	4648000	4562600	4649400	4649500		
Hi-K A	T	D	BA	BA	Hi-K BA	D	R	**Standard**	
Lava	Lava	Lava	Block	Block	Lava	Block	Block	**Errors**	
Aphy	Aphy	Porph	Porph	Porph	Porph	Porph	Porph		
–	–	Plg	Plg Px	Plg Amp	Plg Px	Plg Amp	Qtz Plg	(±%)	
61.15	65.35	70.37	49.90	54.45	60.26	65.78	77.04	0.36	
0.852	0.747	0.479	0.747	0.689	0.649	0.489	0.214	0.004	
16.37	14.51	15.79	14.51	18.73	17.71	16.46	11.91	0.07	
6.96	5.88	2.08	10.00	8.43	5.65	4.20	1.54	0.07	
0.125	0.113	0.061	0.187	0.144	0.071	0.082	0.031	0.002	
2.25	0.91	0.39	8.30	3.49	2.54	1.70	0.17	0.03	
4.43	4.30	2.00	12.78	10.41	6.53	5.26	2.81	0.05	
3.43	2.98	3.64	2.10	2.65	4.24	3.33	6.16	0.04	
3.99	4.79	5.11	1.31	0.93	2.08	2.55	0.12	0.02	
0.45	0.42	0.09	0.18	0.08	0.26	0.14	0.00	0.02	
3.40	2.20	1.40	3.30	2.60	1.40	1.60	1.80	(±ppm)	
17	15	7	52	38	13	18	4	0.5	
4	3	4	188	23	53	6	5	1.7	
99	81	24	299	305	109	127	26	3.1	
			57	12	29	4		1.4	
9	4	4	43	25	15	8		0.8	
28.9	3.2	18.2	107.9	42.7	89.8	49.2	9.6	0.2	
95	69	41	78	84	56	57	19	2.2	
102	186	66	29	23	55	64	2	0.7	
370	367	267	455	294	619	287	67	4.2	
30	25	24	18	19	16	17	17	0.5	
155	137	175	44	47	139	86	119	3.4	
10	7	9	2	2	12	5	5	0.8	
741	872	3469	251	245	465	735	32	11.8	
27.0	24.9	28.5	8.8	6.1	21.4	14.6	3.4	0.4	
54	49	52	19	13	40	26	10	1.5	
7	6	6	3	2	4	3	1	1.2	
28.2	26.2	24.5	13.3	8.4	15.8	12.2	5.7	0.7	
6.06	5.59	4.86	3.42	2.42	3.06	2.64	1.60	0.09	
1.49	1.39	1.60	0.93	0.70	0.92	0.82	0.36	0.06	
5.76	5.52	4.68	3.23	2.58	3.05	2.82	1.71	0.24	
0.86	0.78	0.70	0.50	0.41	0.39	0.42	0.32	0.05	
4.77	4.66	4.86	3.16	2.80	2.54	2.73	2.44	0.13	
1.06	0.98	0.88	0.64	0.62	0.46	0.57	0.58	0.04	
2.73	2.44	2.24	1.56	1.65	1.42	1.58	1.75	0.05	
0.48	0.41	0.39	0.23	0.25	0.13	0.28	0.30	0.04	
2.48	2.33	2.34	1.48	1.65	1.54	1.62	1.72	0.07	
0.32	0.27	0.27	0.16	0.18	0.07	0.21	0.19	0.02	
5.09	3.51	4.71	1.38	1.41	3.20	2.32	3.10	0.04	
0.68	0.42	0.65	0.17	0.09	0.70	0.33	0.24	0.04	
28	11	21	8	11	12	14	7	3.5	
8.5	7.6	9.5	2.4	1.2	5.2	5.4	1.8	0.4	
6.5	5.8	6.9	1.0	0.5	1.4	2.0	0.8	0.06	

are given in the table, below the sample numbers. Abbreviations: Hi-K BA, high-K (potassium) basaltic-andesite; BA, basaltic andesite;

First Stage: Lavas of the Dereköy Formation

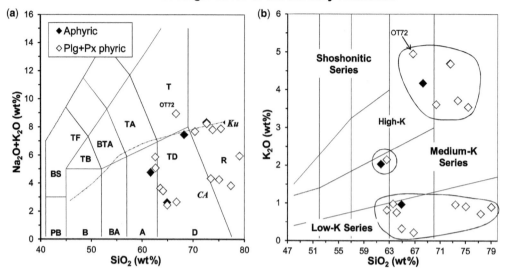

Second Stage: Lavas of the Cambu Formation

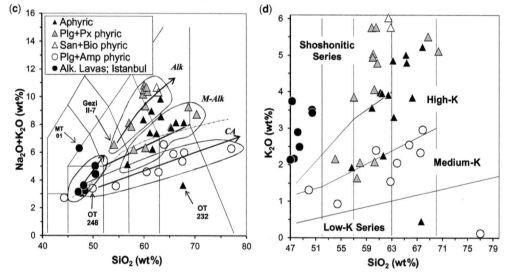

Fig. 6. Classification of the lavas of the first and second stages from the study area. (**a**) & (**b**) TAS (total alkali v. silica) diagram of Le Bas *et al.* (1986) from the early and second stage units (i.e. the Dereköy and Cambu formations, respectively). Abbreviations: PB, picritic basalt; B, basalt; BA, basaltic andesite; A, andesite; D, dacite; R rhyolite; TB, trachybasalt; BTA, basaltic trachyandesite; TA, trachyandesite; TD, trachydacite; T, trachite. K is the alkaline–subalkaline divide of Kuno (1966). (**c**) & (**d**) K_2O v. SiO_2 diagram of Peccerillo & Taylor (1976) displaying the magma series. Note that the Upper Cretaceous alkaline lavas from the north of Istanbul are also plotted on the diagrams for comparison.

one sample (OT72) which plots into the trachydacite field (Fig. 6a). They may be divided into three series on the basis of the Peccerillo & Taylor (1976) K_2O v. SiO_2 diagram: the Low-, Medium- and High-K Series (Fig. 6b). On the AFM diagram of Irvine & Baragar (1971), all samples fall into the calc-alkaline field except for two samples that plot into the tholeiitic field (Fig. 7).

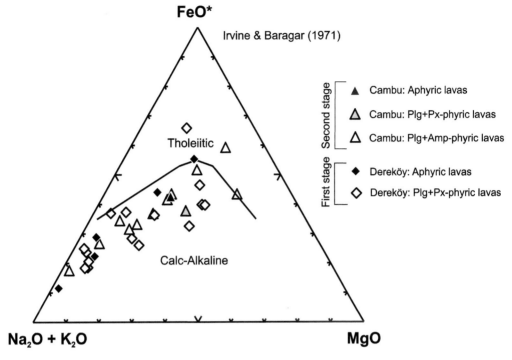

Fig. 7. (**a**) AFM diagram of Irvine & Baragar (1971) showing the calc-alkaline character of the subalkaline lavas (see Fig. 6a, c). Alkaline or mildly alkaline lavas of the second stage are not plotted on this diagram.

Lavas of the Cambu Formation may be divided into three series on the basis of their total alkalis (TA) content: (1) calc-alkaline lavas (TA <6% for intermediate and <8% for acid lavas); (2) alkaline lavas (TA >6% for intermediate and >8% for acid ones); and (3) transitional or mildly alkaline lavas (see the fields in Fig. 6c). Calc-alkaline (CA) lavas correspond to Plg + Amp- and Plg + Px-phyric lavas. They span a broad compositional range from basalt to rhyolite. Alkaline and mildly alkaline lavas, on the other hand, coincide with aphyric and Plg + Px-phyric pillow lavas. They plot into the High-K, shoshonitic and banakite fields on the Peccerillo & Taylor (1976) diagram (Fig. 6d). Results from the Winchester & Floyd (1977) classification diagram, constructed using immobile elements, are in good agreement with those of the TAS and K_2O v. SiO_2 diagrams (Fig. 8). This implies that alteration did not have a substantial effect on classification schemes based on major element geochemistry.

Tectonic discrimination

Basalts and basaltic andesites, containing less than 60 wt% SiO_2, have been plotted onto the tectono-magmatic discrimination diagrams of Pearce & Cann (1973), Pearce (1982), Meschede (1986) and Wood (1980) (Fig. 9). Note that none of the samples

of the Dereköy Formation has been plotted in this figure, since all of them have silica contents greater than 60%. We also plotted alkaline basalts of the Kısırkaya Formation (Keskin *et al.* 2003, 2010) from the north of Istanbul, displaying a within-plate signature, to be able to make a comparison with the coeval alkaline lavas of the Cambu Formation.

Lavas of the Cambu Formation plot into the calc-alkaline basalt field, while the majority of the alkaline lavas from the Kısırkaya Formation fall into the within-plate field on the Pearce & Cann (1973) diagram (Fig. 9a). On the Meschede (1986) diagram, these lavas are basically confined to the volcanic-arc field, with the exception of one alkaline sample (Gezi II-7) from the Cambu Formation that plots into the within-plate field (Fig. 9b). Similarly, all Cambu samples are located in the volcanic arc field on the Pearce (1982) Zr v. Ti diagram, except for that particular alkaline sample (Gezi II-7) (Fig. 9d). Lavas of the Kısırkaya Formation fall consistently into the within plate field on these diagrams, except for three samples that plot on the border dividing volcanic arc field from the within-plate field. In summary, the distribution of data points on the tectonomagmatic discrimination diagrams implies that the majority of calc-alkaline lavas show geochemical affinities with island-arc basalts. In contrast, alkaline lavas of the Kısırkaya Formation

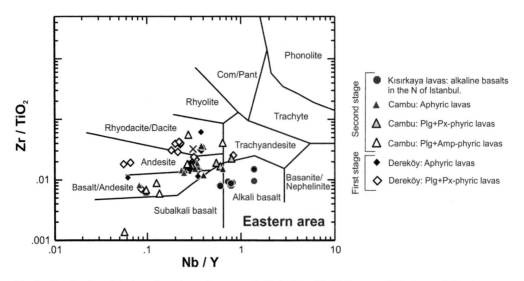

Fig. 8. Classification of the lavas from the study area on the Zr/TiO_2 v. Nb/Y diagram of Winchester & Floyd (1977). Upper Cretaceous alkaline lavas from the north of Istanbul (i.e. Kısırkaya lavas: Keskin *et al.* 2010) are also plotted on the diagrams for comparison.

and sample Gezi II-7 bear a close resemblance to within-plate (WP) basalts. A subset of samples classifying as alkaline or mildly alkaline from the higher stratigraphic levels of the Cambu Formation also plot close to the within-plate basalt fields. Data points of both alkaline and calc-alkaline lavas seem to align along the trends between the most basic samples of the alkaline and calc-alkaline lavas (OT248 and MT-01, respectively: Fig. 9a).

Variations of major and trace elements with silica

Harker diagrams for major and trace elements plotted against silica demonstrate some significant differences between alkaline-WP lavas and the calc-alkaline series (Figs 10 & 11). In order to make tracking of fractionation trends easier, we have plotted the data onto two sets of diagrams side-by-side, corresponding to the first and second stages of the volcanism, on the left and right, respectively (Fig. 10).

It is evident from the behaviour of some major (e.g. TiO_2, CaO and P_2O_5) and trace elements (e.g. Sc, Y, Zr, Nb, La, Hf, Ta and Pb) that the fractionation trends formed by the calc-alkaline (CA) lava series of the Dereköy and Cambu formations, on the one hand, and those formed by alkaline lavas of the Cambu Formation, on the other, appear to start from two different basic end-member compositions (the data points shown in circles in Figs 10 & 11). This may be regarded as an indication of the existence of two diverse parental magma

compositions from which calc-alkaline and alkaline magmas have started to evolve separately. For ease of exposition, the following abbreviations will be used throughout the rest of the paper: calc-alkaline lava series of Dereköy (henceforth, De_{CA}), Cambu (C_{CA}), and alkaline lava series of the Cambu (C_A).

Determination of basaltic end members. The most primitive calc-alkaline basalt sample OT248 from the C_{CA} series, containing the lowest SiO_2 and the highest MgO concentrations (49 and 8.3%, respectively), might be regarded as the parental magma for calc-alkaline lava series of both the first and second stages. The trends formed by data points of the CA lava series either start from this sample or their backwards projection intersects with this composition on both major and trace element Harker diagrams. It is termed here BEM-SC (BEM stands for Basaltic End Member) because it has a very clear subduction component (SC), consistently falling into 'arc' fields on tectonomagmatic discrimination diagrams (Fig. 9).

Similarly, parental magma for alkaline lava series corresponds to the most basic alkaline basaltic lava sample (i.e. MT-01 with $SiO_2 = 47.32$ wt%) from the Western Pontides, collected from the Kısırkaya Formation (Keskin *et al.* 2010). None of the alkaline samples classify as basalt in the study area, even the most basic alkaline sample (Gezi II-7) being basaltic trachyandesite. Although there are no basaltic samples available for the lavas of De_{CA}, backwards projections of the main liquid line of descent trends on the Harker diagrams of this series appear to head

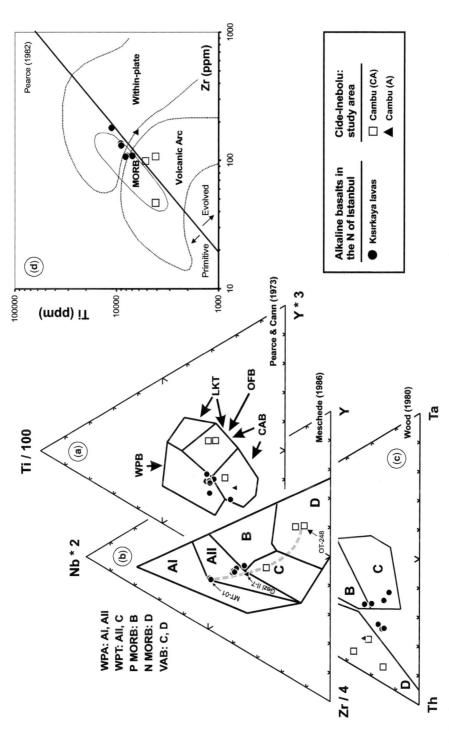

Fig. 9. Discrimination diagrams reflecting possible palaeo-tectonic setting(s) of the volcanic units from the study area: (**a**) Pearce & Cann (1973), (**b**) Meschede (1986), (**c**) Wood (1980) and (**d**) Pearce (1982). Only basaltic and basaltic andesitic lavas (SiO₂, <60 wt%) are plotted on these diagrams. The Upper Cretaceous alkaline basalts from the north of Istanbul (i.e. Kısırkaya lavas) are also plotted on the diagrams for comparison. Abbreviations: WPA, within-plate alkaline basalts; WPT, within-plate tholeiites; P MORB, P-type mid-ocean ridge basalts (MORB); N MORB, N-type MORB; VAB, volcanic arc basalt; WPB, within-plate basalts; LKT, low-K (potassium) tholeiites; OFB, ocean floor basalts; CAB, island arc (calc-alkaline) basalts; B, enriched-MORB and within-plate tholeiites; C, alkaline within-plate basalts; D, calc-alkaline basalts.

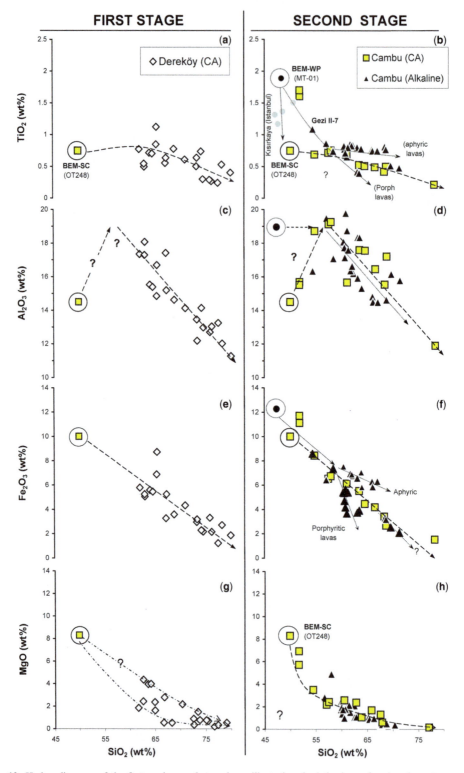

Fig. 10. Harker diagrams of the first- and second-stage lavas illustrating the behaviour of major element

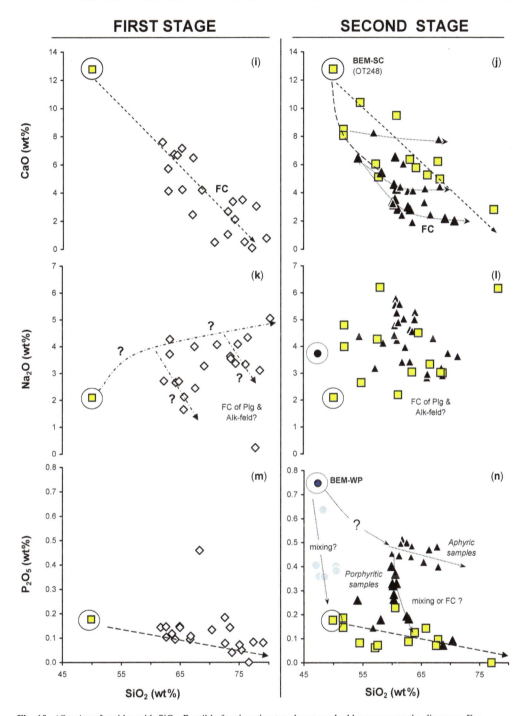

Fig. 10. (*Continued*) oxides with SiO₂. Possible fractionation trends are marked by arrows on the diagrams. For comparison, the Upper Cretaceous alkaline basalts from the north of Istanbul (i.e. Kısırkaya lavas) are also plotted on the diagrams. Abbreviations: FC, fractional crystallization; BEM-WP, basaltic end-member composition displaying a clear within-plate signature; BEM-SC; basaltic end-member composition displaying a clear subduction component.

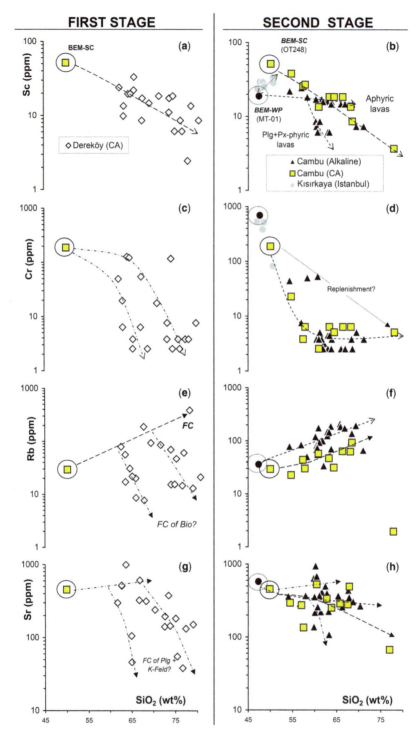

Fig. 11. Harker diagrams of the first- and second-stage lavas illustrating the behaviour of trace elements with SiO$_2$. Possible fractionation trends are marked by arrows on the diagrams. For comparison, the most primitive Upper Cretaceous alkaline basalts from the north of Istanbul (i.e. Kısırkaya lavas) are also plotted on the diagrams. FC, BEM-WP and BEM-SC are as explained in Figure 12.

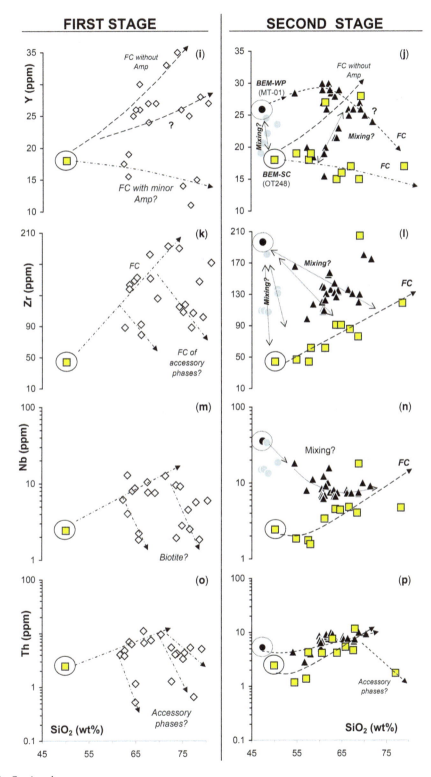

Fig. 11. *Continued.*

towards the BEM-SC composition on almost all major and trace element diagrams (Figs 10 & 11). Therefore, the BEM-SC may also be considered as the parental composition for the De_{CA} lavas of the first stage. Figures 10 and 11 show that the BEM-WP (Basaltic End Member of Within-Plate lavas: i.e. MT-01) is depleted in CaO and Sc but enriched in almost all other elements, especially in TiO_2, MgO, P_2O_5, Cr, Ni (not shown), Rb, Y, Zr (c. 197 ppm), Nb (c. 36 ppm), La, Hf (3.6 ppm), Ta (3.1 ppm) and Th with respect to BEM-SC (i.e. OT248).

Major element fractionation trends. In all lava series, irrespective of their stratigraphic position and alkalinity, major oxides decrease with increasing silica, except for Al_2O_3 at basic and Na_2O and K_2O at intermediate compositions (Fig. 10). There is a major inflection in Al_2O_3 at around 56–57% SiO_2 in the calc-alkaline lava series. A similar inflection is also present in the fractional crystallization (FC) trend of the alkaline lava series (i.e. C_A). This may be interpreted as a turning point during which the early ferromagnesian-dominated crystallization possibly gave way to feldspar-dominated crystallization. Scatter in Na_2O in the lavas of the C_{CA} series may be due to weathering and, possibly to a lesser extent, crystal accumulation. Behaviour of MgO appears to be consistent with this interpretation. It falls rapidly at the basic end in the lavas of C_A and C_{CA} series, possibly due to fractionation dominated by mafic mineral phases, and then starts to decrease relatively more slowly after about 58 wt% SiO_2 as fractionation proceeds, probably in response to incoming plagioclase crystallization. A similar pattern is also visible in Cr and Ni v. SiO_2 diagrams (Ni is not shown in Fig. 11). These findings indicate that fractionation of olivine was more important at basic compositions in the lavas of the C_A and C_{CA} series with respect to the De_{CA} series.

Fractionation trends of the C_A series are diverged into two or three trends on TiO_2, Fe_2O_3 and P_2O_5 v. SiO_2 diagrams at about 60% SiO_2. On each diagram, the lower trend displaying more depletion corresponds to the porphyritic lavas, while the upper trend showing more enrichment for equivalent silica contents is represented by aphyric lavas (Fig. 10).

Trace element fractionation trends. Although behaviours of the major elements in the first and second stage calc-alkaline lava series (i.e. De_{CA} and C_{CA} series) are fairly similar, there are significant differences in trace element behaviours (Fig. 11). Lavas of the De_{CA} series, in general, are characterized by higher values of Zr, Nb, La and Hf with respect to the lavas of the C_{CA} series at a given silica value.

Data points of lavas of the De_{CA} series display two major inflections at around 62 and 72% SiO_2 from the main liquid line of descent in a number of trace elements, including Rb, Sr, Zr, Nb, Ba, La (not shown) and Th (Fig. 11). Note that a similar inflection is seen in Na_2O v. SiO_2 diagram at almost the same silica values. This can be attributed to the onset of alkali feldspar (for Ba and Sr), biotite (for Rb, Ba and Nb), Th-bearing accessory phases (e.g. zircon and allanite) and plagioclase (for Na_2O and Sr) at intermediate and acid compositions of the De_{CA} series.

On the Y v. SiO_2 diagram, data points of the De_{CA} lavas seem to display a divergence at intermediate composition (c. 55% SiO_2) forming two or possibly three trends (Fig. 11i). The first trend displaying a slight negative correlation with silica may reflect crystallization of minor amounts of amphibole in the mafic phase, while the second and third trends with positive gradients necessitate fractionation of anhydrous mineral phases. Behaviour of Rb and Nb (Fig. 11e, m) implies biotite crystallization for a subset of De_{CA} lavas. Hence, lavas of the De_{CA} series can be divided into two fractionation series: (1) the one possibly containing minor amphibole ± clinopyroxene ± biotite; and (2) the other dominated by clinopyroxene + olivine. Lavas of the De_{CA} series are enriched in Cr and Ni (not shown), as well as in MgO with respect to the lavas of the C_A and C_{CA} series at intermediate composition. This may indicate that fractionation of Mg-bearing mafic phases are more important in the lavas of the second stage relative to that of the first stage at intermediate composition.

Notably, a subset of data points of the C_A series diverges from the main trend heading towards the fractionation trend of the C_{CA} series on Y, Zr, Nb, Hf and Ta v. SiO_2 diagrams (Fig. 11). This brings in the question of whether these divergences are a sign of interaction between the alkaline and calc-alkaline magma series during the second stage of the volcanism. When FC trends are closely examined in Figures 10 and 11, a number of points draw one's attention: (1) FC trends of the alkaline and calc-alkaline series are strongly diverged on most of the diagrams; and (2) there are significant temporal differences in the calc-alkaline lava series, namely between the lavas of the first and second stages. On account of these findings, we argue that there may be three distinct fractionation trends in the Upper Cretaceous volcanic units exposed throughout the Western Pontides:

- calc-alkaline lavas of the first stage, represented by the lavas of the De_{CA} series;
- calc-alkaline lavas of the earlier period of the second stage (i.e. C_{CA} series);
- alkaline or mildly-alkaline lavas of the final period of the second stage (i.e. C_A series).

Calc-alkaline lavas of the first and second stages appear to have fractionated both anhydrous and hydrous mineral assemblages basically consisting of plagioclase, clino- and ortho-pyroxenes, amphibole with occasional biotite, and oxides, whereas alkaline and mildly alkaline lavas of the second stage underwent a fractionation dominated by an anhydrous assemblage composed basically of plagioclase, clinopyroxene and olivine. Before examining the nature and extent of any possible interaction between alkaline and calc-alkaline lavas of the second-stage lavas, we would like to further examine the fractionation history of the magma series by interpreting

chondrite-normalized REE patterns in the next subsection.

Interpretation of REE patterns

Figure 12 illustrates chondrite-normalized REE patterns for a subset of representative samples from first- and second-stage lava series. Note that we plotted BEM-SC and BEM-WP compositions on these diagrams for comparison. These two basaltic end-member compositions form two distinct patterns

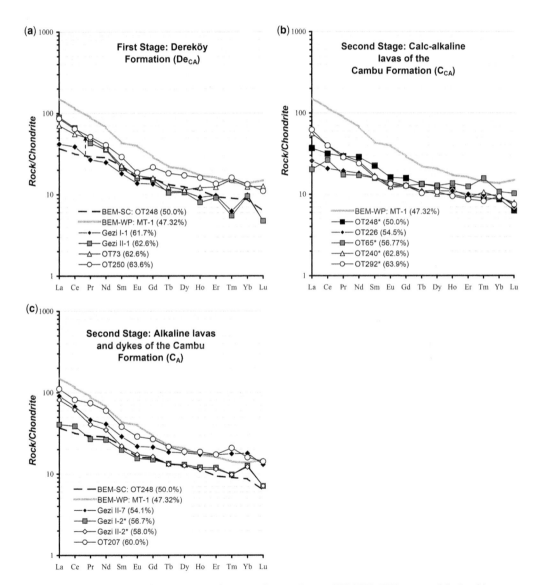

Fig. 12. Chondrite-normalized REE patterns of the lavas from study area. BEM-WP, REE pattern of the basaltic end-member composition displaying a within-plate signature; BEM-SC, REE pattern of the basaltic end-member composition displaying a clear subduction component (Sun & McDonough 1989). For an explanation, see the text.

plotted away from each other, while the rest of the samples fall mostly between these two patterns. BEM-WP shows a much steeper LREE/MREE (light REE/medium REE) profile relative to that of the BEM-SC. This is compatible with the enriched nature of the alkaline BEM-WP lava.

The C_{CA} lava series displays the flattest REE patterns in contrast to the alkaline lavas of the same stage (i.e. C_A) displaying marked LREE enrichments. The most primitive alkaline second-stage lavas (i.e. BEM-WP, Gezi II-2 and Gezi II-7) exhibit higher LREE and MREE concentrations, slightly higher HREE (heavy REE) concentrations, and much steeper LREE/MREE profiles relative to the calc-alkaline series. These profiles are also consistent with crystallization of an anhydrous mineral assemblage and the more profound effect of pyroxenes in the C_{CA} lava series.

Petrological modelling of magma chamber processes

In this section, we will focus on the magmatic processes that might have been operational in magma chamber evolution of the Upper Cretaceous lavas by constructing petrological models utilizing trace elements. For this aim, we will concentrate on crystal fractionation and magma-mixing processes for the second-stage lavas, as suggested by the behaviour of trace elements (i.e. Fig. 11).

Fractional crystallization v. magma mixing

In order to test the feasibility of a magma mixing process, we plotted some highly compatible elements V, Ti, Fe and Sc (not shown) against Zr, an immobile and highly incompatible element at basic to intermediate magma compositions (Fig. 13). Notably, BEM-SC and BEM-WP compositions plot far away from each other on these diagrams; BEM-WP being enriched in Zr, Ti and Fe, and depleted in V and Sc with respect to BEM-SC. The majority of calc-alkaline lavas follow a curve with some scatter, while most of the alkaline lavas align along one or two linear trends between BEM-WP and the curved line. Hf v. Fe, Ti and V plots (not shown) display almost the same relationship between the alkaline and calc-alkaline magma series. Among these diagrams, Fe v. Zr plot displays the most coherent trends (Fig. 13c). We modelled Rayleigh fractionation curves that best-fit to the distribution of the data points and then plotted alongside the data (Fig. 13b, c), taking the BEM-SC as the starting composition and using bulk partition coefficient (D) values given in the inset of the diagrams. We then marked possible straight mixing lines between BEM-WP and the FC curve, by taking a linear

arrangement of the data points of the alkaline lavas into consideration.

Lavas of the C_{CA} series appear to follow a fractionation curve, while lavas of the C_A series follow one or two linear mixing trends between the BEM-WP composition and the fractionation curve (Fig. 13c). In general, the mixing trend on this diagram intersects with the fractionation curve at an intermediate composition. What all these relationships may imply is that only the samples following the modelled curve might have evolved through fractional crystallization or the Assimilation Combined with Fractional Crystallization (i.e. AFC) processes, while the ones aligning along the inferred mixing trends were probably produced as a result of the episodic injection of primitive alkaline magmas into the magma chambers in which calc-alkaline magmas had been evolved via fractional crystallization. Similar relationships are present in the Late Cretaceous units exposed to the north of Istanbul (Keskin *et al.* 2010).

As is known, Nb concentrations are lower in the lavas containing the subduction-signature compared to those erupted in a within-plate environment. The aforementioned contrast between the alkaline and calc-alkaline magma series provides a good opportunity for testing the mixing model that we have proposed. With this aim, we have constructed a 3D graph whose axes are represented by Fe, Zr and Nb (Fig. 14), and then plotted the data series on it together with the Upper Cretaceous lava samples from north of Istanbul (Keskin *et al.* 2003, 2010). Also plotted are a modelled theoretical Rayleigh FC curve and hypothetical mixing lines. Note that data points fit fairly well to the modelled fractionation curve and mixing lines (Fig. 14). These observations support our model involving the injection of primitive alkaline magmas into fractionally evolved calc-alkaline magma chambers and the mixing of these two compositionally diverse magmas derived from contrasting mantle sources.

AFC models (e.g. Keskin *et al.* 2010), based on the formulations of both DePaolo (1981) and Aitcheson & Forrest (1994), indicate that the calc-alkaline lavas of the first stage have variable crustal assimilation rates (i.e. r, ratio of the rate of assimilation to the rate of fractionation), varying from 0.03 to 0.7, in contrast to the alkaline lavas of the second stage which contain almost no crustal assimilation. Hence, there is a time-integrated decrease in crustal contribution to magma genesis.

Nature of the mantle source regions

Multi-element patterns

N-type mid-ocean ridge basalt (MORB)-normalized patterns of a subset of representative basic samples

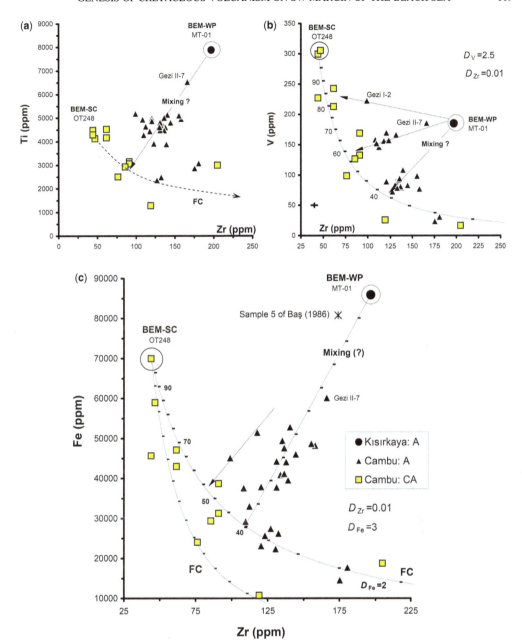

Fig. 13. Diagrams displaying the petrological model of magma mixing v. fractional crystallization processes for the calc-alkaline and alkaline lavas of the second stage. FC, BEM-WP and BEM-SC are as explained in Figure 12. D_V and D_{Zr}, bulk partition coefficients of the elements V and Zr, respectively, used in the FC model. For an explanation, see the text.

(i.e. SiO_2 <60%) from the alkaline and calc-alkaline lava series are displayed in Figure 15 with the aim of assessing variations in source compositions across the WPMB across the Western Pontides. We have to plot a couple of more evolved samples to make sure that the samples represent the whole evolutionary sequence.

The calc-alkaline lavas of both the first and second stages (i.e. De_{CA} and C_{CA}), including the most mafic ones, display a clear subduction signature

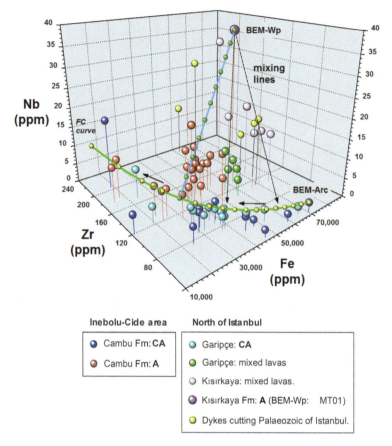

Fig. 14. Petrological model involving fractional crystallization and magma mixing processes on a 3D diagram whose axes are represented by a highly incompatible trace element (Zr), a highly compatible one (Fe) and a subduction sensitive trace (Nb). FC, BEM-WP and BEM-SC are explained in Figure 12. For an explanation, see the text.

represented by a selective enrichment in large ion lithophile elements (LILEs) and LREE relative to their high field strength elements (HFSEs) and MORB (Fig. 15). Concentrations of LILEs, LREEs and HFSEs are elevated with increasing silica in contrast to those of Tb, Ti, Y and Yb. This can be attributed to the FC process of mafic minerals (e.g. pyroxenes and amphiboles).

Calc-alkaline lavas of the first stage (i.e. De$_{CA}$) have elevated Hf, Zr and Sm concentrations, and a P depletion with respect to those of the second stage (i.e. C$_{CA}$). This may be attributed to differences in the AFC histories of the lavas from these two stages. The patterns for calc-alkaline lavas of the De$_{CA}$ and C$_{CA}$ series across the Western Pontides closely resemble those of the active continental margin patterns of Pearce (1983). Assimilation of continental crust may also cause a significant depletion in Nb and Ta concentrations combined with enrichment in LILEs and LREEs in evolved magmas. However, this could not be the case for the basic lavas. We

argue that existence of the subduction signature in calc-alkaline lavas can be explained by their derivation from a depleted mantle source enriched by a subduction component.

The most primitive alkaline basalt sample (i.e. MT-01 from the Kısırkaya Formation, north of Istanbul) shows a humped trace element pattern without Nb–Ta depletion (Fig. 15c). Its pattern resembles those of the ocean island basalts (OIBs), especially alkaline within-plate lavas from the Southern Atlantic (i.e. Ascension and Bouvet: Weaver *et al.* 1987), and Red Sea islands (Rogers 1993) (Fig. 15d) and Cenozoic mafic lavas from West Antarctica (Hole & LeMasurier 1994) (Fig. 15e). What is common for all these lavas is that they are thought to have been derived from an asthenospheric source.

Th/Yb v. Ta/Yb diagram

The Ta/Yb values of the lavas containing less than 60 wt% silica have been plotted against their

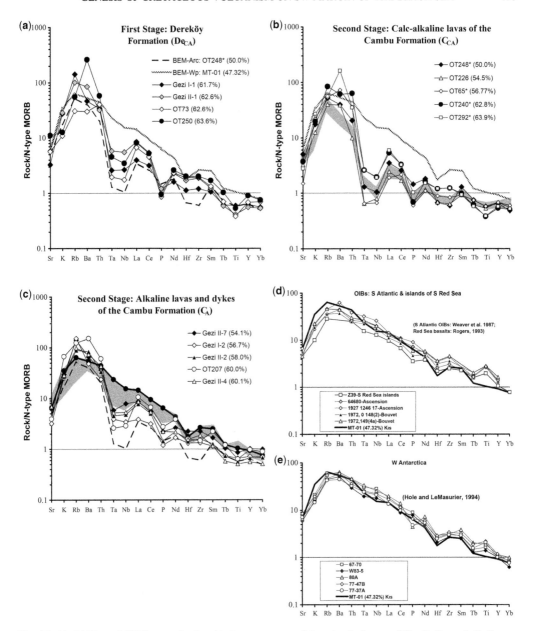

Fig. 15. (**a**)–(**c**) N-type MORB-normalized multi-element patterns of the calc-alkaline v. mildly alkaline and alkaline lavas in the study area. N-type MORB composition is that of Sun & McDonough (1989). Abbreviations: L, lava; D, dyke; Int, sub-volcanic intrusion. Samples with no abbreviations are lava blocks in volcaniclastic sequence. The silica percentages of the samples are given in parentheses in the legend. Shaded areas correspond to the fields of the Upper Cretaceous lavas from the north of Istanbul (Keskin *et al.* 2003, 2010). (**d**) & (**e**) Comparison of the multi-element patterns of the well-known OIB-type within plate alkaline lavas with those of the Kısırkaya basalt (i.e. Sample MT-01 D: the basaltic end-member of the alkaline series) from the north of Istanbul.

Ta/Yb values in Figure 16. Note that we also plotted the Upper Cretaceous alkaline and calc-alkaline samples from north of Istanbul (i.e. Kısırkaya and Garipçe lavas, respectively: Keskin *et al.* 2010) on this figure for comparison. Note that BEM-WP plots very close to the mantle metasomatism (MM) array with elevated Th/Yb and Ta/Yb ratios. In contrast, BEM-SC exhibits a consistent displacement from

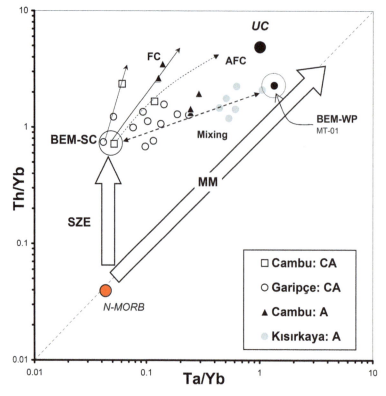

Fig. 16. Ta/Yb v. Th/Yb diagram (after Pearce 1983) for Upper Cretaceous lavas of the Western Pontides. Only basaltic and basaltic andesitic lavas with <60% SiO_2 are plotted on the diagram. The Upper Cretaceous second-stage lavas from the north of Istanbul are also plotted on the figure (i.e. calc-alkaline lavas of the Garipçe Formation and alkaline lavas of the Kısırkaya Formation) for comparison. MM, mantle metasomatism trend; SZE, subduction zone enrichment; UC, average upper crust; N-MORB, average N-type MORB of Sun & McDonough (1989).

the MM array towards a higher Th/Yb value. The location of BEM-WP in Figure 16 requires derivation of this lava from an enriched mantle source, comparable with those of OIBs, with no subduction component and no crustal assimilation. In contrast, the location of BEM-SC suggests derivation of this particular lava composition from a depleted 'N-type MORB-like source' on which a subduction component (i.e. SZE) has been overprinted (Fig. 16).

The key observation is that, except for a few samples from C_{CA} and C_A series, the rest plot along a line between the BEM-SC and BEM-WP compositions. The presence of the aforementioned trend supports the magma mixing model inferred from trace element behaviours. A few samples from the C_{CA} and C_A series plot away from the mixing trend towards higher Th/Yb values. As the trend does not head towards the composition of the upper crust (e.g. UC in Fig. 16), these relationships may imply the presence of fractional crystallization of hydrous phases, especially those containing amphibole.

Because fractionation of amphibole holds back Yb but not Th and Ta, it consequently causes an increase both in Th/Yb and Ta/Yb values.

Discussion

A number of issues revealed in this study need explanation, including: (1) why there was an extensional regime throughout the Western Pontides during the Late Cretaceous, continuously stretching the Pontide arc (Fig. 17); (2) why the volcanism occurred in two periods, divided by a pause period during which the region experienced sudden subsidence; (3) why two different magma series derived possibly from contrasting lithospheric and asthenospheric mantle sources coexisted during the second stage; (4) why the input of the alkaline lavas increased towards the end of the second stage (i.e. Campanian); and (5), finally, why volcanism suddenly ceased at the end of Campanian. Before moving onto these subjects, we will briefly examine a number of extensional arc settings in the literature where arc-type calc-alkaline and within-plate alkaline magma series coexist and discuss the models proposed for their magma genesis.

Back-arc basins of the Western Pacific are good present- day examples of arc settings (e.g. Uyeda & Kanamori 1979; Tamaki & Honza 1991) in which extensional tectonics and volcanism coexist with voluminous volcanoclastic material intercalated with sediments (e.g. Kano et al. 1993). Some arcs also contain OIB-type alkaline lavas alongside the calc-alkaline ones (e.g. Bacon et al. 1997; Kita et al. 2001; Mueller et al. 2002; Scott et al. 2002), similar to what we observe in the Western Pontides. Researchers have proposed a number of hypotheses in order to explain why OIB-type alkaline lavas erupt in the arc settings. Some researchers suggest that the diversities in lava chemistry in arcs could be related to an abrupt change in the tectonic evolution of a subduction system that causes upwelling the asthenosphere to shallow depths. For example, Stein et al. (1992) argued that OIB-type plutonic rocks of SW Japan could have been related to a 'slab window' in the subducted plate. In a similar way, Scarrow et al. (1997) suggested that the OIB signature in mafic dykes cutting through the Antarctic Peninsula continental margin batholith should be related to an oceanic spreading ridge-trench collision. For the generation of OIB-type volcanism in the Western Trans-Mexican volcanic belt, Ferrari et al. (2001) proposed a model involving slab rollback, asthenosphere infiltration, and variable flux melting. The most widely accepted models for the coexistence of calc-alkaline and OIB lavas involve the upwelling of asthenospheric mantle due to intra-arc (e.g. Cascade Range of Western USA: Bacon et al. 1997) or back-arc extension (e.g. Kyushu Island, Japan: Kita et al. 2001), followed by magma generation by adiabatic decompression in the asthenosphere. The basic rationale behind these models is that OIB-type basalts represent deeper and smaller-degree melts of enriched asthenospheric mantle, unaffected by subduction.

Geodynamic model proposed for the Western Pontides

Pre-Upper Cretaceous geology of the Western Pontides (i.e. the Istanbul Zone) indicates an extensional setting that prevailed between the Late Barremian and Aptian (Tüysüz 2017). During that period, the Istanbul Zone was affected by normal faults and, as a result, the Zonguldak and Ulus basins were rifted, deepened and filled by siliciclastic sediments (Tüysüz 1999). This extensional period, which possibly caused the thinning of continental crust, is generally accepted as the opening phase of the Western

Black Sea (Görür 1988; Schleder et al. 2015). Although this period of rifting was attributed to a back-arc setting in older studies (Letouzey et al. 1977; Zonenshain & Le Pichon 1986; Finetti et al. 1988; Görür 1988; Manetti et al. 1988), the absence of any evidence for an Aptian–Albian arc-volcanism contradicts this interpretation. Tüysüz (1999) and recently Tari (2015) proposed two stages of opening: thinning of continental lithosphere due to wide rifting without volcanism during the Aptian–Albian followed by narrow back-arc/intra-arc rifting during the Turonian–Santonian.

As has been discussed throughout this paper, the Turonian–Campanian magmatism of the WPMB consists of two periods of volcanism separated by a non-volcanic period. Magmatism during the second period is geochemically different from what is expected in the case of a classical arc magmatism as lavas with clear subduction and OIB signatures coexist. Possible reasons for this coexistence are discussed in the following paragraphs.

A ridge–trench collision may result in a pause in the magmatism across an arc system and can generate OIB-like alkaline magmatism along with the calc-alkaline magmas in a subduction system due to slab-window opening. The mid-ocean ridge represents the youngest, hottest, thinnest and the most buoyant part of an oceanic crust. Therefore, it either clogs or considerably slows down the subduction system by decreasing the subduction angle. Such an event may also uplift the overriding plate, deform the active margin by compression, result in migration of an alkaline pulse of volcanism from the trench towards the back-arc and create transform faults (e.g. the San Andreas Fault in California and related extensional regions, e.g. Basin and Range in western USA). Moreover, ridge subduction creates high-temperature metamorphism and crustal melting along the overriding plate. Because we have no evidence for such variations (e.g. migration of alkaline volcanism from south to north, the presence of high-temperature metamorphism) in the Western Pontides, a ridge–trench collision event (discussed in Keskin et al. 2010) cannot be regarded as a valid model.

Our data indicate that the initial stage of the arc volcanism produced calc-alkaline lavas with a typical subduction signature, possibly in response to northwards subduction of the Tethys Ocean beneath the southern margin of the Laurasia (Fig. 17a) (see fig. 10 of Tüysüz et al. 2016). The extensional nature of this magmatic arc is evidenced by the presence of synsedimentary growth faults, as well as by the fining- and deepening-upwards nature of the associating sediments, containing debris flows and olistoliths (Tüysüz 1999; Tüysüz et al. 2012). We argue that the extensional setting was created by the rollback of the northwards subducting oceanic

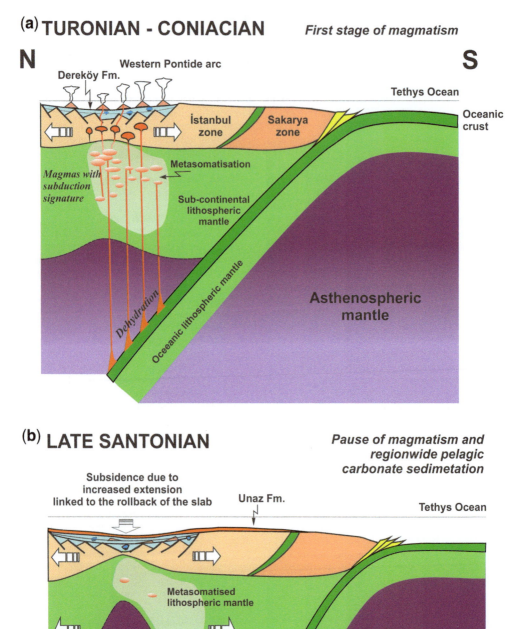

Fig. 17. Cartoons illustrating our geodynamic model for Western Pontides in a period between the Turonian and Maastrichtian. Abbreviations: MOR, mid-ocean ridge; MORB, mid-ocean ridge basalts. For a thorough discussion, see the text.

(c)

CAMPANIAN

Second stage of magmatism

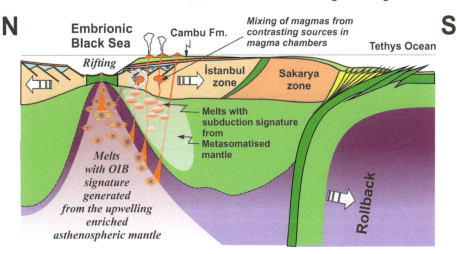

(d)

MAASTRICHTIAN

*Cesseation of magmatism
on the passive margin*

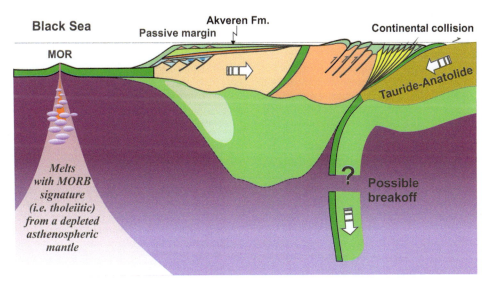

Fig. 17. *Continued.*

lithosphere (Çinku *et al.* 2010). This extensional period possibly corresponds to back-arc or intra-arc opening phase of the Western Black Sea. During this stage of rifting, between the Turonian and Early Santonian, subduction of the Tethyan Oceanic lithosphere was possibly generating melts, producing calc-alkaline lavas with a subduction signature on the continental crust of the southern Laurasian margin (Fig. 17a). These melts were also metasomatizing the subcontinental lithospheric mantle beneath the region, imprinting a pronounced arc signature to the mantle via veins. We argue that the reason why an arc massif never emerged as a positive topography across the Western Pontides was due to the extensional tectonic regime.

As rifting proceeded, the southern Laurasian margin should have thinned and finally ruptured (Fig. 17b, c). Because gradually older and, hence, denser parts of the oceanic lithosphere arrived to the trench and subducted beneath the Western Pontides in the south, the slab possibly further steepened and rolled back, intensifying the stretching of the rift system in the north along the axis of the arc. Slab rollback possibly also caused the arc volcanism to pose during the Late Santonian in response to gradual steepening, deepening and, hence, retreat of the slab away beneath the Istanbul Zone (Fig. 17b). The increased stretching probably resulted in a region-wide subsidence, giving way to the deposition of deep-marine pelagic carbonates (i.e. Unaz Formation) during the Late Santonian (Fig. 17b).

Volcanism resumed in Campanian (i.e. the second stage) during the deposition of the Cambu Formation. Because the rolled-back slab was no longer capable of generating an arc magmatism across the rift, we suggest that the subduction component in the second-stage lavas was not genuine, instead possibly coming from the abovementioned metasomatized subcontinental mantle source in a rift setting (Fig. 17a). Hence, the subduction signature in calc-alkaline lavas of the second stage might have been inherited from the previous subduction event. A similar inheritance model was proposed previously to explain the presence of subduction signature seen in the collision-related lavas of the NE Anatolia (i.e. Cenozoic volcanics on the Erzurum-Kars Plateau) in the absence of any active arc magmatism (Pearce et al. 1990; Keskin et al. 1998). The heat needed to melt the metasomatized lithospheric mantle was supplied possibly by passively upwelled asthenosphere in response to the extension (Fig. 17c). The presence of the asthenospheric source is evidenced by the presence of OIB-type alkaline within-plate lavas during the second stage. All these findings indicate that the arc setting almost completely disappeared before the second stage during the Campanian.

As rifting proceeded, the enriched asthenospheric source became gradually dominant in magma genesis, producing lavas with OIB-like signature along the Pontides and generating an embryonic oceanic crust (i.e. Black Sea) along the rift axis. Therefore, alkaline magma generation towards the end of the second stage could be attributed to adiabatic decompression melting controlled by the passive upwelling of asthenospheric mantle in response to rifting. Magmas derived from two contrasting mantle sources (i.e. asthenosphere v. metasomatized subcontinental lithosphere: see the melting models given in fig. 14 of Aysal et al. 2017) coexisted, especially during the final period of the second stage (i.e. Campanian). Primitive alkaline magmas injected into the crustal magma chambers in which calc-alkaline magmas were evolving and mixed with them, as evidenced by the geochemical data discussed in this paper. All of these volcanosedimentary products were deposited on the passive margins of the embryonic Black Sea Basin (Fig. 17c). Subduction associated with rollback caused the Laurasian continental sliver, known at present as the Istanbul Zone (corresponding to Western Pontides), to move to the south towards the Sakarya Continent. This southwards movement widened the Western Black Sea Basin behind the Istanbul Zone. Magmatism diminished at the end of the Campanian by the continental collision of the Istanbul Zone with the Sakarya Zone to the south (Fig. 17d).

The character of the magmas changed from OIB to MORB over time, possibly along the mid-Black Sea ridge, because the asthenospheric source might have become exhausted due to extensive melt generation to form the oceanic lithosphere while the ridge was moving away from the Black Sea passive margin (Fig. 17d). However, the presumed MORB-type crust is covered by an exceptionally thick (c. 12 km) sedimentary sequence at present. On the basis of deep seismic profiles, the stratigraphy of this sequence was recently revised by Nikishin et al. (2015b). They proposed that the aforementioned sedimentary sequence, ranging in age from Late Santonian to Present, rests on oceanic crust.

In the onshore, post-volcanic sediments, representing the southern passive margin sedimentary prism of the Black Sea (Tüysüz 1999), consist of Maastrichtian–Paleocene calciturbidites and overlying Eocene siliciclastic turbidites. In contrast to the undeformed nature of basinal sediments, the passive margin sediments along the Pontides were intensively deformed by a compressional tectonic regime. Collision between the İstanbul and the Sakarya zones at the end of the Campanian caused uplifting in the southern areas. These uplifted areas were the main source of the southern margin of the Black Sea Basin. Post-collisional compression progressed towards the north and has caused the development of a fold and thrust belt along the whole length of the Pontides since the middle Eocene (Sunal & Tüysüz 2002). This fold and thrust belt consists of north- and south-vergent thrusts and intense folding of Eocene and older sediments. Therefore, the Oligo-Miocene was a period of uplift and erosion that sourced sediments for hydrocarbon-prone Maykop deposits within the Black Sea Basin. Finetti *et al.* (1988) showed occurrences of these compressional structures along the southern Black Sea margin. The steep topography of the Pontides and the recent compressional events, such as the Bartın 1968 M_s 6.8 earthquake (Alptekin *et al.* 1986), implies that this compressional regime is possibly still active.

Data presented in this paper support that the Western Black Sea oceanic basin was opened during the Turonian–Late Santonian as a back-arc/intra-arc basin along the southern margin of Laurasia. We propose an extensional arc model associated with steepening and rollback of the northwards-subducting Tethyan slab. Our data also show that Upper Cretaceous sediments and the associated volcanic rocks of the Pontides are far from being regarded as a reservoir and source for petroleum occurrences.

The fieldwork and laboratory analyses in this study were financially supported by two Turkish foundations in Turkey: (1) The Istanbul University Research Fund (project number: 1145/010598); and (2) The Scientific and Technological Research Council of Turkey (YDABCAG-590/6, TÜBİTAK 197Y045 and TÜBİTAK 104Y215). Field studies were supported by the Turkish Petroleum Company (TPAO). We thank Sabri Kirici of Turkish Petroleum Co. for his careful study on the sedimentary material we supplied. We particularly thank Dr Gabor Tari and Dr Prelević Dejan for their constructive comments and suggestions that definitely improved the quality of our paper. We also appreciate suggestions made by the volume Editors Dr Mike Simmons and Dr Aral Okay. An earlier version of this study was presented at the 100th Anniversary of Geology at Istanbul University symposium, which was partly supported by Istanbul University Research Fund (project No. IRP-51703), coordinated by the first author. We appreciate their support. We are also grateful to Dr Judy Monthie-Doyum for proofreading an earlier version of the manuscript.

References

AITCHESON, S.J. & FORREST, A.H. 1994. Quantification of crustal contamination in open magmatic systems. *Journal of Petrology*, **35**, 461–488.

AKYOL, Z., ARPAT, E. ET AL. 1974. Cide-Kurucaşile dolayının jeoloji haritası [Geology Map of Cide-Kurucaşile and Surroundings]. Series of 1:50 000 Scale Geology Map of Turkey. General Directorate of Mineral Research and Exploration, Ankara [explanation in Turkish].

ALPTEKIN, Ö., NÁBĚLEK, J.L. & TOKSÖZ, M.N. 1986. Source mechanism of the Bartın earthquake of September 3, 1968 in northvvestern Turkey: evidence for active thrust faulting at the southern Black Sea margin. *Tectonophysics*, **122**, 73–88.

AYKOL, A. & TOKEL, S. 1991. The geochemistry and tectonic setting of the Demirköy-Istranca granitoid chain, NW Turkey. *Mineralogical Magazine*, **55**, 249–256.

AYSAL, N., KESKIN, M., PEYTCHEVA, I. & DURU, O. 2017. Geochronology, geochemistry and isotope systematics

of a mafic–intermediate dyke complex in the İstanbul Zone. New constraints on the evolution of the Black Sea in NW Turkey. *In:* SIMMONS, M.D., TARI, G.C. & OKAY, A.I. (eds) *Petroleum Geology of the Black Sea.* Geological Society, London, Special Publications, **464**, https://doi.org/10.1144/SP464.4

BACON, C.R., BRUGGMAN, P.E., CHRISTIANSEN, R.L., CLYNNE, M.A., DONNELLYNOLAN, J.M. & HILDRETH, W. 1997. Primitive magmas at five Cascade volcanic fields: melts from hot, heterogeneous sub-arc mantle. *Canadian Mineralogist*, **35**, 397–423.

BOCCALETTI, M., MANETTI, P. & PECCERILLO, A. 1974. Hypothesis on the plate tectonic evolution of the Carpatho-Balkan arcs. *Earth and Planetary Science Letters*, **23**, 193–198.

BOCCALETTI, M., MANETTI, P., PECCERILLO, A. & STANISHEVA-VASSILEVA, G. 1978. Late Cretaceous high-potassium volcanism in Eastern Srednogorie, Bulgaria. *Geological Society of American Bulletin*, **89**, 439–447.

ÇINKU, M.C., USTAÖMER, T., HIRT, A.M., HISARLI, Z.M., HELLER, F. & ORBAY, N. 2010. Southward migration of arc magmatism during latest Cretaceous associated with slab-steepening, East Pontides, N Turkey: new paleomagnetic data from the Amasya region. *Physics of the Earth and Planetary Interiors*, **182**, 18–29.

DEPAOLO, D.J. 1981. Trace element and isotopic effects of combined wall-rock assimilation and fractional crystallisation. *Earth and Planetary Science Letters*, **53**, 189–202.

FERRARI, L., PETRONE, C.M. & FRANCALANCI, L. 2001. Generation of oceanic-island basalt-type volcanism in the Western Trans-Mexican volcanic belt by slab rollback, asthenosphere infiltration, and variable flux melting. *Geology*, **29**, 507–510.

FINETTI, I., BRICCHI, G., DEL BEN, A., PIPAN, M. & XUAN, Z. 1988. Geophysical study of the Black Sea: bolletino di Geofisica Teorica ed Applicata. *Monograph on the Black Sea*, **30/117-118**, 197–324.

GÖRÜR, N. 1988. Timing and opening of the Black Sea Basin. *Tectonophysics*, **147**, 247–262.

HIPPOLYTE, J.-C., MÜLLER, C., KAYMAKCI, N. & SANGU, E. 2010. Dating of the Black Sea Basin: new nannoplankton ages from its inverted margin in the Central Pontides (Turkey). *In:* STEPHENSON, R.A., KAYMAKCI, N., SOSSON, M., STAROSTENKO, V. & BERGERAT, F. (eds) *Sedimentary Basin Tectonics from the Black Sea and Caucasus to the Arabian Platform.* Geological Society, London, Special Publications, **340**, 113–136, https://doi.org/10.1144/SP340.7

HIPPOLYTE, J.-C., MÜLLER, C., SANGU, E & KAYMAKCI, N. 2015. Stratigraphic comparisons along the Pontides (Turkey) based on new nannoplankton age determinations in the Eastern Pontides: geodynamic implications. *In:* SOSSON, M., STEPHENSON, R.A. & ADAMIA, S.A. (eds) *Tectonic Evolution of the Eastern Black Sea and Caucasus.* Geological Society, London, Special Publications, **428**. First published online 27 October 2015, https://doi.org/10.1144/SP428.9

HOLE, M.J. & LEMASURIER, W.E. 1994. Tectonic controls on the geochemical composition of Cenozoic, mafic alkaline volcanic rocks from West Antarctica. *Contributions to Mineralogy and Petrology*, **117**, 187–202.

IRVINE, T.N. & BARAGAR, W.R.A. 1971. A guide to the chemical classification of the common volcanic rocks. *Canadian Journal of Earth Sciences*, **8**, 523–548.

KANO, K., YAMAMOTO, T. & TAKEUCHI, K. 1993. A Miocene island-arc volcanic seamount – the Takashibiyama formation, Shimane Peninsula, SW Japan. *Journal of Volcanology and Geothermal Research*, **59**, 101–119.

KARACIK, Z. & TÜYSÜZ, O. 2010. Petrogenesis of the Late Cretaceous Demirköy Igneous Complex in NW Turkey: implications for magma genesis in Strandja Zone. *Lithos*, **114**, 369–384.

KESKİN, M., PEARCE, J.A. & MITCHELL, J.G. 1998. Volcano-stratigraphy and geochemistry of collision-related volcanism on the Erzurum-Kars Plateau, North Eastern Turkey. *Journal of Volcanology and Geothermal Research*, **85**, 355–404.

KESKİN, M., USTAÖMER, T. & YENIYOL, M. 2003. İstanbul kuzeyinde yüzeylenen Üst Kretase yaşlı volkanosedimenter birimlerin stratigrafisi, petrolojisi ve tektonik ortamı [Stratigraphy, petrology and tectonic setting of the Upper Cretaceous volcano-sedimentary units in the north of Istanbul]. *İstanbul'un Jeolojisi Sempozyumu [Symposium on the Geology of Istanbul]*, **1**, 23–35 [in Turkish].

KESKİN, M., USTAÖMER, T. & YENIYOL, M. 2010. İstanbul kuzeyindeki Üst Kretase volkanojenik istiflerinin magmatik evrimi ve jeodinamik ortamı [Magmatic evolution and geodynamic setting of the Upper Cretaceous sequence in the north of İstanbul]. *İstanbul'un Jeolojisi Sempozyumu III Kitabı [Symposium on the Geology of Istanbul]*, **3**, 130–180 [in Turkish].

KETIN, İ. 1966. Anadolu'nun tektonik birlikleri (Tectonic Units of Asia minor). *Bulletin of Directorate of the Mineral Research and Exploration*, **66**, 20–34 [in Turkish].

KITA, I., YAMAMOTO, M., ASAKAWA, Y., NAKAGAWA, M., TAGUCHI, S. & HASEGAWA, H. 2001. Contemporaneous ascent of within-plate and island-arc type magmas in the Beppu-Shimabara graben system, Kyushu Island, Japan. *Journal of Volcanology and Geothermal Research*, **111**, 99–109.

KUNO, H. 1966. Lateral variation of basalt magma types across continental margins and island arcs. *Bulletin of Volcanology*, **29**, 195–222.

LE BAS, M.J., LEMAITRE, R.W., STRECKEISEN, A. & ZANETTIN, B. 1986. A chemical classification of volcanic rocks based on the total alkali-silica diagram. *Journal of Petrology*, **27**, 745–750.

LETOUZEY, J., BIJU-DUVAL, B., DORKEL, A., GONNARD, R., KRISTCHEV, K., MONTADERT, L. & SUNGURLU, O. 1977. The Black Sea: a marginal basin; geophysical and geological data. *In*: BIJU-DUVAL, B. & MONTADERT, L. (eds) *International Symposium on the Structural History of the Mediterranean Basins*. Editions Technip, Paris, 363–376.

MANETTI, P., PECCERILLO, A. & POLI, G. 1979. REE distribution in Upper Cretaceous calc-alkaline and shoshonitic volcanic rocks from Eastern Srednogorie (Bulgaria). *Chemical Geology*, **26**, 51–63.

MANETTI, P., BOCCALETTI, M. & PECCERILLO, A. 1988. The Black Sea: remnant of a marginal basin behind the Srednogorie-Pontides island arc system during Upper Cretaceous–Eocene time. *Bolletino di Geofisica Teorica ed Applicata*, **30**, 39–51.

MASSE, J.P., TÜYSÜZ, O., FENERCI-MASSE, M., ÖZER, S. & SARI, B. 2009. Stratigraphic organisation, spatial distribution, palaeoenvironmental reconstruction, and demise of Lower Cretaceous (Barremian-lower Aptian) carbonate platforms of the Western Pontides (Black Sea region, Turkey). *Cretaceous Research*, **30**, 1170–1180.

MESCHEDE, M. 1986. A method of discriminating between different types of mid-ocean ridge basalts and continental tholeiites with the Nb–Zr–Y diagram. *Chemical Geology*, **56**, 207–218.

MUELLER, W.U., DOSTAL, J. & STENDAL, H. 2002. Inferred Palaeoproterozoic arc rifting along a consuming plate margin: insights from the stratigraphy, volcanology and geochemistry of the Kangerluluk sequence, southeast Greenland. *International Journal of Earth Sciences (Geologische Rundschau)*, **91**, 209–230.

MUNTEANU, I., MATENCO, L., DINU, C. & CLOETINGH, S. 2011. Kinematics of back-arc inversion of the Western Black Sea Basin, *Tectonics*, **30**, TC5004, https://doi.org/10.1029/2011TC002865

NIKISHIN, A.M., OKAY, A.I., TÜYSÜZ, O., DEMIRER, A., AMELIN, N. & PETROV, E. 2015a. The Black Sea basins structure and history: new model based on new deep penetration regional seismic data. Part 1: basins structure and fill. *Marine and Petroleum Geology*, **59**, 638–655.

NIKISHIN, A.M., OKAY, A.I., TÜYSÜZ, O., DEMIRER, A., WANNIER, M., AMELIN, N. & PETROV, E. 2015b. The Black Sea basins structure and history: new model based on new deep penetration regional seismic data. Part 2: tectonic history and paleogeography. *Marine and Petroleum Geology*, **59**, 656–670.

OKAY, A.İ. 1989. Tectonic units and sutures in the Pontides, northern Turkey. *In*: ŞENGÖR, A.M.C. (ed.) *Tectonic Evolution of the Tethyan Region*. Kluwer Academic, Dordrecht, The Netherlands, 109–116.

OKAY, A.İ. & TÜYSÜZ, O. 1999. Tethyan sutures of northern Turkey. *In*: DURAND, B., JOLIVET, L., HOVARTH, F. & SÉRANNE, M. (eds) *The Mediterranean Basins: Tertiary Extension within the Alpine Orogen*. Geological Society, London, Special Publications, **156**, 475–515, https://doi.org/10.1144/GSL.SP.1999.156.01.22

OKAY, A.İ., ŞENGÖR, A.M.C. & GÖRÜR, N. 1994. Kinematic history of the opening of the Black Sea and its effect on the surrounding regions. *Geology*, **22**, 267–270.

OKAY, A.İ., TANSEL, İ. & TÜYSÜZ, O. 2001. Obduction, subduction and collision as reflected in the Upper Cretaceous–Lower Eocene sedimentary record of Western Turkey. *Geological Magazine*, **138**, 117–142.

OKAY, A.I., TÜYSÜZ, O., SATIR, M., ÖZKAN-ALTINER, S., ALTINER, D., SHERLOCK, S., & EREN, R.H. 2006. Cretaceous and Triassic subduction-accretion, HP/LT metamorphism and continental growth in the Central Pontides, Turkey. *Geological Society of America Bulletin*, **118**, 1247–1269.

OKAY, A.İ., SUNAL, G., SHERLOCK, S., ALTINER, D., TÜYSÜZ, O., KYLANDER-CLARK, A.R.C. & AYGÜL, M. 2013. Early Cretaceous sedimentation and orogeny: an active margin of Eurasia: Southern Central Pontides, Turkey. *Tectonics*, **32**, 1–25.

PEARCE, J.A. 1982. Trace element characteristics of lavas from destructive plate boundaries. *In*: THORP, R.S. (ed.) *Andesites: Orogenic Andesites and Related Rocks*. John Wiley, New York, 525–548.

PEARCE, J.A. 1983. Role of the sub-continental lithosphere in magma genesis at active continental margins. *In*: HAWKESWORTH, C.J. & NORRY, M.J. (eds) *Continental Basalts and Mantle Xenolites*. Shiva, Nanthwich, UK, 230–249.

PEARCE, J.A. & CANN, J.R. 1973. Tectonic setting of basic volcanic rocks determined using trace element analyses. *Earth and Planetary Science Letters*, **19**, 290–300.

PEARCE, J.A., BENDER, J.F. *ET AL.* 1990. Genesis of collision volcanism in Eastern Anatolia, Turkey. *Journal of Volcanology and Geothermal Research*, **44**, 189–229.

PECCERILLO, A. & TAYLOR, S.R. 1975. Geochemistry of Upper Cretaceous volcanic rocks from the Pontide chain, northern Turkey. *Bulletin Volcanologique*, **39**, 1–13.

PECCERILLO, A. & TAYLOR, S.R. 1976, Geochemistry of Eocene calc-alkaline volcanic rocks from the Kastamonu area, northern Turkey. *Contributions to Mineralogy and Petrology*, **58**, 63–81.

ROBINSON, A.G., RUDAT, J.H., BANKS, C.J. & WILES, R.L.F. 1996. Petroleum geology of the Black Sea. *Marine and Petroleum Geology*, **13**, 195–223.

ROGERS, N.W. 1993. The isotope and trace element geochemistry of basalts from the volcanic island of the southern Red Sea. *In*: PRICHARD, H.M., ALABASTER, T., HARRIS, N.B.W. & NEARY, C.R. (eds) *Magmatic Processes and Plate Tectonics*. Geological Society, London, Special Publications, **76**, 455–467, https://doi.org/10.1144/GSL.SP.1993.076.01.24

ROSS, D.A. 1974. The Black Sea. *In*: BURK, C.A. & DRAKE, C.L. (eds) *The Geology of Continental Margins*. Springer, Berlin, 669–682.

ROSS, D.A. 1978. Summary of results of Black Sea drilling. *In*: ROSS, D.A. & NEPROCHNOV, Y.P. (eds) *Initial Reports on the Deep Sea Drilling Project, Volume 42, Part 2*. United States Government Printing Office, Washington, DC, 1149–1177.

ROSS, D.A. & DEGENS, E.T. (eds). 1974. *The Black Sea – Geology, Chemistry, and Biology*. American Association of Petroleum Geologists Memoirs, **20**.

ROSS, D.A., STOFFERS, P. & TRIMONIS, E.S. 1978. Black Sea sedimentary framework. *In*: ROSS, D.A. & NEPROCHNOV, Y.P. (eds) *Initial Reports on the Deep Sea Drilling Project, Volume 42, Part 2*. United States Government Printing Office, Washington, DC, 359–363.

ŞAHIN, S.Y., AYSAL, N. & GÜNGÖR, Y. 2012. Petrogenesis of late Cretaceous adakitic magmatism in the İstanbul zone (Çavuşbaşı Granodiorite, NW Turkey). *Turkish Journal of Earth Sciences*, **21**, 1029–1045.

ŞAHINTÜRK, Ö. & ÖZÇELIK, Y. 1983. *Zonguldak-Bartın-Amasra-Kurucaşile-Cide dolaylarının jeolojisi ve petrol olanakları [Geology and petroleum potential of the area among Zonguldak, Bartın, Amasra, Kurucaşile and Cide]*. TPAO Arama Grubu Arşivi, Rapor [Report in the archive of Turkish Petroleum]. [in Turkish].

SCARROW, J.H., LEAT, P.T., WAREHAM, C.D. & MILLAR, I.L. 1997. Mantle sources for Cretaceous–Tertiary mafic magmatism in the Antarctic Peninsula. *In*: RICCI, C.A. (ed.) *The Antarctic Region: Geological Evolution and Processes*. Terra Antartica Publications, Siena, Italy, 327–332.

SHILLINGTON, D.J., WHITE, N., MINSHULL, T.A., EDWARDS, G. R.H., JONES, S.M., EDWARDS, R.A. & SCOTT, C.L. 2008.

Cenozoic evolution of the Eastern Black Sea: a test of depth dependent stretching models. *Earth Planetary Science Letters*, **265**, 360–378.

SHILLINGTON, D.J., SCOTT, C.L., MINSHULL, T.A., EDWARDS, R.A., BROWN, P.J. & WHITE, N. 2009. Abrupt transition from magma-starved to magma-rich rifting in the Eastern Black Sea. *Geology*, **37**, 7–10.

SCHLEDER, Z., KRESZEK, C., TURI, V., TARI, G., KOSI, W. & FALLAH, M. 2015. Regional structure of the Western Black Sea Basin: constraints from cross-section balancing. *In*: POST, P.J., COLEMAN JR., J.L., NORMAN, C.R., BROWN, D.E., ROBERTS-ASHBY, T., KAHN, P. & ROWAN, M. (eds) *Transactions of the GCSSEPM Foundation Perkins-Rosen 34th Annual Research Conference 'Petroleum Systems in Rift Basins', 13–16 December 2015, Houston, Texas*, 396–411.

SCOTT, C.R., MUELLER, W.U. & PILOTE, P. 2002. Physical volcanology, stratigraphy, and lithogeochemistry of an Archean volcanic arc: evolution from plume-related volcanism to arc rifting of SE Abitibi Greenstone Belt, Val d'Or, Canada. *Precambrian Research*, **115**, 223–260.

ŞENGÖR, A.M.C. & YILMAZ, Y. 1981. Tethyan evolution of Turkey: a plate tectonic approach. *Tectonophysics*, **75**, 181–241.

SPADINI, G., ROBINSON, A. & CLOETINGH, S. 1996. Western v. Eastern Black Sea tectonic evolution: pre-rift lithospheric controls on basin formation. *Tectonophysics*, **266**, 139–154.

SPADINI, G., ROBINSON, A.G. & CLOETINGH, S.A.P.L. 1997. Thermomechanical modeling of Black Sea Basin formation, subsidence, and sedimentation. *In*: ROBINSON, A.G. (ed.) *Regional and Petroleum Geology of the Black Sea and Surrounding Region*. American Association of Petroleum Geologists Memoirs, **68**, 19–38.

STEIN, G., LAPIERRE, H. & CHARVET, J. 1992. Within-plate alkaline magmatism in an island-arc environment – the Ashizuri plutonic body (SW Japan). *Comptes Rendus de l'Academie des Sciences Serie II*, **315**, 1501–1508.

STEPHENSON, R. & SCHELLART, W.P. 2010. The Black Sea back-arc basin: insights to its origin from geodynamic models of modern analogues. *In*: SOSSON, M., KAYMAKCI, N., STEPHENSON, R.A., BERGERAT, F. & STAROSTENKO, V. (eds) *Sedimentary Basin Tectonics from the Black Sea and Caucasus to the Arabian Platform*. Geological Society, London, Special Publications, **340**, 11–21.

SUN, S.-S. & MCDONOUGH, W.F. 1989. Chemical and isotopic systematics of oceanic basalts: implications for mantle composition and processes. *In*: SAUNDERS, A. D. & NORRY, M.J. (eds) *Magmatism in Ocean Basins*. Geological Society, London, Special Publications, **42**, 313–345, https://doi.org/10.1144/GSL.SP.1989.042.01.19

SUNAL, G. & TÜYSÜZ, O. 2002. Palaeostress analysis of Tertiary post-collisional structures in the Western Pontides, Northern Turkey. *Geological Magazine*, **139**, 343–359.

TAMAKI, K. & HONZA, H. 1991. Review of global marginal basins. *Episodes*, **14**, 224–230.

TARI, G. 2015. Is the Black Sea really a back-arc basin? *In*: POST, P.J., COLEMAN JR., J.L., NORMAN, C.R., BROWN,

D.E., ROBERTS-ASHBY, T., KAHN, P. & ROWAN, M. (eds) *Transactions of the GCSSEPM Foundation Perkins-Rosen 34th Annual Research Conference 'Petroleum Systems in Rift Basins', 13–16 December 2015, Houston, Texas*, 509–520.

TÜYSÜZ, O. 1999. Geology of the Cretaceous sedimentary basins of the Western Pontides. *Geological Journal*, **34**, 75–93.

TÜYSÜZ, O. 2017. Cretaceous geological evolution of the Pontides. *In*: SIMMONS, M.D., TARI, G.C. & OKAY, A.I. (eds) *Petroleum Geology of the Black Sea*. Geological Society, London, Special Publications, **464**, https://doi.org/10.1144/SP464.9.

TÜYSÜZ, O., AKSAY, A. & YİĞİTBAŞ, E. 2004. *Batı Karadeniz Bölgesi Litostratigrafi Birimleri [Lithostratigraphic Units of the Western Pontide Region]*. Maden Tetkik ve Arama Genel Müdürlüğü, Stratigrafi Komitesi, Litostratigrafi Birimleri Serisi [Mineral Research & Exploration Institute of Turkey, Stratigraphy Committee, Lithostratigraphy Series], **1** [in Turkish].

TÜYSÜZ, O., YILMAZ, İ.Ö., SVABENICKA, L. & KIRICI, S. 2012. The Unaz formation: a key unit in the Western Black Sea region, N Turkey. *Turkish Journal of Earth Sciences*, **21**, 1009–1028, https://doi.org/10.3906/yer-1006-30

TÜYSÜZ, O., MELINTE-DOBRINESCU, M.C., YILMAZ, İ.Ö., KIRICI, S., ŠVABENICKÁ, L. & SKUPIEN, P. 2016. The Kapanboğazı formation: a key unit for understanding Late Cretaceous evolution of the Pontides, N Turkey. *Palaeogeography, Palaeoclimatology, Palaeoecology*, **441**, 565–581, https://doi.org/10.1016/j.palaeo.2015.06.028

UYEDA, S. & KANAMORI, H. 1979. Backarc opening and the mode of subduction. *Journal of Geophysical Research*, **84**, 1049–1061.

WEAVER, B.L., WOOD, D.A., TARNEY, J. & JORON, J.L. 1987. Geochemistry of ocean island basalts from the South Atlantic: ascension, St. Helena, Gough and Tristan da Cunha. *In*: FITTON, J.G. & UPTON, B.G.J. (eds) *Alkaline Igneous Rocks*. Geological Society, London, Special Publications, **30**, 253–267, https://doi.org/10.1144/GSL.SP.1987.030.01.11

WINCHESTER, J.A. & FLOYD, P.A. 1977. Geochemical classification of different magma series and their differentiation products using immobile elements. *Chemical Geology*, **20**, 325–343.

WOOD, D.A. 1980. The application of a Th–Hf/Ta diagram to problems of tectonomagmatic classification and to establishing the nature of crustal contamination of basaltic lavas of the British Tertiary volcanic province. *Earth and Planetary Science Letters*, **50**, 11–30.

YILMAZ, Y., TÜYSÜZ, O., YİĞİTBAŞ, E., GENÇ, Ş.C. & ŞENGÖR, A.M.C. 1997. Geology and tectonic evolution of the Pontides. *In*: ROBINSON, A. (ed.) *Regional and Petroleum geology of the Black Sea and Surrounding Region*. American Association of Petroleum Geologists Memoirs, **68**, 183–226.

ZONENSHAIN, L.P. & LE PICHON, X. 1986. Deep basins of the Black Sea and Caspian Sea as remnants of Mesozoic backarc basins. *Tectonophysics*, **123**, 181–211.

Geochronology, geochemistry and isotope systematics of a mafic–intermediate dyke complex in the İstanbul Zone. New constraints on the evolution of the Black Sea in NW Turkey

NAMIK AYSAL[1]*, MEHMET KESKIN[1], IRENA PEYTCHEVA[2] & OLGUN DURU[3]

[1]*Department of Geological Engineering, Faculty of Engineering, İstanbul University, 34320 Avcılar, İstanbul, Turkey*

[2]*Geological Institute, Bulgarian Academy of Sciences, Sofia, Bulgaria*

[3]*Graduate School of Science and Engineering, İstanbul University, 34126 Vezneciler, İstanbul, Turkey*

*Correspondence: aysal@istanbul.edu.tr

Abstract: We report new U–Pb zircon ages, major and trace element data, mineral chemistry, and Sr–Nd isotopic analyses of the mafic–intermediate dykes and intrusions in the İstanbul Zone. Mafic dykes are represented by calc-alkaline to alkaline lamprophyre and diabase. Intermediate dykes and subvolcanics are andesitic to dacitic in composition and calc-alkaline in character, while intrusive rocks (stocks and small plutons) are granodioritic and dioritic in composition. New zircon U–Pb laser ablation inductively coupled plasma mass spectrometry (LA-ICP-MS) dating yielded ages from 72.49 ± 0.79 (Upper Cretaceous–Campanian) to 65.44 ± 0.93 Ma (Lower Paleocene–Danian) for the intermediate dykes, and 58.9 ± 1.8 Ma (Upper Paleocene–Thanetian) for a small granodiorite stock. $^{87}Sr/^{86}Sr_{(i)}$ values of the mafic and intermediate dykes and small stocks span a range from 0.703508 to 0.706311, while their $^{143}Nd/^{144}Nd_{(i)}$ values vary from 0.512614 to 0.512812 and eNd$_{(i)}$ values from 5.09 to 1.24. Nd$_{TDM}$ model ages range between 0.46 and 0.77 Ga.

Dykes are enriched in large ion lithophile elements (LILEs) and light rare earth elements (LREEs) relative to high field strength elements (HFSEs). Normal-type mid-ocean ridge basalt (N-MORB)-normalized multi-element spidergrams of the majority of the mafic and intermediate dykes display a clear subduction signature, except a subset, which cut the Palaeozoic of İstanbul and the upper part of the Upper Cretaceous volcanics in the north of İstanbul (i.e. feeder dykes of the Kısırkaya Formation) and show a clear ocean island basalt (OIB) signature indicating that the melts feeding the dyke system during the Upper Cretaceous–Paleocene period were derived from two contrasting mantle sources: (1) initially a lithospheric mantle modified by subducted slab-derived melts which sourced the magmas with a clear subduction signature; and (2) followed by an asthenospheric mantle from which basic magmas with OIB signature. Petrological models indicate the interaction of these two discrete magma series via magma-mixing processes.

Geothermometric calculations based on the composition of amphiboles are in the range of 769–953 and 938–994°C. Geobarometric calculations indicate crystallization depths ranging over an interval between 3.0 and 20.2 km, implying a polybaric crystallization. The oxygen fugacity ($\log fO_2$) values vary between −10.10 and −13.07 bar in the dykes cutting the Upper Cretaceous volcanics, and from −8.71 to −10.33 bar in intermediate dykes cutting the İstanbul Palaeozoic unit. H$_2$O$_{melt}$ contents change between 4.91–6.89 and 4.82–7.51%, respectively implying that the dykes were emplaced at mid to shallow crustal levels. Dyke complexes of the İstanbul zone are interpreted to have been emplaced in a rifted volcanic arc margin related to the opening of the Black Sea during the Late Cretaceous–Paleocene period.

Supplementary material: Tables of representative analyses are available at https://doi.org/10.6084/m9.figshare.c.3841276

Magmatic rocks play a key role in understanding the tectonic settings of past periods. Modern analytic techniques used in geochemistry, geochronology, isotope geochemistry and mineral chemistry have enabled researchers to make more reliable interpretations about geodynamic settings of the magmatic rocks.

A series of dyke and plutonic systems related to various magmatic activities are observed on both sides of the Bosphorus (İstanbul) in the south of the Black Sea. Extrusive and intrusive products of these magmatic activities show contrasting geochemical characteristics and display an age range from Late Cretaceous to Paleocene, coinciding with

From: SIMMONS, M. D., TARI, G. C. & OKAY, A. I. (eds) 2018. *Petroleum Geology of the Black Sea.* Geological Society, London, Special Publications, **464**, 131–168. First published online September 25, 2017, https://doi.org/10.1144/SP464.4

the period during which the Western Black Sea Basin was opened. Therefore, they can be regarded as an important data source for understanding the opening and the tectonic evolution of the Black Sea. The above-mentioned magmatic association is represented by dyke systems and granitoid intrusions (e.g. Çavuşbaşı granodiorite: Öztunalı & Satır 1975; Yılmaz Şahin *et al.* 2012), cutting the İstanbul Palaeozoic unit which is exposed widely in and around İstanbul (Fig. 1). The internal stratigraphy of this unit is presented in Özgül (2012).

There is a volcanic/volcaniclastic sequence exposed widely in the north of İstanbul along the Black Sea coast, which is coeval with the aforementioned dykes and plutonic bodies of Late Cretaceous–Paleocene age (Fig. 2). That particular unit is named by Keskin *et al.* (2003) as the Kavaklar Group. Keskin *et al.* (2003, 2010) divided this unit into three sub-units. From bottom to top these are: (1) the Bozhane, (2) the Garipçe and (3) the Kısırkaya formations. The Bozhane Formation consists predominantly of siliciclastic sediments, whereas the Garipçe Formation is composed of epiclastic sandstone, volcanic breccia, hyaloclastics and volcanoclastic debris-flow deposits. Individual beds of these debris-flow deposits may reach up to 50 m in thickness, while the total thickness of the Kavaklar Group is over 2 km. Lava flows are scarce in the Garipçe Formation, but feeder dyke swarms and sills are abundant, especially along the Black Sea coast (Keskin *et al.* 2010). The uppermost unit of the Kavaklar Group namely the Kısırkaya Formation is made up of alkaline olivine basalt lavas. It also contains basaltic to intermediate epiclastic and pyroclastic intercalations. Basic dykes, which are similar in composition to the Kısırkaya lavas, cut the underlying Garipçe Formation along the Black Sea coast and hence they are interpreted as the feeder dykes of the Kısırkaya Formation (Keskin *et al.* 2003). Because it is beyond the scope of this paper to discuss the Kavaklar Group, we concentrate solely on dyke systems in the İstanbul Zone.

Yavuz & Yılmaz (2009) obtained an K–Ar biotite age of 75 ± 2 Ma (Maastrichtian) from an andesitic lava sample and a K–Ar whole-rock age of 67 ± 2 Ma (Maastrichtian) from an olivine basalt sample in the Upper Cretaceous Volcanics. The olivine basalt possibly belongs to the Kısırkaya Formation. The published Rb–Sr whole-rock age from the Çavuşbaşı granodiorite is 63 ± 13 Ma (Öztunalı & Satır 1975), the K–Ar biotite age is 63.5 ± 1.6 Ma and the zircon U–Pb SHRIMP-II ages are 67.91 ± 0.63 and 67.59 ± 0.5 Ma (Yılmaz Şahin *et al.* 2012). These results indicate that the volcanics in the north and the Çavuşbaşı granodiorite cutting the İstanbul Palaeozoic unit in the south are coeval (i.e. they span a period from the Late Cretaceous to the Paleocene) (see Fig. 2).

The mafic–intermediate dykes and small intrusions cutting the basement rocks (i.e. the İstanbul Palaeozoic unit) are exposed almost everywhere in İstanbul (Özgörüş & Okay 2005). In this paper, we present new geochemical, mineral chemistry, geochronology and isotope data from the mafic–intermediate dykes (a few are felsic) and a subset of samples from the plutonic rocks cutting the İstanbul Palaeozoic unit, as well as the dykes cutting the Upper Cretaceous volcanic rocks in the north of İstanbul (i.e. the Kavaklar Group). In the light of our new data, the source characteristics, melting parameters and magma chamber processes (e.g. crystallization, assimilation and mixing) of these magmatic rocks, with special reference to the physicochemical conditions of fractionation, are evaluated in this paper. We conclude the paper with a geodynamic model connecting the genesis and evolution of the Late Cretaceous–Paleocene dyke and plutonic systems to the opening of the Western Black Sea Basin.

Geological background and petrography

The study area is located in the İstanbul Zone, comprising the western part of the Pontides, in the north of the Sakarya Zone (Fig. 1) (Şengör & Yılmaz 1981; Okay & Tüysüz 1999). The İstanbul Zone consists of Ordovician–Lower Carboniferous sedimentary rocks that overlay a metamorphic basement ranging in age from Late Precambrian to Early Palaeozoic. These metamorphic units are observed around the Bolu region, east of the İstanbul Zone (Ustaömer *et al.* 2005). The first magmatic activity that cut the İstanbul Palaeozoic unit is represented by the Permian Sancaktepe granite. The Sancaktepe granite gave a Rb–Sr age of 255 ± 5 Ma and a K–Ar age of 254 Ma (Yılmaz 1977). Yılmaz Şahin *et al.* (2015) obtained U–Pb SHRIMP-II zircon ages of between 257.3 ± 1.5 and 253.7 + 1.75 Ma from the same pluton. This sequence is unconformably overlain by the Permo-Triassic Kapaklı Formation and then by the Triassic Gebze Group (Özgül 2012 and references therein). These are represented by clastic and carbonate beds that are unconformably overlain by the Late Cretaceous–Paleocene Kavaklar Group. The İstanbul Palaeozoic unit thrusts over the Kavaklar Group in the north of İstanbul from south to the north (Fig. 2). Both the İstanbul Palaeozoic unit and the Kavaklar Group are cut by mafic–intermediate dykes and small stocks ranging in age from Upper Cretaceous to Paleocene. All these units are unconformably overlain by Neogene continental sedimentary rocks.

Dykes and small stocks cutting the İstanbul Palaeozoic Unit

Petrographic features of the dykes and small stocks are presented in Table 1 based on the classification

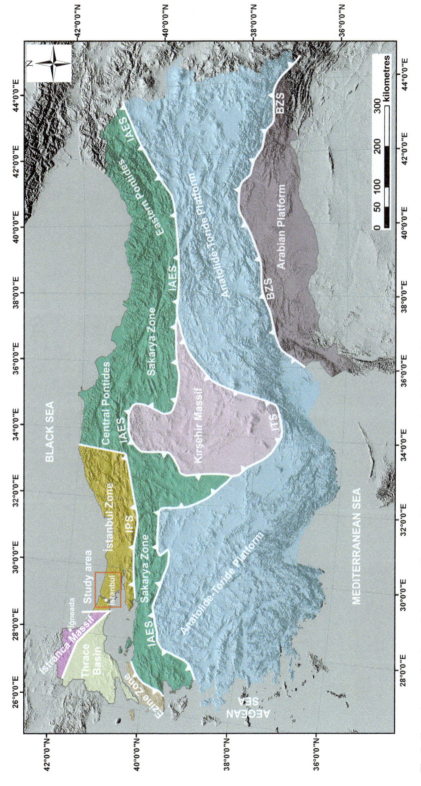

Fig. 1. Tectonic map of Turkey and the surrounding area with major suture zones (modified from Okay & Tüysüz 1999). IAES, İzmir–Ankara–Erzincan Suture; IPS, Intra-Pontide Suture; ITS, Intra-Tauride Suture; BZS, Bitlis–Zagros Suture.

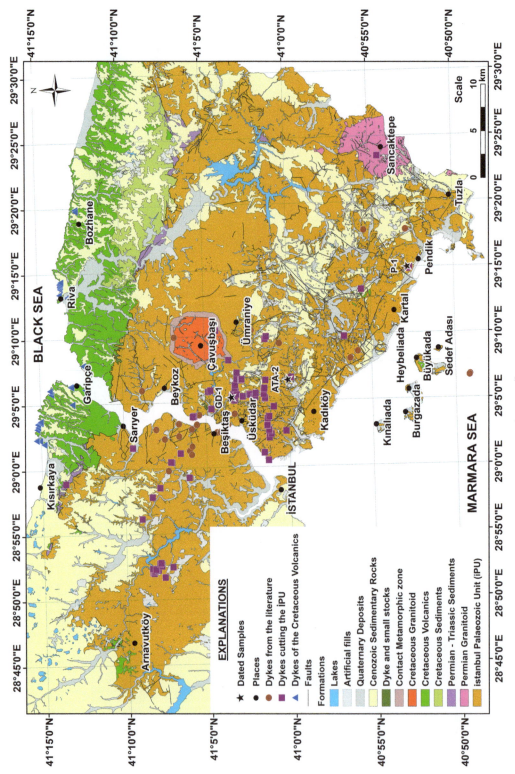

Fig. 2. Geological map of the study area (simplified from the Geological Map of İstanbul: Özgül *et al.* 2011).

Table 1. *Petrographical and geochemical properties, and age of the dykes and small stocks exposed around Bosphorus, Istanbul*

Group	Formation	Type	Geochemical tendency	Serial	Lithology	Texture	Mineral composition	Age
Upper Cretaceous volcanics in northern Istanbul	Kısırkaya	Lava	Within-plate alkaline	Dry (POAM)	Alkaline Olivine basalt	Aphyric, microporphyric, microlithic	Plg + Cpx + Ol ± Opx	67 ± 2 K–Ar (1)
		Dyke						
	Garipçe and Bozhane	Lava and blocks	Calc-alkaline with SC	Wet (PAm)	Bazaltic andesite, andesite, rarely basalt and dacite	Porphyric, microlithic	Plg + Amp + Cpx + Opx	75 ± 2 K–Ar (1)
				Dry	Andesite and dacite	Porphyric, microlithic	Plg + Cpx + Opx	–
		Dyke	Calc-alkaline with SC	Wet	Andesite	Porphyric,	Plg + Amp + Cpx + Opx	–
				Dry		Vitrophyric porphyric,	Plg + Cpx	
			Within-plate alkaline	Dry	Olivine basalt (Like as Kısırkaya)	Aphyric Vitrophyric, hyaloplitic and porphyric	Plg + Ol + Cpx	–
Dykes and small stocks in Istanbul Palaeozoic units	Tavşantepe	Small Stock	Calc-alkaline with SC	Wet	Microdiorite	Microgranular	Plg + Pg + Cpx + Kf + Q	58.9 ± 1.8 U–Pb LA-ICP-MS (this study)
	Çavuşbaşı	Small Pluton	Calc-alkaline with SC	Wet	Granodiorite	Equigranular hypidiomorphic	Plg + Hb + Kf + Q	67.91 ± 0.63–67.59 ± 0.5 U–Pb SHRIMP-II (2)
		Dyke	Calc-alkaline with SC	Wet	Bazaltic andesite, andesite, dacite	Porphyric	Plg + Hb + Kf	65.44 ± 0.93 U–Pb LA-ICP-MS (this study)
					Microgranodiorit (Like as Çavuşbaşı granodiorite)	Porphyric	Plg + Hb + Kf + Q + Bi	72.49 ± 0.79 U–Pb LA-ICP-MS (this study)
		Dyke	Calc-alkaline with SC – weakly alkaline	Wet	Lamprofirler (spessartite, vogesite)	Intergranular, subophitic	Plg + Amp + Cpx ± Phl ± Kf	–
		Dyke	Within-plate alkaline	Dry	Diabase	Intergranular, ophitic	Plg + Cpx + Ol	–

Abbreviations: Plg, plagioclase; Amp, amphibole; Pg, Pargasite; Cpx, clinopyroxene; Opx, orthopyroxene; Ol, olivine; PAm, Plg + Olv + Augite + Magnetite fractionation assemblage; POAM, Plg + Olv + Augite + Magnetite fractionation assemblage; SC, subduction component (modified from Keskin *et al.* 2003).

of Keskin et al. (2003). The ones cutting the İstanbul Palaeozoic unit can be divided into four sub-groups based on their chemical compositions and petrographic features: (1) diabase dykes, (2) lamprophyre dykes, (3) intermediate to felsic dykes and (4) microdioritic to granodioritic stocks and sub-volcanic bodies (Keskin et al. 2003).

The diabase dykes are typically aphyric in texture and green to dark green in colour. Igneous textures are mostly well preserved and predominantly sub-ophitic, intergranular and intersertal. Main mineral assemblage is represented by plagioclase, clinopyroxene and rarely olivine pheno- and microcrystals. The main mineral assemblages have suffered variable degrees of alteration. Secondary mineral phase is composed of albite, epidote, chlorite, calcite, sericite, clay minerals and large amount of opaque minerals. Pyroxenes have been partly altered to a fine-grained mass of chlorite, while plagioclase (mainly labradorite) is replaced partly by albite, epidote and calcite. The groundmass has been turned into chlorite, albite and quartz. Their crystal sizes range between 1 and 3 mm (i.e. medium to fine). Note that because many diabase dykes were observed and sampled along the tunnel excavations around the town of Sarıyer and rock quarries, they are much fresher than their exposures on the surface.

Lamprophyre dykes are green, dark green and greyish (Fig. 3a–f). They are spessartitic and/or vogesitic in composition, displaying fine to moderate grained intergranular and sub-ophitic texture (Fig. 4a–c). They are composed of plagioclase, pargasitic amphibole, clinopyroxene phenocrysts and microcrysts. They rarely include quartz, phlogopite and K-feldspar. Secondary mineral phases are composed of epidote, chlorite and calcite. Their grain sizes are between 1 and 3 mm. Pargasitic amphiboles display acicular to prismatic shapes and are brown. Plagioclase crystals are euhedral, subhedral and rarely anhedral in shape. The albitization in plagioclase, chloritization and epidotization in amphiboles and phlogopites are observed as alteration products. Albite is commonly observed in the chloritized groundmass. Clinopyroxene crystals are colorless, subhedral and the smaller ones are occasionally anhedral. They are generally augitic in composition and show partial chloritization. Quartz, phlogopite and K-feldspar have not been observed under the microscope, but they were determined by the XRD analysis. Thickness of the lamprophyre dykes varies from 0.2 to 5 m (Fig. 3a–d). The lamprophyre dykes display intrusive contacts with the İstanbul Palaeozoic unit (Fig. 3e). A narrow contact metamorphic zone (1–10 mm wide) is occasionally observed at their contacts with the wall rocks. In some locations they also include xenoliths of the host rock close to contacts.

Intermediate to felsic dykes (Fig. 3g) are composed of basaltic andesite, andesite, dacite and rhyolite. They have medium to fine crystal size (2–5 mm), porphyric texture and felsic groundmass. Intermediate dykes consist of quartz, plagioclase, K-feldspar, hornblende and biotite crystals. Titanite, apatite, zircon and magnetite can be found as accessory phases, while epidote, chlorite, sericite, clay minerals and calcite are secondary minerals. Plagioclase and amphibole crystals have euhedral, subhedral and rarely anhedral shapes (Fig. 4d, e). Thickness of the intermediate dykes varies from 0.5 to 3 m (Fig. 3g). They are form small intrusions. They are microdiorite to microgranodiorite in compositions, similar to the Çavuşbaşı granodiorite hence they may be the porphyric equivalent of that pluton.

The Çavuşbaşı granodiorite covers an area of c. 20 km^2 (Ketin 1941). This pluton displays equigranular and hypidiomorphic textures with medium to large crystals of quartz, plagioclase, orthoclase, amphibole and biotite. Secondary mineral phase consists of epidote, chlorite, calcite and kaolinite. In fact, the Çavuşbaşı granodiorite is not a homogeneous body as it contains a range of lithologies from granodiorite to quartz-monzodiorite and granite (Fig. 4f), granodiorite being the most abundant one. It contains microdioritic enclaves with aplite veins (Yılmaz Şahin et al. 2012). The Çavuşbaşı granodiorite intruded into the İstanbul Palaeozoic unit and formed a contact metamorphic zone in the wall rocks (Ketin 1941).

The smallest and the youngest intrusion is the Tavşantepe diorite. It covers an area of approximately 0.5 km^2 in the vicinity of the town of Pendik, SE of İstanbul. This stock display equigranular, hypidiomorphic and porphyric textures (Fig. 4g). The Tavşantepe diorite consists of quartz, plagioclase, orthoclase, pargasitic amphibole, clinopyroxene (diopside) and primary epidote minerals in the endoskarn zone along its contacts with the Devonian limestones.

Dykes cutting the Upper Cretaceous volcanic rocks

In the north of İstanbul, the Garipçe and the Bozhane formations are cut by numerous dykes, which display a wide compositional range from alkaline olivine basalt to calc-alkaline basaltic andesite, andesite and dacite (Figs 3h–j & 4h, i). The olivine-bearing dykes display reddish to scarlet alteration colours due to iddingsitization of olivine crystals and can be easily distinguished from the plagioclase- and pyroxene-bearing intermediate dykes, which are generally dark-grey to black in colour and porphyritic in texture. The olivine-bearing dykes are similar to the lavas of the Kısırkaya Formation, in terms of

Fig. 3. Field photographs showing the details of the contact zone between the wall rocks and the cross-cutting mafic–intermediate dykes. (**a**)–(**d**) Fine-grained lamprophyre dykes intruded into the Carboniferous greywacke beds (Trakya Formation) in the Cebeci and Ayazağa quarry areas. (**e**) An intermediate dyke displaying an intrusive contact with the Carboniferous greywacke close to Bosphorus. (**f**) A lamprophyre dyke cutting the Çavuşbaşı granodiorite. (**g**) An intermediate dyke cutting Devonian shale in the vicinity of the town of Anadoluhisarı. (**h**) & (**i**) An alkali olivine basalt dyke cropping out nearby the Black Sea coast, cutting the Garipçe Formation and extending along the southern margin of a small island called Atlamataşı. It displays columnar jointing. (**j**) An andesite dyke cutting the volcaniclastics of the Garipçe Formation near Yenidalyan Bay.

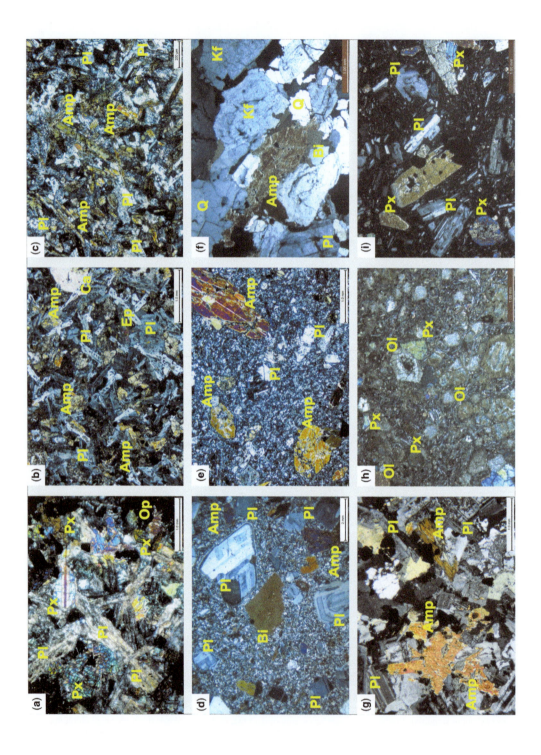

Table 2. *GPS determined geographical coordinates of the dated samples*

Sample No.	Rock type	Coordinates	Age
GD-1	Microgranodiorite	29° 5′ 17.91″ E–41° 3′ 33.38″ N	72.49 ± 0.79 Ma
ATA-2	Andesite	29° 6′ 34.82″ E–41° 0′ 8.41″ N	65.44 ± 0.93 Ma
P-1	Microdiorite	29° 15′ 0.15″ E–40° 52′ 51.47″ N	58.9 ± 1.8 Ma

their mineral content and texture (Keskin *et al.* 2010). On the other hand, plagioclase- and pyroxene-bearing porphyritic dykes are petrographically similar to the lavas of the Garipçe Formation. Dykes of various thicknesses and shapes are abundant in certain zones forming dyke swarms, especially close and parallel to the Black Sea coastline (Keskin *et al.* 2003, 2010).

Analytical techniques

A total of 46 rock samples were selected from the mafic–intermediate dykes for geochemical analyses. Whole-rock chemical analyses of 46 samples were conducted at Acme Analytical Laboratories Ltd in Vancouver, Canada. The analytical package included inductively coupled plasma atomic emission spectrometer (ICP-AES) analyses after $LiBO_2$ fusion for all major oxides (SiO_2, TiO_2, Al_2O_3, MnO, MgO, CaO, K_2O, Na_2O, P_2O_5) and loss on ignition (LOI). The trace elements were analysed by inductively coupled plasma mass spectrometer (ICP-MS), with rare earth and incompatible elements determined from a $LiBO_2$ fusion, and precious and base metals determined from an aqua regia digestion. LOI was calculated as the weight difference after ignition at 1000°C. The total iron concentration was determined as Fe_2O_3. All samples are analysed together with the STD-SO-17 international standard.

Mineral chemistry analyses were performed using the Cameca SX-100 instrument at the University of Edinburgh, School of Geosciences (Grant Institute). The wavelength-dispersive mode with 15 kV acceleration potential, 300 nA beam current and a 10 µm beam diameter was used during the analyses. Results of representative feldspar, pyroxene and amphibole analyses are given in the Supplementary material.

Sr and Nd isotope analyses were performed at the Radiogenic Isotope Laboratory of the Middle East Technical University, Central Laboratory, Ankara, Turkey. Analytically details are given in Aysal *et al.* (2012). For zircon analysis, the method we followed is described in detail in Aysal *et al.* (2012).

Results

LA-ICP-MS zircon U–Pb geochronology

Zircon crystals suitable for U–Pb laser ablation inductively coupled plasma mass spectrometry (LA-ICP-MS) dating were found in three samples: two intermediate dyke samples (GD-1 and ATA-2) and in a sample from a small stock (P-1) cutting the İstanbul Palaeozoic unit (Fig. 2; Table 2). The zircon grains were primarily colourless or pale brown, transparent, euhedral to subhedral, mostly elongated to prismatic grains (50–250 µm) with magmatic oscillatory zoning in cathodoluminesce (CL) images (Fig. 5a–c). For sample GD-1, 23 analyses yielded apparent $^{206}Pb/^{238}U$ ages ranging from 66.8 ± 3.2 to 82.6 ± 5.3 Ma (Table 3), and a concordia $^{206}Pb/^{238}U$ age of 72.49 ± 0.79 Ma (mean square weighted deviation (MSWD) = 1.5: Fig. 5a). Analyses of zircons from sample ATA-2 yielded $^{206}Pb/^{238}U$ ages of 62.0 ± 4.2–88.7 ± 7.5 Ma (Table 4). The 12 analyses yielded a concordia $^{206}Pb/^{238}U$ age of 65.44 ± 0.93 Ma (MSWD = 1.4: Fig. 5b). Inherited cores of zircons from sample ATA-2 yielded $^{206}Pb/^{238}U$ ages of 114.5 ± 4.2–584.8 ± 15.2 Ma (Table 4). Analyses of zircons from sample P-1 yielded $^{206}Pb/^{238}U$ ages of 54.8 ± 6.4–64.0 ± 9.7 Ma (Table 5). The 12 analyses yielded a concordia $^{206}Pb/^{238}U$ age of 58.9 ± 1.8 Ma (MSWD = 3.2: Fig. 5c). An inherited zircon or

Fig. 4. Microphotographs of studied dykes. (**a**) Plagioclase and clinopyroxene phenocrysts in a diabase dyke with ophitic texture. (**b**) & (**c**) Acicular, brown pargasitic amphibole and plagioclase phenocrysts in a lamprophyre dyke. (**d**) & (**e**) Plagioclase phenocrysts and green amphiboles set in a fine-grained groundmass in an intermediate dyke. Note that (a)–(e) are microphotographs of the dykes cutting the İstanbul Palaeozoic Unit. (**f**) Microphotograph of the Çavuşbaşı granodiorite. (**g**) Brown pargasitic amphibole and plagioclase crystals in the Tavşantepe diorite. (**h**) Corroded olivine phenocrysts in a basaltic dyke belonging to the Kısırkaya Formation. (**i**) Plagioclase and clinopyroxene phenocrysts in a glassy groundmass, displaying vitrophyric and porphyritic texture in an andesite dyke from the Garipçe Formation (i.e. Upper Cretaceous Kavaklar Group). Abbreviations: Pl, plagioclase; Px, pyroxene; Amp, amphibole; Ca, calcite; Ep, epidote; Q, quartz; Ol, olivine.

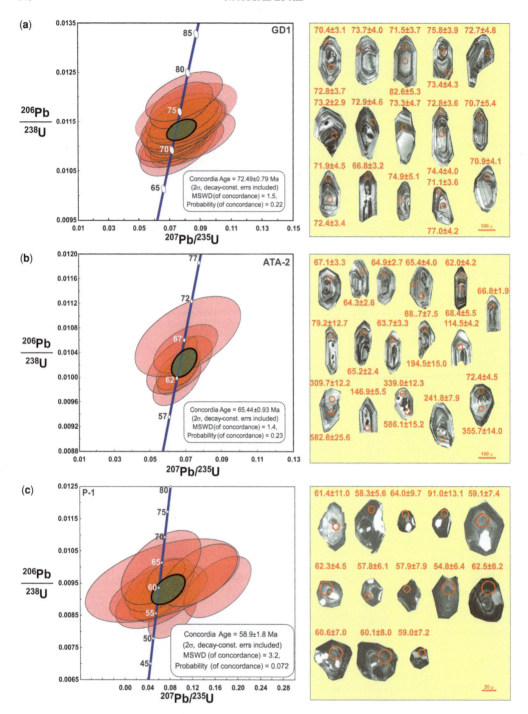

Fig. 5. Concordia diagrams and CL images of the zircon grains from the mafic–intermediate dykes (GD-1 and ATA-2) and Tavşantepe diorite (P-1). The red circles show spots of U–Pb LA-ICP-MS analyses in zircon grains.

zircon antecryst (according to Miller *et al.* 2007) from sample P-1 yielded $^{206}Pb/^{238}U$ ages of 91.0 ± 13.1 Ma (Table 5). Th/U ratios of the zircons varied between 0.201 and 1.096, which is consistent with a magmatic origin (Teipel *et al.* 2004; Linnemann *et al.* 2011).

Table 3. LA-ICP-MS radiometric age determination from intermediate dyke cutting İstanbul Palaeozoic unit (GD-1)

GD-1 Zircon	Isotope ratios									Age (Ma)		Th/U ratio
	$^{206}Pb/^{238}U$	1SE	$^{207}Pb/^{235}U$	1SE	$^{208}Pb/^{232}Th$	1SE	$^{207}Pb/^{206}Pb$	1SE	Rho	$^{206}Pb/^{238}U$	2σ	
1	0.010981984	0.00024	0.0779	0.0105	0.0044	0.0004	0.0514	0.0070	0.54	70.4	3.1	0.801
2	0.011503694	0.00031	0.0803	0.0156	0.0040	0.0004	0.0506	0.0099	0.53	73.7	4.0	0.550
3	0.012893901	0.00042	0.0882	0.0249	0.0057	0.0006	0.0496	0.0141	0.53	82.6	5.3	0.522
4	0.011154213	0.00029	0.0768	0.0151	0.0043	0.0005	0.0499	0.0099	0.53	71.5	3.7	0.472
5	0.011832833	0.00031	0.0775	0.0135	0.0043	0.0004	0.0475	0.0083	0.53	75.8	3.9	0.677
6	0.011454923	0.00034	0.0784	0.0179	0.0045	0.0006	0.0496	0.0114	0.53	73.4	4.3	0.369
7	0.011334411	0.00038	0.0778	0.0147	0.0049	0.0006	0.0498	0.0095	0.54	72.7	4.8	0.444
8	0.011422656	0.00023	0.0748	0.0096	0.0036	0.0004	0.0475	0.0061	0.53	73.2	2.9	0.702
9	0.01137459	0.00036	0.0779	0.0149	0.0033	0.0005	0.0497	0.0096	0.54	72.9	4.6	0.417
10	0.011434754	0.00037	0.0794	0.0180	0.0033	0.0005	0.0504	0.0115	0.53	73.3	4.7	0.393
11	0.011613975	0.00031	0.0780	0.0156	0.0033	0.0005	0.0487	0.0098	0.53	74.4	4.0	0.509
12	0.011351789	0.00028	0.0782	0.0119	0.0037	0.0004	0.0499	0.0077	0.54	72.8	3.6	0.830
13	0.011034343	0.00042	0.0757	0.0165	0.0027	0.0005	0.0497	0.0110	0.54	70.7	5.4	0.453
14	0.011214445	0.00035	0.0735	0.0168	0.0042	0.0006	0.0475	0.0110	0.53	71.9	4.5	0.438
15	0.011293156	0.00027	0.0739	0.0112	0.0039	0.0005	0.0475	0.0073	0.53	72.4	3.4	0.629
16	0.010421353	0.00025	0.0736	0.0095	0.0032	0.0003	0.0512	0.0067	0.54	66.8	3.2	0.893
17	0.011685002	0.00040	0.0808	0.0219	0.0027	0.0006	0.0501	0.0137	0.53	74.9	5.1	0.357
18	0.011214015	0.00031	0.0910	0.0143	0.0040	0.0005	0.0588	0.0094	0.54	71.9	4.0	0.502
19	0.01107382	0.00033	0.0772	0.0183	0.0027	0.0004	0.0506	0.0121	0.53	71.0	4.2	0.530
20	0.012012493	0.00033	0.0793	0.0165	0.0036	0.0004	0.0479	0.0100	0.53	77.0	4.2	0.727
21	0.01108325	0.00028	0.0758	0.0122	0.0039	0.0005	0.0496	0.0081	0.53	71.1	3.6	0.614
22	0.011063662	0.00032	0.0761	0.0124	0.0037	0.0005	0.0499	0.0083	0.54	70.9	4.1	0.554
23	0.011362586	0.00029	0.0748	0.0132	0.0037	0.0005	0.0478	0.0085	0.53	72.8	3.7	0.713

Table 4. *LA-ICP-MS radiometric age determination from an intermediate dyke cutting the İstanbul Palaeozoic unit (ATA-2)*

ATA-2 Zircon	Isotope ratios									Age (Ma)		Th/U ratio
	$^{206}Pb/^{238}U$	1SE	$^{207}Pb/^{235}U$	1SE	$^{208}Pb/^{232}Th$	1SE	$^{207}Pb/^{206}Pb$	1SE	Rho	$^{206}Pb/^{238}U$	2σ	
1	0.01129	0.00035	0.0819	0.0161	0.0039	0.0004	0.0526	0.0105	0.53	72.4	4.5	0.670
2	0.01046	0.00026	0.0701	0.0118	0.0033	0.0003	0.0486	0.0083	0.53	67.1	3.3	0.714
3	0.01002	0.00020	0.0648	0.0070	0.0028	0.0002	0.0469	0.0051	0.54	64.3	2.6	0.767
4	0.00966	0.00033	0.0770	0.0173	0.0025	0.0003	0.0578	0.0131	0.53	62.0	4.2	0.721
5	0.01066	0.00043	0.0713	0.0195	0.0030	0.0005	0.0485	0.0134	0.53	68.4	5.5	0.484
6	0.03823	0.00064	0.2759	0.0168	0.0166	0.0015	0.0524	0.0032	0.56	241.8	7.9	0.346
7	0.01793	0.00033	0.1315	0.0096	0.0061	0.0007	0.0532	0.0040	0.55	114.5	4.2	0.201
8	0.05672	0.00115	0.5131	0.0374	0.0233	0.0025	0.0656	0.0049	0.56	355.7	14.0	0.489
9	0.01236	0.00100	0.0831	0.0748	0.0056	0.0018	0.0487	0.0441	0.52	79.2	12.7	0.418
10	0.01016	0.00019	0.0703	0.0055	0.0032	0.0004	0.0501	0.0040	0.55	65.2	2.4	0.464
11	0.04922	0.00099	0.3567	0.0349	0.0180	0.0023	0.0526	0.0052	0.54	309.7	12.2	0.288
12	0.09458	0.00218	0.7732	0.0746	0.0284	0.0035	0.0593	0.0058	0.55	582.6	25.6	0.658
13	0.01011	0.00021	0.0656	0.0080	0.0031	0.0004	0.0470	0.0058	0.54	64.9	2.7	0.629
14	0.01020	0.00031	0.0667	0.0121	0.0037	0.0005	0.0474	0.0087	0.54	65.4	4.0	0.642
15	0.01385	0.00059	0.0917	0.0279	0.0042	0.0006	0.0480	0.0148	0.53	88.7	7.5	0.600
16	0.00993	0.00026	0.0653	0.0089	0.0027	0.0002	0.0477	0.0066	0.54	63.7	3.3	1.058
17	0.01042	0.00015	0.0716	0.0033	0.0030	0.0002	0.0498	0.0024	0.56	66.8	1.9	0.599
18	0.02304	0.00044	0.1582	0.0127	0.0057	0.0005	0.0498	0.0041	0.55	146.9	5.5	0.510
19	0.09496	0.00129	0.7811	0.0284	0.0272	0.0019	0.0597	0.0022	0.57	584.8	15.2	0.843
20	0.05399	0.00101	0.3976	0.0280	0.0191	0.0015	0.0534	0.0038	0.55	339.0	12.3	0.613
21	0.03063	0.00120	0.2721	0.0453	0.0072	0.0009	0.0644	0.0110	0.55	194.5	15.0	0.582

Table 5. *LA-ICP-MS radiometric age determination from the Tavşantepe diorite (P-1)*

P-1	Isotope ratios								Rho	Age (Ma)		Th/U ratio
Zircon	$^{206}Pb/^{238}U$	1SE	$^{207}Pb/^{235}U$	1SE	$^{208}Pb/^{232}Th$	1SE	$^{207}Pb/^{206}Pb$	1SE		$^{206}Pb/^{238}U$	2σ	
1	0.0657	0.0903	0.3599	0.0126	0.0030	0.0011	0.0498	0.0687	0.52	61.4	11.0	0.462
2	0.0624	0.0229	0.3620	0.0163	0.0027	0.0003	0.0498	0.0185	0.53	58.3	5.6	0.831
3	0.0643	0.0469	0.3612	0.0142	0.0019	0.0005	0.0467	0.0344	0.52	64.0	9.7	0.546
4	0.5615	0.0799	0.3581	0.0113	0.0106	0.0011	0.2865	0.0456	0.57	91.0	13.1	0.835
5	0.0595	0.0573	0.3296	0.0395	0.0036	0.0006	0.0469	0.0453	0.52	59.1	7.4	0.548
6	0.0727	0.0224	0.3807	0.0142	0.0026	0.0002	0.0543	0.0169	0.53	62.3	4.5	1.096
7	0.0592	0.0289	0.4401	0.0301	0.0023	0.0003	0.0476	0.0235	0.52	57.8	6.1	0.672
8	0.0589	0.0500	0.3817	0.0146	0.0041	0.0008	0.0473	0.0404	0.52	57.9	7.9	0.498
9	0.0846	0.0284	0.3613	0.0201	0.0029	0.0003	0.0719	0.0245	0.54	54.8	6.4	1.086
10	0.1321	0.0443	0.3859	0.0143	0.0010	0.0005	0.0983	0.0336	0.54	62.5	8.2	0.701
11	0.0613	0.0404	0.3568	0.0159	0.0033	0.0005	0.0471	0.0312	0.52	60.6	7.0	0.579
12	0.1639	0.0393	0.3647	0.0194	0.0040	0.0005	0.1269	0.0316	0.55	60.1	8.0	0.797
13	0.0600	0.0352	0.3605	0.0124	0.0027	0.0004	0.0473	0.0279	0.52	59.0	7.2	0.860

Whole-rock geochemistry

Whole-rock geochemical data from the mafic–intermediate dykes and small stocks are given in Table 6. LOI values ranging from 1.28 to 9.1 wt% (9% LOI is very high and indicates a high degree of alteration) imply that most of the samples have weak to strong alteration, which is consistent with the secondary mineralogy. Therefore, we consider mainly relatively immobile elements (high field strength elements (HFSEs) and rare earth elements (REEs)) in our petrogenetic interpretations.

The chemical index of alteration (CIA = molar $[(Al_2O_3)/(Al_2O_3 + CaO + Na_2O + K_2O)] \times 100$) values range between 49.29 and 70.66, and fall in the 'low degree of chemical weathering' field of Nesbitt & Young's (1984) ternary diagram (Fig. 6a). This observation implies that the element systematics of the dyke samples have not been considerably affected by the alteration. Note that the dyke samples mainly fall on the left side of the plagioclase–K-feldspar connection line (CIA of *c.* 50–70) and the alteration trends are parallel to the A-CN side on Figure 6a. These observations indicate that some of the plagioclase crystals changed partly into clay minerals (Fig. 6a).

Geochemical data from the İğneada volcanic rocks (i.e. calc-alkaline lavas from western part of the Black Sea, possibly related to an arc setting: unpublished data of the first author) and from the Çavuşbaşı granodiorite (taken from Yılmaz Şahin *et al.* 2012) are also plotted on the diagrams for comparison.

The wide compositional range of the dykes, determined petrographically, is also shown by their heterogeneous geochemical compositions. On the Zr/TiO_2 v. SiO_2 wt% diagram (Fig. 6b) of Winchester & Floyd (1977), the lamprophyric rocks (Bergman 1987; Rock 1987; Mitchell & Bergman 1991; Woolley *et al.* 1996) and diabase dykes cutting the İstanbul Palaeozoic unit plot into the sub-alkali basalt field (Fig. 6b). The SiO_2 content of lamprophyric rocks and diabase span a narrow range from 48.25 to 51.93 wt%, while their total alkali contents ($Na_2O + K_2O$) vary between 3.47 and 6.14 wt%, indicating that they are either alkaline or transitional between the alkaline and sub-alkaline series. Magnesium numbers (Mg# = $(100 \times MgO/(MgO + FeO_t))$) range from 49.42 to 67.41.

The mafic–intermediate dykes (some of felsic) cutting the İstanbul Palaeozoic unit fall into the andesite, rhyodacite and dacite fields (Fig. 6b). The mafic–intermediate dykes have SiO_2 contents between 53.85 and 70.74 wt%, and Mg# values ranging from 35.66 to 66.75.

The mafic–intermediate dykes cutting the Upper Cretaceous volcanics are divided into two compositionally different groups: (1) the first group is

Table 6. The results of a whole-rock major (wt%), trace (ppm) and rare earth elements (REE) (ppm) geochemical analysis of mafic–intermediate dykes and small stocks

Lamprophyres and diabasis cutting the Istanbul Palaeozoic unit

	SU-1	SU-3	HT-6	HT-7	HT-8	HT-11	KB-1	KB-2	KB-3	KB-5	KB-7	CS-4	MT42	MT60	MT22	MT40
SiO_2	50.22	48.74	50.37	51.14	51.93	50.25	50.84	51.08	48.72	49.97	50.91	51.89	48.25	49.26	49.60	49.66
TiO_2	1.57	1.32	1.86	1.25	1.82	1.24	1.89	1.84	1.69	1.58	1.63	2.40	1.68	1.51	2.24	1.58
Al_2O_3	17.84	14.80	17.61	18.54	18.44	18.11	18.35	18.62	17.32	16.70	16.80	17.06	15.74	15.65	16.77	16.39
$Fe_2O_3^T$	9.23	8.61	10.82	10.50	10.49	9.98	10.64	10.52	9.54	9.90	9.82	12.93	10.32	9.79	11.37	9.30
MnO	0.18	0.16	0.18	0.17	0.16	0.17	0.17	0.16	0.17	0.16	0.16	0.21	0.16	0.15	0.18	0.15
MgO	4.55	8.99	5.67	6.16	5.51	6.34	6.30	6.04	6.96	6.76	6.76	8.12	9.14	10.20	7.16	7.26
CaO	11.64	11.38	8.30	7.30	5.98	8.72	6.26	6.08	10.29	9.87	8.91	4.41	9.96	9.18	8.58	11.06
Na_2O	5.03	3.62	4.48	4.28	5.64	4.29	5.49	5.56	4.27	4.98	4.98	2.85	3.53	3.07	3.66	3.84
K_2O	0.15	0.22	0.77	0.94	0.50	1.08	0.44	0.45	0.85	0.41	0.37	0.61	0.85	0.76	0.14	0.41
P_2O_5	0.59	0.40	0.44	0.29	0.52	0.27	0.44	0.47	0.56	0.42	0.41	0.39	0.38	0.42	0.31	0.35
LOI	3.20	3.80	3.90	3.60	4.40	4.00	4.80	4.90	3.30	4.60	4.80	4.50	4.80	7.90	9.10	5.50
Sum	100.99	98.25	100.49	100.58	101.00	100.46	100.82	100.82	100.37	100.76	100.75	100.87	100.00	100.00	100.00	100.00
Sc	15	27	24	28	20	28	22	20	25	24	23	29	35	29	38.6	35
V	240	223	260	251	257	252	249	235	237	207	204	247	243	214	255.5	234
Cr	13.68	458.42	41.05	34.21	13.68	41.05	13.68	20.53	95.79	184.74	177.89	136.84	232.63	280.52	112.60	109.47
Co	27	34.9	31.3	33.5	27.6	31.1	29.2	29	30.9	30.4	29.2	39	45	43.5	29.2	38.5
Ni	28.6	210.5	33	43.7	16.8	49.2	25	27.9	33.6	79.4	74.9	97.8	93	165	74.9	52
Cu	74.8	61.1	49.6	75	24.8	99.7	41.6	39.7	49	61.8	54.5	42.3	101	70	47	66
Zn	73	45	83	68	82	60	72	82	64	65	62	81	59	55	69	58
Ga	17	13.8	16.3	14.9	17.3	15.1	17.5	16.3	16.5	16.5	16.2	17.8	18.4	18.3	18.4	19.4
Rb	1.4	3.6	10.1	21.8	12.9	22.7	9.4	10.1	10	8.3	7.4	24.7	17.1	25.1	7.3	7.8
Sr	1173.4	959.9	465.8	506	643.7	502.2	699.3	697.3	624.5	531.4	535.3	292.6	692.8	775.7	453.4	710.4
Y	24.1	17.4	24.3	20	36.8	20.7	21	21	23	19.8	20.4	34.9	21.5	19.6	33.6	22
Zr	159.9	116.6	142.5	99.4	176.9	101.3	149.1	149.4	136	130.2	128.4	223.2	125.8	111.2	173.6	127.5
Nb	24.7	16.9	18	10.8	22	10.6	14.1	15	19.7	14.9	14.9	10.3	15.1	17.9	6.4	13.1
Ba	42	74	163	252	179	304	286	182	146	166	167	90	1303	174	33	117
Cs	1.4	6.3	1.6	1.7	1.1	1.8	3.1	4.5	0.1	0.6	0.6	1	0.5	4.5	0.5	0.5
La	34.7	21.2	24.8	16.8	33.9	16.3	23.3	24.5	26.7	24.2	22.8	20.9	25.6	23.9	10.9	22.9
Ce	71	43.4	51.3	35.9	71.4	34.7	49.5	52.2	53.5	46.9	47.1	47.2	54.1	47.9	28.3	46.6
Pr	8.61	5.29	6.5	4.47	9.01	4.67	6.26	6.53	6.7	5.95	5.8	6.08	7	6.08	4.32	6.11
Nd	34	21	26.9	19.4	36.3	20.2	25.5	26.6	28.2	23.9	23.5	25.3	29	24.8	20.4	25.1
Sm	6.52	4.08	5.88	3.95	7.45	4.32	5.2	5.25	5.43	4.6	4.71	5.96	6.4	5	5.9	5.3
Eu	1.95	1.29	1.76	1.32	2.24	1.38	1.68	1.64	1.84	1.53	1.5	1.88	1.89	1.73	2.13	1.78
Gd	5.71	3.82	5.44	4.02	7.47	4.17	4.89	5.07	5.43	4.61	4.58	6.36	5.48	4.09	6.32	5.13
Tb	0.84	0.56	0.82	0.62	1.11	0.68	0.74	0.77	0.81	0.71	0.69	1.07	0.73	0.63	0.9	0.71
Dy	4.67	3.23	4.53	3.63	6.19	3.88	4.09	4.27	4.47	3.97	3.94	6.49	4.24	3.61	6.18	4.47
Ho	0.84	0.61	0.86	0.73	1.19	0.76	0.8	0.8	0.82	0.77	0.76	1.27	0.91	0.78	1.28	0.92
Er	2.33	1.79	2.27	2.04	3.11	2.24	2.05	2.26	2.34	1.96	2.09	3.6	2.39	2.13	3.69	2.47

Tm	0.33	0.26	0.34	0.31	0.43	0.32	0.32	0.34	0.32	0.3	0.29	0.56	0.29	0.28	0.47	0.31
Yb	2.01	1.69	2.04	2.01	2.54	2.04	1.96	1.93	1.94	1.8	1.71	3.51	1.85	1.94	3.55	2.18
Lu	0.33	0.25	0.3	0.32	0.39	0.32	0.31	0.3	0.29	0.28	0.28	0.55	0.3	0.29	0.5	0.34
Hf	3.6	2.6	3.4	2.5	4.1	2.5	3.5	3.7	3.3	3.2	3.1	5.1	3.2	2.9	3.9	3.2
Ta	1.4	1	1.1	0.6	1.3	0.6	0.9	0.9	1.1	0.9	0.9	0.7	1.2	1.2	0.4	0.9
Pb	3	2.3	1.4	1.5	3.3	2.2	2.1	2.5	1.1	1.1	0.9	3.5				
Th	4.9	3.3	3.6	2.7	4.6	2.8	3.1	3.6	3.5	3.6	3.4	2.7	4.3	3.9	0.6	3.4
U	1.2	0.8	1	0.6	1	0.7	0.7	0.8	0.9	0.9	0.9	0.6	1.4	1.3	0.3	1
W	5.1	75.4	1.4	8.7	2	80.5	0.9	1.9	2.2	6.9	11	17.7				
Mg#	49.42	67.41	50.95	53.75	51.0	55.72	54.0	53.22	59.11	57.51	57.69	55.43	63.69	67.37	55.51	60.74

					Intermediate dykes cutting the Istanbul Palaeozoic unit									Tavşantepe diorite	
	YMK-1	GD-1	KN-1	KN-1B	CB-1	MT64	MT61	MT63	MT41-b	GD-2	ATA-2	SU-4	CS-6	P-1	P-2
SiO$_2$	65.43	66.40	61.12	60.97	55.66	57.70	58.49	56.81	53.85	67.63	68.78	70.23	70.74	56.24	60.87
TiO$_2$	0.58	0.56	0.63	0.63	1.10	1.65	0.78	3.10	1.30	0.56	0.39	0.91	0.85	0.93	0.90
Al$_2$O$_3$	19.00	17.22	18.57	18.61	15.16	16.79	19.32	15.85	15.10	16.65	17.18	13.73	13.50	18.26	18.22
Fe$_2$O$_3^T$	3.98	3.65	5.71	5.69	8.71	8.84	6.74	12.87	7.07	3.68	2.93	5.88	5.69	6.70	4.12
MnO	0.05	0.06	0.10	0.10	0.15	0.16	0.11	0.13	0.14	0.06	0.05	0.14	0.13	0.15	0.08
MgO	1.20	2.30	2.76	2.79	6.37	3.60	2.97	3.58	7.16	2.15	1.40	2.98	3.03	3.63	2.31
CaO	1.64	4.02	5.40	5.48	8.03	6.17	5.86	2.31	9.51	3.61	3.52	1.81	1.76	9.96	7.52
Na$_2$O	6.12	4.75	4.77	4.77	3.95	3.60	4.34	4.01	3.99	4.43	5.88	3.56	3.54	4.19	4.53
K$_2$O	2.66	1.70	0.93	0.93	0.74	1.16	1.14	0.90	1.50	1.85	0.69	1.03	1.00	0.81	2.35
P$_2$O$_5$	0.17	0.17	0.20	0.21	0.25	0.33	0.25	0.44	0.38	0.19	0.12	0.16	0.16	0.30	0.27
LOI	4.72	4.81	2.70	2.70	2.80	3.30	7.70	3.70	5.30	4.90	4.99	3.60	3.80	6.16	6.25
Sum	100.82	100.83	100.18	100.17	100.10	100.00	100.00	100.00	100.00	100.81	100.95	100.43	100.39	101.17	101.18
Sc	3	8	10	10	30	22	9	38	18	7	5	15	14	20	9
V	68	75	109	108	290	173	93	322	143	84	46	110	107	155	136
Cr	13.68	27.37	13.68	13.68	171.05	20.53		13.68	218.94	20.53	13.68	95.79	88.95	13.68	13.68
Co	5.6	9.5	13.9	13.8	41.6	23.8	15.7	31.8	31.9	9	5.3	15.3	13.5	16.6	6.7
Ni	23.8	13.9	9.9	10.1	90.3	12			123	14.9	3.6	48.1	104.3	4.9	3.9
Cu	13.2	3	45.6	45.5	49.4	16	10	11	36	39.6	0.4	22.6	68.3	66.8	3.7
Zn	75	31	48	51	69	61	62	91	31	31	25	64	65	26	23
Ga	17.6	14.6	15.1	15.1	16.7	19.8	19.8	22.7	16.1	14.3	14.4	13.3	13.1	20.5	18.5
Rb	67.8	32.6	18.6	18.9	10.4	32.1	25.6	16	31.4	35.3	14.3	36	32	28.3	60.9
Sr	685.8	631.1	636.1	633.7	423.1	391	502.4	133.6	710.3	534.9	536.2	132.1	127.2	610.2	687
Y	16.9	11.4	13.3	12.8	19.9	34.4	15.7	45.9	19.9	9.6	8	22.8	18.4	23.8	22.9
Zr	162.6	125.5	121.7	122	144.2	229.6	128.1	278	155.3	101.8	92.9	170.7	154.7	124.5	137.5
Nb	25.3	11.2	10.3	9.9	28.4	13.5	9.7	13.6	31.3	11.3	3.8	8.5	7.5	12.5	11.6
Ba	516	396	472	457	211	334	248	328	498	402	95	203	230	275	427
Cs	1.1	0.2	0.6	0.5	0.1	0.9	1.3	0.5	1	0.3	1.4	1.2	1.3	3	1

(Continued)

Table 6. *The results of a whole-rock major (wt%), trace (ppm) and rare earth elements (REE) (ppm) geochemical analysis of mafic–intermediate dykes and small stocks (Continued)*

	Intermediate dykes cutting the Istanbul Palaeozoic unit													Tavşantepe diorite	
	YMK-1	GD-1	KN-1	KN-1B	CB-1	MT64	MT61	MT63	MT41-b	GD-2	ATA-2	SU-4	CS-6	P-1	P-2
La	29.2	22.9	19.6	18.2	21.1	33.7	19.3	27.1	32.1	21.1	11.8	24.8	20.8	22.6	17.2
Ce	45.9	37.7	35.7	33.7	43	69.3	39.1	61.2	58.9	36.5	22.4	51.7	44.6	44.9	34.4
Pr	4.95	4.69	4.01	3.87	5.44	8.3	4.79	8.11	6.84	4.21	2.49	6.1	5.39	6.84	4.57
Nd	17.8	16.7	15.2	15.5	22.8	35.2	18.3	36.1	27.3	14.8	9.6	23.2	21.3	25.8	20.1
Sm	3.18	3	2.88	2.88	5.1	7.9	3.8	8.3	5.2	2.5	1.89	4.73	4.38	5.38	3.99
Eu	0.98	0.88	0.9	0.91	1.76	1.84	1.23	2.41	1.66	0.8	0.62	1.12	1.11	1.58	1.22
Gd	3.11	2.71	2.65	2.72	5.17	6.8	3.31	8.21	4.55	2.2	1.64	4.39	3.96	5.26	4.2
Tb	0.49	0.39	0.41	0.42	0.78	0.91	0.48	1.26	0.61	0.32	0.26	0.68	0.61	0.88	0.68
Dy	2.97	2.17	2.26	2.38	4.22	6.21	3.29	8.67	4.04	1.71	1.42	4	3.59	4.81	3.75
Ho	0.57	0.38	0.46	0.46	0.78	1.23	0.67	1.79	0.71	0.35	0.26	0.81	0.71	0.92	0.79
Er	1.63	1.08	1.35	1.29	1.92	3.85	1.71	5.13	2.05	0.91	0.7	2.27	1.89	2.59	2.32
Tm	0.26	0.16	0.19	0.2	0.25	0.45	0.21	0.68	0.29	0.14	0.11	0.32	0.28	0.37	0.37
Yb	1.76	1.05	1.26	1.24	1.54	3.48	1.69	4.71	1.85	0.91	0.73	2.16	1.85	2.35	2.24
Lu	0.29	0.17	0.22	0.2	0.22	0.54	0.25	0.73	0.29	0.14	0.12	0.33	0.29	0.33	0.35
Hf	3.8	3.3	3	2.9	4	5.4	3	6.4	3.5	2.7	2.3	4.6	4.2	4.3	3.8
Ta	1.6	0.8	0.7	0.6	1.8	0.9	0.6	1.1	2.4	0.9	0.2	0.5	0.4	0.8	0.6
Pb	4.3	1	2.1	2.2	2.1					0.9	1.2	3.6	6.4	5.8	8.6
Th	7.6	5.4	3.2	3.1	2.8	6.2	3.1	4.9	7.7	6.1	2.1	5.1	4.3	3	2.4
U	2	1.2	0.9	1	0.6	1.8	1.3	1.9	2.5	1.3	0.5	1.2	1.4	1.1	0.5
W	26.5	17.8	1.5	1.4	3.4					3.3	1.3	3.3	205.1	1.8	1.3
Mg#	37.38	55.52	58.89	49.26	59.16	44.64	46.63	35.56	66.75	53.60	48.67	50.11	51.36	51.81	52.64

	Mafic-intermediate dykes and small stocks cutting Upper Cretaceous volcanics														
	SL-1	SL-2	MT25	MT09	MT7	MT8	MT23	MT27	MT53	MT46	MT24*	MT50	MT3-b	MT1	MT5
SiO_2	56.65	64.62	58.64	57.86	58.61	58.51	58.69	58.42	58.43	58.95	60.06	61.65	65.03	47.32	47.04
TiO_2	0.71	0.52	0.61	0.66	0.60	0.65	0.62	0.70	0.74	0.70	0.66	0.64	0.49	1.31	1.31
Al_2O_3	16.90	15.56	17.78	16.84	16.70	17.04	18.06	16.98	17.02	16.49	18.34	15.61	16.00	18.96	14.89
$Fe_2O_3^T$	9.13	6.03	8.01	7.46	7.99	7.76	7.96	6.63	7.88	7.26	6.08	6.46	5.67	12.27	9.91
MnO	0.13	0.08	0.11	0.23	0.15	0.16	0.11	0.21	0.15	0.15	0.10	0.17	0.11	0.25	0.23
MgO	4.32	2.77	3.07	2.96	3.98	2.87	3.52	2.47	4.18	5.35	3.06	4.49	2.32	5.77	12.67
CaO	7.33	4.95	7.70	9.56	8.18	8.55	5.77	10.66	6.71	7.73	5.67	6.27	4.74	7.09	10.40
Na_2O	3.45	3.29	2.74	2.68	2.92	2.97	3.25	2.39	3.45	2.16	4.81	2.10	3.29	3.73	2.13

K$_2$O	1.01	2.55	2.24	2.50	1.00	1.05	1.31	1.38	1.92	1.36	0.77	1.63	1.23	2.91	2.07
P$_2$O$_5$	0.41	0.75	0.11	0.12	0.20	0.15	0.11	0.17	0.12	0.12	0.10	0.13	0.11	0.14	0.19
LOI	4.39	5.64	1.70	2.50	1.87	6.30	2.70	3.04	1.28	4.04	1.79	6.99	1.53	5.17	5.08
Sum	100.00	100.00	100.00	100.00	100.00	100.00	100.00	100.00	100.00	100.00	100.00	100.00	100.00	100.87	100.86
Sc	23.7	18.9	14	26	13	30	26	32.7	36.4	25.9	29.9	31.3	39.2	16	22
V	210.4	185	121	176	116.5	214	199	279	183.4	210.2	213.7	224.9	203.1	141	229
Cr	528.70	82.20	20.53	88.95	22.00	61.58	34.21	25.20	15.00	27.70	30.30		16.10	27.37	20.53
Co	48.9	33	14.6	19.8		22.6	23.4							13.1	20.4
Ni	320.1	38.3	4	18	14.1	14	11	10.6	7.1	10.9	12.1		7.7	144.3	184.6
Cu	58.6	49.6	56	71	32.3	23	80	36.9	75.7	19.9	62	24.8	54.4	201.7	727.3
Zn	68.6	89.5	45	52	54.4	59	45	63.2	71.3	66	66.4	62.8	71.5	37	150
Ga	16.9	19.7	15.2	14.9		15.3	16							11.3	13.4
Rb	16.7	35.7	48.4	44.6	19.8	18.2	31	22.1	26.3	23.1	23.2	31.4	31	42.7	33.6
Sr	471.6	576.5	324.6	316.5	667.4	403.6	272.6	378.1	261.5	293.4	294.5	317	236.3	273.8	377.1
Y	19	25.9	16.4	18.6	14.5	19.9	20.7	23.4	18.7	20.8	20.6	18.3	17.3	18	18.6
Zr	108.7	196.8	94.7	66.8	127.2	76	87.4	75.1	79.8	80.1	87.2	79.8	81.6	115.7	81.3
Nb	14.9	35.7	4.1	3.1	12.1	3.3	4.1	3.1	3.7	3.2	3	3.4	4	4.1	3.3
Ba	294	337.7	342	417	304.4	274	265	215.3	206.4	239.1	242.9	270.5	211.4	618	413
Cs	1.3	1.8	1.3	0.6		2	1.6							0.2	0.3
La	22.8	36	14.8	10.4	12.2	12.8	12.8	9.7	10.2	9.5	9.3	12.1	6	15	14
Ce	46.5	70.5	28.7	22.4	34.1	26.5	26.5	25.9	24.1	26.3	24.7	28.8	23.8	29.2	29.1
Pr	6.1	8.44	3.49	2.93		3.37	3.46							3.44	3.78
Nd	24.3	31.5	13	11.7	14.6	14.4	14	15.2	13.1	14.4	11.7	11.7	12.3	13.4	15.7
Sm	4.7	6.6	2.7	3.4		3.8	3.4							2.88	3.57
Eu	1.63	2.31	0.9	1		1.04	1.01							0.86	1.01
Gd	4.61	6	3.06	3.28		3.71	3.58							2.91	3.48
Tb	0.62	0.82	0.42	0.45		0.52	0.53							0.48	0.55
Dy	3.57	5.22	2.82	3.19		3.67	3.86							2.98	3.17
Ho	0.8	0.97	0.63	0.59		0.76	0.76							0.56	0.64
Er	2.14	2.68	1.94	2.08		2.4	2.47							1.73	1.8
Tm	0.24	0.36	0.26	0.27		0.32	0.34							0.27	0.29
Yb	1.6	2.32	1.91	2.09		2.15	2.11							1.79	1.88
Lu	0.24	0.38	0.32	0.33		0.38	0.32							0.3	0.31
Hf	2.5	3.6	2.6	1.9		2.2	2.2							3.1	2.1
Ta	1	3.1	0.3	0.3		0.2	0.3							0.4	0.3
Pb		3		3										2.7	2.1
Th	3.6	5.3	4.6	2.3		2.9	3.3					5.8		6.7	3.8
U	1.2	1.8	1.3	0.7		1.2	1.2					1.7		1.6	1.2
W														294.6	207.9
Mg#	71.70	48.22	44.74	57.92	49.93	59.36	51.25	42.46	46.67	42.30	49.70	43.99	43.17	47.64	48.38

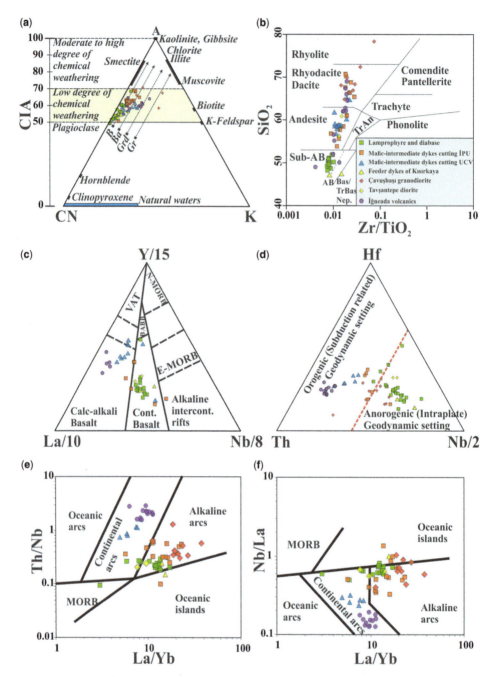

Fig. 6. (a) A–CN–K ternary diagram (Nesbitt & Young 1984) reflecting the alteration degree of dykes. Classification and tectonomagmatic discrimination diagrams of the mafic–intermediate dykes (CIA, chemical index of alteration (Yan *et al.* 2010); A, Al_2O_3; CN, $CaO + Na_2O$; K, K_2O). **(b)** SiO_2 v. Zr/TiO_2 diagram (Winchester & Floyd 1977) for the mafic–intermediate dykes. **(c)** Cabanis & Lecolle (1989). **(d)** Krmíček *et al.* (2011). **(e)** & **(f)** Hollocher *et al.* (2012).

represented by the dykes with alkali basaltic and basanite–trachybasanite composition (i.e. feeder dykes of the Kısırkaya lavas) (Fig. 6b); and (2) the second group is represented by the dykes and infrequent sills with andesitic and dacitic compositions (Fig. 6b), and are calc-alkaline in character. The second group corresponds to the feeder dykes of the Garipçe and Bozhane formations (Keskin *et al.* 2010). The mafic–intermediate dykes have SiO_2 contents between 47.04 and 65.03 wt%, and Mg# values ranging from 42.30 to 71.70.

On the La/10–Y/15–Nb/8 tectonomagmatic discrimination diagram of Cabanis & Lecolle (1989), mafic dykes cutting the Upper Cretaceous volcanics and the İğneada volcanic rocks predominantly fall into 'calc-alkaline basalt' field, in contrast to the lamprophyre and alkali olivine dykes plotting mainly into the 'continental basalt' field (Fig. 6c). One sample plots in the alkaline intercontinental rift field.

Th–Hf–Nb/2 diagram of Krmíček *et al.* (2011) discriminates alkaline and calc-alkaline dyke series more clearly (Fig. 6d). Almost all lamprophyre dykes and alkali olivine basalt dykes fall into the anorogenic (intraplate) geodynamic setting. In contrast, the mafic–intermediate dykes cutting the Upper Cretaceous volcanics and lavas and the İğneada volcanics plot into the orogenic (i.e. subduction-related) setting.

La/Yb v. Th/Nb (Fig. 6e) and La/Yb v. Nb/La (Fig. 6f) diagrams (Hollocher *et al.* 2012) clearly discriminate two series: (1) mafic–intermediate dykes cutting the Upper Cretaceous volcanics and İğneada volcanics falling into the 'continental arc' field; and (2) lamprophyres, mafic–intermediate dykes, Çavuşbaşı granodiorite and the alkali olivine basalts, falling into the 'alkaline arc' field.

To sum up, on the basis of the discrimination diagrams in Figure 6, dykes around İstanbul can be divided into two geochemically distinct groups: (1) the first group includes the İğneada volcanic rocks and the mafic–intermediate dykes cutting the Upper Cretaceous volcanics; these are calc-alkaline in character and possibly related to a subduction-related orogenic setting; and (2) the second group includes the olivine basaltic dykes (i.e. the feeder dykes of the Kısırkaya Formation), most of the lamprophyre dykes and a few mafic dykes cutting the İstanbul Palaeozoic unit. These are alkaline in character and possibly genetically related to an anorogenic or intraplate setting.

On the Harker diagrams (Fig. 7), the mafic–intermediate dyke samples show distinct magma evolution trends. Among them, lamprophyres and the feeder dykes of the Kısırkaya Formation form discrete trends that are clearly diverged from other dyke series on TiO_2, AL_2O_3, CaO, Na_2O and Nb v. SiO_2 diagrams. In lamprophyre rocks, TiO_2, Al_2O_3, F_2O_3, Na_2O and K_2O are positively correlated with SiO_2: however, CaO and MgO are negatively correlated with SiO_2. In these rocks, K_2O concentrations span a much narrower range relative to those of other dyke series and intrusions. The incompatible trace elements (i.e. Rb) and HFS

elements, such as Zr and Nb, slightly increase with increasing SiO_2, in contrast to Sr which slightly decreases with increasing SiO_2. TiO_2, Al_2O_3, F_2O_3, MgO, CaO and P_2O_5 concentrations of the mafic–intermediate dyke samples cutting the İstanbul Palaeozoic unit are positively correlated with SiO_2. In contrast, Na_2O and K_2O are scattered against silica.

Normal-type mid-ocean ridge basalt (N-MORB)-normalized multi-element spidergrams (Fig. 8) indicate that the majority of the mafic–intermediate dyke samples and small intrusions display selective enrichment in large ion lithophile elements (LILEs) (Sr, K, Rb, Ba, Th) and, to a lesser extent, in light REEs (LREEs) (La, Ce and Nd) relative to the HFS elements (Tb, Ti, Y and Yb). They contain notable depletions in Ta and Nb relative to adjacent LILEs and LREEs. The calc-alkaline dykes cutting the Upper Cretaceous volcanics and the İğneada volcanics show a clear subduction signature reflected by a profound depletion in Nb and Ta relative to the adjacent LILEs and LREEs. Note that compared to the above-mentioned dykes, Ta–Nb depletions are less profound in lamprophyre dykes and two plutonic intrusions (i.e. the Tavşantepe microdiorite and the Çavuşbaşı granodiorite). In contrast, alkaline olivine basalt dykes display ocean island basalt (OIB)-like patterns with very little or no Ta and Nb depletion. The findings presented above imply that two compositionally contrasting magma series coexisted during the Late Cretaceous–Paleocene.

On chondrite-normalized REE spidergrams (Fig. 9), the majority of dyke samples show an average OIB composition with REE concentrations, but are consistently lower than that of the OIB. A small subset of alkaline basaltic and lamprophyric dykes display broadly parallel and similar patterns to that of the average OIB. Note that that particular subset of samples fall into the anorogenic (intraplate) field in Figure 6d. Among the dykes, REE patterns of the mafic–intermediate calc-alkaline dykes cutting the Upper Cretaceous volcanics display the flattest patterns, with a notable depletion in contrast to the patterns of the Çavuşbaşı granodiorite which show a greater depletion in heavy REEs (HREEs) relative to that of the OIB.

Sr–Nd isotope geochemistry

The whole-rock Sr–Nd isotopic compositions of the representative samples from the dykes and small stocks are listed in Table 7. The initial Sr and Nd isotopic data were calculated by taking new geochronological data into consideration. The results of the calculations indicate that the initial $^{87}Sr/^{86}Sr_{(i)}$ values of the lamprophyres span a range from 0.704809 to 0.705451, while their initial

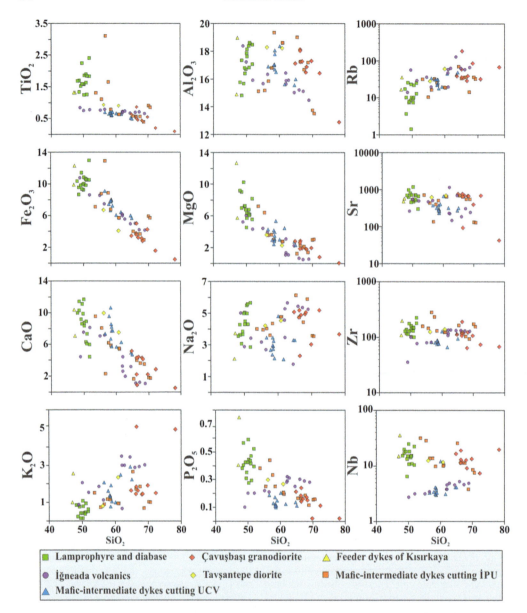

Fig. 7. Variations of selected major and trace elements against SiO_2 for the studied dykes and small stocks cutting the İstanbul Palaeozoic units and the Upper Cretaceous volcanics (the Çavuşbaşı granodiorite data are from Yılmaz Şahin *et al.* 2012; data from the İğneada volcanics are unpublished).

$^{143}Nd/^{144}Nd_{(i)}$ values vary between 0.512752 and 0.512812 (Fig. 10a), and their $\varepsilon Nd_{(i)}$ values fall between 3.9 and 5.1 (Fig. 10b). The depleted mantle model ages (T_{DM}) vary between 0.46 and 0.55 Ga (after Liew & Hofmann 1988). The $^{87}Sr/^{86}Sr_{(i)}$ values of the mafic–intermediate dykes cutting the İstanbul Palaeozoic unit span a range from 0.703633 to 0.704673, while their $^{143}Nd/^{144}Nd_{(i)}$

values vary between 0.512614 and 0.512699 (Fig. 10a), and $\varepsilon Nd_{(i)}$ values between 1.2 and 2.9 (Fig. 10b). The T_{DM} model ages vary between 0.64 and 0.77 Ga. An intermediate dyke cutting the Upper Cretaceous volcanics has an $^{87}Sr/^{86}Sr_{(i)}$ value of 0.704551, $^{143}Nd/^{144}Nd_{(i)}$ value of 0.51252 and $\varepsilon Nd_{(i)}$ value of 2.0. The T_{DM} model age of that sample is 0.71 Ga. The $^{87}Sr/^{86}Sr_{(i)}$ values of the

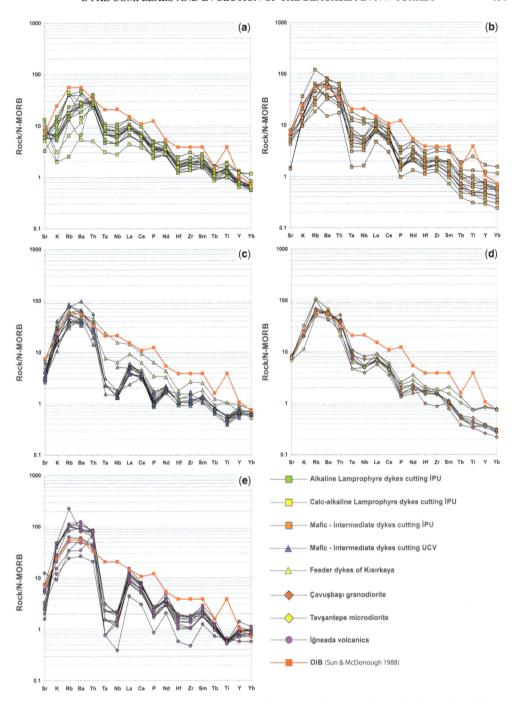

Fig. 8. N-MORB-normalized (Sun & McDonough 1989) multi-element spidergrams of the mafic–intermediate dykes and small stocks cutting the Palaeozoic units and the dykes cutting the Upper Cretaceous volcanics around İstanbul.

Çavuşbaşı granodiorite and the Tavşantepe diorite span the range from 0.703508 to 0.706344, while their $^{143}Nd/^{144}Nd_{(i)}$ values vary between 0.512713 and 0.512740 (Fig. 10a), and their $\varepsilon Nd_{(i)}$ values lie between 3.0 and 3.7 (Fig. 10b). The T_{DM} model ages vary between 0.57 and 0.62 Ga.

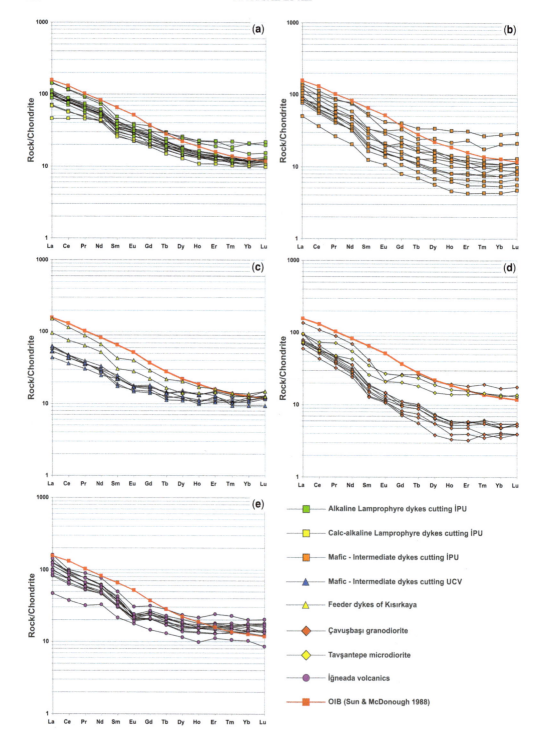

Fig. 9. Chondrite-normalized (Sun & McDonough 1989) spidergrams of the mafic–intermediate dykes and small stocks.

Table 7. *Sr and Nd isotopic compositions of the mafic–intermediate dykes and small stocks cutting the İstanbul Palaeozoic unit (İPU) and Upper Cretaceous volcanics (UCV; analyses of the Çavuşbaşı granodiorite are taken from Yılmaz Şahin et al. 2012)*

Sample	Age	SiO_2	Sr	Rb	Sm	Nd	$^{87}Rb/^{86}Sr$	$^{87}Sr/^{86}Sr$	$^{87}Sr/^{86}Sr$ (i)	$^{147}Sm/^{144}Nd$	$^{143}/^{144}Nd$	$^{143}/^{144}Nd$ (i)	εNd (0)	εNd (i)	T_{DM}	Series
KB-2	68	51.08	697	10.1	5.25	26.6	0.0419	0.705491	0.705451	0.1199	0.512805	0.512752	3.3	3.9	0.55	Lamprophyre
KB-5	68	49.97	531	8.3	4.60	23.9	0.0452	0.704929	0.704885	0.1169	0.512809	0.512757	3.3	4.0	0.54	Lamprophyre
SU-3	68	48.74	959	3.6	4.08	21.0	0.0109	0.705163	0.705153	0.1180	0.512864	0.512812	4.4	5.1	0.46	Lamprophyre
HT-6	68	50.37	466	10.1	5.88	26.9	0.0627	0.704870	0.704809	0.1327	0.512816	0.512757	3.5	4.0	0.54	Lamprophyre
ATA-2	68	68.78	536	14.3	1.89	9.6	0.0771	0.704348	0.704273	0.1195	0.512667	0.512614	0.6	1.2	0.77	Intermediate (IPU)
KN-1	68	61.12	636	18.6	2.88	15.2	0.0846	0.704755	0.704673	0.1151	0.512680	0.512629	0.8	1.5	0.75	Intermediate (IPU)
GD-1	68	66.40	631	32.6	3.00	16.7	0.1494	0.703777	0.703633	0.1091	0.512733	0.512684	1.9	2.6	0.66	Intermediate (IPU)
GD-2	68	67.63	535	35.3	2.50	14.8	0.1909	0.703891	0.703707	0.1026	0.512745	0.512699	2.1	2.9	0.64	Intermediate (IPU)
SL-1	68	56.65	377	33.6	3.57	15.7	0.2577	0.704800	0.704551	0.1381	0.512713	0.512652	1.5	2.0	0.71	Intermediate (UCV)
P-2	59	60.87	687	61	3.99	20.1	0.2569	0.706559	0.706344	0.1205	0.512760	0.512713	2.4	3.0	0.62	Tavşantepe diorite
ÇG-4	68	66.92	692	34.5	2.38	13.2	0.1442	0.703694	0.703555	0.1095	0.512765	0.512716	2.5	3.2	0.61	Çavuşbaşı granodiorite
ÇG-16	68	68.35	675	39	2.01	11.4	0.1671	0.703669	0.703508	0.1071	0.512788	0.512740	2.9	3.7	0.57	Çavuşbaşı granodiorite
UT-1	320	61.82	66.3	117	8.84	46.2	5.1037	0.729957	0.706713	0.1162	0.512520	0.512277	-2.3	1.0	1.00	Trakya Formation

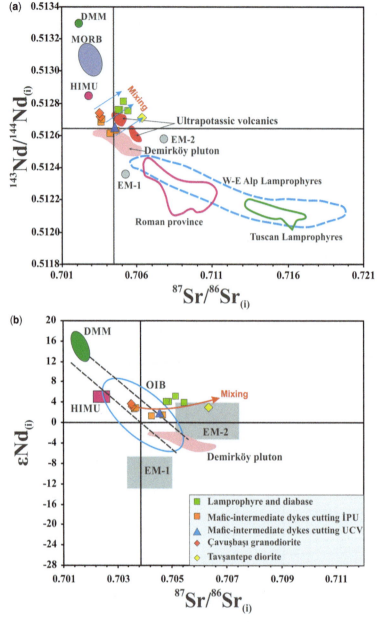

Fig. 10. (a) ^{143}Nd/^{144}Nd v. ^{87}Sr/^{86}Sr ratios for the studied dykes compared with the mantle components HIMU, EM-1, EM-2, DMM (after Zindler & Hart 1986). The Eastern and Western Alps, Tuscan lamprophyres, and Roman Province fields are from Bell *et al.* (2013) and Moghadam *et al.* (2014). Demirköy pluton data (71–84 Ma) are from Karacık & Tüysüz (2010); ultrapotassic volcanics (74–78 Ma) are from Gülmez *et al.* (2016). (b) Sr–Nd isotope plot demonstrating the composition of the mafic–intermediate dykes from İstanbul Palaeozoic unit (İPU) and Upper Cretaceous volcanic (UCV) rocks. DMM, depleted mantle; EM-I, enriched mantle I; EM-2, enriched mantle 2; HIMU-high μ (Zindler & Hart 1986; Shellnutt *et al.* 2014 and references therein).

Mineral chemistry

Representative and selected analyses are given in the Supplementary material.

Plagioclases. Forty-five point analyses on feldspar minerals from six samples were obtained from the mafic–intermediate dykes. Plagioclase is a common phenocryst in both mafic and intermediate dykes, and occurs as euhedral or subhedral phenocrysts (0.1–7 mm), showing polysynthetic twinning and compositional zoning in intermediate dykes and stocks. The composition of the plagioclases ranges from An_0 to An_{71} (albite–bytownite: Fig. 11a). Note that plagioclase crystals in samples VY-5, VY-9, VY-21, MT-3b and MT-21 display albitic composition (Ab_{96-99}). Therefore, albite crystals in that sample should have formed by the alteration of Ca-rich plagioclases. Albitization is not important for the rest of the samples.

Pyroxenes. Based on the classification of Morimoto *et al.* (1988), all clinopyroxenes from the mafic–intermediate dykes are augite. In the Di–Hd–En–Fs classification of Morimoto *et al.* (1988), all of the studied clinopyroxenes ($Wo_{42-44}En_{39-43}Fs_{13-18}$) fall into the augite field (Fig. 11b). Orthopyroxene phenocrysts of the mafic–intermediate dykes display a

corona texture and are mantled by clinopyroxenes. All of the analysed orthopyroxenes ($Wo_{3-4}En_{65-69}$ Fs_{28-32}) fall into the clinoenstatite field (Fig. 11b).

Amphiboles. Amphiboles from mafic–intermediate dykes have magnesiohornblende, magnesiohastingsite and pargasite compositions. According to the IMA classification and Fe^{3+} calculation proposed by Leake *et al.* (1997), all the amphiboles from the mafic–intermediate dykes are in the calcic (BCa >1 apfu (atom per formula unit) group, and mostly contain a high Si content (TSi = 5.5–6.5 apfu: Fig. 11c, d).

Crystallization conditions

Mineral chemistry data are used to constrain the crystallization conditions (i.e. geothermometry (*T*), geobarometry (*P*), oxygen fugacity ($\log fO_2$) and

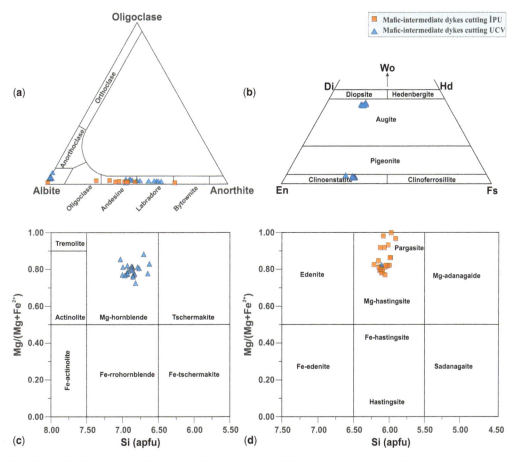

Fig. 11. (**a**) Or–Ab–An ternary diagram showing compositions of the plagioclase phenocrysts from the studied dykes (after Deer *et al.* 1992). (**b**) Pyroxenes from mafic–intermediate dykes from the Upper Cretaceous volcanic units plotted in a conventional Di–Hd–En–Fs diagram (Morimoto *et al.* 1988). (**c**) & (**d**) The amphibole compositions on $Mg/(Mg + Fe^{2+})$ v. Si classification diagrams (after Leake *et al.* 1997). For an explanation, see the text.

water content wt% H_2O_{melt}) of the mafic–intermediate dyke complex cutting both the İstanbul Palaeozoic unit and the Upper Cretaceous volcanics rocks in the study area (please refer to the Supplementary material).

Pyroxene geothermobarometry. Crystallization temperatures of pyroxenes were calculated using the pyroxene geothermometer of Lindsley (1983). The maximum temperature estimated for the clinopyroxenes of the mafic–intermediate dykes using the Lindsley (1983) graphical geothermometer is 900°C at 5 kbar. Single clinopyroxene temperature calculations were performed by using the formulae of Dal Negro *et al.* (1982) and Molin & Zanazzi (1991). According to the formulations of Dal Negro *et al.* (1982), single clinopyroxene geothermometers give similar cooling temperatures ranging from 854 to 882°C (mean 866 ± 18.5°C). Based on Molin & Zanazzi's (1991) formulae, cooling temperatures vary between 895 and 901°C (mean 898 ± 2°C) for diopside crystals from the studied dykes.

Compositions of clinopyroxenes and orthopyroxenes are used for the estimation of temperature (T) and pressure (P) calculations in magmatic rocks (Putirka 2008). Calculated temperatures for studied rocks vary between 1135 and 1187°C (mean = 1155 ± 19°C). The pressure of crystallization for the studied dykes range from 5.28 to 10.37 kbar, corresponding to a depth interval of 14–28 km from the palaeosurface. Calculated pressures and temperatures reflect the conditions of pyroxene crystallization, which possibly occurred at two different levels: initial crystallization in a mid- to upper-crustal magma chamber, prior to emplacement; and crystallization at shallower crustal levels.

Amphibole geothermobarometry. The calculated temperatures for the mafic–intermediate dyke complex (using the formulations of Ridolfi *et al.* 2010) cutting the Upper Cretaceous volcanics vary between 769 and 953°C, whereas those for the dykes cross-cutting the İstanbul Palaeozoic unit range from 938 to 994°C. The calculated barometry results for the mafic–intermediate dykes cross-cutting the Upper Cretaceous volcanics range from 0.8 to 3.83 kbar, while those for the dykes cross-cutting the İstanbul Palaeozoic unit varies between 3.30 and 5.35 kbar, respectively.

In the *P–T* diagram (Fig. 12a), the amphiboles from the dyke samples cutting the Upper Cretaceous volcanics mainly plot within domain 1, in contrast to one pargasite plot falling within domain 3. The dyke complex cutting the Upper Cretaceous volcanics include Mg-hornblende, plagioclase, clinopyroexene, ortopyroxene, olivine, magnetite and ilmenite. The amphiboles from the dyke samples cutting the İstanbul Palaeozoic unit mainly plot within domains 2 and 3. The dyke complex cutting the Upper Cretaceous volcanics include pargasite, plagioclase, clinopyroexene, magnetite and ilmenite (Fig. 12a).

Oxygen fugacity (log fO_2) and hygrometric estimations. The oxygen fugacity values of calc-alkaline magmas (relative NNO buffer) were calculated from the Mg content of amphiboles (using the formulations of Ridolfi *et al.* 2010; Ridolfi & Renzulli 2012). Relative oxygen fugacity values from the mafic–intermediate dykes vary between 0.81 and 1.82, and range from 0.49 to 1.61 in the intermediate dykes cross-cutting the İstanbul Palaeozoic unit. These values are consistent with the intervals of relative oxygen fugacity values of the calc-alkaline magmas (ΔNNO from −1 to +3: see references in Ridolfi *et al.* 2010). The oxygen fugacity (log fO_2) values vary between −10.10 and −13.07 bar in the dykes cutting the Upper Cretaceous volcanics, and from −8.71 to −10.33 bar in intermediate dykes cutting the İstanbul Palaeozoic unit, and all of the samples fell into the arc magma field (Fig. 12b) described by Harald & Galliard (2006).

The Al^{VI} content of amphibole is sensitive to the water content (H_2O_{melt}) in the melt and can be used to estimate the stability field of amphibole crystallization. The H_2O_{melt} concentrations of the studied dyke complex were calculated using Ridolfi *et al.*'s (2010) hygrometric formulation for Mg-hornblende and pargasite crystals. In the mafic–intermediate dykes cutting the Upper Cretaceous volcanics, the H_2O_{melt} concentrations ranged from 4.91 to 6.89%. The H_2O_{melt} concentrations of intermediate dykes cutting the İstanbul Palaeozoic unit varied between 4.82 and 7.51%. These values are within the typical limits for calc-alkaline magma crystallization in arc-related settings (Fig. 12c).

Discussion

Timing of emplacement

Upper Cretaceous volcanic rocks and plutonic equivalents are not only observed in and around İstanbul but also in the Pontides along the Black Sea coastal area. The aforementioned belt reaches the Georgian border in the east, and the town of İğneada and Bulgaria in the west.

For the first time, we dated the dykes and small intrusions cutting the İstanbul Palaeozoic unit. U–Pb LA–ICP-MS zircon dating of the intermediate dykes and small stocks yielded ages between 72.49 ± 0.79 (Upper Cretaceous–Campanian) and 65.44 ± 0.93 Ma (Lower Paleocene–Danian): however, U–Pb LA–ICP-MS zircon dating of the Tavşantepe diorite yielded an age of 58.9 ± 1.8 Ma (Middle Paleocene–Thanetian). The dating result from the

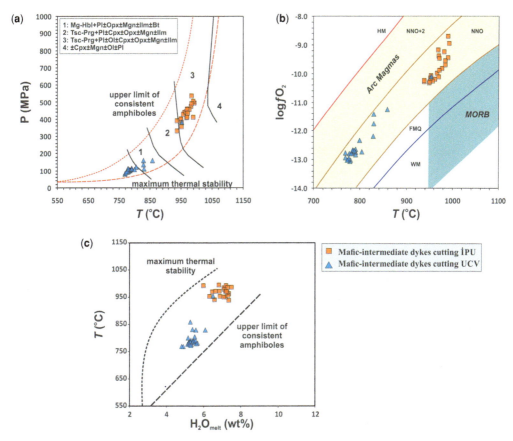

Fig. 12. The amphibole compositions from the studied dykes plotted on: (**a**) a *P–T* diagram (after Ridolfi *et al.* 2010); (**b**) a log *f*O$_2$–*T* diagram (after Ridolfi *et al.* 2010; 'MORB' and 'Arc magmas' areas from Harald & Galliard 2006); and (**c**) a *T*–H$_2$O melt diagram (after Ridolfi *et al.* 2010).

Tavşantepe diorite is the youngest crystallization age from the Upper Cretaceous–Paleocene magmatism in the region. All the age dating implies that these magmatic products were coeval and hence formed in the same geodynamic setting. Lamprophyre dykes cutting the Çavuşbaşı granodirite should be younger than 68 Ma.

Petrogenesis

Nature of the mantle source region. Most of the dykes, stocks and plutons emplaced into the İstanbul Palaeozoic unit and the Upper Cretaceous volcanic, as well as the lavas from İğneada, show multi-element patterns similar to that of a continental arc basalt with selective enrichment in LILEs and LREEs, and depletion in Ta and Nb. Although less profound, the Çavuşbaşı granodiorite and the Tavşantepe diorite also display subduction signature. Note that among the calc-alkaline magmatic series, even the most primitive basalts display the

aforementioned subduction signature. Therefore, we argue that this signature is genuine, unrelated to any other magmatic processes (e.g. assimilation).

The fact that all of the samples of dykes, stocks and plutons plot within the mantle array and enriched quadrants on the Sr–Nd isotope diagram, partly overlapping the OIB field (Fig. 10a), can be explained by their derivation from a subduction-modified lithospheric mantle source. Lamprophyres, Tavşantepe diorite and the alkali olivine basalt dykes have ^{143}Nd/^{144}Nd$_{(i)}$ values ranging between 0.512713 and 0512812 (i.e. initial εNd = +3.0 to +4.1), plotting within the OIB field (Fig. 10b). This can be explained by the derivation of melts from an enriched mantle source, possibly an asthenospheric one (Keskin *et al.* 2010; Keskin & Tüysüz 2017). The trend formed by the data points of the lamprophyre and diabase dykes, aligning towards the second quadrant (see the orange arrow in Fig. 10b), can be explained by the mixing of the primitive alkaline magmas with evolved calc-alkaline magmas.

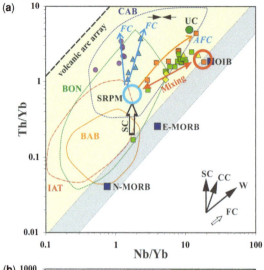

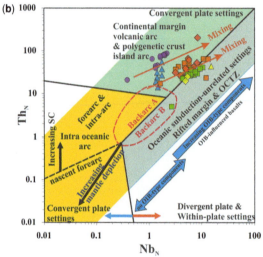

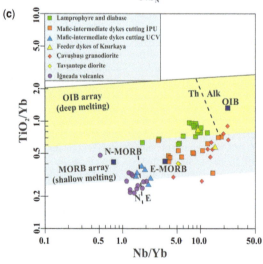

The depleted mantle model ages (T_{DM}) of the dykes, stocks and plutons emplaced into the İstanbul Paleozoic unit and the Upper Cretaceous volcanics range from 0.57 to 0.70 Ga, indicating that they were probably derived from an ancient lithospheric mantle source or an enriched arc mantle. The Nd-isotope model ages (T_{DM}) of the lamprophyres and Tavşantepe diorite (0.46–0.62 Ga) imply that the source was probably the OIB-type mantle.

On the Th/Yb v. Nb/Yb diagram (Pearce 2008), data points of the dykes follow three or four trends (Fig. 13a). A lamprophyre dyke plots close to E-MORB (enriched-MORB), while two alkaline olivine basic dyke samples fall very close to the enriched side of the mantle metasomatism trend close to OIB (Fig. 13a). Data points of the lamprophyric and mafic dykes cutting the İstanbul Palaeozoic unit and alkaline olivine basalt dykes cutting the Garipçe formation align along an almost linear trend between the inferred basalt composition derived from a subduction-related mantle source (i.e. calc-alkaline lavas from the Subduction-Related Primitive Magma composition: SRPM) and the alkaline OIB composition. We interpret this line as a magma-mixing trend between magmas derived from two contrasting sources.

Although the lamprophyre dykes plot close to the enriched portion of the mantle metasomatism trend, they still seem to display a weak subduction signature in multi-element spidergrams (Fig. 8a). This may be explained by mixing between arc-type calk-alkaline magmas and OIB-type within plate magmas (red arrow in Fig. 13a). The mafic dykes cutting the İstanbul Palaeozoic unit display a small shift towards the average upper-crust (i.e. UC) composition. This may be explained by the effect of AFC associated with magma mixing in these magmas (orange arrow in Fig. 13a).

The İğneada lavas and the mafic dykes cutting the Upper Cretaceous volcanics fall into the calc-alkaline basalt field (i.e. CAB) in Figure 13a. Data points of these two magmatic series align along two separate trends with higher gradients due to a consistent shift from the mantle metasomatism array towards higher Th/Yb values, starting from the inferred SRPM composition (the two blue arrows

in Fig. 13a). These trends may be explained by the evolution of a calc-alkaline magma derived from a subduction-modified mantle source (SRPM), followed possibly by combined effects of fractional crystallization (i.e. FC) with or without assimilation with no interactions with the OIB type magmas.

The Th_N v. Nb_N diagram (Saccani 2015) enables the researchers to compare the composition of the magmas derived from different source regions with their geodynamic settings. In this diagram, lamprophyre dykes and primitive alkaline olivine basalt dykes fall into the field of continental rift and/or ocean–continental transition zone. They display a moderate to high OIB-type within-plate component (Fig. 13b). In contrast, most of the mafic–intermediate dykes cutting the İstanbul Palaeozoic unit and the Upper Cretaceous volcanic, as well as the İğneada volcanic, fall in the Cordilleran type arc region and/or poligenetic crustal island arc field. These findings indicate that they may be linked to a convergent plate-margin setting. Only one sample plots on the border of the back-arc basin basalt field. On this diagram, the mixing of arc-type and OIB-type magmas and the transition from convergent plate to divergent plate settings are clearly observed (Fig. 13b).

The TiO_2/Yb v. Nb/Yb diagram of Pearce (2008) markedly separates the dykes and lavas derived from OIB-type deep magma sources from the ones derived from MORB-type shallow magma sources (Fig. 13c). Calc-alkaline dykes cutting the Upper Cretaceous volcanics in the north of İstanbul and İğneada fall into the MORB mantle array with a shallow melting zone. In contrast, lamprophyre dykes, mafic–intermediate dykes, small stock and plutonic rocks plot in a field that spans a range from a OIB-type deep melting array to a shallow melted mantle array. All of the above-mentioned findings that we obtained from dykes, stocks and volcanic rocks support a model involving the mixing of magmas derived from contrasting mantle sources (Fig. 13c).

In order to examine the source mineralogy and melting parameters of the dyke samples, we constructed a set of melting models presented in Figure 14, using the equations of Shaw (1970) for the

Fig. 13. (a) Th/Yb v. Nb/Yb diagram (Pearce & Peate 1995) of the studied dykes and small stocks displaying a mantle source enriched by a subduction component. (b) N-MORB-normalized (Sun & McDonough 1989) Nb v. Th diagram (after Saccani 2015) displaying possible tectonic settings for the mafic–intermediate dykes. (c) TiO_2/Yb v. Nb/Yb diagram (Pearce 2008) clearly separating the lavas derived from OIB-type deep sources from the ones derived from MORB-type shallow sources for the studied dykes. *Abbreviations*: BABB, back-arc basin basalt; CAB, continental arc basalt; BON, boninite; IAT, island arc tholeiite; N-MORB, N-type mid-ocean ridge basalt (high-Ti basalts); E-MORB, enriched mid-ocean ridge basalt; OIB, ocean island basalt; UC, average upper crust (Taylor & McLennan 1985); SC, subduction component; SRPM, Subduction-Related Primitive Magma composition; CC, crustal contamination; W, within-plate enrichment; FC, fractional crystallization; AFC, assimilation combined fractional crystallization; OCTZ, ocean–continent transition zone; Th, tholeiitic; Alk, alkaline.

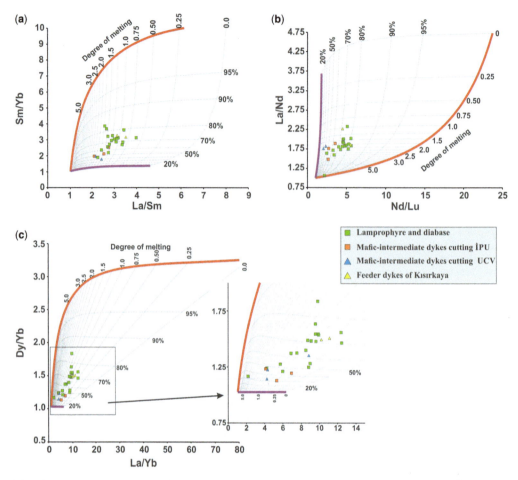

Fig. 14. (a)–(c) Non-modal batch melting models for the primitive mantle composition utilizing the LREE/HREE v. MREE/HREE plots. Red, melting curve of the garnet peridotite; purple, melting curve of the spinel peridotite. Numbers on the sides of the curves represent the percentage of the garnet and spinel peridotites in the source. Parameters used for the models are given in Table 8. For an explanation, see the text.

non-modal batch melting process, employing the REE ratios. Only the samples with $SiO_2 < 60\%$ and MgO > 3.5 have been plotted in Figure 14 to avoid the possible effects of fractionation processes. The parameters we used in the model are presented in Table 8. The choice of La/Sm v. Sm/Yb, Nd/Lu v. La/Nd, and La/Yb v. Dy/Yb plots was intentional because LREE/HREE ratios (i.e. La/Yb) along the horizontal axis are sensitive to the degree of melting in contrast to the MREE/HREE ratios (i.e. Sm/Yb, Dy/Yb), which are sensitive to the percentages of spinel and garnet. Diagrams imply that lamprophyre dykes cutting the İstanbul Palaeozoic unit and the alkaline olivine basalt dykes cutting the Upper Cretaceous volcanics possibly had garnet in their mantle source in contrast to the calc-alkaline dykes and microdioritic intrusions which contained

spinel in their source (Fig. 14). There also appears to be a marked difference in the melting degrees of these two group of rocks: lamprophyre dykes cutting the İstanbul Palaeozoic unit and the alkaline olivine basalt dykes cutting the Upper Cretaceous volcanics formation display lower degrees of melting (between *c.* 0.6 and 1.75%) in comparison with the calc-alkaline dykes and microdioritic intrusions which range from approximately 1.3 to 2.25%. What these findings may indicate is that magmas that fed the lamprophyre dykes and alkali olivine basalt dykes might have been generated by relatively lower-degree melting of a garnet-bearing deeper (possibly asthenospheric) mantle source relative to those of the calc-alkaline dykes and microdioritic intrusions which contained a spinel-bearing shallower source.

Table 8. *Non-modal batch melting parameters and partition coefficients used in the melting models. The source mode and melt modes for garnet peridotite and spinel peridotite mantle (i.e. primitive mantle) sources are from McKenzie & O'Nions (1991) and Thirlwall et al. (1994), respectively. Distribution coefficients used in this model are from McKenzie & O'Nions (1991)*

Source	Mode	Mineral percentage					
		Olivine	Orthopyroxene	Clinopyroxene	Spinel	Garnet	Total
Garnet peridotite	Source mode	0.598	0.211	0.076	–	0.115	1
	Melt mode	0.05	0.20	0.30	–	0.45	1
Spinel peridotite	Source mode	0.578	0.27	0.119	0.033	–	1
	Melt mode	0.10	0.27	0.50	0.13	–	1
	K_d values						
	La	0.0004	0.002	0.054	0.01	0.01	
	Nd	0.001	0.0068	0.21	0.01	0.087	
	Lu	0.0015	0.06	0.28	0.01	5.5	
	Yb	0.0015	0.049	0.28	0.01	4.03	
	Dy	0.0017	0.022	0.33	0.01	1.06	
	Sm	0.0013	0.01	0.26	0.01	0.217	

Magma chamber processes (FC, AFC, mixing). We have selected two highly compatible elements (i.e. Fe and V) and a highly incompatible element (i.e. Zr), and plotted them against each other to examine the possible effects of fractional crystallization (FC) and magma-mixing processes during the magma evolution (Fig. 15a, b). These two processes were modelled based on two different end-member compositions and then plotted on this diagram following the approach discussed in Keskin et al. (2010) and Keskin & Tüysüz (2017). Possible fractional crystallization curves (i.e. FC1 for the volcanics and dykes, and FC2 for the plutonic rocks; as the latter requires higher K_d values of the elements to generate a curve marked by FC2), linear magma mixing and replenishment trends are marked on these two diagrams. The most primitive basaltic end members of the subduction-related calc-alkaline magma series and the alkaline within-plate magma series are marked on the diagrams as BEM-Arc (i.e. sample İğneada - 7) and BEM-WP (i.e. alkaline sample CS-4: a lamprophyre dyke), respectively. Figure 15a, b indicates that a small subset of calc-alkaline dykes and plutonic rocks possibly follow the pure fractionation trend (i.e. mafic–intermediate dykes cutting the İstanbul Palaeozoic unit and the Upper Cretaceous volcanics, the İğneada volcanics and Çavuşbaşı granodiorite). The rest of the calc-alkaline samples plot along a set of hypothetical replenishment trends, aligning between the BEM-Arc composition and the evolved part of the FC curves. Distribution of the data points imply the importance of the replenishment of evolved magmas by primitive magmas during the magma-chamber evolution of the calc-alkaline series. These results are consistent with the textural properties of the samples (e.g. sieve texture, corona texture,

glass inclusions in plagioclase and pyroxene phenocrysts etc.).

In contrast to the calc-alkaline series, the alkaline magma series (i.e. feeder dykes of the Kısırkaya Formation and lamprophyres plus two mafic samples cutting the İstanbul Palaeozoic unit) follow a series of radiating linear trends between the BEM-WP composition (i.e. lamprophyre sample CS-4) and different parts of the FC1 curve. These may be related to the injection of OIB-type alkaline primitive magmas into the magma chambers of variously evolved calc-alkaline magmas.

In this study, we constructed an energy-constrained AFC model using our Sr–Nd isotopic data from the studied mafic–intermediate dykes and small stocks (Fig. 15c). We used the thermodynamic framework of the energy-constrained assimilation fractional crystallization (EC-AFC) equations of Bohrson & Spera (2001) and Spera & Bohrson (2001). Thermal and compositional parameters used for our modelling are given in Table 9. The geochemical dataset is used in combination with published partitioning data, and an energy-constrained modelling approach (EC-AFC: Spera & Bohrson 2001) is implied to predict the evolution of a variety of Sr–Nd isotopes during crustal assimilation and/or fractional crystallization, considering the lower crust (LC Model 1), upper crust (UC Model 2) and the Trakya Formation (Model 3) as assimilants. Note that the Trakya Formation belongs to the İstanbul Palaeozoic unit. It is Carboniferous in age and made up of greywacke beds. The reason why we selected the composition of the Trakya Formation is that most of the dykes are emplaced in this formation and they contain partly absorbed xenoliths of the lithologies of that formation.

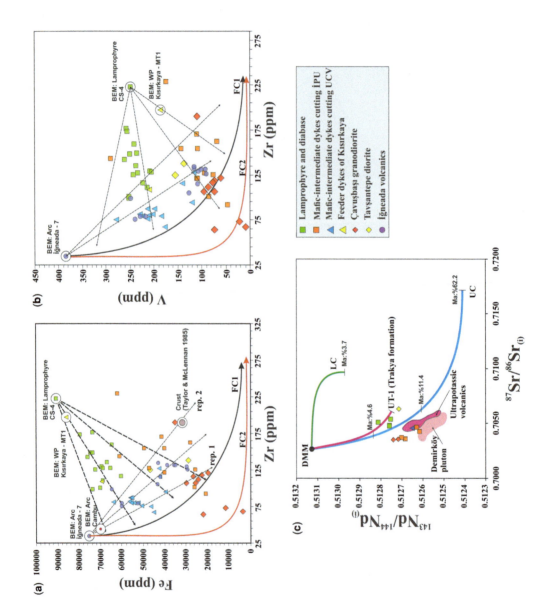

Table 9. *Thermal and compositional parameters used for the EC-AFC model calculations of mafic–intermediate dykes cutting the İstanbul Palaeozoic unit (İPU) and Upper Cretaceous volcanics (UCV)*

Thermal and compositional parameters	Lower crust (Model 1)	Upper crust (Model 2)	Trakya Formation (Model 3)
Magma liquidus temperature (T_{lm})	1320°C	1280°C	1320°C
Magma temperature (T_{m0})	1320°C	1280°C	1320°C
Assimilant liquidus temperature (T_{la})	1100°C	1000°C	1100°C
Country rock temperature (T_{a0})	600°C	400°C	400°C
Solidus temperature (T_s)	950°C	900°C	900°C
Magma specific heat capacity (C_{pm})	1484 J kg^{-1} K^{-1}	1484 J kg^{-1} K^{-1}	1484 J kg^{-1} K^{-1}
Assimilant specific heat capacity (C_{pa})	1388 J kg^{-1} K^{-1}	1370 J kg^{-1} K^{-1}	1388 J kg^{-1} K^{-1}
Crystallization enthalpy (ΔH_{cry})	396 000 J kg^{-1}	396 000 J kg^{-1}	396 000 J kg^{-1}
Fusion enthalpy (ΔH_{fus})	354 000 J kg^{-1}	270 000 J kg^{-1}	270 000 J kg^{-1}
Equilibration temperature (T_{eq})	980°C	980°C	980°C
Magma isotope ratio (ε_m)	Sr 0.70263; Nd 0.51313	Sr 0.70263; Nd 0.51313	Sr 0.70263; Nd 0.51313
Asimilant isotope ratio (ε_a)	Sr 0.7100; Nd 0.5122	Sr: 0.7220 – Nd: 0.5118	Sr 0.7006713; Nd 0.512277

For the upper- and lower-crustal assimilants, the thermal parameters for the 'standard upper crustal' and the 'standard lower crustal' case, respectively, were taken from Bohrson & Spera (2001).

Trends on our EC-AFC modelling graph (Fig. 15c) indicate clear signatures which can be explained by variable fractionation and assimilation of magmas derived from a common parental magma (DMM). We plotted fractionation and assimilation paths of lower and upper crusts and of the Trakya Formation. Our EC-AFC modelling results (Fig. 15c) suggest that all the samples, including mafic–intermediate dykes and small stocks, assimilated upper-crustal rocks. The assimilation (Ma*/Mc) values vary between 4.6 and 11.4%. Ma*/Mc = the amount of assimilant partial melt (Ma*)/the mass of cumulates (Mc).

Geodynamics

Late Cretaceous arc magmatism can be followed from the Apuseni–Banat–Timok–Srednogorie (ABTS) Belt in Bulgaria (von Quadt et al. 2005; Marchev et al. 2009; Georgiev et al. 2012), to İğneada (Turkish–Bulgarian border) to the Eastern Pontides. In the İstanbul region, the arc magmatism is represented by a volcanic succession (e.g. the Kavaklar Group), small plutons (i.e. Çavuşbaşı granodiorite and Tavşantepe diorite) and mafic–

intermediate dykes. It has been suggested in earlier studies that this arc magmatism could be ascribed to the north-dipping subduction of the Neo-Tethyan Ocean that closed along the İzmir–Ankara–Erzincan Suture (Şengör & Yılmaz 1981; Okay & Şahintürk 1997; Keskin & Tüysüz 1999; Yılmaz Şahin et al. 2012). A number of researchers (Keskin & Tüysüz 1999, 2001; Keskin et al. 1999; Keskin & Ustaömer 2001) argued that the Late Cretaceous volcanic units along the Black Sea coast in Western Pontides displayed arc-related characteristics formed in a rift setting related to the opening of the Western Black Sea Basin.

According to Yavuz & Yılmaz (2009), Upper Cretaceous volcanism in the north of İstanbul may be related to the İzmir–Ankara Suture or the Intra-Pontide Suture. Also, Yılmaz Şahin et al. (2012) inferred that the Upper Cretaceous andesitic lavas, dykes and small acidic intrusions were the products of the arc magmatism related to the northwards subduction of either the Intra-Pontide or İzmir–Ankara–Erzincan oceans (Yılmaz Şahin et al. 2012 and references therein). They also argued that adakitic Çavuşbaşı granodiorite was most probably generated

Fig. 15. (**a**) & (**b**) Bivariate diagrams on which a highly incompatible and immobile element (i.e. Zr) is plotted against a subset of highly compatible ones (i.e. Zr v. Fe and Zr v. V diagrams are adopted from Keskin et al. 2010; Keskin & Tüysüz 2017). Possible fractionation curves (FC1 and FC2: FC2 for highly evolved plutonic rocks; BEM, basic end-member composition), replenishment and mixing trends are marked on the diagrams. (**c**) ^{87}Sr/^{86}Sr$_{(i)}$ v. ^{143}Nd/^{144}Nd$_{(i)}$ plot displaying our EC-AFC model using the energy-constrained AFC equations of Bohrson & Spera (2001) and Spera & Bohrson (2001). The compositional and thermal parameters used in the model are shown in Table 8. Standard LC and UC data are from Bohrson & Spera (2001); DMM (starting end-member) isotope data from Workman & Hart (2005). The red curve represents the model in which the average composition of the Trakya Formation (sample UT-1, isotope values are given in Table 6) has been used as the contaminant.

(a)   **TURONIAN - CONIACIAN: Arc-rifting along the W Pontides.**
Formation of calc-alkaline magmas with subduction signature

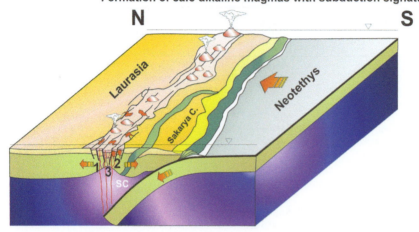

(b)   **CAMPANIAN TO PALEOCENE: Streching & thinning of the lithosphere,**
asthenospheric melting, formation of OIB type alkaline lavas and dykes

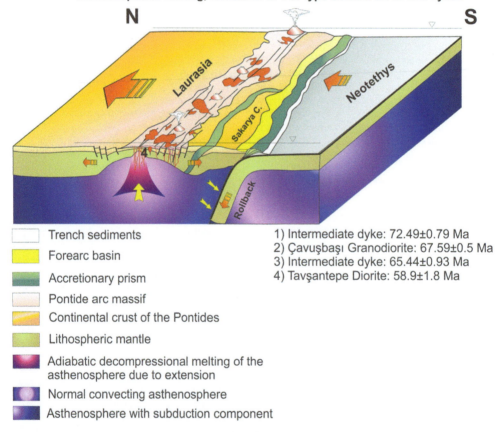

Trench sediments	1) Intermediate dyke: 72.49±0.79 Ma
Forearc basin	2) Çavuşbaşı Granodiorite: 67.59±0.5 Ma
Accretionary prism	3) Intermediate dyke: 65.44±0.93 Ma
Pontide arc massif	4) Tavşantepe Diorite: 58.9±1.8 Ma
Continental crust of the Pontides	
Lithospheric mantle	
Adiabatic decompressional melting of the asthenosphere due to extension	
Normal convecting asthenosphere	
Asthenosphere with subduction component	

Fig. 16. Block diagrams showing the geodynamic evolution of the Late Cretaceous–Paleocene magmatism in the Western Pontides in this study (modified from Keskin *et al.* 2010; see also the model in Keskin & Tüysüz 2017). *Abbreviations*: RZ, rift zone; SC, subduction component; MOR, mid-ocean ridge; Sakarya C., Sakarya continent.

from the partial melting of a subducted oceanic slab of the northern branch of the Neo-Tethyan Ocean along the İzmir–Ankara–Erzincan Suture Zone (Yılmaz Şahin *et al.* 2012).

Keskin *et al.* (2010) divided the Late Cretaceous volcanic activity in the Western Pontides into two stages: the early stage is represented by the calc-alkaline volcanic rocks that display a clear subduction signature (i.e. the Bozhane Formation in the north of İstanbul); and in the late stage, calc-alkaline lavas dominated at the beginning of the volcanic succession. On the other hand, around the middle of the late stage, alkaline and mildly alkaline lavas coexisted, intercalated with calc-alkaline lavas (Keskin *et al.* 2010). According to Keskin *et al.* (2010) and Keskin & Tüysüz (2017), these two contrasting magma series interacted intensively via magma mixing during magma-chamber evolutions in an extensional arc setting. Mafic–intermediate dykes, lamprophyres and small stocks are intruded into both the Upper Cretaceous volcanic sequence and the İstanbul Palaeozoic unit. Mafic dykes are represented by calc-alkaline to alkaline basalts, lamprophyres and diabases. Intermediate dykes are represented by the calc-alkaline andesitic to dacitic subvolcanic rocks and small stocks. The olivine-bearing alkaline basalt dykes cutting the İstanbul Palaeozoic unit are very similar to the lavas of the Kısırkaya Formation in terms of their mineral content, texture and general appearance. Keskin *et al.* (2010) interpreted the olivine-bearing alkaline basalt dykes as the feeder dykes of the lavas making up the Kısırkaya Formation. They formed during the late stage of the Upper Cretaceous volcanism. The İstanbul Palaeozoic unit is cut by numerous dykes and subvolcanic intrusions. There are three different types of magmatic intrusions cutting through this basement unit: (1) lamprophyre and diabase dykes; (2) intermediate subvolcanic intrusions with porphyiritic texture; and (3) small microdioritic and microgranitic intrusions. Lamprophyre dykes, which cut the İstanbul Palaeozoic unit, display a similar geochemical tendency to the feeder dykes of the Kısırkaya Formation with a slight Ta–Nb depletion, while the rest display a subduction signature.

The data and findings we present in this paper fit well with the rifted-arc model related to the opening of the Black Sea proposed by Keskin & Tüysüz (2017). For example, older calc-alkaline dykes cutting the Garipçe Formation and microdioritic intrusions with a clear calc-alkaline character were probably derived from a shallower mantle source, such as a metasomatized lithospheric mantle wedge, that contained a subduction signature during the initial stages of rifting. This in contrast to the younger alkaline feeder dykes of the Kısırkaya Formation and lamprophyre dykes which were derived from a possibly deeper, and hence presumably asthenospheric,

mantle source that was richer in garnet and with relatively lower degrees of melting. The aforementioned temporal change may be interpreted as thinning of the lithosphere by rifting of an arc massif during the opening of the Black Sea and the subsequent upwelling of the asthenospheric mantle that generated OIB-type magmas towards the final stages of the magmatism, resulting in extensive interaction with the calc-alkaline magmas, as shown in this paper (and as also proposed by Keskin & Tüysüz 2017). We present our geodynamic model in Figure 16.

Conclusions

Results of our study indicate that the melts feeding the dyke system during the Late Cretaceous–Paleocene were derived from two contrasting mantle sources: (1) a lithospheric mantle source modified by subducted-slab-derived melts which sourced the calc-alkaline magmas with a clear subduction signature; and (2) an asthenospheric mantle from which alkaline basic magmas with an OIB signature were derived towards the end of the magmatism. Our data also imply a widespread interaction of magmas derived from these contrasting sources. We propose a geodynamic model for the emplacement of dyke complexes that involves a rifted volcanic arc margin related to the opening of the Black Sea during the Late Cretaceous–Paleocene period.

This work was supported by three research projects: numbers 40841 and 1145/010598, granted by the İstanbul University Research Fund and number 104Y215, granted by the Scientific and Technological Research Council of Turkey. We sincerely thank Timur Ustaömer for EPMA analyses. We are grateful to Hamdi Çinal from Akçin Engineering Co. (İstanbul) for assistance and logistical support during our fieldwork. We thank Volume Editors Mike Simmons and Aral Okay and two anonymous referees for their thoughtful reviews and constructive comments on our manuscript.

References

Aysal, N., Ustaömer, T., Öngen, S., Keskin, M., Köksal, S., Peytcheva, I. & Fanning, M. 2012. Origin of the Early–Middle Devonian magmatism in the Sakarya Zone, NW Turkey: geochronology, geochemistry and isotope systematics. *Journal of Asian Earth Sciences*, **45**, 201–222.

Bell, K., Lavecchia, G. & Rosatelli, G. 2013. Cenozoic Italian magmatism – isotope constraints for possible plume-related activity. *Journal of South American Earth Sciences*, **41**, 22–40.

Bergman, S.C. 1987. Lamproites and other potassium rich rocks: a review of their occurrence, mineralogy and geochemistry. *In*: Fitton, J. & Upton, B.G.J. (eds) *Alkaline Igneous Rocks*. Geological Society, London,

Special Publications, **30**, 103–190, https://doi.org/10.1144/GSL.SP.1987.030.01.08

BOHRSON, W.A. & SPERA, F.J. 2001. Energy-constrained open-system magmatic processes II: application of energy-constrained assimilation–fractional crystallization (EC-AFC) model to magmatic systems. *Journal of Petrology*, **42**, 1019–1041.

CABANIS, B. & LECOLLE, M. 1989. Le diagramme La/10-Y/15-Nb/8: un outil pour la discrimination des séries volcanique et la mise en evidence des processus de mélange et/ou de contamination crustale [The La/10-Y/15-Nb/8 diagram: a tool for the discrimination of the volcanic series and the identification of crustal mixing and/or contamination processes]. *Comptes Rendus de l'Académie des Sciences*, **309**, 2023–2029.

DAL NEGRO, A., CARBONIN, S., MOLIN, G.M., CUNDARI, A. & PICCIRILLO, E.M. 1982. Intracrystalline cation distribution in natural clinopyroxenes of tholeiitic, transitional, and alkaline basaltic rocks. *In*: SAXENA, S.K. (ed.) *Advances in Physical Geochemistry, Vol. 2*, Springer, New York, 117–150.

DEER, W.A., HOWIE, R.A. & ZUSSMAN, J. 1992. *An Introduction to the Rock-Forming Minerals*. 2nd edn. Longman, London.

GEORGIEV, S., VON QUADT, A., HEINRICH, C.A., PEYTCHEVA, I. & MARCHEV, P. 2012. Time evolution of a rifted continental arc: integrated ID-TIMS and LA-ICPMS study of magmatic zircons from the Eastern Srednogorie, Bulgaria. *Lithos*, **154**, 53–67.

GÜLMEZ, F., GENÇ, Ş.C., PRELEVIC, D., TÜYSÜZ, O., KARACIK, Z., RODEN, M.F. & BILLOR, Z. 2016. Ultrapotassic volcanism from the waning stage of the Neotethyan subduction: a key study from the Izmir–Ankara–Erzincan Suture Belt, Central Northern Turkey. *Journal of Petrology*, **57**, 561–593.

HARALD, B. & GALLIARD, F. 2006. Geochemical aspects of melts: volatiles and redox behavior. *Elements*, **2**, 275–280.

HOLLOCHER, K., ROBINSON, P., WALSH, E. & ROBERTS, D. 2012. Geochemistry of amphibolite facies volcanics and gabbros of the Støren Nappe in extensions west and southwest of Trondheim, Western Gneiss Region, Norway: a key to correlations and paleotectonic settings. *American Journal of Science*, **312**, 357–416.

KARACIK, Z. & TÜYSÜZ, O. 2010. Petrogenesis of the Late Cretaceous Demirköy igneous complex in NW Turkey: implications for magma genesis in the Strandja Zone. *Lithos*, **114**, 369–384.

KESKIN, M. & TÜYSÜZ, O. 1999. Geochemical evidence for nature and evolution of the rift volcanism related to the opening of the Black Sea, Central Pontides, Turkey. *EUG10 in Strasbourg, France, Journal of Conference Abstracts*, **4**, 816.

KESKIN, M. & TÜYSÜZ, O. 2001. Genesis of the rift volcanism and basin formation related to the opening of the Black Sea during Late Cretaceous, Western Pontides, Turkey. *EUG11 in Strasbourg, France, Journal of Conference Abstracts*, **5**, 732.

KESKIN, M. & TÜYSÜZ, O. 2017. Stratigraphy, petrogenesis and geodynamic setting of the Late Cretaceous volcanism on the SW margin of the Black Sea, Turkey. *In*: SIMMONS, M.D., TARI, G.C. & OKAY, A.I. (eds) *Petroleum Geology of the Black Sea*. Geological Society, London, Special Publications, **464**, https://doi.org/10.1144/SP464.5

KESKIN, M. & USTAÖMER, T. 2001. Stratigraphy and geochemistry of Cretaceous volcano-sedimentary units in the North of Istanbul: development of a volcanic rifted margin, Western Pontides, Turkey. *EUG11 in Strasbourg, France, Journal of Conference Abstracts*, **5**, 732.

KESKIN, M., USTAÖMER, T. & YENIYOL, M. 1999. Volcanism associated with extension at a consuming margin, W Pontides, N Turkey. *EUG10 in Strasbourg, France, Journal of Conference Abstracts*, **4**, 841.

KESKIN, M., USTAÖMER, T. & YENIYOL, M. 2003. Stratigraphy, petrology and tectonic setting of Upper Cretaceous volcano-sedimentary units, North of İstanbul, Turkey. *In*: MERIÇ, E. (ed.) *Symposium on the Geology of İstanbul Symposium, Chamber of Geological Engineers of Turkey*. Kadir Has University, Istanbul, 23–35.

KESKIN, M., USTAÖMER, T. & YENIYOL, M. 2010. Magmatic evolution and geodynamic setting of the Late Cretaceous volcanogenic sequences in the north of İstanbul, NW Turkey. *In*: ÖRGÜN, Y., YILMAZ ŞAHIN, S. (eds) *Symposium on the Geology of İstanbul Symposium, Chamber of Geological Engineers of Turkey*. Kadir Has University, Istanbul, 130–180.

KETIN, İ. 1941. *Das Granitmassiv westlich von Alemdağ [The Granite Massif West of Alemdağ]*. İstanbul University Institute of Geology Publications, **7**.

KRMÍČEK, L., CEMPÍREK, J., HAVLÍN, A., PŘICHYSTAL, A., HOUZAR, S., KRMÍČKOVÁ, M. & GADAS, P. 2011. Mineralogy and petrogenesis of a Ba–Ti–Zr-rich peralkaline dyke from Šebkovice (Czech Republic): recognition of the most lamproitic Variscan intrusion. *Lithos*, **121**, 74–86.

LEAKE, B.E., WOOLLEY, A.R. *ET AL.* 1997. Nomenclature of amphiboles: report of the subcommittee on amphiboles of the International Mineralogical Association Commission on New Minerals and Mineral Names. *Canadian Mineralogist*, **35**, 219–246.

LIEW, T.C. & HOFMANN, A. 1988. Precambrian crustal components, plutonic associations, plate enviroment of the Hercynian fold belt of Central Europe; indications from Nd and Sr isotopic study. *Contributions to Mineralogy and Petrology*, **98**, 129–138.

LINDSLEY, D.H. 1983. Pyroxene thermometry. *American Mineralogist*, **68**, 477–493.

LINNEMANN, U., OUZEGANE, K., DRARENI, A., HOFMANN, M., BECKER, S., GÄRTNER, A. & SAGAWE, A. 2011. Sands of West Gondwana: an archive of secular magmatism and plate interactions – a case study from the Cambro-Ordovician section of the Tassili Ouan Ahaggar (Algerian Sahara) using U–Pb- LA-ICP-MS detrital zircon ages. *Lithos*, **123**, 188–203.

MARCHEV, P., GEORGIEV, S., ZAJACZ, Z., RAYCHEVA, R., MANETTI, P., VON QUADT, A. & TOMASSINI, S. 2009. High-K ankaramitic melt inclusions and lavas in the Upper Cretaceous Eastern Srednogorie continental arc, Bulgaria: implications for the genesis of arc shoshonites. *Lithos*, **113**, 228–245.

MCKENZIE, D.P. & O'NIONS, R.K. 1991. Partial melt distributions from inversion of rare earth element concentrations. *Journal of Petrology*, **32**, 1021–1091.

MILLER, J.S., MATZEL, J.E.P., MILLER, C.F., BURGESS, S.D. & MILLER, R.B. 2007. Zircon growth and recycling during the assembly of large, composite arc plutons. *Journal of Volcanology and Geothermal Research*, **167**, 282–299.

MITCHELL, R.H. & BERGMAN, S.C. 1991. *Petrology of Lamproites*. Plenum Press, New York.

MOGHADAM, H.S., GHORBANI, G. ET AL. 2014. Late Miocene K-rich volcanism in the Eslamieh Peninsula (Saray), NW Iran: implications for geodynamic evolution of the Turkish–Iranian High Plateau. *Gondwana Research*, **26**, 1028–1050.

MOLIN, G. & ZANAZZI, P.F. 1991. Intracrystalline Fe^{2+}–Mg ordering in augite: experimental study and geothermometric applications. *European Journal of Mineralogy*, **3**, 863–875.

MORIMOTO, N., FABRIES, J., FERGUSON, A.K., GINZBURG, I.V., ROSS, M., SEIFERT, F.A. & ZUSSMAN, J. 1988. Nomenclature of pyroxenes. *Mineralogical Magazine*, **52**, 535–550.

NESBITT, H.W. & YOUNG, G.M. 1984. Prediction of some weathering trends of plutonic and volcanic rocks based on thermodynamic and kinetic considerations. *Geochimica et Cosmochimica Acta*, **48**, 1523–1534.

OKAY, A.I. & ŞAHINTÜRK, Ö. 1997. Geology of the Eastern Pontides. *In*: ROBINSON, A.G. (ed.) *Regional and Petroleum Geology of the Black Sea and Surrounding Region*. American Association of Petroleum Geologists Memoirs, **68**, 291–311.

OKAY, A.I. & TÜYSÜZ, O. 1999. Tethyan sutures of northern Turkey. *In*: DURAND, B., JOLIVET, L., HORVÁTH, F. & SÉRANNE, M. (eds) *The Mediterranean Basins: Tertiary Extension within the Alpine Orogen*. Geological Society, London, Special Publications, **156**, 475–515, https://doi.org/10.1144/GSL.SP.1999.156.01.22

ÖZGÖRÜŞ, Z. & OKAY, A.İ. 2005. Orientation of the andesitic dykes in the İstanbul region: an approach to the Cretaceous stress distribution. *Mineral Research and Exploration Bulletin*, **130**, 17–27.

ÖZGÜL, N. 2012. Stratigraphy and some structural features of the İstanbul Palaeozoic. *Turkish Journal of Earth Sciences*, **21**, 817–866.

ÖZGÜL, N., ÖZCAN, İ. ET AL. 2011. *Geology of the İstanbul City*. İstanbul Metropolitan Municipality, İstanbul.

ÖZTUNALI, Ö. & SATIR, M. 1975. Rb and Sr Altersbestimungen an Tiefengesteinen aus Çavuşbaşı (İstanbul). *İstanbul Üniversitesi, Fen Fakültesi Mecmuası Seri B*, **40**, 1–7.

PEARCE, J.A. 2008. Geochemical fingerprinting of oceanic basalts with applications to ophiolite classification and the search for Archean oceanic crust. *Lithos*, **100**, 14–48.

PEARCE, J.A. & PEATE, D.W. 1995. Tectonic implications of the composition of volcanic arc magmas. *Annual Review of Earth and Planetary Sciences*, **23**, 251–285.

PUTIRKA, K.D. 2008. Thermometers and barometers for volcanic systems. *In*: PUTIRKA, K.D. & TEPLEY, F. (eds) *Mineral Inclusions and Volcanic Processes. Reviews in Mineralogy and Geochemistry*, **69**, 61–120.

RIDOLFI, F. & RENZULLI, A. 2012. Calcic amphiboles in calc-alkaline and alkaline magmas: thermobarometric and chemometric empirical equations valid up to 1130°C and 2.2 GPa. *Contributions to Mineralogy and Petrology*, **163**, 877–895.

RIDOLFI, F., RENZULLI, A. & PUERINI, M. 2010. Stability and chemical equilibrium of amphibole in calc-alkaline

magmas: an overview, new thermobarometric formulations and application to subduction-related volcanoes. *Contributions to Mineralogy and Petrology*, **160**, 45–66.

ROCK, N.M.S. 1987. The nature and origin of lamprophyres: an overview. *In*: FITTON, J.G. & UPTON, B.G.J. (eds) *Alkaline Igneous Rocks*. Geological Society, London, Special Publications, **30**, 191–226, https://doi.org/10.1144/GSL.SP.1987.030.01.09

SACCANI, E. 2015. A new method of discriminating different types of post-Archean ophiolitic basalts and their tectonic significance using Th–Nb and Ce–Dy–Yb systematic. *Geoscience Frontiers*, **6**, 481–501.

ŞENGÖR, A.M.C. & YILMAZ, Y. 1981. Tethyan evolution of Turkey: a plate tectonic approach. *Tectonophysics*, **75**, 181–241.

SHAW, D.M. 1970. Trace element fractionation during anatexis. *Geochimica et Cosmochimica Acta*, **34**, 237–243.

SHELLNUTT, J.G., BHAT, G.M., WANG, K.L., BROOKFIELD, M.E., JAHN, B.M. & DOSTAL, J. 2014. Petrogenesis of the flood basalts from the Early Permian Panjal Traps, Kashmir, India: geochemical evidence for shallow melting of the mantle. *Lithos*, **204**, 159–171.

SPERA, F.J. & BOHRSON, W.A. 2001. Energy-constrained open-system magmatic processes I: general model and energy-constrained assimilation and fractional crystallization (EC-AFC) formulation. *Journal of Petrology*, **42**, 999–1018.

SUN, S.-s. & MCDONOUGH, W.F. 1989. Chemical and isotopic systematics of oceanic basalts: Implications for mantle composition and processes. *In*: SAUNDERS, A.D. & NORRY, M.J. (eds) *Magmatism in the Ocean Basins*. Geological Society, London, Special Publications, **42**, 313–345, https://doi.org/10.1144/GSL.SP.1989.042.01.19

TAYLOR, S.R. & MCLENNAN, S.M. 1985. *The Continental Crust: Its composition and Evolution. Geoscience Texts*. Blackwell Scientific, London.

TEIPEL, U., EICHHORN, R., LOTH, G., ROHRMÜLLER, J., HÖLL, R. & KENNEDY, A. 2004. U–Pb SHRIMP and Nd isotopic data from the western Bohemian Massif (Bayerischer Wald, Germany): implications for Upper Vendian and Lower Ordovician magmatism. *International Journal of Earth Sciences*, **93**, 782–801.

THIRLWALL, M.F., UPTON, B.G.J. & JENKINS, C. 1994. Interaction between continental lithosphere and the Iceland plume-Sr–Nd–Pb isotope geochemistry of Tertiary basalts, NE Greenland. *Journal of Petrology*, **35**, 839–879.

USTAÖMER, P.A., MUNDIL, R. & RENNE, P.R. 2005. U/Pb and Pb/Pb zircon ages for arc-related intrusions of the Bolu Massif (W Pontides, NW Turkey): evidence for Late Precambrian (Cadomian) age. *Terra Nova*, **17**, 215–224.

VON QUADT, A., MORITZ, R., PEYTCHEVA, I. & HEINRICH, C.A. 2005. Geochronology and geodynamics of Late Cretaceous magmatism and Cu–Au mineralization in the Panagyurishte region of the Apuseni–Banat–Timok–Srednogorie belt, Bulgaria. *Ore Geology Reviews*, **27**, 95–126.

WINCHESTER, J.A. & FLOYD, P.A. 1977. Geochemical discrimination of different magma series and their differentiation products using immobile elements. *Chemical Geology*, **20**, 325–343.

WOOLLEY, A.R., BERGMAN, S.C., EDGAR, A.D., LE BAS, M.J., MITCHELL, R.H., ROCK, N.M.S. & SCOTT SMITH, B.H. 1996. Classification of lamprophyres, lamproites, kimberlites, and the kalsilitic, melilitic and leucitic rocks. *Canadian Mineralogist*, **34**, 175–186.

WORKMAN, R.K. & HART, S.R. 2005. Major and trace element composition of the depleted MORB mantle (DMM). *Earth and Planetary Science Letters*, **231**, 53–72.

YAN, D., CHEN, D., WANG, Q. & WANG, J. 2010. Large-scale climatic fluctuations in the latest Ordovician on the Yangtze block, south China. *Geology*, **38**, 599–602.

YAVUZ, O. & YILMAZ, Y. 2009. Geochemical characteristics of upper cretaceous volcanism in the north of İstanbul. *In*: *62nd Geological Congress of Turkey*, *13–17 April 2009, MTA–Ankara, Türkiye, Abstract Book*, 622–623.

YILMAZ ŞAHIN, S., AYSAL, N. & GÜNGÖR, Y. 2012. Petrogenesis of late cretaceous adakitic magmatism in the İstanbul zone (Çavuşbaşı Granodiorite, NW Turkey). *Turkish Journal of Earth Sciences*, **21**, 1029–1045.

YILMAZ ŞAHIN, S., AYSAL, N., GÜNGÖR, Y. & PEYTCHEVA, I. 2015. Geochemical, geochronological and isotopic data from Permo-Triassic plutons in Western Pontides, NW Turkey. *Goldschmidt Abstracts*, **2015**, 3527.

YILMAZ, İ. 1977. Absolute age and genesis of the Sancaktepe granite (Kocaeli peninsula). *Bulletin of the Geological Society of Turkey*, **20**, 11–20.

ZINDLER, A. & HART, S. 1986. Chemical geodynamics. *Annual Review of Earth and Planetary Sciences*, **14**, 493–571.

The geological history of the Istria 'Depression', Romanian Black Sea shelf: tectonic controls on second-/third-order sequence architecture

DAVID R. D. BOOTE

12 Elsynge Road, London SW18 2HN, UK
drdboote@gmail.com

Abstract: The Istria 'Depression' or sub-basin of offshore Romania lies at the intersection of the trans-European Tornquist–Teisseyre 'Zone' and the Black Sea back-arc basin, just outboard of the East Carpathian orogenic welt. Its Late Mesozoic–Cenozoic succession records an extraordinary polyphase history of subsidence and sedimentation, interrupted by several quite spectacular second-/third-order erosional unconformities, reflecting the interplay between these tectonic domains. The unconformities divide the succession into a number of stratigraphic sequences.

The sub-basin first developed as a transtensional rift in the Triassic–Early Jurassic, evolving into a narrow oceanized trough in the later Jurassic. This was tilted west during the Early Cretaceous, and the residual Late Jurassic topography was filled and buried by a west-facing clastic–evaporite wedge. Following Late Aptian–Albian(?) rifting, post-rift subsidence and spreading in the Western Black Sea imposed a strong easterly tilt, encouraging the partial evacuation of its Early Cretaceous sedimentary fill by gravity-driven mass wastage. The incised valley topography was subsequently infilled and buried during the later Cretaceous and Early Cenozoic. During the mid-Late Cenozoic, the Black Sea Basin experienced intermittent periods of partial to complete isolation from the world ocean and significant base-level drawdown. The first major sea-level fall occurred in the Eocene when the Istria 'Depression' was deeply incised, to be healed by Oligocene shales during the subsequent rise. Yet another period of drawdown and exposure occurred in the mid-Miocene, with extensive shelf-margin mass wastage and erosion, followed by re-flooding and deposition of a transgressive backstepping sequence in the middle-late Miocene. Messinian drawdown in the Mediterranean caused a further period of isolation and falling base level. The shelf margin was again exposed, and experienced widespread mass wastage and slumping. Rising sea level eroded the earlier slumped sequence and the margin was healed by a lowstand prograding wedge in the late Miocene–early Pliocene. This was followed by shelf sedimentation in the Plio-Pleistocene periodically interrupted by canyon-incision events, testifying to continued climatically or tectonically imposed base-level fluctuations.

Several direct and indirect tectonic factors were responsible for valley/canyon incision within the Istria Depression and erosion of the Romanian Black Sea shelf margin. These include: (1) the local structural framework; (2) direct tectonic uplift and tilting; and (3) more indirect tectonically imposed isolation encouraging significant base-level falls.

The Istria 'Depression' or sub-basin is located on the Romanian Black Sea shelf (Figs 1 & 2). Its Late Mesozoic–Cenozoic succession is interrupted by several pronounced unconformities of near base Albian, intra-Eocene, top Eocene/base Oligocene, ?Badenian–Sarmatian and intra-Pontian age, dividing the complex fill sequence into several second-/third-order stratigraphic sequences (Fig. 3). Their scale and intensity, sometimes cutting down deeply into the underlying section, suggests a direct or indirect tectonic rather than a global eustatic cause. This review attempts to assess how this local stratigraphic architecture was controlled by regional tectonostratigraphic drivers.

Regional overview

Catuneanu (1992, 1994) first defined the Istria Depression as an easterly-thickening Albian–Recent depocentre bounded by the so-called 'Euxinic Threshold', extending from the deep-water Black Sea across the continental shelf to the onshore 'Babadag Basin'. This embayed depocentre lies directly above the offshore extension of the North Dobrogea Orogen, constrained to the south by the Peceneaga-Camena Fault and Midia Platform, and by the Pelikan Platform in the north (Fig. 2). The Istria Depression lies at the southern terminus of the transcontinental Tornquist–Teisseyre Lineament on the margin of the Western Black Sea back-arc basin, and just outboard of the eastern Carpathians and Moesian Platform (Fig. 1). The present configuration of these several structural elements reflects a long history of deformation across the northern margin of Tethys. Late Palaeozoic consolidation of the Pangean supercontinent left a triangular Tethyan oceanic remnant separating Laurasia to the north and Gondwana in the south (Barrier & Vrielynck 2008). Early Mesozoic subduction along the northern

From: Simmons, M. D., Tari, G. C. & Okay, A. I. (eds) 2018. *Petroleum Geology of the Black Sea*.
Geological Society, London, Special Publications, **464**, 169–209.
First published online October 9, 2017, https://doi.org/10.1144/SP464.8

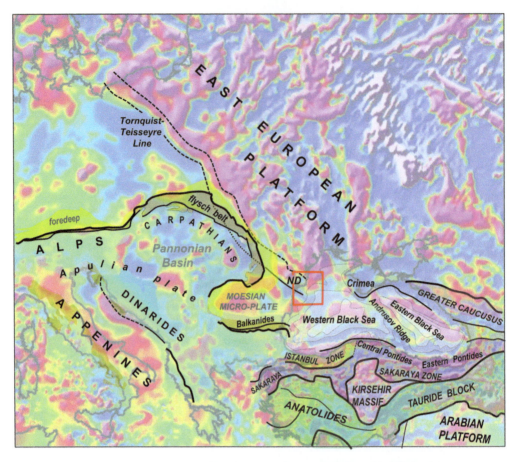

Fig. 1. Central European structural elements: highlighting the major tectonic elements that influenced the Cretaceous–Cenozoic sequence architecture of the Istria Depression, offshore Romania, including the Tornquist–Teisseyre 'Line', Moesian Platform, Carpathian fold belt, and the western Black Sea (adapted from Pharaoh 1999; Hippolyte 2002; Picha *et al.* 2006; Korhonen *et al.* 2007 Stephensen & Schellart 2010; Tari *et al.* 2014).

flank of this oceanic basin gave way to slab rollback and opening of small back-arc basins ('Palaeotethys') in the Triassic–Early Jurassic. These closed in the mid-Jurassic ('Cimmerian' event) as the Moesian Platform broke away from Europe and moved SE along the Tornquist–Teisseyre Lineament/ Peceneaga-Camena fault system during the later Jurassic, leaving the North Ligurian seaway in its wake (Tari 2005; Barrier & Vrielynck 2008). Early Cretaceous extension and rifting was followed by back-arc spreading in the western Black Sea in the mid-Cretaceous and in the eastern Black Sea some time later (Okay *et al.* 1994; Banks & Robinson 1997; Okay & Gorur 2007). Several large continental blocks broke away from the Gondwanan margin of the Tethys during the Late Triassic–Early Jurassic and, followed by the Afro-Arabian Plate, moved north across Tethys to collide with Eurasia (Laurasia) in the Late Cretaceous/Early Cenozoic (Barrier & Vrielynck 2008). The northern arm of 'Neotethys' finally closed in the early–mid Eocene, followed by a collision with Afro-Arabia and closure of the southern arm in the mid–late Eocene (Robertson *et al.* 2006; Barrier & Vrielynck 2008), while the Carpathian orogenic front advanced around the Moesian Platform and up the North Ligurian seaway to collide with the Tornquist–Teisseyre Lineament and Eurasian Platform during the Miocene (Linzer *et al.* 1998; Hippolyte *et al.* 1999). This regional crustal consolidation created a semi-isolated Parathethyan seaway stretching across central Europe and the Carpathian foredeep (Central Paratethys) to the Black and Caspian seas (Eastern Paratethys). Its complex topography with narrow connecting passages made it particularly susceptible to climatic and tectonically enforced base-level changes. As a result, it experienced complete isolation from the global ocean several times during the mid-Late Cenozoic.

The tectonostratigraphic architecture of the Istria Depression and the second-/third-order sequence boundaries dividing its Mesozoic–Cenozoic succession reflect this complex tectonic evolution. The following review explores this relationship by comparing a comprehensive sequence analysis of the basin fill with the tectonostratigraphic evolution of the larger region.

Database and analytical methodology

Stratigraphic sequence analysis of the Istria succession utilized a 6700 km 2D seismic dataset acquired by Enterprise–CanOxy–OGY consortia in 1992 (Boote 1994). This comprised two surveys (E92RO-17 and ER92RO-16) covering the Midia and Pelikan concessions and extending locally across the intervening Istria Depression. The dataset also included three additional Petrom (P93) lines and composite logs from 14 wells. Stratal geometries and reflector amplitudes, locally constrained by well lithologies, were used to identify the erosional unconformities and correlatable conformities defining the main second-/third-order sequence boundaries within the basin fill. These were then correlated around the Istria Depression and onto the flanking platforms (Boote 1994, 2007, 2014a, b). Data quality and lithological control were generally insufficient to describe higher-frequency sequence architecture, but enough to identify the second-/third-order depositional systems within each sequence. Biostratigraphic data constraining the age of the key sequence boundaries were of uncertain quality and often ambiguous. This ambiguity was compounded by the difficulty in correlating local Paratethyan chronostratigraphy with other regional and global schemes. As a result, correlations between local unconformities and the structural or stratigraphic events responsible for their development were sometimes rather tenuous. However, published analyses focused on both the local and regional geology provided some very useful constraints. The more important of these included:

- North Dobrogea Basin Frame: the classic 1984 and 1988 articles by Gradinaru supplemented and expanded by Seghedi (2001) and Hippolyte (2002). Plate tectonic reconstructions by Banks & Robinson (1997), Tari (2005) and Barrier & Vrielynck (2008) provided a robust regional perspective.
- Sequence 1 – Neocomian–Aptian: the detailed stratigraphic analysis of the onshore Dobrogea succession by Avram et al. (1993) and Swidrowska et al. (2008) proved critical in reconstructing the original extent of the basal Cretaceous evaporite systems tract in the North Dobrogea and offshore. The regional Lower Cretaceous

palaeogeography was constrained by Barrier & Vrielynck (2008) and Okay et al. (2014).
- Sequence 2 – Albian–Upper Cretaceous/Paleocene: analysis of the near base Albian erosional evacuation and mid-Late Cretaceous transgressive systems tracts was constrained by descriptions of the equivalent section onshore (Avram et al. 1993; Gradinaru 1995; Swidrowska et al. 2008) supported by plate tectonic reconstructions of the Western Black Sea back-arc basin by Okay et al. (1994), Banks & Robinson (1997), Kazmin et al. (2000), Gorur & Tuysuz (2001), Okay & Gorur (2007) and a robust biostratigraphic analysis of Cretaceous coastal outcrops in northern Turkey by Hippolyte et al. (2010).
- Sequence 3 – Eocene–mid-Miocene: regional evidence for a late Eocene–basal Oligocene drawdown of the Black Sea Basin by Tari et al. (2013) provided useful support for the composite intra-Eocene/near base Oligocene valley-incision event observed offshore Romania. An insight into the possible drivers responsible for the event came from descriptions of the Balkanide orogenic event by Doglioni et al. (1996), Harbury & Cohen (1997), Sinclair et al. (1997) and Stuart et al. (2011), mid-Tertiary Pontide deformation by Okay & Sahinturk (1997), Yilmaz et al. (1997) and Robertson et al. (2006), and more regional plate tectonic reconstructions by Linzer et al. (1998), Hippolyte (2002) and Barrier & Vrielynck (2008).
- Sequence 4 – Badenian–Meotian: the mid-Miocene drawdown event and evaporite deposition in Central Paratethys described by Rogl et al. (1978), Baldi (2006), Oszczypko et al. (2006), Peryt (2006) and Harzhauser & Piller (2007), with evidence of falling base level and canyon incision from Paraschiv (1979) and Picha et al. (2006), provided useful constraints for Badenian–Sarmatian erosion of the Romanian Black Sea shelf. Geodynamic reconstructions of Carpathian orogenesis by Linzer et al. (1998) and Hippolyte et al. (1999) offered a tectonic explanation for this event.
- Sequence 5 – Messinian–Pontian: sequence analysis of the intra-Pontian erosional event and lowstand fill was complemented by independent descriptions of the same sequence from Gillet et al. (2007), Konerding et al. (2010) and Munteanu et al. (2012). Local biostratigraphic control was imprecise and correlation with the Messinian event in the Mediterranean was based on a regional synthesis of information from Clauzon et al. (2005), Snel et al. (2006), Leever et al. (2009), Lofi et al. (2011), Bache et al. (2012) and Suc et al. (2015b), tightly constrained by high-resolution chronostratigraphic control from the equivalent succession encountered in the DSDP 380A well, offshore NW Turkey, Black

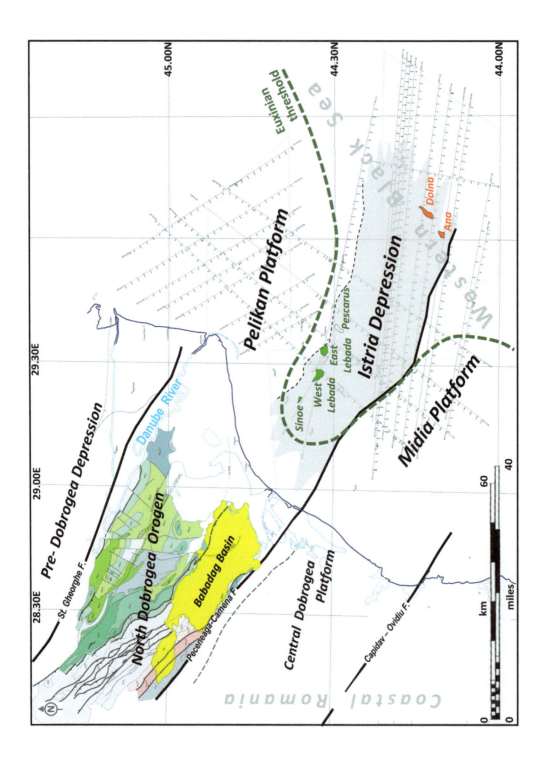

Sea (after Popescu 2006; Popescu *et al.* 2010; Suc *et al.* 2015*a*) and molecular biomarker analysis (Vasiliev *et al.* 2015).

- Sequence 5 – Plio-Pleistocene: descriptions of the Danube deep-water fan–channel system and feeder canyon by Winguth *et al.* (2000), Popescu *et al.* (2001), Popescu *et al.* (2004) and Lericolais *et al.* (2009) provided a useful regional context for the incised Plio-Pleistocene (palaeo-Danube) canyon systems crossing the Romanian shelf. Kinematic models of Carpathian orogenesis proposed by Linzer *et al.* (1998) and Hippolyte *et al.* (1999) helped to explain the late diversion of the Danube River.

Ultimately, these several published reviews provided a critical regional perspective for identifying and assessing the more likely regional geological factors responsible for the second-/third-order sequence boundaries observed on the Romanian Black Sea shelf.

Basin frame: Palaeozoic platforms and Triassic–Jurassic suture zone

The Palaeozoic–Early Mesozoic basin frame of the offshore Istria Depression is blanketed by a thick sequence of Cretaceous–Tertiary sediments, and poorly constrained by existing seismic and well control. However, because of its apparent stratigraphic and structural continuity with the adjacent Dobrogea region, onshore-based palaeo-tectonic reconstructions provide a very useful perspective for understanding the early tectonostratigraphic evolution of the offshore (Boote 2014*b*).

Coastal onshore Dobrogea structural fabric and extension offshore

The NW onshore margin of the Black Sea comprised several structural zones, separated by major linear crustal faults (Fig. 2):

- Scythian Platform and the Pre-Dobrogea Depression: forming the western margin of the East European Platform with flat-lying Mesozoic–Tertiary sediments resting upon a late Permian rift sequence.
- North Dobrogea 'Orogen' or Fold-Thrust Zone: is a narrow zone (40–55 km wide) of highly deformed Palaeozoic and Mesozoic igneous and sedimentary rocks bounded by the Sfantu-Gheorghe Fault (SGF) and Pre-Dobrogea Depression to the NE, and by the Peceneaga-Camena Fault (PCF) and Central–South Dobrogea in the SW. The Palaeozoic basement was consolidated by Hercynian magnetism and structuring, followed by Late Permian–Jurassic extension, interrupted by episodic compressional events during the Late Triassic and Jurassic (early Cimmerian). The age of the final paroxysmal phase of Cimmerian orogenesis is controversial, but flat-lying late Albian–Cenomanian rocks rest unconformably upon the deformed Kimmeridgian and older succession, and bounding Peceneaga-Camena Fault, constrain it to Tithonian–early Albian time.
- Central Dobrogea: Neoproterozoic basement outcrops across much of the Central Dobrogea with erosional remnants of Bathonian–Kimmeridgian carbonates preserved locally.
- South Dobrogea: contrasting with the Central Dobrogea, the South Dobrogea is blanketed by weakly to undeformed Ordovician–recent sediments.

The North Dobrogea 'Orogen' and flanking Pre-Dobrogea and Central Dobrogea zones extend offshore where they underlie the Cretaceous–Tertiary Istria Depression and adjacent Pelikan and Midia Platforms.

North Dobrogea Fold-Thrust Zone

The North Dobrogea Orogen and offshore Istria Trough lie along the SE terminus of the Tornquist–Teisseyre Suture Zone separating the ancient Baltic Shield/East European Platform from the younger crustal collage of Western Europe (Zeigler 1984, 1988). It is structurally complex in outcrop, with highly deformed pre-Albian Mesozoic and

Fig. 2. Major structural elements and location map, Coastal Dobrogea and offshore Romanian Black Sea shelf. The Istria 'Depression' is shown in grey, bounded by the Midia and Pelikan platforms, continuing onshore as the Central, North and Pre-Dobrogea. The outcropping geology of the North Dobrogea is highlighted in colour. Highly structured Carboniferous–Early Permian (PM_3^{rot}), late Permian rhyolites, granites and syenites (PM_3^{int}/PZ^{int}), Early Scythian Cemurlia Formation (TR^1), Late Scythian Somova and Bogza formations (TR^{2-1}), Late Scythian–Anisian Niculitel Formation (TR^{2-2}), Norian–Anisian Murighiol and Poppina formations (TR^{2-3}), Agighiol, Congaz and Uspenia formations (TR^{2-3}), Ladinian–Norian Cataloi Formation (TR^{3-1}), Norian–Carnian Alba Formation (TR^{3-2}), Early–Middle Jurassic Nalbant, Dunavat and Aiorman formations (J^{1-2}), Late Jurassic Carjelari, Baspunar and Carabair formations (J^3), and the bounding Peceneaga-Camena Fault are sealed by flat-lying Middle Cretaceous (K^2) clastics and carbonates in the Babadag 'Basin'. This sequence can be traced offshore into the subsurface, where structured Palaeozoic–Jurassic clastics and volcanics have been encountered in wells below an intra-Albian unconformity. Oil and gas fields are highlighted in green and red. The database used to constrain the sequence stratigraphic analysis comprised 14 wells and 6700 km 2D seismic (Boote 1994, 2014*a*, *b*). The key seismic lines are shown on the map.

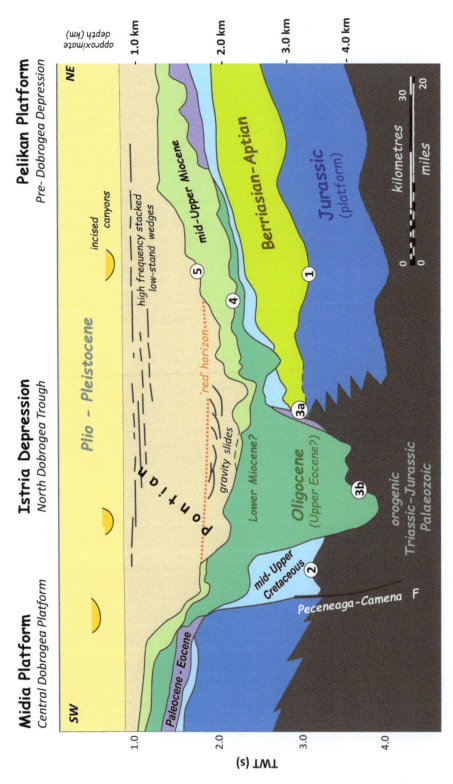

Fig. 3. Sequence architecture of the Istria Depression (schematic). Cretaceous–Cenozoic second-/third-order sequences and sequence boundaries are numbered: Sequence 1, west-facing (synrift?) Neocomian–Aptian wedge; Sequence 2, valley incision and east-facing mid-Upper Cretaceous? transgressive infill; Sequence 3, composite intra-Eocene valley incision and late Eocene?–Oligocene (Lower Miocene?) valley fill; Sequence 4, intra-Badenian (c. Langhian–Serravalian) valley incision/shelf-margin erosion and mid–late Miocene infill; Sequence 5, intra-Pontian (Messinian) shelf-margin erosion, lowstand wedge and late Pontian–Recent infill/healing sequence (interrupted by multiple fourth-order canyon-incision events). Numbers 1–5 refer to the basal unconformity bounding each sequence. TWT, two-way time.

Palaeozoic sedimentary and igneous rocks plunging SE to form the structural substrate of the Istria Depression offshore (Gradinaru 1984, 1988, 1995; Seghedi & Szakacs 1990; Stefan *et al.* 1992; Banks 1997; Banks & Robinson 1997; Seghedi 2001, 2012; Hippolyte 2002). Where observed in outcrop, it comprises two major zones, organized obliquely with respect to the bounding Sfantu-Gheorghe and Peceneaga-Camena faults (Fig. 2):

- Macin Zone: the western Macin Zone is dominated by Lower Cambrian–Vendian metamorphic basement, shallow-marine Palaeozoics and Permo-Carboniferous continental clastics, rhyolitic volcanic and volcano-clastics, and calc-alkaline intrusives. Marginal- to deep-marine Triassic–Jurassic rocks outcrop sporadically along the Peceneaga-Camena Fault (Cirjelari and Baspunar-Camena areas) and the entire zone is partially blanketed by undeformed late Albian (?)–Coniacian transgressive marine clastics and carbonates of the Babadag 'Basin'.
- Tulcea Zone: the eastern Tulcea Zone is separated from the Macin Zone by the high-angle Luncavita-Consul (LCF) reverse fault and comprises three thrust subzones – the Consul unit dominated by Scythian carbonates and clastics; the Niculitel unit with marked oceanic affinities; and the Tulcea unit comprised folded Triassic–Jurassic turbidites and deep-marine Palaeozoic clastics.

Its Palaeozoic evolution has been summarized by Seghedi (2012), after Mosar & Seghedi (1999), in a speculative plate tectonic reconstruction highlighting the Early Palaeozoic(?) closure of the 'Tornquist' seaway, and leading to Variscan collision and suturing of Baltica and East Moesia (Zeigler 1988). This was followed by Late Carboniferous–Early Permian wrenching and (? later Permian–) Early Triassic rifting across the Pre-Dobrogea and extending into the North Dobrogea during the early Scythian associated with continental fan-glomerates, fluvial clastics and shales, evaporites and local calc-alkaline intrusives, alkaline volcanic, and volcano-clastics. The North Dobrogea began to subside more rapidly in the late Scythian with continental red beds (Bogza and Tulcea Veche formations) and marginal-marine marls and clays (Ceamurlia Formation) passing laterally and upwards into shallow- to deeper-water slope carbonates and calci-turbidites (Somova Formation) confined to narrow troughs with tuffs, ignimbrites and basalt flows locally. By the Anisian, a carbonate platform (Murighiol Formation and equivalents) had extended across the Pre-Dobrogea and eastern North Dobrogea area, passing laterally into a westerly-deepening basin with deeper-water nodular and cherty carbonates (Agighiol Formation and equivalents) interbedded with thick pillow basalts (Niculitel Formation). These display a mid-ocean ridge basalt

(MORB)-like tholeiitic character thought to represent submarine extrusion during a period of rifting (Seghedi 2001; Tari 2005) and extension between the Moesian microplate and Eurasia. Carbonate sedimentation continued into the Ladinian (Uspenia, Agighiol and Murighiol formations) and may have extended across into the Central Dobrogea, only to be eroded in the later Triassic. This was interrupted by the sudden appearance of a deep-water clastic sequence (Alba Formation) of sandstones, marls and conglomeritic intercalations with Carnian-Ladinian-aged carbonate clasts, during the late Carnian–Norian. The Alba sequence is confined to the western Tulcea Zone passing laterally into marls (Cataloi Formation) and limestones (Congaz Formation and equivalents) towards the east, implying a westerly source. Carbonate deposition continued to dominate the North Dobrogea (Cataloi and Trestenic formations) into the Rhaetic, with more marginal clastic equivalents in the adjacent Pre-Dobrogea. By mid-Rhaetic time, clastic input had increased dramatically with deep-water turbidite fan systems building out across the North Dobrogea during the Early Jurassic (Fig. 4: Telita Formation, Denis Tepe Formation and Nalbant Beds). Although these systems may have been shed locally from the western 'Macin Range' (Seghedi 2001), facies organization (Gradinaru 1984) suggests that they more probably reflect erosional unroofing of the flanking Pre-Dobrogea and Moesian Platform highs (Seghedi & Oaie 1995), with earlier Mesozoic and Palaeozoic cover stripped away and recycled into the adjacent basin.

The Moesian microplate began to separate from central Europe in the mid-Jurassic, moving SE along the Tornquist–Teisseyre Lineament during the later Jurassic to leave the North Ligurian/'Magaru' oceanic salient in its wake (Tari 2005; Barrier & Vrielynck 2008). The tectonostratigraphic history of the North Dobrogea trough is poorly constrained at this time because of later erosion. Deep-water Bathonian sandstones have been described locally along the southern margin of the basin at Cirjelari (Aiorman Formation) and Camena (Mavila Goala Formation with Neoproterozoic clasts recycled from the adjacent Central Dobrogea) and the eastern part of the Tulcea Zone at Dunavatu (Dunavatu Formation) perhaps representing the erosional remnants of a once more-continuous deep-marine clastic system. Relative sea level began to rise during the Bathonian–Callovian, shutting off sand input into the basin and flooding the adjacent platforms with shallow-marine clastics and Oxfordian–Kimmeridgian (?Tithonian–Berriasian) platform carbonates (Casimcea Formation and equivalents) above. This transgression appears regional in character as time-equivalent shallow-marine clastics and carbonates are present across much of the Eastern

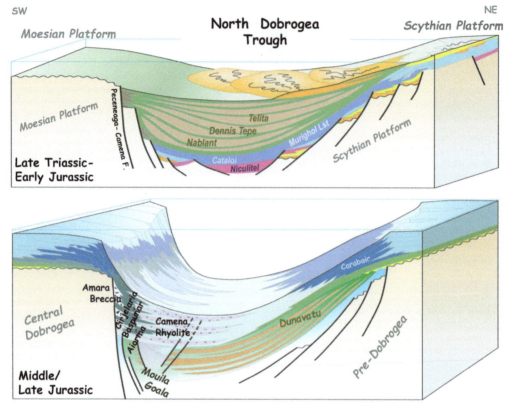

Fig. 4. Schematic model: Late Triassic–Early Jurassic and Middle–Late Jurassic depositional systems, North Dobrogea Trough. Lithostratigraphic descriptions and chronostratigraphy of the Late Triassic–Late Jurassic North Dobrogea basin-fill sequence are summarized from Gradinaru (1984, 1988) and Seghedi (2001).

European Platform, the Crimea, and the Western and Central Pontides, resting unconformably on sometimes highly structured Triassic and older rocks (Robinson & Kerusov 1997; Yilmaz *et al.* 1997; Ustaömer & Robertson 2010). Oxfordian–Kimmeridgian (?Tithonian) sediments within the North Dobrogea Trough are largely limited to small scattered outcrops along the Peceneaga-Camena Fault near Cirjelari and Camena. These represent a very complex fault-controlled facies assemblage described in some detail by Gradinaru (1984, 1988, 1995) comprising (Fig. 4):

- Amara Breccia: slope breccias of angular, cobble- to boulder-sized 'Schistes Vertes' clasts shed directly from the adjacent Central Dobrogea Platform;
- Cirjelari Formation (Amara and Sfinta facies): oo-clastic and bio-clastic calcarenites and calcirudites locally with dispersed to concentrated conglomerate interbeds comprising 'Schistes Vertes' (up to 7 m diameter), Palaeozoic marble and quartzite, red Permian Carapelit sands and silts, Upper Jurassic limestone and Cirjelari rhyolite

clasts interbedded with air-fall rhyolitic tuffs/tephra, and siliceous spongo-radiolarian muds passing laterally into;
- Baspunar Formation: deep-water siliceous muds (gaizes), marls, thin rhyolitic tuffs/tephra and quartz sandstone to micro-conglomerate interbeds with common bioclastic detritus. These in turn interfinger with;
- Camena Rhyolites: rhyolites, rhyolitic ignimbrites, tuffs and related hydroclastics capped by;
- Baspunar Spilites: spongo-radiolarian cherts, marly shales interlayered with pillow basalts displaying a tholeiitic character typical of intra-continental rifting; and
- Baspunar Mélange: comprising chaotically organized blocks of Lower Palaeozoic marble and quartzite, and Jurassic metabasic rocks (metabasalts and metadolerites) with a MORB geochemical signature suggestive of incipient oceanic crust (Gradinaru 1984, 1988).

Because of the linear organization of the outcrops and shallow-water aspect of the associated Cirjelari carbonate detritus, Gradinaru (1984, 1988) suggested that this assemblage represented deposition in a narrow (<5 km) transtensional graben system bounded by the Peceneaga-Camena Fault and a conjectural intra-basin 'Macin High'. Facies architecture with extra-formational Palaeozoic clasts and shallow-water carbonate detritus shed directly from the adjacent Central Dobrogea Platform certainly points to a deeper-water submarine fault-apron sequence backed up against the Peceneaga-Camena fault scarp. However, its regional extent is poorly constrained because of later erosion and may once have been part of a far more extensive deeper-water sequence (Oxfordian–Kimmeridgian), stretching across the North Dobrogea Trough to merge with the pelagic marls and limestones of the Carabair Formation (Oxfordian) in the eastern part of the Tulcea Zone at Dunavatu.

Tithonian–mid-Albian sediments are absent in the North Dobrogea, with flat-lying late Albian–Coniacian clastics and carbonates (Iancila and Dolojman formations) of the 'Babadag Basin' resting with pronounced unconformity on highly structured Jurassic, Triassic, Palaeozoic and basement rocks (Fig. 2). Because of the long gap in time, the age of deformation is necessarily ambiguous and controversial (Gradinaru 1984, 1988, 1995; Banks & Robinson 1997; Seghedi 2001; Hippolyte 2002). Tithonian and Berriasian platform carbonates are preserved locally in the Pre-Dobrogea and Birlad Depression to the north, subcropping an intra-Berriasian unconformity with a mixed evaporate–clastic sequence above (Swidrowska et al. 2008). Upper Kimmeridgian–?Tithonian platform carbonates are also preserved in the southern part of the Central Dobrogea (Gradinaru & Barbulescu 1994; Gradinaru et al. 1995) and South Dobrogea (Casimcea Formation and equivalents) unconformably overlain by Berriasian–Valanginian evaporites and carbonates. It is possible these erosional remnants may represent once more-continuous platform sequences flanking a still underfilled pre-tectonic North Dobrogea Trough.

The Dobrogea structural fabric observed onshore continues directly offshore where it underlies the Istria Depression and the flanking Midia and Pelikan platforms (Dinu et al. 2002). A highly deformed sequence observed seismically in the more proximal part of the depression (Line P93-16, see later) has been tagged locally by several wells. These encountered Jurassic, Triassic and Palaeozoic volcanic and sedimentary rocks similar to those of the North Dobrogea, while Jurassic platform carbonates have been found on the adjacent platforms comparable to those of the Central and Pre-Dobrogea. Because of the apparent stratigraphic and structural continuity with the adjacent onshore, the offshore basin frame of the Istria Depression is assumed to have shared a similar early Mesozoic tectonostratigraphic evolution comprising:

- Variscan consolidation in the Late Carboniferous followed by Kasimovian–Sakmarian intercontinental wrenching along the Tornquist–Teisseyre crustal boundary zone.
- Early Triassic transtensional (dextral?) rifting and volcanism followed by rapid subsidence in the Middle Triassic.
- Deep-water Middle Triassic carbonate deposition, giving way to easterly-sourced axial (?) fan systems in the later Triassic and Early Jurassic (Fig. 4).
- Separation of the Moesian microplate from the Bohemian margin of the European crustal collage towards the end of the Bathonian/early Callovian was followed by the opening of the North Ligurian/'Magaru' oceanic seaway and southwards movement of the Moesian microplate. Initial rift-margin uplift was followed by subsidence and transgression of the bounding Midia and Pelikan platforms with thin basal clastics passing up into platform carbonates.
- As Moesia moved southwards during the later Jurassic, stratigraphic projections from the onshore North Dobrogea suggest that the offshore Istria extension probably experienced increased crustal extension, subsidence and incipient development of oceanic crust.
- Regional projections suggest that the trough remained underfilled into the Cretaceous with significant residual topography.

Sequence 1: west-facing Berriasian–Aptian wedge

Berriasian–Aptian rocks are largely absent in the North Dobrogea and Istria Depression because of later erosion. Consequently, stratigraphic reconstructions of this period must rely on extrapolation from the adjacent platforms where the section is locally still preserved (Fig. 5).

Seismic, well and outcrop control in the onshore Dobrogea, Midia and Pelikan platforms all suggest the Istria–North Dobrogea Trough faced westwards towards the Magara oceanic embayment during the Berriasian–Aptian. The offshore depression appears to have remained underfilled at this time and its faulted margins were locally eroded. A near base Cretaceous unconformity can be traced north onto the Pelikan Platform where the Jurassic substrate is deeply incised (Fig. 5, R016-226) towards the west and NW. The corrugated erosional topography was gradually transgressed and onlapped by a younger terrestrial to shallow-marine clastic and carbonate sequence of Valanginian–Aptian age, thinning and lapping out towards the east. This sequence is

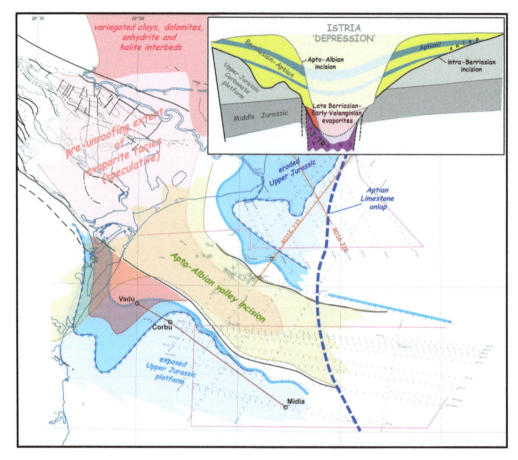

Fig. 5. Sequence 1 – Upper Berriasian–Valanginian? incision and infill sequence, Istria Depression. Sequence architecture and bounding unconformities of Sequence 1 are summarized on the map and schematic cross-section. Intra-Berriasian erosion of the (underfilled) Istria Depression and flanking highs incised a west-facing palaeo-valley system into the underlying Jurassic. The incision surface and eroded Jurassic substrate are illustrated by RO16-226 and highlighted in darker blue on the map. Evaporites encountered in the Vadu well (correlation section) and Pre-Dobrogea (Kongaz Beds after Swidrowska *et al.* 2008) are shown in red on the map. These may be part of a once more-laterally continuous lowstand drawdown sequence, confined by the incised topography, later transgressed and infilled by ?Valanginian–Aptian clastics and carbonates onlapping high ground towards the east (RO16-226 and map). Easterly tilting in late Aptian–Albian encouraged a new phase of erosional incision, evacuating the Istria Trough (RO16-233) and leaving only remnants of Sequence 1 on the flanking platforms.

truncated by pre-Albian erosion along the northern margin of the Istria 'Trough' (Fig. 5, R016-233), but an age-equivalent section was penetrated by the Corbu and Vadu wells on the Midia Platform. The Vadu borehole encountered a thick (*c.* 1200 m) sequence of stromatolitic dolomites, anhydrite and halite not present in any other well nearby. The interval is poorly dated, resting unconformably upon Kimmeridgian limestones with late Berriasian–early Valanginian clastics and thin carbonates directly above (Bancila *et al.* 1997). Rather ambiguous seismic data (Tambrea *et al.* 2002*a*, *b*), supported by a stratigraphic correlation between the Vadu and Corbu wells, suggests Vadu evaporites

infilled an erosional valley deeply incised into Late Jurassic (Kimmeridgian–Tithonian) platform carbonates, later buried by Neocomian–?early Aptian clastics. These are very similar to those on the Pelikan Platform and are probably the remnants of an originally more continuous wedge infilling the Istria Trough and extending onshore towards the North Ligurian seaway.

Regional significance of the Sequence 1 unconformity

Although no longer preserved in the Istria–North Dobrogea Trough because of Apto-Albian incision,

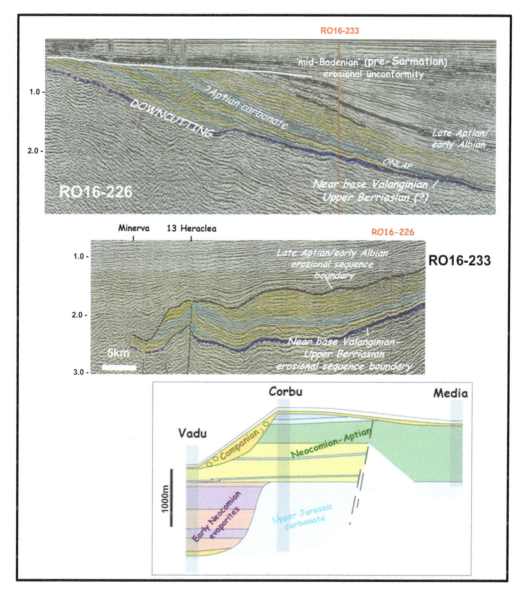

Fig. 5. *Continued.*

the erosional remnants still preserved offshore appear to be part of a more extensive sequence extending onshore across the North, Central and Pre-Dobrogea:

- Once more-extensive Tithonian and Berriasian platform carbonates preserved locally in the Pre-Dobrogea and Birlad Depression to the north are overlain unconformably by Kongaz Beds of basal Valanginian (plus possible uppermost Berriasian) age. These comprise variegated clays, anhydrite/gypsum, dolomite interbeds and halite

locally in the south (Swidrowska *et al.* 2008), passing up into variegated shales, sandstones and thin carbonates of the Hauterivian–Valanginian Chadyr-Lunga and Kormat beds. This Early Cretaceous sequence is partially truncated by an intra-Albian unconformity with mid to late Albian–Cenomanian clastics and carbonates directly above (Fig. 5).

- In addition, further south in the South Dobrogea region, locally preserved Upper Jurassic platform carbonates are unconformably overlain by a

180   D. R. D. BOOTE

sequence of evaporitic marls, gypsum and anhydrite (Amara Member, basal Cernavoda Formation), lithostratigraphically similar to the Kongaz Beds of the Pre-Dobrogea (Swidrowska *et al.* 2008). They are generally considered to be of late Berriasian–early Valanginian age, although there is some biostratigraphic evidence suggesting they might extend back to the late Tithonian (Avram *et al.* 1993). The evaporitic sequence passes up into late Berriasian–Valanginian marginal-marine carbonates (Poarta Alba, Medgidia and Alimanu members of the Cernavoda Formation) with late Barremian–Aptian calcareous sands and gravels (Ramadan and Gherghina formations) above.

The presence of early Cretaceous evaporites at three such widely spaced localities implies some general equivalence and, combined with the surprisingly thick halite section at Vadu, highlights the possibility of a laterally continuous lowstand evaporite sequence facing west towards the North Ligurian ('Magaru') seaway, partially confined by the underfilled Istria palaeotopography (Fig. 6). Where still preserved, these evaporitic facies pass up into transgressive late Valanginian–Aptian clastics and carbonates, onlapping rising ground towards the east. Stratigraphic and petrological constraints indicate this larger region was relatively positive at the time, traversed by fluvial systems extending south from the Ukrainian shield to the Tethyan

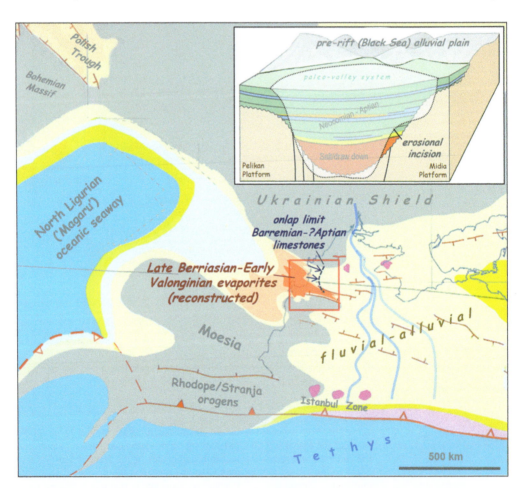

Fig. 6. Middle Aptian palaeostratigraphic reconstruction, Istria Depression and northern Tethyan margin. The remnant Berriasian–Aptian succession of the larger Istria region infills west-facing palaeotopography, onlapping rising ground towards the east. Its stratal geometry (summarized by the west-facing block diagram) appears to reflect subsidence of the mid–late Jurassic North Ligurian ('Magara') basin margin to the west and/or early pre-rift doming of the western Black Sea. The mid-Aptian palaeo-facies reconstruction is from Barrier & Vrielynck (2008) modified to accommodate the south-flowing drainage system implied by provenance analyses of Tethyan margin turbidite systems of the Central Pontides (Okay *et al.* 2014), suggesting a northerly source from the Ukrainian Shield.

margin (Fig. 6), prior to significant rifting and subsidence of the Western Black Sea (Okay *et al.* 2014). The original stratigraphic continuity, implied by this reconstruction, suggests the sequence predates the 'late Cimmerian' deformation responsible for the highly structured character of the Palaeozoic–Jurassic succession in the North Dobrogea and adjacent offshore. Instead, this event may be related to the opening of the western Black Sea back-arc basin (Gradinaru 1984, 1988). The relationship between transpressional deformation of the North Dobrogea–Istria Trough and Black Sea extension is unclear, but may have occurred in response to an initial oblique rather than orthogonal direction of opening to the east.

Sequence 2: Middle–Late Cretaceous (late Aptian?–early Paleocene) east-facing valley incision (evacuation) and infill

The Istria Depression was deeply incised in the late Aptian?–early Albian and subsequently infilled and buried by a later Cretaceous transgressive system before renewed incision during the Eocene (Fig. 7).

Late Aptian–early Albian incised valley unconformity

The major mid-Cretaceous unconformity observed in the Babadag Basin and adjacent offshore shelf marks a dramatic shift from Early Cretaceous depositional systems facing west towards the North Ligurian oceanic embayment to east-facing systems fringing the western Black Sea during the later Cretaceous and Cenozoic. Regional projections suggest the fault-bounded Istria Trough was erosionally evacuated some time in the late Aptian–early Albian. The incised unconformity can be observed truncating a highly structured Palaeozoic–Triassic–Jurassic succession in the proximal part of the trough and merging with erosional fault scarps along its southern and northern flanks (P93-16, see later; Fig. 7, R017-13/R016-233). The deeply incised valley topography of the trough was subsequently onlapped and infilled by a thick sequence of transgressive (?late Aptian) Albian–Late Cretaceous sands, shales and limestones, and then partially re-incised during the Eocene.

The unconformity can be traced onshore where undeformed flat-lying ?late Albian (Vraconian)–Cenomanian sandstones, limestones and marls of the Iancila Formation (Gradinaru 1995) overlie highly deformed Macine and Tulcea zones in the Babadag area of the North Dobrogea (Fig. 2). These extend southwards locally across the Peceneaga-Camena Fault and onto the Central Dobrogea to rest upon Precambrian basement. The unconformity appears to continue north into the Pre-Dobrogea where mid-Upper Albian sediments rest directly on Aptian and older Cretaceous sediments (Swidrowska *et al.* 2008). A similar transgressive facies (Cochirleni Formation) of early Albian to possible latest Aptian age overlaps earlier Cretaceous in the South Dobrogea (Avram *et al.* 1993) and south into eastern Bulgaria (Harbury & Cohen 1997).

Regional significance of the Sequence 2 unconformity

The unconformity appears to reflect significant regional uplift and erosion associated with 'late Cimmerian' deformation in the North Dobrogea–Istria Trough and the opening of the Western Black Sea back-arc basin (Fig. 8). Undercompacted Lower Cretaceous (and older) sediment fill within the Istria Trough probably become unstable at this time and, perhaps encouraged by plastic basal evaporite layers, collapsed and slid away to the east. Significant disagreement about the precise time of opening remains. Several kinematic models have been proposed to describe its evolution. The more comprehensive interpretations include those by Banks & Robinson (1997), Kazmin *et al.* (2000), Okay & Gorur (2007), Barrier & Vrielynck (2008) and Krezsek *et al.* (2017) variously suggesting initial crustal separation in the late Albian–Cenomanian, late Albian (Fig. 7), Santonian–Campanian and Cenomanian.

The stratigraphic evidence for all these several estimates come from outcrops along the coastal margins of the Black Sea supported by limited offshore well control and seismic. A number of key onshore sections have been described by Yudin & Kitchka (2010), Harbury & Cohen (1997) and Hippolyte *et al.* (2010).

Southern Crimea. A striking Early Cretaceous unconformity is well exposed along the northern flank of the Crimean Mountains (Muratov 1969; Yudin 2009) separating a complex Triassic–Jurassic (–Valanginian) succession from gently dipping early-mid Cretaceous–Neogene sediments above. Field observations suggest the unconformity surface is deeply incised, with quite pronounced erosional valley systems locally. The age of incision is weakly constrained because of limited and sometimes conflicting biostratigraphic control with poorly dated basal sandstones (?Hauterivian/Upper Albian) and Albian marine shales both resting directly upon the incised unconformity locally (Popadyuk *et al.* 2010). This valley system is assumed to have faced south towards the opening Black Sea Basin during the Cretaceous only to be tilted northwards in response to late uplift of the Crimean Mountains.

SE Moesian Platform. A pronounced late Aptian–Albian unconformity has been described along

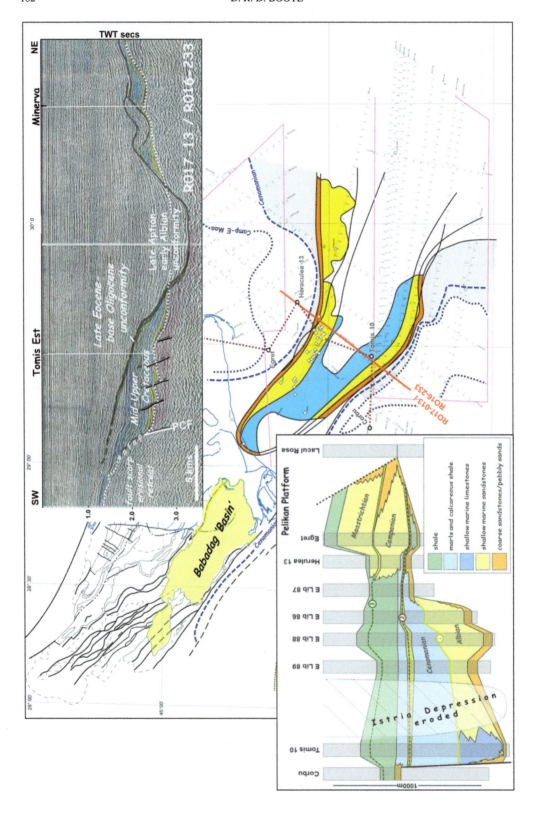

the SE margin of the Moesian Platform flanking the Kamchia Depression (Balkanide Foredeep) by Harbury & Cohen (1997). They noted that erosion increased east and NE, with older rocks subcropping the unconformity towards the Black Sea margin, transgressed and onlapped from the south (and east?) by shallow-marine sands and carbonates of Cenomanian–Campanian age.

Central Pontides. A Cretaceous–Early Tertiary syn- to post-rift sequence is exposed in scattered outcrops along the coastal margin of the central Black Sea, and Zonguldak and Ulus basins of northern Turkey. This has been described in some detail by Gorur (1997), Yilmaz *et al.* (1997), Tuysuz (1999) and others, but Hippolyte *et al.* (2010) have since revised its chronostratigraphy very significantly. The succession comprises a south-facing deep-water Barremian–early Albian sequence (Caglayan Formation) locally capped unconformably (Tuysuz 1999) by a shallow-water clastic unit (Cemaller Formation) of late Albian age (Hippolyte *et al.* 2010) with reworked Jurassic limestone clasts (olistoliths). This interval is overlain by transgressive marine Coniacian sandstones with later Cretaceous–early Cenozoic pelagic limestones, deep-water clastics and volcanoclastics above. The Coniacian transgression is generally considered to reflect the transition from rifting to backarc spreading in the western Black Sea (Hippolyte *et al.* 2010). However, the intra-Albian unconformity may be more significant than previously assumed. Recent analysis suggests that the Caglayan turbidite system was sourced from the north and so must have preceded Black Sea rifting (Okay *et al.* 2014). The unconformity directly above, with a dramatic switch from deep- to shallow-water sediments, may therefore represent its initial subsidence and collapse, deflecting more regional depositional systems to the north.

Although information from these several marginal successions is enough to constrain the age of rifting and break-up to some time in the 'mid' Cretaceous, it is too imprecise and conflicting to provide anything more exact. The Istria Depression sequence provides additional evidence refining this estimate, with slow post-break-up subsidence recorded by the gradual onlap and burial of the incised topography and flanking platforms during the later Cretaceous and early Cenozoic.

Sequence 3: composite intra-Eocene incised valley and Eocene–Miocene valley-fill sequence

A quite spectacular composite valley system was deeply incised into the Romanian Black Sea shelf during the Eocene, partially constrained by the underlying Istria Trough and adjacent platforms, and subsequently infilled by Oligocene shales (Fig. 9).

Composite intra-Eocene valley incision and fill

Two, or possibly three, intra-Eocene incision events can be distinguished in the inner part of the shelf, but merge into one deep valley further outboard (Fig. 9, P93-16). These comprise:

- an initial incision event eroding down through the mid-Upper Cretaceous into the basin frame below and healed by an onlapping 'middle– upper' Eocene sequence;
- a second phase of erosion incising through the earlier valley fill further outboard with an onlapping 'upper' Eocene sequence resting directly on the Cretaceous;
- a final phase of incision, cutting down through the earlier valley-fill sequences into the pre-rift section. This was later filled by thick Oligocene– early Miocene shales, bounded by a mid-Badenian/Sarmatian unconformity above.

Mid/late Eocene incision and valley-fill sequences. Only the proximal part of the earlier fill sequences are preserved because of later near top Eocene/base Oligocene erosion. These are represented by more basinal facies in the main intra-Eocene valley system and truncated shelf facies on the adjacent platforms (Tambrea *et al.* 2002*b*; Dinu *et al.* 2005*a*). Stratal geometries of the fill facies are locally very complex with multiple cut and fill packages amalgamated together into two intra-Eocene sequences of re-sedimented skeletal packstones–mudstones and

---

Fig. 7. Sequence 2 – Mid–late Albian valley incision and infill sequence, Istria Depression. Evacuation of the Istria Trough in late Aptian?–early Albian time was followed by subsidence and flooding from the east. Transgressive marine clastics and carbonates began to infill the incised topography (well correlation section and RO17-13/ RO16-233) and extend west into the North Dobrogrea (latest Albian–Cenomanian Iancila Formation). These pass up into deeper-marine marls and shales onlapping the valley walls and flooding the adjacent platforms, interrupted by high-frequency delta systems prograding SE across the Pelikan Platform during the Campanian and Maastrichtian. Only the more prominent parts of the adjacent platforms remained exposed by the end of the Cretaceous. The westerly (onshore) extent of the sequence is unknown because of subsequent unroofing. Cenomanian and Caampanian–Maastrichtian onlap limits are highlighted by blue dashed lines. Areas eroded by later Eocene incision are uncoloured (time lines 1, 2 and 3 on the section = tectono-stratigraphic maps illustrated on Figure 8).

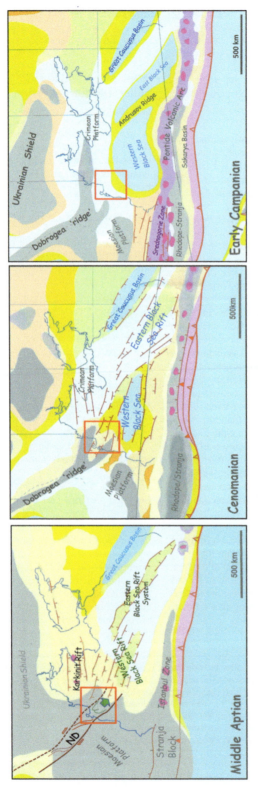

Fig. 8. Mid–Late Cretaceous tectonostratigraphic evolution of the Istria Depression and northern Tethyan margin. Late Aptian–early Albian paroxysmal deformation of the North Dobrogea has been inferred from outcropping and subsurface stratigraphy of the coastal Dobrogea onshore and adjacent offshore, causally linked to rifting and opening of the western Black Sea. The time of opening remains controversial (Krezsek *et al.* 2017). In this reconstruction adapted from Barrier & Vrielynck (2008), back-arc spreading in the western Black Sea and basin margin subsidence is tentatively assumed to have started during the late Albian and Campanian (or later) in the eastern Black Sea. This finds support from the well-dated Albian transgressive sequence in the Istria Depression.

marls. Further outboard, Eocene argillaceous lime-stones are locally preserved, draped over a palaeo-topographical high at East Lebada with thick Oligocene valley-fill shales above. Although gener-ally impermeable, the crestal section immediately below the base Oligocene unconformity is micro-karstified and porous (Boote 1994), testifying to a brief period of subaerial exposure at the time of valley incision.

Near top Eocene/base Oligocene incision and valley-fill sequence. The following near top Eocene/base Oligocene incision cuts down increas-ingly deeply towards the east, removing most of the early Eocene valley fill and exposing flat-lying Creta-ceous along the valley walls (Fig. 9, R017-13/R016-233). The initial fill facies within the valley thalweg is characterized by a distinctive high-amplitude reflector package suggestive of a hetero-lithic sand–shale lag (Fig. 9). This can be traced west and updip past the West Lebada structure, where deep-water subarkosic sands with reworked carbon-ate detritus, have been encountered along the flanks of a mid–upper Eocene palaeotopographical high (Ionescu *et al.* 2002; Tambrea *et al.* 2002a, b). Although there is some uncertainty about their age, regional constraints and seismic correlation with the thalweg package favours the late Eocene/basal Oli-gocene, marking the transition from valley incision to infill. The heterogeneous thalweg facies pass up into a thick, seismically more homogeneous silt and shale succession dominated by flat-lying strata onlap-ping the valley walls, consistent with deep-water dep-osition from suspension and dilute turbidite flows. The upper part of this silty shale sequence spills out of the incised valley onto the eroded platform mar-gins and is finally terminated by a mid-Miocene (Badenian–Sarmatian) base-level fall.

Regional significance of the Sequence 3 unconfor-mity. Stratal geometries of the incision surfaces and fill architecture of the incised valley system highlight its sequential development, with progressively deeper down-cutting towards the Black Sea Basin. Its position on the shelf appears to be constrained by the underlying structural grain, with the valley thalweg centred on the Istria Trough, bounded by the Pelikan and Midia platforms. Frequent large-scale base-level alternations appear unlikely, and the earlier incision events were probably driven by retrograde mass wasting and slumping in a marine environment, triggered by relatively minor base-level falls. Together, the three incision events appear to reflect a fluctuating, but falling, base level, culmi-nating with a dramatic fall of some 2000–2500 m in latest Eocene–near base Oligocene.

A general coincidence of timing suggests this composite event may reflect the closure of the Neotethys, and the initial collision between the Afro-Arabian Plate and the Eurasian crustal collage, variously dated as 'late Eocene' (Hempton 1987), 'late Eocene–Oligocene' (Yilmaz 1993), 'post-Eocene/probably Oligocene' (Robertson *et al.* 2006) and 'earliest Oligocene' (Vincent *et al.* 2007). The Para-tethyan seaway formed at this time (Fig. 10) and oro-genic uplift associated with the collision may have briefly isolated it from the global ocean, encouraging rapid drawdown of the palaeo-Black Sea Basin. This was marked by a rapid switch from carbonate-to clastic-dominated sedimentation, canyon/valley incision and shelf-margin erosion around the proto-Black Sea at that time, implying a significant base-level fall. Tari *et al.* (2013) correlated this event with a lowstand prograding wedge observed on ION SeaSPAN seismic data in the central part of the basin. They argued this wedge indicated deposition in water depths of only 60–80 m and con-sequently must reflect a base-level fall of 2000–3000 m.

Erosional incision of the Istria Trough may also have been encouraged by Balkanide orogenesis (Fig. 10) (Munteanu *et al.* 2012). The north-verging Balkanide thrust sequence and foreland plunges east towards the Black Sea coast and offshore where the leading thrust wedge and retro-arc foredeep succes-sion are still preserved in the Kamchia Trough. Excellent seismic, well and outcrop control confirm that the proximal thrust wedge was emplaced onto the platform margin in the mid-Eocene, roughly coincident with the time of the initial intra-Eocene valley incision to the north (Harbury & Cohen 1997; Sinclair *et al.* 1997). The age of thrust emplacement further west is less well constrained because of late unroofing (Sinclair *et al.* 1997; Tari *et al.* 2011), but regional constraints suggest it may have been diachronous, extending up into the later Eocene and early Oligocene further along the platform margin. Thrust loading at this time may have tilted the platform and acted as a tectonic trig-ger. With falling base level, the undercompacted sediment fill of the Istria Depression might have become susceptible to retrograde slumping, enhanced by fluvial erosion cutting down <2000 m into the earlier Cenozoic–Cretaceous substrate. The subse-quent base-level rise appears to have been quite rapid, as the incised topography is largely unaffected by transgressive shoreface erosion and the valley system was subsequently filled by marine shales deposited in a deeper-water environment.

Sequence 4: Badenian–Meotian (Middle–Late Miocene) shelf-margin erosion and healing

The Dobrogea shelf suffered yet another episode of incision during the mid-Miocene. Intra-Badenian

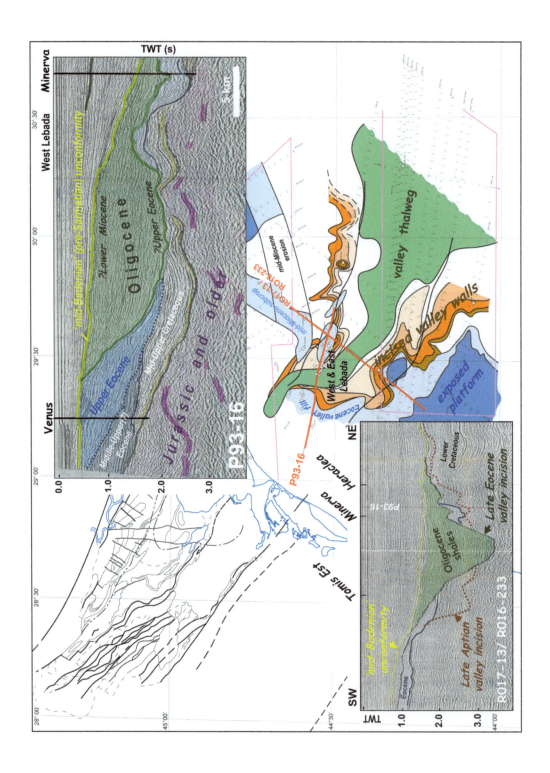

(late Langhian) erosion cut down through the underlying Tertiary, to leave a shallow scallop-shaped unconformity, directly above the Istria Trough and extending laterally along the shelf margin (Fig. 11). Part of the unconformity and overlying sequence was re-incised during the Pontian (Fig. 11, R017-01/R016-227 & R017-25/R016-237). However, enough remains to suggest a rather scallop-shaped erosional surface comprising several entrenched channels separated by low ridges onlapped and infilled by flat-lying muds and local channel sands. Subaerial exposure, dissolution and karstification of the southern platform is implied by the rather wormy reflector character of the section immediately underlying the unconformity surface (Fig. 11, R017-25). Further offshore, a lenticular mounded package, with disorganized high-amplitude reflector segments floating in an otherwise seismically diffuse matrix, appears to rest directly upon the correlative conformity (Fig. 11, R017-20). These high-amplitude 'segments' correlate with large blocks of displaced Eocene limestone and Oligocene sandstone and shale encountered by the Ovidiu Est well, embedded in deep-water mid-Miocene shales. This package represents an extensive debris-flow apron shed from the adjacent shelf, triggered by a significant base-level fall.

Regional significance of the Sequence 4 unconformity

Biostratigraphic control is too imprecise to assign anything more than a general 'Badenian–Sarmatian' (Meotian?) age, both to the sequence overlying the unconformity surface along the eroded shelf margin and the mudstone matrix of the deeper-water debris flow system further outboard. However, the unconformity can be traced west across the inner part of the shelf, where it erosionally truncates an easterly-dipping Early Cenozoic–retaceous succession with Sarmatian sediments directly above (Fig. 5, R016-226; Fig. 9, P93-16), appearing to mirror a similar truncational unconformity observed along the western flank of the Dobrogea (Paraschiv 1979; Pene et al. 2009). This palaeo-high appears to have developed as a peripheral arch or bulge following the advance of a Carpathian orogenic front up the North Ligurian embayment and collision with the East European Platform (Fig. 12). The foreland basin of the advancing orogen was briefly isolated from the global ocean, experiencing a significant sea-level fall(s) associated with mid-Badenian (14.2–13.6 Ma) drawdown evaporites (Wielician crisis after Baldi 2006; Oszczypko et al. 2006; Peryt 2006, after Dudek et al. 2004; Harzhauser & Piller 2007), shelf-margin erosion, and incised palaeo-valley systems in Poland (Krzywiec et al. 2009), Ukraine (Oszczypko et al. 2006) and the 'West Bucharest' region of the Moesian Platform of Romania (Paraschiv 1979). A correlative mid-Badenian (late Chorakian–Karaganian) base-level fall has also been observed in outcrop along the southern margin of the Black Sea. Gorur et al. (2000) traced it across Eastern Paratethys, where it is associated with basin margin emergence, de-salinization and the appearance of endemic brackish–marine faunas (Gorur et al. 2000), followed by marine floods in the late Badenian and again in the early Sarmatian, becoming increasingly brackish with time.

While significant ambiguity remains about the precise age of these several events, there is enough to suggest that their general equivalence reflects a paroxysmal phase of regional compression, crustal consolidation and isolation of both the Central and Eastern Paratethys seaways during mid-Badenian time. The possible coincidence in timing implies the (relative) base-level fall and erosion observed offshore Dobrogea may reflect both local and regional tectonism. Regional emergence and

---

Fig. 9. Sequence 3 – Middle–Late Eocene composite valley incision and infill sequence, Istria Depression. The mid–late Eocene stratigraphy of the Istria Depression and flanking platforms records several cut and fill events culminating with deep incision in late Eocene–basal Oligocene time. An axial view of the proximal-fill sequence is provided by line P93-16, highlighting three incision events, tentatively dated as: (1) base 'middle–late' Eocene with subcropping Upper Cretaceous; (2) base 'late' Eocene; and (3) near top Eocene? bounded by a mid-Badenian (pre-Sarmatian) unconformity above. Stratal geometries demonstrate that each incision event was healed by complex infill sequences dominated by basinal lime mudstones and marls with shelf facies equivalents on the adjacent platforms, before re-incision by the next event. The transverse seismic profile (R017-13/R016-233) highlights the deeply incised geometry of the final event, cutting down more deeply towards the east and testifying to a major fall in base level. This final event removed much of the earlier intra-Eocene fill sequence in more distal parts of the valley system, exposing Middle–Upper Cretaceous strata along the valley walls (highlighted in light brown). The valley thalweg fill displays a distinctive high-amplitude reflector character suggestive of very heterogenous interbedded lithologies (dark green on the map and line R017-13/R016-233). This package was intersected at West Lebada where it is represented by mixed clastics and carbonates. The incised valley was subsequently filled by a thick (2000+ m) Oligocene–early Miocene shale/silty shale-dominated sequence. Reflector geometries of the fill sequence display a generally consistent onlapping character (with local chaotic slump packages) indicative of deeper-water marine deposition (R017-13/R016-233).

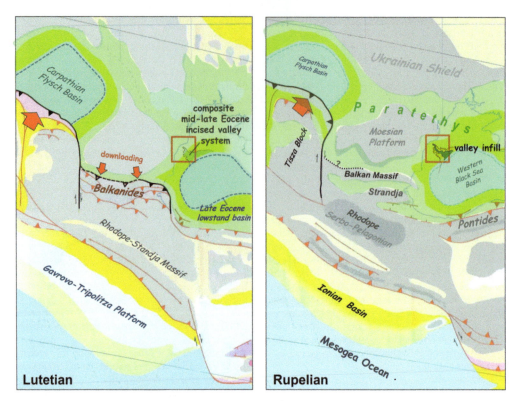

Fig. 10. Regional tectonic controls on Eo-Oligocene sequence architecture, Istria Depression and 'birth' of Paratethys. Late Cretaceous–Eocene collision of the Pontide and Tauride–Anatolide microplates and closure of 'northern' Neotethys (Sengor & Yilmaz 1981; Okay & Sahinturk 1997; Gorur & Tuysuz 2001; Robertson & Ustaömer 2004) was followed by closure of the 'southern Neotethyan' seaway and initial collison between the north-moving Arabian Plate and the southern accretionary margin of Eurasia some time during the later Eocene–early Oligocene (Hempton 1987; Yilmaz 1993; Robertson *et al.* 2006). These several suturing events led to isolation of the Paratethyan seaway with a brief but dramatic base-level fall (Tari *et al.* 2013) followed by a switch from carbonate- to clastic-dominated depositional systems and widespread appearance of anoxic ('Maikop' and equivalent) shales (Schultz *et al.* 2005; Vincent *et al.* 2007). Progressive Cenozoic crustal consolidation around the margins of the Moesian Platform started with mid-Eocene Balkanide orogenesis followed by the northerly advance of the Alcapa/ Tisza terrains up the North Ligurian seaway and collision with the East European Platform in the middle Miocene. While any correlation between Mid-Eocene Balkanide thrust emplacement (Harbury & Cohen 1997; Sinclair *et al.* 1997) and initiation of valley incision across the Romanian shelf must be speculative, the apparent coincidence in timing suggests a possible causal link between the two, with thrust loading and depression/tilting of the southern Moesian Platform margin enhancing the effect of the Paratethyan base-level fall by encouraging local sediment instability, headward erosion and valley incision in the Istria Depression (reconstruction from Barrier & Vrielynck 2008).

exposure of the Romanian shelf could have been enhanced by local uplift and easterly tilting, encouraging widespread gravitational collapse and slumping along the shelf margin. The subsequent transgression across the Dobrogea Arch reflects intra-Sarmatian subsidence, perhaps in response to a pause in the easterly advance of the Carpathian Orogen. The eroded shelf margin was buried and healed during the later Miocene by a transgressive sequence set. This was partially incised in mid–late Pontian (intra-Messinian), but is locally preserved updip, where at least three high-frequency progradational sequences

can be distinguished backstepping towards the west separated by truncational unconformities and their correlative conformity surfaces (Boote 1994; Boote 2014*a*, *b*)

Sequence 5: Pontian ('Messinian') shelf-margin erosion and healing

The Dobrogea shelf suffered a further period of mass wastage and erosion some time during the Pontian (Fig. 13), described in some detail recently by Krezsek *et al.* (2016*a*, *b*). Intra-Pontian erosion appears to be confined to the mid–outer part of the shelf, with little evidence of inboard canyon incision. The erosional surface (basal intra-Pontian unconformity (bIPu)) is steep and scallop-shaped, terminating within the Pontian section to the west and cutting down through older Miocene sediments towards the east, continuing laterally along the Miocene shelf margin into the western Ukraine. The eroded margin was subsequently healed by a complex lowstand systems tract (Fig. 13, P93-16B/16C/16), which comprised two sequences separated by a planar truncational unconformity (IPrvs). The lower sequence (Pontian Unit B) onlaps and is confined to the deeper part of the erosional scallop. Internal stratal geometries are complex. Poorly imaged clinoforms can be observed in some places, but it is dominated by two distinctive erosionally truncated packages of rotated, folded and faulted strata, interpreted as large (Midia and Rapsodia) gravity slides (Fig. 13). Stratification/bed cohesion is preserved in the proximal parts of both slides, with sometimes extreme bed rotation and fold rippling on shallow listric faults, soleing out on a lower Miocene–intra Oligocene detachment surface. However, primary stratification is less obvious in the more distal part of the central (Midia) slide, with scattered chaotic blocks of stratified sediment (higher-amplitude reflector segments) locally floating within an otherwise seismically homogeneous matrix.

The highly structured Pontian Unit B is truncated by a strikingly uniform and regular erosional surface (intra-Pontian ravinement surface (IPrs)) which merges updip with the bIPu. Its regular peneplaned character is characteristic of a shoreface ravinement surface, eroded during a comparatively slow base-level rise. This is overlain by a complex high-frequency lowstand prograding wedge or apron sequence (Pontian Unit A), comprising thinly developed topset strata and a thick clinoforming package incised by multiple intercutting channels/canyons, passing laterally into aggrading fan–channel complexes. Where penetrated by wells, the interval appears to be dominated by silts and shales, although the frequency of erosional channels suggests efficient sand bypass, with the possibility of outer-shelf fan systems further to the east. Clinoform geometries indicate that initial progradation was towards the south (prograding lowstand wedge 1 (PLSW 1)), later changing to the SE (prograding lowstand wedge 2 (PLSW 2)). A similar southerly progradation direction can be observed in the equivalent late Miocene (–early Pliocene?) interval, on trend to the east in western Ukraine (Khriachtchevskaia *et al.* 2010). The generally consistent progradation direction towards the south supports a northerly sediment source. The clinoforming sequence passes laterally and upwards into uniformly flat-lying strata typical of aggrading shelf sedimentation. This interval was incised by at least three distinct groups of canyon systems. The first of these incision events ('C' system) represents the upper sequence boundary of Pontian Unit A.

Intra-Pontian base-level fall – local stratigraphic models

During the final Oligo-Miocene closure of Tethys, the residual Paratethyan seaway was fragmented into several semi-isolated basins. These evolved into variably marine–brackish and freshwater lakes often with endemic faunas (Jones & Simmons 1997; Harzhauser & Piller 2007; Jipa & Olariu 2009). Several local biostratigraphic schemes were developed to accommodate these differences, but inter-basin correlations have proved difficult, imprecise and often ambiguous. Over the last few years, increasing amounts of biostratigraphic, palynological and magnetostratigraphic data (Vasiliev *et al.* 2005, 2011) have provided more robust constraints, but such correlations remain tenuous. This is especially apparent for the stratigraphically complex Pontian sequence of the Romanian shelf with at least four conflicting chronostratigraphic schemes proposed over the last 10 years:

- Dinu (2009) and Dinu *et al.* (2005*b*) provided a rather generalized description of the offshore Neogene succession illustrating the Pontian sequence architecture on supporting seismic illustrations with a tentative age of between ±5.8 and 4.8 Ma.
- Gillet *et al.* (2007) focused upon the more proximal part of the Pontian incision event and upper clinoforming unit, and apparently did not observe the intra-Pontian ravinement surface or lower Pontian B Unit. They correlated the basal intra-Pontian unconformity with the Messinian erosional event of the Mediterranean with an estimated age of ±6.0–5.5 Ma (after Snel *et al.* 2006).
- Konerding *et al.* (2010) presented a very detailed description of Pontian sequence architecture subdividing it into four unconformity bounded packages suggesting a tentative correlation between their IPU2 unconformity (=intra-Pontian ravinement surface) and the Messinian erosional event with an approximate age of ±6.3–6.0 Ma.
- More recently, Munteanu *et al.* (2012) provided a well-constrained description of the Pontian, highlighting an intra-Pontian unconformity surface (SB3 = intra-Pontian ravinement surface) representing a possible base-level fall of 1.6–2.0 km at approximately 5.6 Ma, directly correlative with the Messinian drawdown event of the Mediterranean.

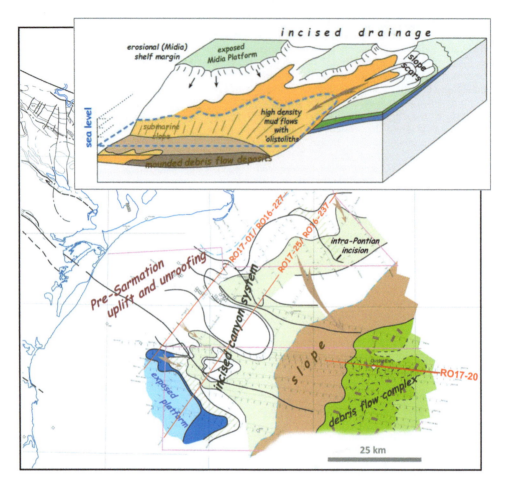

Fig. 11. Sequence 4 – 'Badenian–Sarmatian' unconformity and lowstand systems tract. A significant intra-'Badenian–Sarmatian' base-level fall is suggested by inboard uplift (Fig. 4, RO16-226; Fig. 8, P93-16) shelf-margin erosion, incised valleys and canyons. The inboard transect (R017-01/R016-237) highlights a shallow incised valley system overlying the Istria Trough, onlapped and infilled by flat-lying sediment. Further outboard (R017-25/R016-227) the unconformity surface has been more deeply incised by later intra-Pontian erosion, removing evidence of its original seawards extent. The 'wormy' seismic reflector character observed on the Midia Platform margin subcropping the unconformity (R017-25) may reflect subaerial exposure and karstification. Further outboard, a seismically homogeneous mound form has been observed with short irregularly orientated high-amplitude reflectors (R017-20). This package was intersected by the Ovidiu well, where it comprised ?Badenian–Sarmatian mudstones with shallow-water Eocene, Oligocene sandstone and shale(?) clasts of varying size, representing a debris-flow complex shed from the inboard margin during a period of subaerial exposure. The lowstand systems tract was subsequently buried by several high-frequency back-stepping shelf-margin sequences, partially truncated by later intra-Pontian erosion.

Because of inter-basin chronostratigraphic ambiguity, correlation between the Pontian lowstand sequence, offshore Romania and the Messinian salinity crisis of the Mediterranean is poorly constrained. However, recent stratigraphic descriptions of key sections supported by new biostratigraphic data provide a more robust framework with which to assess the relationship between the dramatic late Miocene sea-level fall and rise in the Mediterranean and Euxinic (Black Sea) basins (Figs 14 & 15).

Regional significance of the Sequence 5 base-level fall

The first evidence of a significant late Miocene base-level fall in the Black Sea Basin came from two DSDP holes (DSDP sites 380 and 381) drilled offshore Turkey in 1975 (Fig. 14, Section 3 & Fig. 15). A thin interval of pebbly mudstone with clasts of shallow-water supratidal stromatolitic dolomites (Unit IVd) was encountered at DSDP Site 380

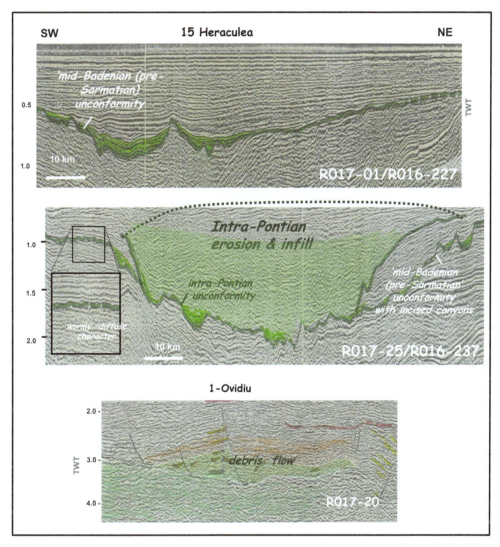

Fig. 11. *Continued.*

with deeper-water brackish–marine (>22‰) laminated aragonitic shale (IVc) above, and laminated marine carbonate and marls (Unit IVe) below (Ross *et al.* 1978; Stoffers & Müller 1979). Hsu & Giovanoli (1979) first proposed this as evidence of a dramatic fall in the Black Sea water level (c. 1600 m below global sea level) towards the end of the Messinian, interpreting the overlying aragonitic layer (Unit Vc) as marine transgressive facies recording the re-flooding of the Black Sea Basin at the beginning of the Pliocene ('Zanclean Flood').

Since that time the deep-desiccated basin model for the Messinian salinity crisis and stratigraphic correlations between the Black Sea and Mediterranean base-level fall events have evolved very significantly. More recently, Bache *et al.* (2012) synthesized earlier outcrop-based interpretations (Hsu *et al.* 1973; Clauzon *et al.* 1996, 2005; Lofi *et al.* 2005, 2011; Pierre *et al.* 2006; Roveri *et al.* 2008; Bache *et al.* 2009; Krijgsman *et al.* 2010 and others) around the Mediterranean Basin with offshore seismic data to provide a robust interpretation of the Messinian salinity crisis (Fig. 14, Section 1 & Fig. 15). This interpretation provides a critical model for stratigraphic and chronological correlations with the Black Sea base-level fall event. Their analysis supported and amplified the

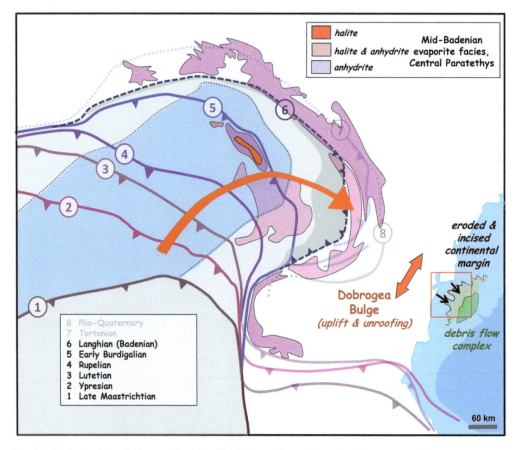

Fig. 12. 'Badenian–Sarmatian' event, Carpathian Foreland and Central Paratethys. The presence of lowstand evaporites in the Carpathian foredeep and Panonnian Basin (highlighted schematically after Baldi 2006; Oszczypko *et al.* 2006; Peryt 2006; Harzhauser & Piller 2007), basin-margin emergence and valley incision flanking the Central and Eastern Paratethyan basins (Paraschiv 1979; Gorur *et al.* 2000; Oszczypko *et al.* 2006; Krzywiec *et al.* 2009), and desalinization across Eastern Paratethys (Gorur *et al.* 2000) suggest the Paratethyan seaway experienced a significant base-level fall during the Badenian. Its impact upon the Romanian shelf margin may have been enhanced by crustal loading during the advance of the Eastern Carpathians (Hippolyte *et al.* 1999). This was responsible for rapid subsidence of the Carpathian foredeep and complementary uplift of the Dobrogea Arch or peripheral bulge, where pronounced pre-Sarmatian erosional truncation is observed both along its eastern (R016-226; P93-16) and western flank (Paraschiv 1979; Stefanescu *et al.* 2000; Pene *et al.* 2009). The coincidence in timing between the Sequence 4 unconformity and these more regional events suggests a causal link. Uplift and unroofing of the Dobrogea Arch may have tilted the Romanian shelf seawards, encouraging sediment instability, mass wastage and erosion of the shelf margin during falling base level. Cenozoic orogenic evolution of the Carpathians was synthesized from Harbury & Cohen (1997), Sinclair *et al.* (1997), Linzer *et al.* (1998) and Hippolyte (2002).

two-phase model proposed earlier by Clauzon *et al.* (1996) comprising:

- an initial modest sea-level fall (<100 m) associated with basin-margin evaporites (5.96–*c.* 5.6 Ma) followed by a brief rise (Lago Mare 1);
- and a dramatic base-level fall of 1500–00 m at 5.6 Ma.

This fall event was associated with severe basin-margin erosion, canyon incision and deposition

of thick clastic aprons (detrital unit (DU)) passing laterally and upwards into basinal evaporites. Re-flooding occurred in three distinct steps:

- an initial slow rise between approximately 5.53 and 5.46 Ma from >1100 to 600 m, responsible for a remarkably even peneplaned ravinement surface (MESrvs);
- followed by a very rapid rise at 5.46 Ma triggered by the sudden collapse of the Gibraltar Channel drowning the incised basin-margin topography;

- and a final slow rise extending into the marginal basins and depressions around the Mediterranean (Lago Mare 3) re-establishing a marine connection with Eastern Parathethys. As this slowed in the early Zanclean, clinoforming sequences ('Gilbert' deltas) began to build back out and bury the incised basin-margin topography.

High sea-level exchanges between the Mediterranean and the Eastern Parathethys occurred immediately before the 5.6 Ma fall (LM 1) and then following the rapid sea-level rise (LM 3), marked by floods of brackish-water Parathethyan fauna and flora into the Mediterranean and Mediterranean nannoplankton influxes into the Eastern Parathethys. The location of the connecting strait between the two provinces is uncertain. Clauzon et al. (2005), Bache et al. (2012) and Suc et al. (2015b) discounted the possibility of a link through the Mamara Sea region and argued for a trans-Balkans gateway (Fig. 14) allowing two-way exchanges (Bache et al. 2012) with the Dacic Basin by 5.45 Ma, later reconnecting with the Euxinic Basin (proto-Black Sea) at the beginning of the Pliocene (5.31 Ma).

Because of its uncertain chronostratigraphy, the Pontian lowstand sequence of the Istria Depression is best constrained indirectly by a detailed stratigraphic analysis of better-dated upper Miocene–lower Pliocene sections in the Dacic and Euxinic basins:

- Messinian re-flooding of the Dacic Basin, Turnu Severin section (Fig. 14, Section 2 & Fig. 15): a pronounced Late Miocene unconformity has been observed in the Gura Vaii/Turna Severin area of the western Dacic Basin just outboard of the 'Iron Gates' gorge (Clauzon et al. 2005; Suc et al. 2011, 2015b). This unconformity cuts down through older Tertiary and Mesozoic rocks into metamorphic basement, representing a very significant base-level fall. It is locally overlain by a poorly dated polygenic breccia at Gura Vaii with reworked clasts of Sarmatian limestones, while a distinctive 'Gilbert-type' fan-delta sequence is more widely developed at Turna Severin nearby. This is described in some detail by Clauzon et al. (2005), although it has recently been criticized. A marine (nannofossil and dynocyst) fauna found in the lower ('bottomset') part of the sequence is indicative of a significant base-level rise and re-flooding from the Mediterranean (Clauzon et al. 2005). The interval has been assigned to the nannoplankton NN12b Subzone by Popescu et al. (2006, 2009) and Suc et al. (2011) with an estimated age of ±5.279–5.345 Ma. However, evidence of an Eastern Parathethyan flood into the Mediterranean at 5.45 Ma (Popescu et al. 2009; Suc et al. 2011, 2015b; Bache et al. 2012) supports a somewhat earlier

reconnection with the erosional downcutting event responsible for the basal unconformity occurring prior to that time. The 'foreset' beds of the fan delta above pass up into a 'topset' facies of interbedded clays and lignites. The transition between C3r chron and C3n–4n (Threva) subchron at 5.23 Ma occurs in the lower part of this interval just above the A and B lignite horizons (Clauzon et al. 2005; Popescu et al. 2006) constraining the 'Gilbert' fan delta to the basal Zanclean stage.

- Upper Miocene base-level fall and rise, DSDP 380, western Black Sea (Fig. 14, Section 3 & Fig. 15): the DSDP 380A and 381 holes drilled on the SW margin of the Black Sea record a significant Late Miocene base-level fall and rise (Hsu 1978; Ross et al. 1978; Hsu & Giovanoli 1979). DSDP 380A terminated at −3190 m (1075 m below mean level (m bml)) in laminated black shale (Unit V) with a marine dinoflagellate fauna of Sarmatian age (Suc et al. 2015a). This unit appears equivalent to a lithologically similar interval (Unit 7) in DSDP 381 (Ross et al. 1978; Hsu 1978), and is overlain by a sequence of laminated carbonates and marls (Unit IVe) in DSDP 380A represented by a distinctive, laterally continuous seismic reflector package (Tari et al. 2015). Although poorly constrained biostratigraphically, a flood of freshwater Pediastrum algae in the lower part and unusual hydrogen isotopic character of molecular biomarkers above (Vasiliev et al. 2015) support palynological evidence (Popescu et al. 2006) for a significant climate change. An equivalent unit has not been recognized in DSDP 381, although its apparent absence might reflect the very poor core recovery in that well. The overlying unit in DSDP 380A (Unit IVd) of pebbly mudstones with angular clasts of shallow-water stromatolitic dolomite was previously offered as evidence of a dramatic fall in base level and desiccation (Hsu & Giovanoli 1979). However, its fabric is more suggestive of a synsedimentary debris flow displaced from a shallower part of the margin and does not constrain the magnitude of the base-level fall. However, the equivalent unit (Unit 6) in DSDP 381 rests upon a very pronounced unconformity which can be tentatively traced updip to the deeply incised Karadeniz shelf-margin canyon (Fig. 15: 900–3800 m wide and 900–1000 m deep) infilled with Pliocene sediment (Gillet et al. 2007; Suc et al. 2015a). The equivalent conformable(?) surface may be represented downdip by the apparent hiatus separating Unit IVe and Unit V in DSDP 380A The overlying 'aragonitic' unit (Unit IVc and Unit 5) was encountered in both wells with a very similar brackish–marine fauna. Popescu et al. (2010) assigned an age of 5.45 Ma to the lower part of

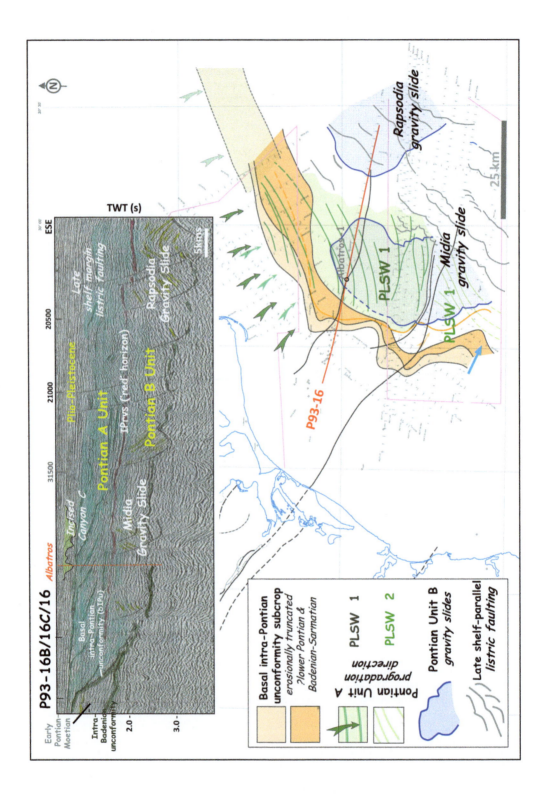

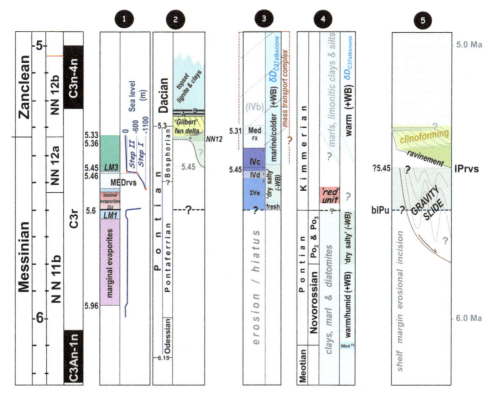

Fig. 14. Tentative late Miocene–early Pliocene chronostratigraphic summary of the Istria Depression, and correlation with the Mediterranean and Eastern Paratethyan (Dacic and Euxinic basins) realms. The several chronostratigraphic schemes are summarized from: (1) Clauzon *et al.* (2005), Roveri *et al.* (2008) and Bache *et al.* (2012); (2) Clauzon *et al.* (2005) and Snel *et al.* (2006); (3) Popescu (2006), Popescu *et al.* (2010), Bache *et al.* (2012) and Vasiliev *et al.* (2015); and (4) Krijgsman *et al.* (2010) and Vasiliev *et al.* (2011, 2013, 2015). The locations of these sections are highlighted in Figure 15. In the absence of any definitive biostratigraphic, palynological or magnetostratigraphic data, the correlation proposed here between the intra-Pontian interval, offshore Romania (Section 5) and better constrained sections elsewhere (sections 1–4) remains tentative, and is based on the assumption that a linked causal relationship existed between the late Miocene base-level fall events of the Mediterranean and Eastern Paratethys. LM, Lago Mare events; DU, detrital unit; MEDrws, shoreface ravinement surface, Western Mediterranean; +WB/−WB, positive/ negative water budget; MedFX, (fully) marine influx from Mediterranean; IPrvs, intra-Pontian ravinement surface, offshore Romania.

this interval in DSDP 380A, while a dramatic hydrogen isotopic shift ($\delta D_{C37alkenone}$) observed in terrestrial and aquatic biomarkers (Vasiliev *et al.* 2015) reflects a return to normal marine conditions in the upper part. Tari *et al.* (2015) have presented convincing seismic evidence of a thick

(*c.* 370 m) mass-transport complex in DSDP 380A just above and, perhaps, including units IVd and IVc (Fig. 14, Section 3). However, the position of the basal detachment surface is rather ambiguous and the apparent stratigraphic continuity of the 'Aragonite' Unit (IVc and 5) between the

Fig. 13. Sequence 5 – intra-Pontian unconformity and lowstand infill. A scallop-shaped intra-Pontian unconformity (bIPu) can be observed seismically cutting down into older Miocene sediment across the central and outer part of the Romanian and western Ukrainian shelf (Khriachtchevskaia *et al.* 2010; Boote 2014*b*). The erosional surface with subcropping early Pontian–Sarmatian section is highlighted and its complex lowstand infill geometry is summarized by line P93-16B/16C/16. This comprises a lower Pontian Unit B characterized by two large (Midia and Rapsodia) gravity slides (blue) and upper prograding wedge Unit A separated by a distinctive erosional ravinement suface (IPrvs – 'red horizon') locally displaced by shelf-parallel listric faults. Both the gravity slides and clinoform surfaces above face south and SE, reflecting the prevailing depositional slope. The lowstand wedge passes up into rather uniform transgressive highstand shelf strata terminated by the first of several Plio-Pleistocene canyon-incision events.

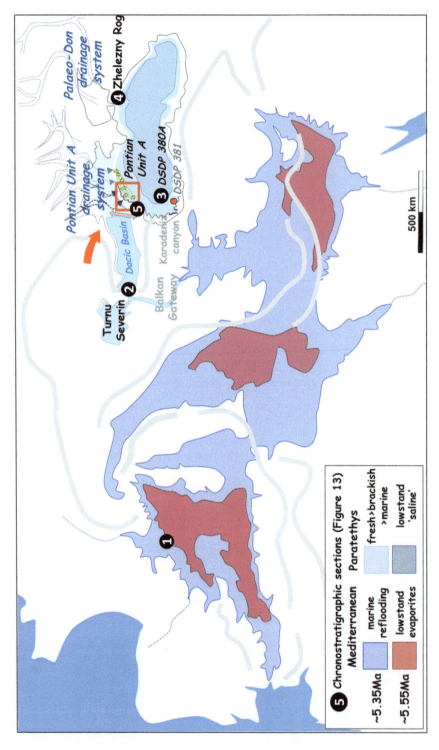

Fig. 15. Late Miocene–Early Pliocene drawdown and re-flooding of the Mediterranean and Eastern Paratethys (Dacian and Euxinic basins). Tentative reconstruction of the Messinian event at time of maximum drawdown (5.6–?5.3 Ma) and following re-flooding at approximately 5.35 Ma (from Bache *et al.* 2012, modified after Popov *et al.* 2010; Suc *et al.* 2015*a*, *b*). Bache *et al.* (2012) suggested the Euxinic Basin (ancestral Black Sea) was only reconnected with the Mediterranean around 5.31 Ma via the Dacian Basin and Balkans valley system. However, recalibration of the DSDP 380 section (Popescu *et al.* 2010; Suc *et al.* 2015*a*, *b*) suggests the (brackish–marine) base level began to rise shortly after Stage II re-flooding of the Mediterranean at approximately at the same time as the Dacian Basin (*c.* 5.45 Ma), although fully marine conditions were only re-established at approximately 5.31 Ma.

two DSDP holes suggests it might be located a short distance above. Marine diatom and dinoflagellate cysts found in the lower part of the displaced Unit IVb (DSDP 380A) just above have been assigned an age of 5.31 Ma, and a possibly equivalent calcareous-nannoplankton-bearing horizon (Sub-zone NN12b) has been observed in the Crimea (Popescu *et al.* 2009). Popescu *et al.* (2010) and Bache *et al.* (2012) have argued that this influx (Med^{FX}) marks the final reconnection with the Mediterranean through the Dacic Basin and across the so-called Scythian sill. However, the isotopic evidence of marine flooding during deposition of the 'Aragonitic' Unit (units IVc and 5) suggests the Euxinic (Black Sea) base level began to rise somewhat earlier, at approximately 5.46 Ma.

- Late Miocene base level fall, Zheleznyi Section, Taman Peninsular, NE Black Sea margin (Fig. 14, Section 4 & Fig. 15). Evidence of a late Miocene base-level fall and rise has been observed in the classic Zheleznyi Rog (Iron Cape) section, outcropping on the NE margin of the Black Sea (Taman Peninsular). This section includes a late Miocene–Pliocene interval comprising deep-water lacustrine brackish marls and diatomites interrupted by a thin (2.5 m) red oolithic/pisolithic sandstone layer ('Red Unit') with gypsum- and jarosite-bearing marl, overlain by late Miocene?–Pliocene dark grey limonitic marls (Krijgsman *et al.* 2010). The 'red layer' has been assigned to the basal 'Kimmerian' (equivalent to the upper part of the Pontafarian in the Dacic Basin) on the basis of its shallow-water molluscan fauna with an estimated age of 5.6–5.5 Ma (Krijgsman *et al.* 2010; Vasiliev *et al.* 2015). Contacts with the deeper-water Pontian marls below and Kimmerian above are abrupt, and Vasiliev *et al.* (2015) has suggested both represent significant hiatuses. Vasiliev *et al.* (2013) identified a very significant hydrogen isotope (δD$_{C37alkenone}$) shift in the Pontian marl sequence some 40 m below the 'red layer' representing a dramatic shift in the hydrological budget (+WB to −WB) and change from warm to cold arid climate at approximately 5.8 Ma. This appears to have reversed during the Kimmerian (Vasiliev *et al.* 2013), returning to a warmer climate, increased river inflow, positive water budget (+WB) and brackish water to freshwater taxa (Filippova 2002). Both Krijgsman *et al.* (2010) and Vasiliev *et al.* (2013) correlate the 'red layer' with the 5.6 Ma event in the Mediterranean, while the earlier hydrological shift may be related to the initial (5.96–5.6 Ma) base-level fall. Based upon lithofacies and faunal content, they suggest the 'red layer' represents a comparatively small fall of some 50–100 m. However, further north

in the West Kuban Depression, Popov *et al.* (2010) identified an extensive palaeo-Don incised drainage system representing a late 'Pontian' base-level fall of 500 m or more (Fig. 14). The precise age of incision is uncertain but may be correlative with the Zheleznyi Rog 'red layer' event.

Regional significance of the Sequence 5 unconformity

These several sections record a significant base-level fall and incision in both the Dacic and Euxinic basins during the late Miocene age. The time of incision is poorly constrained by the age of the overlying sediments, but is widely considered to be correlative with the 5.6 Ma event in the Mediterranean. The time of re-flooding is more uncertain. Faunal links between the Mediterranean and Dacic basins appear to have been established by 5.45 Ma (Bache *et al.* 2012), although a fully marine interchange may have occurred somewhat later at around 5.31 Ma (5.345–5.279 Ma) in both the Dacic (Clauzon *et al.* 2005; Popescu *et al.* 2006, 2009) and Euxinic basins (Popescu *et al.* 2010; Bache *et al.* 2012). The position and character of the connection between the Dacic and Euxinic basins is conventionally considered to have been a distinct, silled gateway (Galati or Scythian Sill) located north of the Dobrogea. However, this area suffered significant Quaternary uplift and unroofing, and may have been far more subdued during the late Miocene–Pliocene with a fairly open connection between the two basins. Certainly, isotopic and faunal evidence from the DSDP 380A hole suggests that re-flooding started at around 5.45 Ma with freshwater to brackish water, but becoming fully marine shortly thereafter.

These sometimes tentative correlations provide a relatively consistent regional chronological framework with which to reconstruct the history of the intra-Pontian lowstand sequence of the Istria Depression and to assess its relationship with the Messinian salinity crisis, commencing with:

- Stage 1 (*c.* 5.95–>5.6 Ma): the dramatic change in water chemistry at 5.8 Ma identified by Vasiliev *et al.* (2015) at Zheleznyi Rog may be linked to the initial base-level fall in the Mediterranean.
- Stage II (*c.* 5.6 Ma): a laterally extensive episode of late Miocene (Pontian) shelf-margin collapse and erosion offshore Romania reflects a pronounced sea-level fall of the proto-Black (Euxinic) Sea, perhaps correlative with the 5.6 Ma event in the Mediterranean. The magnitude of the fall is uncertain, although stratal geometries of the Pontian Unit A lowstand sequence suggest at least 1100 m. This compares with a maximum of 500–600 m suggested by Krezsek *et al.* (2016a, b).

The pebbly breccias (units 6 and IVd) with shallow-water clasts in the DSDP holes further south are clearly displaced and so offer little constraint, but evidence of erosion and canyon incision further updip (Suc *et al.* 2015a; Tari *et al.* 2015) supports a similar magnitude of 1000 m up to, perhaps, as much as 2000 m.

- Stage III (*c.* 5.6–5.45 Ma): as base level fell, sediments eroded from the exposed shelf margins were re-deposited further outboard (Pontian Unit B) analogous to the DU in the Mediterranean. Although not observed in the Istria Depression area, these may pass basinwards into shallow-water/supratidal lowstand carbonates equivalent to those observed in the DSDP holes, although there is no evidence of basin dessication and drawdown evaporites (Fig. 15).
- Stage IV (*c.* 5.45 Ma): following full reconnection with the global ocean (5.46 Ma), rising sea level in Mediterranean eventually broke through the Balkan 'sill' and re-flooded the Dacic and Euxinic basins As the water level rose, some of the earlier re-deposited sediment became unstable and slid downslope as mass-transport complexes (Midia and Rapsodia gravity slides).
- Stage V (<5.45 Ma): with rising water level, shoreface erosion cut a remarkably even peneplaned ravinement surface (IPrvs), truncating the sometimes strongly deformed sequence beneath.
- Stage VI (*c.* <5.45 Ma – ?5.31 Ma): preservation of the marginal intra-Pontian erosional scarp (bIPu) may reflect an increasingly rapid water-level rise. As it rose, northerly-sourced sediment was trapped along the shelf margin to form lowstand prograding wedges (Figs 13 & 15).
- Stage VII (<5.33 Ma/early Dacian): as rising water level crested and flooded back across the adjacent shelf, clastic systems were pushed back and the shelf-margin lowstand wedges abandoned to be buried beneath flat-lying shelf sediments.

Sequence 5: Pliocene–Holocene incised canyon systems

The intra-Pontian clinoforming sequence passes upwards into uniformly flat-lying stratal geometries typical of aggrading shelf sedimentation. This interval was incised during at least three distinct periods (C, B and A), each represented by two or more canyons appearing in a mid-shelf position and trending SE for more than 60 km towards the shelf break (Fig. 16). Jipa & Olariu (2009), after Tambrea (2007), and Konerding *et al.* (2010) suggested the earliest canyon system (C) is of basal or early Dacian age (PDU after Konerding *et al.* 2010) representing a higher-frequency sequence boundary, separating the intra-Pontian healing sequence from the

Dacian–Holocene aggrading system above. However, regional control favours a mid–late Dacian age (SB4 after Munteanu *et al.* 2012) for this boundary, coincident with infilling of the Dacian Basin and advance of the Danube palaeo-drainage system across the Dobrogea into the western Black Sea. The age of the two later canyons (B and A) systems are less well constrained to the late Pliocene and/or Pleistocene (Jipa & Olariu 2009, after Popescu 2002; Konerding *et al.* 2010).

The same flat-lying sequence is cut by a series of listric fault trains along the outer shelf paralleling the shelf margin (Fig. 13). These terminate within the shallower part of the interval, displacing the underlying Mio-Pliocene section, and soleing out within the lower Miocene and upper Oligocene. It was not possible to establish the relative ages of canyon incision and fault displacement with any precision. However, from the ambiguous data at hand, they appear to be approximately synchronous and it is possible that the faulting may represent periods of incipient mass wasting during periods of shelf exposure.

Regional significance of the Sequence 5 incised canyon systems

During much of the later Miocene, the onshore Dacic Basin acted as a trap for clastics shed from the Carpathian Orogen, starving the deep-water Black Sea Basin of terrestrial sediment. Following a brief marine influx at the end of the Miocene/early Dacian, water salinities of the Black (or 'Euxinic) Sea dropped rapidly and deposition was initially dominated by lacustrine chalks ('seekreide') and sideritic muds (Hsu 1978). However, by mid–late Dacian time, the Dacic Basin had been filled and the palaeo-Danube drainage system (Figs 15 & 16) was able to extend east, transporting clastics across the Dobrogea shelf and beyond to the deep-water Danube fan (Jipa 2009; Jipa & Olariu 2009).

Global cooling and advance of continental ice sheets across the northern hemisphere had an increasing influence on the Black Sea base level during the late Pliocene–early Pleistocene (2.0–1.5 Ma), fluctuating as the drainage catchment area expanded or contracted in response to regional ice movement. Short intermittent marine incursions from the Mediterranean during global eustatic highstands alternated with lowstands. By the early Pleistocene, the Black (Euxinic) Sea had become a freshwater lake (Hsu 1978) with a changing lake level reflecting more local (wet–dry) climatic cycles. Falling lake levels exposed the shelf and increased the volume of mud-rich sediment transported from shelf-margin delta systems directly to the lake floor (Popescu *et al.* 2001, 2004). Rising lake levels pushed the delta systems back, trapping

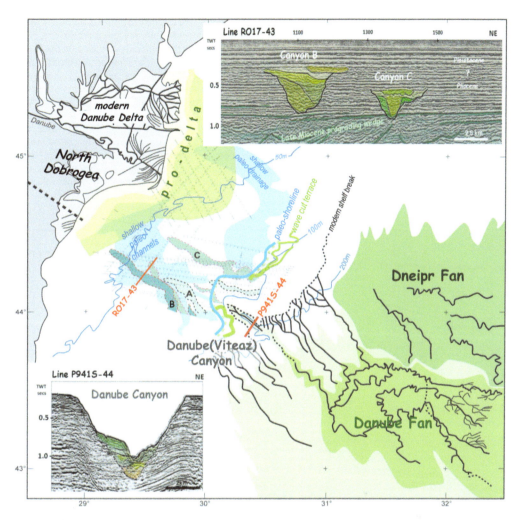

Fig. 16. Sequence 5 – Plio-Pleistocene shelf sedimentation and incised canyon systems, offshore Romania. The Romanian shelf was incised several times in the Pliocene–early Pleistocene, during brief base-level fall events. At least three discrete palaeo-canyon systems (C, B and A) can be identified seismically of very similar size, shape and internal stratal architecture to the late Pleistocene–Holocene Danube or Viteaz shelf-margin canyon (R017-43 v. P941S-44 from Popescu *et al.* 2004) further outboard. The four canyon systems cluster together some distance south of the modern Danube delta (Panin 2009). Their position and orientation suggests incision by a palaeo-fluvial system perhaps constrained by the tectonic fabric of the North Dobrogea, transporting mud-rich sediments across the Dobrogea (Istria) shelf to the deep-water Danube fan beyond (Popescu *et al.* 2001, 2004; Lericolais *et al.* 2009, 2013; Panin 2009).

sediment on the shelf. The last major rise occurred at approximately 7150 years BP, reflecting a flood of marine water via the Bosphorus Channel, which Ryan & Pitman (1998) linked to the biblical 'Noah's Flood' and extraordinary human migration from the drowned Black Sea shelf across Europe, Asia, Arabia and Egypt.

The three Plio-Pleistocene canyon systems record intermittent falls in base level and exposure of the Romanian shelf. Stratal geometries suggest

that the base-level falls responsible for their incision occurred quite rapidly over short periods of time. Their age is poorly constrained, but may predate significant Pleistocene glaciations (Popescu 2002; Tambrea 2007; Konerding *et al.* 2010). Nevertheless, a comparison of their size, shape and fill geometry with the late Pleistocene–Holocene Danube (Viteaz) canyon suggests that they formed in a similar way, in response to significant base-level falls and shelf exposure, followed by cross-shelf and shelf-margin erosional incision (Fig. 16, R016-226 and P9415-44).

Although the immediate cause(s) for the base-level fluctuations is unclear, the location of the canyon systems may have been tectonically controlled. The Dacic Basin had been filled by mid–late Dacian time, allowing the palaeo-Danube drainage system to extend east towards the deep-water Black Sea Basin. Although the Danube River is now deflected to the north around the Dobrogea High, the three shelf canyon systems (C, B & A) face NW directly towards the North Dobrogea (Figs 16 & 17). This suggests that the elevated topography of the modern

Dobrogrea 'High' and northwards diversion of the modern river was imposed quite recently, perhaps in response to the southerly advance of the Carpathian orogenic front. The Dobrogea 'barrier' might, in fact, have been more subdued in the past, allowing the palaeo-Danube drainage system to extend directly across to the Black Sea shelf, weakly constrained by the underlying tectonic fabric. This drainage system would have been particularly susceptible to base-level fluctuations with significant cross-shelf canyon incision occurring during low sea stands caused by tectonically imposed isolation from the world ocean enhanced by regional climatic changes.

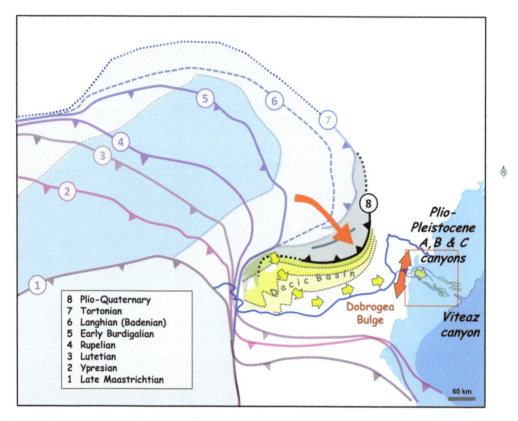

Fig. 17. Late Cenozoic incised canyon systems. The Dacic foreland basin acted as a sediment sink in the later Miocene–basal Pliocene trapping clastics shed from the adjacent Carpathian uplands. By early–middle Dacian time this had been filled by a distinctive regressive package of brackish-water shales, shoreline sands, and freshwater coastal plain shales and coals (Lower Pliocene regressive fill sequence of the Dacic Basin adapted and redrawn from Jipa & Olariu 2009) allowing the palaeo-Danube drainage system to extend eastwards across the Dobrogea to the western Black Sea. As the advancing Carpathian orogenic front was deflected south by the rigid East European Platform in late Cenozoic time (reconstruction synthesized from Harbury & Cohen 1997; Sinclair *et al.* 1997; Linzer *et al.* 1998; Hippolyte 2002), crustal loading may have depressed the northern Moesian Platform with compensatory arching and uplift of the Dobrogea. This may have tilted the Romanian shelf, encouraging brief periods of shelf incision and eventually forcing the Danube River northwards to its modern location. Some support for late uplift of the Dobrogea comes from the Murfatlar quarry west of Constanta where flat-lying Santonian–Campanian chalks appear to have been incised and infilled/buried by a surficial ?Quaternary clinoform package comprised almost entirely of reworked chalk fragments (Krezsek *et al.* 2016*a, b*).

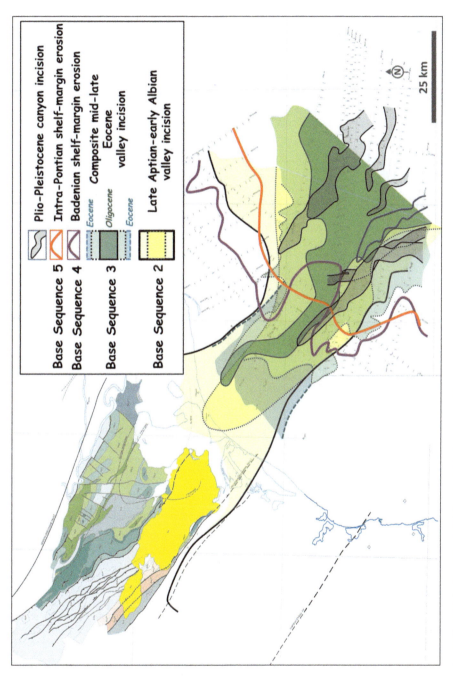

Fig. 18. Local tectonic controls on sequence architecture, Istria Depression. The approximate erosional limits of sequences 2–5 basal unconformities are highlighted in colour. Sequences 2 (yellow) and 3 (blue green/green) are bounded by deeply incised erosional surfaces constrained by the underlying structural fabric, contrasting with the more laterally extensive shelf-margin erosion characteristic of sequences 4 (purple line) and 5 (red line) above. The Plio-Pleistocene canyons (grey) appear to cluster together above the basin frame and face NW, suggesting the palaeo-river system responsible for their incision may have been weakly constrained by the North Dobrogea, prior to uplift of the Dobrogea Arch.

Concluding synthesis

The second-/third-order sequence boundaries inter-
rupting the Late Mesozoic–Cenozoic succession of
the Romanian Black Sea shelf are characterized by
deep erosional incision, cut during short periods of
basin isolation and significant base-level fall. The
location and geometry of each incised sequence
boundary (Figs 3 & 18) was controlled in varying
degrees by the underlying North Dobrogea basin
frame, regional tectonic events and indirect tectonically imposed base-level fluctuations. The evolution
and more likely mechanisms responsible for each
boundary surface include:

- Base Sequence 1 boundary (?Upper Berriasian–
  Lower Valanginian): the base Sequence 1 unconformity is poorly constrained because of later
  erosional unroofing. However, the reconstruction
  presented in this review suggests it was controlled
  by the underfilled topography of the North Dobrogea Trough and flanking platforms facing west
  towards the North Ligurian seaway. Erosion of
  the trough margins and intra-trough lowstand
  evaporites reflect a significant base-level fall, perhaps driven by subsidence to the west and pre-rift
  uplift of the greater Black Sea region.
- Base Sequence 2 boundary (?late Aptian–early
  Albian): the deep valley incision marking the
  base of Sequence 2 reflects rifting and subsidence
  of the western Black Sea. As this tilted the Romanian shelf to the east, the earlier Cretaceous infill
  within the Istria Depression became unstable,
  and perhaps augmented by Valanginian evaporites, slumped into the opening seaway. Headwards and flank erosion enhanced this process
  and, guided by the underlying basin frame, the
  valley system was extended far back to the west.
  Continued post-rift subsidence was followed by
  transgression and infill, gradually onlapping the
  valley walls and flanking platforms during the
  later Cretaceous.
- Base Sequence 3 boundary (intra-Eocene): intra-
  and late Eocene valley incision appears to have
  been a multiple event, related to the collision of
  Eurasia with the central Turkey/Balkanide
  crustal collage in mid-Eocene and Afro-Arabia
  during the late Eocene. The Paratethyan seaway
  created by this episode of crustal consolidation
  was briefly cut off from the world ocean, leading
  to a dramatic fall in base level and exposure of
  the proto-Black Sea margins. Once again, the earlier infill of the Istria Depression became unstable
  and, enhanced by headward erosion, an incised
  valley system cut back across the shelf to be partially filled and then re-incised during the late
  Eocene. Erosional incision and valley-head
  retreat appears to have been guided by the earlier

mid-Cretaceous valley fill and underlying basin
frame. Contemporaneous Balkanide orogensis
to the south may have enhanced this process by
tilting the Moesian micro-plate and increasing
sediment instability along the Istria proto-Black
Sea margin.

- Base Sequence 4 boundary ('mid-Badenian'/pre-
  Sarmatian): continuing crustal consolidation and
  uplift led to yet another base-level fall in the mid-
  Miocene ('mid-Badenian') when the deeper parts
  of the Paratethyan seaway became partially, to
  completely, isolated from the world ocean. This
  is reflected by drawdown evaporites in the Carpathian foredeep and restricted brackish conditions
  in the proto-Black Sea. Carpathian orogensis to
  the north arched up the coastal Dobrogea area
  and tilted the outboard shelf to the east, encouraging mass wasting, shelf-margin collapse and erosion, largely unconstrained by the underlying
  valley systems or underlying basin frame.
- Base Sequence 5 boundary (intra-Pontian/Messi-
  nian): a further period of intra-Pontian (late Miocene) erosion reflects yet another base-level fall
  isolating the Black Sea from the Mediterranean
  and the world ocean beyond. The shelf-margin
  erosional surface appears unconstrained by the
  earlier incised valley systems or basin frame,
  and can be traced laterally NE and eastwards
  across the Odessa shelf. This base-level fall can
  be attributed indirectly to the tectonism responsible for closing the Gibraltar straits and the Messinian drawdown event. Late lowstand slumping
  and gravity sliding towards the south and SE
  demonstrate gravity-driven sediment instability,
  perhaps weakly constrained by uplift to the
  north. Late Messinian re-flooding of the Mediterranean, followed by reconnection and flooding of
  the Black Sea Basin was marked by a distinctive
  shoreface ravinement surface and shelf-margin
  clinoforming wedge along its NW margin.
- Plio-Pleistocene canyon systems: the several
  incised canyons crossing the Romanian shelf
  display a distinct NW–SE orientation clustered
  above the earlier Cenozoic and Cretaceous valley
  systems. This association may be coincidental, but
  the consistent orientation of the canyons suggests
  the proto-Danube drainage system responsible for
  their incision once lay some distance to the south
  of its modern location, perhaps constrained by the
  tectonic fabric of the North Dobrogea prior to
  Holocene uplift and unroofing. These canyons
  reflect intermittent, oscillatory base-level changes
  within the Black Sea Basin related to Plio-
  Pleistocene climatic cycles or local tectonism.
  Seawards tilting driven by uplift of the Dobrogea
  Arch was probably also responsible for the late
  shelf-parallel listric faulting, auguring the future
  collapse of the shelf margin.

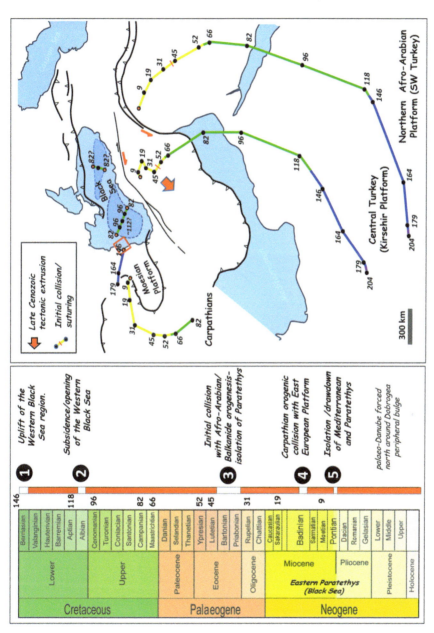

Fig. 19. Regional tectonic controls on sequence architecture, Istria Depression. The influence of regional plate kinematics upon the Romanian shelf margin is illustrated by several relative motion trackways between Afro-Arabia (SW Turkey), the central Turkey Kiresehir (Platform), the Moesian Platform and the western Black Sea (constructed from Barrier & Vrielynck 2008). Their relative position at particular times is shown in million years during the Triassic (mauve), Jurassic (blue), Cretaceous (green) and Cenozoic (yellow). The trackways highlight the SE movement of the Moesian Platform during the Jurassic, the opening of the western Black Sea in the middle–late Cretaceous and the closure of the Tethys to form the Paraethyan seaway in the Cenozoic. While correlations between regional tectonic events and erosional incision offshore Romania remain speculative, the apparent coincidence in timing supports a significant causal link summarized by the chronostratigraphic synthesis.

The three particular tectonic factors highlighted in this analysis:

- local structural framework – influencing the location and size of trans-shelf depocentres (Fig. 18);
- local tectonic uplift – tilting such depocentres basinwards;
- and more indirect regional tectonism – responsible for increasing basin isolation and encouraging significant base-level falls (Fig. 19)

have all controlled the local sequence architecture of the offshore Romanian shelf by increasing its susceptibility to mass wasting, valley/canyon incision and headwards erosion. These processes are observed on passive continental margins elsewhere in the world but their frequency and scale appears to be unique to the Black Sea. This close relationship between local stratigraphic and regional tectonism observed offshore Romania (Figs 18 & 19) can be expected to have particular significance to other parts of the Black Sea; thus constraining interpretations of its tectonostratigraphic history and shelf-margin architecture elsewhere.

This article grew out of an inhouse sequence stratigraphic analysis for Canadian Occidental in late 1994/early 1995. Since that time, additional data, ideas and encouragement have come from many friends and colleagues, without which it would never have been completed. I am very grateful for their help. Sterling Resources and OMV gave their approval to publish an earlier version of this work, and Dr Alastair Baird of Kingston University and Diana Monckton acted as critical sounding boards over the several years it took to complete. Their help and advice is much appreciated. I would also like to thank Gabriel Ionescu and Csaba Krezsak for their reviews. Csaba's perceptive and very knowledgeable critique was especially useful in forcing me to reconsider some of my more speculative projections. With time, I know some of these will fall by the wayside, but hope they will at least move things along just a little in a generally forward direction.

References

AVRAM, E., SZASZ, L. ET AL. 1993. Cretaceous terrestrial and shallow marine deposits in northern South Dobrogea (SE Romania). *Cretaceous Research*, **14**, 265–305.

BACHE, F., OLIVET, J.L., GORINI, C., RABINEAU, M., BAZTAN, J., ASLANIAN, D. & SUC, J.-P. 2009. Messinian erosional and salinity crisis: view from the Provence Basin (Gulf of Lions, Western Mediterranean). *Earth and Planetary Science Letters*, **286**, 139–157.

BACHE, F., POPESCU, S.-M. ET AL. 2012. A two step process for the reflooding of the Mediterranean after the Messinian Salinity crisis. *Basin Research*, **24**, 125–153.

BALDI, K. 2006. Paleoceanography and climate of the Badinian (Middle Miocene, 16.4–13.0 Ma) in the Central Paratethys based on foraminifera and stable isotope ($\delta^{18}O$ and $\delta^{13}C$) evidence. *International Journal Earth Science (Geologische Rundschau)*, **95**, 119–142.

BANCILA, I., NEAGU, Th., MUTIIU, R. & DRAGASTAN, O. 1997. Jurassic–Cretaceous stratigraphy and tectonic framework of the Romanian Black Sea offshore. *Revue Roumaine de Geologie*, **41**, 65–76.

BANKS, C.J. 1997. Basins and thrust belts of the Balkan Coast of the Black Sea. *In*: ROBINSON, A.G. (ed.) *Regional and Petroleum Geology of the Black Sea and Surrounding Region*. American Association of Petroleum Geologists, Memoirs, **68**, 115–128.

BANKS, C.J. & ROBINSON, A.G. 1997. Mesozoic strike-slip back-arc basins of the Western Black Sea. *In*: ROBINSON, A.G. (ed.) *Regional and Petroleum Geology of the Black Sea and surrounding Region*. American Association of Petroleum Geologists, Memoirs, **68**, 53–62.

BARRIER, E. & VRIELYNCK, B. 2008. *Paleotectonic Maps of the Middle East (Scale 1/18 500 000)*. Middle East Basins Evolution Programme 2003–2007 Commission for the Geological Map of the World (CGMW/CCGM)/UNESCO, Paris, http://www.ccgm.org.

BOOTE, D.R.D. 1994. *Reconnaissance Stratigraphic Analysis, Midia and Pelikan Blocks, Offshore Romanian Black Sea*. Canadian Occidental Internal Report, March 1994

BOOTE, D.R.D. 2007. The geological history of the Istria Depression, offshore Romania – tectonic controls on 2nd order sequence architecture. Oral presentation at the *AAPG Europe Energy Conference & Exhibition, November 2007*, Athens.

BOOTE, D.R.B. 2014a. The geological history of the Istria 'Depression', tectonic controls on 2nd/3rd order sequence architecture. *Doctoral thesis*, Kingston University, London.

BOOTE, D.R.D. 2014b. Tectonic controls on 2nd/3rd order sequence architecture istria 'Depression', Romanian Black Sea Shelf. Oral presentation at the *AAPG International Conference & Exhibition*, 14–17 September 2014, Istanbul, Turkey.

CATUNEANU, O. 1992. The geology of the Black Sea Romanian Shelf of Central Dobrogea Type. *Revue Roumaine de Geologie*, **36**, 73–81.

CATUNEANU, O. 1994. The geology of the Black Sea Romanian Shelf of North Dobrogean type: useful resources. *Revue Roumaine de Geologie*, **38**, 53–64.

CLAUZON, G., SUC, J.P., GAUTIER, F. & LOUTRE, A. 1996. Alternate interpretation of the Messinian salinity crisis: controversy resolved? *Geology*, **24**, 363–366.

CLAUZON, G., SUC, J.P., POPESCU, S.M., MARANTEANU, M., RUBINO, J.-L., MARINESCU, F. & MELINTE, M.C. 2005. Influence of Mediterranean sea-level changes on the Dacic Basin (Eastern Paratethys) during the late Neogene: the Mediterranean Lago Mare facies deciphered. *Basin Research*, **17**, 437–462.

DINU, C. 2009. Neogene and Quaternary tectonic evolution of the western Black Sea. *In*: *Topo-Europe Summer School on Carpathian–Danube Delta–Black Sea Sedimentary System*. European Science Foundation (ESF), Strasbourg, 39–69.

DINU, C., WONG, H.K. & TAMBREA, D. 2002. Stratigraphic and tectonic synthesis of the Romanian Black Sea Shelf and correlation with major land structures. *In*: DINU, C. & MORCANU, V. (eds) *Geology and Tectonics of the Romanian Black Sea Shelf and its Hydrocarbon Potential*. Bucharest Geoscience Forum, Special Volume, **2**, 101–117.

DINU, C., TAMBREA, D. & RAILEANU, A. 2005a. Structural
characteristics, seismic facies and depositional frame-
work of Eocene deposits in central Romanian Black
Sea offshore. In: DIMITRIU, R.G., PROCA, A. & IOANE,
D. (eds) 4th Congress of the Balkan Geophysical Soci-
ety, Bucharest 2005, Conference Volume. Romanian
Society of Geophysics, Bucharest, O18-02.
DINU, C., WONG, H.K., TAMBREA, D. & MATENCO, L. 2005b.
Stratigraphic and structural characteristics of the Roma-
nian Black Sea shelf. Tectonophysics, 410, 417–435.
DOGLIONI, C., BUSATTA, C., BOLIS, G., MARIANINI, L. &
ZANELLA, M. 1996. Structural evolution of the eastern
Balkans (Bulgaria). Marine and Petroleum Geology,
13, 225–251.
DUDEK, K., BUKOWSKI, K. & WIEWIORKA, J. 2004. Datowa-
nia radiometryczne badenskich osadow piroklastycz-
nych z okolic Wieliczki i Bochni [The radiometric
dating of malarial pyroclastic deposits from Wieliczka
and Bochnia]. In: MICHALIK, M., JACHER-SLIWCZYNSKA,
K., SKIBA, M. & MICHALIK, J. (eds) VIII Ogolnopolska
Sesja Naukowa Datowanie Mineralow i Skal, 19–26.
FILIPPOVA, N.Yu. 2002. Spores, pollen and organic walled
phytoplankton from Neogene deposits of the Zhelezez-
nyi Rog reference section (Taman Peninsular). Stratigraphy and Geological Correlation, 10, 176–188.
GILLET, H., LERICOLAIS, G. & REHAULT, J.-P. 2007. Messinian event in the Black Sea: evidence of a Messinian
erosional surface. Marine Geology, 244, 142–165.
GORUR, N. 1997. Cretaceous syn- to post-rift sedimentation
on the southern continental margin of the Western
Black Sea Basin. In: ROBINSON, A.G. (ed.) Regional
and Petroleum Geology of the Black Sea and Surrounding Region. American Association of Petroleum Geologists, Memoirs, 68, 227–240.
GORUR, N. & TUYSUZ, O. 2001. Cretaceous to Miocene
palaeogeographic evolution of Turkey: implications
for hydrocarbon potential. Journal of Petroleum Geology, 24, 119–146.
GORUR, N., CAGATAY, N., SAKINC, N., AKKOK, R., TCHAPA-
LYGA, A. & NATALIN, B. 2000. Neogene Paratethyan
succession in Turkey and its implications for the palaeogeography of and Eastern Paratethys. In: BOZKURT, E.,
WINCHESTER, J.A. & PIPER, J.D.A. (eds) Tectonics and
Magnetism in Turkey and the Surrounding Area. Geological Society, London, Special Publications, 173,
251–269, https://doi.org/10.1144/GSL.SP.2000.173.
01.13
GRADINARU, E. 1984. Jurassic rocks of north Dobrogea: a
depositional–tectonic approach. Revue Roumaine Geologie, Geophysique et Geographie, Geologie, 28, 61–72.
GRADINARU, E. 1988. Jurassic sedimentary rocks and bimodal volcanics of the Cirjelari-Camena Outcrop Belt: evi-
dence for a transtensile regime of the Peceneaga-Camena
Fault. Studii si cercetari de Geologie, 33, 97–121.
GRADINARU, E. 1995. Mesozoic rocks in North Dobrogea:
an overview. In: Field Guidebook, Central and North
Dobrogea, Romania, Comparative Evolution of Peri-
Tethyan Rift Basins. IGCP Project No. 369. Geological
Institute of Romania, Bucharest, 17–26.
GRADINARU, E. & BARBULESCU, A. 1994. Upper Jurassic
brachiopod faunas of Central and North Dobrogea
(Romania): biostratigraphy, paleoecology and paleo-
biogeography. Jahrbuch Geologischen Bundesanstalt,
137, 43–84.

GRADINARU, E., SEGHEDI, A., OAIE, G. & RADAN, S. 1995.
Field trip in Central and North Dobrogea: description
of itinerary and stops. In: Field Guidebook, Central
and North Dobrogea, Comparative evolution of Peri-
Tethyan Rift Basins. IGCP Project No. 369. Geological
Institute of Romania, Bucharest, 29–70.
HARBURY, N. & COHEN, M. 1997. Sedimentary history of the
Late Jurassic Paleogene of northeast Bulgaria and the
Bulgarian Black Sea. In: ROBINSON, A.G. (ed.) Regional
and Petroleum Geology of the Black Sea and the Sur-
rounding Region. American Association of Petroleum
Geologists, Memoirs, 68, 129–168.
HARZHAUSER, M. & PILLER, W.E. 2007. Benchmark data of a
changing sea – paleogeography, paleobiogeography
and events in the Central Paratethys during the Miocene. Palaeogeography, Palaeoclimatology, Palaeoecology, 253, 8–31.
HEMPTON, M.R. 1987. Constraints on Arabian plate motion
and extensional history of the Red Sea. Tectonics, 6,
687–705.
HIPPOLYTE, J.-C. 2002. Geodynamics of Dobrogea
(Romanian): new constraints on the evolution of the
Tornquist–Teisseyre Line, the Black Sea and the Carpathians. Tectonophysics, 357, 33–53.
HIPPOLYTE, J.-C., BADESCU, C. & CONSTANTIN, P. 1999. Evolution of the transport direction of the Carpathian belt
during its collision with the east European Platform.
Tectonics, 18, 1120–1138.
HIPPOLYTE, J.-C., MULLER, C., KAYMAKCI, N. & SANGU, E.
2010. Dating of the Black Sea Basin: new nannoplankton ages from its inverted margin in the Central Pontides (Turkey). In: SOSSON, M., KAYMAKCI, N.,
STEPHENSON, R.A., BERGERAT, F. & STAROSTENKO, V.
(eds) Sedimentary Basin Tectonics from the Black Sea
and Caucasus to the Arabian Platform. Geological
Society, London, Special Publications, 340, 113–136,
https://doi.org/10.1144/SP340.7
HSU, K.J. 1978. Stratigraphy of the lacustrine sedimentation
in the Black Sea. In: ROSS, D.A., NEPROCHNOV, Y.P.
ET AL. (eds) Initial Reports of the Deep Sea Drilling Pro-
ject Leg 42, Istanbul (Turkey) to Istanbul (Turkey)
May–June 1975, Volume 42. United States Government
Printing Office, Washington, DC, 509–524.
HSU, K.J. & GIOVANOLI, F. 1979. Messinian event in the
Black Sea. Palaeogeography, Palaeoclimatology,
Palaeoecology, 29, 75–93.
HSU, K.J., CITA, M.B. & RYAN, W.B.F. 1973. The origin of
the Mediterranean evaporites. In: RYAN, W.B.F. & HSU,
K.J. (eds) Initial Reports Deep Sea Drilling Project,
Volume 13, Part 1. United States Government Printing
Office, Washington, DC, 1203–1231.
IONESCU, G., SISMAN, M. & CATARAIANI, R. 2002. Source and
reservoir rocks and trapping mechanisms on the Roma-
nian Black Sea shelf. In: DINU, C. & MORCANU, V. (eds)
Geology and Tectonics of the Romanian Black Sea
Shelf and its Hydrocarbon Potential. Bucharest Geosci-
ence Forum, Special Volume, 2, 67–83.
JIPA, D.C. 2009. The Dacian Basin source to sink system.
In: Topo-Europe Summer School on Carpathian–Dan-
ube Delta–Black Sea Sedimentary System. European
Science Foundation (ESF), Strasbourg, 71–88.
JIPA, D. & OLARIU, C. 2009. Dacian Basin: Depositional
Architecture and Sedimentary History of a Paratethys
Sea. Geo-Eco-Marina, Special Publications, 3.

JONES, E.W. & SIMMONS, M.D. 1997. A review of the stratigraphy of Eastern Paratethys (Oligocene–Holocene)
with particular emphasis on the Black Sea. In: ROBIN-
SON, A.G. (ed.) Regional and Petroleum Geology of
the Black Sea and Surrounding Regions. American
Association of Petroleum Geologists, Memoirs, 68,
39–52.

KAZMIN, V.G., SHREIDER, A.A. & BULYCHEV, A.A. 2000.
Early stages of evolution of the Black Sea. In: BOZKURT,
E., WINCHESTER, J.A. & PIPER, J.D.A. (ed.) Tectonics
and Magnetism in Turkey and the Surrounding Area.
Geological Society, London, Special Publications,
173, 235–249, https://doi.org/10.1144/GSL.SP.2000.
173.01.12

KHRIACHTCHEVSKAIA, O., STOVBA, S. & STEPHENSON, R.
2010. Cretaceous –Neogene tectonic evolution of the
northern margin of the Black Sea from seismic reflection data and tectonic subsidence analysis. In: SOSSON,
M., KAYMAKCI, N., STEPHENSON, R.A., BERGERAT, F. &
STAROTENKO, V. (eds) Sedimentary Basin Tectonics
from the Black Sea and Caucasus to the Arabian Platform. Geological Society, London, Special Publications, 340, 137–157, https://doi.org/10.1144/SP340.8

KONERDING, C., DINU, C. & WONG, H.K. 2010. Seismic
sequence stratigraphy, structure and subsidence history
of the Romanian Black Sea shelf. In: SOSSAN, M., KAY-
MAKCI, N., STEPHENSON, R.A., BERGERAT, F. & STAROS-
TENKO, V. (eds) Sedimentary Basin Tectonics from the
Black Sea and Caucasus to the Arabian Platform. Geological Society, London, Special Publications, 340,
159–180, https://doi.org/10.1144/SP340.9

KORHONEN, J.V., FAIRHEAD, J.D. ET AL. 2007. Magnetic
Anomaly Map of the World. Commission for the Geo-
logical Map of the World (CGMW/CCGM)/
UNESCO, Paris.

KREZSEK, C., BERCEA, R.I., SEGHEDI, A. & TARI, G. 2016a. Cretaceous depositional systems and tectonic evolution of the
Romanian Black Sea: guide for the post-conference field
trip, 21st May 2016. AAPG European Regional Conference and Exhibition, 19–20 May 2016, Bucharest.

KREZSEK, C.M., SCHLEDER, Z., BEGA, Z., IONESCU, G. & TARI,
G. 2016b. The Messinian sea-level fall in the western
Black Sea: small or large? Insights from offshore
Romania. Petroleum Geoscience, 23, 392–399,
https://doi.org/10.1144/petgeo2015-093

KREZSEK, C., BERCEAL, R.-I., TARI, G. & IONESCUL, G. 2017.
Cretaceous sedimentation along the Romanian margin
of the Black Sea: inferences from onshore to offshore
correlations. In: SIMMONS, M.D., TARI, G.C. & OKAY,
A.I. (eds) Petroleum Geology of the Black Sea. Geological Society, London, Special Publications, 464,
https://doi.org/10.1144/SP464.10

KRIJGSMAN, W., STOICA, M., VASILIEV, I. & POPOV, V.V.
2010. Rise and fall of the Paratethys Sea during the
Messinian Salinity Crisis. Earth and Planetary Science
Letters, 290, 183–191.

KRZYWIEC, P., FLOREK, R. & POPADYUK, I. 2009. Polish
Ukrainian Carpathian subthrust Prospects – selected
problems. Extended abstract presented at the AAPG
European Region Annual Conference, 23–24 November 2009, Paris-Malmaison, France.

LEEVER, K.A., MANTENCO, L., RABAGIA, T., CLOETINGH, S.,
KRIJGSMAN, W. & STOICA, M. 2009. Messinian sea
level fall in the Dacic Basin (Eastern Paratethys):

paleogeographical implications from seismic sequence
stratigraphy. Terra Nova, 22, 12–17.

LERICOLAIS, G., BULOIS, C., GILLET, H. & GUICHARD, F. 2009.
High frequency sea level fluctuations in the Black Sea
since the LGM. Global and Planetary Change, 66,
65–75.

LERICOLAIS, G., BOURGET, J., POPESCU, I., JERMANNAUD, P.,
MULDER, T., JORRY, S. & PANIN, N. 2013. Late Quaternary
deep-sea sedimentation in the western Black Sea: new
insights from recent coring and seismic data in the deep
basin. Global and Planetary Change, 103, 232–247.

LINZER, H.-G., FRISCH, W., ZWEIGEL, P., GIRBACEA, R., HANN,
H.-P. & MOSER, F. 1998. Kinematic evolution of the
Romanian Carpathians. Tectonophysics, 297, 133–156.

LOFI, J., GORINI, C., BERNE, S., CLAUZON, G., TADEU DOS
REIS, A., RYAN, W.B.F. & STECKLER, M.S. 2005. Erosional processes and paleo-environmental changes in
the Western Gulf of Lions (SW France) during the Messinian Salinity Crisis. Marine Geology, 217, 1–30.

LOFI, J., DEVERCHERE, J. ET AL. 2011. Seismic Atlas of the
'Messinian Salinity Crisis' Markers in the Mediterranean and Black Seas. Memoires de la Societe Geologique de France, 179.

MOSAR, J. & SEGHEDI, A. 1999. North Dobrogea and the
Paleozoic plate tectonics. EGS, XXIV General Assem-
bly, The Hague, Netherlands, 19–23 April 1999,
Abstracts, Annales Geophysicae 17 Supplement.

MUNTEANU, I., MATENCO, L., DINU, C. & CLOETINGH, C.
2012. Effects of large sea-level variations in connected
basins: the Dacian–Black Sea system of the Eastern
Paratethys. Basin Research, 24, 583–597.

MURATOV, M.V. (ed.). 1969. Geology of the USSR. Volume
VIII, Crimea, Part 1. Geological Description [in
Russian].

OKAY, A. & GORUR, N. 2007. Tectonic evolution models for
the Black Sea. In: YILMAZ, P.O. & ISAKSEN, G.H. (eds)
Oil and Gas of the Greater Caspian Area. American
Association of Petroleum Geologists, Studies in Geology, 55, 13–16.

OKAY, A. & SAHINTURK, O. 1997. Geology of the Eastern
Pontides. In: ROBINSON, A.G. (ed.) Regional and Petroleum Geology of the Black Sea and Surrounding
Regions. American Association of Petroleum Geologists, Memoirs, 68, 291–311.

OKAY, A.I., SENGOR, A.M.C. & GORUR, N. 1994. Kinematic
history of the opening of the Black Sea and its effects on
the surrounding region. Geology, 13, 267–270.

OKAY, A.I., SUNAI, G., TUYSUZ, O., ALTINER, D., KYLANDER-
CLARCK, A.R. & AKDOGAN, R. 2014. Lower Cretaceous
turbidites of the Poontides and the opening the Black
Sea. Abstract presented at the AAPG International Conference & Exhibition, September 2014, Istanbul, Turkey.

OSZCZYPKO, N., KRZYWIEC, P., POPADYUK, I. & PERYT, T.
2006. Carpathian Foredeep Basion (Poland and
Ukraine): its sedimentary, structural and geodynamic
evolution. In: GOLONKA, J. & PICHA, F.J. (eds) The Carpathians and their Foreland: Geology and Hydrocarbon Resources. American Association of Petroleum
Geologists, Memoirs, 84, 261–318.

PANIN, N. 2009. The Danube Delta: the mid term of the geo-
system Danube River Danube Delta-Black Sea geological setting, sedimentology and Holocene present day
evolution. In: Topo-Europe Summer School on Carpathian–Danube Delta–Black Sea Sedimentary System.

European Science Foundation (ESF), Strasbourg, 11–38.

PARASCHIV, D. 1979. *Romanian Oil and Gas Fields.* Technical and Economical Studies Series, **A/13**. Institute of Geology and Geophysics, Bucharest.

PENE, C., NICHESCU, B. & COLTOI, O. 2009. Hydrocarbon entrapment in the east of the Moesian Platform. Abstract presented at the *71st EAGE Conference & Exhibition, 8–11 June*, Amsterdam, The Netherlands.

PERYT, T.M. 2006. The beginning, development and termination of the middle Miocene Badinian salinity crisis in Central Paratethys. *Sedimentary Geology*, **188–189**, 379–396.

PHARAOH, T.C. 1999. Palaeozoic terranes and their lithospheric boundaries within the Trans-European Suture Zone. *Tectonophysics*, **314**, 17–41.

PIERRE, C., CURUSO, A., BLANC-VALLERON, M.-M., ROUCHY, J. & ORSZAG-SPERBER, F. 2006. Reconstruction of the paleo-environmental changes around the Miocene–Pliocene boundary along a west–east transect across the Mediterranean. *Sedimentary Geology*, **188–189**, 339–340.

PICHA, F.J., STRANIK, Z. & KREJCI, O. 2006. Geology and hydrocarbon resources of the outer Western Carpathians and their foreland, Czech Republic. *In*: GOLONKA, J. & PICHA, F.J. (eds) *The Carpathians and their Foreland: Geology and Hydrocarbon Resources.* American Association of Petroleum Geologists, Memoirs, **84**, 49–175.

POPADYUK, I., KHRIACHTCHEVSKAIA, O. & STOVBA, S. 2010. Geology of the Crimea Mountains in the context of petroleum exploration in the Black Sea. *AAPG European Region Conference & Exhibition, 15–16 October*, Kiev Ukraine. Field Trip Guide Book, Trip #1.

POPESCU, I.S. 2002. Analyse des processus sedimentaires recents dans l'evantail profund du Danube (Mere Noire) [Analysis of the recent sedimentary processes in the Danube (Black Mother)]. *These de doctorat*, Universite de Bretagne Occidentale – Université de Bucarest, Brest.

POPESCU, I., LERICOLAIS, G.L., PANIN, N., WONG, H.K. & DROZ, L.D. 2001. Late Quaternary channel avulsions on the Danube deep-sea fan, Black Sea. *Marine Geology*, **179**, 25–37.

POPESCU, I., LERICOLAIS, G.L., PANIN, N., NORMAND, A., DINU, C. & LE DREZEN, E. 2004. The Danube submarine canyon (Black Sea): morphology and sedimentary processes. *Marine Geology*, **206**, 249–265.

POPESCU, S.-M. 2006. Late Miocene and early Pliocene environments in the southwestern Black Sea region from high resolution palynology of DSDP Site 380A (Leg 42B). *Palaeogeography,Palaeoclimatology, Palaeoecology*, **238**, 64–77.

POPESCU, S.-M., KRIJGSMAN, W., SUV, J.-P., CLAUZON, G., MARUNTEANU, M. & NICA, T. 2006. Pollen record and integrated high-resolution chronology of the early Pliocene Dacic Basin (southwestern Romania). *Palaeogeography, Palaeoclimatology, Palaeoecology*, **238**, 78–90.

POPESCU, S.-M., DALESME, F. *ET AL.* 2009. Galaecysta etrusca complex: dinoflagellate cyst marker of Parathethyan influxes to the Mediterranean Sea before and after the peak of the Messinian Salinity Crisis. *Palynology*, **33**, 105–134.

POPESCU, S.-M., BILTEKIN, D. *ET AL.* 2010. Pliocene and Lower Pleistocene vegetation and climate changes at

the European scale: long pollen records and climatostratigraphy. *Quaternary International*, **219**, 152–167.

POPOV, S.V., ANTIPOV, M.P., ZASTROZHNOV, A.S., KURINA, E. E. & PINCHUK, T.N. 2010. Sea-level fluctuations on the northern shelf of the Eastern Paratethys in the Oligocene–Neogene. *Stratigraphy and Geological Correlation*, **18**, 200–224.

ROBERTSON, A.H.F. & USTAÖMER, T. 2004. Tectonic evolution of the intra-Pontide suture zone in the Armutlu Peninsular, NW Turkey. *Tectonophysics*, **381**, 175–209.

ROBERTSON, A.H.F., USTAOMER, T., PARLAK, O., UNLUGENC, U.C. & INAN, N. 2006. The Berit transect of the Tauride thrust belt, South Turkey: Late Cretaceous –Early Cenozoic accretionary/collisional processes related to closure of the southern Neotethys. *Journal of Asian Earth Sciences*, **27**, 108–145.

ROBINSON, A G. & KERUSOV, E. 1997. Stratigraphic and structural development of the Gulf of Odessa, Ukrainian Black Sea: implications for petroleum exploration. *In*: ROBINSON, A.G. (ed.) *Regional and Petroleum Geology of the Black Sea and Surrounding Region.* American Association of Petroleum Geologists, Memoirs, **68**, 369–380.

ROGL, F., STEININGER, F. & MULLER, C. 1978. Middle Miocene salinity crisis and paleogeography of the Paratethys (Middle and Eastern Europe. *In*: ROSS, D.A., NEPROCHNOV, Y.P. *ET AL.* (eds) *Initial Reports of the Deep Sea Drilling Project Leg 42, Istanbul (Turkey) to Istanbul (Turkey) May–June 1975, Volume 42.* United States Government Printing Office, Washington, DC, 985–990.

ROSS, D.A., NEPROCHNOV, Y.P. *ET AL.* (eds) 1978. *Initial Reports of the Deep Sea Drilling Project Leg 42, Istanbul (Turkey) to Istanbul (Turkey) May– June 1975, Volume 42.* United States Government Printing Office, Washington, DC.

ROVERI, M., MANZI, V., GENNARI, R., IACCARINO, S. & LUGLI, S. 2008. Recent advancements in the Messinian stratigraphy of Italy and their Mediterranean-scale implications. *Bollettino Societa Paleontologia Italiana*, **47**, 71–85.

RYAN, W. & PITMAN, W. 1998. *Noah's Flood – The New Scientific Discoveries about the Event that Changed History.* Simon & Schuster, New York.

SCHULTZ, H.M., BECHTEL, A. & SACHSENHOFER, R.F. 2005. The birth of Paratethys during the early Oligocene: from Tethys to an ancient Black Sea Analogue? *Global and Planetary Change*, **49**, 163–176.

SEGHEDI, A. 2001. The North Dobrogea orogenic belt (Romania): a review. *In*: ZEIGLER, P.A., CAVAZZA, W., ROBERTSON, A.H.F. & CRASQUIN- SOLEAU, S. (eds) *Peri-Tethys Memoir 6: Peri-Tethyan Rift/Wrench Basins and Passive Margins.* Memoires Museum National d'Historie Naturelle, **186**, 237–257.

SEGHEDI, A. 2012. Palaeozoic Formations from Dobrogea and Pre-Dobrogea – an Overview. *Turkish Journal of Earth Science*, **21**, 669–721.

SEGHEDI, A. & OAIE, G. 1995. Paleozoic evolution of North Dobrogea. *In*: *Comparative Evolution of Peri-Tethyan Rift Basins, Field Guidebook, Central and North Dobrogea, Romania.* IGCP Project No. 369. Geological Institute of Romania, Bucharest, 5–16.

SEGHEDI, I. & SZAKACS, A. 1990. A model for Mesozoic volcanism and tectonic evolution in North Dobrogea, Romania. *Studii si Cercetari de Geologie*, **39**, 21–34.

SENGOR, A.M.C. & YILMAZ, Y. 1981. Tethyan evolution of Turkey: a plate tectonic approach. *Tectonophysics*, **75**, 181–241.

SINCLAIR, H.D., JURANOV, S.G., GEORGIEV, G., BYRNE, P. & MOUNTNEY, N.P. 1997. The Balkan Thrust Wedge and foreland basin of Eastern Bulgaria: structural and stratigraphic development. *In*: ROBINSON, A.G. (ed.) *Regional and Petroleum Geology of the Black Sea and Surrounding Region*. American Association of Petroleum Geologists, Memoirs, **68**, 91–114.

SNEL, E., MARUNTEANU, M., MACALET, R., MEULENKAMP, J.E. & VAN VUGT, N. 2006. Miocene to early Pliocene chronostratigraphic framework for the Dacic Basin, Romania. *Palaeogeography, Palaeoclimatology, Palaeoecology*, **238**, 107–124.

STEFAN, A., ROSU, E., BRATOSIN, I., VAJDEA, E., GRABARI, G. & STOIAN, M. 1992. Camena Rhyolites (North Dobrogea). *Romanian Journal of Petrology*, **75**, 39–52.

STEFANESCU, M., DICEA, O. & TARI, G. 2000. Influence of extension and compression on salt diapirism in its type area, East Carpathians Bend area, Romania. *In*: VENDEVILLE, B., MART, Y. & VIGNERESSE, J.-L. (eds) *Salt, Shale and Igneous Diapirs in and around Europe*. Geological Society, London, Special Publications, **174**, 131–147, https://doi.org/10.1144/GSL.SP.1999.174.01.08

STEPHENSEN, R. & SCHELLART, W.P. 2010. The Black Sea back-arc basin: insights to its origin from geodynamic models of modern analogues. *In*: SOSSON, M., KAYMAKCI, N., STEPHENSON, R.A., BERGERAT, F. & STAROSTENKO, V. (eds) *Sedimentary Basin Tectonics from the Black Sea and Caucasus to the Arabian Platform*. Geological Society, London, Special Publications, **340**, 11–21, https://doi.org/10.1144/SP340.2

STOFFERS, P. & MÜLLER, G. 1979. Carbonate rocks in the Black Sea basin: indicators for shallow water and subaerial exposure during Miocene–Pliocene time. *Sedimentary Geology*, **23**, 137–147.

STUART, C.J., NEMCOK, M., VANGELOV, D., HIGGINS, E.R., WELKER, C. & MEAUX, D.P. 2011. Structural and depositional evolution of the East Balkan thrust belt, Bulgaria. *AAPG Bulletin*, **95**, 649–673.

SUC, J.-P., COUTO, D.D. ET AL. 2011. The Messinian Salinity Crisis in the Dacic Basin (SW Romania) and early Zanclean Mediterranean–Eastern Paratethys high sea level connection. *Palaeogeography, Palaeoclimatology, Palaeoecology*, **310**, 256–272.

SUC, J.-P., GILLET, H. ET AL. 2015a. The region of the Strandja (North Turkey) and the Messinian events. *Marine and Petroleum Geology*, **66**, 149–164.

SUC, J.-P., POPESCU, S.M. ET AL. 2015b. Marine gateway v. fluvial stream within the Balkans from 6 to 5 Ma. *Marine and Petroleum Geology*, **66**, 231–145.

SWIDROWSKA, J., HAKENBERG, M., POLUHTOVIC, B., SEGHEDI, A. & VISNAKOV, I. 2008. *Evolution of the Mesozoic Basins on the south-western edge of the East European Craton (Poland, Ukraine, Moldova, Romania). Part I & Part II*. Studia Geologica Polonica, **130**.

TAMBREA, D. 2007. *Analiza de subsidenta si evolutia tectonica-termica a depresiunii Istria (Marea Neagra). Implicatii in generarea hidrocarburilor [Analysis of substation and tectonic-thermal evolution of the Istria (Black Sea) depression. Implications in hydrocarbon generation]*. PhD thesis, Bucharest University, Faculty of Geology and Geophysics.

TAMBREA, D., DINU, C. & SAMPETREAN, E. 2002a. Characteristics of the tectonics and lithostratigraphy and the Black Sea Shelf, offshore Romania. *In*: DINU, C. & MORCANU, V. (eds) *Geology and Tectonics of the Romanian Black Sea Shelf and its Hydrocarbon Potential*. Bucharest Geoscience Forum, Special Volume, **2**, 29–42.

TAMBREA, D., RAILEANU, A. & BOROSI, V. 2002b. Seismic facies and depositional framework of Eocene deposits. *In*: DINU, C. & MORCANU, V. (eds) *Geology and Tectonics of the Romanian Black Sea Shelf and its Hydrocarbon Potential*. Bucharest Geoscience Forum, Special Volume, **2**, 85–100.

TARI, G. 2005. The divergent continental margins of the Jurassic proto-Pannonian basin: implications for the petroleum systems of the Vienna Basin and the Moesian Platform. *In*: POST, P.J., ROSEN, N., OLSON, D.L., PALMES, S.L., LYONS, K.T. & NEWTON, G.B. (eds) *Transactions of the GCSSEPM Foundation Perkins-Rosen 34th Annual Research Conference 'Petroleum Systems in Rift Basins', Houston, Texas*, 955–986.

TARI, G., CIUDIN, D., KOSTNER, A., RAILEANU, A., TULUCAN, A., VACARESCU, G. & VANGELOV, D. 2011. Play types of the Moesian Platform of Romania and Bulgaria. AAPG Search & Discovery Article 10311, *AAPG European Region Annual Conference*, 17–19 October 2010, Kiev, Ukraine.

TARI, G., KOSI, W. ET AL. 2013. The end-Eocene drawdown in the Black Sea Basin. Oral presentation at the *AAPG Europe Conference, 26–27 September 2013*, Tbilisi, Georgia.

TARI, G., SCHLEDER, Z., KOSI, W., KREZEK, C., FALLAH, M. & TURI, V. 2014. Regional structure of the western Black Sea: map-view kinematics. Oral presentation at the *AAPG International Conference and Exhibition*, Istanbul, Turkey.

TARI, G., FALLAH, M., KOSI, W., FLOODPAGE, J., BAUR, J., BATI, Z. & SIPAHIOGLU, N.O. 2015. Is the impact of the Messinian Salinity Crisis in the Black Sea comparable to that of the Mediterranean. *Marine and Petroleum Geology*, **66**, 135–148.

TUYSUZ, O. 1999. Geology of the Cretaceous sedimentary basins of the Western Pontides. *Geological Journal*, **34**, 75–93.

USTAÖMER, T & ROBERTSON, A.H.F. 2010. Late Paleozoic–Early Cenozoic tectonic development of the Eastern Pontides (Artvin area) Turkey: stages of closure of Tethys along the southern margin of Europe. *In*: SOSSON, M., KAYMAKCI, N., STEPHENSON, R.A., BERGERAT, F. & STAROSTENKO, V. (eds) *Sedimentary Basin Tectonics from the Black Sea and Caucasus to the Arabian Platform*. Geological Society, London, Special Publications, **340**, 281–327, https://doi.org/10.1144/SP340.13

VASILIEV, I., KRIJGSMAN, W., STOICA, M. & LANGEREIS, C.G. 2005. Mid- Pliocene magnetostratigraphy in the southern Carpathian Foredeep and Mediterranean–Paratethys correlations. *Terra Nova*, **17**, 376–384.

VASILIEV, I., IOSIFIDI, A.G. ET AL. 2011. Magnetostratigraphy and radio-isotope dating of the upper Miocene–lower Pliocene sedimentary successions of the Black Sea Basin (Taman Peninsular, Russia).

Palaeogeography, Palaeoclimatology, Palaeoecology, **310**, 163–175.

VASILIEV, I., REICHART, G.-J. & KRIJGSMAN, W. 2013. Impact of the Messinian Salinity Crisis on Black Sea hydrology – insights from hydrogen isotopes analysis on biomarkers. *Earth and Planetary Science Letters*, **362**, 272–282.

VASILIEV, I., REICHART, G.-J. *ET AL.* 2015. Recurrent phases of draught in the upper Miocene of the Black Sea region. *Palaeogeography, Palaeoclimatology, Palaeoecology*, **423**, 18–31.

VINCENT, S.J., MORTON, A.C., CARTER, A., GIBBS, S. & BARABADZE, T.G. 2007. Oligocene uplift of the Western Greater Caucasus: an effect of initial Arabia Eurasia collision. *Terra Nova*, **19**, 160–166.

WINGUTH, C., WONG, H.K. *ET AL.* 2000. Upper Quaternary water level history and sedimentation in the north western Black Sea. *Marine Geology*, **169**, 127–146.

YILMAZ, Y. 1993. New evidence and model on the evolution of the southeast Anatolian orogen. *Geological Society America Bulletin*, **105**, 251–271.

YILMAZ, Y., TUYSUZ, O., YIGITBAS, E., CANGENC, S. & SENGOR, A.M. 1997. Geology and tectonic evolution of the Pontides. *In*: ROBINSON, A.G. (ed.) *Regional and Petroleum Geology of the Black Sea and Surrounding Regions*. American Association of Petroleum Geologists, Memoirs, **68**, 183–226.

YUDIN, V.V. 2009. *Geological Map of the Mountains and Foothills of Crimea*. Crimea Academy of Sciences, Simferopol, Crimea.

YUDIN, V.V. & KITCHKA, A.A. 2010. Understanding tectonic evolution of the Crimea to ensure strategy for hydrocarbon prospecting, offshore Ukraine. P021 Extended abstract at the *72th EAGE Conference & Exhibition*, 14 June 2010, Barcelona, Spain.

ZEIGLER, P.A. 1984. Caledonian and Hercynian crustal consolidation of Western Europe and Central Europe – a working hypothesis. *Geologisch Mijnbouwkundige*, **65**, 93–108.

ZEIGLER, P.A. (ed.) 1988. *Evolution of the Arctic-North Atlantic and Western Tethys*. American Association of Petroleum Geologists, Memoirs, **43**.

Cretaceous sedimentation along the Romanian margin of the Black Sea: inferences from onshore to offshore correlations

C. KREZSEK[1]*, R.-I. BERCEA[1], G. TARI[2] & G. IONESCU[1]

[1]*OMV Petrom, Petrom City, 22 Coralilor Street, 013329 Bucharest, Romania*

[2]*OMV Exploration and Production GmbH, Trabrennstraße 6–8, 1020 Vienna, Austria*

**Correspondence: csaba.krezsek@petrom.com*

Abstract: It is generally believed that the western part of the Black Sea opened during the Early Cretaceous. However, recent data and interpretation from the Turkish margin suggest rifting continued into the Coniacian or Santonian. In this review, the evidence related to the Black Sea rifting on the conjugate Romanian margin is reassessed. Our integrated interpretation of this region, supported by outcrop observations, core and detrital zircon data, suggests that rifting started during the Aptian and continued intermittently until the mid-Turonian in two distinct stages. These stages are bounded by significant unconformities and reflect the progressive widening of the rift system. The first synrift stage started in the Aptian with the deposition of fluvial and lacustrine clastic successions, and locally marine carbonates in semi-isolated depocentres. These sinks began to coalesce during the latest Aptian–Albian with shallow-marine transgression from the east, and deposition of coastal swamp, deltaic and littoral facies. The second phase of rifting during the Cenomanian was marked by transgressive shallow-marine deposits overstepping the earlier Albian depocentres. Continental break-up followed in the mid-Turonian associated with regional uplift and erosion of the basin margin and the local deposition of fluvial conglomerates.

The western Black Sea is one of the largest and relatively unexplored hydrocarbon basins in Europe (Robinson *et al.* 1996). Significant oil and gas reserves have been discovered in the syn- and post-rift Cretaceous section of offshore Romania (Ionescu 2000; Ionescu *et al.* 2002) and Bulgaria (Georgiev 2012). Therefore, understanding facies variability and its relationship to the tectonics is very important in predicting the reservoir potential.

Most authors agree that the rifting in the Black Sea began during the Early Cretaceous (Letouzey *et al.* 1977; Kazmin *et al.* 1986; Zonenshain & Le Pichon 1986; Görür 1988; Okay *et al.* 1994; Banks & Robinson 1997) and reached an oceanic stage during the early part of the Late Cretaceous (Görür 1997; Tüysüz *et al.* 2012, 2016; Okay *et al.* 2017). The age of rifting along the Romanian margin has been variously interpreted as Barremian and/or Aptian–Cenomanian (Ionescu 2000; Dinu *et al.* 2005; Țambrea 2007). There is less certainty about the age of break-up unconformity and it is often referred to rather vaguely as having occurred at some time during the Turonian–Maastrichtian period (Munteanu *et al.* 2011).

The largest synrift section offshore Romania is about 2 km thick. Onshore, the thickness of the synrift is generally less than a 1 km (Bucur & Baltres 2002). These depocentres are certainly smaller than the ones exposed in the Pontides (Hippolyte *et al.* 2010), but still provide important constraints about

the basin opening. Therefore, our objectives in this paper are: (1) to revisit the lines of evidence for synrift tectonics along the Romanian margin; and (2) to illustrate the evolution of the synrift sedimentary environments. We believe our observations will help to better understand the timing and the architecture of opening in the wider Black Sea region.

Data and methods

Our analysis is based on a diverse dataset that includes 2D and 3D seismic data, well logs and cores from the OMV Petrom Exploration licences offshore (Fig. 1) supported by published onshore and offshore data, and interpretations from Ionescu (2000), Dinu *et al.* (2005), Țambrea (2007) and Munteanu *et al.* (2011).

Several seismic horizons tied to wells were mapped using seismic interpretation software. Particular attention was given to mapping the syn-extensional growth packages and the unconformities. Interpretation was difficult in some areas due to the limited number of wells, the post-Cretaceous inversion events and associated unconformities, and the large-scale Cenozoic incisions, which have removed a significant amount of the Cretaceous succession locally (Dinu *et al.* 2005).

The fault network has been regionally mapped on seismic data and tied to onshore geology. Defining

From: SIMMONS, M. D., TARI, G. C. & OKAY, A. I. (eds) 2018. *Petroleum Geology of the Black Sea.*
Geological Society, London, Special Publications, **464**, 211–245.
First published online September 28, 2017, https://doi.org/10.1144/SP464.10

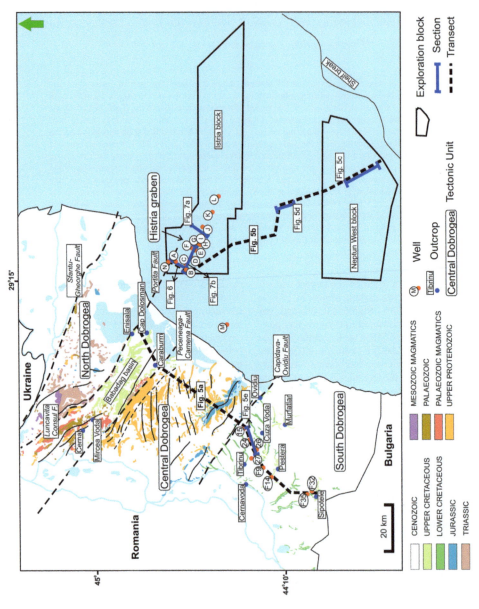

Fig. 1. The database shown over a simplified geological map of the Romanian margin (Chiriac & Mînzatu 1967; Chiriac *et al.* 1968; Mîrăuță *et al.* 1968; Săndulescu *et al.* 1978).

the amount of extension along the faults was often challenging because of late inversion (Munteanu *et al.* 2011). Nevertheless, we were able to construct time–thickness maps (isochores) illustrating the extensional growth packages within the main depocentres.

Most Cretaceous offshore data come from the hydrocarbon-producing fields of the Histria half-graben (Fig. 1). Tens of metres of cores were described using industry software. Sedimentary structures and biofacies data were then utilized to calibrate the depositional facies of well logs and well log motifs to infer the evolution of the synrift depositional systems.

Additional geological constraints came from the nearby onshore areas. The Cretaceous often crops out close to the Black Sea coast, most importantly in the onshore extension of the Histria half-graben (often referred as the 'Babadag Basin': e.g. Szász & Ion 1986), and further to the south in South Dobrogea (Chiriac & Mînzatu 1967, Chiriac *et al.* 1968; Avram *et al.* 1993; Antoniade 2016) (Fig. 1). The Cretaceous has been also intensively studied along the northern Turkish margin (e.g. Hippolyte *et al.* 2010; Tüysüz *et al.* 2012; Okay *et al.* 2017) providing some useful constraints on the conjugate margin. To facilitate comparison between the onshore and offshore, we have compiled a stratigraphic chart illustrating the key Cretaceous sections and the most significant unconformities discussed in this review (Fig. 2).

Previous U–Pb detrital zircon geochronology analysis in the broader Dobrogea area was focused on Palaeozoic and older sequences (e.g. Balintoni *et al.* 2010, 2011; Balintoni & Balica 2016). Here we report, for the first time, on detrital zircon data derived from the Cretaceous outcrops of South Dobrogea and cores from offshore wells. Only five samples have been analysed to date, therefore our preliminary conclusions will need confirmation by additional samples in the near future. However, the results are sufficient to provide some insights about provenance areas and to attempt a rough comparison with the Turkish margin.

Overview of the geology of the Romanian part of the Black Sea

Most authorities agree the western Black Sea is an extensional basin developed in the back-arc region of the Pontides during the Cretaceous (e.g. Görür 1988; Okay *et al.* 1994; Munteanu *et al.* 2011; Nikishin *et al.* 2015; Schleder *et al.* 2015; Boote 2017). However, although previously considered to reflect simple back-arc extension, additional mechanisms are now considered to be at least partially responsible for the extension in the basin (Tari 2015). Although different basin-opening models

have been proposed by various authors (Görür 1988; Okay *et al.* 1994; Robinson *et al.* 1996; Nikishin *et al.* 2015; Okay *et al.* 2017), most of them agree that during the Early Cretaceous (Aptian?–Albian), a few narrow, possibly interconnected, rift basins have developed in the NW Black Sea (Fig. 3). Extension progressed to the south as the western Black Sea opened either by: (1) clockwise rotation as the Pontides broke away from the Bulgarian, Romanian and Ukrainian margins along a major transform fault zone (Fig. 3) (Schleder *et al.* 2015; Tari *et al.* 2015); or (2) extension accommodated by a major strike-slip fault zone located west of the Istanbul zone (e.g. Okay *et al.* 2017). In any case, the amount of rift associated with the extension is estimated to have been at least 250 km (Schleder *et al.* 2015).

In contrast to the more locally developed Lower Cretaceous, Upper Cretaceous and younger sequences are widespread across the Western Black Sea Basin (Fig. 4). The Senonian–Eocene interval is dominated by deep-marine sediments reflecting a gradual and steady deepening of the basin margin (Fig. 2) (e.g. Ţambrea 2007; Boote 2017). This was interrupted by a number of sea-level falls (Dinu *et al.* 2005), most importantly during the latest Eocene (Tari *et al.* 2014) incising deep canyons into the underlying Cretaceous deposits. Munteanu *et al.* (2011) suggested that this erosion occurred in response to inversion tectonics of the Balkanides and Moesian Platform. The incised valleys were quickly flooded, and during the Oligocene were filled by deep-marine Maykop shales (e.g. Sachsenhofer *et al.* 2013).

The modern shelf of the Romanian Black Sea extends more than 100 km from the coastline to the slope break (Lericolais *et al.* 2013), reflecting large-scale progradation of the basin margin, from the Mid-Miocene to the present (Bega & Ionescu 2009). Progradation was interrupted several times by major sea-level falls, notably in the Mid-Sarmatian, intra-Pontian and intra-Dacian. At these times, the shelf was exposed and sedimentation was restricted to the outboard part of the shelf and slope. The Mid-Sarmatian and intra-Dacian unconformities are more probably related to the Carpathians compressional tectonics ('Styrian' and 'Vallachian' phases: Sǎndulescu 1988), while the intra-Pontian unconformity was triggered by a sea-level drop related to the Messinian salinity crisis in the Mediterranean (Krezsek *et al.* 2016).

Regional interpretation

A number of key Cretaceous horizons have been identified on the Romanian shelf and correlated with time-equivalent events onshore. These include the base of the Aptian, the top of the Albian and the top of the Lower Turonian.

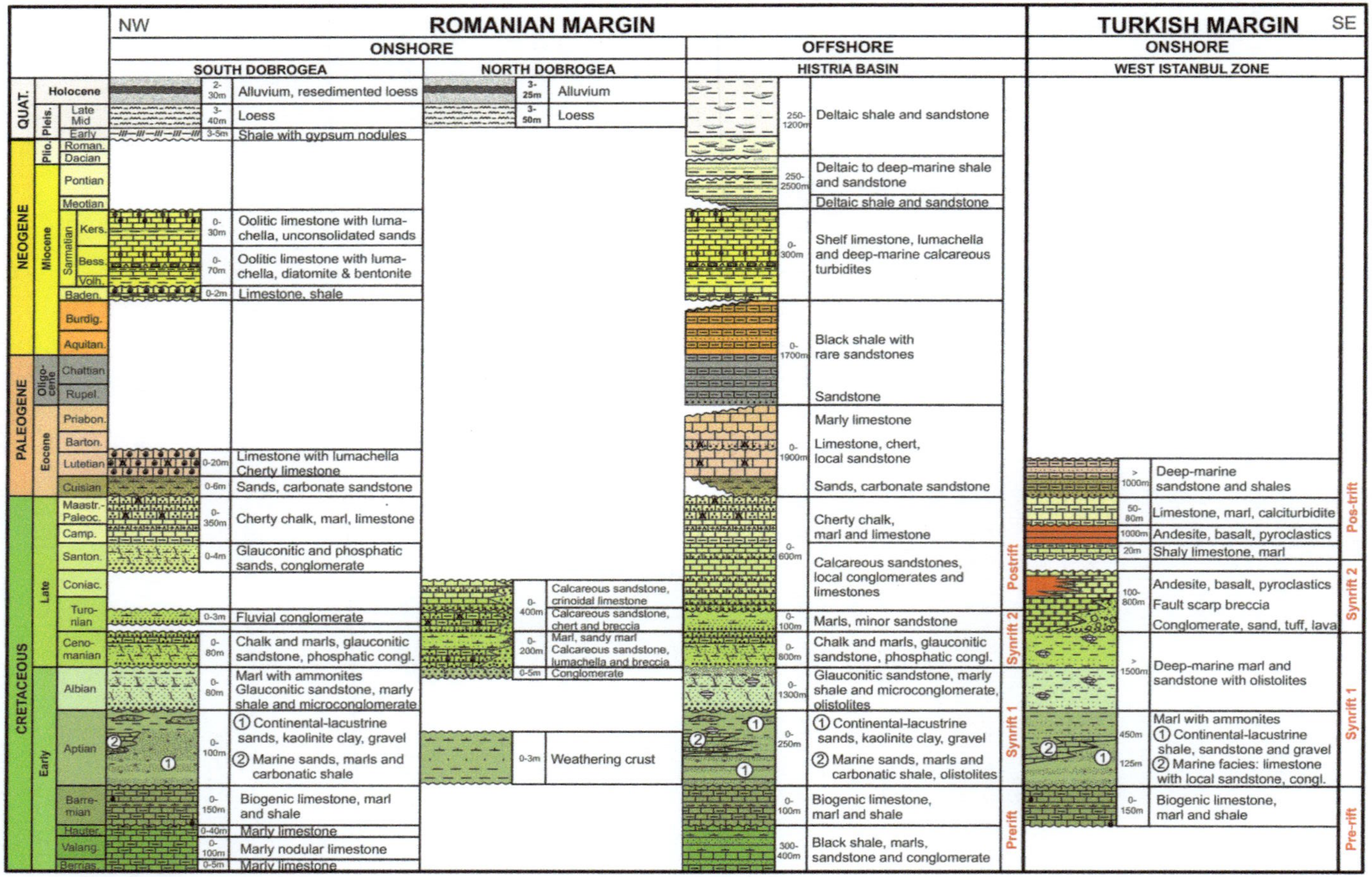

Fig. 2. Stratigraphic charts of onshore (South and North Dobrogea) and offshore (Histria half-graben) Romania correlated with the onshore Turkey (Istanbul Zone). The stratigraphy of the Romanian margin was compiled based on Chiriac *et al.* (1968), Szász & Ion (1986), Dinu *et al.* (2005), Țambrea (2007) and Munteanu *et al.* (2011), and the Turkish margin based on Okay *et al.* (2017).

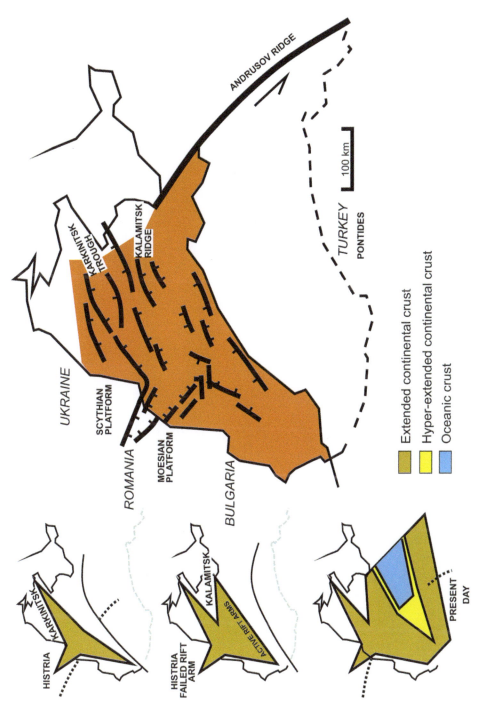

Fig. 3. The map shows the restored Early Cretaceous location of the Turkish coastline with the present-day rift-related normal faults. This opening model is based on Schleder *et al.* (2015) and Tari (2015).

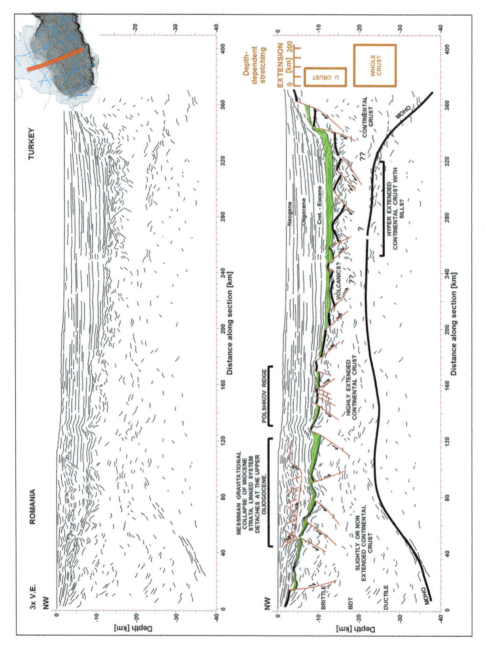

Fig. 4. Large-scale architecture of the western Black Sea between offshore Romania and Turkey (based in Schleder *et al.* 2015). The Lower Cretaceous synrift section (in green) is much thinner compared to the post-rift succession. V.E., vertical exaggeration.

Pre-Aptian

The base of the Aptian is an unconformity that delimits undeformed Aptian above from the thrusted and folded older sediments and basement below (Ionescu 2000; Dinu et al. 2005; Ţambrea 2007) (Fig. 5a–e). Locally, the Aptian is missing and the Albian, or even the Cenomanian, covers the deformed deposits below. The youngest deformation phase of the pre-Aptian units is related to the Cimmerian Orogeny (Nikishin et al. 2011).

The pre-Aptian geology offshore is broadly comparable to the stratigraphy exposed onshore (Fig. 1) where four NW–SE-trending tectonic units (South Dobrogea, Central Dobrogea, North Dobrogea Orogen and the Scythian margin/pre-Dobrogea) are bounded by the Capidava-Ovidiu, Peceneaga-Camena and Sfântu-Gheorghe strike-slip fault systems (Seghedi et al. 2010). The Peceneaga-Camena fault system has been variously interpreted as a transtensional (Gradinaru 1988) and compressional (Ianovici et al. 1961) fault system during the Late Jurassic related to the Cimmerian Orogeny (e.g. Seghedi et al. 2010; Nikishin et al. 2011). This period of deformation is responsible for the highly structured Jurassic and older sequence now outcropping in North Dobrogea and partially covered unconformably by flat-lying Upper Cretaceous (Săndulescu et al. 1978; Săndulescu 1988). The equivalent Late Cimmerian units offshore are interpreted as pre-rift 'basement' on our sections (e.g. Fig. 5b).

Aptian

Onshore, the largest, approximately 200 m-thick, section of the Aptian depocentre in South Dobrogea is exposed along the Capidava-Ovidiu Fault (Fig. 5a–e). Shallow coal exploration wells drilled in this area suggest the Aptian is thick just south of the Capidava-Ovidiu Fault and wedges out further to the south, and is virtually missing to the north (Chiriac et al. 1968). This we have interpreted to reflect a relatively small normal fault-bounded extensional depocentre, the 'Ţibrinu graben'. Alternatively, it may represent a down-faulted block near the Capidava-Ovidiu Fault where the Aptian has been preserved from later erosion. Elsewhere across Central and North Dobrogea, the Aptian is absent either by erosion or non-deposition, reappearing locally as erosional remnants in the pre-Dobrogea (Rădan 2000).

Albian

The Albian is regionally much more extensive than the Aptian and was deposited unconformably over older rock units (Fig. 5a, b). It overlies the Aptian near the Capidava-Ovidiu Fault and the folded

'pre-rift' Neocomian carbonates further to the south (Avram et al. 1993). Offshore, on seismic sections, the Albian is wedge shaped and is thickening into extensional faults (Fig. 6). In the Histria half-graben (offshore), it reaches several hundreds of metres in thickness (Figs 5b & 6); while, in contrast, it is represented by conglomerates a few metres thick lying discordantly on folded Triassic carbonates onshore at Enisala (Fig. 1) (Gradinaru 2002).

The top of the Albian sequence is eroded, as is frequently observed offshore on seismic sections (Fig. 5d) and also onshore in outcrops (Szász & Ion 1986; Avram et al. 1993). This minor erosional unconformity is transgressively covered by Lower Cenomanian deposits.

Cenomanian

The Cenomanian is typically thicker than the underlying Albian sequence, ranging from about 200–800 m onshore (Fig. 5a) (Szász & Ion 1986; Bucur & Baltres 2002) to more than 1 km offshore (Fig. 5b). It is typically wedge shaped, thickening into major extensional faults in a synkinematic manner (Figs 5 & 6). Onshore, in the 'Babadag Basin', the Cenomanian thickens into the Luncavita-Consul Fault. Near the fault, the Cenomanian facies consists of conglomerates (Szász & Ion 1986). The depositional thickness of the Cenomanian in South Dobrogea is difficult to assess because of later erosion (Fig. 5a).

Turonian

The Turonian is tens of metres thick offshore (Fig. 5b–d) and it is exposed onshore only in the 'Babadag Basin' (Fig. 5a). It features an important, but poorly dated, intra-Turonian (Mid-Turonian?) unconformity (Fig. 2) (Szász & Ion 1986). This erosional surface is much more pronounced offshore, where the Lower Turonian fine-grained deposits are partly eroded and abruptly overlain by Mid?–Upper Turonian conglomerates (Fig. 7a, b). In contrast to the 'Babadag Basin', the Turonian is practically missing from South Dobrogea apart from some conglomerates and sandstones interpreted as Mid-Turonian in age (Fig. 2) (Avram et al. 1993; Chiriac 1981 in Dragastan et al. 1998).

Senonian

The Senonian deposits are regionally developed across the offshore not only in the extensional depocentres, but extending well over pre-rift structural highs blanketing older deposits without any evidence of extensional growth (Figs 5b–d, 6 & 7). Similar architecture is observed locally onshore. In

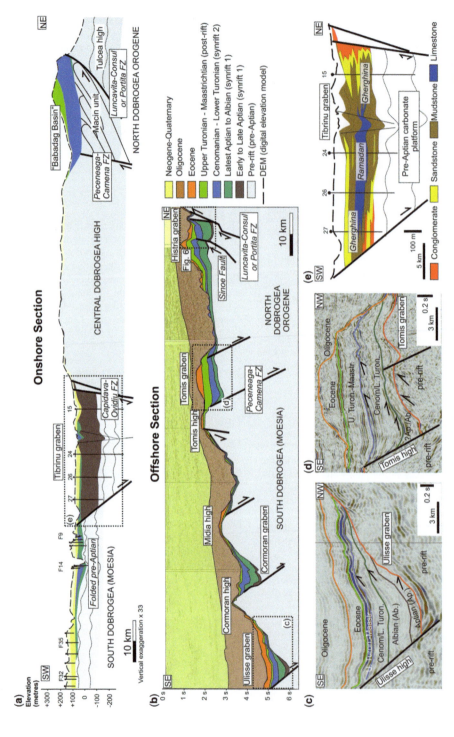

Fig. 5. (**a**) This onshore section illustrates the western end of the Black Sea Basin on the Romanian margin and features two Cretaceous depocentres: the 'Tibrinu graben' and the 'Babadag Basin'. (**b**) The most important offshore extensional depocentres and intra-basinal highs along the Romanian margin. (**c**) & (**d**) Seismic details of the Ulisse and Tomis half-graben (shown in Fig. 5a) illustrating synrift architecture and unconformities. (**e**) Aptian facies details of the 'Tibrinu graben' (Fig. 5a) based on shallow well data (Avram *et al.* 1993).

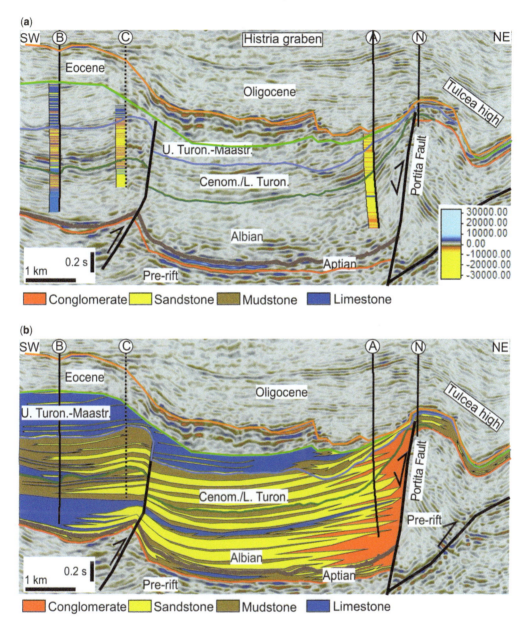

Fig. 6. (a) Seismic detail of the Histria half-graben (Fig. 1) with inverted synrift faults. The facies drilled by the wells illustrate fining and more calcareous depositional trends away from the extensional faults. (b) Overlay of the facies interpretation of the Cretaceous fill of the Histria half-graben. The synrift sedimentary architecture is interpreted related to the activity of the Portita Fault. The post-rift (upper Turonian–Maastrichtian) succession is dominated by limestones that unconformably cover the break-up unconformity.

addition, an important Upper Turonian–Coniacian unconformity may be observed in South Dobrogea (Dragastan *et al.* 1998) (Fig. 2). In turn, in the 'Babadag Basin' (Szász & Ion 1986), the Santonian–Maastrichtian strata are missing due to erosion occurring some time during the Cenozoic.

Extensional depocentres

The top Albian and top Mid-Turonian unconformities split the Cretaceous stratigraphy into three packages: (1) Aptian–Albian; (2) Cenomanian– Mid-Turonian; and (3) Upper Turonian–Senonian. Of these, only

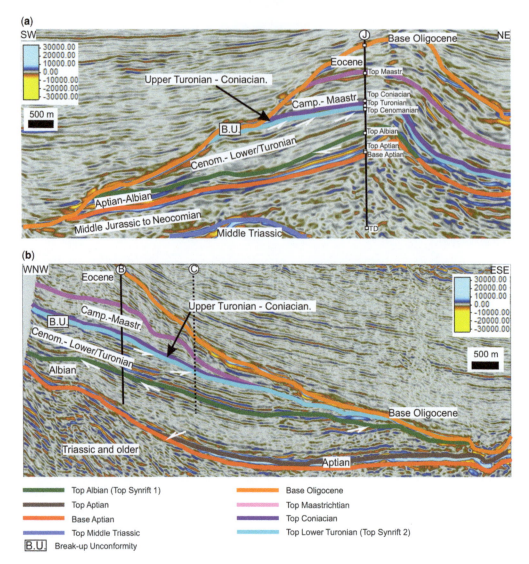

Fig. 7. (**a**) & (**b**) Seismic details of the Histria half-graben featuring the key stratigraphic horizons. The lines are located in Figure 1. The top Albian and in the mid-Turonian unconformities are interpreted to mark two different phases of rifting (Tari 2015). The mid-Turonian unconformity is interpreted as the break-up unconformity of the western Black Sea.

the first two packages show evidence of syn-extensional growth.

Time–thickness maps illustrate the distribution and evolution of the synrift depocentres (Fig. 8) with synrift faults superimposed. These were identified by their association with synsedimentary growth packages combined with isochore changes on the maps (Figs 5, 6 & 8).

It is easy to observe that the onshore structural trends continue offshore and most of the faults appear to have been active during rifting (Fig. 8).

However, the fault network onshore predates Cretaceous extension of the Black Sea (Hippolyte 2002; Seghedi *et al.* 2010). Consequently, it appears that the pre-Cretaceous basement fabric offshore must have been reactivated as the western Black Sea began to open. This extension is progressively dying out onshore, as indicated by only weak NW–SE-orientated deformation measured in Dobrogea (Hippolyte 2002) (Fig. 8).

The most prominent extensional depocentres are located offshore next to the most important

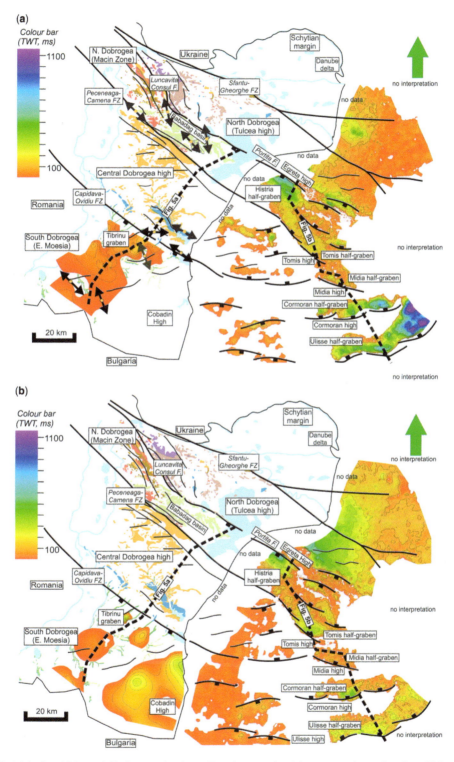

Fig. 8. (**a**) Aptian–Albian and (**b**) Cenomanian–lower Turonian extensional depocentres observed on time–thickness isochore maps (two-way time (TWT) in ms). For the discussion, refer to the text.

222   C. KREZSEK ET AL.

pre-Cretaceous Luncavita-Consul, the Peceneaga-Camena and the Capidava-Ovidiu fault systems (Fig. 8). The reactivation of these faults controlled two major, rather broadly defined, extensional depocentres represented by half-graben systems: the Histria–Tomis and the Ulisse–Cormoran (Fig. 8). These depositional areas are delimited by significant horst blocks with no or limited extension: the Tulcea high, the Central Dobrogea–Tomis high, and the Midia, Cormoran and Ulisse highs (Fig. 8).

Histria and Tomis half-graben and the 'Babadag Basin'

The offshore Histria half-graben is confined to the south by the Peceneaga-Camena Fault Zone and the Portita Fault to the north (Figs 6–8) (e.g. Dinu et al. 2005; Munteanu et al. 2011). In map view, the trace of the offshore Portita Fault lines up with the Luncavita-Consul Fault (Figs 1 & 8). This separates the highly deformed and thrusted Macin Unit and the folded Tulcea Unit within the North Dobrogea Orogen (Mutihac 1964; Seghedi 2001). Therefore, we interpret the Portita Fault as the offshore segment of the Luncavita-Consul Fault that has been reactivated extensionally during the Early Cretaceous. This is also supported by the seismic character of the pre-rift section on the north side of the Portita Fault which is gently folded and dominated by coherent seismic reflections, and is folded similar to reminiscent of the Tulcea Unit, while to the south they are much more chaotic and, perhaps, characteristic of highly structured metamorphic rocks like those in the Macin Unit.

Onshore, in the 'Babadag Basin', minor mid-Cretaceous NW- to SE-orientated extension has been described near and between the Luncavita-Consul and Peceneaga-Camena fault systems (Fig. 8a) (Hippolyte 2002). Wells drilled in this area encountered Cenomanian–Coniacian about 1600 m thick (Bucur & Baltres 2002), but virtually no wells outside of the 'Babadag Basin' intersected this interval (Figs 5a & 8) (Szász & Ion 1986). Although this thickness difference may reflect later erosion, it may also be interpreted as being due to extension, albeit much weaker than in the offshore.

Offshore, the Aptian–Albian is thickening into the Portita Fault (Figs 6 & 8a). Much less sediment is recorded along the Peceneaga-Camena Fault Zone. These indicate active Aptian and Albian tectonics along the Portita Fault and less along Peceneaga-Camena. In turn, by the Cenomanian, sediments are equally thick along both faults. Therefore, it seems that the extensional activity along the offshore Peceneaga-Camena fault systems is increasing from the Aptian–Albian to the Cenomanian.

The isochore patterns suggest that the polarity of the fault-controlled subsidence changed across a relay zone from the Histria half-graben during the Aptian–Albian in the north to the Tomis half-graben in the Cenomanian–Lower Turonian (south). Further offshore, beyond the limit of the mapped area (Fig. 8a, b), the orientation of the Histria half-graben changes to a more east–west trend, on strike with the structural fabric observed in the Odessa shelf (Fig. 3). Schleder et al. (2015) suggested that this change of orientation may reflect a triple-junction organization of rifting along the Romanian margin.

Ulisse and Cormoran half-graben

The Capidava-Ovidiu Fault can be traced offshore where it turns east to merge with the Peceneaga-Camena on the Midia high (Fig. 8). Thus, Central Dobrogea is interpreted to wedge out to the east between North and South Dobrogea (Figs 5 & 8) (Munteanu et al. 2011).

Two-east facing (Ulisse and Cormoran) half-graben can be observed (Fig. 8a, b) on the outer part of the continental shelf just south of the merged Capidava-Ovidiu–Peceneaga-Camena fault system with a different, more west–east, structural orientation than the Histria and Tomis half-graben (Fig. 8). This orientation reflects the basement fabric of Moesia rather than that of Central and North Dobrogea (Tari et al. 1997; Seghedi 2001; Hippolyte 2002). As they are undrilled, the stratigraphy of their fill sequences is uncertain. However, their synrift fill architecture is very clear (Fig. 5c, d) with a similar seismic character to the Histria half-graben. This similarity suggests a generally equivalence in age and facies.

'Ţibrinu graben'

Despite the widespread Cretaceous extension observed offshore, extension onshore South Dobrogea has rarely been discussed, in part because earlier studies have focused on the outcropping stratigraphy rather than structure (e.g. Chiriac et al. 1968; Avram et al. 1993; Antoniade 2016). The only significant exceptions are those presented by Ciulavu in Seghedi et al. (1999) and Hippolyte (2002). Both authors measured mid-Cretaceous east–west deformation along the Capidava-Ovidiu Fault in outcrops near Ovidiu (Fig. 1). We believe this extension may have continued further west along the fault, as suggested by Hippolyte (2002), and controlled the deposition of the Aptian–Albian (Fig. 7a) within an extensional depocentre, the 'Ţibrinu graben'. Additional evidence includes: (a) the thickening pattern of the Aptian into the Capidava-Ovidiu Fault (Chiriac et al. 1968); (b) the absence of Aptian deposits north of the fault (onshore Central Dobrogea); and (c) the Aptian facies distribution encountered by wells illustrating synsedimentary tectonics (Fig. 5e).

In addition, the Aptian–Albian and the Cenomanian thickness maps (Fig. 8) constructed from well data recorded by Avram *et al.* (1993) display a number of SW- to NE-trending depocentres orientated at 60° to the Capidava-Ovidiu Fault Zone suggesting the possibility that fault movement may have had a strike-slip component during the deposition of the Cretaceous sequence. Yet, given the paucity of the extensional deformation observed in the area (Hippolyte 2002), we cannot rule out that an alternative explanation for at least some of the features presented above is that an originally more continuous Neocomian–Aptian section which once covered most of Dobrogea sealing the Late Cimmerian structures (Săndulescu *et al.* 1978; Săndulescu 1988) may have been preserved here in a faulted depression south of the Capidava-Ovidiu Fault. Nevertheless, the Cretaceous facies exposed here and the sedimentary cycles are very similar to the offshore, as will be discussed in the next section.

Depositional facies

A comprehensive description of mid-Cretaceous synrift sedimentary facies and depositional environments has been synthesized from outcrops and offshore core data.

Aptian–Albian sequence

The depositional facies of the Aptian and Albian is constrained by several offshore well core samples, outcrops in the 'Ţibrinu graben' and red weathering crusts of Aptian age (Rădan 2000) exposed in Central and North Dobrogea.

Early–Mid? Aptian. The Lower Aptian interval encountered in Well M (Fig. 9d) offshore Central Dobrogea comprises red conglomerates, whereas the equivalent section in the Histria half-graben consists of grey, green, and black mudstones interbedded with thin quartz-rich glauconite sandstones and sandy limestones (Well F: Fig. 10a). The stratigraphic relationship between the two facies is not well understood, but their age was confirmed by the presence of Aptian foraminifera and palynomorphs. Their lithologies and fauna suggest continental setting with marine incursions (Dragastan 2008; Świdrowska *et al.* 2008).

Simiar deposits crop out onshore. Coarse-grained red fluvial conglomerates are well known on Southern Dobrogea (Dragastan *et al.* 1998; Antoniade 2016). In the classical outcrop at Cernavoda (Fig. 1), the reddish cross-bedded fluvial conglomerates erode deeply into the Valanginian platform limestones (Fig. 9a–c). Similar red conglomerates have been encountered offshore in Well M (Fig. 9d). These conglomerates were not encountered in wells in the 'Babadag Basin', where instead a red pre-Albian weathering crust (Fig. 9d) has been described (Fig. 9e) (Rădan 2000). This crust may be at least partially coeval with the red conglomerates.

The red conglomerates are overlain by continental–lacustrine deposits (Gherghina Formation: Avram *et al.* 1993). The package contains kaolinite clays and quartz-rich sandstones that are intensely mined all along the Capidava-Ovidiu Fault. In one of the quarries, at Ţibrinu (Figs 1 & 11a), the grey kaolinite clay, locally interbedded with coal layers (Fig. 11b), is exploited in the lower part of the quarry. Locally, the clays are yellow and red due to palaeo-soil levels (Fig. 11d). The kaolinite clays were deposited in a freshwater lake during the Aptian, as suggested by the charophyta assemblages (Avram *et al.* 1993) and ostracod fauna (Antoniade 2016). The ostracod assemblages and the acritarchs from the coaly intercalations suggest that some marine ingressions occurred during the overall continental–lacustrine deposition (Avram *et al.* 1993). Other evidence of the marine ingressions is the less than 1 m-thick limestones with marine fauna (Ramadan Formation: Avram *et al.* 1993), which may be locally found interbedded in the continental–lacustrine succession (Fig. 9b). The presence of coals and palaeo-soils also suggest frequent lake-level changes, which led to eutrophication and pedogenesis. Regional studies indicate that the kaolinite levels have variable thickness and often pass laterally into quartz-rich fluvial deposits (Avram *et al.* 1993; Antoniade 2016). These fluvial deposits are observed in the upper part of the Ţibrinu quarry, where cross-bedded, quartz-rich, yellow fluvial sandstone a few metres thick marks the end of the continental–lacustrine succession (Fig. 11a). They differ in facies from the reddish fluvial conglomerates of the Cernavodă Bridge outcrop (Fig. 9b, c). The sandstones are poorly consolidated subarkose with a high content of well-rounded quartz grains, and exhibit large-scale cross-bedding, with several concave-upwards erosional surfaces (Fig. 11c). No fauna has been found so far, but continental pollen has been described from this location. These deposits are interpreted as gravel-rich braided fluvial deposits.

Late Aptian–Early Albian. Marine waters invaded the Histria half-graben and South Dobrogea (Cochirleni Formation: Avram *et al.* 1993) starting with the Late Aptian. Both offshore and onshore, this is expressed by glauconite-rich shallow-marine sandstones overstepping the former continental–lacustrine depocentres (Figs 10 & 11).

Offshore, in Well E, the Upper Aptian consists of 1 m-thick, fine to medium, well- to poorly-sorted cross-bedded glauconitic sandstones (Fig. 10b–d) intercalated with thin conglomerates (Fig. 10e),

Fig. 9. (**a**) Outcrop of Cernavoda (Fig. 1) with folded and eroded Valanginian platform carbonates (Cernavoda Formation) overlain by Lower Aptian red fluvial conglomerates (Gherghina Formation) and marine limestones (Ramadan Formation). (**b**) & (**c**) Sedimentary details of the red fluvial cross-bedded conglomerates. (**d**) Red fluvial conglomerate in offshore well core (Well M, Fig. 1), similar in facies to Figure 9b, c. (**e**) Red weathering crust, several metres thick, developed over the Palaeozoic basement of North Dobrogea near Cerna (Fig. 1).

including occasional bioclasts and coal clasts. The sandstone beds sometimes display normal grading (Fig. 10b) and are locally cross-bedded with mud drapes (Fig. 10c, d). Vertical and horizontal bioturbation (Fig. 10b, d) and cryptobioturbation is quite pervasive (the bioturbation index (BI) ranges from 2 to 4: Taylor & Goldring 1993) with Skolithos-type ichnofacies (Fig. 10d), including *Paleophycus* isp., *Ophiomorpha* isp. and *Macaronichnus* isp. The sedimentary succession is interpreted to represent tidal channels and tide-modified mouth bars.

Onshore, the Late Aptian transgression may be nicely observed in the upper part of the quarry at Ţibrinu (Fig. 11a). There, the Mid?-Aptian quartz-rich

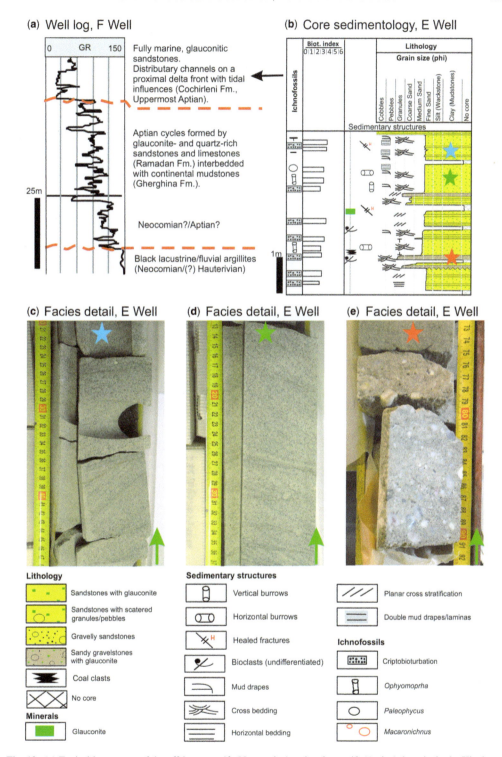

Fig. 10. (**a**) Typical log pattern of the offshore pre-rift (Neocomian) and early synrift (Aptian) deposits in the Histria half-graben. (**b**) Core description of the latest Aptian delta/shoreface deposits. (**c**) & (**d**) Cross-bedded and bioturbated upper shoreface sandstones with tidal mud drapes. (**e**). Gravelly distributary channel.

Fig. 11. (**a**) Overview of the kaolinite quarry at Ţibrinu (Fig. 1) with Aptian–Albian synrift deposits overlain by Mid-Miocene limestones and then by Quaternary loess. (**b**) The lower part of the quarry exposes whitish and reddish kaolinite clays with coals (Gherghina Formation). (**c**) Quartz-rich Aptian fluvial cross-bedded sandstones located above the kaolinite clays. (**d**) Transgressive glauconite-rich lower shoreface sandstones of Late Aptian–Early Albian age (Cochirleni Formation) above the fluvial sandstones. (**e**) & (**f**) Sedimentary features of the transgressive sandstones (Fig. 11d): bioturbations, frequent bioclasts, glauconite and diagenetic concretions.

fluvial sandstones are overlain by a transgressive lag of a heterogeneous conglomerate, a few centimetres thick, with frequent shale rip-up clasts (Fig. 11d), and then by glauconitic, intensely bioturbated sandstones with a Late Aptian–Early Albian shallowmarine fauna (Fig. 11e, f).

Mid-Albian–Late Albian. The shallow-marine sedimentation persisted on the Romanian margin of the Black Sea during the rest of the Albian. Offshore, the deposits became more calcareous, shown by the occurrence of limestones and sandy limestones forming several well-organized high-frequency depositional cycles (Figs 12 & 13). Typically, each cycle starts with bioturbated fine-grained glauconitic sand and mud (e.g. wells H and E: Fig. 12a), sometimes with pyrite, grading up into coarser often crossbedded sandstones (Fig. 12a–d), sandy conglomerates and calcarenites (Fig. 12e). Glauconite becomes less common upwards, and the calcarenite lithofacies is better developed away from the active normal faults (Figs 6 & 13a) and the main clastic inputs. The finer-grained mudstones, also muddy/silty fine to very fine sandstones, are typically bioturbated and rather cyclic in character (Fig. 13b–d) with variable amounts of normal graded thin and fine gravelly sandstones (Fig. 13d). Recognizable burrows include, for example, *Thalassinoides* isp., *Planolites* isp., *Asterosoma* isp., *Ophiomorpha* isp., *Phycosiphon* isp., *Rhizocorallium* isp. and *Macaronichnus* isp., and escape traces (fugichnia) and criptobioturbation were observed (Fig. 13b–e). Much of primary stratification has been homogenized by burrowing, but hummocky-like structures, flames/load casts, planar parallel laminations and asymmetrical ripples could be identified. Mud drapes (often double) and mudstones rip-up clasts are also very common.

The coarsening-up profiles and associated sedimentary structures represent high-frequency parasequences with offshore/prodelta facies passing up into upper shoreface/delta-front sediments cut locally by distributary channels (Figs 10 & 12b, d, e). Thin normally graded and hummocky sandstones (Fig. 13b, d, e) suggest occasional storm events, while a tidal component is inferred from frequent mud drapes and mud-sand laminations (Fig. 13b, e). Very often, the primary depositional structures are disrupted, often completely wiped out by the intense bioturbation.

Similar sedimentary cycles are observed onshore South Dobrogea, where sediments are grouped into two different facies associations. The first facies unit is coarse grained, represented by glauconite-rich (often over 30% volume content), intensely bioturbated, quartz-rich or muddy sandstones with strings of microconglomerates. Locally, the sandstones feature large-scale planar or through cross-bedding. The second facies unit is finer grained with grey siltstones

and marls. Both are rich in ammonites and other fauna, pointing to a Late Aptian–Mid-Albian age (Avram *et al.* 1993). The succession ends with a major flooding event that produced about 1 m of shales rich in phosphates, glauconitites and lumachelles, with pelocypods, gasteropods and ammonites of Late Albian age (Seimenii Mari in Avram *et al.* 1993).

Interpretation. The Early Aptian red conglomerates are interpreted as high-energy braided fluvial deposits and alluvial fans. These depositional environments were progressively replaced during the Mid-Aptian by continental lowlands where ephemeral lakes coexisted with major fluvial channel belts (Gherghina Formation: Avram *et al.* 1993). It is most likely that, by this time, the Early Aptian depocentres must have been interconnected, as illustrated by the presence of large fluvial systems carrying well-sorted quartz-rich sands. Other evidence for the connection is the marine limestones (Ramadan Formation) locally found interbedded in the continental–lacustrine succession.

The Late Aptian transgression coalesced most of the extensional depocentres on the Romanian margin into a single large shallow-marine basin with local subaerial highs where coastal/brackish swamps, deltaic deposits and offshore shales have been deposited. This basin maintained its overall character during the Albian as well. The depositional trend remained overall transgressive, which seems to have had a peak around the Late Albian, as indicated by the condensed shales in Southern Dobrogea and also the first marine deposits observed in the 'Babadag Basin' at Enisala (Fig. 1).

Metre-thick olistoliths of Mid-Jurassic (Fig. 14) and Neocomian black mudstones have been encountered in the Aptian–Albian section of the Histria half-graben. Similarly, Hippolyte *et al.* (2010) and Tüysüz *et al.* (2012) described – albeit larger – olistoliths along the Turkish margin, embedded in deepwater Aptian sediments. These authors attributed the olistoliths to fault-scarp erosion during extensional rifting.

An erosional event was recorded by the top Albian, followed by Early Cenomanian transgression. This event was observed both in outcrops and on seismic data (Fig. 5c, d). This minor erosion may be interpreted as being related to the change in extensional style in the Black Sea rift (Tari 2015). As such, we consider it to mark the end of the first (Aptian–Albian) phase of synrift evolution of the Romanian margin.

Cenomanian–Lower Turonian sequence

We have examined the Cenomanian and Turonian deposits in outcrop and in offshore cores, but have

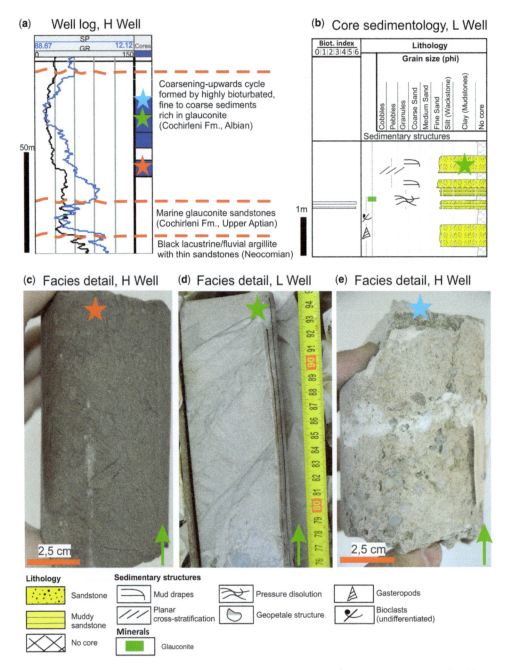

Fig. 12. (**a**) Characteristic coarsening-upwards log pattern of the Albian deltaic/shoreface deposits in Well H, Histria half-graben (Fig. 1). (**b**) Detailed core description of middle shoreface/delta facies. (**c**) Highly bioturbated glauconitic lower shoreface sandstones similar to onshore (Fig. 11d). (**d**) Upper shoreface sandstones with large-scale cross-bedding. (**e**) Sandy conglomerate lag at the base of a deltaic distributary channel.

relied on the interpretations provided by Szász & Ion (1986) and Bucur & Baltres (2002) for the 'Babadag Basin'.

In the 'Babadag Basin', the Cenomanian–Lower Turonian comprises 200–800 m of conglomerates, calcareous sandstones, sandy limestones and marls

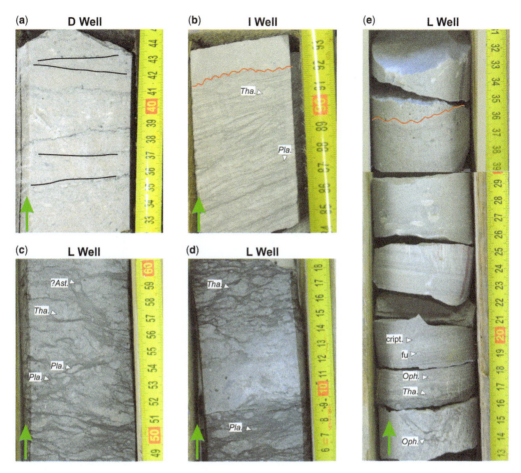

Fig. 13. Representative Albian offshore well cores from the Histria half-graben (Fig. 1) interpreted as shallow-marine tidal deltaic/shoreface sandstones. Abbreviations: *Tha.*, *Thalassinoides* isp.; *Pla.*, *Planolites* isp.; *?Ast.*, *Asterosoma?* isp.; *Oph.*, *Ophiomorpha* isp.; cript., criptobitorbations; fu, fugichnia. (**a**) Cross-bedding, normal grading and early-stage pressure-dissolution structures in sandy calcarenites. (**b**) Moderately bioturbated sandstones and muddy/silty sandstones erosionally overlain by normal-graded sandstone. (**c**) & (**d**) Highly bioturbated mudstones, muddy-silty sandstones repetitions with Cruziana ichonofacies (MacEachern & Bann 2008). (**e**) Low to highly bioturbated sandstones with thin mudstone intercalations. Locally, planar parallel lamination and erosional normal-graded sandstones may be observed.

organized into a general upwards-fining pattern (Iancila Formation: Szász & Ion 1986; Bucur & Baltres 2002). The entire succession features frequent shallow-marine fauna.

The Lower Cenomanian is well exposed at Cerna (Fig. 1), and comprises calcareous and glauconitic sandstones with cherty concretions (Fig. 15a). The sandstones feature trough cross-bedding with thin erosional normal-graded beds often rich in bioclasts (Fig. 15b). Also, horizontal and vertical bioturbations may occur (Fig. 15c). This is interpreted in terms of a shallow-marine environment in a shoreface/delta-front setting.

The Lower Cenomanian calcareous sandstones at Cerna are gradually replaced by limestones upwards into the Upper Cenomanian (Bucur & Baltres 2002). A similar change is observed in the 'Tibrinu graben', where the Cenomanian has about a 50 m-thick section of conglomerates, sandstones and limestones (the Pestera Formation of Avram *et al.* 1993).

The contact between the Albian and the overlying Cenomanian is best observed at the Sipotele outcrop (Fig. 16a–c). At this location, the Cenomanian erodes into the structurally tilted Albian cross-bedded sandstones. This erosional unconformity is overlain by quite constant and thin (about 1 m)

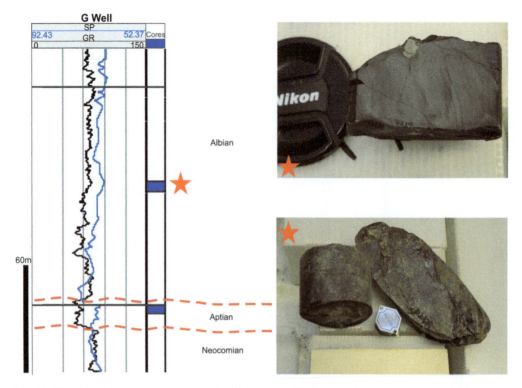

Fig. 14. Olistoliths of Middle Jurassic black mudstones encountered in 5 m cores in the Albian by Well G, Histria half-graben (Fig. 1).

conglomerate beds with reworked Albian clasts. The basal conglomerate grade upwards into cross-bedded sandstones with lens-shaped conglomerates interpreted as deltaic in origin. Higher up, the succession becomes overall chalky with frequent interbedding of finer-grained quartz- and glauconite-rich calcareous sandstones (Fig. 16c). This is similar to the facies encountered in cores, offshore Histria half-graben. There, the Upper Cenomanian comprises calcareous silty sandstones, fine gravelly sandstones and sandstones (Well A, Fig. 17a). The very-fine- to medium-grained sandstones and gravelly sandstones may be massive or feature normal grading with internal erosional surfaces and mudstones rip-up clasts (Fig. 17a). Shallow-marine bioclasts (i.e. shell hash) are frequent. The traction-related structures include asymmetrical ripples, cross-bedding, planar parallel lamination and flames. Some glauconite, horizontal burrows (BI 1–4) and centimetre-size coal clasts appear. All in all, this well core sample is interpreted as having been deposited in a shallow-marine depositional setting, possibly in a delta-front channel affected by storms.

The depositional pattern is also maintained in the Early Turonian. In Well J (Fig. 17b), the Lower Turonian is represented by alternation of mudstones,

muddy siltstones, silty/muddy sandstones and rare, fine gravely sandstones. The sandstones are fine to very fine (rarely medium) and typically form coarsening-upwards successions. The observed sedimentary structures include symmetrical ripples, load casts, flames structures, mud drapes and hydroplastic deformations (e.g. slumps, Fig. 17b). In addition, mudstone clasts, glauconite, pyrite, coal clasts and bioclasts tend to occur (Fig. 18a, b). The entire core is bioturbated (BI 3–4) with, for example, *Thalassinoides* isp., *Planolites* isp. and *Phycosiphon* isp. borrows (Fig. 18a, b). This core interval is interpreted to indicate a distal delta-front/lower shoreface to offshore/(?) prodelta environment. A tidal component may also be inferred due to the frequent mudstone–sand repetitions.

Interpretation. The Lower Cenomanian lies transgressively above the top Albian unconformity (Fig. 7a, b). Overall, the Cenomanian–Lower Turonian comprises a large-scale fining- and deepening-upwards section accompanied by an increase of the carbonate content. Most of the deposits are interpreted as being related to tidal deltas. The succession ends abruptly with a major unconformity observed on seismic data, logs offshore and in outcrops

Fig. 15. (**a**) Lower Cenomanian shallow-marine calcareous and glauconitic sandstones. (**b**) Cross-bedded sandstones. (**c**) Close-up example of vertical and horizontal burrows (*Ophiomorpha* isp.).

onshore. This unconformity is best expressed in South Dobrogea, where the Lower Turonian deposits are missing (probably eroded), and the Cenomanian is overlain by very localized Mid to Late (?) Turonian conglomerates or, more frequently, by regionally widespread Santonian–Campanian chalk.

Mid-Turonian–Maastrichtian sequence

The offshore wells encountered a complete Mid-Turonian–Maastrichtian section lying above the Mid-Turonian unconformity (Figs 7a & 17a, b). Well J has cored this unconformity and found mid- to upper Turonian cross-bedded and normal-graded conglomerates and gravelly sandstones (Fig. 17b). The entire succession has a flat (erosional) base and fining-upwards well-log pattern (Fig. 17a).

Overall, it is interpreted as a shallow-marine transgressional sequence deposited over a major unconformity. In the 'Babadag Basin', this unconformity is attributed to a relatively poorly dated mid-Turonian erosional event (Szász & Ion 1986). Above the unconformity, Upper Turonian and Coniacian shallow-marine calcareous sediments may be found (Bucur & Baltres 2002).

The age-equivalent deposits from South Dobrogea are missing, but rare conglomerates and coarse-grained sandstones are present (Cuza Vodă Formation of Dragastan *et al.* 1998). These deposits include reworked Albian, Cenomanian (?) and Mid-Turonian fauna, and their age is considered Mid- to Late Turonian. A very characteristic reworked fauna association is with Mid-Turonian echinoids (*Conulus subrotundus, C. spheroidalis, C. rhotomagensis*

(a)

Fig. 16. (a) The abandoned quarry outcrop at Sipotele (Fig. 1) exposes structurally tilted Albian glauconitic sandstones (Cochirleni Formation) erosionally overlain by Cenomanian calcareous deposits (Peştera Formation). (b) Detail of the cross-bedded Upper Albian glauconite sandstone (Fig. 16a) interpreted as a delta-front deposit. The D03 sample (Fig. 21e) from this sandstone level was used for the U–Pb age dating of detrital zircons. (c) Detail of the Upper Cenomanian mollusk-rich sandy limestones located at the uppermost part of the quarry (Fig. 16a).

elevates and *C. nucula*: Dragastan *et al.* 1998) found in only two locations (Fig. 1) in the entire South Dobrogea: at Cuza Voda (Avram *et al.* 1993) and Pestera (Dragastan *et al.* 1998).

We have studied in detail the sedimentology of the outcrop at Peştera, where the conglomerates and sandstones reach 15 m in thickness (Fig. 19a). The sandstones feature spectacular metre-scale low-angle cross-stratification alternating with high-angle cross-bedding (Fig. 19b). The bedding surfaces are erosional, orientated concave upwards. A hierarchy of large erosional surfaces and several minor ones between can be identified by careful mapping (Fig. 19a). The sandstones are reported to contain marine (or brackish?) burrows (Avram *et al.* 1993). The coarse-grained nature of the sediments in the Pestera outcrop and the frequent erosional surfaces suggests high-energy depositional settings represented by amalgamated channel successions. Thus, the high-angle cross-beds are interpreted as

migrating point-bar deposits, parts of a large channel-fill sequence. The marine influence is possibly indicated by the marine fauna and burrows of *Paleophycus?* isp., *Macaronichnus* isp. and *Conichnus* isp. (Fig. 19b). The lithology and stratigraphic architecture of the Pestera Formation at this location suggests that this sequence was deposited in distributary channels of a larger deltaic system transporting sands into the Black Sea.

In South Dobrogea, the much localized deltaic distributary channels are cut by a major unconformity followed by a depositional hiatus that spans the entire Coniacian (Dragastan *et al.* 1998). Sedimentation resumes only in the Santonian, with transgressive conglomerates formed by well-rounded quartzite fragments that grade upwards into regionally extensive white massive chalks dated as Santonian–Early Campanian based on a rich planktonic and benthonic foraminifer fauna, *Inoceramus* isp., echinoids and siliceous sponges (Dragastan *et al.*

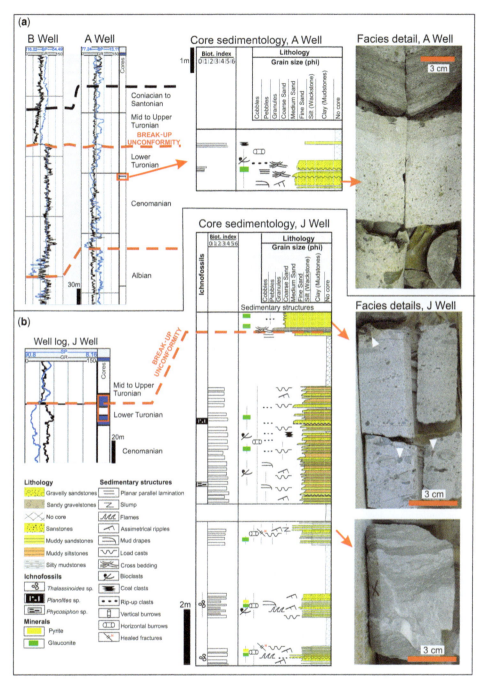

Fig. 17. (**a**) Log patterns and sedimentology of the Upper Cretaceous in the Histria half-graben (Fig. 1). The top Albian and mid-Turonian unconformities (break-up unconformity) are characterized by a flat erosional base followed by a fining-upwards log pattern. The Cenomanian core may be interpreted as part of a deltaic distributary channel. (**b**) Break-up unconformity cored in Well J, Histria half-graben (Fig. 1). Below the unconformity, the Lower Turonian features fine-grained mudstones and sandstones interpreted as distal delta-front/lower shoreface to prodelta deposits. The conglomerates above the break-up unconformity are full of reworked green clasts of the Histria Formation (Seghedi 2001) from Central Dobrogea, and reddish fragments typical of Aptian fluvial deposits (Fig. 9c, d). Overall, the conglomerates are interpreted as coarse-grained channel lags, likely to be part of deltaic distributary channels.

(a) **J Well**

(b) **J Well**

Fig. 18. (**a**) Upper Cenomanian–lower Turonian highly bioturbated (e.g. *Pla.*, *Planolites* isp.; *Phy.*, *Phycosiphon* isp.) muddy siltstones, silty-muddy sandstones and sandstone repetitions encountered by Well J (Histria half-graben, Fig. 1). This may be interpreted as a lower shoreface setting influenced by tides. (**b**) Lower Turonian highly bioturbated (e.g. *Tha.*, *Thalassinoides* isp.) mudstones, muddy and silty sandstones with mudstone rip-up clasts in Well J (Histria half-graben, Fig. 1) interpreted as tidal deposits.

1998) (Fig. 20a). The chalks often display intense bioturbation (Fig. 20b) and contain siliceous levels (flintstones, Fig. 20c) (Crandell 2013), mostly in their lower part. More importantly, the white chalks are quite constant in thickness and cover unconformably the underlying deposits in South Dobrogea, including Upper Jurassic, Aptian and Mid?–Upper Turonian successions (Dragastan *et al.* 1998).

Interpretation. The abrupt facies rearrangement, combined with the sand-rich and highly erosional character of the Middle–Upper Turonian deposits, suggest that a major change in depositional style took place on the Romanian margin starting with the Mid-Turonian. Uplift and erosion observed in the 'Babadag Basin' and offshore Histria graben caused massive reworking and widespread (Coniacian) erosion in South Dobrogea. Such a major event could only be explained by a tectonically enhanced relative sea-level fall. Importantly, this uplift was coeval with the termination of the extensional activity along the Romanian margin, as illustrated by our assessment of the seismic data (Figs 5–7). Therefore, this long-lived erosional event, which started in the mid-Turonian, could be interpreted as the expression of the final continental break-up and initiation of the

oceanic seafloor spreading the western Black Sea (Fig. 3).

The uplift of the basin margin restricted the Upper Turonian and Coniacian sedimentation mostly to the deeper parts of the basin and in the deepest depocentres (e.g. the 'Babadag Basin'). As discussed, the early post-rift sediments following a short period of lowstand have a clear transgressive pattern that is interpreted due to post-rift thermal subsidence. This thermal subsidence-driven transgression reached onshore South Dobrogea by the Santonian and resulted in deposition of thick sections of chalk in the basin.

U–PB age dating of detrital zircons

Whereas there are a growing number of detrital zircon studies on the conjugate margin of the Romanian Black Sea, in the Pontides of Turkey (Ustaömer *et al.* 2005, 2011, 2012; Okay *et al.* 2006, 2013; Sunal *et al.* 2008; Karslıoğlu *et al.* 2012; Yılmaz-Şahin *et al.* 2014; Akdoğan *et al.* 2017), only limited amount of comparable information has been published regarding North and Central Dobrogea (Balintoni *et al.* 2010, 2011; Balintoni & Balica 2016).

Fig. 19. (**a**) The outcrop at Peştera exposes mid-Turonian fluvial microconglomerates (Cuza Vodă Formation) with frequent reworked marine echinoids overlain by Mid-Miocene deposits. The mid-Turonian section is interpreted to indicate uplift and erosion of the basin margin, which occurred due to the continental break-up in the western Black Sea. (**b**) Detail of the upper part of the outcrop with spectacular cross-bedded sandstones interpreted as point-bar deposits. (**c**) Close-up of a stacked distributary channel. (**d**) A possible marine influence indicated by vertical (*Conichnus* isp.) and horizontal (*Paleophycus* isp., *Macaronichnus* isp.) bioturbations.

Data and observations

In the following subsection, the detrital zircon age spectra of the five Cretaceous samples, generally from oldest to youngest, are described.

The oldest sample (D04) among these Cretaceous sandstones was collected in a large semi-active sand quarry near the village of Cuza Vodă (Fig. 21a). Here, Aptian alluvial to fluvial sandstones crop out; the sample was taken from a fluvial sand-bar element (Fig. 21a). The U–Pb age spectrum of the 100 detrital zircons displays several prominent peaks. The oldest of these peaks is a Neoarchaean one, around 2.7 Ga.

Another three important peaks are located in the Meso-to Palaeoproterozoic, with approximate ages of 1.5, 1.8 and 2.1 Ga. Among the youngest peaks, the most important one has an age of about 600 Ma (Ediacaran). A subordinate double peak has an approximate age of 300 Ma (Carboniferous). Note that the depositional age of the sandstone sample is about 120 Ma (intra-Aptian).

The second sample (D07) was a drill core from the offshore Lebada area (Well J, Fig. 1) dated as Albian (Fig. 21b). The sandstone provided a very different age spectrum from that of the Aptian sample, based on 99 zircon grains. In this core sample,

(a)

?Quaternary fan deltas with reworked chalks

3m

Santonian-Lower Campanian in situ chalks
with silex (chert) nodules:
Murfatlar Formation

(b)   **(c)**   **(d)**

⊢ 5cm ⊣

2,5 cm

Fig. 20. (**a**) The abandoned chalk quarry at Murfatlar exposes massive Santonian–Campanian chalks with silex (chert) nodules overlain by Quaternary (?) fan-delta deposits. (**b**) Detail of intensely bioturbated chalks. (**c**) Chert nodule from the chalk (Crandell 2013). (**d**) Bioturbated chalk from an offshore well core.

by far the most important peaks were in the Phanerozoic, with ages of about 260 Ma (Late Permian) and 320 Ma (Carboniferous). In addition, much smaller peaks had approximate ages of 170 Ma (Middle–Late Jurassic) and 2.1 Ga (Palaeoproterozoic).

The third sample (D06) was obtained from the same offshore well (Well J, Figs 1 & 17b), but 300 m higher, in the Mid-Turonian early post-rift sequence (Fig. 21c). The zircon spectrum of the pebbly to coarse-grained sandstone, based on 100 grains, has a dominant peak with a Phanerozoic age of about 270 Ma (Late Permian). Additional peaks of lesser importance include a 350 Ma (Carboniferous) and, similarly to sample D04, a latest Neoproterozoic population of about 600 Ma (Ediacaran) in age. Note that this later peak has a small spread between 620 and 540 Ma.

The fourth sample (D01) was collected from an abandoned quarry near the village of Sipotele (Figs 16 & 21d), from loose, glauconitic Albian sands.

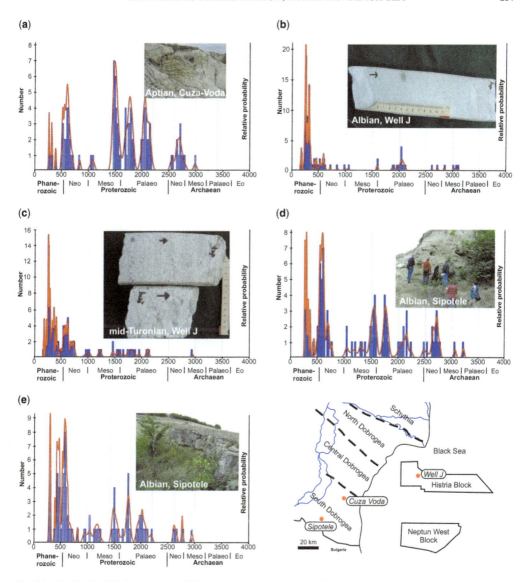

Fig. 21. (a)–(e). Detrital zircon spectra of Cretaceous sandstone outcrops in South Dobrogea and offshore cores displayed as relative probability/zircon grains v. Pb–U age per zircon grains. For an interpretation, see the text.

In this sample, the two most dominant peaks have a 280 Ma (Permian) and, similarly to samples D04 and D06, a latest Neoproterozoic age of about 600 Ma (Ediacaran). The later population has a small spread to younger ages, between 620 and 540 Ma. In addition, moderately important peaks show a pattern very similar to that of D04 (i.e. a Neoarchaean one) around 2.7 Ga and another three peaks in the Meso- to Palaeoproterozoic, with approximate ages of 2.15, 1.75 and 1.55 Ga. Note that the depositional age of the sandstone sample itself is about 105 Ma (latest Albian).

The fifth sample (D03) was collected in the same abandoned quarry near the village of Sipotele. This coarser, microconglomeratic Albian sandstone (Fig. 21e) is located on the other side of the quarry, just by a few metres below the base Cenomanian unconformity. The age spectrum is consistent with the other sample (D01) collected in the same quarry (Fig. 21d, e). Here, again, the most significant peaks are at 290 Ma (Permian) and 600 Ma (Ediacaran). Interestingly, there is also a prominent 450 Ma (Ordovician) peak. Subordinate peaks have Neo-archaean ages clustering at 2.95, 2.75 and 2.6 Ga,

and another three peaks in the Meso- to Palaeoproterozoic, with approximate ages of 2.0, 1.75 and
1.5 Ga.

Interpretation

Although the sample number is clearly too small to
draw far-reaching conclusions, the nearly 500 zircons grains collected from five samples do indicate
certain trends. Our preliminary interpretation of the
results is as follows:

• One of the common signals among the samples is
  the presence of an Ediacaran peak, clustering
  around 600 Ma. This zircon population may be
  derived from the Histria Formation, exposed in
  Central Dobrogea (Żelaźniewicz *et al.* 2009).
  These flyschoid, very-low-grade metamorphosed
  siliciclastics (Oaie *et al.* 2005) provided youngest
  zircon ages of 587 and 584 Ma, suggesting a maximum Late Ediacaran depositional age for the Histria Formation (Balintoni *et al.* 2011). In addition,
  it is also possible that some of these zircons were
  derived from late Neoproterozoic granites common in the Balkanides and in the Strandja Massif
  (Natal'in *et al.* 2016). However, these sources
  would have been located further away compared
  to the Central Dobrogea provenance area.
  The U–Pb detrital zircon ages in our Cretaceous
  sediments provide indirect evidence for a late Ediacaran, possibly earliest Cambrian, age of the fine-
  grained turbidites of the Histria Formation (Oaie
  et al. 2005). The detrital zircon signal documented
  in this Ediacaran formation by Żelaźniewicz *et al.*
  (2009) and Balintoni *et al.* (2011) can be seen
  clearly in our Cretaceous samples. We envision
  the Histria Formation, currently exposed only in
  Central Dobrogea (e.g. Seghedi 2012), being
  exposed on the surface in a much larger area during the Early and Mid-Cretaceous. The dominance of these anchimetamorphosed siliciclastics
  in the Early Cretaceous provenance area explains
  their common occurrence in all the zircon spectra
  (Fig. 21a–e). We also speculate that there might
  have been an important element of recycling of
  the 600 Ma zircons into other intermediate clastic
  formations (i.e. Permian and Triassic) before they
  were re-deposited into the Cretaceous synrift and
  early post-rift formations.
• The other outstanding age signature in our samples is the presence of earliest Triassic–Permian
  and Carboniferous peaks, scattered between 350
  and 250 Ma. Permian and Carboniferous granites
  are widespread in the Black Sea region (Okay &
  Topuz 2016). The closest provenance area for
  these zircons could be the alkali granites and syenites, associated with alkali rhyolites, described
  from the southern margin of North Dobrogea

(Seghedi 2001), from the pre-Dobrogea Basin further to the north (Seghedi 2012; Seghedi & Neaga
2016). In contrast, the Carboniferous zircons
should be derived from Hercynian granitic plutons
in the broader area. There are, indeed, Carboniferous granites in the Tulcea zone and their ages were
confirmed by detrital zircons of Balintoni &
Balica (2016). There are granite clasts found in
the red deposits on the northern edge of the 'Babadag Basin', ascribed either to the Permian or to
lowermost Triassic. Regardless of their origin,
both of these lithologies were eroded from the
exposed footwall blocks of the Early to Mid-
Cretaceous synrift extensional terrain, which
could have a wide-rift style, basin-and-range land-
scape (Tari 2015).
• With varying relative importance in the overall
  zircon spectrum, in all samples there are the
  same populations present for the Mesoprotero-
  zoic–Neoarchaean period, namely the ones with
  approximate ages of 2.7, 2.1, 1.8 and 1.5 Ga.
  These peaks are the most prevalent in the oldest
  Aptian fluvial sample (Fig. 21a). We interpret
  this as the reflection of a relatively flat, not yet
  structurally differentiated, palaeotopography at
  the beginning of the Early Cretaceous synrift
  period. During this time, river systems were still
  transporting clastics from a considerable distance,
  from the nearby cratonic areas. It is tempting to
  draw some conclusions regarding the overall terrain provenance, building on the results of a recent
  paper by Balintoni & Balica (2016), but given the
  current small sample size we opted not to do so at
  this point of time.
• Only the two offshore samples (Fig. 21b, c) record
  smaller peaks at around 170–150 Ma (Middle–
  Late Jurassic). These are attributed to the Jurassic
  volcanism described from many parts of the
  broader Black Sea area (Okay & Nikishin 2015),
  including North Dobrogea (e.g. Seghedi 2001),
  Crimea (e.g. Meijers *et al.* 2010*b*) and the Pontides (e.g. Genç & Tüysüz 2010).

Discussion

In order to illustrate the changes of the structural
style from the pre-rift (Late Jurassic–Neocomian)
to synrift 1 (Aptian–Albian) configuration, we drew
a regional section (Fig. 22a) based on Seghedi (2001)
and Cavazza *et al.* (2004). Our conceptual transect
runs north to south, following the Black Sea coast.
The Neocomian reconstruction (Fig. 22a) illustrates
a number of thrusts related to the Late Triassic
(Tari *et al.* 1997) to Neocomian Cimmerian transpressional episodes and to the emergence of the
North Dobrogea Orogen (Seghedi 2001). Note that
in our interpretation, the Sfântu-Gheorghe and

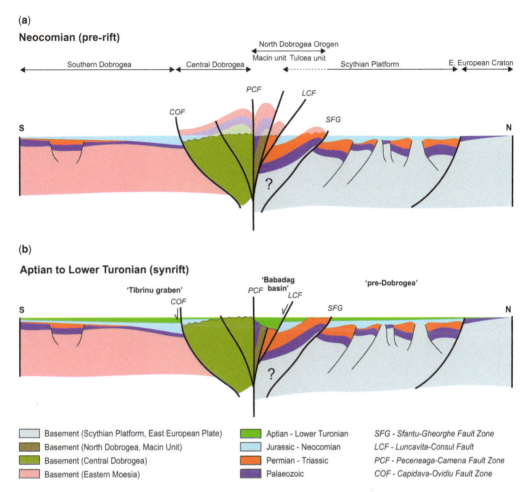

Fig. 22. Conceptual reconstructions for the Cretaceous, along a regional transect onshore Romania. (**a**) The Neocomian transect illustrates the exposed and eroded North Dobrogea Orogen surrounded by the Neocomian platform carbonates. (**b**) The Aptian–early Turonian reconstruction features the 'Babadag Basin' and the 'Țibrinu graben' developed along reactivated basement faults and filled with shallow-marine sediments.

Capidava-Ovidiu fault zones are only branches of a much larger strike-slip deformation belt developed along the Peceneaga-Camena Fault Zone. Also, the Tulcea Unit, part of the North Dobrogea Orogen, is interpreted as a piece of the Scythian Platform involved in the transpressional deformation.

Although the seismic architecture of the Late Aptian–Albian marine deposits clearly indicates their syn-extensional character, the tectonic setting of the Early–Mid-Aptian continental deposits is uncertain. This is because this continental package is typically very thin on seismic sections, and therefore it is difficult to assess their syn-extensional character and its relationship to the 'Neocomian' pre-rift deposits below. Some authors (Săndulescu 1988; Seghedi 2001; Hippolyte 2002) have considered

them as related to the 'late Cimmerian' deformations of North Dobrogea. Others considered them as the expression of the incipient rifting in the Black Sea (Robinson et al. 1996; Nikishin et al. 2011). We prefer the latter interpretation at least for the Mid-Aptian continental lowlands with limestone intercalations, because similar facies has been described as syn-extensional on the Schytian (Nikishin et al. 2011) and the Turkish conjugate margin (Görür 1988; Hippolyte et al. 2010). These continental lowlands may have been developed in relatively small and separated individual graben that typically characterize the incipient stages of wide rifting (Tari 2015). On the other hand, we cannot exclude the possibility at present that the coarse-grained Early Aptian conglomerates may be related to the post-tectonic

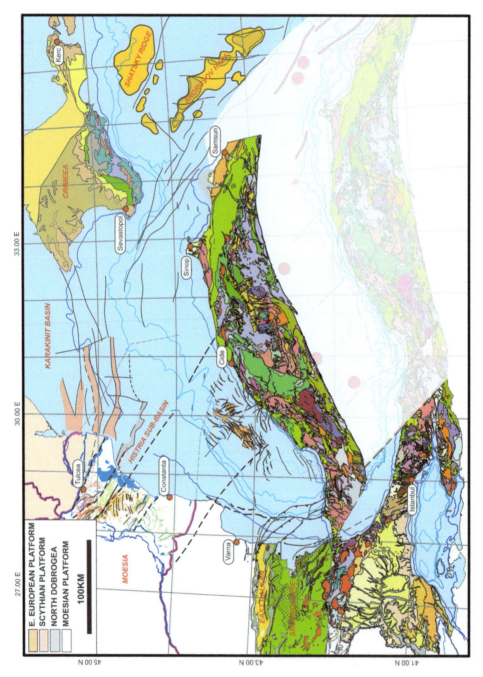

Fig. 23. Pre-rift planispastic reconstruction of the Pontides back to the Bulgarian, Romanian and Ukrainian margin based on Tari (2015). Dashed red lines highlight the hinge zones on both margins.

cover of the 'late Cimmerian' deformations in North Dobrogea (Săndulescu 1988).

In contrast to the Neocomian configuration, the Late Aptian–Albian reconstruction (Fig. 22b) illustrates the synrift depocentres developing along the most important strike-slip faults of the North and Central Dobrogea. Thus, in a way, the extensional opening of the Black Sea contributed to the extensional collapse of the North Dobrogea Orogen. The Histria half-graben, one of the largest extensional depocentres, opened over the weakest part of the North Dobrogea Orogen (i.e. the Macin Unit) (Seghedi 2001). The Mid-Aptian–Albian extension (synrift 1) along the Romanian margin generated localized extensional depocentres in a wide-rift-style manner (Fig. 8a). The Cenomanian–Early Turonian (synrift 2) (Fig. 8b) episode corresponds to a truly back-arc rifting stage in the western Black Sea with the appearance of arc volcanics on the conjugate Turkish margin (e.g. Tari 2015).

The map-view reconstruction of the Western Black Sea Basin (Tari et al. 2015) prior to its mid-Cretaceous opening brings the onshore South Dobrogea within 250 km of the Central Pontides of Turkey (Fig. 23). The pre-rift geometry of the western Black Sea was obtained by a 12.5° counterclockwise rotation of the Pontides in relation to a fixed Eastern European and Moesian Platform margin (see also Schleder et al. 2015). The pole of rotation for this reconstruction was chosen a few hundred kilometres to the SW of the Black Sea, located in the present-day Aegean Sea. The primary constraint on this reconstruction is the match of the 'hinge zones' on the conjugate margins. Note that the rotation used is a minimum, as the reconstruction does not take into account the extension associated with several Cretaceous basins offshore and onshore (e.g. the Ulus Basin) to the south of the hinge zone on the Turkish margin. Moreover, the post-rift oroclinal bending of the Pontides (Meijers et al. 2010a) still needs to be accounted for in the kinematic reconstruction shown (Fig. 23).

The importance of this reconstruction is that the corresponding segment of South Dobrogea on the conjugate margin in the Central Pontides is the area between Zonguldak and Cide (Fig. 23). Therefore, the growing detrital zircon database in the Cretaceous sequence of the Central Pontides (Okay et al. 2013; Akdoğan et al. 2017) can be compared with our preliminary results obtained from the synrift and early post-rift strata on the conjugate margin in South Dobrogea (Fig. 21).

The Phanerozoic–Neoproterozoic stratigraphy of the Western and Central Pontides (Akdoğan et al. 2017) shows multiple potential sources to provide zircons, by direct or multiple recycling into the Early Cretaceous siliciclastic sequence. Akdoğan et al. (2017) analysed 1194 detrital zircon ages

from 13 Lower Cretaceous sandstone samples collected in the Central Pontides. Their detrital zircon ages show different patterns in the eastern, central and western parts of the Pontides. In the eastern part of the Çağlayan Basin, the majority of the detrital zircon ages are Archaean and Palaeoproterozoic. Rocks of these ages are absent in the Pontides but are present in the Ukranian Shield (e.g. Bogdanova et al. 2010; Safonova et al. 2010; Bibikova et al. 2015), which indicates a source north of the Black Sea. In the western part of the basin, the majority of the zircons are Carboniferous and late Neoproterozoic, which implies sources located within the Pontides. The detrital zircons from the central part show a mixture of western and eastern parts. Significantly, Jurassic and Early Cretaceous zircons make up less than 2% of the total zircon population, just like in the samples from Dobrogea discussed above, which implies a lack of a coeval magmatic arc in the region. This is compatible with the absence of Lower Cretaceous granites in the Pontides (e.g. Okay et al. 2006; Aygül et al. 2016). Thus, although the Çağlayan Basin occupied a forearc position above the subduction zone, the arc was missing, probably due to flat subduction, and the basin was largely fed from the Ukranian Shield in the north (Akdoğan et al. 2017; Okay et al. 2017).

Conclusions

The Romanian dataset presented here places the onset of the western part of the Black Sea rifting to the Early–Mid Aptian and the break-up unconformity to be by the Mid-Turonian. This suggests that the synrift period spanned a time interval of more than 30 myr, an unusually long time for a single synrift period in any basin (Tari 2015). Therefore, we consider, instead, that the western Black Sea has opened over several rifting phases (e.g. Dinu et al. 2005; Tari 2015). Again, a hint for the different synrift episodes is the presence of various unconformities observed on seismic data and in South Dobrogea. Among these, the most important is the top Albian one. This unconformity divides the synrift strata into two distinct units: Mid-Aptian–Albian (synrift 1) and Cenomanian–Early Turonian (synrift 2). Tari (2015) described these stages as wide-rift- and narrow-rift-style extensional periods, respectively.

On the Romanian margin, the Aptian–Albian (synrift 1) depositional environments are represented by an early phase of continental lacustrine settings followed by shallow-marine deltaic deposits. The continental influence seems to diminish further offshore, where more marine limestones appear interbedded in the continental succession.

242 C. KREZSEK *ET AL.*

During the Cenomanian–Early Turonian (synrift 2) the role of continental deposits diminishes. The Romanian margin deepens, but still remains in shallow-marine depositional conditions. Deeper marine settings are likely to have prevailed SE of the Histria half-graben, as suggested by Akdoğan *et al.* (2017).

The depositional systems markedly change across a regional unconformity dated as Mid-Turonian in age. By that time, the whole of South Dobrogea is exposed, and coarse-grained Upper Turonian–Coniacian sediments are deposited offshore. Sedimentation returns to South Dobrogea with deep-marine chalks that blanket the older unconformities. We propose this Mid-Turonian unconformity as the break-up unconformity of the western Black Sea. In addition to the sedimentological evidence, the Mid-Turonian age of opening for the western part of the Black Sea is supported by the seismic architecture of the Aptian–Mid-Turonian packages that show clear evidence of syn-extensional growth, in contrast to the Senonian strata.

We acknowledge OMV Petrom's management and the Romanian National Agency for Mineral Resources for allowing us to publish proprietary data. The detailed reviews by David Boote and an anonymous reviewer helped to make our interpretations clearer. We thank Antoneta Seghedi (Geoecomar) and Silviu Rădan (Geoecomar) for guidance in the field and for the constructive discussions related to the geology of Dobrogea. The core interpretations would not have been possible without the help of ICPT Campina, especially Claudia Antoniade. Cornel Olariu (University of Texas at Austin) is acknowledged for helping with the sedimentological interpretation of the cores, and Maria Fotu (Petromar) with the offshore well database. Zircon age dating was performed by Mihai Ducea (University of Arizona). This work has been developed working together with many colleagues from Petrom, in particular Zamir Bega, Zsolt Schleder, Dan Stefaniuc, Roxana Dudus, Anas Al Khder, Roxana Sbingu and John Anderson.

References

AKDOĞAN, R., OKAY, A.I., SUNAL, G., TARI, G., MENHOLD, G. & KYLANDER-CLARK, A.R.C. 2017. Provenance of a large Lower Cretaceous turbidite submarine fan complex on the active Laurasian margin: central Pontides, northern Turkey. *Journal of Asian Earth Sciences*, **134**, 309–329.

ANTONIADE, C.-G. 2016. *Studiul lito-biostratigrafic al depozitelor de varsta Cretacic inferior din zona Cernavoda, Dobrogea de Sud [Litho-biostratigraphic study of the Lower Cretaceous deposits in the Cernavoda area, South Dobrogea].* PhD thesis, University of Bucharest.

AVRAM, E., SZÁSZ, L. *ET AL.* 1993. Cretaceous terrestrial and shallow marine deposits in northern South Dobrogea (SE Romania). *Cretaceous Research*, **14**, 265–305.

AYGÜL, M., OKAY, A.I., OBERHÄNSLI, R. & SUDO, M. 2016. Pre-collisional accretionary growth of the Southern Laurasian active margin, Central Pontides, Turkey. *Tectonophysics*, **671**, 218–234.

BALINTONI, I. & BALICA, C. 2016. Peri-Amazonian provenance of the Euxinic Craton components in Dobrogea and of the North Dobrogean Orogen components (Romania): a detrital zircon study. *Precambrian Research*, **278**, 34–51.

BALINTONI, I., BALICA, C., SEGHEDI, A. & DUCEA, M.N. 2010. Avalonian and Cadomian terranes in north Dobrogea, Romania. *Precambrian Research*, **182**, 217–229.

BALINTONI, I., BALICA, C., SEGHEDI, A. & DUCEA, M. 2011. Peri-Amazonian provenance of the Central Dobrogea terrane (Romania) attested by U/Pb detrital zircon age patterns. *Geologica Carpathica*, **62**, 299–307.

BANKS, C.J. & ROBINSON, A.G. 1997. Mesozoic strike-slip back-arc basins of the western Black Sea region. *In:* ROBINSON, A.G. (ed.) *Regional and Petroleum Geology of the Black Sea and Surrounding Region.* American Association of Petroleum Geologists, Memoirs, **68**, 53–62.

BEGA, Z. & IONESCU, G. 2009. Neogene structural styles of the NW Black Sea Region, offshore Romania. *The Leading Edge*, **28**, 1082–1089.

BIBIKOVA, E.V., BOGDANOVA, S.V. *ET AL.* 2015. The early crust of the Volgo-Uralian segment of the East European Craton: isotope-geochronological zirconology of metasedimentary rocks of the Bolshecheremshanskaya Formation and their Sm–Nd model ages. *Stratigraphy and Geological Correlation*, **23/1**, 1–23.

BOGDANOVA, S.V., DE WAELE, B., BIBIKOVA, E.V., BELOUSOVA, E.A., POSTNIKOV, A.V., FEDOTOVA, A.A. & POPOVA, L.P. 2010. Volgo-Uralia: the first U–Pb, Lu–Hf, and Sm–Nd isotopic evidence of preserved Paleoarchean crust. *American Journal of Science*, **310**, 1345–1383.

BOOTE, D.R.D. 2017. The geological history of the Istria 'Depression', Romanian Black Sea shelf: tectonic controls on second-/third-order sequence architecture. *In:* SIMMONS, M.D., TARI, G.C. & OKAY, A.I. (eds) *Petroleum Geology of the Black Sea.* Geological Society, London, Special Publications, **464**, https://doi.org/10.1144/SP464.8

BUCUR, I. & BALTRES, A. 2002. Cenomanian microfossil in the shallow water limestones from Babadag Basin: biostratigraphic significance. *Studia Universitatis Babes-Bolyai, Geologia, Special Issue*, **1**, 79–95.

CAVAZZA, Z., ROURE, F.M., SPAKMAN, W., STAMPFLI, G.M. & ZIEGLER, P.A. (eds). 2004. *The TRANSMED Atlas. The Mediterranean Region from Crust to Mantle.* Springer, Berlin.

CHIRIAC, M. & MÎNZATU, S. 1967. *Geological Map of Romania: Scale 1:200 000. Mangalia Sheet.* Geological Institute of Romania, Bucharest [in Romanian].

CHIRIAC, M., MUTIHAC, V., MIRĂUŢĂ, O. & MÎNZATU, S. 1968. *Geological Map of Romania: Scale 1:200 000. Constanta Sheet.* Geological Institute of Romania, Bucharest [in Romanian].

CRANDELL, O. 2013. The provenance of Neolithic and chalcolithic stone tools from sites in Teleorman county, Romania. *Buletinul Muzeului Judetean Teleorman, Seria Arheologie*, **5**, 125–142.

DINU, C., WONG, H.K., ȚAMBREA, D. & MATENCO, L. 2005. Stratigraphic and structural characteristics of the Romanian Black Sea shelf. *Tectonophysics*, **410**, 417–435.

DRAGASTAN, O.N. 2008. Lithostratigraphy of the Upper Jurassic–Cretaceous deposits and hydrocarbon perspective in the Romanian Shelf of the Black Sea. Search and Discovery Article 10144, adapted from a poster presentation at *AAPG and AAPG European Region Energy Conference*, 18–21 November 2007, Athens, Greece.

DRAGASTAN, O.N., NEAGU, T., BĂRBULESCU, A. & PANĂ, I. 1998. *The Jurassic and Cretaceous from Central and South Dobrogea – Paleontology and Stratigraphy*. Editura Supergraph, Bucharest [in Romanian].

GENÇ, S.C. & TÜYSÜZ, O. 2010. Tectonic setting of the Jurassic bimodal magmatism in the Sakarya Zone (Central and Western Pontides), Northern Turkey: a geochemical and isotopic approach. *Lithos*, **118**, 95–111.

GEORGIEV, G. 2012. Geology and Hydrocarbon system of the Western Black Sea. *Turkish Journal of Earth Sciences*, **21**, 723–754.

GÖRÜR, N. 1988. Timing of opening of the Black Sea basin. *Tectonophysics*, **147**, 247–262.

GÖRÜR, N. 1997. Cretaceous syn- to post-rift sedimentation on the southern continental margin of the western Black Sea Basin. *In*: ROBINSON, A.G. (ed.) *Regional and Petroleum Geology of the Black Sea and Surrounding Region*. American Association of Petroleum Geologists, Memoirs, **68**, 227–240.

GRADINARU, E. 1988. Jurassic sedimentary rocks and bimodal volcanics of the Carjelari-Camena outcrop belt: evidence for transtensile regime of the Peceneaga-Camena Fault. *Studii si Cercetari de Geologie, Geofizica si Geografie, Seria Geologie*, **33**, 97–121.

GRADINARU, E. 2002. Vraconian age of the Enisala limestone from the Babadag (North Dobrogea Orogene): *Lepthoplites Enisalaensis* new species. *Studii si cercetari de Geologie*, **47**, 55–63.

HIPPOLYTE, J.-C. 2002. Geodynamics of Dobrogea (Romania): new constraints on the evolution of the Tornquist-Teisseyre Line, the Black Sea and the Carpathians. *Tectonophysics*, **357**, 33–53.

HIPPOLYTE, J.-C., MÜLLER, C., KAYMAKCI, N. & SANGU, E. 2010. Dating of the Black Sea Basin: new nannoplankton ages from its inverted margin in the Central Pontides (Turkey). *In*: STEPHENSON, R.A., KAYMAKCI, N., SOSSON, M., STAROSTENKO, V. & BERGERAT, F. (eds) *Sedimentary Basin Tectonics from the Black Sea and Caucasus to the Arabian Platform*. Geological Society, London, Special Publications, **340**, 113–136, https://doi.org/10.1144/SP340.7

IANOVICI, V., GIUSCA, D., MUTIHAC, V., MIRAUTA, O. & CHIRIAC, M. 1961. General view over Dobrogea geology. *In*: ILIE, M. (ed.) *Excursion Guide, D, V*[th]* Congress of the Carpathian-Balkan Geological Association*, 4–19 September 1961. CBGA, Bucharest [in Romanian].

IONESCU, G. 2000. *Modele faciale ale formațiunilor Paleogene pe șelful Nord-Vestic al Mării Negre [Facies models of the Paleogene Formations on the NW shelf of the Black Sea]*. PhD thesis, University of Bucharest.

IONESCU, G., SISMAN, M. & CATARAIANI, R. 2002. Source and reservoir rocks and trapping mechanisms of the Romanin Black Sea Shelf. *In*: DINU, C. & MOCANU, V. (eds) *Geology and Tectonics of the Romanian Black Sea Shelf and its Hydrocarbon Potential*. Bucharest Geoscience Forum, Special Volume, **2**, 67–83.

KARSLIOĞLU, Ö., USTAÖMER, T., ROBERTSON, A.H.F. & PEYTCHEVA, I. 2012. Age and provenance of detrital zircons from a sandstone turbidite of the Triassic–Early Jurassic Küre Complex, Central Pontides. *In*: AKAL, C., ERSOY, E.Y. *ET AL.* (eds) *Abstracts of the International Earth Science Colloquium on the Aegean Region, IAESCA-2012*, 1–5 October 2012, Dokuz Eyül University, Izmir, 57.

KAZMIN, V., RICOU, L.-E. & SBORTSHIKOV, I.M. 1986. Structure and evolution of the passive margin of the eastern tethys. *Tectonophysics*, **123**, 153–179.

KREZSEK, C., SCHLEDER, Z., BEGA, Z., IONESCU, G. & TARI, G. 2016. The Messinian sea-level fall in the western Black Sea: small or large? Insights from offshore Romania. *Petroleum Geoscience*, **22**, 392–399, https://doi.org/10.1144/petgeo2015-093

LERICOLAIS, G., BOURGET, J., POPESCU, I., JERMANNAUD, P., MULDER, T., JORRY, S. & PANIN, N. 2013. Late Quaternary deep-sea sedimentation in the western Black Sea: new insights from recent coring and seismic data in the deep basin. *Global and Planetary Change*, **103**, 232–247.

LETOUZEY, J., BIJU-DUVAL, B., DORKEL, A., GONNARD, R., KRISTCHEV, K., MONTADERT, L. & SUNGURLU, O. 1977. The Black Sea: a marginal basin – Geophysical and geological data. *In*: BIJU-DUVAL, B. & MONTADERT, L. (eds) *Structural History of the Mediterranean Basins*. Editions Technip, Paris, 363–376.

MACEACHERN, J. & BANN, K. 2008. The role of ichnology in refining shallow marine facies models. *In*: HAMPSON, G.J., STEEL, R.J., BURGESS, P.M. & DALRYMPLE, R.W. (eds) *Recent Advances in Models of Siliciclastic Shallow-Marine Stratigraphy*. SEPM, Special Publications, **90**, 73–116.

MEIJERS, M.J.M., KAYMAKCI, N., VAN HINSBERGEN, D.J.J., LANGEREIS, C.G., STEPHENSON, R.A. & HIPPOLYTE, J.-C. 2010a. Late Cretaceous to Paleocene oroclinal bending in the central Pontides (Turkey). *Tectonics*, **29**, TC4016, https://doi.org/10.1029/2009TC002620

MEIJERS, M.J.M., VROUWE, B. *ET AL.* 2010b. Jurassic arc volcanism on Crimea (Ukraine): implications for the paleosubduction zone configuration of the Black Sea region. *Lithos*, **119**, 412–426.

MIRĂUȚĂ, O., MUTIHAC, V., BANDRABUR, T. & DRĂGULESCU, A. 1968. *Geological Map of Romania: Scale 1:200 000. Tulcea Sheet*. Geological Institute of Romania, Bucharest [in Romanian].

MUNTEANU, I., MATENCO, L., DINU, C. & CLOETINGH, S. 2011. Kinematics of back-arc inversion of the Western Black Sea Basin. *Tectonics*, **30**, TC5004, https://doi.org/10.1029/2011TC002865

MUTIHAC, V. 1964. The Tulcea zone and its position within the structural framework of Dobrogea. *Anuarul Comitetului Geologic*, **XXXIV**, 215–253 [in Romanian with French abstract].

NATAL'IN, B.A., SUNAL, G., GÜN, E., WANG, B. & ZHIQING, Y. 2016. Precambrian to Early Cretaceous rocks of the Strandja Massif (northwestern Turkey): evolution of a long lasting magmatic arc. *Canadian Journal of Earth Sciences*, **53/11**, 1312–1335.

NIKISHIN, A.M., ZIEGLER, P.A., BOLOTOV, S.N. & FOKIN, P.A. 2011. Late Paleozoic to Cenozoic evolution of the

Black Sea–southern Eastern Europe region: a view from the Russion Platform. *Turkish Journal of Earth Sciences*, **20**, 571–634.

NIKISHIN, A.M., OKAY, A., TÜYSÜZ, O., DEMIRER, A., WANNIER, M., AMELIN, N. & PETROV, E. 2015. The Black Sea basins structure and history: new model based on new deep penetration regional seismic data. Part 2: tectonic history and paleogeography. *Marine and Petroleum Geology*, **59**, 656–670.

OAIE, G., SEGHEDI, A., RĂDAN, S. & VAIDA, M. 2005. Sedimentology and source area composition for the Neoproterozoic-Eocambrian turbidites from East Moesia. *Geologica Belgica*, **8/4**, 78–105.

OKAY, A.I. & NIKISHIN, A.M. 2015. Tectonic evolution of the southern margin of Laurasia in the Black Sea region. *International Geology Review*, **57**, 1051–1076.

OKAY, A.I. & TOPUZ, G. 2016. Variscan Orogeny in the Black Sea region. *International Journal of Earth Sciences*, **106/2**, 569–592.

OKAY, A.I., ŞENGÖR, A.M.C. & GÖRÜR, N. 1994. Kinematic history of the opening of the Black Sea and its effects on the surrounding regions. *Geology*, **22**, 267–270.

OKAY, A.I., TÜYSÜZ, O., SATIR, M., ÖZKAN-ALTINER, S., ALTINER, D., SHERLOCK, S. & EREN, R.H. 2006. Cretaceous and Triassic subduction–accretion, HP/LT metamorphism and continental growth in the Central Pontides, Turkey. *Geological Society of America Bulletin*, **118**, 1247–1269.

OKAY, A.I., SUNAL, G., SHERLOCK, S., ALTINER, D., TÜYSÜZ, O., KYLENDER-CLARK, A.R.C. & AYGÜL, M. 2013. Early Cretaceous sedimentation and orogeny on the southern active margin of Eurasia; Central Pontides, Turkey. *Tectonics*, **32**, 1247–1271.

OKAY, A.I., ALTINER, D., SUNAL, G., AYGÜL, M., AKDOĞAN, R., ALTINER, S. & SIMMONS, M.D. 2017. Geological evolution of the Central Pontides. *In*: SIMMONS, M.D., TARI, G.C. & OKAY, A.I. (eds) *Petroleum Geology of the Black Sea*. Geological Society, London, Special Publications, **464**, https://doi.org/10.1144/SP464.3

RĂDAN, S. 2000. Lateritic paleoweathering crusts in Central and North Dobrogea. *Academia Română, Studii şi cercetări de geologie*, **45**, 51–70.

ROBINSON, A.G., RUDAT, J.H., BANKS, C.J. & WILES, R.L.F. 1996. Petroleum geology of the Black Sea. *Marine and Petroleum Geology*, **13/2**, 195–223.

SACHSENHOFER, R., BECHTEL, A., FRANCU, J. & MAYER, J. 2013. Oligocene and Miocene source rocks in the Central and Eastern Paratethys. Abstract presented at the *AAPG Conference, Petroleum Systems of the Paratethys*, 26–27 September 2013, Tbilisi, Georgia.

SAFONOVA, I., MARUYAMA, S., HIRATA, T., KON, Y. & RINO, S. 2010. LA ICP MS U–Pb ages of detrital zircons from Russia largest rivers: implications for major granitoid events in Eurasia and global episodes of supercontinent formation. *Journal of Geodynamics*, **50**, 134–153.

SĂNDULESCU, M. 1988. Cenozoic tectonic history of the Carpathians. *In*: ROYDEN, L.H. & HORVÁTH, F. (eds) *The Pannonian Basin, A Study in Basin Evolution*. American Association of Petroleum Geologists, Memoirs, **45**, 17–25.

SĂNDULESCU, M., KRÄUTNER, H. *ET AL.* 1978. *Geological Atlas of Romania, Sheet 1. Geological Map of Romania: Scale 1:1 000 000*. Geological Institute of Romania, Bucharest.

SCHLEDER, Z., TARI, G., KREZSEK, C., KOSI, W., TURI, V. & FALLAH, M. 2015. Regional structure of the Western Black Sea Basin: constraints from cross-section balancing. *In*: POST, P.J., COLEMAN JR., J.L., ROSEN, N.C., BROWN, D.E., ROBERTS-ASHBY, T., KAHN, P. & ROWAN, M. (eds) *Petroleum Systems in Rift Basins*. 34th Annual GCSSEPM Foundation Perkins-Rosen Research Conference, 13–16 December 2015, Houston, Texas, Golf Coast Section SEPM Foundation, 396–411.

SEGHEDI, A. 2001. The North Dobrogea orogenic belt (Romania): a review. *In*: ZIEGLER, P.A., CAVAZZA, W., ROBERTSON, A.F.H. & CRASQUIN-SOLEAU, S. (eds) *Peri-Tethys Memoir 6: PeriTethyan Rift/Wrench Basins and Passive Margins*. Memoires du Museum National d'Histoire Naturelle, **186**, 237–257.

SEGHEDI, A. 2012. Palaeozoic formations from Dobrogea and Pre-Dobrogea – an overview. *Turkish Journal of Earth Sciences*, **21**, 669–721.

SEGHEDI, A. & NEAGA, V. 2016. Permian volcanism and rifting in the basement of Pre-Dobrogea depression (Scythian Platform). *In*: *AAPG Europe Conference, 19–20 May 2016, Bucharest, Romania*. Abstract Book, 94.

SEGHEDI, A., OAIE, G. *ET AL.* 1999. Geology and structure of the Precambrian and Paleozoic basement of North and Central Dobrogea. Mesozoic history of North and Central Dobrogea. *Romanian Journal of Tectonics and Regional Geology*, **77/2**, 72.

SEGHEDI, A., BARBU, V. *ET AL.* 2010. The Romanian segment of the Black Sea margin: an overview of the stratigraphy and structure of North Dobrogea and East Moesia. *In*: *AAPG Europe Regional Annual Conference & Exhibition*, 17–19 October 2010, Kiev, Ukraine, Field Trip 3, Field trip guide, 54.

SUNAL, G., SATIR, M., NATAL'IN, B.A. & TORAMAN, E. 2008. Paleotectonic position of the Strandja Massif and surrounding continental blocks based on zircon Pb–Pb age studies. *International Geology Review*, **50**, 519–545.

ŚWIDROWSKA, J., HAKENBERG, M., POLUHTOVIČ, B., SEGHEDI, A. & VISNĂKOV, I. 2008. Evolution of the Mesozoic basins on the Southwestern edge of the East European Craton (Poland, Ukraine, Moldova, Romania). *Studia Geologica Polonica*, **130**, 3–130.

SZÁSZ, L. & ION, J. 1986. Crétacé supérieur du bassin de Babadag (Roumanie). Biostratigraphie intégrée (ammonites, inoceramés, foraminifères planctoniques). *Mémoires de l'institut de Géologie et de Géophysique, Bucharest*, **33**, 91–149.

ŢAMBREA, D. 2007. *Subsidence analysis and tectonic-thermal evolution of the Istria Depression. (Black Sea). Implications for the hydrocarbon generations*. PhD thesis, Bucharest University [in Romanian].

TARI, G. 2015. Is the Black Sea really a back-arc basin? *In*: POST, P.J., ROSEN, N., OLSON, D.L., PALMES, S.L., LYONS, K.T. & NEWTON, G.B. (eds) *Transactions of the GCSSEPM Foundation Perkins-Rosen 34th Annual Research Conference 'Petroleum Systems in Rift Basins'*, Houston, Texas, Golf Coast Section SEPM Foundation, 510–520.

TARI, G., DICEA, O., FAULKERSON, J., GEORGIEV, G., POPOV, M., STEFANESCU, M. & WEIR, G. 1997. Cimmerian and Alpine stratigraphy and structural evolution of the Moesian Platform (Romania, Bulgaria), Regional and petroleum geology of the Black Sea and surrounding region. *AAPG Memoir*, **68**, 63–90.

TARI, G., KOSI, W. *ET AL.* 2014. Messinian-style drawdown in the Black Sea at the End Eocene. Search and Discovery Article 90194, *International Conference & Exhibition*, 14–17 September 2014, Istanbul, Turkey.

TARI, G., SCHLEDER, Z., KREZSEK, C., KOSI, W., TURI, V. & FALLAH, M. 2015. Regional structure of the Western Black Sea Basin: map-view kinematics. *In*: POST, P.J., ROSEN, N., OLSON, D.L., PALMES, S.L., LYONS, K.T. & NEWTON, G.B. (eds) *Petroleum Systems in Rift Basins*, 34th Annual GCSSEPM Foundation Perkins-Rosen Research Conference, 13–16 December 2015, Houston, Texas, Golf Coast Section SEPM Foundation, 372–395.

TAYLOR, A.M. & GOLDRING, R. 1993. Description and analysis of bioturbation and ichno-fabric. *Journal of the Geological Society, London*, **150**, 141–148, https://doi.org/10.1144/gsjgs.150.1.0141

TÜYSÜZ, O., YILMAZ, I.Ö., SVABENICKA, L. & KIRICI, S. 2012. The Unaz Formation: a key unit in the Western Black Sea Region, N Turkey. *Turkish Journal of Earth Sciences*, **21**, 1009–1028.

TÜYSÜZ, O., MELINTE-DOBRESCU, M.C., YILMAZ, I.Ö., KIRICI, S., ŠVABENICKÁ, L. & SKUPIEN, P. 2016. The Kapanboğazı formation: a key unit for understanding Late Cretaceous evolution of the Pontides, N Turkey. *Palaeogeography, Palaeoclimatology, Palaeoecology*, **441**, 565–581.

USTAÖMER, P.A., MUNDIL, R. & RENNE, P.R. 2005. U/Pb and Pb/Pb zircon ages for arc-related intrusions of the Bolu Massif (W Pontides NW Turkey): evidence for Late Precambrian (Cadomian) age. *Terra Nova*, **17**, 215–223.

USTAÖMER, P.A., USTAÖMER, T., GERDES, A. & ZULAUF, G. 2011. Detrital zircon ages from a lower Ordovician quartzite of the İstanbul exotic terrane (NW Turkey): evidence for Amazonian affinity. *International Journal of Earth Sciences*, **100**, 23–41.

USTAÖMER, T., ROBERTSON, A.H.F., USTAÖMER, P.A., GERDES, A. & PEYTCHEVA, I. 2012. Constraints on Variscan and Cimmerian magmatism and metamorphism in the Pontides (Yusufeli–Artvin area), NE Turkey from U–Pb dating and granite geochemistry. *In*: ROBERTSON, A.H.F., PARLAK, O. & ÜNLÜGENÇ, U.C. (eds) *Geological Development of Anatolia and the Easternmost Mediterranean Region*. Geological Society, London, Special Publications, **372**, 49–74, https://doi.org/10.1144/SP372.13

YILMAZ-ŞAHIN, S., AYSAL, N., GÜNGÖR, Y., PEYTCHEVA, I. & NEUBAUER, F. 2014. Geochemistry and U–Pb zircon geochronolgy of metagranites in Istranca (Strandja) Zone, NW Pontides Turkey: implications for the geodynamic evolution of Cadomian Orogeny. *Gondwana Research*, **26**, 755–771.

ŻELAŹNIEWICZ, A., BUŁA, Z., FANNING, M., SEGHEDI, A. & ŻABA, J. 2009. More evidence on Neoproterozoic terranes in Southern Poland and southeastern Romania. *Geological Quarterly*, **53**, 93–124.

ZONENSHAIN, L.P. & LE PICHON, X. 1986. Deep basins of the Black Sea and Caspian Sea as remnants of Mesozoic back-arc basins. *Tectonophysics*, **123**, 181–211.

Deep-water plays in the western Black Sea: insights into sediment supply within the Maykop depositional system

E. V. L. REES, M. D. SIMMONS* & J. W. P. WILSON

Halliburton, 97 Jubilee Avenue, Milton Park, Abingdon OX14 4RW, UK

Correspondence: mike.simmons@Halliburton.com

Abstract: The Oligocene–Early Miocene Maykop depositional system of the Western Black Sea Basin is investigated in terms of sediment supply and provenance. Potential sediment source regions and conduits for sediment supply into the deep-water portion of the basin are evaluated based on the tectonic history and framework of the region, and are supported by observations from published well, reflection seismic and isopach data. The outcrop geology of the present-day land areas adjacent to the basin is used as a guide to the likely provenance and, hence, quality of potential siliciclastic reservoirs. Reservoir presence and reservoir quality are key subsurface risks for exploration in deep-water plays involving Maykop turbidite sandstones and charge from the well-known Maykop organic carbon-rich mudstones that are widespread across the basin.

Sediments sourced from the NE Moesian Platform and Dobrogea, channelled into the off-shore Black Sea via the Histria Trough, are considered moderate risk in terms of primary res-ervoir quality, as evidenced by thick packages of fine-grained sediment. In contrast, sediments derived from the southern Strandja Massif fed into the Burgas Basin, and potentially into the deeper-water Turkish Black Sea, are relatively low risk in terms of reservoir quality, given the abundance of acidic intrusions within the massif. Sediment derived from parts of the northern Strandja Massif, especially the volcaniclastics of the Srednogornie region, are likely to have poorer reservoir quality characteristics. Sediments derived from the granitic Bolu Massif within the Pontides might be of good reservoir quality but are likely to be ponded behind the offshore Kozlu Ridge. An important sediment source-to-sink system was derived from the Balkanides and entered the deeper-water western Black Sea via the Kamchia Trough. The present-day Kam-chia river is a relatively minor sediment supplier to the Black Sea, but the palaeo-Kamchia river of the Oligocene–Early Miocene would have exploited a much greater drainage area consisting of an axial trunk stream, occupying the newly formed Kamchia Foredeep to the north of the Balkanides, and transverse rivers sourcing sediment from the granitic and gneissic bodies of the Balkan Moun-tains and from Early Cretaceous and Palaeogene sandstones. These would provide reasonable res-ervoir quality, and it is estimated from reference source-to-sink relationships that offshore sediment flux via this system was probably at least eight times greater than at present. Known shelf-edge canyons in offshore Bulgaria facilitated this sediment reaching the deep water offshore, where a sedimentary fan with a length in excess of 150 km is likely to have developed. This sug-gests that the potential is good for encountering good-quality reservoir sands in the Maykop suc-cession deep water of the western Black Sea, and this aspect of regional play risk could be of less concern than was previously considered.

The Maykop Suite (or Maykop Group) is the name given to distinctive, often organic carbon-rich, sed-iments deposited during the Oligocene–Early Mio-cene (Fig. 1) that occur within a region spanning the Black Sea and its margins, the Greater Cauca-sus, and the South Caspian Sea (Bazhenova *et al.* 2003). The name is derived from the town of May-kop in the Russian Caucasus, nearby to which sev-eral hundred metres of dominantly dark, partly high total organic carbon (TOC) shale occur, inter-bedded with thin sandstones (Zaporozhets 1999; Saint-Germes *et al.* 2000; Sachsenhofer *et al.* 2017*a*). There are local lithostratigraphic equiva-lents for this unit, including, for example, the Rus-lar Formation onshore and offshore Bulgaria

(Sachsenhofer *et al.* 2009), and the Histria Forma-tion offshore Romania (Dinu *et al.* 2005). The term 'Maykopian' also has chronostratigraphic con-notations in Russian language literature (Voronina *et al.* 1988; Jones & Simmons 1996).

The Oligocene–Miocene period encompasses several eustatic (Haq *et al.* 1988; Miller *et al.* 2005) and regional (Popov *et al.* 2010) changes in sea level, which are recorded within the Maykop Suite by cyclic deposition of fine-grained organic-rich sediments and sandstone packages (Mateo *et al.* 2013). It is the organic carbon-rich horizons that have been most studied historically because of their importance as potential source rocks – they are the known source for oil and gas accumulations

From: SIMMONS, M. D., TARI, G. C. & OKAY, A. I. (eds) 2018. *Petroleum Geology of the Black Sea.*
Geological Society, London, Special Publications, **464**, 247–265.
First published online September 15, 2017, https://doi.org/10.1144/SP464.13

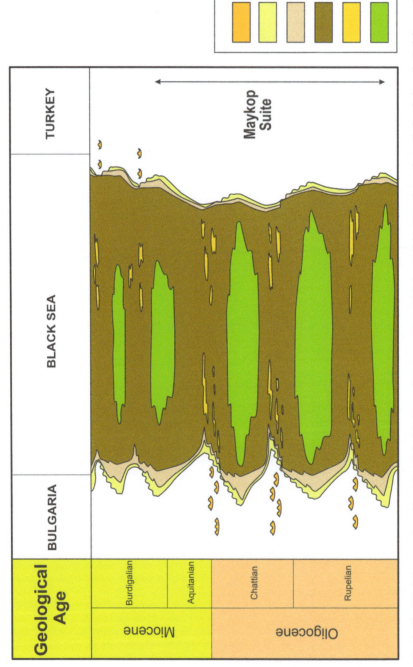

Fig. 1. Simplified chronostratigraphic setting of the Maykop sedimentary system and facies cyclicity. Note that as the vertical scale is geological time, sedimentary thickness is not implied (i.e. lowstand deep-marine sands might be deposited during a short duration but could, nonetheless, be thick). The coastal onlap indicated on the left-hand side of the figure provides an indication of the regional relative sea-level change.

in the South Caspian Basin, Azov-Kuban Basin, and the East and West Black Sea basins (Bazhenova *et al.* 2003; Sachsenhofer *et al.* 2009, 2017*b*; Cranganu & Şaramet 2011; Georgiev 2012; Moroşanu 2012; Vincent & Kaye 2017). This paper focuses on the coarse clastic depositional systems of the Maykop, with the aim of assessing the reservoir potential of any deep-water plays that might exist within the formation in the western Black Sea (WBS).

Commercial hydrocarbon discoveries in rock units other than the Maykop have been made along the coast and in the shallow waters of the WBS over the last several decades (Tari *et al.* 2009; Georgiev 2012). Examples include the Tyulenovo (in 1951) and Galata (in 1993) fields in Bulgaria, the East Lebada Field (in 1979) discovered offshore Romania (Ionescu 2002), and the Akçakoca gas field (in 2004) off the coast of Turkey (Menlikli *et al.* 2009). More recently, the deeper-water sector of this region has experienced some success (Mateo *et al.* 2013), with the Domino (in 2012) and associated discoveries (1–3 TCF (trillion cubic feet) of gas) made offshore Romania by OMV Petrom and ExxonMobil in Late Miocene sandstones with a biogenic source of gas (Moroşanu 2012). This has drawn attention to the exploration possibilities of the deep-water WBS and a consideration of plays with a thermogenic source: for example, within Cretaceous synrift clastics or carbonate build-ups, or turbidite (slope) fans within post-rift Tertiary clastic systems (Tari *et al.* 2009, 2011). The Oligocene–Miocene Maykop (and equivalents) depositional system contains several such potential turbidite horizons (Dinu *et al.* 2005; Menlikli *et al.* 2009; Mityukov *et al.* 2011, 2012; Mateo *et al.* 2013; Sipahioğlu *et al.* 2013) and is the focus of this review.

The potential for exploiting plays in the WBS with Maykop sandstones as a reservoir has gathered momentum during the last decade. Wells, such as Istranca-1 drilled by TPAO offshore northwest Turkey (Korucu *et al.* 2013), have proven that a potential thermogenic hydrocarbon source (in this case, gas charging a Middle Miocene sandstone reservoir) exists within the WBS. This is further supported by discoveries, such as Galata, offshore Bulgaria, and gas chimneys and gas-related velocity push-downs on high-quality 3D seismic reflection data (Tari *et al.* 2009). The Subbotina discovery offshore Crimea demonstrated that Maykop Suite sandstones can form effective reservoirs with associated mudstones creating an effective seal (Stovba *et al.* 2009). There are also small discoveries in reservoirs within the upper Maykop in the Gulf of Odessa (Palii & Tochkov 1994). Although the main reservoir target was Mesozoic carbonates, gas has been encountered in Maykop-aged sandstones at Tyulenovo and nearby fields (e.g. Bulgarevo) in coastal Bulgaria (Georgiev 2012). Reservoir-quality sands

have been encountered at the base of the Oligocene section in the Portiţa Field, offshore Romania (Moroşanu 2012). The Samotino Melrose-1 well drilled offshore Bulgaria encountered Maykop-equivalent arkosic sandstones with good reservoir characteristics (Tari pers. comm. 2016). The Sile-1 well drilled by Shell and TPAO offshore Turkey (in 2016) is expected to have targeted a Maykop turbiditic sandstone play. Most importantly, in 2016, the OMV–Total–Repsol consortium tested a Maykop slope-fan play in the Han Asparuh block, offshore Bulgaria, and the well was announced a success by the operator.

Reservoir presence and reservoir quality remain the key risks for the success of plays involving Maykop deep-water sandstones. To assess the prospectivity of plays within the Maykop coarse clastic horizons, the provenance of the sediment needs to be established, as this will be a primary control on reservoir quality (Menlikli *et al.* 2009; Tari *et al.* 2011; Maynard *et al.* 2012). By taking into account the geodynamic history, the topography of the hinterland surrounding the WBS can be constrained and sediment provenance areas identified. Data from seismic reflection surveys, wells and outcrops can then be used to further recognize sediment pathways within the basin. These elements (provenance and pathway) combined can then be used to assess the prospectivity of deep-water turbidite plays within the Maykop Suite and its equivalents.

Tectonic setting of the Maykop suite

The structure of the Western Black Sea Basin is dominated by NW–SE-trending faults inherited from the Permo-Triassic rifting of Dobrogea, as well as troughs and depressions (Fig. 2) formed during and after the opening of the WBS ocean basin in the Late Cretaceous and Tertiary (Dinu *et al.* 2005; Bergerat *et al.* 2010; Konerding *et al.* 2010). Figure 3 shows a tectonic events chart summarizing the tectonic evolution of the WBS region.

Rifting in the WBS began in the Barremian (Munteanu *et al.* 2011), with the deposition of synrift clastics and carbonates identified in the Pontide Mountains (Hippolyte *et al.* 2010, 2016; Tari 2015; Okay *et al.* 2017), Crimea (Nikishin *et al.* 2015*b*) and offshore Romania (Ionescu 2002). The process involved the Istanbul Terrane (modern-day Western Pontides) splitting away from Moesia as a consequence of the subduction of the Neotethyan Ocean to the south (Okay *et al.* 1994; Banks & Robinson 1997). Seafloor spreading then commenced during the Coniacian–Santonian, with deep-water sequences well established in the Western Black Sea Basin by the Coniacian and an associated island-arc contributing volcaniclastic material (Görür 1988;

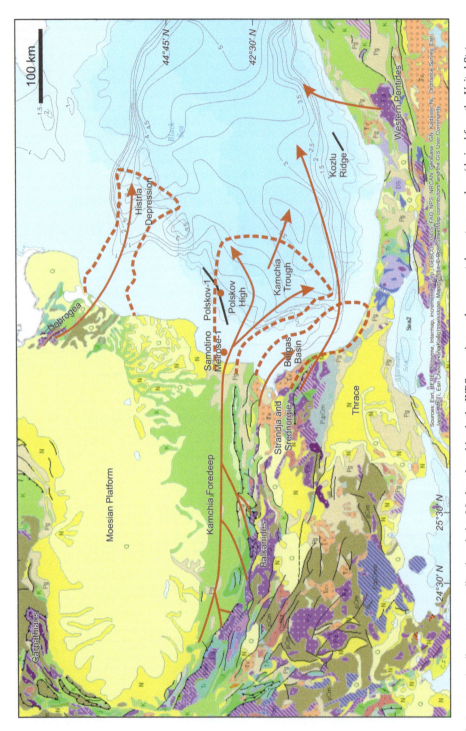

Fig. 2. Interpreted sediment transport directions during Maykop deposition in the WBS, superimposed on present-day outcrop geology (derived from the United States Geological Survey World Energy Project: ArcGIS 2011). The offshore sedimentary thickness of the Maykop Suite (Gorshkov *et al.* 1989; Meisner & Tugolesov 2003) is shown (isopachs at 500 m intervals) and contributes to the understanding of the sedimentary pathways. Red dashed lines represent the main depocentres. Key locations and tectonic units mentioned in the text are indicated.

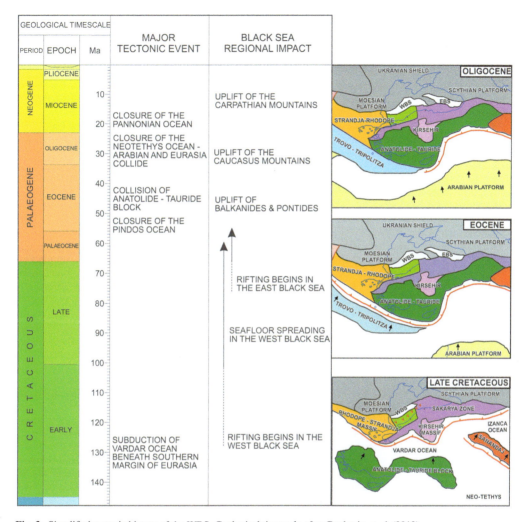

Fig. 3. Simplified tectonic history of the WBS. Geological timescale after Gradstein *et al.* (2012).

Georgiev *et al.* 2001; Hippolyte *et al.* 2010; Nikishin *et al.* 2015a, b; Okay & Nikishin 2015; Okay *et al.* 2017). In the Early Tertiary, ocean spreading ceased and the rate of sedimentation increased, with sediment ultimately sourced from erosion of the surrounding orogens (Mateo *et al.* 2013).

From the Eocene onwards, the progressive convergence of Arabia towards Europe and the successive collision of terranes (e.g. the Anatolia–Tauride block) with the southern margin of Eurasia (Fig. 3) closed the Neotethys Ocean, and uplifted the Caucasus, Pontide and Alborz mountain ranges (Jolivet & Faccenna 2000; Meulenkamp & Sissingh 2003; Cavazza *et al.* 2012; Espurt *et al.* 2014). This isolated the Black Sea and Caspian Sea basins from the closing Tethys to the south, and created the Paratethys realm in Early Oligocene times (Baldi 1980;

Popov *et al.* 1993, 2010; Jones & Simmons 1996; Popov & Stolyarov 1996; Rögl 1999; Schulz *et al.* 2005). Runoff from the emerging highlands surrounding Paratethys introduced both terrestrially derived organic matter and nutrients that led to high phytoplankton production (Bazhenova *et al.* 2003).

The Paratethys region experienced restricted and episodic connection to the open ocean (mostly via the Dnieper–Donets Basin and Pripyat Strait to the North Sea Basin or the Birladsky Strait to the Carpathian Basin: Popov *et al.* 2004), which led to the development of a stratified water column during times of high global sea level, and the deposition and preservation of the organic matter (Popov & Stolyarov 1996; Dinu *et al.* 2005; Schulz *et al.* 2005) within the Maykop Suite. At times of low global sea level, Paratethys became even more isolated from the open ocean and coarser clastic sediments were derived from the emergent hinterland, and associated nutrient influxes resulted in blooms of calcareous nannoplankton adapted to brackish surface waters. One such event occurred at the beginning of the Solenovian regional stage (i.e. within the Early Oligocene at the NP22–NP23 nannofossil zone boundary) (Popov & Stolyarov 1996; Rögl 1999; Schulz *et al.* 2005; Sachsenhofer *et al.* 2009). Overall, the Oligocene climate of the WBS was warm (mean annual temperature of 15–17°C) and wet (mean annual precipitation of 800–1300 mm) based on palynological data (Ivanov *et al.* 2007). This would have assured regularity of sediment transport from local fluvial systems.

As the Paratethys domain developed, so did the hinterland surrounding the WBS (Jones & Simmons 1996). During the Oligocene, the Scythian and Moesian platforms were exposed to the north and NW, and the Balkanides were a topographical feature to the west. The Pontides were a growing mountain range to the south, as evidenced by a region-wide peneplain (Yilmaz *et al.* 1997), and a number of other small highs, such as the Strandja Massif in the SW, were also topographical features. With regard to the Western Black Sea Basin, the shelf and seafloor were dominated by fault structures inherited from the opening of the basin in the Cretaceous (Banks & Robinson 1997; Hippolyte *et al.* 2010; Konerding *et al.* 2010). Features, such as the Kozlu Ridge, the Polshkov High, and the Histria, Kamchia and Burgas basins, all heavily influenced sediment distribution patterns during deposition of the Maykop (Fig. 2), with previous structures remaining from extension in the Dobrogea region during the Permo-Triassic (Dinu *et al.* 2005) also influencing bathymetry and, therefore, sediment transport pathways.

Sediment source regions

The quality of siliciclastic petroleum reservoirs is tied closely to sediment provenance because different mineralogical composition directly controls the effects of diagenesis and, ultimately, porosity and permeability characteristics. In the WBS, multiple sediment sources exist that introduced sediment from a variety of exposed lithotypes. Subsequently, an understanding of where sediment within the deep Western Black Sea Basin has been derived is key to assessing the quality of any potential deep-water turbidite reservoirs. Figure 2 summarizes the findings from this review, highlighting sediment provenance regions and likely sediment pathways into the basin. The following discussion deals first with the aspect of sediment source region, followed by the bathymetric variation of the seafloor and the

routes that sediment might take to the basin, and then drawing conclusions on the overall potential reservoir quality of Maykopian turbidites in the WBS.

NE Moesian Platform/Dobrogea

The Moesian Platform is a Precambrian basement block bordered to the south by the Balkanide Mountains, to the north and west by the Carpathian Mountains and the Dobrogea Orogen, and extends eastward into the WBS (Dinu *et al.* 2005) (Fig. 2). It is separated from the rest of the East European Craton by North Dobrogea, a region which records several accretionary events thought to represent the suturing of the Moesian Platform to the East European Craton during the Varsican Orogeny in the Late Palaeozoic (Dinu *et al.* 2005; Carrigan *et al.* 2006; Golonka *et al.* 2006). The sedimentary succession contains deposits ranging from the Late Cambrian to the Cenozoic, across which unconformities can be recognized representing several tectonic phases of the platform's history (Tari *et al.* 1997; Seghedi *et al.* 2005). One such notable unconformity is from the Paleocene to the Lower Miocene, where a large erosional surface has been identified not only in core and outcrop but also in seismic reflection profiles (Tari *et al.* 1997; Mirea 2009). During this erosional period on the platform, sediment was being deposited in the offshore Histria Depression, as observed in seismic reflection data (Dinu *et al.* 2005), with up to 4900 m of sediment being deposited in the Ovidiu region offshore.

The portion of the Moesian Platform of interest to this study is the NE section bordering the Dobrogean Terrane. This part of the Moesian Platform was, as with the rest of the platform, subaerially exposed during the Oligocene following the uplift of the Balkanides to the south in the Eocene (Tari *et al.* 1997; Jolivet & Brun 2010). As the NE platform borders a significant part of the WBS coastline, drainage from this region and the adjacent Dobrogea will have a significant impact on the clastic input to the basin through the Maykopian period, with some sedimentation focused into the Histria Depression (Dinu *et al.* 2005; Moroşanu 2012).

The main drainage pathways across the northern Moesian Platform in the Oligocene have been shown to flow north, towards the developing Carpathian Foredeep, also known as the Getic Depression (Tari *et al.* 1997; Matenco *et al.* 2003; Agalareva 2012). This would suggest that any rivers flowing into the WBS off the NE Moesian Platform/Dobrogea would have had short drainage pathways with localized sources (Fig. 2).

Outcrops in the Măcin Mountains of North Dobrogea include Precambrian amphibolites, schist and quartzite overlain by a weakly metamorphosed succession of Palaeozoic shales, phyllites, greywackes and limestones intruded by Variscan granitoids (Burchfiel 1976; Seghedi 2001). This was, in turn, unconformably overlain by a Mesozoic succession, including Cretaceous sandstones and chalks. In Central Dobrogea, there are extensive outcrops of Late Precambrian schist and gneiss overlain by a thick succession of greywackes and chloritic phyllites (Seghedi *et al.* 1999) called the 'Schistes Vertes' (Burchfiel 1976). This is overlain by Late Jurassic shallow-marine limestones and a synrift succession of Cretaceous sandstones, followed by chalks. Sediment derived from this potential source material is likely to have limitations with regard to reservoir quality. Nonetheless, investigation of sediments within the Carpathian Foredeep can therefore provide an indication of the quality of the source areas on the northern Moesian Platform.

The Fusaru and Kliwa formations are flysch deposits shown to be sourced from southerly (Moesian Platform) and easterly (Scythian Platform) provinces (Sylvester 2002; Panaiotu *et al.* 2007). Analysis of data from these formations shows that the clastic component has a high percentage of quartz, with minimal lithic content. This would suggest that the Moesian and Scythian platforms in the proximity of the NW margin of the Black Sea have a source of clean quartz-rich sediment, which could also be the source region for rivers draining into the WBS through the Histria Depression. These rivers could have deposited sediment derived from the Precambrian granite gneiss and quartzite basement of the Moesian Platform (Burchfiel 1976; Tari *et al.* 1997; Seghedi *et al.* 1999) or overlying Palaeozoic quartz-rich sandstones (Dinu *et al.* 2005). Even though the river systems draining into the Histria Depression might have been small, well data show that significant thicknesses of sediment were deposited on this margin (Moroşanu 2012), especially during the Oligocene (Dinu *et al.* 2005). The track record of exploration within the Histria Depression channels suggests that deep-water fans developed within this system tend to be fine-grained (Dinu *et al.* 2005), with grain size and resultant porosity–permeability issues being the main factors in reservoir risk evaluation.

Balkanides

The Balkanides are an east–west-trending thrust belt generated by sinistral transpression and compression during the Palaeogene (Doglioni *et al.* 1996; Banks 1997; Sinclair *et al.* 1997; Bergerat *et al.* 1998, 2010; Stuart *et al.* 2011; Mateo *et al.* 2013). The terrane on which the Balkanides sit is floored by a Neoproterozoic ophiolite and a Cambrian–Ordovician island-arc complex of Gondwanan affinity (Carrigan *et al.* 2005). This underlying terrane was accreted to the Moesian Platform during the

Variscan Orogeny and contains many Variscan-age granitoid bodies intruded into the metamorphosed basement (Carrigan *et al.* 2005). After the Variscan Orogeny, the region was part of a passive margin, on which platform carbonates accumulated through the Triassic (Sinclair *et al.* 1997). Early Jurassic shallow-marine siliciclastics were then followed by deeper-water Late Jurassic and Cretaceous carbonate and clastic flysch deposits (Georgiev *et al.* 2001). During the Early Cretaceous, shelfal siliciclastic sediments prograded into this flysch basin (Sinclair *et al.* 1997). The Late Cretaceous and Early Tertiary succession includes platformal carbonates and clastics deposited during and after rifting of the WBS in the Cretaceous, as well as numerous Late Cretaceous granitoid bodies similar to those emplaced within the Srednogorie region to the south (Von Quadt *et al.* 2005).

Thrusting within the Balkanides occurred as a result of closure of the Pindos Ocean and collision of the Trovo-Tripolitza Terrane with Eurasia in the early Eocene (Fig. 3), with orogenesis continuing through to the Priabonian (Jolivet & Brun 2010). Basins in the Balkanides were inverted during this time, with syn- and post-rift sediments being eroded and deposited as flysch in the Balkanides (Kamchia) Foredeep through to the Middle Eocene (Sinclair *et al.* 1997; Bergerat *et al.* 1998, 2010). Following uplift, the mountains were then a prominent topographical feature, thus forming the drainage divide from which transverse rivers issued north into the Kamchia Foredeep that developed to the north and east of the Balkanides, and which form a major depocentre for the sedimentary products of Balkanides erosion (Sinclair *et al.* 1997; Suttill 2009; Mateo *et al.* 2013). The Kamchia Foredeep first formed in the early Middle Eocene ('Illyrian') phase of Eastern Balkanide uplift (Bonchev 1986; Doglioni *et al.* 1996; Marinov 1997; Dimitrov & Georgiev 2011). In particular, a major axial river system can be envisaged within the Kamchia Foredeep to the north of the Balkanides. Such axial rivers are typical of the overfilled stage of retro-arc foreland basin development (Suttill 2009; Garcia-Castellanos & Cloetingh 2012), with drainage patterns changing to transverse as the basin fills and reaches an overfilled state. This axial river would have been fed by small, but erosionally vigorous, transverse tributaries draining the area of granite bodies in the Balkanides. In its current overfilled state, drainage in the Kamchia foreland system is dominantly to the north and joins the Danube; in the Oligocene, however, drainage would have been west to east and would have flown out into the Black Sea (Suttill 2009; Stuart *et al.* 2011) via canyon systems identified in seismic data (Mateo *et al.* 2013).

By considering the geological history and surface geology exposed within the Balkanides present day (Fig. 4), it is possible to infer that the trunk stream

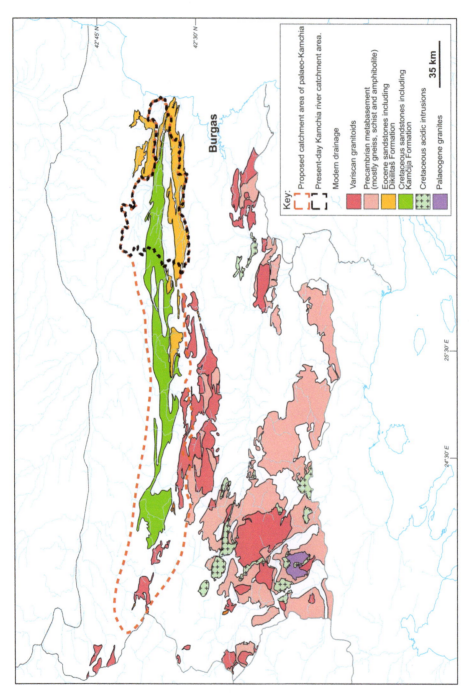

Fig. 4. Catchment area of the palaeo-Kamchia river compared with the extent of the modern Kamchia. Rock units likely to yield quartz-rich good-quality reservoir sediment are highlighted. Geological features are based on Cheshitev & Kănčev (1989). The southern limit of the palaeo-drainage basin area uses the drainage divide in the Balkanides, and the northern limit uses likely points of capture and inflection in the drainage planform.

1672.38 m   **1673.38 m**

1673.86 m

1673.38 m

of the Oligocene palaeo-Kamchia river occupied an axial position in the Kamchia Foredeep, and was sourced by transverse tributary rivers with headwater regions actively eroding areas of Precambrian metabasement (e.g. the gneisses and granite gneisses of the Arda Group), Variscan (e.g. Stara Planina granodiorite–granite complex) and Cretaceous granites, and thick Early Cretaceous and Palaeogene sandstones of the Kamčija and Dikilitaš formations. This latter formation is quartz-rich and easily eroded (Dimitrov *et al.* 1997), and forms part of the drainage area of the modern Kamchia river. Today, this river has a small drainage area, occupied mostly by Late Cretaceous flysch (Fig. 4). Its sediment composition includes quartz (60–70%), feldspar (10–15%), rock fragments (15–20%) and accessory minerals (5–6%) (Dimitrov *et al.* 1997).

The Samotino Melrose-1 well drilled offshore of the Balkanides, in the southern part of the Kamchia Trough offshore Bulgaria, encountered reasonable reservoir quality Oligocene Maykop-equivalent sandstones with porosities ranging between 14 and 28% (Tari pers. comm. 2016). These are arkosic or lithic arkosic arenites with a significant proportion of rock fragments derived from crystalline, metamorphic and volcanic sources (Mateo *et al.* 2013). Clays present are mostly kaolinite. Poor cementation by ankerite and the leaching of grains maintains porosity and permeability. A notable feature is the presence of conglomeratic and pebbly sandstone horizons (Fig. 5), also observed in the onshore outcrops of the Ruslar Formation south of Varna on the Bulgarian coast (Fig. 6) (see Suttill 2009). These sediments are considered canyon-fills in a partial shelf-bypass zone that potentially feeds a slope fan system farther to the east (Mateo *et al.* 2013).

Strandja Massif

The Strandja Massif is the name given to a region situated between the Balkanide Mountains of Bulgaria and the Pontide Mountains of northern Turkey (Fig. 7). As such, it has geological associations with both mountain ranges (Sunal *et al.* 2008), although the geodynamic evolution of the massif is likely to be linked to the Istanbul Terrane of Turkey (Okay *et al.* 2006; Okay 2008). In contrast

Fig. 5. Poorly sorted pebbly sandstone within the Maykop Suite equivalent, Samotino Melrose-1 well, offshore Bulgaria. Diverse lithologies are present, including granite, gneiss, quartzite, schist, shales and volcanic rock fragments. A Balkanides source supplied by the palaeo-Kamchia appears likely. The Samotino Melrose-1 location is with an offshore channel supplying sediment into the offshore basin. Photograph courtesy of Dr Gabor Tari, OMV.

Fig. 6. Poorly sorted conglomerate within the Maykop Suite equivalent at an outcrop south of Varna on the Bulgarian coast (42° 55′ 32.58″ N; 27° 54′ 14.28″ E). Note the erosive base. This can be interpreted as a mass-transport deposit feeding into the offshore channel observed around Samotino Melrose (Fig. 5). Photograph courtesy of Dr Gabor Tari, OMV.

to the Istanbul Zone and Pontides, the massif is composed mainly of metamorphic and granitic rocks. The basement of the Strandja Massif comprises mostly Late Palaeozoic rocks, heavily metamorphosed during the Variscan Orogeny, to quartzo-feldsphathic amphibolite gneisses and schists (Okay *et al.* 2001; Okay 2008; Sunal *et al.* 2008; Natal'in *et al.* 2012, 2016). The basement also contains undeformed Late Variscan granitoids that cut through the Palaeozoic strata (Carrigan *et al.* 2006), and which have been dated to between 260 and 270 Ma by zircon Pb/Pb evaporation mass spectrometry (Okay *et al.* 2001; Sunal *et al.* 2008). The Variscan crystalline basement is overlain by a Triassic–Jurassic sedimentary cover (Okay *et al.* 2001; Elmas *et al.* 2011; Natal'in *et al.* 2012, 2016). Fluviatile sediments were deposited during the Early Triassic, with Middle/Late Triassic–Jurassic transgressions resulting in the deposition of shallow-marine clastics and carbonates (Banks 1997; Okay 2008). A metamorphism event in the Late Jurassic altered these sedimentary rocks to meta-sandstones and conglomerates, phyllites, and marbles. The

metamorphic rocks of the Strandja Massif are unconformably overlain by Cenomanian shallow-marine sandstones (Okay *et al.* 2001).

In the northern part of the Strandja region (Srednogornie), a thick series of Late Cretaceous volcanic and volcanogenic rocks exists, the products of arc magmatism within the present-day southern Black Sea relating to the Neotethys subduction south of the present-day Pontides (Görür 1988; Georgiev *et al.* 2001; Nikishin *et al.* 2015*a*, *b*; Okay & Nikishin 2015). This magmatic belt can be traced into the Pontides (Okay 2008; Okay *et al.* 2017). In the Strandja Massif, this event also produced a large number of andesitic dykes and sills, and the intrusion of the Demiriciköy granitic pluton (Moore *et al.* 1980).

In the Oligocene, the Strandja Massif was a topographical high, having not been transgressed since the Early Eocene and the uplift of the Balkans. As such, the massif would be a likely source of sediment for the WBS through the Burgas Basin. The granites and metamorphic rocks of the massif exposed present day (Fig. 7) provide an indication of the nature

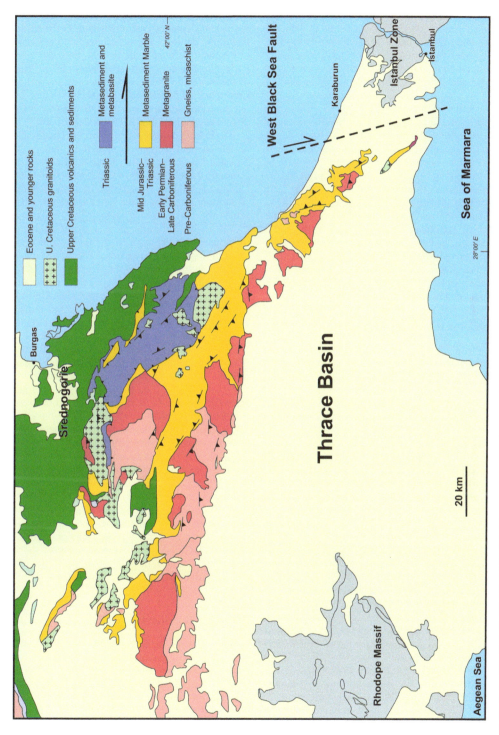

Fig. 7. Geology of the Strandja Massif and Srednogorie. Adapted from Okay *et al.* (2001).

of exposed rocks in the Oligocene. With a high proportion of granitic bodies potentially exposed, the Strandja Massif can be considered as a low-risk sediment source region for clastic input into the WBS. Okay (pers. comm. 2017) reports that apatite fission-track ages from the Strandja Massif range from the Maastrichtian to the Early Paleocene (70–60 Ma) and indicates modest levels of erosion of approximately 3 km.

Even though the exposed granitic bodies of the Strandja Massif suggest a high-quality reservoir sediment source, the presence of the aforementioned Late Cretaceous volcanic and volcaniclastic sediments within the Srednogorie region to the north of the massif (Fig. 7) does add an element of risk to sediments shed from this region. However, Late Cretaceous granite intrusions, also observed in the Balkanides (Von Quadt *et al.* 2005), offer better sediment provenance characteristics with regard to reservoir quality. Nonetheless, modern-day sediments derived from the Srednogorie region are rich in lithic grains derived from the volcaniclastics (Dimitrov *et al.* 1997). Magnetite and zircon are conspicuous components of heavy mineral assemblages.

Pontides and Bolu Massif

The Pontide Mountains of northern Turkey were a prominent feature of the Paratethys realm, providing the southern barrier to open-ocean conditions and, hence, helping to create the restricted basin in which the Maykop was deposited. The mountains were uplifted towards their present elevation during the Eocene–Miocene, initially in response to the amalgamation of the Anatolide–Tauride Terrane (Yilmaz *et al.* 1997) and subsequently as a result of the collision of the Arabian peninsula with Europe (Fig. 3). Consequently, their stratigraphy documents a varied geodynamic history. Apatite fission-track thermochronology shows that initial uplift of the Western Pontides occurred between the Late Lutetian and the Early Rupelian (Cavazza *et al.* 2012) or as old as the Ypresian (Espurt *et al.* 2014). The northern part of the Western Pontides relevant to this review was originally part of the Istanbul Terrane (Okay 2008). Within this terrane, the Bolu and Sünnice massifs expose the late Precambrian crystalline basement characterized by gneiss, amphibolite, and metavolcanic rocks and large grantoid intrusions (Ustaömer & Rogers 1999; Chen *et al.* 2002; Okay 2008). Also present are Ordovician–Carboniferous sediments (Ustaömer & Rogers 1999; Özgül 2012), including Ordovician red beds, Upper Silurian–Devonian limestones and a thick succession of Early Carboniferous siliciclastic turbidites, although marked differences exist in the facies of the Palaeozoic section between the eastern and western parts of the terrane (Okay 2008). Following Variscan

deformation and the intrusion of Late Permian granitoids, deposition recommenced in the Triassic with a thick succession of sandstones, shales and limestones deposited east of Istanbul. In the eastern part of the terrane, Late Jurassic shallow carbonates of the Inalti Formation are present, followed by an Early Cretaceous synrift succession. In all parts of the Istanbul Terrane, a thick succession of Late Cretaceous deep-water carbonates, clastics and arc-related volcaniclastics are present (Özcan *et al.* 2012; Okay & Nikishin 2015; Okay *et al.* 2017).

The Palaeozoic–pre-Late Cretaceous stratigraphy of the Istanbul Terrane is similar to that of the Moesian Platform and prior to the Late Cretaceous opening of the WBS, it was located south of the Odessa Shelf (Okay *et al.* 1994). With the inception of back-arc spreading in the Late Cretaceous, it was rifted away and translated south.

The major Bolu and Sunnice massifs acted as topographical highs through the Early Cretaceous, with the Inpiri and Cengellidere formations onlapping the flanks of these highs. The region was not submerged again until the Albian, in connection with rifting and the break-up of the WBS. Uplift of the Pontides in the Eocene exposed the massifs once again (Yilmaz *et al.* 1997) and the region remains subaerially exposed to the present day.

The limited time that the Bolu and Sunnice massifs were submerged in the Mesozoic means that the sedimentary cover there was likely to be thin. Once uplifted in the Eocene, initial eroded material would be lithic-rich (as observed in Kusuri Formation turbidites near Ayancik: Janbu *et al.* 2007) but would quickly evolve in composition as the quartz-rich basement of the Bolu Massif was exposed. These erosional products would have been transported northwards by fluvial systems draining towards the southern WBS. Sediments sourced from the Palaeozoic granitic complexes of the Western Pontides could therefore contribute to high-quality reservoir sandstones and can be considered a low-risk source region, although there are risks involved with the timing of sediment supply and submarine sediment transport pathways (discussed further below).

Sediment transport pathways

Major pathways for sediments to reach the basin floor of the WBS are limited to only a few key conduits, namely the Histria, Kamchia and Burgas troughs (Fig. 2). Sediment thickness and depth contour maps within the WBS (Gorshkov *et al.* 1989; Meisner & Tugolesov 2003) highlight these features as sediment conduits during Maykopian deposition. Additionally, topographical features on the seafloor acted as barriers, trapping or diverting sediment along specific pathways, such as the Kozlu Ridge. During Maykopian lowstands, sediments were channelled through the troughs, bypassing the shelves and creating submarine fan complexes, as observed on seismic data from the Bulgarian and Turkish offshore (Tari *et al.* 2009; Stuart *et al.* 2011; Mateo *et al.* 2013; Sipahioğlu *et al.* 2013; Nikishin *et al.* 2015*a*).

The surface drainage pattern to supply sediment into these conduits is, in the absence of detailed studies of sediment provenance using detrital zircon and heavy mineral studies techniques, speculative. Likewise, the uplift and denudation histories of the mountain belts acting as sediment sources for the Maykopian WBS are imprecise because of the lack of a thermochronological constraint. Modern-day drainage patterns of rivers entering the WBS provide limited but important clues about the likely pattern and extent of drainage during Maykopian times, and can be used to reconstruct the catchment area of some major river systems. The modern-day Danube, which is a significant feature of the current drainage pattern, reached its present configuration in the Pleistocene (Miklós & Neppel 2010). Furthermore, the fill of the Kamchia Foredeep to an overfilled state has allowed drainage from the Balkanides to be captured by the Danube such that drainage is prominently northwards. Drainage from the southern side of the Balkanides and Strandja Massif is currently mostly captured by the Evros, which drains into the northern Aegean Sea, although it probably flowed into the Marmara Sea until approximately 1.5 Ma (Okay & Okay 2002). Other modern rivers (e.g. the Kamchia) are small (the Kamchia length is 245 km) and have a relatively limited sediment flux, although, as discussed by Milliman & Syvitski (1992), small mountainous rivers can supply large volumes of sediment to receiving basins. Before dam construction in the 1950s, the three largest Turkish rivers emptying into the Black Sea discharged an estimated 50 Mt (million tonnes) of sediment annually. This includes the Sakarya and Filyos entering the WBS.

Maynard *et al.* (2012) suggested that during Maykop Suite deposition, most of the rivers entering the WBS would have had short lengths and local sediment sources, providing a limited (and poor reservoir quality) sediment flux. This view can be challenged both from observations on the sediment fill of the main depocentres, as previously discussed, and from estimates of sediment flux from reference source-to-sink relationships (e.g. Sømme *et al.* 2009).

For example, the present-day Kamchia river has a drainage basin area of 5338 km^2, with a maximum relief of 1100 m and an average annual temperature of 11°C (Skoulikidis *et al.* 2009). Using the empirical scaling relationships from the regression analysis of Sømme *et al.* (2009), and the BQART analytical

model of Syvitski & Milliman (2007), which takes into account geomorphic and climatic factors, a long-term sediment flux of 0.68 Mt a^{-1} is estimated. The predicted long-term value is lower than the pre-irrigation channel construction value of 1.2 Mt a^{-1} reported by Dimitrov *et al.* (1997) and Jaoshvili (2002), but can be explained by the larger time average (*c.* 30 year) adopted in the approach of Syvitski & Milliman (2007). This would create a sediment fan area of 2132 km^2 with a length of 52 km. However, as previously discussed, the drainage basin of the palaeo-Kamchia river was probably much larger, comprising an axially situated river with the west–east-trending Kamchia Foredeep fed by a system of transverse tributaries issuing from the Balkanides to the south (Fig. 4). A drainage basin area of approximately 20 000 km^2 can be estimated from clues preserved within the present-day drainage planform. Temperatures would have averaged 15–17°C a^{-1} (Ivanov *et al.* 2007). The maximum relief within the palaeo-drainage basin is difficult to estimate, but a lower boundary can be inferred from the present-day relief, which is a maximum of 2300 m. This would yield a sediment flux of 5.1–5.87 Mt a^{-1} and a sediment fan area of 16 697–18 437 km^2, with a length of 155–163 km, subject to further consideration of shelf bypass, slope gradient and sediment composition. If relief was greater, then sediment flux would increase accordingly. For example, a maximum relief of 3 km yields a sediment flux of 6.75–7.65 Mt a^{-1}, and a maximum relief of 4 km yields a sediment flux of 9.01–10.21 Mt a^{-1}. These results match seismic observations that show Maykopian fan systems extending up to 200 km into the basin (Sipahioğlu *et al.* 2013). These figures are estimates but serve to demonstrate that the sediment flux into the WBS from a single river system during the Maykopian was probably much greater than in the present day.

The notion that the Kamchia Trough was a major conduit for Maykopian sedimentation into the deeper WBS is supported by seismic and well observations. Georgiev (2004) noted a major thickening of Oligocene sediments within the trough, as did Sachsenhofer *et al.* (2009). Maykop-age strata vary from 30 to 40 m in thickness on the shelf between the Histria and Kamchia troughs up to almost 500 m within the Kamchia Trough itself (Sachsenhofer *et al.* 2009).

A west–east-trending 'Oligocene Channel' has been identified to the west of the Galata gas field, with other channel features close to the Samotino More and Samotino Melrose wells (Melrose Resources 2002; Mateo *et al.* 2013), and Oligocene channel-like features have been imaged on attribute expressions of 3D seismic data volumes from deep water offshore Bulgaria to the west of the Polshkov High (Tari *et al.* 2009), suggesting a major sediment routing to slope fans in deep water offshore Bulgaria (e.g. the plays over the Polshkov High).

Seismic reflection survey results from the southern margin of the WBS recently provided further insight into deep-water sediment pathways during Maykopian deposition. Analysis of 3D seismic reflection data showed that turbidite channels along the southern margins of the WBS have a distinct west to east flow direction (Sipahioğlu *et al.* 2013). The systems can be observed to evolve from channelized sand deposits (see also Nikishin *et al.* 2015*a*) through to unconfined sinuous channels and eventually to basin-floor lobes that onlap bathymetric highs. This study also shows an evolution of the sediment source area through the Oligocene–Miocene. Turbidite systems in the Oligocene can be observed to be sourced from the west, flowing along the southern margin of the Turkish WBS, with little input from southerly sediment sources. This then switches to a dual source by the early Miocene, with sediments derived from both the west and south. By the middle Miocene, sediments are only originating from southerly sources.

With regard to sediment provenance, it can be assumed that river systems flowing from the Strandja Massif would have drained into the Burgas Basin. Data from Sipahioğlu *et al.* (2013) and Nikishin *et al.* (2015*a*) suggest that the sediment being deposited in the deep part of the southern Western Black Sea Basin during the Oligocene was derived from a westerly source, which from analysis of the likely seafloor bathymetry in the Oligocene (Gorshkov *et al.* 1989), suggests a conduit through the Burgas Basin. The authors are doubtful of the conjecture of Nikishin *et al.* (2015*a, b*) that such sediment was derived from the Thrace Basin via a 'palaeo-Bosporus'. Sedimentological studies (Islamoglu *et al.* 2010) show that the Thrace Basin was infilling during the Maykopian with a delta prograding from the NE off the Strandja Massif. It might therefore be expected that sediments of Oligocene–Miocene age within the southern WBS would be of high quality if sourced from the Strandja Massif and represent a low-risk reservoir target. An additional factor to consider when assessing these turbidite systems is the impact of the long run-out paths on maintaining grain size within the deposits.

The modern-day Sakarya river is the third longest river in Turkey (824 km) and drains a catchment area of 58 160 km^2 (Akbulut *et al.* 2009). The catchment area includes Cretaceous and Early Tertiary deep-water shales and sandstones, as well as volcaniclastics. It probably drained into the Marmara Sea until relatively recently in geological history. Modern rivers transporting sediment sourced from the granitic Bolu Massif are relatively small (e.g. the Filyos, which is 228 km long and has a catchment area of 13 156 km^2) but contribute a significant sediment flux (approximately 3.8 Mt a^{-1}: Algan *et al.* 1997; Jaoshvili 2002). However, when considering the influence of the Western Pontides as a sediment source for Maykopian sediments, the analysis by Sipahioğlu *et al.* (2013) suggested that there was only a southerly sediment source for deep-water turbidites after the early Miocene. This might, in part, result from the influence of the Kozlu Ridge, a high offshore Turkey, which formed during the rifting of the WBS. Interpreted seismic reflection profiles transecting the Kozlu Ridge (Menlikli *et al.* 2009) suggest that there is an absence of Oligocene sediments on the shelf bordering the Turkish coast. Syn- and post-rift Cretaceous–Eocene sediments fill the sub-basin behind the ridge, with Oligocene sediments identified on the section in the deep basin only, onlapping the high (Menlikli *et al.* 2009; Sipahioğlu *et al.* 2013). Sediment thickness maps also suggest a distinct thinning of Maykopian-age strata over the Kozlu Ridge (Gorshkov *et al.* 1989). This would indicate that sediments sourced from granitic highs, such as the Bolu Massif, would not have reached the deep water of the WBS until the Miocene and thus would play only a minor role in the deep-water Maykopian plays of the WBS.

Conclusions

Maykopian-aged turbidite channels and fans have been identified by 3D seismic evaluations of the WBS (e.g. Tari *et al.* 2009; Mateo *et al.* 2013; Sipahioğlu *et al.* 2013), and their association with thermally mature organic-rich mudstones means that they might form viable plays in stratigraphic/anticlinal traps. The recent discovery of hydrocarbons in the Han Aspuah block, offshore Bulgaria, is an indication that deep-water exploration of Maykop horizons in the WBS is potentially of great economic value and will only serve to push investigation into these plays further.

Although many deep-water fan plays exist in the WBS within the Maykop Formation, it is clear when considering possible sediment source areas that not all regions will provide the same quality of reservoir. Maynard *et al.* (2012) concluded that local drainage from orogens and volcanic arcs would provide poor-quality sediment in limited volumes. Nonetheless, areas such as parts of the Balkanides, the Strandja Massif and the Western Pontides are the most likely to have provided clean quartz-rich sediments to the basin, and represent low-risk sources; whereas regions such as Srednogorie and Dobrogea were more likely to provide lithic-rich sediments to the WBS, and are considered higher-risk sediment source areas.

Drainage patterns in the land areas surrounding the WBS during deposition of the Maykop Suite were very different from the present-day configuration. Although the rivers were relatively small in length, their mountainous nature meant they could yield significant sediment flux (Milliman & Syvitski 1992). In particular, the palaeo-Kamchia river is considered to have occupied a much larger catchment area than that at present, and was fed by tributaries draining from the granitic regions of the Balkanides and the quartz-rich Cretaceous and Palaeogene sandstones in the Kamchia Foredeep. This would have yielded a sediment flux more than eight times that of the river in the recent past and led to the development of a fan system in excess of 150 km in length after bypassing the shelf through a submarine channel/canyon identified on seismic data (Mateo et al. 2013). In support of this, sediment passing through the Burgas Basin has been shown to travel up to 200 km along the southern margin of the WBS (Sipahioğlu et al. 2013). Conduits, such as the Kamchia and Histria depressions and the Burgas Basin, channel material away from the shelf into the deep parts of the northern and western WBS, and overall reduce the risk of encountering good-quality reservoirs within the Maykop deposition system of the WBS.

The authors are grateful to their colleagues at Halliburton for discussions about many of the geological concepts presented herein. Allie Nawell and Helen Ding are thanked for their help with drafting. Dr Gabor Tari (OMV) and Prof. Aral Okay (İstanbul Technical University) have been an inspiration to our efforts, although they may not agree with all of our conclusions.

References

AGALAREVA, M.N. 2012. Palaeo-valleys and reconstruction of river valley network in the Moesian Platform, northern Bulgaria. Comptes Rendus de l'Académie Bulgare des Sciences: Sciences Mathematiques et Naturelles, 65, 1725–1730.

AKBULUT, N., BAYARI, S., AKBULUT, A. & ŞAHIN, Y. 2009. Rivers of Turkey. In: TOCKNER, K., ROBINSON, C.T. & UEHLINGER, U. (eds) Rivers of Europe. Academic Press, New York, 643–672.

ALGAN, O., GAZYOĐLU, C., YÜCEL, Z., ÇAĐATAY, N. & GÖNENÇGYL, B. 1997. Sediment and freshwater discharges of the Anatolian rivers into the Black Sea. In: CAGATAY, N. (ed.) Proceedings of the IOC/BSRC Workshop on Black Sea Fluxes, Istanbul, Turkey, 10–12 June 1997. IOC/UNESCO, Geneva, 38–50.

ArcGIS. 2011. Geology of Europe including Turkey (WMS). http://www.arcgis.com/home/item.html?id= 9a304e471f6846c4bd3ccba1c121fe77 [last accessed 19 June 2017].

BALDI, T. 1980. The early history of the Paratethys. Földtani Közlöny, 110, 456–471.

BANKS, C.J. 1997. Basins and thrust belts of the Balkan coast of the Black Sea. In: ROBINSON, A.G. (ed.) Regional and Petroleum Geology of the Black Sea and Surrounding Region. American Association of Petroleum Geologists, Memoirs, 68, 115–128.

BANKS, C.J. & ROBINSON, A.G. 1997. Mesozoic strike-slip back-arc basins of the western Black Sea region. In: ROBINSON, A.G. (ed.) Regional and Petroleum Geology of the Black Sea and Surrounding Region. American Association of Petroleum Geologists, Memoirs, 68, 53–62.

BAZHENOVA, O.K., FADEEVA, N.P., SAINT-GERMES, M.L. & TIKHOMIROVA, E.E. 2003. Sedimentation conditions in the eastern Paratethys ocean in the Oligocene–Early Miocene. Moscow University Geology Bulletin, 58, 11–21.

BERGERAT, F., MARTIN, P. & DIMOV, D. 1998. The Moesian Platform as a key for understanding the geodynamical evolution of the Carpatho-Balkan Alpine system. In: CRASQUIN-SOLEAU, S. & BARRIER, E. (eds) Peri-Tethys Memoir 3: Stratigraphy and Evolution of Peri-Tethyan Platforms. Mémoires du Muséum National d'Histoire Naturelle, 177, 129–150.

BERGERAT, F., VANGELOV, D. & DIMOV, D. 2010. Brittle deformation, palaeostress field reconstruction and tectonic evolution of the Eastern Balkanides (Bulgaria) during Mesozoic and Cenozoic times. In: SOSSON, M., KAYMAKCI, N., STEPHENSON, R., BERGERAT, F. & STAROSTENKO, V. (eds) Sedimentary Basin Tectonics from the Black Sea and Caucasus to the Arabian Platform. Geological Society, London, Special Publications, 340, 77–111, https://doi.org/10.1144/SP340.6

BONCHEV, E. 1986. The Balkanides – Geotectonic Position and Evolution. BAS, Sofia.

BURCHFIEL, B.C. 1976. Geology of Romania. Geological Society of America, Special Papers, 158.

CARRIGAN, C.W., MUSKASA, S.B., HAYDOUTOV, I. & KOLCHEVA, K. 2005. Age of Variscan magmatism from the Balkan sector of the orogen, central Bulgaria. Lithos, 82, 125–147.

CARRIGAN, C.W., MUKASA, S.B., HAYDOUTOV, I. & KOLCHEVA, K. 2006. Neoproterozoic magmatism and Carboniferous high-grade metamorphism in the Sredna Gora Zone, Bulgaria: an extension of the Gondwana-derived Avalonian–Cadomian belt? Precambrian Research, 147, 404–416.

CAVAZZA, W., FEDERICI, I., OKAY, A.I. & ZATTIN, M. 2012. Apatite fission-track thermochronology of the Western Pontides (NW Turkey). Geological Magazine, 149, 133–140.

CHEN, F., SIEBEL, W., SATIR, M., TERZIOGLU, M. & SAKA, K. 2002. Geochronology of the Karadere basement (NW Turkey) and implications for the geological evolution of the Istanbul zone. International Journal of Earth Sciences (Geologische Rundschau), 91, 469–481.

CHESHITEV, G. & KĂNČEV, I. 1989. Geological Map of Bulgaria, 1:500 000. Committee of Geology, Sofia.

CRANGANU, C. & ŞARAMET, M. 2011. Hydrocarbon generation and accumulation in the Histria Basin of the western Black Sea. In: RYANN, A.L. & PERKINS, N.J. (eds) The Black Sea: Dynamics, Ecology and Conservation. Nova Publishers, New York, 243–263.

DIMITROV, H.B. & GEORGIEV, G.V. 2011. Correlation between main seismic sequence boundaries in Kamchia Basin (offshore Bulgaria) and Western Black Sea Basin. In: 73rd EAGE Conference & Exhibition incorporating SPE EUROPEC, Vienna, Austria, Extended Abstracts, P295.

DIMITROV, P., SOLAKOV, D., PEJCHEV, V. & DIMITROV, D. 1997. The source provinces in the western Black Sea. *In*: CAGATAY, N. (ed.) *Proceedings of the IOC/BSRC Workshop on Black Sea Fluxes*, Istanbul, Turkey, 10–12 June 1997. IOC/UNESCO, Geneva, 51–58.

DINU, C., WONG, H.K., TAMBREA, D. & MATENCO, L. 2005. Stratigraphic and structural characteristics of the Romanian Black Sea shelf. *In*: CLOETINGH, S., MAŢENCO, L., BADA, G., DINU, C. & MOCANU, V. (eds) *The Carpathians–Pannonian Basin System: Natural Laboratory for Coupled Lithospheric–Surface Processes*. *Tectonophysics*, **410**, 417–435.

DOGLIONI, C., BUSATTA, C., BOLIS, G., MARIANINI, L. & ZANELLA, M. 1996. Structural evolution of the eastern Balkans (Bulgaria). *Marine and Petroleum Geology*, **13**, 225–251.

ELMAS, A., YILMAZ, I., YIGITBAS, E. & ULLRICH, T. 2011. A Late Jurassic–Early Cretaceous metamorphic core complex, Strandja Massif, NW Turkey. *International Journal of Earth Sciences (Geologische Rundschau)*, **100**, 1251–1263.

ESPURT, N., HIPPOLYTE, J.-C., KAYMAKCI, N. & SANGU, E. 2014. Lithospheric structural control on inversion of the southern margin of the Black Sea Basin, Central Pontides, Turkey. *Lithosphere*, **6**, 26–34.

GARCIA-CASTELLANOS, G. & CLOETINGH, S. 2012. Modeling the interaction between lithospheric and surface processes in foreland basins. *In*: BUSBY, C. & AZOR, A. (eds) *Tectonics of Sedimentary Basins: Recent Advances*. Blackwell, Oxford, 152–181.

GEORGIEV, G. 2004. Geological structure of Western Black Sea region. *In*: *EAGE 66th Conference and Exhibition*, 7–10 June 2004, Paris, Extended Abstracts, B040.

GEORGIEV, G. 2012. Geology and hydrocarbon systems in the Western Black Sea. *Turkish Journal of Earth Sciences*, **21**, 723–754.

GEORGIEV, G., DABOVSKI, C. & STANISHEVA-VASSILEVA, G. 2001. East Srednogornie–Balkan Rift Zone. *In*: ZIEGLER, P.A., CAVAZZA, W., ROBERTSON, A.H.F. & CRASQUIN-SOLEAU, S. (eds) *Peri-Tethys Memoir 6: Peri-Tethyan Rift/Wrench Basins and Passive Margins*. Mémoires du Muséum National d'Histoire Naturelle, **186**, 259–293.

GOLONKA, J., MARKO, F., GAHAGAN, L., OSZCZYPKO, N., KROBICKI, M. & SLACZKA, A. 2006. Plate-tectonic evolution and paleogeography of the Circum-Carpathian region. *In*: GOLONKA, J. & PICHA, F.J. (eds) *The Carpathians and Their Foreland: Geology and Hydrocarbon Resources*. American Association of Petroleum Geologists, Memoirs, **84**, 11–46.

GORSHKOV, A.S., MEISNER, L.B., SOLOVIEV, V.V., TUGOLESOV, D.A. & KHAKHALEV, E.M. 1989. *Album of Structural and Thickness Maps of Black Sea Basin Cenozoic Sediments*. Main Department for Geodesy and Cartography (GUGK),Moscow.

GÖRÜR, N. 1988. Timing of opening of the Black Sea basin. *Tectonophysics*, **147**, 247–262.

GRADSTEIN, F.M., OGG, J.G., SCHMITZ, M. & OGG, G. (eds). 2012. *The Geologic Time Scale 2012*. Elsevier, Amsterdam.

HAQ, B.U., HARDENBOL, J. & VAIL, P.R. 1988. Mesozoic and Cenozoic chronostratigraphy and cycles of sea-level change. *In*: WILGUS, C.K., HASTINGS, B.S., POSAMENTIER, H., VAN WAGONER, J., ROSS, C.A. & KENDALL, C.G. (eds)

Sea-Level Changes: An Integrated Approach. Society of Economic Palaeontologists and Mineralogists (SEPM), Special Publications, **42**, 71–108.

HIPPOLYTE, J.-C., MÜLLER, C., KAYMAKCI, N. & SANGU, E. 2010. Dating of the Black Sea Basin: new nannoplankton ages from its inverted margin in the Central Pontides (Turkey). *In*: SOSSON, M., KAYMAKCI, N., STEPHENSON, R., BERGERAT, F. & STAROSTENKO, V. (eds) *Sedimentary Basin Tectonics from the Black Sea and Caucasus to the Arabian Platform*. Geological Society, London, Special Publications, **340**, 113–157, https://doi.org/10.1144/SP340.7

HIPPOLYTE, J.-C., ESPURT, N., KAYMACKCI, N., SANGU, E. & MÜLLER, C. 2016. Cross-section anatomy and geodynamic evolution of the Central Pontide orogenic belt (northern Turkey). *International Journal of Earth Sciences (Geologische Rundschau)*, **105**, 81–106.

IONESCU, G. 2002. *Facies Architecture and Sequence Stratigraphy of the Black Sea Offshore Romania*. Bucharest Geoscience Forum, Special Volume, **2**.

ISLAMOGLU, Y., HARZHAUSER, M. *ET AL.* 2010. From Tethys to eastern Paratethys: oligocene depositional environments, paleoecology and palaeobiogeography of the Thrace Basin (NW Turkey). *International Journal of Earth Sciences (Geologische Rundschau)*, **99**, 183–200.

IVANOV, D.A., ASHRAF, A.R. & MOSBRUGGER, V. 2007. Late Oligocene and Miocene climate and vegetation in the Eastern Paratethys area (northeast Bulgaria), based on pollen data. *Paleogeography, Paleoclimatology, Paleoecology*, **255**, 342–360.

JANBU, N.E., NEMEC, W., KIRMAN, E. & OZAKSOY, V. 2007. Facies anatomy of a sand-rich channelised turbidite system: the Eocene Kusuri Formation in the Sinop Basin, north-central Turkey. *In*: NICHOLS, G., WILLIAMS, E. & PAOLA, C. (eds) *Sedimentary Processes, Environments and Basins: A Tribute to Peter Friend*. International Association of Sedimentologists (IAS), Special Publications, **38**, 457–511.

JAOSHVILI, S. 2002. *Rivers of the Black Sea*. European Environment Agency, Technical Report 71.

JOLIVET, L. & BRUN, J.-P. 2010. Cenozoic geodynamic evolution of the Aegean. *International Journal of Earth Sciences (Geologische Rundschau)*, **99**, 109–138.

JOLIVET, L. & FACCENNA, C. 2000. Mediterranean extension and the Africa–Eurasia collision. *Tectonics*, **19**, 1095–1106.

JONES, R.W. & SIMMONS, M.D. 1996. A review of the stratigraphy of Eastern Paratethys (Oligocene–Holocene). *Bulletin of the Natural History Museum, London (Geology Series)*, **52**, 25–49.

KONERDING, C., DINU, C. & WONG, H.K. 2010. Seismic sequence stratigraphy, structure and subsidence history of the Romanian Black Sea shelf. *In*: SOSSON, M., KAYMAKCI, N., STEPHENSON, R., BERGERAT, F. & STAROSTENKO, V. (eds) *Sedimentary Basin Tectonics from the Black Sea and Caucasus to the Arabian Platform*. Geological Society, London, Special Publications, **340**, 159–180, https://doi.org/10.1144/SP340.9

KORUCU, Ö., SIPAHIOĞLU, N.Ö., AKTEPE, S. & BENG, E. 2013. Ultra-Derin Deniz Kuyu Verileri ile Orta-Batı Karadeniz Neojen Istifinin Korelasyonu [Correlation of Neogene sequence in western-central part of Turkish Black Sea based on ultra-deep well data]. *In*: *19th*

International Petroleum and Natural Gas Congress and Exhibition of Turkey, Extended Abstracts, TAPG/EAGE.

MARINOV, E. 1997. The Alpine structural complex in Lower Kamchia Foredeep and the adjacent parts of East-Balkan Zone. *Geology and Mineral Resources*, **5**, 3–9.

MATENCO, L., BERTOTTI, G., CLOETINGH, S. & DINU, C. 2003. Subsidence analysis and tectonic evolution of the external Carpathian–Moesian Platform region during Neogene times. *Sedimentary Geology*, **156**, 71–94.

MATEO, P., FLOODPAGE, J., POPA, C., MORICE, M., PHILIPPE, Y. & VARELA SUSINA, C. 2013. Tertiary structural evolution, palaeogeography and sedimentation in the SW Black Sea deep offshore. *75th EAGE Conference and Exhibition*, London, Extended Abstracts WE 16 16.

MAYNARD, J.R., ARDIC, C. & MCALLISTER, N. 2012. Source to sink assessment of Oligocene to Pleistocene sediment supply in the Black Sea. *Gulf Coast Section Society of Economic Palaeontologists and Mineralogists Conference Transactions*, **32**, 664–700.

MEISNER, L.B. & TUGOLESOV, D.A. 2003. Key reflecting horizons in sedimentary fill seismic records of the Black Sea Basin (correlation and stratigraphic position). *Stratigraphy and Geological Correlation*, **11**, 606–619.

MELROSE RESOURCES PLC 2002. Annual Report and Accounts 2002.

MENLIKLI, C., DEMIRER, A., SIPAHIOLU, Ö., KÖRPE, L. & AYDEMIR, V. 2009. Exploration plays in the Turkish Black Sea. *The Leading Edge*, **28**, 1066–1075.

MEULENKAMP, J.E. & SISSINGH, W. 2003. Tertiary palaeogeography and tectonostratigraphic evolution of the Northern and Southern Peri-Tethys platforms and the intermediate domains of the African-Eurasian convergent plate boundary zone. *Paleogeography, Paleoclimatology, Paleoecology*, **196**, 209–228.

MIKLÓS, D. & NEPPEL, F. 2010. Palaeogeography of the Danube and its catchment. *In*: BRILLY, M. (ed.) *Hydrological Processes of the Danube River Basin*. Springer, Dordrecht, The Netherlands, 79–124.

MILLER, K.G., KOMINZ, M.A. ET AL. 2005. The Phanerozoic record of global sea-level change. *Science*, **310**, 1293–1298.

MILLIMAN, J.D. & SYVITSKI, J.P.M. 1992. Geomorphic/tectonic control of sediment discharge to the ocean: the importance of small mountainous rivers. *Journal of Geology*, **100**, 525–544.

MIREA, A. 2009. Sedimentary and structural features of the Upper Tertiary from the Western Moesian Platform. *In*: *71st EAGE Conference and Exhibition, Amsterdam*, Extended Abstracts.

MITYUKOV, A.V., AL'MENDINGER, O.A., MYASOEDOV, N.K., NIKISHIN, A.M. & GAIDUK, V.V. 2011. The sedimentation model of the Tuapse Trough (Black Sea). *Doklady Earth Sciences*, **440**, 1245–1248.

MITYUKOV, A.V., NIKISHIN, A.M., ALMENDINGER, O.A., BOLOTOV, S.N., LAVRISHCHEV, V.A., MYASOEDOV, N.K. & RUBTSOVA, E.V. 2012. A sedimentation model of the Maikop deposits of the Tuapse Basin in the Black Sea according to the results of 2-D and 3-D seismic surveys and field works in the Western Caucasus and Crimea. *Moscow University Geology Bulletin*, **67**, 81–92.

MOORE, W.J., MCKEE, E.H. & AKINCI, Ö. 1980. Chemistry and chronology of plutonic rocks in the Pontoid Mountains, northern Turkey. *In*: JANKOVIC, S. & SILLITOE, R.

(eds) *European Copper Deposits*. Society for Geology Applied to Mineral Deposits, Special Publications, **1**, 209–216.

MOROŞANU, I. 2012. The hydrocarbon potential of the Black Sea continental plateau in Romania. *Romanian Journal of Earth Sciences*, **86**, 91–109.

MUNTEANU, I., MATENCO, L., DINU, C. & CLOETINGH, S. 2011. Kinematics of back-arc inversion of the Western Black Sea Basin. *Tectonics*, **30**, TC5004.

NATAL'IN, B.A., SUNAL, G., SATIR, M. & TORAMAN, E. 2012. Tectonics of the Strandja Massif, NW Turkey: history of a long-lived arc at the northern margin of Palaeo-Tethys. *Turkish Journal of Earth Sciences*, **21**, 755–798.

NATAL'IN, B.A., SUNAL, G., GÜN, E., WANG, B. & ZHIQING, Y. 2016. Precambrian to Early Cretaceous rocks of the Strandja Massif (northwestern Turkey): evolution of a long lasting magmatic arc. *Canadian Journal of Earth Sciences*, **53**, 1312–1335.

NIKISHIN, A.M., OKAY, A.I., TÜYSÜZ, O., DEMIER, A., AMELIN, N. & PETROV, E. 2015a. The Black Sea basins structure and history: new model based on new deep penetration regional seismic data. Part 1: basins structure and fill. *Marine and Petroleum Geology*, **59**, 638–655.

NIKISHIN, A.M., OKAY, A.I., TÜYSÜZ, O., DEMIER, A., WANNIER, M., AMELIN, N. & PETROV, E. 2015b. The Black Sea basins structure and history: new model based on new deep penetration regional seismic data. Part 2: tectonic history and paleogeography. *Marine and Petroleum Geology*, **59**, 656–670.

OKAY, A.I. 2008. Geology of Turkey: a synopsis. *Anschnitt*, **21**, 19–42.

OKAY, A.I. & NIKISHIN, A.M. 2015. Tectonic evolution of the southern margin of Laurasia in the Black Sea region. *International Geology Review*, **57**, 1051–1076.

OKAY, A.I., ŞENGÖR, A.M. & GÖRÜR, N. 1994. Kinematic history of the opening of the Black Sea and its effect on the surrounding regions. *Geology*, **22**, 267–270.

OKAY, A.I., SATIR, M., TUYSUZ, O., AKYUZ, S. & CHEN, F. 2001. The tectonics of the Strandja Massif: Late-Variscan and mid-Mesozoic deformation and metamorphism in the Northern Aegean. *International Journal of Earth Sciences (Geologische Rundschau)*, **90**, 217–233.

OKAY, A.I., SATIR, M. & SIEBEL, W. 2006. Pre-Alpine Palaeozoic and Mesozoic orogenic events in the Eastern Mediterranean region. *In*: GEE, D.G. & STEPHENSON, R.A. (eds) *European Lithosphere Dynamics*. Geological Society, London, Memoirs, **32**, 389–405, https://doi.org/10.1144/GSL.MEM.2006.032.01.23

OKAY, A.I., ALTINER, D., SUNAL, G., AYGÜL, M., AKDOĞAN, R., ALTINER, S. & SIMMONS, M. 2017. Geological evolution of the Central Pontides. *In*: SIMMONS, M.D., TARI, G.C. & OKAY, A.I. (eds) *Petroleum Geology of the Black Sea*. Geological Society, London, Special Publications, **464**, https://doi.org/10.1144/SP464.3

OKAY, N. & OKAY, A.I. 2002. Tectonically induced Quaternary drainage diversion in the northeastern Aegean. *Journal of the Geological Society, London*, **159**, 393–399, https://doi.org/10.1144/0016-764901-065

ÖZCAN, Z., OKAY, A.I., ÖZCAN, E., HAKYEMEX, A. & ÖZKAN-ALTINER, S. 2012. Late Cretaceous–Eocene geological evolution of the Pontides in northwest Turkey between the Black Sea coast and Bursa. *Turkish Journal of Earth Sciences*, **21**, 933–960.

ÖZGÜL, N. 2012. Stratigraphy and some structural features of the Istanbul Palaeozoic. *Turkish Journal of Earth Sciences*, **21**, 933–960.

PALII, A. & TOCHKOV, D. 1994. Oil and gas in the Cenozoic deposits of the Black Sea Basin. *In: Proceedings of the Conference on Petroleum Geology and Hydrocarbon Potential of the Black Sea Area*, 16–18 October, Varna, 16–18.

PANAIOTU, C.E., VASILIEV, I., PANAIOTU, C.G., KRIJGSMAN, W. & LANGEREIS, C.G. 2007. Provenance analysis as a key to orogenic exhumation: a case study from the East Carpathians (Romania). *Terra Nova*, **19**, 120–126.

POPOV, S.V. & STOLYAROV, A.S. 1996. Palaeogeography and anoxic environments of the Oligocene – Early Miocene Eastern Paratethys. *Israel Journal of Earth Sciences*, **45**, 161–167.

POPOV, S.V., AKHETIEV, M.A., ZAPOROZHETS, N.I., VORONINA, A.A. & STOLYAROV, A.S. 1993. Evolution of the Eastern Paratethys in the Late Eocene–Early Miocene. *Stratigraphy and Geological Correlation*, **1**, 572–600.

POPOV, S.V., RÖGL, F., ROZANOV, A.Y., STEININGER, F.F., SHCHERBA, I.G. & KOVAC, M. (eds). 2004. *Lithological–Paleogeographic Maps of Paratethys: 10 Maps Late Eocene to Pliocene*. Courier Forschungsinstitut Senckenberg, Frankfurt, **250**.

POPOV, S.V., ANTIPOV, M.P., ZASTROZHNOV, A.S., KURINA, E.E. & PINCHUK, T.N. 2010. Sea-level fluctuations on the north shelf of the Eastern Paratethys in the Oligocene–Neogene. *Stratigraphy and Geological Correlation*, **18**, 200–224.

RÖGL, F. 1999. Mediterranean and Paratethys. Facts and hypotheses of an Oligocene to Miocene paleogeography (short overview). *Geologica Carpathica*, **50**, 339–349.

SACHSENHOFER, R.F., STUMMER, B., GEORGIEV, G., DELLMOUR, R., BECHTEL, A., GRATZER, R. & CORIĆ, S. 2009. Depositional environment and hydrocarbon source potential of the Oligocene Ruslar Formation (Kamchia Depression; Western Black Sea). *Marine and Petroleum Geology*, **26**, 57–84.

SACHSENHOFER, R.F., POPOV, S.V. *ET AL.* 2017a. The type section of the Maikop Group (Oligocene–Lower Miocene) at the Belaya River (North Caucasus): depositional environment and hydrocarbon potential. *AAPG Bulletin*, **101**, 289–319.

SACHSENHOFER, R.F., POPOV, S.V. *ET AL.* 2017b. Oligocene and Lower Miocene source rocks in the Paratethys: palaeogeographical and stratigraphic controls. *In*: SIMMONS, M.D., TARI, G.C. & OKAY, A.I. (eds) *Petroleum Geology of the Black Sea*. Geological Society, London, Special Publications, **464**, https://doi.org/10.1144/SP464.1

SAINT-GERMES, M.L., BAZHENOVA, O.K., BAUDIN, F., ZAPOROZHETS, N.I. & FADEEVA, N.P. 2000. Organic matter in Oligocene Maikop sequence of the North Caucasus. *Lithology and Mineral Resources*, **35**, 47–62.

SCHULZ, H.-M., BECHTEL, A. & SACHSENHOFER, R.F. 2005. The birth of the Paratethys during the Early Oligocene: from Tethys to an ancient Black Sea analogue? *Global and Planetary Change*, **49**, 163–176.

SEGHEDI, A. 2001. The North Dobrogea orogenic belt (Romania): a review. *In*: ZIEGLER, P.A., CAVAZZA, W.,

ROBERTSON, A.H.F. & CRASQUIN-SOLEAU, S. (eds) *Peri-Tethys Memoir 6: Peri-Tethyan Rift/Wrench Basins and Passive Margins*. Mémoires du Muséum National d'Histoire Naturelle, **186**, 237–257.

SEGHEDI, A., OAIE, G. *ET AL.* 1999. Geology and structure of the Precambrian and Paleozoic basement of North and Central Dobrogea. *Romanian Journal of Tectonics and Regional Geology*, **77**, (Suppl. 2), 1–72.

SEGHEDI, A., VAIDA, M., IORDAN, M. & VERNIERS, J. 2005. Paleozoic evolution of the Romanian part of the Moesian Platform: an overview. *Geologica Belgica*, **8**, 99–120.

SINCLAIR, H.D., JURANOV, S.G., GEORGIEV, G., BYRNE, P. & MOUNTNEY, N.P. 1997. The Balkan thrust wedge and foreland basin of Eastern Bulgaria: structural and stratigraphic development. *In*: ROBINSON, A.G. (ed.) *Regional and Petroleum Geology of the Black Sea and Surrounding Regions*. American Association of Petroleum Geologists, Memoirs, **68**, 91–114.

SIPAHIOĞLU, O., KORUCU, Ö., AKTEPE, S. & BENGÜ, E. 2013. Westerly-sourced Late Oligocene–Middle Miocene axial sediment dispersal system in Turkish Western Black Sea: myth or reality? *Paper presented at the 19th International Petroleum and Natural Gas Congress and Exhibition of Turkey*, 15–17 May 2013, Ankara, Turkey.

SKOULIKIDIS, N.T., ECONOMOU, A.N., GRITZALIS, K.C. & ZOGARIS, S. 2009. Rivers of the Balkans. *In*: TOCKNER, K., ROBINSON, C.T. & UEHLINGER, U. (eds) *Rivers of Europe*. Academic Press, New York, 421–466.

SØMME, T.O., HELLAND-HANSEN, W., MARTINSEN, O.J. & THURMOND, J.B. 2009. Relationship between morphological and sedimentological parameters in source-to-sink systems: a basis for predicting semi-quantitative characteristics in subsurface systems. *Basin Research*, **21**, 361–387.

STOVBA, S., KHARIACHTCHEVSKAIA, O. & POPADYUK, I. 2009. Hydrocarbon-bearing areas in the eastern part of the Ukranian Black Sea. *The Leading Edge*, **28**, 1042–1045.

STUART, C.J., NEMČOK, M., VANGELOV, D., HIGGINS, E.R., WELKER, C. & MEAUX, D.P. 2011. Structural and depositional evolution of the East Balkan thrust belt, Bulgaria. *AAPG Bulletin*, **95**, 649–673.

SUNAL, G., SATIR, M., NATAL'IN, B.A. & TORAMAN, E. 2008. Paleotectonic position of the Strandja Massif and surrounding continental blocks based on zircon Pb–Pb age studies. *International Geology Review*, **50**, 519–545.

SUTTILL, H.L. 2009. *Sedimentological Evolution of the Emine and Kamchia Basins, Eastern Bulgaria*. MPhil thesis, University of Edinburgh.

SYLVESTER, Z. 2002. *Facies, architecture, and bed-thickness structure of turbidite systems: examples from the East Carpathian Flysch, Romania, and the Great Valley Group, California*. Doctoral dissertation, Stanford University.

SYVITSKI, J.P.M. & MILLIMAN, J.D. 2007. Geology, geography and humans battle for dominance over the delivery of fluvial sediment to the coastal ocean. *Journal of Geology*, **115**, 1–19.

TARI, G. 2015. Is the Black Sea really a back-arc basin? *In*: POST, P.J., COLEMAN, J.L., JR, ROSEN, N.C., BROWN, D.E., ROBERT-ASHBY, T., KAHN, P. & ROWAN, N. (eds) *Proceedings of the 34th Annual GCSSEPM Foundation*

Bob F Perkins Research Conference 2015: Petroleum Systems in Rift Basins. Gulf Coast Section of the SEPM (GCSSEPM), Houston, TX, 509–520.

TARI, G., DICEA, O., FAULKERSON, J., GEORGIEV, G., POPOV, S., STEFANESCU, M. & WEIR, G. 1997. Cimmerian and Alpine stratigraphy and structural evolution of the Moesian Platform (Romania/Bulgaria). *In*: ROBINSON, A.G. (ed.) *Regional and Petroleum Geology of the Black Sea and Surrounding Region.* American Association of Petroleum Geologists, Memoirs, **68**, 63–90.

TARI, G., DAVIES, J., DELLMOUR, R., LARRATT, E., NOVOTNY, B. & KOZHUHAROV, E. 2009. Play types and hydrocarbon potential of the deepwater Black Sea, NE Bulgaria. *The Leading Edge*, **28**, 1076–1081.

TARI, G., MENLIKLI, C. & DERMAN, S. 2011. Deepwater play types of the Black Sea: a brief overview. *Search and Discovery Article #10310 presented at the AAPG European Region Annual Conference*, 17–19 October 2010, Kiev, Ukraine.

USTAÖMER, P.A. & ROGERS, G. 1999. The Bolu Massif; remnant of a pre-Early Ordovician active margin in the West Pontides, northern Turkey. *Geological Magazine*, **136**, 579–592.

VINCENT, S.J. & KAYE, M.N.D. 2017. Source rock evaluation of Middle Eocene–Early Miocene mudstones from the NE margin of the Black Sea. *In*: SIMMONS, M.D., TARI, G.C. & OKAY, A.I. (eds) *Petroleum Geology of the Black Sea.* Geological Society, London, Special Publications, **464**, https://doi.org/10.1144/SP464.7

VON QUADT, A., MORITZ, R., PEYTCHEVA, I. & HEINRICH, C.A. 2005. Geochronology and geodynamics of Late Cretaceous magmatism and Cu–Au mineralization in the Panagyurishte region of the Apuseni–Banat–Timok–Srednogorie belt, Bulgaria. *Ore Geology Reviews*, **27**, 95–126.

VORONINA, A.A., KURGALIMOVA, G.G., POPOV, S.V., SEMENOV, G.J. & STROLJAROV, A.S. 1988. Biostratigraphy and facial features of the Maykopian beds in the Volga-Don region. *Izvestiya Akademii Nauk SSSR, Seriya Geologicheskaya*, **9**, 39–50.

YILMAZ, Y., TUYSUZ, O., YIGITBAS, E., GENC, S.C. & ŞENGÖR, A.M.C. 1997. Geology and tectonic evolution of the pontides. *In*: ROBINSON, A.G. (ed.) *Regional and Petroleum Geology of the Black Sea and Surrounding Region.* American Association of Petroleum Geologists, Memoirs, **68**, 183–226.

ZAPOROZHETS, N.I. 1999. Palynostratigraphy and dinocyst zonation of the Middle Eocene–Lower Miocene deposits at the Belata River (Northern Caucasus). *Stratigraphy and Geological Correlation*, **7**, 161–178.

Oligocene and Lower Miocene source rocks in the Paratethys: palaeogeographical and stratigraphic controls

R. F. SACHSENHOFER[1]*, S. V. POPOV[2], A. BECHTEL[1], S. CORIC[3], J. FRANCU[4], R. GRATZER[1], P. GRUNERT[5], M. KOTARBA[6], J. MAYER[1,7], M. PUPP[1], B. J. RUPPRECHT[1] & S. J. VINCENT[8]

[1]*Chair of Petroleum Geology, Montanuniversitaet Leoben, Peter-Tunner-Strasse 5, 8700 Leoben, Austria*

[2]*Paleontological Institute, Russian Academy of Sciences, ul. Profsoyuznaya 123, Moscow 117997, Russia*

[3]*Geological Survey of Austria, Neulinggasse 38, 1030 Vienna, Austria*

[4]*Czech Geological Survey, Branch Brno, Leitnerova 22, 65869 Brno, Czech Republic*

[5]*Institute for Earth Sciences, University of Graz, Heinrichstrasse 26, 8010 Graz, Austria*

[6]*Faculty of Geology, Geophysics and Environmental Protection, AGH-University of Science and Technology, Al. Mickiewicza 30, 30–059 Kraków, Poland*

[7]*OMV Exploration and Production GmbH, Trabrennstraße 6–8, 1020 Vienna, Austria*

[8]*Cambridge Arctic Shelf Programme (CASP), 181a Huntingdon Road, Cambridge CB3 0DH, UK*

Correspondence: reinhard.sachsenhofer@unileoben.ac.at

Abstract: Oligocene and Lower Miocene deposits in the Paratethys are important source rocks, but reveal major stratigraphic and regional differences. As a consequence of the first Paratethys isolation, source rocks with very good oil potential accumulated during Early Oligocene time in the Central Paratethys. Coeval source rocks in the Eastern Paratethys are characterized by a lower source potential. With the exception of the Carpathian Basin and the eastern Kura Basin, the source potential of Upper Oligocene and Lower Miocene units is low. In general, this is also valid for rocks formed during the second (Kozakhurian) isolation of the Eastern Paratethys. However, upwelling along a shelf-break canyon caused deposition of prolific diatomaceous source rocks in the western Black Sea.

Overall, Oligocene–Lower Miocene sediments in the Carpathian Basin (Menilite Formation) can generate up to 10 t HC m^{-2}. Its high petroleum potential is a consequence of the interplay of very high productivity of siliceous organisms and excellent preservation in a deep silled basin. In contrast, the petroleum potential of Oligocene–Lower Miocene (Maikopian) sediments in the Eastern Paratethys is surprisingly low (often <2 t HC m^{-2}). It is, therefore, questionable whether these sediments are the only source rocks in the Eastern Paratethys.

The hydrological regime of silled intracratonic basins is highly sensitive to changes of internal (e.g. geodynamics, tectonic uplift and subsidence, basin infill) and external (e.g. climate, sea-level fluctuations) factors (e.g. van Baak 2015). Frequently, these changes result in anoxic conditions and the accumulation of organic matter. Hence, prolific hydrocarbon source rocks may accumulate in silled basins (e.g. Allen & Allen 2013). The huge epicontinental Oligo-Miocene Paratethys Sea, stretching from western Europe to Kopetdagh (Fig. 1), offers a unique possibility to study the spatial and temporal distribution of source rocks in (semi-) silled basins.

The Paratethys Sea originated from a sea-level drop at the Eocene–Oligocene boundary (Zachos *et al.* 2001; Popov *et al.* 2010) and strong Alpine tectonic activity (Allen & Armstrong 2008), which resulted in the separation of the Tethys realm into a southern 'Mediterranean' and a northern 'Parate-thyian' domain (Rögl 1998, 1999; Popov *et al.* 2004*b*). Based on different environmental histories due to differently timed geotectonic events, the

From: SIMMONS, M. D., TARI, G. C. & OKAY, A. I. (eds) 2018. *Petroleum Geology of the Black Sea.*
Geological Society, London, Special Publications, **464**, 267–306.
First published online September 7, 2017, https://doi.org/10.1144/SP464.1

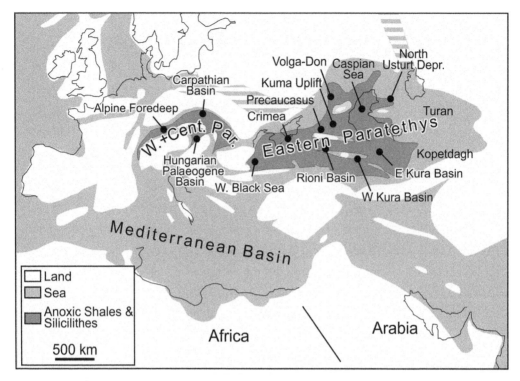

Fig. 1. Palaeogeography of the Paratethyan realm during early Oligocene (early Rupelian; Pshekhian) time. The distribution of oxygen-depleted environments is shown after Popov *et al.* (2004*b*). W. + Cent. Par., Western and Central Paratethys.

Paratethys is subdivided into the small Western (Rhône Basin; Alpine Foreland Basin west of Munich), the Central (remaining Alpine Foreland Basin, Carpathian Basin, Hungarian Palaeogene Basin) and the larger Eastern Paratethys (Fig. 1).

During its evolution, the Paratethys was frequently disconnected from the world ocean (Rögl 1998, 1999). The isolation created conditions favourable to the development of anoxia (Popov *et al.* 2004*b*; Schulz *et al.* 2005) (Fig. 2), and allowing the deposition of fine-grained Oligocene and Lower Miocene rocks with a maximum total organic carbon (TOC) content exceeding 10 wt%, which are proven source rocks in the aforementioned basins (e.g. Menilite Formation, Maikop Group: Robinson *et al.* 1996; Katz *et al.* 2000; Ulmishek 2001; Kotarba *et al.* 2007; Gratzer *et al.* 2011; Badics & Vetö 2012).

The isolation of the Paratethys favoured the development of rapidly evolving endemic benthic faunas that are used for correlation within the Paratethys. Correlation with the global stratigraphic framework is achieved mainly through calcareous nannoplankton. However, benthic groups are often absent in deep anoxic basins and calcareous plankton have frequently been dissolved, making stratigraphic

correlations of sedimentary successions challenging on both the regional and global levels. An overview of the correlation of standard and Paratethys stages, as well as sedimentary successions in different parts of the Central and Western Paratethys, is provided in Figure 3.

Typical lithologies of the Oligocene–Lower Miocene succession comprise carbonate-free shales, cherts, diatomaceous shales, marls and coccolith limestones. Successions show high vertical variability, whereas lateral continuity of specific units is high. For example, light-coloured, oligospecific nannomarls (Dynow and Polbian marls) form a main marker horizon within the entire Paratethys ('Solenovian Event': e.g. Voronina & Popov 1984; Nagymarosy & Voronina 1993; Schulz *et al.* 2004; Popov & Studencka 2015). The high lateral continuity reflects both deposition in a deep-water environment and control by basin-wide processes, including variations in salinity, redox potential and nutrient supply.

The present contribution reviews a number of high-resolution source-rock studies of key sections from different parts of the Paratethys and explores the impact of basin evolution on lateral and vertical variations in source-rock quality.

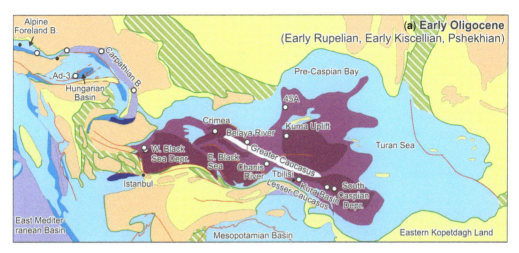

Fig. 2. Palaeogeographical sketches of the Paratethys for (**a**) Early Oligocene, (**b**) Late Oligocene and (**c**) Early Miocene time after Popov *et al.* (2004*a*, *b*). Deep-water zones with permanent anoxia (dark red) and zones with temporary anoxia (light red) are shown for the Eastern Paratethys (Popov & Stolyarov 1996). Locations of sections discussed in the text are labelled.

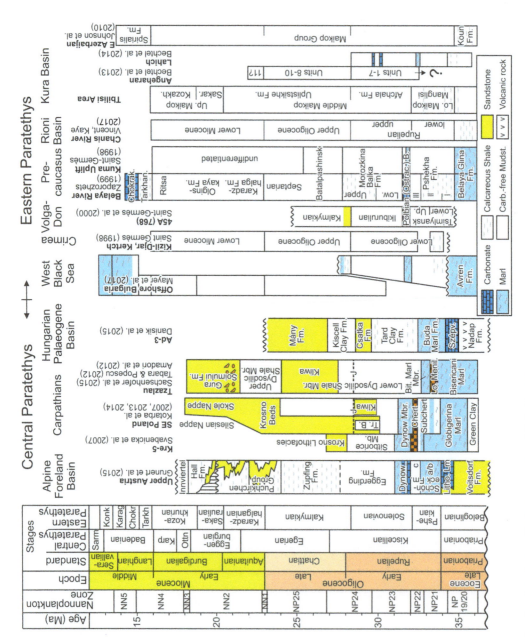

Fig. 3. Stratigraphy of sections discussed in the present paper. J, Jaslo Limestone.

Oligo-Miocene evolution of the Paratethys

Following Rögl (1999) and Popov *et al.* (2004*a*, *b*), the evolution of the Paratethys is briefly described with special attention given to periods with anoxic environments. Popov & Stolyarov (1996) distinguished two types of anoxic regimes: (1) environments with oxygen-deficiency within the uppermost sediment layers resulting in a reduction of benthic organisms; and (2) anoxic settings with free hydrogen sulphide in the water column prohibiting deep-water life.

In the Late Eocene, Europe formed an archipelago and was covered by a vast oxygenated, semi-open, subtropical sea. Around the Eocene–Oligocene boundary, strong tectonic activity separated the Paratethys from the Mediterranean domain (Fig. 2a) and led to an increase in water depth in depocentres of the Paratethys to >1000 m. Basin isolation resulted in oxygen-depleted conditions during Early Oligocene time (nannoplankton zones NP21–NP22; dinoflagellate zones D12 and D13) and deposition of organic matter-rich rocks in deep-water settings from the Western Alps to the North Usturt Depression (Fig. 1). The presence of benthic fauna and abundant deep-water fish with photophores testifies to the absence of free hydrogen sulphide in the water column, except for a short anoxic event near the Eocene–Oligocene boundary (Sachsenhofer *et al.* 2017). Consequently, shallow areas like the eastern Turanian and the southern Transcaucasian shelves remained oxygenated. Palaeobotanical data show a gradual change from a xerophytic subtropical to a temperate mesophilic climate around the NP21–NP22 boundary. *Micrhystridium* became dominant in phytoplankton assemblages in the Volga-Don region, which implies intense runoff (Akhmetev & Zaporozhets data in Popov *et al.* 1993*a*). These climatic changes led to salinity and thermal stratification of the water column, and stagnation in the deepest zones of the Paratethys at the beginning of the NP22 (Fig. 2a).

The closure of seaways culminated with the onset of nannoplankton zone NP23. The Paratethys lost its connection with the world ocean and became inhabited by a peculiar endemic fauna and phytoplankton (Solenovian Event, NP23; D14a *Wetzeliella gochtii*: e.g. Voronina & Popov 1984; Čtyroký 1991; Rusu 1999). A positive water balance caused changes in water circulation patterns. Initially anoxia intensified, but very soon shallowing of the sea and hydrological changes led to the restoration of aerobic conditions on the outer shelf where dark, carbonate-free, clayey sediments and diatomites were replaced by light-coloured, oligospecific nannomarls (Dynow and Polbian marls) with ostracods and mollusks. These sediments, forming a widespread synchronous marker horizon in the Paratethys, accumulated in a basin with strongly reduced salinity (12–14‰) and an altered ion composition (Popov *et al.* 1993*b*). During the upper part of nannoplankton zone NP23 and nannoplankton zone NP24 (late Solenovian), communication with the open sea was partly restored. This is indicated by the appearance of benthic foraminifers, ostracods, fish, mollusks and dinoflagellates of marine origin. Offshore of the shelf zone, upper Solenovian deposits are represented by dark-coloured, largely carbonate-free clays without benthic fauna. Salinity was probably in the range 13–15‰ and occasionally reached 20‰ (Popov *et al.* 1993*b*).

In the Late Oligocene (NP24–NP25; D14b–D15 *Chiropteridium partispinatum*), seaways to the ocean had broadened and the entire Paratethys was inhabited by marine fauna. Palaeontological data indicate a renewed connection with the North Sea Basin. In contrast, the sea regressed from the western part of the Alpine Foreland Basin. The restoration of marine environments caused alternating aerobic–anaerobic conditions and periodic development of benthic microfaunas (*Virgulinella* Beds NP23–NP24; D14b). Later, the basin became contaminated by hydrogen sulphide, and typical dark, carbonate-free muds, rich in organic matter and fish remains ('fish facies', D14b), were deposited. These rocks, widespread in the Eastern Paratethys (Fig. 2b), even extend into the inner shelf zone, restricting benthic organisms to the shallowest margins of the basin. The fish fauna became less diverse and pelagic forms disappeared. The salinity of basinal parts of the Eastern Paratethys reached about 30‰ in the early Kalmykian, while in marginal areas, and in basinal areas during late Kalmykian time, it fell to 15–20‰, and in northern parts possibly as far as 10‰ (Popov *et al.* 1993*a*). Stagnant and unstable environments in the late Kalmykian (NP25, D15) are evidenced by the diversity of green algae, acritarchs and dinocysts (Akhmetiev and Zaporozhets data in Sachsenhofer *et al.* 2017). A few euryhaline mollusk and foraminiferal species dominated along the northern shelf of the Eastern Paratethys (Popov *et al.* 1993*b*). Faunistic data support a fundamental change of seaways. Predominant biogeographical relationships with the North Sea Basin ceased and the proportion of Mediterranean taxa increased during the latest Oligocene (Popov *et al.* 2004*a*).

At the beginning of the Early Miocene (Karadzhalganian; NN1–NN2; D16), the sedimentation pattern in the Eastern Paratethys was similar to the early Late Oligocene: organic matter-rich mud with abundant fish remains ('fish facies') accumulated in hydrogen-sulphide-contaminated deep-water zones (Fig. 2c). Despite a regressive trend, the seaways connecting the Paratethys with the Mediterranean had broadened during the early Burdigalian (Sakaraulian). The Lesser Caucasus passage connected the Eastern Paratethys with the eastern Turkey and Iranian part of the Burdigalian Sea (Fig. 2c) (Popov *et al.* 2004*b*). Oxygenation improved and benthos inhabited new areas. Fossil data suggest a pronounced warming, especially in the Transcaucasian part of the basin (Popov *et al.* 1993*a*).

A major palaeogeographical rearrangement occurred in the late Burdigalian (Kozakhurian), when the counterclockwise rotation of Africa and Arabia separated the Mediterranean Sea from the Indian Ocean. At the same time, the Eastern Paratethys became isolated, forming the Kozakhurian Sea with strongly reduced salinity and endemic brackish faunas which appear simultaneously in the (late Ottnangian) Central Paratethys (Popov *et al.* 1993*b*). The sea was still regressing and sandy sediments were prevalent in shallow-water zones. At the same time, the lithology of sediments deposited in hydrogen-sulphide-contaminated deep-water zones remained similar to that of the underlying units.

A Middle Miocene (early Badenian) transgression flooded the entire Central Paratethys, except the Alpine Foreland Basin, and for a short time a seaway connecting the Eastern Paratethys with the Indian Ocean opened again. The establishment of aerobic conditions enabled the benthic fauna to inhabit most parts of the Eastern Paratethys during Tarkhanian time.

Amount and type of organic matter in key sections

Studies that focus on the stratigraphic variability of amount and type of organic matter are reviewed in this section. These data are supplemented by new data from the authors. With few exceptions (the Grybów Subunit in the Northern Carpathians; Tbilisi and Lahich sections), only successions with immature organic matter ($T_{max} \leq 430°C$; vitrinite reflectance (Rr) $\leq 0.5\%$) are considered to determine the original hydrocarbon potential. Detailed maturity information is provided for sections reaching oil window maturity. Regarding maturity parameters in immature successions, the reader is referred to the cited papers. For most sections carbonate and total organic carbon (TOC) contents are reported together with the hydrogen index (HI), which determines the kerogen type (e.g. Espitalié *et al.* 1977). Total organic carbon/total sulphur (TOC/S) ratios are used as simple indicator for sulphate- or oxygen-limited settings (e.g. Berner 1984). Where possible, biomarker ratios, as well as stable carbon ($\delta^{13}C$) and nitrogen ($\delta^{15}N$) isotope data, are added. A short explanation of the applied biomarker ratios is provided in the following, but for a thorough discussion of the applied biomarker and isotope proxies, the reader is referred to Peters *et al.* (2007). Pristane/phytane

(Pr/Ph) ratios are commonly used as a redox indicator. According to Didyk *et al.* (1978), Pr/Ph ratios <1.0 indicate anaerobic conditions, whereas values >1.0 reflect suboxic to oxic environments. The presence of aryl-isoprenoids may indicate photic zone anoxia (Summons & Powell 1987). Methyltrimethyltridecylchroman (MTTC) ratios are proxies for salinity conditions (de Leeuw & Sinninghe Damsté 1990; Barakat & Rullkötter 1997). Because different MTTC ratios (di-/tri-MTTC; tri-/[mono-+di- + tri-] MTTC) are in usage, an increase in salinity is indicated by an arrow in the relevant figures. Gammacerane is considered a proxy for a stratified water column, often related to salinity stratification (Fu *et al.* 1986). C_{27} and C_{28} steranes are produced by phytoplankton and photosynthetic bacteria, whereas C_{29} steranes are often associated with land plants (Volkman 1986). The sum of di- and triterpenoids and the di-/(di- + triterpenoids) ratio are well-established proxies for land plant input and the ratio of gymnosperms to angiosperms, respectively (e.g. Bechtel *et al.* 2008). Highly branched isoprenoids (HBIs) are considered to be biomarkers for diatoms (Sinninghe Damsté *et al.* 1989; Grossi *et al.* 2004).

Alpine Foreland Basin

Pelitic rocks, some with very high TOC contents, accumulated in the Alpine Foreland Basin between Oligocene and Early Miocene time. A high number of source-rock data are available from the Lower Oligocene succession, the main source for oil in the Austrian sector of the basin (Gratzer *et al.* 2011). In contrast, geochemical logs for Upper Oligocene and Lower Miocene units are available only from borehole Hochburg 1 (Grunert *et al.* 2013, 2015). In Figure 4, data from this borehole are complemented by data from other wells to provide a continuous log. A detailed view of Lower Oligocene units is presented in Figure 5.

Accumulation of fine-grained rocks started around the Eocene–Oligocene boundary (NP21 or NP19–NP20: see Schulz *et al.* 2002) with deposition of the Schöneck Formation (Figs 4 & 5). The Schöneck Formation, 10–25 m thick, overlies shallow-marine Priabonian sandstone or limestone. It is composed of marly members 'a' and 'b' containing globigerinoid planktonic foraminifera, and black shale member 'c' (Schulz *et al.* 2002). Member 'c' (NP21–NP22) is typically carbonate-free, but contains a few micritic limestone layers. Average TOC contents are about 2.5 wt% in members 'a' and 'b', and about 5.5 wt% in the shale member 'c' (max. 12%). The HI values reveal the presence of type II kerogen and display a general upwards increasing trend from 400 to 600 mg HC/g TOC. Whereas phytoplankton productivity was probably low, very high TOC and HI values in member 'c' result from

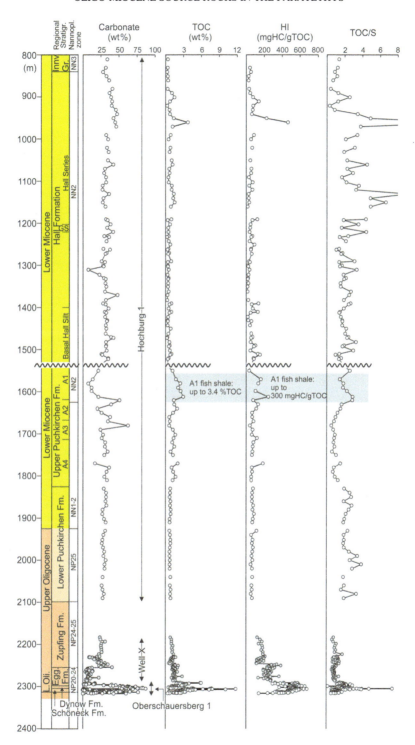

Fig. 4. Compiled bulk rock and geochemical data in the Oligo-Miocene rocks in borehole Hochburg 1 (Austrian part of the Alpine Foreland Basin). Stratigraphic data are from Grunert *et al.* (2015). Bulk geochemical data are compiled from different boreholes according to Schulz *et al.* (2002), Belaed (2007), Sachsenhofer *et al.* (2010) and Grunert *et al.* (2013, 2015).

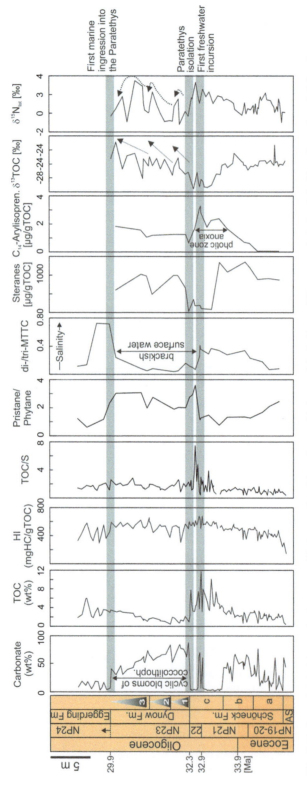

Fig. 5. Bulk geochemical data and palaeogeographical proxy data (biomarkers, stable isotopes) in the Lower Oligocene rocks in borehole Oberschauersberg 1. Inferred palaeogeographical changes are shown after Schulz *et al.* (2002). AS, Ampfing Sandstone.

photic zone anoxia. MTTC ratios indicate that salinity dropped significantly during deposition of member 'c'.

Light-coloured coccolith limestones and marls (Dynow Formation; NP23: Schulz et al. 2004) follow with a sharp boundary above the Schöneck Formation. Nannoplankton assemblages and MTTC ratios indicate continued low salinity during deposition of the Dynow Formation. High sterane concentrations confirm enhanced primary productivity. Limestones with low TOC contents (min. 0.5 wt%) were deposited during blooms of calcareous nannoplankton, whereas organic-rich marls (max. 2.0 wt%) accumulated during periods of low production of calcareous nannoplankton (Schulz et al. 2004). HI values in the order of 500–600 mg HC/g TOC reflect excellent preservation conditions due to prevailing anoxia. Subsequently, an increase in salinity, caused by the reconnection of the Paratethys, worsened the environment for calcareous nannoplankton (Schulz et al. 2005).

Oxygen-deficient conditions continued during deposition of the Eggerding Formation (NP23–NP24), typically about 45 m thick. Salinity variations are recorded in the lower part of the formation, consisting of dark grey laminated shaly marlstone with white, coccolith-bearing bands. TOC contents (1.9–6.0 wt%) and HI values (up to 600 mg HC/g TOC) are very high in the lower part. The upper part consists of homogenous mudstones with low carbonate content (c. 10 wt%) and a moderate amount of organic matter (c. 1.5 wt% TOC) dominated by type II/III kerogen (HI 200–400 mg HC/g TOC). Slope instabilities are indicated by slumps and extensive submarine slides culminating at the transition from the Eggerding to the Zupfing formations, when locally 70 m-thick successions were removed from the northern slope of the basin (Sachsenhofer & Schulz 2006; Sachsenhofer et al. 2010).

The Zupfing Formation (NP24–NP25) consists of calcareous mudstone, up to 450 m thick. Oxygen-depleted conditions continued during deposition of the Zupfing Formation, but only its base, a few metres thick, contains good amounts of type II/III kerogen (1.5 wt% TOC; HI 200–400 mg HC/g TOC: Fig. 4). The main part of the Zupfing Formation contains moderate amounts of type III kerogen (1.1 wt% TOC; HI 100–250 mg HC/g TOC).

The overlying Puchkirchen Group (upper NP25–upper NN2: Grunert et al. 2015) thickens towards the Alps and reaches a maximum thickness of approximately 2500 m. It includes coarse-grained siliciclastic rocks deposited in a deep-marine basin axial channel system encased by fine-grained rocks. Traditionally, a Lower and an Upper Puchkirchen Formation are distinguished. TOC contents are below 1.0 wt% in the Lower Puchkirchen Formation (HI <100 mg HC/g TOC) and slightly higher in the lower units (A4–A2) of the Upper Puchkirchen Formation (c. 1.5 wt%: Fig. 4). In contrast, organic matter-rich 'fish shale' with TOC contents up to 3.4 wt% and HI values up to 300 mg HC/g TOC are observed in its uppermost unit A1. To the north, the Puchkirchen Group passes into laminated slope deposits with TOC contents varying between 1.5 and 1.9 wt% (Ebelsberg Formation, NN2: Grunert et al. 2010a; Rupp & Coric 2012). Micropalaeontological and geochemical proxies indicate oxygen-depleted conditions caused by intense upwelling. In addition, increased runoff provided large amounts of nutrients stimulating productivity. Nevertheless, HI values are only in the range of 170–200 mg HC/g TOC (Grunert et al. 2010a).

The Burdigalian Hall Formation (NN2–NN3), about 750 m thick, follows above a basin-wide submarine erosional unconformity. It is composed of sands at the base, which record the reactivation of the basin axial channel. During an early Burdigalian sea-level rise, the channel was cut off from its western sediment sources (Grunert et al. 2013). Although palaeontological data prove temporal suboxic environments, TOC contents are generally low and only exceed 1.0 wt% in a prograding unit with strong terrestrial input (990–1150 m depth) and at 890–970 m depth. Foraminiferal assemblages suggest that the latter two samples are from an interval characterized by high nutrient supply. HI values are generally low (10–140 mg HC/g TOC), reflecting the predominance of terrestrial organic matter (type III kerogen), but reach 464 mg HC/g TOC at 960 m depth.

Carpathian Basin

Oligocene–Lower Miocene rocks, rich in organic matter, occur in different thrusts of the Carpathian orogenic belt. The term Menilite Formation is used for the clayey and siliceous rocks in the Czech, Polish and Ukrainian Carpathians (e.g. Kotarba & Koltun 2006; Picha et al. 2006) and is used here for the Romanian sector as well.

The Menilite Formation, up to 550 m thick (Oszczypko 2006), is the main source rock in the Carpathian Basin and the focus of the following discussion. It overlies Upper Eocene Green Clays and pelagic Globigerina Marls (Fig. 3; Leszczynski 1997). Oxic conditions prevailed during deposition of the Globigerina Marls, but benthic foraminiferal assemblages and low bioturbation imply an upwards decrease in bottom oxygenation (Krhovský 1995). The upper boundary of the Menilite Formation is diachronous (NP24–NN3) and marked by a change to flysch- ('Krosno-type') or molasse-type deposits (e.g. Svabenicka et al. 2007; Kuśmierek et al. 2013).

The Menilite Formation includes calcareous and non-calcareous black shales and siliceous sediments, which are often rich in fish remains (e.g. Brzobohaty 1981; Kotlarczyk *et al.* 2006). Chert, interpreted as diagenetically altered diatomite (e.g. Krhovský *et al.* 1992), is a prominent lithology within the Oligocene–Lower Miocene succession in the Carpathian Basin. Its origin is debated. According to Picha & Stranik (1999), cherty rocks were deposited due to upwelling resulting in high siliceous bioproductivity. Based on fish and trace fossil assemblages, Kotlarczyk & Uchman (2012) argued that cherts deposited during NP23 accumulated in a silled, anoxic basin with a stratified water column. Thin bands of coccolith limestones formed during algal blooms and are used as stratigraphic markers (Haczewski 1989). The Jasło Limestone (middle part of NP24) is used here to distinguish Lower and Upper Oligocene units. Deep-marine turbiditic sequences (e.g. Kliwa sandstones) occur within the Menilite Formation with varying thicknesses.

Western Carpathians (Czech part). The Menilite Formation in the Western Carpathians is subdivided from base to top into the Subchert, Chert, Dynow and Sitborice members. It is discussed using the example of borehole Kre-5, where the Menilite Formation has an apparent thickness of about 90 m (Francu & Feyzullayev 2010) (Fig. 6).

The Subchert Member (NP22) overlies the Green Clay Formation and *Globigerina* Marls, which contain moderate amounts of organic matter (0.5–2.7 wt% TOC) but with very low HI (<70 mg HC/g TOC: Fig. 6). The Subchert Member consists of laminated marls and shales, and is about 10 m thick across the basin. In Kre-5 it is thinner, but contains abundant organic matter (2.5–5.9 wt% TOC) with moderately high HI values (250–345 mg HC/g TOC).

The Chert Member is up to 4 m thick. Data from equivalent diatomites suggest that its upper part accumulated in an environment with strongly reduced salinity (Krhovský *et al.* 1992). Hence, in both the Carpathian Basin and the Alpine Foreland Basin, the major salinity drop characteristic for the first Paratethys isolation (Solenovian Event) occurred shortly before deposition of the Dynow Marlstone (NP23). The latter is characterized by low-diversity calcareous nannofossils and brackish-water mollusks typical for the Solenovian fauna (e.g. *Korobkoviella*, Lenticorbula: Čtyroký 1991; Popov & Studencka 2015). Silica from diatom frustules resulted in local silicification (Krhovský *et al.* 1992). In Kre-5, both cherts (3.8–5.3 wt% TOC; HI 580–680 mg HC/g TOC) and marlstones (5.1–5.9 wt% TOC; HI 460–560 mg HC/g TOC) contain abundant organic matter with very high HI values (type II kerogen).

The Sitborice Member is the uppermost unit of the Menilite Formation in the Western Carpathians. It is composed of brown-grey clays with thin layers of nanno-chalk. According to palaeontological evidence, oxygen-depleted conditions prevailed during deposition (Krhovský *et al.* 1992). Salinity was variable and often low (Krhovský *et al.* 1992). Debris-flow deposits occur frequently in the lower part of the Sitborice Member (e.g. Picha *et al.* 2006), and may represent the same phase of slumping and submarine erosion described by Sachsenhofer & Schulz (2006) in the Alpine Foreland Basin. In Kre-5, the Sitborice Member includes mainly carbonate-free shales with moderate TOC contents (average 1.5 wt%), but low HI values (*c.* 70 mg HC/g TOC). Only the lowermost sample at 92 m depth contains 4.6 wt% TOC with a moderately high HI (315 mg HC/g TOC). Abundant highly branched isoprenoid (HBI) alkanes indicate a significant contribution of diatom biomass (e.g. Grossi *et al.* 2004). Nannoplankton blooms and the occurrence of prasinophytes and diatoms in the upper part (*c.* 40–45 m) reflect environmental perturbations, such as decreased salinity, water stratification and high nutrient input (Svabenicka *et al.* 2007). Nevertheless, TOC and HI values remain low.

The Menilite Formation is overlain by light grey, mostly calcareous mudstones and clay with turbiditic fine-grained 'Krosno-type' sandstones (Zdanice-Hustopece Formation). In Kre-5, the boundary is located within NP24 (Svabenicka *et al.* 2007).

Northern Carpathians (Polish part). In the Polish Outer (Flysch) Carpathians, the Oligocene Menilite Formation and the overlying Oligo-Miocene Krosno Beds mainly occur within the Skole, Silesian and Dukla nappes (Figs 7 & 8). The total original thickness of the Menilite Formation ranges from 550 m in the Skole Nappe to 140 m in the Silesian Nappe (Kuśmierek 1990). The differences in thickness reflect different conditions of sedimentation within individual tectonic nappes. Partly, the higher thickness in the Skole Nappe is due to the fact that deposition of organic-rich rocks continued into NP25 in this nappe, but had terminated in NP23 in the Silesian Nappe (Köster *et al.* 1998*a*) (Fig. 7). In the central part of the Silesian Nappe and the NW part of Dukla Nappe, the Menilite Formation is overlain by Lower Oligocene 'Transitional Beds'. The Jasło Limestones subdivide the Oligo-Miocene succession into the Sub-Jasło complex and the Supra-Jasło complex (Jucha 1969; Haczewski 1989) (Fig. 7). Drastic changes in the thickness of the Krosno Beds in the limbs of folds and overthrusts (Kuśmierek *et al.* 2013) prove that deposition of the Krosno Beds was coeval with early fold-and-thrust deformation (Kuśmierek 2010; Kuśmierek *et al.* 2001).

Due to the importance of the Menilite Formation as a petroleum source rock in Poland, a wealth of source-rock data are available (e.g. Köster *et al.* 1998*a*, *b*; Kotarba & Koltun 2006; Kotarba *et al.*

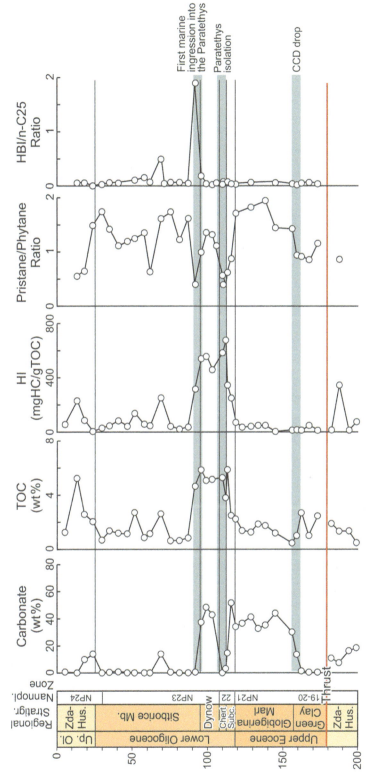

Fig. 6. Bulk rock and geochemical (including biomarker) data in the Oligocene rocks in borehole Kre-5 (Western Carpathians; Moravia) after Francu & Feyzullayev (2010). Formation boundaries and nannoplankton zones after Krhovský *et al.* (2001) and Svabenicka *et al.* (2007). Zda-Hus., Zdanice-Hustopece Formation; HBI, highly branched isoprenoids; *n*-C25, normal alkanes with 25 carbon atoms.

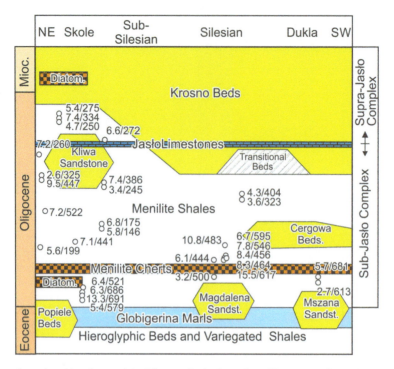

Fig. 7. Schematic stratigraphic columns of the Oligocene in the Outer Carpathian nappes of SE Poland together with TOC and hydrogen index values of samples after Curtis *et al.* (2004) modified by authors. All data are from immature rocks. Dia, intercalated diatomites.

2007, 2013, 2014), but the attribution to specific units within the Menilite Formation is often difficult. Figure 7 shows schematic stratigraphic columns in different nappes of the Polish Carpathians together with selected TOC and HI data according to Curtis *et al.* (2004). This compilation shows that HI values above 500 mg HC/g TOC are restricted to rocks underlying or directly overlying the Menilite Cherts. This agrees with the observation that maximum TOC (12.6 and 20.2 wt%) and HI (731 and 784 mg HC/g TOC), both in the Skole and in the Silesian nappes, occur just above the Menilite Cherts (Kotarba *et al.* 2007). It indicates that organic matter preservation was enhanced in NP22 and especially in the lower NP23, a period of time characterized by brackish environments. A diverse brackish bivalve fauna was described from the Dynow Marls (lower NP23) overlying chert in Zone B of the Silesian Nappe by Popov & Studencka (2015) and Studencka *et al.* (2016). Hence, organic matter preservation during early NP23 may have been promoted by salinity stratification. In contrast, Kotlarczyk & Uchman (2012) suggested thermo-stratification for the middle part of NP23.

Figure 8 presents the average thickness of source-rock complexes for the Menilite Formation

(excluding sandstone beds), Transitional and Krosno beds in different nappes together with source-rock data compiled after Kotarba *et al.* (2013) for the western part (Zone A), Kotarba *et al.* (2014) for the central part (Zone B) and Kosakowski *et al.* (2009) for the eastern part (Zone C). T_{max} values show that, with the exception of the western part of the Dukla Nappe (Grybów Subunit; southern part of Zone A), organic matter is immature. In the central and eastern parts of the Skole Nappe, the median TOC contents of the Menilite Formation are 5.0 and 4.7 wt%, respectively. The maximum TOC is 18.1% and the maximum HI value is 769 mg HC/g TOC. Average HI values are significantly lower (Zones B 363 mg HC/g TOC; Zone C 289 mg HC/g TOC). In the Silesian Nappe, average TOC contents in the Menilite Formation decrease from west (4.4 wt%) to east (3.6 wt%). Mean HI reaches a maximum in the central zone B (500 mg HC/g TOC). In the Dukla Nappe, average TOC contents of the Menilite Formation are 3.0 wt% in the thermally mature western part (Grybów Subunit, Zone A) and 2.7 wt% in the thermally immature central part (Zone B). Due to enhanced maturity, average HI in the western part is only 198 mg HC/g TOC, whereas it is 349 mg HC/g TOC in Zone B.

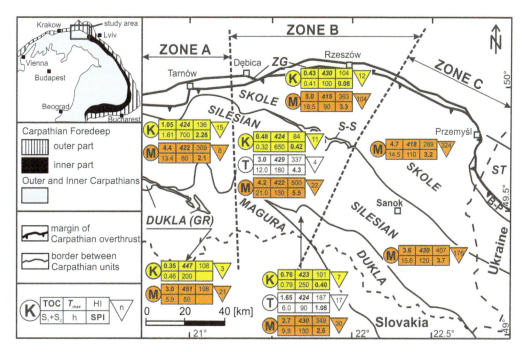

Fig. 8. Distribution of Rock-Eval data, SPI and average thickness of source-rock complex for the Oligocene Menilite, Oligocene Transitional and Upper Oligocene–Lower Miocene Krosno Beds in different nappes of the Polish Outer Carpathians. Rock-Eval pyrolysis data after Kotarba *et al.* (2013) for Zone A, Kotarba *et al.* (2014) for Zone B and Kosakowski *et al.* (2009) for Zone C. SPI has not been calculated for mature rocks in the Grybów Subunit. GR, Grybów Subunit of the Dukla Nappe; S-S, Subsilesian Nappe; ST, Stebnik Unit; ZG, Zgłobice Unit; TOC, total organic carbon (wt%); T_{max}, T_{max} temperature (°C); HI, hydrogen index (mg HC/g TOC); petroleum potential (S1 + S2) (mg HC/g r ock); h, thickness (m); SPI, source potential index (t HC m^{-2}); n, number of samples. In circle: K, Krosno Beds; T, Transitional Beds; M, Menilite Formation.

Geochemical analysis (ten Haven *et al.* 1993; Bessereau *et al.* 1996; Köster *et al.* 1998*b*; Kotarba *et al.* 2007, 2013, 2014) and hydrous pyrolysis experiments (Curtis *et al.* 2004; Lewan *et al.* 2006) reveal that the Menilite Formation contains varying mixtures of type II, type IIS and type III kerogen. Whereas type III kerogen dominates in Zone C of the Skole Nappe, type IIS kerogen is found in Zone B. According to Lewan *et al.* (2006), type IIS kerogen accumulated in areas with low siliciclastic input and a greater dominance of siliceous rocks.

Median TOC contents in the Transition Beds vary between 3.0 wt% (Silesian Nappe) and 1.65 wt% (Dukla Nappe: Fig. 8). Average (337 mg HC/g TOC) and maximum (513 mg HC/g TOC) HI values are higher in the Silesian Nappe than in the Dukla Nappe (187 and 360 mg HC/g TOC, respectively: Kotarba *et al.* 2014). The Krosno Beds typically contain low amounts of organic matter (<1 wt%) with low HI values (*c.* 100 mg HC/g TOC: Fig. 8). Maximum values of TOC and HI are 3.7 wt% and 248 mg HC/g TOC (Kotarba *et al.* 2013, 2014). Hence, the Transition Beds may be

considered as potential source rocks, whereas the Krosno Beds have only a limited source potential.

Eastern Carpathians (Romanian part). The Tazlau section (Fig. 9), located in the Vrancea Nappe, is used to represent the Menilite Formation in the Eastern Carpathians. Amadori *et al.* (2012) and Guerrera *et al.* (2012) assumed a Late Oligocene–Early Miocene age of the Menilite Formation in this section. In contrast, an Early Oligocene age has been attributed to the lower part of the succession by Melinte (2005) and Melinte-Dobrinescu & Brustur (2008). Based on additional nannoplankton data, Sachsenhofer *et al.* (2015) returned to the traditional age assignment, but the precise age remains debatable. Adopting the sedimentological model of Miclăus *et al.* (2009), organic matter accumulation in the Tazlau section has been described by Sachsenhofer *et al.* (2015). The following overview follows this paper.

The Menilite Formation, about 430 m thick, comprises, from bottom to top, Lower Menilites, Bituminous Marls, Lower Dysodilic Shales, which include

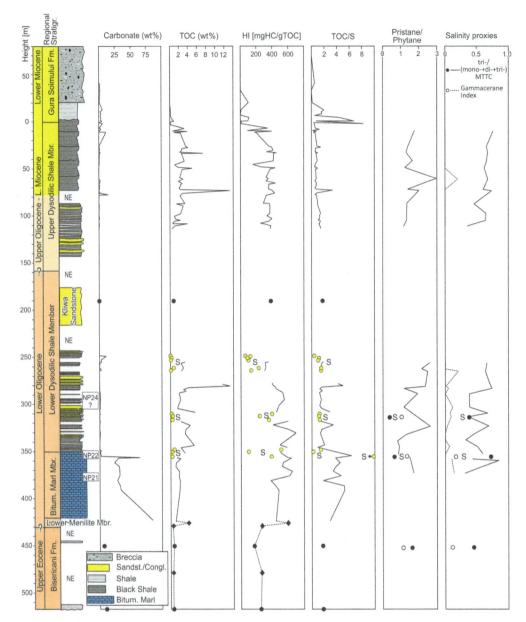

Fig. 9. Lithological log of the Tazlau section (Eastern Carpathians, Romania). Bulk geochemical parameters are shown together with some biomarker proxies. Sandy samples from the Lower Dysodilic Shale Member are labelled 'S'. TOC, total organic carbon; HI, hydrogen index; NE, not exposed (after Sachsenhofer *et al.* 2015).

the turbiditic Kliwa sandstone in their upper part, and Upper Dysodilic Shales (e.g. Miclăus *et al.* 2009). They overlie Eocene mudstones (Bisericani Formation) with low TOC contents (*c.* 0.7 wt%), and type III and mixed II–III kerogens (HI 170–270 mg HC/g TOC). Deposition of the Lower Menilite Member marks a change towards strongly

oxygen-deficient conditions. These, together with high siliceous bioproductivity, caused the accumulation of organic matter-rich rocks (4.0 wt% TOC; HI 570 mg HC/g TOC).

Calcareous nannoplankton constrains the overlying Bituminous Marl Member to NP21–NP22. According to Miclăus & Schieber (2014), the

Bituminous Marl Member accumulated in a deep-marine depositional environment characterized by high levels of bottom current activity. Mollusk assemblages from Piatra Neamţ (25 km N of Tazlau) include typical Solenovian endemic forms (*Korobkoviella lipoldii* (Rolle), *Lenticorbula sokolovi* (Karlov) and *Janschinella garetzkii* (Merklin)), supporting an Early Oligocene age (Rusu 1999). The organic matter is mainly derived from autochthonous marine organisms including bacterial biomass. TOC contents are moderate (average 1.7 wt%; HI 400–650 mg HC/g TOC) because of dilution by carbonate minerals. According to MTTC and Pr/Ph ratios (Fig. 9), salinity and redox conditions varied from reduced to slightly enhanced and from strictly anoxic to dysoxic, respectively (Sachsenhofer *et al.* 2015).

The lower part of the Lower Dysodilic Shale Member contains black shale and a significant number of sandstone beds deposited in a depositional lobe. Channel-fill sediments (Kliwa Sandstone) form the top of the Lower Dysodilic Shale Member. Bivalve assemblages with Solenovian-type mollusks (*Cerastoderma* cf. *serogosicum* (Nossovsky), *Korobkoviella lipoldii* (Rolle), *Urbnisia lata* Goncharova, *Janschinella garetzkii* (Merklin)) predominate, but euryhaline boreal taxa (*Nucula compta* Goldfuss, *Saccella westendorpi* (Nyst) and *Glossus subtrans* v. Orbigny) also occur suggesting episodic connections with the North Sea (Rusu 1999). Anoxic environments caused accumulation of abundant organic matter (average TOC 2.6 wt%) with hydrogen-rich type II kerogen (HI 500–650 mg HC/g TOC). Despite high HI values, maceral composition and palynological data show that land plants form a significant part of the organic matter (Sachsenhofer *et al.* 2015). Pr/Ph ratios (Fig. 9) and the abundance of land-plant-derived biomarkers suggest that a decrease in HI in the upper part of the Lower Dysodilic Shale Member to 300 mg HC/g TOC is due to the combined effect of increasing contributions of terrestrial organic matter and increased oxygen contents. Similar to the Bituminous Marl Member, salinity varied significantly during the deposition of this member.

The Upper Dysodilic Shale Member shows a fining-upwards trend and represents the transition from a depositional lobe to a basin plain setting. MTTC ratios increase upwards (Fig. 9) and reflect a trend from slightly enhanced to slightly decreased normal marine salinity (Sachsenhofer *et al.* 2015). Whereas oxygen-depleted but not strictly anoxic conditions are indicated by Pr/Ph ratios (Fig. 9), the presence of aryl-isoprenoids suggests a temporary photic zone anoxia controlled by salinity variations (Sachsenhofer *et al.* 2015). Both autochthonous marine biomass and land plants contributed to the organic matter (average TOC in lower part 2.2 wt%;

in the upper part 3.2 wt%), classified as type II kerogen, but with a lower HI (300–400 mg HC/g TOC) than in the Lower Dysoldilic Shale Member.

A major change towards oxic conditions occurred at the boundary between the Upper Dysoldilic Shale Member and the overlying Lower Burdigalian Gura Soimului Formation (NN2–NN3: Tabara & Popescu 2012), which is characterized by megaolistoliths in the Tazlau section (Popescu & Popescu 2002; Popescu 2005).

Hungarian Palaeogene Basin

The Hungarian Palaeogene Basin, a precursor of the Pannonian Basin, was formed as a wrench basin (Nagymarosy 1990) or a retro-arc foreland basin (Tari *et al.* 1993). The Lower Oligocene succession is represented by the pelagic, bathyal Buda Marl Formation (NP20–NP21) and the organic-matter-rich Tard Clay Formation (NP21/NP22–NP23) (Fig. 3). Previous papers showed the great impact of salinity and redox variations on organic matter accumulation (Vetö 1987; Brukner-Wein *et al.* 1990; Vetö & Hetényi 1991; Vetö *et al.* 1995; Vetö & Hertelendi 1996). More recently, Badics & Vetö (2012) and Bechtel *et al.* (2012) studied the vertical and lateral variation of its source rock and shale gas/shale oil potential.

In the present paper, data from borehole Ad-3 are used to characterize the Lower Oligocene succession (Fig. 10). In Ad-3, the top of the Buda Marl Formation is at 720 m (NP21: e.g. Vetö & Hertelendi 1996) or 698 m depth (middle NP22: Danisik *et al.* 2015) and is marked at 698 m depth in Figure 10. TOC contents of the Buda Marl Formation are typically below 1.0 wt% (max. 2.3 wt%). The presence of type III kerogen (HI <150 mg HC/g TOC) indicates high contributions of land plants or the oxygenation of marine organic matter.

The Tard Clay Formation is, on average, 68 m thick, but may reach a maximum thickness of 200 m in the NE part of the basin (Badics & Vetö 2012). In Ad-3, it is 86 m thick. Its non-laminated to weakly laminated lower part (679–698 m depth; uppermost NP22) was deposited in a semi-marine environment, and displays high TOC contents up to 5.0 wt% and low HI values (<185 mg HC/g TOC). Significantly higher HI values (210–440 mg HC/g TOC) occur between 640 and 679 m depth (lower NP23); this coincides roughly with the strongly laminated, low-salinity middle part of the formation. Brukner-Wein *et al.* (1990) assumed that alteration of volcanic tuff induced high bioproductivity due to an increased supply of nutrients in this interval. Very low Pr/Ph ratios and high contents in C_{14}-arylisoprenoids (Fig. 10) suggest photic zone anoxia (Bechtel *et al.* 2012). The upper part of the Tard Clay (610–640 m depth; upper part of NP23) is characterized by an increase in salinity,

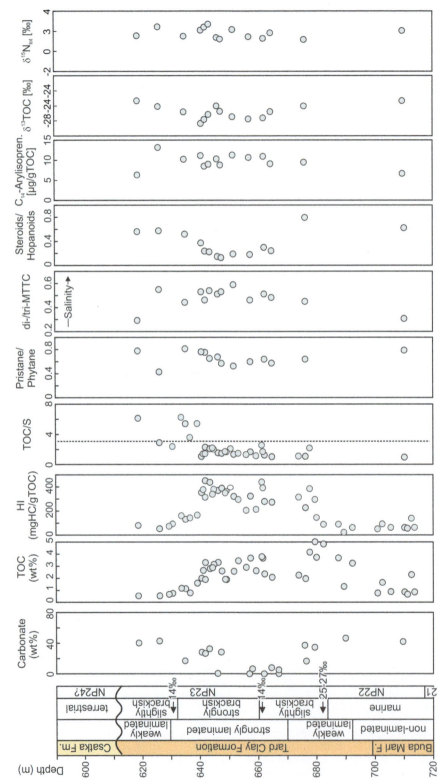

Fig. 10. Bulk and biomarker parameters of Lower Oligocene rocks in borehole Ad-3 (Hungarian Palaeogene Basin) after Bechtel *et al.* (2012 and references therein). The formation boundaries and nannoplankton zones after Danisik *et al.* (2015). Salinity estimates are after Nagymarosy (1983).

and a gradual decrease in TOC and HI to values of about 0.5 wt% TOC and 75 mg HC/g TOC, respectively. Interestingly, di-/tri-MTTC ratios are not as low as expected in view of the very low salinity indicated by calcareous nannoplankton and brackish mollusks (e.g. Nagymarosy 2000).

At 612 m, the Tard Clay Formation is truncated by the terrestrial Csatka Formation (NP23–NP24), changing upwards from terrestrial to pelagic and bathyal clays of the Kiscell Formation (NP24). The latter was deposited under normal marine conditions and is up to 800 m thick (Haas 2001). Data from Milota et al. (1995) show that TOC contents (<1.0 wt%) and HI values (<200 mg HC/g TOC) are low in the Kiscell Formation. This is also supported by data from Ad-3 (average TOC 0.71 wt%; HI 23 mg HC/g TOC: not shown in Fig. 10).

Western Black Sea

Fine-grained Oligocene sediments in Bulgaria have been called the Ruslar Formation (e.g. Sachsenhofer et al. 2009). However, following previous authors (e.g. Popov et al. 2004a; Stolyarov & Ivleva 2006), the name Maikop Group is used in the present contribution for all fine-grained Oligocene–Lower Miocene rocks in the Eastern Paratethys. With respect to stratigraphy, the Maikop Group comprises the Pshekhian, Solenovian, Kalmykian, Karadgalganian, Sakaraulian and Kozakhurian regiostages (Popov et al. 1993a) (Fig. 3).

Maikopian sediments offshore Bulgaria have been studied by Sachsenhofer et al. (2009) and Mayer et al. (2017). Using age data based on calcareous nannoplankton and applying sequence stratigraphic concepts, Mayer et al. (2017) proposed a new age model for the Oligocene–Lower Miocene rocks in the western Black Sea area, which is used in this paper.

The architecture of the Maikop Group offshore Bulgaria is controlled by a regional erosional unconformity ('Top Eocene') and the incision of the west–east-trending Kaliakra Canyon during late Solenovian time; the latter cut deeply into underlying Oligocene and Eocene rocks (Fig. 11a). In the following, Oligocene rocks outside the canyon and the Solenovian–Middle Miocene canyon fill are discussed separately on the basis of boreholes Samotino More (Sachsenhofer et al. 2009) (Fig. 11b) and Samotino Melrose 1 (Mayer et al. 2017) (Fig. 12). Data from both wells are largely based on cutting samples.

Oligocene rocks outside the Kaliakra Canyon. In Samotino More, Pshekhian rocks overlie calcareous shales of the Eocene Avren Formation with a minor erosional unconformity and are characterized by an upwards decrease in carbonate content (Fig. 11b).

Their average TOC is 1.7 wt%. The average HI in Samotino More is only 130 mg HC/g TOC but is significantly higher (c. 300 mg HC/g TOC) in Pshekhian rocks in some other wells, suggesting the presence of type III–II kerogen. Marls overlying the Pshekhian interval have been cored in Samotino More (Fig. 11c). Based on a low-diversity nannoplankton assemblage dominated by *Reticulofenestra ornata* and very low di-/tri-MTTC ratios, this layer is correlated with the Dynow Marl (NP23). It contains moderate amounts of kerogen classified as type II/III (1.4 wt% TOC; HI 290 mg HC/g TOC). The upper part of the Solenovian succession is dominated by largely carbonate-free shales with similar TOC contents (c. 1.5 wt%) but significantly lower HI (70 mg HC/g TOC), suggesting low-quality type III/IV kerogen. Sediments with upwards-increasing carbonate contents are attributed to the Upper Oligocene. Its lower part contains up to 2.0 wt% TOC: however, HI values remain below 200 mg HC/g TOC. In Samotino More, the Oligocene rocks are overlain with an erosional unconformity by diatom-rich rocks, which are discussed within the frame of the canyon fill.

Canyon fill. Borehole Samotino Melrose 1 is located near the axis of the Kaliakra Canyon, where the canyon fill is more than 1000 m thick (Fig. 11a). According to Mayer et al. (2017), the canyon fill comprises upper Solenovian–Middle Miocene rocks. With the exception of the Middle Miocene rocks, the succession is largely carbonate-free (Fig. 12). Upper Solenovian and Kalmykian deposits include a high percentage of sandstone layers.

Pelitic intervals within the upper Solenovian–Sakaraulian succession contain moderate amounts of type III kerogen (c. 1.5 wt% TOC; HI c. 240 mg HC/g TOC). TOC/S ratios suggest a normal marine environment. A major change in organic matter input occurred during Kozakhurian time, when diatomaceous shales, more than 200 m thick, with high TOC contents (c. 2.5 wt%) and HI values up to 530 mg HC/g TOC were deposited. TOC/S ratios confirm an environment with low salinity, which fits well with the second isolation of the Eastern Paratethys. The content of biogenic opal reaches 50 wt%, suggesting a large contribution of diatoms to the biomass. Similar sediments, but with significantly lower thickness, also occur south of the Kaliakra Canyon (Mayer et al. 2017). Middle Miocene sediments have been dated as NN5 and NN6 in neighbouring wells. The TOC content (c. 0.8 wt%) and HI value (c. 130 mg HC/g TOC) of these sediments are generally low.

Crimea

Oligocene and Lower Miocene pelitic rocks, several hundred metres thick, are exposed on the Crimean

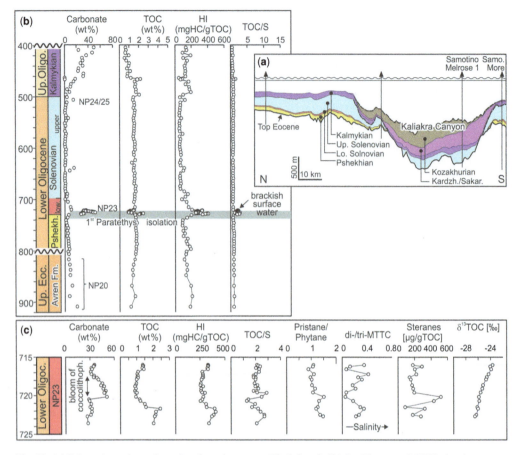

Fig. 11. (**a**) Schematic north–south section along the western Black Sea shelf (after Mayer *et al.* 2017) showing position of boreholes Samotino More and Samotino Melrose 1. (**b**) Bulk parameters of Oligocene rocks in borehole Samotino More. (**c**) Bulk, biomarker and isotope parameters of lower Solenovian rocks in borehole Samotino More. Geochemical data after Sachsenhofer *et al.* (2009). The stratigraphy follows Mayer *et al.* (2017).

Peninsula. Bulk geochemical parameters from two profiles in the eastern part of the peninsula (Kerch) are shown in Figure 13a after Saint-Germès (1998). The greyish or brownish pelitic rocks include varying amounts of silt and frequent fish remains. Apart from large siderite nodules, Oligocene rocks are free of carbonate minerals, whereas Lower Miocene rocks contain low amounts (max. 4.0 wt%). TOC contents up to 6 wt% occur in Upper Oligocene sediments. However, average TOC contents (<1.0 wt%) and HI values (<100 mg HC/g TOC) of all stratigraphic units are low, showing a negligible petroleum potential.

Volga-Don

The Volga-Don region forms the northern shelf of the Indolo-Kuban and Terek-Caspian depressions. Oligocene sediments from this shallow shelf

zone have been studied by Saint-Germès *et al.* (2000) using borehole 45-A (later called 768: Fig. 13b). Palynomorphs from this well have been described by Zaporozhets (1998) and Zaporozhets & Akhmetiev (2015).

A small erosional gap exists in this borehole between the Upper Eocene Belaya Glina marls, 2.5 m thick, and the Lower Oligocene Tsimlyansk Formation (Pshekhian Regiostage). The Lower Tsimlyansk Subformation (222–325 m depth) is composed of alternations of laminated and bioturbated carbonate-free silty clays with a diverse mollusc and foraminifer fauna. The upper subformation (201–222 m depth) is bioturbated and contains a larger amount of silty material. The Solenovian succession is subdivided into the Lower Solenovian Subformation ('Ostracoda Beds') (163–201 m depth) and the Upper Solenovian Subformation (124–163 m depth). The lower one begins with a 5

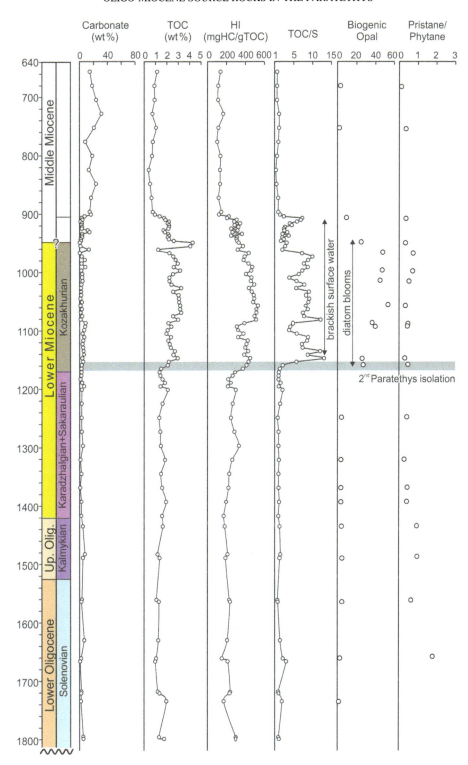

Fig. 12. Bulk rock, bulk geochemical and biomarker parameters of Oligocene and Miocene rocks in borehole Samotino Melrose 1 located within the Kaliakra Canyon (after Mayer *et al.* 2017).

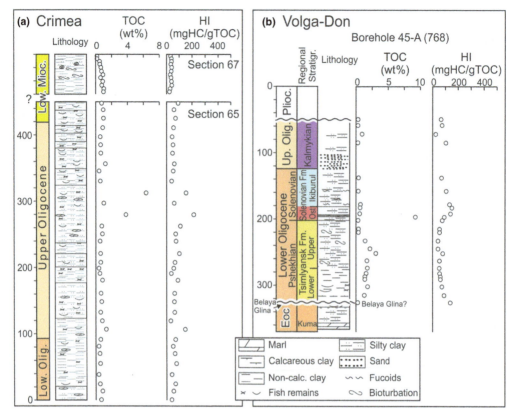

Fig. 13. TOC contents and HI values of Maikopian sediments in (**a**) outcrops on Crimea (Saint-Germès 1998) and (**b**) borehole 45-A in the Volga-Don region (Saint-Germès *et al.* 2000). The Lower–Upper Oligocene boundary in borehole 45-A is shown after Zaporozhets & Akhmetiev (2015).

m-thick light grey clay with nannoplankton with *Transversopontis pax, Reticulofenestra clatrata, R. lockeri* (NP23, Musatov data) and remains of brackish-water mollusks with *Ergenica cimlanica* (Pop.). These are overlain by ostracod-bearing marly clays, silty clays and siltstones. The Ikiburul Subformation is composed of silty clays, silts and clayey sands with very rare brackish and euryhaline marine mollusks (*Janshinella, Nucula*).

The Upper Oligocene Kalmykian Formation (50–124 m depth) comprises sandstones and silty clays with marine mollusks (*Chlamys bifida* Zone; Chatt A) in its lower part and greenish grey massive silty clays with fucoids in its upper part. Very rare mollusks (*Nucula*, cardiids) suggest an environment with reduced salinity. The top of the Kalmykian Formation is truncated by an erosional surface. Erosion removed Early Miocene sediments and cut deeply into the Kalmykian Formation (Stolyarov & Ivleva 2004).

TOC contents are typically around 1.5 wt% in the Tsimlyansk Formation (max. 3.1 wt%), and significantly below 1.0 wt% in Solenovian and Kalmykian

deposits. A very high TOC content (9.2 wt%) is restricted to the base of the Ostracoda Bed. The average HI is 78 mg HC/g TOC (max. HI 150 mg HC/g TOC). In agreement with Saint-Germès (1998), the hydrocarbon potential of Oligocene rocks is considered to be very low.

Precaucasus

The Precaucasus area extends along the northern margin of the Greater Caucasus from the Black Sea to the Caspian Sea. It includes, from west to east, the Indolo-Kuban and Terek-Caspian foredeeps, separated by the Stavropol High (for location see Fig. 2b). This section focuses on the Belaya River section located at the southern slope of the Indolo-Kuban Foredeep. Additional data are provided for the Kuma Uplift, forming the northern slope of the Terek-Caspian Foredeep.

Belaya River section. The Belaya River section (south of Maikop) is considered the type section for the Maikop Group. Sedimentology, fauna and flora of this section have been described in detail by Akhmetiev *et al.* (1995), Zaporozhets (1999) and Zaporozhets & Akhmetiev (2015). Saint-Germès *et al.* (2000) reported bulk geochemical parameters. Recently, the section has been studied by Sachsenhofer *et al.* (2017) using inorganic and organic geochemical proxies. Selected parameters are presented in Figure 14a.

At the Belaya River section, the Pshekha Formation (543.3–604.0 m depth; upper part of NP20–NP22) overlies Eocene marls (Belaya Glina Formation) without an erosional hiatus, but the sea-level drop at the Eocene–Oligocene boundary is reflected by increased runoff and enhanced input of land plants near the base of the Pshekha Formation (Sachsenhofer *et al.* 2017). Thereafter, a sea-level rise caused an increase in water depth, probably to more than 1000 m. The Pshekha Formation is dominated by calcareous shales, with only the uppermost part being carbonate-free. Pr/Ph data indicate that anoxic conditions prevailed during Pshekhian time, with the exception of a short time interval when increased oxygen availability enabled benthic fauna to colonize the seabed. Moreover, the presence of bathypelagic fish remains indicates that the water column was oxygenated. The average TOC content is about 1.7 wt%, but HI is relative low (*c.* 180 mg HC/g TOC). A HI of 365 mg HC/g TOC is reached only in the uppermost part of the Pshekha Formation.

The Polbian ('Ostracoda') Bed, a 30 cm-thick white shaly limestone (base of NP23), follows above a minor hiatus. Calcareous nannoplankton (Akhmetiev *et al.* 1995), endemic mollusks, ostracods and palynomorphs (Zaporozhets & Akhmetiev 2015) prove an environment with strongly reduced salinity, which is further supported by a prominent increase in the tri-/(tri- + di-)MTTC ratio. The top of the 'Ostracoda Bed' is marked by an erosional surface.

Salinity increased gradually during deposition of the Lower Morozkina Balka Formation (508–543 m depth; upper NP23), but remained low. High productivity of aquatic organisms resulted in deposition of organic-matter-rich rocks (up to 3.5 wt% TOC) with high HI values (max. 404 mg HC/g TOC) in the lower part of the formation. The input of land plants, dominated by angiosperms, was moderate and decreased further upwards, despite increasing input of detrital minerals. With the exception of the lowermost 1 m, strictly anoxic conditions prevailed during deposition of this unit.

The Upper Morozkina Balka Formation (460–508 m depth; NP23–NP24: Akhmetiev *et al.* 1995; Nagymarosy & Voronina 1993) contains significant amounts of carbonate. MTTC ratios show that salinity continued to increase and reached a maximum in the middle part of the unit. Pr/Ph ratios suggest

that oxygen availability was relatively high during deposition of its lower part and decreased with time. This is the same for input of detrital minerals. Hence, a maximum flooding zone may be located in the upper part of the unit. TOC (1.5–2.6 wt%) and HI (up to 310 mg HC/g TOC) are moderately high in the lower part of the formation, but decrease in its upper part.

Moderate to moderately high TOC contents (1.2–2.8 wt%) but low HI values (40–240 mg HC/g TOC) are also characteristic of the largely carbonate-free pelitic sediments of the Batalpashinsk (370–460 m depth) and Septarian formations (267–370 m depth; NP25–NN1), which were deposited in anoxic environments. Salinity variations including freshening events are recorded by MTTC ratios (Fig. 14) (Sachsenhofer *et al.* 2017) and dinocyst data (Zaporozhets 1999). Thin sand layers shed from the uprising Greater Caucasus (Vincent *et al.* 2013) are restricted to the upper part of the Septarian Formation.

Uplift and erosion of the Greater Caucasus caused a sharp increase in detrital input into the lowermost part of the Kardzhalga Formation (74–267 m depth; NN1–NN2). Above this level, Zr/K ratios reflect a general Lower Miocene fining-upwards cycle (Sachsenhofer *et al.* 2017). A gradual decrease in salinity is indicated by TOC/S and MTTC data. A lack of bioturbation and low Pr/Ph ratios suggest that, with the exception of the Olginskaya Formation (30–74 m depth), the environment was oxygen depleted. TOC contents (average 1.3 wt%) and HI values (*c.* 95 mg HC/g TOC) are typically low. Maceral data show that relatively high TOC contents (max. 3.0 wt%) in the Olginskaya Formation are due to high amounts of terrestrial organic matter. The Maikop Group is conformably overlain by Tarkhanian pelitic rocks, characterized by shells and relatively high carbonate contents.

Summarizing, TOC contents of the Maikop Group are moderately high, but HI values indicate that the presence of type II kerogen (HI >300 mg HC/g TOC) is restricted to layers overlying the Solenovian Polbian 'Ostracoda' Bed.

Kuma Uplift. Saint-Germès (1998) reported TOC and Rock-Eval pyrolysis data from borehole samples from the Kuma Uplift, where the Maikop Group is about 1 km thick (Table 1). These data show that the Lower Oligocene succession (Pshekha Formation–Morozkina Balka Formation) is thin, but contains large amounts of organic matter with moderately high HI values suggesting type II/III kerogen (Table 1). Highly oil-prone type II kerogen (565–580 mg HC/g TOC) occurs in two samples from the Ostracoda Bed. In contrast, HI is typically low in Upper Oligocene and Lower Miocene units.

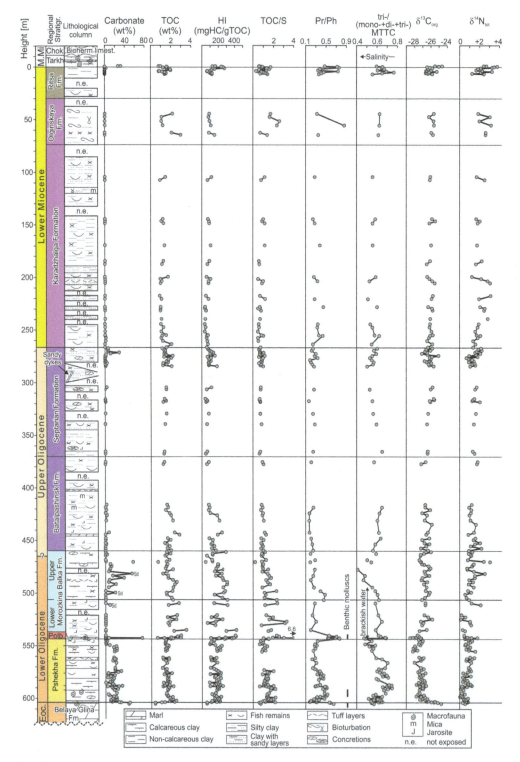

Fig. 14. Bulk and biomarker parameters of Oligocene and Miocene rocks along the Belaya River section (after Sachsenhofer *et al.* 2017).

Table 1. *Bulk geochemical data of the Maikop Group in the Kuma Uplift (from Saint-Germès 1998)*

Stratigraphic unit	Thickness (m)	TOC (wt%)		HI (mg HC/g TOC)		Sample No.
		Range	Average	Range	Average	
Upper Oligocene–Lower Miocene	*c.* 900	0.4–2.9	0.9	30–140	58	18
Batapalshinsk Formation	>60	0.6–3.2	1.9	30–195	84	17
Morozkina Balka Formation	30	0.5–5.6	2.6	40–350	150	29
Ostracoda Bed	0.5–20	2.0–7.2	4.0	50–580	252	7
Pshekha Formation	15	2.0–5.7	2.6	150–350	211	8

Thickness data follow Saint-Germès *et al.* (2000).

Rioni Basin

The NW–SE-trending Rioni Basin is located between the Greater Caucasus and the Lesser Caucasus, and opens westwards into the Eastern Black Sea Depression (Fig. 1). Eocene–Lower Miocene source rocks in the Rioni Basin have been studied recently by Vincent & Kaye (2017). Here, data from Oligocene–Lower Miocene sediments exposed along the Chanis River in the northern part of the Rioni Basin (western Georgia) are presented (Fig. 15).

The Maikopian sediments along the Chanis River section are about 1400 m thick and include potential oil-prone source rocks in their Rupelian part. Maximum TOC contents (4.8 wt%) and HI values (381 mg HC/g TOC) occur in their upper Rupelian part, which may be correlated with the Lower Morozkina Balka Formation in the Precaucasus area (Vincent & Kaye 2017). Lower Rupelian sediments also contain high amounts of organic matter (up to 2.3 wt% TOC) with type II/III kerogen (up to 283 mg HC/g TOC). Because of limited exposures, it is difficult to determine the thickness of the Rupelian succession precisely. Whereas the exposed lower and upper Rupelian sediments are only 40 and 20 m thick, respectively, the entire Rupelian succession is potentially as much as 200 m thick. Similar to the Precaucasus region, TOC (< 1.0 wt%) and HI (*c.* 50 mg HC/g TOC) in Upper Oligocene and Lower Miocene units is low (Fig. 15) (see Vincent & Kaye 2017).

Kura Basin

The Kura Basin is located between the Greater Caucasus and the Lesser Caucasus, and opens eastwards into the South Caspian Depression. Its western part is situated in Georgia, and its eastern part is mainly in Azerbaijan.

Western Kura Basin. Only few source-rock data are available from the Maikop Group in central Georgia. Therefore, new data from a composite profile east of the city of Tbilisi are presented here (Fig. 16). In the Tbilisi area, the Maikop Group is more than 3500 m thick, includes many sandstone beds, and is subdivided from base to top into the Manglisi (Pshekhian), Avchala (Solenovian) and Uplistsikhe formations (Kalmykian–Karadzhalganian), as well as Sakaraulian and Kozakhurian units (Popov *et al.* 1993*b*) (Fig. 16). Eocene and Lower Oligocene rocks contain up to 20 wt% of carbonate minerals. Younger rocks are largely carbonate-free (Fig. 16). Because of their great thickness, Eocene and Lower Oligocene rocks reach oil window maturity (Rr *c.* 0.75%; T_{max} *c.* 450°C). Therefore, their HI might be reduced. However, even if this effect is taken into account, HI values in the entire succession are low (average 67 mg HC/g TOC; max. 230 mg HC/g TOC). This indicates the dominance of type III kerogen, which is corroborated by high amounts of detrital land plants. TOC contents are low (<1.0 wt%) in Eocene, Pshekhian and Kozakhurian units, but moderately high in Solenovian–Sakaraulian units (max. 2.1 wt%). Overall, the low–moderate TOC contents and the dominance of type III kerogen prove a low hydrocarbon potential.

Eastern Kura Basin. A wealth of data has been gathered on the Maikop Group in Azerbaijan (Saint-Germès 1998; Katz *et al.* 2000; Hudson *et al.* 2008; Johnson *et al.* 2010; Bechtel *et al.* 2013, 2014). However, the results do not yield a consistent pattern. Bechtel *et al.* (2013, 2014) observed the best source-rock interval (*c.* 3 wt% TOC; HI *c.* 300 mg HC/g TOC) in Lower Oligocene horizons in the Angeharan and Lahich sections (Fig. 17), whereas Saint-Germès (1998) observed source-rock intervals with elevated TOC contents and HI values also in Upper Oligocene (max. TOC 3.5 wt%; max. HI 350 mg HC/g TOC) and Lower Miocene units (max. TOC 5.8 wt%; max. HI 400 mg HC/g TOC) (Fig. 18a). Data published by Johnson *et al.* (2010) suggested that maximum TOC contents occur in Upper Oligocene rocks, which are characterized by the lowermost average $\delta^{13}C$ and $\delta^{15}N$ ratios (Fig. 18b). In

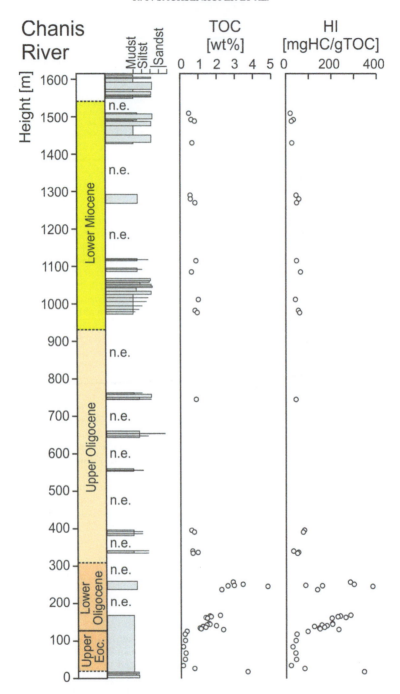

Fig. 15. Bulk rock and bulk geochemical parameters of Oligocene and Miocene rocks along the Chanis River section (after Vincent & Kaye 2017).

contrast, Bechtel *et al.* (2013) observed a trend towards heavier $\delta^{13}C$ and $\delta^{15}N$ ratios across the Lower–Upper Oligocene boundary. These differences either indicate large lateral variability in source-rock parameters, or may reflect problems with age dating.

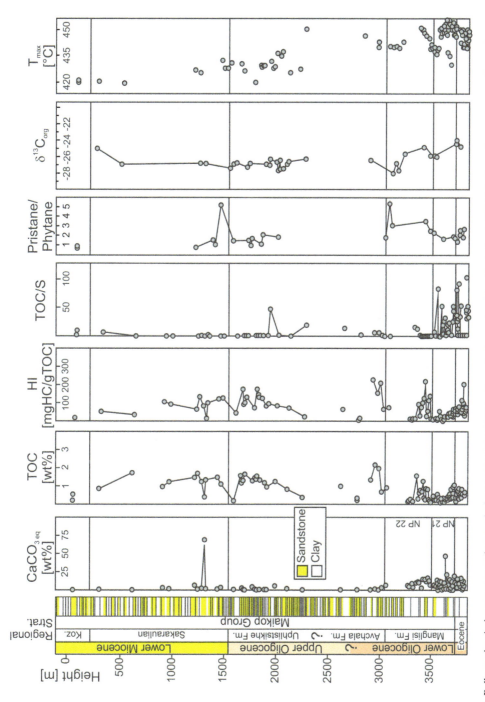

Fig. 16. Bulk geochemical parameters and nannoplankton zones of the Maikop Group in the western Kura Basin (east of Tbilisi). Formation names according to Popov *et al.* (1993*b*). The lithology is taken from Laliev (1964).

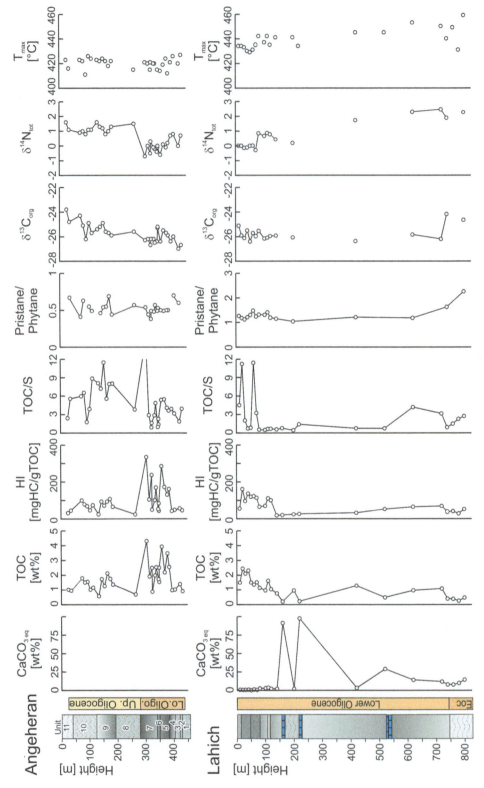

Fig. 17. Bulk and biomarker parameters of Oligocene rocks in Azerbaijan (after Bechtel *et al.* 2013, 2014).

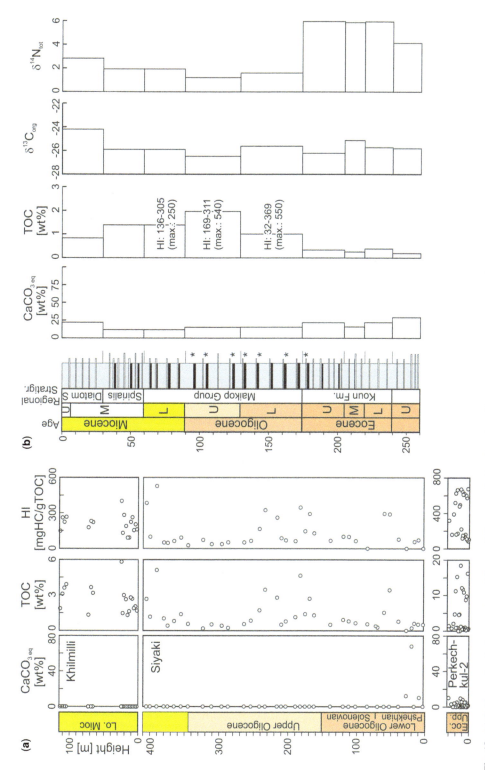

Fig. 18. (a) Bulk geochemical parameters of Eocene–Miocene rocks in selected profiles (after Saint-Germès 1998). The stratigraphy of the Siyaki profile follows Akhmetiev *et al.* (2007). (b) Summary of stratigraphic relationships showing average values for a variety of geochemical parameters (from Johnson *et al.* 2010).

Basin-wide variations of source-rock quality

Petroleum potential

TOC contents and the petroleum potential (S1 + S2 hydrocarbons) are frequently used to classify the quality of potential source rocks (Table 2). Oligocene–Lower Miocene rocks in the Paratethys are often characterized by relative high TOC contents but low petroleum potential, which is therefore considered as the limiting factor. Average values for all stratigraphic units in the Paratethys are shown in Figure 19. Information on the petroleum potential in different nappes in the Polish part of the Carpathians is provided in Figure 8.

Figure 19 shows major vertical and lateral variations of the petroleum potential. With the exception of the Tbilisi and Lahich sections, all data are from immature source rocks (data from mature rocks in the Dukla Nappe are not included in Fig. 19). However, the enhanced maturity of the Tbilisi and Lahich sections has no significant impact on the observed trends. Rocks with a very good petroleum potential (>25 mg HC/g rock) are restricted to Lower Oligocene units in Alpine (Schöneck Formation, unit c) and Carpathian domains (lower part of Menilite Formation), both situated in the Central Paratethys. Rocks with a good (6–25 mg HC/g rock) or fair petroleum potential (3–6 mg HC/g rock) occur both in the Central and Eastern Paratethys. Similar to the best source rocks, rocks with a good to fair petroleum potential are mainly found in Lower Oligocene units. In the case of the Lower (and Upper) Oligocene sediments in the Kaliakra Canyon offshore Bulgaria, characterized by abundant sand layers, the petroleum potential shown in Figure 19 refers only to pelitic intervals.

Whereas Upper Oligocene sediments typically show poor potential, deposition of rocks with good petroleum potential continued into Late Oligocene or even Early Miocene times in the Skole Nappe of the Polish Carpathians and the Eastern Carpathians. However, also in these areas, the petroleum potential typically decreases upwards.

Early Miocene source rocks with a good petroleum potential have also been deposited in the

Table 2. *Parameters describing the petroleum potential (after Peters 1986)*

	TOC (wt%)	S1 + S2 (mg HC/g rock)
Poor	<0.5	<3
Fair	0.5–1.0	3–6
Good	1.0–2.0	6–25
Very good	>2.0	>25

Kaliakra Canyon incised into the western Black Sea shelf (Mayer *et al.* 2017). This setting has not yet been described from any other part of the Paratethys.

The distribution of potential source rocks in the Kura Basin is irregular. In different profiles, the best source rocks are found in Lower Oligocene, Upper Oligocene or even Lower Miocene horizons. Accepting the available age data, this implies that in contrast to the Lower Oligocene source-rock horizons in other Paratethyian basins, their lateral continuity is not very high.

Source potential index

The source potential index (SPI = thickness (S1 + S2) bulk density/1000) integrates source-rock richness and thickness (Demaison & Huizinga 1994). It represents the maximum quantity of hydrocarbon generated within a column of source rock under 1 m^2 of surface area. Taking into account the shallow burial depth of most sections, a density of 2.0 t m^{-3} (shales) or 2.2 t m^{-3} (calcareous shales) has been applied when specific rock densities were not available. Following Demaison & Huizinga (1994), only units with a petroleum potential >2.0 mg HC/g rock have been considered. The resulting SPI values are presented in red numbers in Figure 19 together with the petroleum potential. Note that uncertainties arise from ambiguous age assignments. Because of the uncertain thickness of lower and upper Rupelian rocks, the SPI of the Chanis River section in the Rioni Basin is especially arguable. Accepting a thickness of 115 and 75 m for the lower and upper Rupelian rocks, respectively, the total SPI of the Rupelian succession is 1.9 t m^{-2} (Fig. 19). This value falls with the range suggested by Vincent & Kaye (2017: 0.7–2.5 t HC m^{-2}). The cumulative SPI varies significantly between different (sub-) basins from 0 to 10 t HC m^{-2}, which is discussed below. The calculated SPI values correspond to a low to (moderately) high SPI according to the classification of Demaison & Huizinga (1994).

Notably, high cumulative SPI values indicate the high hydrocarbon potential of the Polish and Romanian Carpathians. For example, the Menilite Formation and the 'Transitional Beds' in the Silesian Nappe in the Northern Carpathians can generate 9.8 t HC m^{-2} (Fig. 8). However, SPI varies considerably along strike of the Silesian Nappe and is only 2.1 and 3.7 t HC m^{-2} in its western and eastern part, respectively (Fig. 8). The cumulative SPI of the Skole Nappe is approximately 3.3 t HC m^{-2}. Significantly higher SPI values were estimated by Matyasik (2006) for the Menilite Formation in wells Paszowa-1 (40.1 t HC m^{-2}) and Żyznów-8 (36.7 t HC m^{-2}) in the Skole Unit. However, these values are too high due to an overestimation of the thickness of the source-rock complexes (394 and

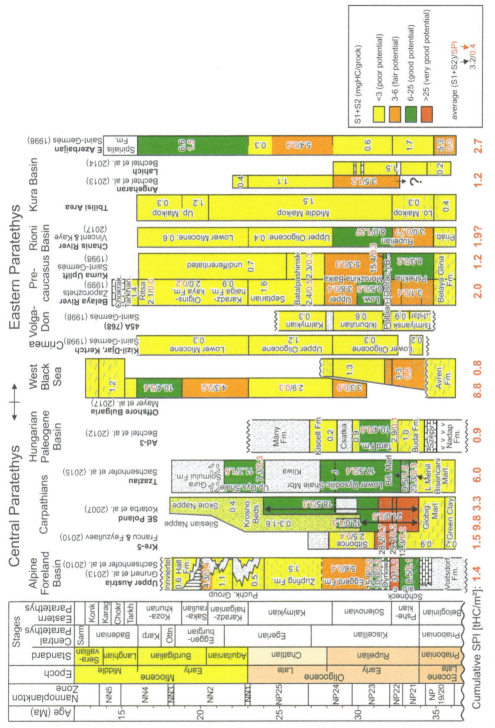

Fig. 19. Average petroleum potential (S1 + S2 (in mg HC/g rock)) for source-rock intervals are shown as black numbers; source potential index (SPI (t HC m^{-2})) is calculated for units with a petroleum potential exceeding 2.0 mg HC/g rock (red numbers).

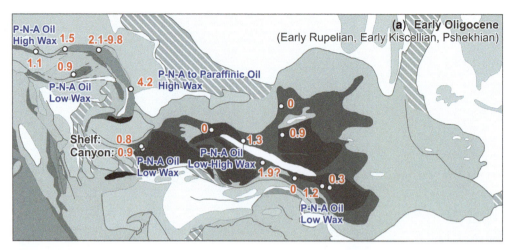

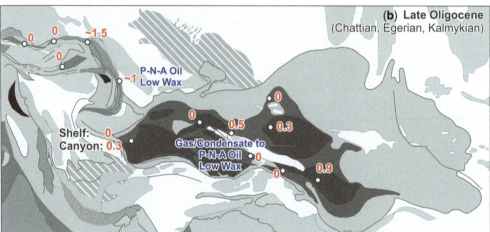

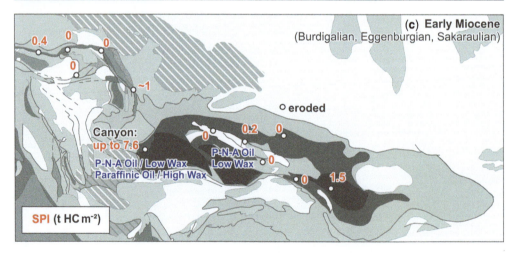

Fig. 20. Map showing the distribution of SPI values and the expected petroleum type for Lower Oligocene, Upper Oligocene and Early Miocene sediments. Note that the age assignment of some units is ambiguous causing some uncertainties.

560 m, respectively). In any case, the fact that Menilite Formation occurs in several tectonic units of the thrust belt additionally increases its hydrocarbon potential.

A very high cumulative SPI is also determined for the axial zone of the Kaliakra Canyon deeply incised into the western Black Sea shelf. However, the lateral continuity of the canyon fill is restricted. Apart from this, the cumulative SPI of Oligocene–Lower Miocene rocks in the Eastern Paratethys is surprisingly low and exceeds 1.5 t HC m^{-2} only in the Belaya River section, in the eastern part of the Kura Basin and potentially in the Rioni Basin. This suggests that in the Eastern Paratethys, an additional prolific hydrocarbon source may be present. The most likely candidate is the Middle–Late Eocene Kuma Formation (e.g. Beniamovski *et al.* 2003; Distanova & Bazhenova 2007), which is an excellent source rock in the Belaya River section (M. Morton pers. comm.), the Kuma Uplift (Saint-Germès 1998) and in western Georgia (Vincent & Kaye 2017).

SPI values for Lower Oligocene, Upper Oligocene and Lower Miocene units are shown in Figure 20 using palaeogeographical maps as a background. Although the delineation of different stratigraphic units is often difficult, Figure 20 shows some notable trends.

The Lower Oligocene part of the Menilite Formation in the Polish and Romanian sectors of the Carpathians rocks has the potential to generate >4 t HC m^{-2}. Coeval rocks in the Western Carpathians, the Alpine Foreland Basin and the Hungarian Palaeogene Basin can generate only 0.9–1.5 t HC m^{-2}. Similar amounts of hydrocarbons (0.8–1.3 t HC m^{-2}) can be generated by the Lower Oligocene part of the Maikop Group in the Precaucasus area, the western Black Sea shelf, and some sections in the Rioni and Kura basins. In contrast, Lower Oligocene rocks deposited on the northern shelf of the Eastern Paratethys (Crimea; Volga-Don region) have a negligible petroleum potential.

Despite prevailing anoxia, the cumulative SPI of Upper Oligocene rocks is typically low. Only the Northern and Eastern Carpathians display a significant SPI (*c.* 1.0–1.5 t m^{-2}), but precise quantification is hindered by the poorly defined upper and lower boundaries of the Upper Oligocene succession. The SPI of Upper Oligocene rocks in other basins in the Central Paratethys is insignificant. In the Eastern Paratethys, SPI values are typically <0.5 t m^{-2}. The highest SPI in the Eastern Paratethys has been reconstructed for the eastern Kura Basin in Azerbaijan (0.9 t HC m^{-2}: Figs 19 & 20).

Lower Miocene rocks in the Central Paratethys typically hold a low SPI. For example, a SPI of 0.4 t HC m^{-2} is calculated for the Alpine Foreland Basin (A1 fish shale). Deposition of these rocks

was limited to the eastern (Austrian) part of the basin. Moreover, Early Miocene erosion (base Hall unconformity) removed large parts of these rocks. The Upper Dysodilic Shale Member of the Menilite Formation in the Eastern Carpathians has been dated as Late Oligocene–Early Miocene and has a SPI of 1.8 t HC m^{-2} (Fig. 19). Although the Oligocene–Miocene boundary is undefined, we assume that about 1 t HC m^{-2} can be generated from Lower Miocene sediments.

The Lower Miocene fill of a shelf-break canyon in the western Black Sea forms a remarkable exception. Its SPI, calculated for the canyon axis, is *c.* 7.6 t m^{-2}. The main contribution (5.4 t HC m^{-2}) comes from brackish water diatomaceous shales, which reach maximum thickness within the canyon, but also extend beyond this structure. Apart from the western Black Sea, Lower Miocene rocks are significant sources for hydrocarbons only in the eastern Kura Basin of Azerbaijan (Figs 19 & 20) (Saint-Germès 1998).

Type of hydrocarbons

Major differences exist between HI values in different basins and different stratigraphic levels. Whereas Lower Oligocene rocks in the Alpine Foreland Basin and the Carpathians often contain type II kerogen with high HI values (500–700 mg HC/g TOC), HI values are typically lower in the Hungarian Palaeogene Basin (100–450 mg HC/g TOC) and the Eastern Paratethys (100–400 mg HC/g TOC). These differences are reflected in the composition of oils generated from the most relevant source-rock units (Fig. 21).

Results from pyrolysis-gas chromatography (Py-GC) show that Lower Oligocene units in the Alpine Foreland Basin and the Carpathians will generate paraffinic-naphthenic-aromatic (P-N-A) mixed oil, often with high wax content, or more rarely high wax paraffinic oil (Fig. 21). A similar oil will be generated from the Upper Oligocene–Lower Miocene Dysodilic Shale Member in the Eastern Carpathians, although it may have a slightly lower wax contents. Sulphur contents will be typically low, but slightly elevated in oils generated from unit 'b' of the Schöneck Formation and in some Tard oils.

Samples from the Eastern Paratethys typically generate a P-N-A mixed oil with low wax content (Fig. 21). Some samples even plot into the gas-condensate field. A high percentage of these samples, including those from Azerbaijan, plot in the sulphur discriminator into the type III kerogen field. In contrast, the lower part of the Lower Morozkina Balka Formation will generate high wax P-N-A mixed oil. A high wax content is also characteristic of some samples from the diatom-rich Lower Miocene (Kozakhurian) unit in the western Black Sea.

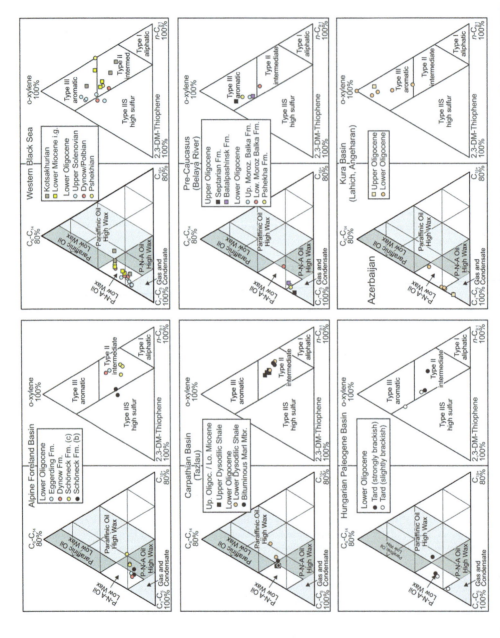

Fig. 21. Screening data from pyrolysis gas chromatography for the most important source-rock units: normal alkyl chain length distribution (Horsfield 1989, 1997) and sulphur discriminator (Eglinton *et al.* 1990). Data points are from Sachsenhofer *et al.* (2010, 2015, 2017) and hitherto unpublished sources.

Paratethys evolution and source-rock distribution

The most prolific source rocks occur in the Lower Oligocene succession (Fig. 20). The relationship of these rocks with the first isolation of the Paratethys is discussed in the first part of this section. It is obvious that the Eastern Carpathians host by far the highest potential. Therefore, factors controlling the pre-eminent position of this basin are considered. Finally, source rocks related to the second (Kozakhurian) Paratethys isolation are discussed briefly.

Source-rock distribution and the first Paratethys isolation (Solenovian Event)

The intensifying isolation of the Paratethys, culminating in the Solenovian Event (lower NP23), caused environmental changes which resulted in deposition of rocks with the highest petroleum potential, both in the Central and in the Eastern Paratethys. In this subsection, lithological profiles from different basins (Alpine Foreland, Western Carpathians, western Black Sea and Precaucasus) are compared and discussed with relation to the source-rock quality.

The lithological successions in the aforementioned basins show remarkable similarities. They include, from base to top: (1) marly rocks overlain by (2) largely carbonate-free shales and (3) marls or limestones exhibiting the typical 'Solenovian fauna', which grade upwards into (4) shales with low carbonate contents.

Calcareous rocks (1). Partial isolation of the Paratethys resulted in deposition of rocks with varying, but often high, carbonate contents during NP20–NP21. In the Alpine Foreland Basin, these rocks (Schöneck Formation units a/b) contain abundant globigerinoids. Thus, the *Globigerina* Marls in the Carpathians may be partial equivalents of the lower part of the Schöneck Formation. In the Alpine Foreland Basin, these rocks contain type II kerogen and have a good hydrocarbon potential. Equivalent rocks in the Belaya section of the Precaucasus area (lower part of the Pshekha Formation) contain a large amount of land plants (type III kerogen) downgrading their hydrocarbon potential. This is likely to be due to the initial uplift of the Caucasus (Vincent *et al.* 2007). The higher petroleum potential of more distal rocks drilled in the Kuma Uplift (type III–II kerogen) may be due to lower land plant input.

Carbonate-free shales (2). Rocks with significantly lower carbonate content accumulated during NP22. These include largely carbonate-free shales (Schöneck Formation unit 'c'; uppermost part of the Pshekha Formation) or chert (Chert Member of the Menilite Formation). Major freshwater incursions

resulted in a significant drop in salinity, recorded by geochemical and palaeontological proxies. The salinity drop occurred during deposition of the carbonate-free interval in the Alpine Foreland Basin and the Carpathians, and near its termination in the Precaucasus. In contrast to the Alpine Foreland Basin, high productivity of siliceous organisms in the Carpathians resulted in deposition of chert. Picha & Stranik (1999), Picha *et al.* (2006) and others assumed that high bioproductivity was due to upwelling. However, based on fish and trace fossil assemblages, Kotlarczyk & Uchman (2012) favoured a stratified water column model, similar to that suggested by Schulz *et al.* (2005) for the Alpine Foreland Basin. Despite differences in productivity, very good oil-prone source rocks were deposited in both the Alpine Foreland Basin and the Carpathians. In the case of the Alpine Foreland Basin, preservation of organic matter was favoured by photic zone anoxia (Fig. 5). In contrast, low bioproductivity resulted in the accumulation of rocks with a low petroleum potential in the Precaucasus area. Carbonate-free Tard Clays in borehole Ad-3 have been dated as occurring in NP23. However, they may form equivalents of the mentioned shales.

Dynow/Polbian Marl (3). Basin isolation and low salinity resulted in the development of a special endemic 'Solenovian-type' fauna and phytoplankton (Voronina & Popov 1984; Zaporozhets & Akhmetiev 2015). Blooms of calcareous nannoplankton (e.g. *Reticulofenestra ornata*) resulted in deposition of limestone or marls, which attain a maximum thickness (>10 m) in the Alpine Foreland Basin. Despite high bioproductivity and excellent preservation conditions, the source potential of the Dynow Formation in the Alpine Foreland Basin is relatively low, an effect of carbonate dilution. In the Carpathians, where nannoplankton blooms and carbonate accumulation was less significant, the Dynow Formation forms a very good source rock. Diatom blooms, recorded by silicification, additionally enhanced the petroleum potential in the Carpathians. Thin, but relatively good, source rocks also accumulated in the western Black Sea. In the Belaya section, the Polbian ('Ostracoda') Bed is only a few decimetres thick. The low petroleum potential of these rocks is a combined effect of carbonate dilution, high input of land plants and a decrease in water depth, which caused oxygenation.

Where available, isotope data suggest that organic matter in the Dynow and Polbian marls is typically characterized by a very light carbon ($\delta^{13}C$ −29‰) and relative heavy nitrogen ($\delta^{15}N$ +2‰). This is also true for the carbonate-rich upper part of the Tard Clay (Ad-3; *c.* 640 m depth: Fig. 10), which therefore may be correlated with the Dynow horizon.

Shales with low carbonate contents (4). Salinity remained reduced during the late Solenovian, but recovered gradually through time. In the Alpine Foreland Basin, salinity varied strongly after deposition of the Dynow Marlstone. Whereas there is a gradual transition from the Dynow Marlstone to the low-carbonate Eggerding Formation in the Alpine Foreland Basin, in most sections the carbonate content decreases sharply above the Dynow and Ostracoda Bed-type limestones and marlstones. Diatomites and cherts are known from the Eastern Carpathians. Rocks with high TOC contents and a type II kerogen with high HI values were deposited during early stages in most basins (e.g. Alpine Foreland Basin, Carpathians). In the Precaucasus area, this interval contains type II kerogen (with HI up to 400 mg HC/g TOC), which represents the best source-rock interval in the entire Maikop Group. The good source-rock quality of these rocks is mainly due to high bioproductivity.

Factors controlling the outstanding petroleum potential of the (Eastern) Carpathians

The Menilite Formation in the Eastern Carpathians hosts by far the greatest petroleum potential of all studied areas. In this subsection, possible reasons are discussed.

The Eastern Carpathians formed a very deep depression within the Paratethys, which was further structured by submarine highs (e.g. the Subsilesian High). According to fish assemblages, the water depth in the basins exceeded 2000 m (Kotlarczyk *et al.* 2006).

Based on fish and trace fossil assemblages, Kotlarczyk & Uchman (2012) showed that anoxic conditions prevailed in the lower part of the water column during Early Oligocene time. Anoxia resulted from a stratified water column (Schulz *et al.* 2005; Kotlarczyk & Uchman 2012). Salinity stratification is probable for the lower part of NP23 (Schulz *et al.* 2005), whereas Kotlarczyk & Uchman (2012) suggested that permanent anoxia, which was established during the middle part of NP23 after deposition of the Dynow Marl Member, was due to thermo-stratification. Upwelling anoxia on the upper slope developed during late Rupelian and Chattian time (NP24–NP25; Kotlarczyk & Uchman 2012). Prevailing anoxic conditions are also recorded by Pr/Ph ratios (Köster *et al.* 1998*b*), which are mostly below 1.0 in the Lower Oligocene section and exceed 1 in the Upper Oligocene part, the presence of biomarkers for green sulphur bacteria indicating photic zone anoxia, and of biomarkers indicative for methanotrophic bacteria (Köster *et al.* 1998*b*). The absence of deep-water fish in some intervals (e.g. NP23) also indicates oxygen-depleted

conditions in the water column (Kotlarczyk *et al.* 2006).

The presence of diatomites, cherts and nanno-marls in different levels proves, together with biomarker data, high bioproductivity of diatoms and calcareous nanoplankton, especially during Early Oligocene time (e.g. Picha *et al.* 2006). Cherts are missing in most other basins and are less widespread in the Western Carpathians.

High bioproductivity of dinoflagellates and diatoms requires the supply of large amounts of nutrients (e.g. Krhovský *et al.* 1992). A significant number of volcanic ash layers (e.g. Kotlarczyk *et al.* 2006) and continental runoff may have triggered eutrophication. As mentioned above, high bioproductivity and anoxic conditions in the Carpathians have been related to upwelling (e.g. Vetö 1987; Picha & Stranik 1999). However, based on the absence of deep-water fish, Kotlarczyk & Uchman (2012) assumed that permanent anoxia during the middle part of NP23 was due to a stratified water column and that upwelling occurred only from the upper part of NP24–NP25.

It is noteworthy that in the Eastern Carpathians, neither fish nor diatom data indicate a strong decrease in salinity of the surface water (Kotlarczyk & Uchman 2012). This agrees with the MTTC trend in the Tazlau section (Sachsenhofer *et al.* 2015). However, brackish mollusks have been described from locations in the Polish (Popov & Studencka 2015) and Romanian sectors of the Eastern Carpathians (Rusu 1999).

Source-rock distribution and the second Paratethys isolation

Very good source rocks occur in a submarine shelf-break canyon offshore Bulgaria and also extend beyond the canyon structure. Because of deposition of diatom-rich rocks in a low-salinity environment, a relationship to the isolation of the Eastern Paratethys in Kotsakhurian time is likely. Based on the relationship with the shelf-break canyon, a model implying locally enhanced upwelling of subsurface water onto the continental shelf (e.g. Kaempf 2007) seems reasonable.

Low-salinity diatomites and diatomaceous shales deposited during Burdigalian time (NN3?–NN4) have also been described from the Skole Nappe of Poland (Leszczawka Member: Kotlarczyk *et al.* 2006) (Fig. 7). They are comparable with diatomites in the Western Carpathians (Pavlovice Formation: Krhovský 1998) and the eastern part of the Alpine Foreland Basin, although Grunert *et al.* (2010*a*) postulated a fully marine environment. Grunert *et al.* (2010*a*, *b*) suggested that upwelling settings were widespread along the northern coast of the

Central Paratethys. Unfortunately, to our knowledge, source-rock data from these units are absent.

Apart from the above described upwelling scenario, prolific source rocks with an Early Miocene age are found only in the eastern Kura Basin. A relationship with the second Paratethys isolation is unclear. In any case, it is obvious that the Kozakhurian isolation of the Eastern Paratethys did not yield such prolific source rocks as the first, Solenovian, one. The reasons for this discrepancy remains poorly understood.

Conclusions

The review of depositional environment, and the amount and type of organic matter in different basins of the Paratethys, proves great vertical and lateral variability of source-rock quality and that a number of different geological settings may result in organic matter accumulation. Some key findings are summarized below:

- Oligocene–Lower Miocene sediments in the Paratethys can generate up to 10 t of hydrocarbons per square metre (t HC m^{-2}), but there is a great variability between different sub-basins.
- The Menilite Formation in the Carpathian Basin hosts the highest petroleum potential, a consequence of the interplay of several favourable factors, including very high bioproductivity of siliceous organisms and excellent preservation, mainly due to water column stratification in a very deep silled basin.
- In most basins, the best source-rock intervals are related to the first isolation of the Paratethys in the Early Oligocene and occur immediately before the isolation (NP22; e.g. Alpine Foreland Basin), during isolation (NP22–NP23 transition; Western Carpathians) or immediately afterwards (lower NP23; Precaucasus area). In the Alpine Foreland Basin, this is despite relatively low bioproductivity; in other basins, bioproductivity was enhanced. In all basins, strictly anoxic conditions favoured organic preservation.
- In general, the petroleum potential of Lower Oligocene rocks is higher in the Central Paratethys than in the Eastern Paratethys.
- With the exception of the Carpathian Basin, the petroleum potential of Upper Oligocene rocks is typically low.
- Lower Miocene rocks include important source rocks in the eastern Kura Basin and in the western Black Sea, where deposition of diatomaceous source rocks was probably related to upwelling along a shelf-break canyon. Upwelling was probably also an important process providing nutrients along the northern margin of the Central Paratethys

- Whereas the first basin isolation in the Early Oligocene resulted in deposition of rich source rocks, no such relationship can be found for the second isolation in the Eastern Paratethys in the Early Miocene (Kozakhurian).

The authors thank M. Morton (Leicester) for providing source rock data from the Kuma Formation in the Belaya River section and E. Gerslova (Brno) for stimulating discussions. The careful reviews by two anonymous reviewers have greatly improved the paper.

References

AKHMETIEV, M.A., POPOV, S.V., KRHOVSKÝ, J., GONCHAROVA, I.A., ZAPOROZHETS, N.I., SYCHEVSKAYA, E.K. & RADIONOVA, E.P. 1995. *Palaeontology and Stratigraphy of the Eocene–Miocene Sections of the Western Pre-Caucasia: Excursion Guidebook*. Russian Committee for IGCP, Moscow.

AKHMETIEV, M.A., ZAPOROZHETS, N.I. ET AL. 2007. New data on stratigraphy of Maikopian deposits in the Central Gobustan. *Stratigraphy and Sedimentology of Oil-Gas Basins*, **1**, 32–52.

ALLEN, M.B. & ARMSTRONG, H.A. 2008. Arabia–Eurasia collision and the forcing of mid-Cenozoic global cooling. *Palaeogeography, Palaeoclimatology, Palaeoecology*, **265**, 52–58.

ALLEN, P.A. & ALLEN, J.R. 2013. *Basin Analysis. Principles and Application to Petroleum Play Assessment*. 3rd edn. Wiley-Blackwell, Chichester.

AMADORI, M.L., BELAYOUNI, H., GUERRERA, F., MARTIN-MARTIN, M., ROJAS, M., MICLĂUS, C. & RAFFAELLI, G. 2012. New data on the Vrancea Nappe (Moldavidian Basin, Outer Carpathian Domain, Romania): paleogeographic and geodynamic reconstructions. *International Journal of Earth Sciences*, **101**, 1599–1623.

BADICS, B. & VETÖ, I. 2012. Source rocks and petroleum systems in the Hungarian part of the Pannonian Basin: The potential for shale gas and shale oil plays. *Marine and Petroleum Geology*, **31**, 53–69.

BARAKAT, A.O. & RULLKÖTTER, J. 1997. A comparative study of molecular paleosalinity indicators: Chromans, tocopherols and C_{20} isoprenoid thiophenes in Miocene lake sediments (Nördlinger Ries, Southern Germany). *Aquatic Geochemistry*, **3**, 169–190.

BECHTEL, A., GRATZER, R., SACHSENHOFER, R.F., GUSTERHUBER, J., LÜCKE, A. & PÜTTMANN, W. 2008. Biomarker and carbon isotope variation in coal and fossil wood of Central Europe through the Cenozoic. *Palaeogeography, Palaeoclimatology, Palaeoecology*, **262**, 166–175.

BECHTEL, A., HÁMOR-VIDÓ, M., GRATZER, R., SACHSENHOFER, R.F. & PÜTTMANN, W. 2012. Facies evolution and stratigraphic correlation in the early Oligocene Tard Clay of Hungary as revealed by maceral, biomarker and stable isotope composition. *Marine and Petroleum Geology*, **35**, 55–74.

BECHTEL, A., GRATZER, R., LINZER, H.-G. & SACHSENHOFER, R.F. 2013. Influence of migration distance, maturity and facies on the stable isotopic composition of alkanes and on carbazole distributions in oils and source rocks of the Alpine Foreland Basin of Austria. *Organic Geochemistry*, **62**, 74–85.

BECHTEL, A., MOVSUMOVA, U., PROSS, J., GRATZER, R., CORIC, S. & SACHSENHOFER, R.F. 2014. The Oligocene Maikop series of Lahich (eastern Azerbaijan): Paleoenvironment and oil-source rock correlation. *Organic Geochemistry*, **71**, 43–59.

BELAED, S. 2007. *Charakterisierung potenzieller Muttergesteine für biogenes Erdgas in der österreichischen Molassezone [Characterization of potential source rocks for biogenic gas in the Austrian Molasse Basin]*. Diploma thesis, Technical University of Clausthal, Clausthal-Zellerfeld, Germany.

BENIAMOVSKI, V.N., ALEKSEEV, A.S., OVECHKINA, M.N. & OBERHÄNSLI, H. 2003. Middle to upper Eocene dysoxic–anoxic Kuma Formation (northeast Peri-Tethys): Biostratigraphy and paleoenvironments. *In*: WING, S.L., GINGERICH, P.D., SCHMITZ, B. & THOMAS, E. (eds) *Causes and Consequences of Globally Warm Climates in the Early Paleogene*. Geological Society of America, Special Papers, **369**, 95–112.

BERNER, R.A. 1984. Sedimentary pyrite formation: an update. *Geochimica et Cosmochimica Acta*, **48**, 605–615.

BESSEREAU, G., ROURE, F., KOTARBA, M., KUŚMIEREK, J. & STRZETELSKI, W. 1996. Structure and hydrocarbon habitat of the Polish Carpathians. *In*: ZIEGLER, P.A. & HORVATH, F. (eds) *Peri-Tethys Memoir 2. Structure and Prospects Alpine Basins and Forelands*. Memoires du Museum National d'Histoire Naturelle, **170**, 343–373.

BRUKNER-WEIN, A., HETÉNYI, M. & VETÖ, I. 1990. Organic geochemistry of an anoxic cycle: a case history from the Oligocene section, Hungary. *Organic Geochemistry*, **15**, 123–130.

BRZOBOHATY, R. 1981. Izolovane rybi zbytky z menilitovych vrstev Ždanicke jednotky na Moravě [Isolated fish remains of the Menilitic Formation in the Ždanice Unit at the Moravia]. *Zemni plyn a nafta*, **26**, 79–87 [in Czech].

ČTYROKÝ, P. 1991. *Mollusk finds in the Dynów Marlstones (Menilitic Formation) of the Ždánice unit*. Geoscience Research Reports for 1991 [Zprávy o geologických výzkumech v roce 1991]. Czech Geological Survey, Prague [in Czech].

CURTIS, J.B., KOTARBA, M.J., LEWAN, M.D. & WIĘCŁAW, D. 2004. Oil/source correlations in the Polish Flysch Carpathians and Mesozoic basement and organic facies of the Oligocene Menilite Shales: insights from hydrous pyrolysis experiments. *Organic Geochemistry*, **35**, 1573–1596.

DANISIK, M., FODOR, L., DUNKL, I., GERDES, A., CSIZMEG, J., HÁMOR-VIDÓ, M. & EVANS, N.J. 2015. A multi-system geochronology in the Ad-3 borehole, Pannonian Basin (Hungary) with implications for dating volcanic rocks by low-temperature thermochronology and for interpretation of (U–Th)/He data. *Terra Nova*, **27**, 258–269.

DE LEEUW, J.W. & SINNINGHE DAMSTÉ, J.S. 1990. Organic sulphur compounds and other biomarkers as indicators of paleosalinity. *ACS Symposium Series*, **219**, 417–443.

DEMAISON, G. & HUIZINGA, B.J. 1994. Genetic classification of petroleum systems using three factors: charge, migration and entrapment. *In*: MAGOON, L.B. & DOW, W.G.

(eds) *The Petroleum System, from Source to Trap*. American Association of Petroleum Geologists Memoirs, **60**, 73–89.

DIDYK, B.M., SIMONEIT, B.R.T., BRASSELL, S.C. & EGLINTON, G. 1978. Organic geochemical indicators of palaeoenvironmental conditions of sedimentation. *Nature*, **272**, 216–222.

DISTANOVA, L. & BAZHENOVA, O.K. 2007. Conditions of source rock potential formation in eocene deposits of Caucasian–Scythian region. *Paper P257 presented at the EAGE 69th Conference & Exhibition*, 11–14 June 2007, London, UK.

EGLINTON, T.I., SINNINGHE-DAMSTÉ, J.S., KOHNEN, M.E.L., DE LEEUW, J.W., LARTER, S.R. & PATIENCE, R.L. 1990. Analysis of maturity-related changes in the organic sulphur composition of kerogens by flash pyrolysis-gas chromatography. *In*: ORR, W.L. & WHITE, C.M. (eds.) *Geochemistry of Sulphur in Fossil Fuels*. American Chemical Society Symposium Series, **429**, 529–565.

ESPITALIÉ, J., LAPORTE, J.L., MADEC, M., MARQUIS, F., LEPLAT, P., PAULET, J. & BOUTEFEU, A. 1977. Méthode rapide de characterisation des roches mères de leur potential pétrolier et de leur degree d'evolution [Rapid method for source rock characterization and for determination of their petroleum potential and degree of evolution]. *Revue de l'Institut Francais du Pétrole*, **32**, 23–42.

FRANCU, J. & FEYZULLAYEV, A. 2010. Molecular evidence of the depositional environment evolution during the Oligocene and Miocene in the early Paratethys and its manifestations in the related petroleum systems. *AAPG Search and Discovery Article 90109 presented at the AAPG European Region Annual Conference*, 17–19 October 2010, Kiev, Ukraine.

FU, J.G., SHENG, P., PENG, S.C., BRASSELL, S.C. & EGLINGTON, G. 1986. Pecularities of salt lake sediments as potential source rocks in China. *Organic Geochemistry*, **10**, 119–127.

GRATZER, R., BECHTEL, A., SACHSENHOFER, R.F., LINZER, H.-G., REISCHENBACHER, D. & SCHULZ, H.-M. 2011. Oil–oil and oil–source rock correlations in the Alpine Foreland basin of Austria: insights from biomarker and stable carbon isotope studies. *Marine and Petroleum Geology*, **28**, 1171–1186.

GROSSI, V., BEKER, B., GEENEVASEN, J.A.J., SCHOUTEN, S., RAPHEL, D., FONTAINE, M.-F. & SINNINGHE DAMSTÉ, J.S. 2004. C25 highly branched isoprenoid alkenes from the marine benthic diatom J.S. Pleurosigma strigosum. *Phytochemistry*, **65**, 3049–3055.

GRUNERT, P., HARZHAUSER, M., RÖGL, F., SACHSENHOFER, R. & GRATZER, R. 2010*a*. Oceanographic conditions as a trigger for the formation of an Early Miocene (Aquitanian) *Konservat-Lagerstätte* in the Central Paratethys Sea. *Palaeogeography, Palaeoclimatology, Palaeoecology*, **292**, 425–442.

GRUNERT, P., SOLIMAN, A., HARZHAUSER, M., MÜLLEGGER, S., PILLER, W.E., ROETZEL, R. & RÖGL, F. 2010*b*. Upwelling conditions in the Early Miocene Central Paratethys Sea. *Geologica Carpathica*, **61**, 129–145.

GRUNERT, P., HINSCH, R. *ET AL.* 2013. Early Burdigalian infill of the Puchkirchen Trough (North Alpine Foreland Basin, Central Paratethys): Facies development and sequence stratigraphy. *Marine and Petroleum Geology*, **39**, 164–186.

GRUNERT, P., AUER, G., HARZHAUSER, M. & PILLER, W.E. 2015. Stratigraphic constraints of the upper Oligocene to lower Miocene Puchkirchen Group (North Alpine Foreland Basin, Central Paratethys). *Newsletter on Stratigraphy*, **48**, 111–133.

GUERRERA, F., MARTIN-MARTIN, M., MARTIN-PEREZ, J.A., MARTIN-ROJAS, I., MICLĂUŞ, C. & SERRANO, F. 2012. Tectonic control on the sedimentary record of the central Moldavidian Basin (Eastern Carpathians, Romania). *Geologica Carpathica*, **63**, 463–479.

HAAS, J. 2001. *Geology of Hungary*. Eötvös University Press, Budapest.

HACZEWSKI, G. 1989. Coccolith limestone horizons in the Menilite-Krosno series (Oligocene, Carpathians) – identification, correlation and origin. *Annales Societatis Geologorum Poloniae*, **59**, 435–523 [in Polish with English abstract].

HORSFIELD, B. 1989. Practical criteria for classifying kerogens: some observations from pyrolysis-gas chromatography. *Geochimica et Cosmochimica Acta*, **53**, 891–901.

HORSFIELD, B. 1997. The bulk composition of first-formed petroleum in source rocks. *In*: WELTE, D.H., HORSFIELD, B. & BACKER, D.R. (eds) *Petroleum and Basin Evolution. Insights from Petroleum Geochemistry, Geology and Basin Modelling*. Springer, Berlin, 335–402.

HUDSON, S.M., JOHNSON, C.L., EFENDIYEVA, M.A., ROWE, H.D., FEYZULLAYEV, A.A. & ALIYEV, C.S. 2008. Stratigraphy and geochemical characterization of the Oligocene–Miocene Maikop series: implications for the paleogeography of Eastern Azerbaijan. *Tectonophysics*, **451**, 40–55.

JOHNSON, C.L., HUDSON, S.M., ROWE, H.D. & EFENDIYEVA, M.A. 2010. Geochemical constraints on the Palaeocene–Miocene evolution of eastern Azerbaijan, with implications for the South Caspian basin and eastern Paratethys. *Basin Research*, **22**, 733–750.

JUCHA, S. 1969. *Les schiestes de Jasło, leur impostance pour la stratigrafie et la sedimentologie de la serie Menilitique et des couches de Krosno (Carpathes flyscheuses) [The Jasło Schists, their importance for the stratigraphy and sedimentology of the Menilitic Series and the Krosno Beds (Flysch Carpathians)]*. Prace Geologiczne PAN, **52** [in Polish with French abstract].

KAEMPF, J. 2007. On the magnitude of upwelling fluxes in shelf-break canyons. *Continental Shelf Research*, **27**, 2211–2223.

KATZ, B., RICHARDS, D., LONG, D. & LAWRENCE, W. 2000. A new look at the components of the petroleum system of the South Caspian Basin. *Journal of Petroleum Science and Engineering*, **28**, 161–182.

KOSAKOWSKI, P., WIĘCŁAW, D. & KOTARBA, M.J. 2009. Evaluation of petroleum potential of the selected source strata in trans-border zone of the Polish Outer Carpathians. *Geologia*, **35**, 155–190 [in Polish with English abstract].

KÖSTER, J., KOTARBA, M., LAFARGUE, E. & KOSAKOWSKI, P. 1998*a*. Source rock habitat and hydrocarbon potential of Oligocene Menilite Formation (Flysch Carpathians, Southeast Poland): an organic geochemical and isotope approach. *Organic Geochemistry*, **29**, 543–558.

KÖSTER, J., ROSPONDEK, M., SCHOUTEN, S., KOTARBA, M., ZUBRZYCKI, A. & SINNINGHE DAMSTÉ, J.S. 1998*b*. Biomarker geochemistry of a foreland basin: the Oligocene Menilite Formation in the Flysch Carpathians of Southeast Poland. *Organic Geochemistry*, **29**, 649–669.

KOTARBA, M.J. & KOLTUN, Y.V. 2006. The origin and habitat of hydrocarbons of the Polish and Ukrainian parts of the Carpathian Province. *In*: GOLONKA, J. & PICHA, F.J. (eds) *The Carpathians and Their Foreland: Geology and Hydrocarbon Resources*. American Association of Petroleum Geologists Memoirs, **84**, 395–442.

KOTARBA, M.J., WIĘCŁAW, D., KOLTUN, Y.V., MARYNOWSKI, L., KUŚMIEREK, J. & DUDOK, I.V. 2007. Organic geochemical study and genetic correlation of natural gas, oil and Menilite source rocks in the area between San and Stryi rivers (Polish and Ukrainian Carpathians). *Organic Geochemistry*, **38**, 1431–1456.

KOTARBA, M.J., WIĘCŁAW, D., DZIADZIO, P., KOWALSKI, A., BILKIEWICZ, E. & KOSAKOWSKI, P. 2013. Organic geochemical study of source rocks and natural gas and their genetic correlation in the central part of the Polish Outer Carpathians. *Marine and Petroleum Geology*, **45**, 106–120.

KOTARBA, M.J., WIĘCŁAW, D., DZIADZIO, P., KOWALSKI, A., KOSAKOWSKI, P. & BILKIEWICZ, E. 2014. Organic geochemical study of source rocks and natural gas and their genetic correlation in the eastern part of the Polish Outer Carpathians and Palaeozoic–Mesozoic basement. *Marine and Petroleum Geology*, **56**, 97–122.

KOTLARCZYK, J. & UCHMAN, A. 2012. Integrated ichnology and ichthyology of the Oligocene Menilite Formation, Skole and Subsilesian nappes, Polish Carpathians: A proxy to oxygenation history. *Palaeogeography, Palaeoclimatology, Palaeoecology*, **331–332**, 104–118.

KOTLARCZYK, J., JERZMANSKA, A., OEWIDNICKA, E. & WISZNIOWSKA, T. 2006. A framework of ichtyofaunal ecostratigraphy of the Oligocene–Early Miocene strata of the Polish Outer Carpathian Basin. *Annales Societatis Geologorum Poloniae*, **76**, 1–111.

KRHOVSKÝ, J. 1995. Early Oligocene palaeoenvironmental changes in the West Carpathian Flysch Belt of southern Moravia. *In*: *Proceedings of the XV Congress of the Carpathian–Balkan Geological Association, September 1995*. Geological Society of Greece, Special Publications, **4**, 209–213.

KRHOVSKÝ, J. 1998. Geology, stratigraphy and palaeoenvironment of the Southern Moravian Flysch Belt. *In*: CICHA, I., RÖGL, F., RUGG, Ch. & CTYROKA, I. (eds) *Oligocene–Miocene Foraminifera of the Central Paratethys*. Abhandlungen der Senckenbergischen Naturforschenden Gesellschaft, **549**, 18–23.

KRHOVSKÝ, J., ADAMOVA, M., HLADIKOVA, J. & MASLOWSKA, H. 1992. Paleoenvironmental changes across the Eocene/Oligocene boundary in the Zdanice et Pouzdrany units (Western Carpathians, Czechoslovakia): long-term trend and orbitally forced changes in calcareous nannofossil assemblages. *In*: HAMRSMID, B. & YOUNG, J. (eds) *Nannoplankton Research. Proceedings 4th INA Conference, Prague 1991, vol. II: Tertiary Biostratigraphy and Paleoecology; Quaternary Coccoliths. Knihovnicka ZPN*, **16**, 73–83.

KRHOVSKÝ, J., RÖGL, F. & HAMRSMID, B. 2001. Stratigraphic correlation of the Late Eocene to Early Miocene of the Waschberg Unit (Lower Austria) with the Zdanice and Pouzdrany Units (South Moravia).

Österreichische Akademie der Wissenschaften Schriftenreihe der Erdwissenschaftlichen Kommissionen, **14**, 225–254.

KUŚMIEREK, J. 1990. *Outline of Geodynamics of the Central Carpathian Petroleum Basin*. Prace Geologiczne PAN, **135** [in Polish with French abstract].

KUŚMIEREK, J. 2010. Subsurface structure and tectonic style of the NE Outer Carpathians (Poland) on the basis of integrated 2D interpretation of geological and geophysical images. *Geologica Carpathica*, **61**, 71–85.

KUŚMIEREK, J., MAĆKOWSKI, T. & ŁAPINKIEWICZ, A.P. 2001. Effects of synsedimentary thrusts and folds on the results of two-dimensional hydrocarbon generation modeling of the eastern Polish Carpathians. *Przegląd geologiczny*, **49**, 412–417 [in Polish with English abstract].

KUŚMIEREK, J., BARAN, U. & GOLONKA, J. 2013. Tectonic and geological characteristic of the eastern part of the Polish Carpathians and transborder zone with Ukraine. *In*: GÓRECKI, W. (ed.) *Geothermal atlas of the Eastern Carpathians*. AGH University of Science and Technology, Kraków, 74–102.

LALIEV, A.G. 1964. *Maykopian Series of Georgia*. Nedra, Moscow [in Russian].

LESZCZYNSKI, S. 1997. Origin of the Sub-Menilite Globigerina Marl (Eocene–Oligocene transition) in the Polish Outer Carpathians. *Annales Societatis Geologorum Poloniae*, **67**, 367–427.

LEWAN, M.D., KOTARBA, M.J., CURTIS, J.B., WIECŁAW, D. & KOSAKOWSKI, P. 2006. Oil-generation kinetics for organic facies with Type-II and -IIS kerogen in the Menilite Shales of the Polish Carpathians. *Geochimica et Cosmochimica Acta*, **70**, 3351–3368.

MATYASIK, I. 2006. *Hydrocarbon Potential of the Skole Unit of the Flysch Carpathians*. Prace INiG, **140** [in Polish with English abstract].

MAYER, J., RUPPRECHT, B.J. *ET AL.* 2017. Source potential and depositional environment of Oligocene and Miocene rocks offshore Bulgaria. *In*: SIMMONS, M.D., TARI, G.C. & OKAY, A.I. (eds) *Petroleum Geology of the Black Sea*. Geological Society, London, Special Publications, **464**, https://doi.org/10.1144/SP464.2

MELINTE, M.C. 2005. Oligocene palaeoenvironmental changes in the Romanian Carpathians, revealed by calcareous nannofossils. *Studia Geologica Polonica*, **124**, 341–352.

MELINTE-DOBRINESCU, M. & BRUSTUR, T. 2008. Oligocene–Lower Miocene events in Romania. *Acta Palaeontologica Romaniae*, **6**, 203–215.

MICLĂUS, C. & SCHIEBER, J. 2014. A hierarchy of current-produced bedforms in a source rock from the Eastern Carpathians points to predominant bedload deposition of an organic-rich mudstone. *Search and Discovery article 51006 presented at the AAPG Annual Convention and Exhibition*, 6–9 April 2014, Houston, Texas, USA.

MICLĂUS, C., LOIACONO, F., PUGLISI, D. & BACIU, D.S. 2009. Eocene–Oligocene sedimentation in the external areas of the Moldavide Basin (Marginal Folds Nappe, Eastern Carpathians, Romania): sedimentological, paleontological and petrographical approaches. *Geologica Carpathica*, **60**, 397–417.

MILOTA, K., KOVACS, A. & GALICZ, Zs. 1995. Petroleum potential of the North Hungarian Oligocene sediments. *Petroleum Geoscience*, **1**, 81–87, https://doi.org/10.1144/petgeo.1.1.81

NAGYMAROSY, A. 1983. Mono- and duospecific nannofloras in Early Oligocene sediments of Hungary. *Proceedings of the Koninlijke Nederlandsche Akademie van Vetenschappen, B*, **86**, 273–283.

NAGYMAROSY, A. 1990. Paleogeographical and paleontological outlines of some intracarpathian Paleogene basins. *Geologica Carpathica*, **41**, 259–274.

NAGYMAROSY, A. 2000. Lower Oligocene nannoplankton in anoxic deposits of the Central Paratethys. *Journal of Nannoplankton Research*, **22**, 128–129.

NAGYMAROSY, A. & VORONINA, A.A. 1993. Calcareous nannoplankton from the lower Maykopian Beds (Early Oligocene, Union of Independent States). *Knihovnicka ZPN*, **2**, 189–223.

OSZCZYPKO, N. 2006. Late Jurassic–Miocene geodynamic evolution of the Outer Carpathian fold and thrust belt and its foredeep (Western Carpathians, Poland). *Geological Quarterly*, **50**, 169–194.

PETERS, K.E. 1986. Guidelines for evaluating petroleum source rocks using programmed pyrolysis. *AAPG Bulletin*, **70**, 318–329.

PETERS, K.E., WALTERS, C.C. & MOLDOWAN, J.M. 2007. *The Biomarker Guide, Biomarkers and Isotopes in Petroleum Exploration and Earth History, Volumes 1 & 2*. Cambridge University Press, New York.

PICHA, F.J. & STRANIK, Z. 1999. Late Cretaceous to early Miocene deposits of the Carpathian foreland basin in southern Moravia. *International Journal Earth Sciences*, **88**, 475–495.

PICHA, F.J., STRANIK, Z. & KREJCI, O. 2006. Geology and hydrocarbon resources of the Outer Western Carpathians and their foreland, Czech Republic. *In*: GOLONKA, J. & PICHA, F.J. (eds) *The Carpathians and Their Foreland: Geology and Hydrocarbon Resources*. American Association of Petroleum Geologists Memoirs, **84**, 49–175.

POPESCU, L.Gh. 2005. *Geological study of the Gura Soimului Formation of Vrancea Nappe (Moldova Valley–Tazlău Valley)*. Editura Sedcom Libris, Iasi [in Romanian].

POPESCU, L.Gh. & POPESCU, D.A. 2002. Geological study of the Gura Soimului Formation of the Bistrita–Râsca Half-window (Vrancea Nappe, East Carpathians). *Analele Universitatii 'Stefan cel Mare' Suceava Sectiunea Geografie*, **XI**, 1–9 [in Romanian].

POPOV, S.V. & STOLYAROV, A.S. 1996. Paleogeography and anoxic environments of the Oligocene–early Miocene Eastern Paratethys. *Israel Journal of Earth Sciences*, **5**, 161–167.

POPOV, S.V. & STUDENCKA, B. 2015. Brackish-water Solenovian Mollusks from the Lower Oligocene of the Polish Carpathians. *Paleontological Journal*, **49**, 342–355.

POPOV, S.V., AKHMETIEV, M.A., ZAPOROZHETS, N.I., VORONINA, A.A. & STOLYAROV, A.S. 1993*a*. Eastern Paratethys evolution during Late Eocene–Early Miocene. *Stratigraphy and Geological Correlation*, **6**, 10–39.

POPOV, S.V., VORONINA, A.A. & GONTSCHAROVA, I.A. 1993*b*. *Stratigraphy and bivalves of the Oligocene–Lower Miocene of the Eastern Paratethys*. Publications of the Paleontological Institute, **256**. Russian Academy of Sciences, Moscow [in Russian].

POPOV, S.V., BUGROVA, E.M. *ET AL.* 2004*a*. Biogeography of the Northern Peri-Tethys from the Late Eocene to the Early Miocene. Part 3. Late Oligocene–Early Miocene.

Marine Basins. *Paleontological Journal*, **38**, (Suppl. Series 6), S653–S716.

POPOV, S.V., RÖGL, F., ROZANOV, A.Y., STEININGER, F.F., SHCHERBA, I.G. & KOVAC, M. 2004*b*. *Lithological– Paleogeographic Maps of Paratethys: 10 Maps Late Eocene to Pliocene*. Courier Forschungsinstitut Senckenberg, Frankfurt, **250**.

POPOV, S.V., ANTIPOV, M.P., ZASTROZHNOV, A.S., KURINA, E.E. & PINCHUK, T.N. 2010. Sea-level fluctuations on the northern shelf of the Eastern Paratethys in the Oligocene–Neogene. *Stratigraphy and Geological Correlation*, **8**, 200–224.

ROBINSON, A.G., RUDAT, J.H., BANKS, C.J. & WILES, R.L.F. 1996. Petroleum geology of the Black Sea. *Marine and Petroleum Geology*, **13**, 195–223.

RÖGL, F. 1998. Palaeogeographic considerations for Mediterranean and Paratethys seaways (Oligocene to Miocene). *Annalen des Naturhistorischen Museums in Wien*, **99**, 279–310.

RÖGL, F. 1999. Mediterranean and Paratethys. Facts and hypotheses of an Oligocene to Miocene paleogeography (short overview). *Geologica Carpathica*, **50**, 339–349.

RUPP, C. & CORIC, S. 2012. Zur Ebelsberg-Formation [On the Ebelsberg Formation]. *Jahrbuch Geologische Bundesanstalt*, **152**, 67–100.

RUSU, A. 1999. Rupelian mollusk fauna of Solenovian type found in Eastern Carpathians (Romania). *Acta Palaeontologica Romaniae*, **2**, 449–452.

SACHSENHOFER, R.F. & SCHULZ, H.-M. 2006. Architecture of Lower Oligocene source rocks in the Alpine Foreland Basin: a model for syn- and post-depositional source-rock features in the Paratethyan realm. *Petroleum Geoscience*, **12**, 363–377, https://doi.org/10.1144/1354-079306-712

SACHSENHOFER, R.F., STUMMER, B., GEORGIEV, G., DELLMOUR, R., BECHTEL, A., GRATZER, R. & CORIC, S. 2009. Depositional environment and hydrocarbon source potential of the Oligocene Ruslar Formation (Kamchia Depression; Western Black Sea). *Marine and Petroleum Geology*, **26**, 57–84.

SACHSENHOFER, R.F., LEITNER, B. *ET AL.* 2010. Deposition, erosion and hydrocarbon source potential of the Oligocene Eggerding Formation (Molasse Basin, Austria). *Austrian Journal Earth Sciences*, **103**, 76–99.

SACHSENHOFER, R.F., HENTSCHKE, J. *ET AL.* 2015. Hydrocarbon potential and depositional environments of Oligo-Miocene rocks in the Eastern Carpathians (Vrancea Nappe, Romania). *Marine and Petroleum Geology*, **68**, 269–290.

SACHSENHOFER, R.F., POPOV, A.V. *ET AL.* 2017. The type section of the Maikop Group (Oligocene–Lower Miocene) at the Belaya River (North Caucasus): Depositional environment and hydrocarbon potential. *AAPG Bulletin*, **101**, 289–319.

SAINT-GERMÈS, M. 1998. *Ètude sédimentologique et géochimique de la matière organique du bassin Maykopien (Oligocène–Miocène Inférieur) de la Crimée a l'Azerbaidjan. Mémoires des Sciences de la Terre [Sedimentological and geochemical study of the organic matter of the Maykop Basin (Oligocene-Lower Miocene) from the Crimea to Azerbaijan]*. PhD thesis, Académie de Paris Université Pierre et Marie Curie, Paris.

SAINT-GERMÈS, M.L., BAZHENOVA, O.K., BAUDIN, F., ZAPOROZHETS, N.I. & FADEEVA, N.P. 2000. Organic

matter in Oligocene Maikop Sequence of the North Caucasus. *Lithology and Mineral Resources*, **35**, 47–62.

SCHULZ, H.-M., SACHSENHOFER, R.F., BECHTEL, A., POLESNY, H. & WAGNER, L. 2002. The origin of hydrocarbon source rocks in the Austrian Molasse Basin (Eocene–Oligocene transition). *Marine and Petroleum Geology*, **19**, 683–709.

SCHULZ, H.-M., BECHTEL, A., RAINER, T., SACHSENHOFER, R.F. & STRUCK, U. 2004. Paleoceanography of the western Central Paratethys during nannoplankton zone NP 23: The Dynow Marlstone in the Austrian Molasse Basin. *Geologica Carpathica*, **55**, 311–323.

SCHULZ, H.-M., BECHTEL, A. & SACHSENHOFER, R.F. 2005. The birth of the Paratethys during the early Oligocene: from Tethys to an ancient Black Sea analogue? *Global and Planetary Change*, **49**, 163–176.

SINNINGHE DAMSTÉ, J.S., VAN KOERT, E.R., KOCK-VAN DALEN, A.C., DE LEEUW, J.W. & & SCHENCK, P.A. 1989. Characterisation of highly branched isoprenoid thiophenes occurring in sediments and immature crude oils. *Organic Geochemistry*, **14**, 555–567.

STOLYAROV, A.S. & IVLEVA, E.I. 2004. Upper Oligocene Sediments of the Ciscaucasus, Volga–Don, and Mangyshlak regions (Central Eastern Paratethys): Communication 1. Main compositional and structural features. *Lithology and Mineral Resources*, **39**, 213–229.

STOLYAROV, A.S. & IVLEVA, E.I. 2006. Lower Miocene sdiments of Central and Western Ciscaucasia. *Lithology and Mineral Resources*, **41**, 174–186.

STUDENCKA, B., POPOV, S.V., BIEŃKOWSKA-WASILUK, M. & WASILUK, R. 2016. Oligocene bivalve faunas from the Silesian Nappe, Polish Outer Carpathians: evidence of the early history of the Paratethys. *Geological Quarterly*, **60**, 317–340.

SUMMONS, R.E. & POWELL, T.G. 1987. Identification of aryl isoprenoids in source rocks and crude oils: biological markers for the green sulphur bacteria. *Geochimica et Cosmochimica Acta*, **51**, 557–566.

SVABENICKA, L., BUBIK, M. & STRANIK, Z. 2007. Biostratigraphy and paleoenvironmental changes on the transition from the Menilite to Krosno lithofacies (Western Carpathians, Czech Republic). *Geologica Carpathica*, **58**, 237–262.

TABARA, D. & POPESCU, L. 2012. Palynological and palynofacies of Gura Soimului Formation from Bistrita-Rasca Half-Window (Eastern Carpathians, Romania). *Acta Palaeontologica Romaniae*, **8**, 23–31.

TARI, G., BÁLDI, T. & BÁLDI-BEKE, M. 1993. Paleogene retroarc flexural basin beneath the Neogene Pannonian Basin: a geodynamic model. *Tectonophysics*, **226**, 433–455.

TEN HAVEN, H.L., LAFARGUE, E. & KOTARBA, M. 1993. Oil/oil and oil/source rock correlations in the Carpathian Foredeep and the Carpathian Overthrust, South-East Poland. *Organic Geochemistry*, **20**, 935–959.

ULMISHEK, G.F. 2001. *Petroleum Geology and Resources of the Middle Caspian Basin, Former Soviet Union*. United States Geological Survey Bulletin, **2201-A**.

VAN BAAK, C. 2015. *Mediterranean–Paratethys Connectivity during Late Miocene to Recent. Unravelling Geodynamic and Paleoclimatic Causes of Sea-Level Change in Semi-Isolated Basins*. Utrecht Studies in Earth Sciences, **87**.

VETÖ, I. 1987. An Oligocene sink for organic carbon: upwelling in the Paratethys? *Palaeogeography, Palaeoclimatology, Palaeoecology*, **60**, 143–153.

VETÖ, I. & HERTELENDI, E. 1996. Sulphur isotope ratios in the laminated Tard Clay (Lower Oligocene of Hungary) reflect a salinity cycle. *Acta Geologica Hungarica*, **39**, (Suppl.), 204–207.

VETÖ, I. & HETÉNYI, M. 1991. Fate of organic carbon and reduced sulphur in dysoxic anoxic Oligocene facies of the Central Paratethys (Carpathian Mountains and Hungary). *In*: TYSON, R.V. & PEARSON, T.H. (eds) *Modern and Ancient Continental Shelf Anoxia*. Geological Society London, Special Publications, **58**, 449–460, https://doi.org/10.1144/GSL.SP.1991.058.01.28

VETÖ, I., HETÉNYI, M., DEMÉNY, A. & HERTELENDI, E. 1995. Hydrogen index, as reflecting intensity of sulfidic diagenesis in non-bioturbated, shaly sediments. *Organic Geochemistry*, **22**, 299–310.

VINCENT, S.J. & KAYE, M.N.D. 2017. Source-rock evaluation of Late Middle Eocene–Early Miocene mudstones from the NE margin of the eastern Black Sea. *In*: SIMMONS, M.D., TARI, G.C. & OKAY, A.I. (eds) *Petroleum Geology of the Black Sea*. Geological Society, London, Special Publications, **464**, https://doi.org/10.1144/SP464.7

VINCENT, S.J., MORTON, A.C., CARTER, A., GIBBS, S. & BARABADZE, T.G. 2007. Oligocene uplift of the Western Greater Caucasus; an effect of initial Arabia–Eurasia collision. *Terra Nova*, **19**, 160–166.

VINCENT, S.J., MORTON, A.C., HYDEN, F. & FANNING, M. 2013. Insights from petrography, mineralogy and U–Pb zircon geochronology into the provenance and reservoir potential of Cenozoic siliciclastic depositional systems supplying the northern margin of the Eastern Black Sea. *Marine and Petroleum Geology*, **45**, 331–348.

VOLKMAN, J.K. 1986. A review of sterol markers for marine and terrigenous organic matter. *Organic Geochemistry*, **9**, 83–99.

VORONINA, A.A. & POPOV, S.V. 1984. Solenovian horizon from Eastern Paratethys. *Bulletin of the Academy of Sciences of the USSR Geologic Series*, **9**, 41–53 [in Russian].

ZACHOS, J.C., PAGANI, M., SLOAN, L., BILLUPS, K. & THOMAS, E. 2001. Trends, rhythms, and aberrations in global climate 65 Ma to present. *Science*, **292**, 686–693.

ZAPOROZHETS, N.I. 1998. New data on Eocene and Oligocene phytostratigraphy of Severnyi Ergeni (the south of the Russian Platform). *Stratigraphy and Geological Correlation*, **6**, 262–279.

ZAPOROZHETS, N.I. 1999. Palynostratigraphy and dinocyst zonation of the Middle Eocene–Lower Miocene deposits at the Belaya River (Northern Caucasus). *Stratigraphy and Geological Correlation*, **7**, 161–178.

ZAPOROZHETS, N.I. & AKHMETIEV, M.A. 2015. Assemblages of organic-walled phytoplankton, pollen, and spores from the Solenovian Horizon (Lower Oligocene) of Western Eurasia. *Stratigraphy and Geological Correlation*, **23**, 326–350.

Source potential and depositional environment of Oligocene and Miocene rocks offshore Bulgaria

J. MAYER[1,2]*, B. J. RUPPRECHT[2], R. F. SACHSENHOFER[2], G. TARI[1], A. BECHTEL[2], S. CORIC[3], W. SIEDL[1], W. KOSI[1] & J. FLOODPAGE[4]

[1]*OMV Exploration and Production GmbH, Trabrennstraße 6–8, A-1020 Vienna, Austria*

[2]*Chair of Petroleum Geology, Montanuniversitaet Leoben, Peter-Tunner-Strasse 5, A-8700 Leoben, Austria*

[3]*Geological Survey of Austria, Neulinggasse 38, A-1030 Vienna, Austria*

[4]*Total E&P, Place Jean Millier, La Défense 6, Paris La Défense Cedex, 92078, France*

*Correspondence: jan.mayer@omv.com

Abstract: Oligo-Miocene ('Maikopian') deposits are considered the main source rocks in the Black Sea area, although only a few source-rock data are available. Geochemical logs from nine wells are used together with age constraints provided by calcareous nannoplankton, well and seismic data to determine vertical and lateral changes of the source potential. Oligocene rocks overlie Eocene deposits with a major unconformity on the western Black Sea shelf in Bulgaria. A west–east-trending erosional structure (the Kaliakra canyon) developed during Lower Oligocene time and was filled with Oligo-Miocene deposits. Potential source rocks are present in different stratigraphic units, but the most prolific intervals accumulated during time intervals when the isolation of the Parathethys resulted in oxygen-depleted, brackish environments with high bioproductivity. These include Lower Solenovian rocks related to blooms of calcareous nannoplankton, which form an extensive layer outside the Kaliakra canyon. This unit hosts a good potential to generate oil and gas. Diatom-rich, very good oil-prone source rocks accumulated during a second isolation event in the Kozakhurian. Thick sections of these diatom-rich rocks occur within the canyon and are present in thin layers outside of it. High productivity of siliceous organisms is attributed to upwelling within the canyon. All studied units are thermally immature on the shelf.

The hydrocarbon prospectivity of the Black Sea depends, amongst other factors, on the distribution and quality of source rocks. Although several source-rock horizons may be present, Oligocene–Lower Miocene rocks are generally considered to be the most important ones. Traditionally, fine-grained Oligocene–Lower Miocene rocks are attributed to the Maikop Group (e.g. Popov *et al.* 1993).

Few data showing the source-rock potential of the Maikop Group in the Black Sea area are currently available. Amongst these are the total organic carbon (TOC) contents and the hydrogen index (HI) values from rocks at the northern Black Sea coast (Crimea: Fig. 1a), which indicate a low petroleum potential (Saint-Germès 1998; Sachsenhofer *et al.* 2017). However, better source rocks have been encountered in the western Black Sea offshore Bulgaria: for example, in the well Samotino More (Fig. 1b) (Sachsenhofer *et al.* 2009).

Results from Sachsenhofer *et al.* (2009) and data from eight additional wells, representing a more than 130 km-long segment of the western Black

Sea shelf (Fig. 1), have been investigated. In addition, 2D and 3D seismic data significantly improved the understanding of Oligocene and Lower Miocene intervals.

The main aims of the present study are to reveal the vertical and lateral variability of the Maikop Group on the western Black Sea shelf, to establish a depositional model and to quantify its source potential.

To determine specific age intervals, a large number of samples were investigated for calcareous nannoplankton. These data provide excellent age constrains for Eocene–Solenovian (NP14–NP23) and Middle Miocene (NN5–NN6) rocks. In contrast, calcareous nannoplankton is largely missing or very rare in Upper Oligocene–Lower Miocene rocks (Fig. 2). Erosional unconformities visible in seismic lines have been correlated with major sea-level drops in the Eastern Parathethys (Eocene–Oligocene boundary, Middle Solenovian and Late Kozakhurian: Popov *et al.* 2010) (Fig. 2) providing additional age constraints. Furthermore, salinity

From: SIMMONS, M. D., TARI, G. C. & OKAY, A. I. (eds) 2018. *Petroleum Geology of the Black Sea.*
Geological Society, London, Special Publications, **464**, 307–328.
First published online February 25, 2017, https://doi.org/10.1144/SP464.2

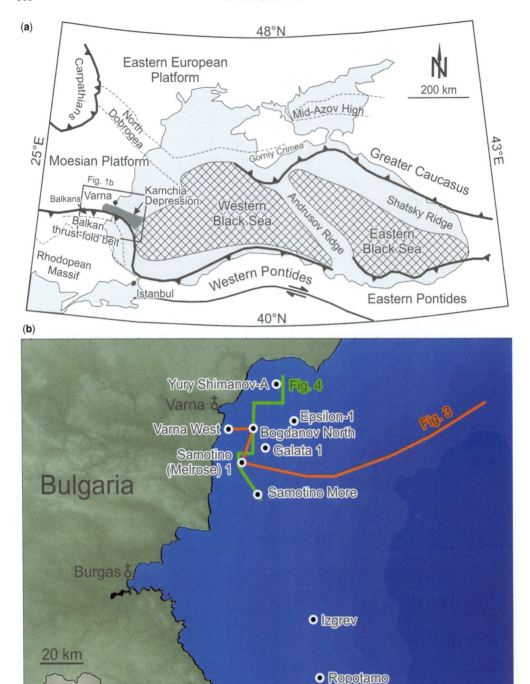

Fig. 1. (a) Overview of the Black Sea and study area outlines. (b) The locations of sampled wells and seismic lines shown in Figures 3 and 4.

proxies (e.g. TOC/S ratios: Berner 1984) have been used together with information on salinity variations in the Eastern Paratethys (e.g. Popov *et al.* 2010) (Fig. 2) to further confine the age of the sediments. Nevertheless, ages of Upper Solenovian–Tarkhanian units remain debatable.

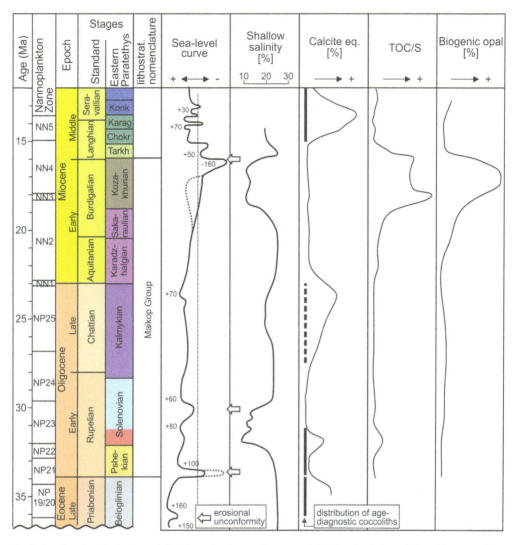

Fig. 2. Stratigraphy (Gradstein *et al.* 2012), sea-level curve (Popov *et al.* 2010) and salinity variations (Popov *et al.* 2001) in the Eastern Paratethys. Geochemical logs are based on the results from this study. The stratigraphic colour code is applied throughout the paper. The age assignment of studied rocks is based on calcareous nannoplankton, the correlation of erosional unconformities with the sea-level curve, and the correlation between salinity proxies (e.g. TOC/S ratios) and evolution of shallow salinity.

Geological overview

The study area is located at the junction of the Moesian Platform, the Balkan thrust-fold belt and the Western Black Sea Basin (Fig. 1b). The Moesian Platform forms the foreland of the Carpathians in the north and the Balkan thrust-fold belt in the south. Its sedimentary cover comprises three main structural sequences: (1) Palaeozoic, (2) Permo-Triassic and (3) Jurassic–Cenozoic, separated by Hercynian and Cimmerian unconformities (Tari *et al.* 1997).

The east–west-trending Balkan thrust-fold belt represents a segment of the Alpine orogen (Boncev 1986). It consists of a stack of dominantly north-verging thrust sheets that developed during multiphase collisional events along a long-lived convergent continental margin. Compression culminated toward the end of the Early Cretaceous and in the Early Middle Eocene (Emery & Georgiev 1993). The Balkan orogen is made up by a southern uplifted overthrust zone (Balkan or Stara Plania) and a north-ern subsided forebalkan thrust-fold zone (Georgiev 2012). The forebalkan zone subsides beneath the

Cenozoic fill of the Kamchia Depression. Offshore, the East Balkan orogen first turns towards the SE, then shifts considerably to the south (Georgiev 2012) (Fig. 1b).

The Black Sea is considered by many authors to be a Late Cretaceous–Palaeogene back-arc extensional basin that developed north of the Pontide magmatic arc, which formed by northwards subduction of the Neo-Tethys Ocean, initiated in the Albian (Tugolesov *et al.* 1985; Finetti *et al.* 1988; Gorur 1988; Okay *et al.* 1994; Dachev & Georgiev 1995; Robinson *et al.* 1995; Banks & Robinson 1997; Nikishin *et al.* 2001, 2003). According to Nikishin *et al.* (2015), the main regional rifting event is Late Barremian–Albian in age and oceanic crust was formed from Cenomanian to Mid-Santonian time.

During Cenozoic time, depositional environments in the western Black Sea area, including the Kamchia Depression, were strongly influenced by the separation of the Paratethys Sea from the Mediterranean Sea. Basin separation was a result of strong tectonic activity (Alpine Orogeny) and a major sea-level drop at the Eocene–Oligocene boundary (Fig. 2) (Zachos *et al.* 2001; Popov *et al.* 2010). It favoured the development of endemic faunas, which hamper correlations with standard time stages. Time stages used in the Eastern Paratethys are shown in Figure 2.

Basin isolation culminated during Early Solenovian time (early NP23) when brackish-water conditions prevailed in the entire Paratethys ('Solenovian Event': e.g. Voronina & Popov 1984; Rögl 1999; Nagymarosy & Voronina 1993; Schulz *et al.* 2004; Popov & Studencka 2015). During this event, marlstones (Dynow Marlstone, Polbian Bed and Ostracoda Bed) were deposited. Today, these beds form a Paratethys-wide marker horizon. The Eastern Paratethys experienced a second phase of basin isolation and strongly reduced salinity in Early Miocene (Kozakhurian) time (e.g. Popov *et al.* 1993). Apart from salinity variations, basin isolation also increased the development of anoxia within the water column (Popov *et al.* 2004a, b; Schulz *et al.* 2005).

Materials and methodology

Owing to the small number of core samples, this paper is based mainly on more than 550 cuttings samples from nine wells (for the locations see Fig. 1b). The spacing between samples varies, but typically ranges from 3 to 12 m. Sample depths are provided in metres measured depth below rotary table (mdbrt).

All samples have been analysed in duplicate for total sulphur (S), total carbon (TC) and total organic carbon (after removal of carbonate minerals: TOC)

using a Leco CS-300 instrument. Assuming calcite to be the only carbonate mineral present, the calcite equivalent percentage was calculated ($calcite_{eq} = (TC - TOC) \times 8.333$). Pyrolysis measurements were carried out in duplicate using a 'Rock-Eval 2+' instrument. The S1 and S2 peaks (mg HC/g rock) were used to calculate the hydrogen index ($HI = S2 \times 100/TOC$ (mg HC/g TOC)) and the production index $PI = S1/(S1 + S2)$: Espitalié *et al.* 1977). T_{max} was measured as a maturity indicator.

For microscopical analysis, polished blocks of selected samples (chosen according to TOC content) were prepared. A LEICA microscope and a point-counting approach (1000 points counted) were used to determine maceral group percentages (vol%) and pyrite semi-quantitatively. The same instrument was used to determine vitrinite reflectance following established procedures (Taylor *et al.* 1998).

Semi-quantitative investigations of calcareous nannoplankton were performed on smear slides, which were prepared using standard techniques. Before preparation, small amounts of sediment were treated by ultrasound in distilled water for a few seconds. Smear slides were analysed with a light microscope (Leica DMLP microscope) under ×1000 magnification (cross and parallel nicols).

The biogenic silica content was quantified following a method described by Zolitschka (1998): 50 mg of dried powdered sample were boiled in 50 ml of 0.5 M KOH for 1 h and left to settle afterwards. Analysis was performed on a Perkin-Elmer 3030 atom-absorption-spectrophotometer. Biogenic opal percentages were calculated (=biogenic silica × 2.4: Pavicevic & Amthauser 2000).

For organic geochemical analyses, representative portions of selected samples were extracted for approximately 1 h using dichloromethane in a Dionex ASE 200 accelerated solvent extractor at 75°C and 50 bars. After evaporation of the solvent to 0.5 ml total solution in a Zymark TurboVap 500 closed-cell concentrator, asphalthenes were precipitated from a hexanedichloromethane solution (80:1) and separated by centrifugation. The fractions of the hexane soluble organic matter were separated into saturated hydrocarbons and aromatic hydrocarbons using medium-pressure liquid chromatography with a Köhnen-Willsch MPLC instrument (Radke *et al.* 1980).

The saturated and aromatic hydrocarbon fractions were analysed with a gas chromatograph equipped with a 30 m DB-1 fused silica capillary column (i.d. 0.25 mm; 0.25 µm film thickness) coupled to a Finnigan MAT GCQ ion trap mass spectrometer. The oven temperature was programmed from 70 to 300°C at a rate of 4°C min^{-1} followed by an isothermal period of 15 min. Helium was used as the carrier gas. The sample was injected splitless with the injector temperature at 275°C. The mass spectrometer was operated in the EI (electron ionization) mode over a mass range from m/z 50 to m/z 650 (0.7 s total scan time). Data were processed with a Finnigan data system. Identification of individual compounds was accomplished on the basis of retention times in the total ion current chromatogram and comparison of mass spectra with published data. Relative percentages and absolute concentrations of different compound groups in the saturated and aromatic hydrocarbon fractions were calculated using peak areas from the gas chromatograms in relation to those of internal standards (deuteriated n-tetracosane and 1,1'-binaphthyl, respectively). The concentrations were normalized to TOC.

The following biomarkers were selected to infer composition and depositional environment of the sampled organic matter:

- Di- and triterpenoids – di- and triterpenoids are biomarker derived from gymnosperms and angiosperms, respectively. Their sum has been used as proxy for the input of terrestrial organic matter. The ration of di-/(di- and triterpenoids) shows the relative abundance of gymnosperms within the land plant input.
- Pristane/phytane – pristane/phytane (Pr/Ph) ratios below 1.0 indicate anaerobic conditions during early diagenesis, and values between 1.0 and 3.0 were interpreted as reflecting dysaerobic environments (Didyk et al. 1978). However, this ratio can be influenced by hypersaline conditions during deposition (ten Haven et al. 1987) or by additional sources (e.g. archaebacterial: Volkman & Maxwell 1986; Texidor et al. 1993).

Pyrolysis gas chromatography was performed on six samples using the Quantum MSSV-2 Thermal Analysis System© in the laboratory of GeoS4 (Michendorf, Germany). The thermally extracted (300°C, 10 min) sample was heated in a flow of helium. Products released over the temperature range 300–600°C (40 K min^{-1}) were focused using a cryogenic trap. The trap was induction-heated ballistically from −196 to 300°C and held at that temperature for the duration of the GC analysis. Pyrolysis products were analyxed using a 50 m × 0.32 mm BP-1 capillary column equipped with a flame ionization detector. The GC oven temperature was programmed from 40 to 320°C at 8°C min^{-1}. Boiling ranges (C_1, C_{2-5}, C_{6-14} and C_{15+}) and individual compounds (n-alkenes, alkylaromatic hydrocarbons, alkylthiophenes) were quantified by external standardization using n-butane. Response factors for all compounds were assumed to be the same, except for methane, whose response factor was 1.1.

Results

Seismic data

The Eocene–Oligocene boundary is a major unconformity on the Black Sea shelf, but erosional features are missing in the deeper parts of the basin (Fig. 3). Furthermore, a significant thickening of Oligocene sediments, onlapping onto Eocene sediments, towards the distal part of the western Black Sea is recognizable. Severe mass-transport complexes within the basin were triggered by the Late Kozakhurian sea-level drop. These mass-transport complexes are clearly visible on 3D seismic surveys available to the authors. A nearly constant thickness of Miocene and Pliocene sediments on the shelf, as well as within the basin, indicates sedimentary bypass.

A roughly north–south-trending composite seismic line along the western Black Sea shelf is shown in Figure 4. Main features relevant for the present paper are: (1) an erosional unconformity at the Eocene–Oligocene boundary (lower stippled line in Fig. 4); (2) an erosional surface, which forms the base of the Kaliakra shelf-break canyon and cuts deeply into Oligocene and Eocene rocks; and (3) another erosional unconformity within the fill of the Kaliakra canyon. These unconformities are related to sea-level drops during Early Pshekian, Late Solenovian and Late Kozakhurian times (see Fig. 2).

A continuous, prominent reflector can be traced from Yury Shimanov-A to Samotino More and is eroded along the canyon axis. It represents the boundary between carbonate-free mudstones and underlying marls. Calcareous nannoplankton dates the reflector as lower NP23 (top of the Lower Solenovian).

Stratigraphy and lithology of Oligo-Miocene rocks. Nannoplankton ages are shown together with depth trends of carbonate, biogenic opal and TOC contents, HI values, TOC/S ratios, and selected biomarkers in Figures 5–7. Oligo-Miocene rocks outside (and underneath) the Kaliakra canyon are discussed separately from the canyon fill.

Oligo-Miocene rocks outside the Kaliakra canyon. Rocks in wells Epsilon 1 and Izgrev contain well-preserved nannoplankton assemblages, with *Coccolithus formosus*, *Isthmolithus recurvus*, *Reticulofenestra umbilicus* and *Chiasmolithus oamaruensis* dating them as Pshekian (NP21: Figs 5–7). Similar ages are likely for samples between 948 and 1014 m depth in the Ropotamo well. Rocks with varying, but typically low, carbonate contents (max. 25%) in-between the Eocene–Oligocene unconformity and carbonatic rocks of Early Solenovian age are also attributed to the Pshekian regiostage. Pshekian

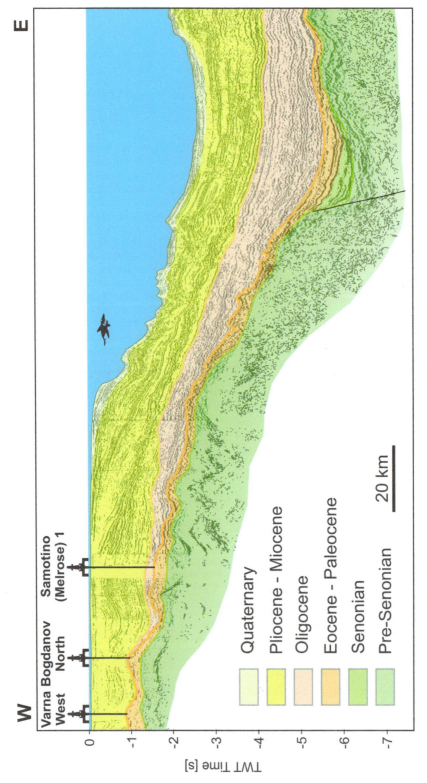

Fig. 3. Line drawing of a seismic composite line with interpretation. Note the increasing thickness of Oligocene and Eocene sediments towards the distal parts of the Western Black Sea Basin. Onlapping of Oligocene sediments is recognizable. All sampled wells are located on the palaeoshelf. TWT Time, two-way travel time.

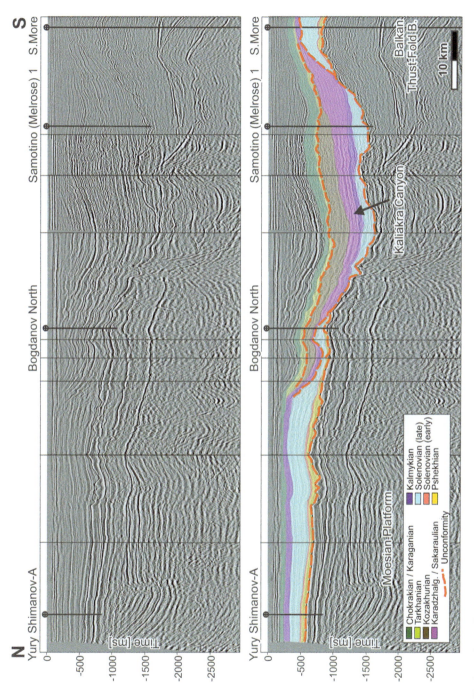

Fig. 4. North–south-trending seismic composite line. Three major erosional events are recognizable. They occurred at the Eocene–Oligocene boundary, in the middle Late Solenovian and in the Late Kozakhurian. Note the different seismic characters of the individual time periods (e.g. chaotic reflectors within the Kozakhurian section). S. More, Samotino More. Colours correspond to the colour scheme introduced in Figure 2.

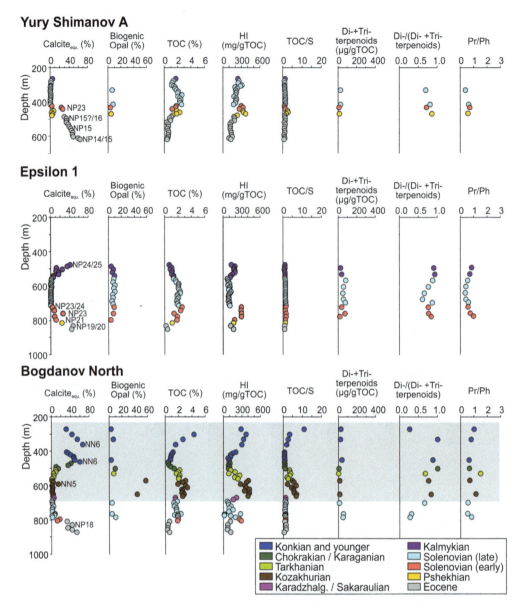

Fig. 5. Bulk geochemical data and selected biomarker parameters of wells in the northern part of the study area. Nannoplankton zones are labelled. The fill of the Kaliakra canyon is highlighted by grey shading.

rocks have been drilled in most wells, but have been eroded along the axis of the Kaliakra canyon (e.g. Samotino (Melrose) 1).

The overlying Solenovian succession is subdivided into rocks with partly high carbonate contents (Lower Solenovian) and largely carbonate-free rocks (Upper Solenovian). Lower Solenovian rocks are often rich in calcareous nannoplankton including *Reticulofenestra ornata*, *R. bisecta*, *Pontosphaera pax* and *P. multipora*. Samples from boreholes Yury

Shimanov-A, Epsilon 1, Varna West, Samotino More and Ropotamo record monospecific blooms of *Reticulofenstra ornata*. These samples represent the brackish-water 'Solenovian Event' dated as lower NP23m (Figs 5–7).

Sediments with upwards-increasing carbonate contents, overlying carbonate-free Upper Solenovian shales, have been dated as NP24-NP25 (Upper Solenovian-Kalmykian) by Sachsenhofer *et al.* (2009).

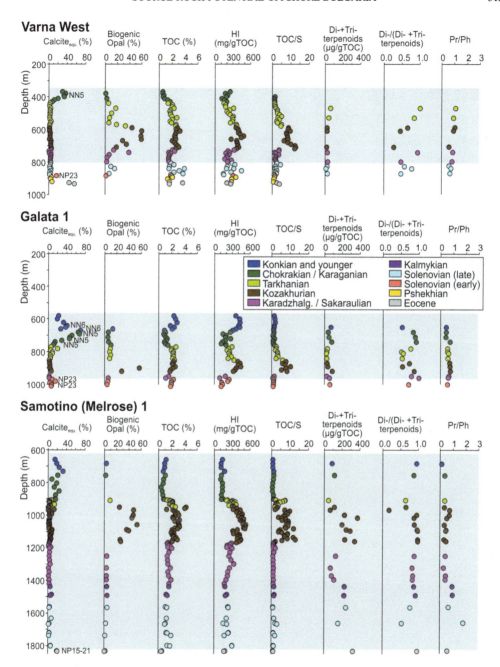

Fig. 6. Bulk geochemical data and selected biomarker parameters of wells in the central part of the study area. Data from Samotino More are partly from Sachsenhofer *et al.* (2009). The fill of the Kaliakra canyon is highlighted by grey shading.

Miocene rocks, outside of the Kaliakra canyon, have been sampled only in boreholes Izgrev and Ropotamo (Fig. 6). They overlie unconformably relatively thin Lower Oligocene deposits and comprise diatomaceous shales with elevated TOC/ S ratios. In analogy to the fill of the Kaliakra canyon (see below), these rocks have been attributed to the Lower Miocene (Kozakhurian, Tarkhanian).

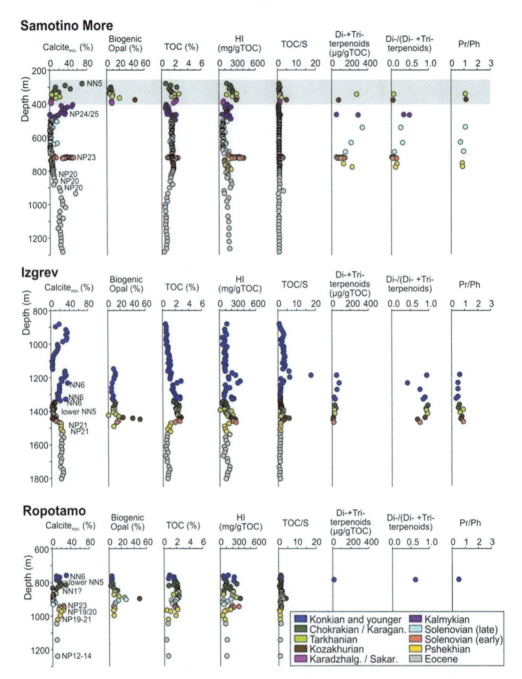

Fig. 7. Bulk geochemical data and selected biomarker parameters of wells in the southern part of the study area. Data from Samotino More are partly from Sachsenhofer *et al.* (2009). The fill of the Kaliakra canyon is highlighted by grey shading.

Overlying carbonate-rich rocks have been dated as lower NN5 (Chokrakian–Konkian) and NN6 (Konkian and younger).

Kaliakra canyon fill. Seismic data show that the uppermost part of the drilled succession in well Samotino More represents the southern part of the

Kaliakra canyon fill. Data from Sachsenhofer *et al.* (2009) were used and assigned with new ages. The canyon fill is largely carbonate-free. Nannoplankton is rare and often contains redeposited material. Considering the presumed Late Solenovian incision age of the canyon, lowermost rocks along the canyon axis, drilled in Samotino (Melrose) 1, are attributed to the Upper Solenovian and Kalmykian. Rocks with a presumed Early Miocene (Karadzhalgian and Sakaraulian) age are present along the northern canyon margin.

A prominent diatomaceous layer with biogenic opal contents up to 60% follows above the Sakaraulian sediments. X-ray diffractograms of samples from the wells Bogdanov North and Samotino (Melrose) 1 show that the primary biogenic opal-A has been transformed into microcrystalline opal-CT. Based on their position beneath the Late Kozakhurian erosional unconformity (Fig. 4), the sediments are attributed to the Kozakhurian regiostage. This age assignment is supported by high TOC/S ratios, indicating a brackish-water environment due to an isolation of the Eastern Paratethys from the Mediterranean during Kozakhurian time. The largely carbonate-free sediments contain a single sample, which yielded a NN5 age. However, this age assignment is probably biased due to caving.

Rocks above the Kozakhurian unconformity are characterized by upwards-decreasing TOC/S ratios and opal contents. As they occur beneath rocks dated as NN5, they are attributed to the Tarkhanian regiostage. Rocks dated as NN5 (Karaganian–Chokrakian) are often characterized by upwards-increasing carbonate contents, which reach maxima near the NN5–NN6 boundary. Konkian (and younger) sediments contain often high and strongly varying carbonate contents. In several wells, the lower part of the Konkian succession is characterized by an upwards decrease in carbonate contents.

Thermal maturity

A detailed investigation of maturity trends is beyond the scope of the present paper. However, in order to test whether geochemical parameters are influenced by thermal maturity, the most important results are summarized here.

T_{max}, vitrinite reflectance and the ratio of $22S/(22S + 22R)$ isomers of the 17α, 21β (H) C_{31} hopane (Seifert & Moldowan 1980) are used as maturity indicators. Average T_{max} values of Eocene–Miocene units in each well are below 430°C. Vitrinite reflectance data are consistently below 0.5%, and the $S/(S + R)$ ratios of C_{31} hopane are below 0.3 and exceed 0.4 only in contaminated samples, often from shallow depth. These results show that the sampled organic matter is thermally immature.

Geochemical parameters

Geochemical logs (Figs 5–7) are discussed combined for the individual time intervals. TOC, HI, TOC/S ratio and pristane/phytane ratios for the individual time periods, as well as for wells, are shown in Table 1. Wells Yury Shimanov-A and Epsilon 1 are located north of the Kaliakra canyon. Wells Bogdanov North, Varna West, Galata 1, Samotino (Melrose) 1 and Samotino More drilled the fill of the Kaliakra canyon, whereas Izgrev and Ropotamo are located south of the canyon.

Pshekian. Pshekian sediments contain a relatively large amount of kerogen type II in the north. The amount of kerogen type III increases towards the south, and reaches a maximum in wells Izgrev and Ropotamo. The average TOC contents range between 1 and 1.8%, and vary in-between wells with no recognizable trend. Carbonate contents are generally low in the north (below 10%) except for a single sample from well Epsilon 1 (22%). Samples from the Izgrev and Ropotamo wells show carbonate contents of 10–20%. TOC/S ratios are generally low.

Early Solenovian. Sediments of Early Solenovian age are characterized by abundant type II (–III) kerogen and high carbonate contents (up to 50%). Average TOC contents are similar or slightly higher than in the underlying Pshekian interval. TOC/S ratios are also within similar ranges to those of the Pshekian sediments. Lower Solenovian rocks directly overlie the Eocene–Oligocene unconformity in wells Bogdanov North and Galata-1.

Late Solenovian. Sediments of Late Solenovian age contain mainly type III kerogen and show similar TOC contents compared with Early Solenovian rocks. The TOC contents of Upper Solenovian rocks decrease with time. TOC/s ratios, as well as carbonate contents, are generally low within this interval. Moderate biogenic opal contents (up to 15%) were measured in samples from wells Varna West and Ropotamo.

Kalmykian. Kalmykian sediments show lower to slightly lower TOC contents than underlying Upper Solenovian samples and decrease further with time. Slightly elevated HI values (kerogen type III), compared to underlying Upper Solenovian sediments, were measured. TOC/S ratios are low and similar to the underlying sediments. Carbonate contents within the Kalmykian section are high (up to 50%).

Karadzhalgian and Sakaraulian. Karadzhalgian and Sakaraulian sediments form the base of the erosional canyon in wells Bogdanov North, Varna West and Galata-1, and contain moderately large amounts of

Table 1. *TOC and HI values combined with TOC/S and pristane/phytane ratios for different stratigraphic units and individual wells*

		Yury Shimanov-A	Epsilon-1	Bogdanov-North	Varna West	Galata 1	Samotino (Melrose) 1	Samotino More	Izgrev	Ropotamo
Konkian and younger	TOC			2.7		2.1	0.9			1.2
	HI			190		390	130			130
	TOC/S			2.02		1.38	0.79			1.06
	Pr/Ph			0.81		0.44	0.12			0.48
Chokrakian–Karaganian	TOC			0.8	0.88	1.06	0.7	1.4		1.8
	HI			90	180	190	130	190		140
	TOC/S			0.8	1.13	1.02	0.85	0.89		1.1
	Pr/Ph			0.65		0.53	0.35			
Tarkhanian	TOC			2	2	2.2	2	2	2.6	2
	HI			170	260	250	290	230	140	160
	TOC/S			2.97	2.9	2.69	3.56	1.48	1.31	1.69
	Pr/Ph			1.48	1	0.57	0.36	1.09		
Kozakhurian	TOC			2.6	2.6	2.2	2.6	2.2	2.6	2
	HI			370	390	380	430	290	210	320
	TOC/S			5.26	9.2	7.31	6.96	4.7	4.24	3.72
	Pr/Ph			0.91	0.94	0.73	0.47	1.14	0.72	
Karadzhalgian–Sakaraulian	TOC			1.5	1.7	1.35	1.6	2		
	HI			180	230	140	250	70		
	TOC/S			1.15	2.26	0.95	1.1	1.13		
	Pr/Ph				0.85	0.63	0.38			
Kalmykian	TOC	1.7	1				1.4	1.2		
	HI	240	160				200	130		
	TOC/S	1.1	0.93				1.31	0.72		
	Pr/Ph		0.78				0.96	0.5		
Late Solenovian	TOC	1.9	1.9	1.7	2.4	2	1.3	1.5		1.5
	HI	220	90	100	260	170	230	70		80
	TOC/S	1.07	1.18	1	2.07	0.92	1.36	0.79		1.14
	Pr/Ph	0.39	0.46	0.63	0.64	0.54	1.24	0.92		
Early Solenovian	TOC	1.5	2	1.8	1.9			1.4	2.2	1.7
	HI	300	270	270	310			290	240	230
	TOC/S	0.94	1.36	1.2	1.25			1.89	1.8	1.23
	Pr/Ph	0.7	0.76						0.88	
Pshekhian	TOC	1.8	1.1		1.7			1.7	1	1.3
	HI	310	170		300			130	100	100
	TOC/S	1.96	1.14		0.84			0.96	1.04	0.74
	Pr/Ph	0.58						0.86		

Please note that the shallowest interval (Konkian and younger) might include contaminated samples.

type III kerogen. TOC/S ratios are similar to underlying sediments and carbonate contents are low.

Kozakhurian. An abrupt increase in TOC/S ratios, TOC contents and HI values marks the base of the Kozakhurian unit, which is characterized by large amounts of kerogen type II and very high biogenic opal contents (up to 87%). TOC/S ratios reach a maxima within this interval (up to 12.8%), which is due to the increased TOC as well as a decrease in sulphur content. A major stratigraphic gap separates Lower Oligocene and Lower Miocene rocks in the Izgrev and Ropotamo wells. Sediments overlying this gap have been attributed to the Kozakhurian regiostage based on high biogenic opal contents and high TOC/S ratios.

Tarkhanian. TOC/S ratios, as well as the biogenic opal contents, of Tarkhanian sediments are lower than in the Kozakhurian succession. TOC contents, HI values and TOC/S ratios gradually decrease upwards in the Tarkhanian section. Tarkhanian rocks comprise mainly type III kerogen and type III–II kerogen in well Samotino (Melrose) 1.

Chokrakian and Karaganian. Sediments from this regiostage contain low to minor amounts of type III kerogen and are characterized by a strong decrease in organic matter in wells located within the Kaliakra canyon. TOC/S ratios are low within all samples. Carbonate contents range from 69% in well Samotino-More down to 1.5% in well Izgrev, and show a high variability within this interval.

An overall decrease of carbonate content towards the south can be observed.

Konkian and younger. Konkian and younger sediments display highly variable carbonate contents and abundant type III to type II (Galata 1) kerogen. The lowest source-rock quality was encountered in well Samotino (Melrose) 1. In well Bogdanov North, S2 peaks with two maxima and high PI values indicate contaminations. Therefore, these data are not further interpreted.

Biomarker data. Pr/Ph ratios for the overall Oligocene and Miocene section are below 1, indicating that anoxic conditions prevailed during the Oligicene and Miocene. Pr/Ph ratios above 1 were only encountered in the Late Solenovian in well Samotino (Melrose) 1 (most likely to have been associated with the major sea-level drop during this time), and in well Samotino More in the Kozakhurian and Tarkhanian. Concentrations of land-plant-derived biomarkers are generally low. Highest terrestrial input occurred during the Late Solenovian and in the Kozakhurian along the Kaliakra canyon axis (Samotino (Melrose) 1). The di-/(di- and triterpenoids) ratio indicates an overall increase in gymnosperm contribution with time.

Microscopy

Semi-quantitative maceral analysis was performed on selected samples (Fig. 8). Framboidal pyrite, sporinite, vitrinite and inertinite can be recognized in

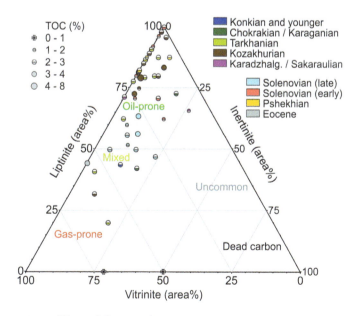

Fig. 8. Maceral percentages of Eocene–Miocene rocks.

most samples. Vitrinite is generally very fine grained. Solid bitumen, interpreted as contamination, was recognized in shallow samples from well Bogdanov North. Furthermore, calcareous shell fragments were found within these samples. Large amounts of diatoms and minor amounts of siliceous sponge needles are observed in shallow Kozakhurian and Tarkhanian sediments.

Maceral percentages show a clear stratigraphic trend. Samples with Eocene, Late Solenovian and Konkian ages typically contain high amounts of terrestrial macerals (vitrinite and inertinite) and plot into the fields for gas- and gas/oil-prone source rocks. In contrast, Pshekian, Lower Solenovian and Kozakhurian deposits contain high amounts of liptinite and plot in areas characteristic for oil-prone rocks. Tarkhanian sediments show mixed percentages of macerals.

Pyrolysis-GC

Twenty-four samples representing different stratigraphic units and a wide range of TOC and HI values were selected for Py-GC. The pyrolysis products contain variable amounts of normal hydrocarbon doublets of alkenes and alkanes extending to large chain lengths. Aromatic hydrocarbons including benzene, toluene and *m*, *p*-oxylenes, and sulphur-bearing compounds are very abundant in the chromatograms. The petroleum-type organofacies has been determined using the ternary diagrams of Horsfield (1989) and Eglinton *et al.* (1990) (Fig. 9). Accordingly, Upper Solenovian samples from well Epsilon 1 and a Tarkhanian sample from well Izgrev produce gas and condensate, while most samples produce paraffinic-naphthenic-aromatic oils with low wax content (Fig. 9a). Kozakhurian samples with high HI values (wells Samotino (Melrose) 1, Bogdanov North) plot into or close to the field of high wax paraffinic oil. In accordance with RockEval data, most samples plot into the type III aromatic or the type II intermediate field in a plot after Eglinton *et al.* (1990). Some sulphur-rich samples plot into the type IIS field (Fig. 9b).

Interpretation

T_{max} v. HI plots for the entire sample set as well as for different stratigraphic units are shown in Figure 10. Cartoons depicting the depositional environments for the different time intervals are presented in Figure 11. Geochemical parameters are summarized in Tables 1 and 2. The applied maturity parameters show a very low thermal overprint of the organic matter and prove that bulk parameters (e.g. HI) are not influenced by maturation.

In the following text main geological events are discussed for each regiostage. Within this context,

it has to be emphasized that the results are based nearly exclusively on data from cuttings. This implies restrictions concerning vertical resolution (3–12 m) and problems like caving.

Depositional environment

Pshekian. A sea-level drop at the Eocene–Oligocene boundary (Zachos *et al.* 2001; Popov *et al.* 2010) (Fig. 11) led to major erosion in the northern part of the study area (e.g. Yury Shimanov A), where Pshe kian sediments rest directly on Middle Eocene (NP16) rocks. Upper Eocene deposits (NP20) are preserved in the southern part of the study area (e.g. Samotino More). The following transgression during the Pshekian resulted in the deposition of rocks with varying and partly moderately high carbonate contents.

TOC and HI values (1.5% TOC; HI 160 mg HC/g TOC) show the presence of moderately large amounts of organic matter, classified as kerogen type III. However, oil-prone type II kerogen with HI values up to 370 mg HC/g TOC (Fig. 10) occurs in Yury Shimanov A in the northern part of the study area.

Biomarker data are characterized by low Pr/Ph ratios, suggesting oxygen-depleted conditions. Land-plant-derived biomarkers suggest larger amounts of terrestrial organic matter and a dominance of angiosperms in the Samotino area. Differences in land plant contribution are also reflected by in varying HI values.

Early Solenovian. The basin isolation of the Paratethys resulted in the development of a peculiar endemic fauna and phytoplankton (Solenovian Event: Voronina & Popov 1984; Čtyroký 1991; Rusu 1999 and others). This event is reflected by the deposition of calcareous rocks rich in calcareous nannoplankton (blooms of *Reticulofenestra ornate; Pontosphaera pax*). Their organic matter contents are similar to those in the underlying Pshekian unit, but the average HI is significantly higher (270 mg HC/g TOC) and can reach up to 410 mg HC/g TOC. High HI values reflect a dominance of aquatic organisms, which is supported by high liptinite contents (Fig. 8) and low concentrations of land-plant biomarkers. Low Pr/Ph ratios (mean 0.7) indicate anoxic conditions during deposition. Low salinity is proven by nannoplankton assemblages.

Late Solenovian. The Late Solenovian time is characterized by a gradual increase in salinity (e.g. Popov *et al.* 1993; Sachsenhofer *et al.* 2017) and a prominent sea-level drop during the middle Late Solenovian (Popov *et al.* 2010) (Fig. 10). The sea-level drop led to the incision of the west–east-trending Kaliakra canyon, clearly visible on seismic sections (e.g. Fig. 4). The sea-level drop is followed

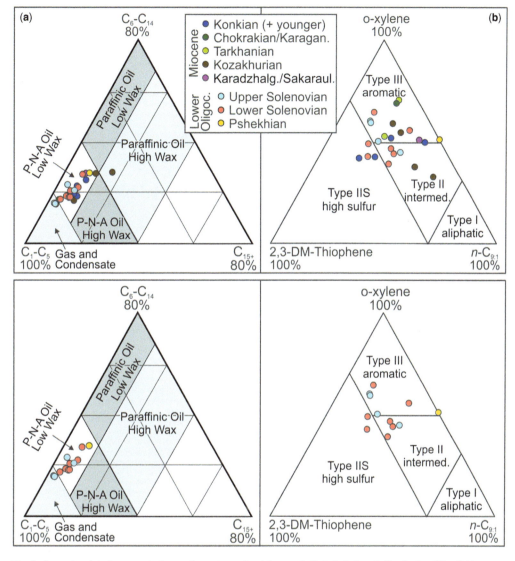

Fig. 9. Screening data from pyrolysis gas chromatography: (**a**) normal alkyl chain length distribution (Horsfield 1989, 1997); and (**b**) sulfur discriminator (Eglinton *et al.* 1990). P-N-A, paraffinic-naphthenic-aromatic.

by a transgression, still within Late Solenovian time. The basal canyon fill, comprising conglomerates and sandstones, is attributed to end Late Solenovian.

Upper Solenovian rocks are largely carbonate-free. They contain, on average, 1.7% TOC, and TOC contents can reach up to 3.9%. An average HI of 130 mg HC/g TOC shows prevailing type III kerogen. This change in kerogen type is also visible in high vitrinite percentages (Fig. 8) and increasing concentrations of land-plant biomarkers. An upwards increase in the ratio between gymnosperms and angiosperms is observed in several wells. Pr/Ph ratios are generally low. The vertical profile in well

Epsilon 1 indicates that the strongest anoxia occurred during the middle part of the Late Solenovian. Higher Pr/Ph ratios (up to 1.8) within samples from Samotino (Melrose) 1 may indicate dysoxic to oxic conditions within the canyon fill.

Kalmykian. Upper Oligocene deposits are found within the Kaliakra canyon, at its southern flank and north of it. They are missing at the northern canyon margin (probably due to erosion) and in the southern study area.

Sediments within the canyon are carbonate-free, whereas rocks deposited on the flanks of the canyon

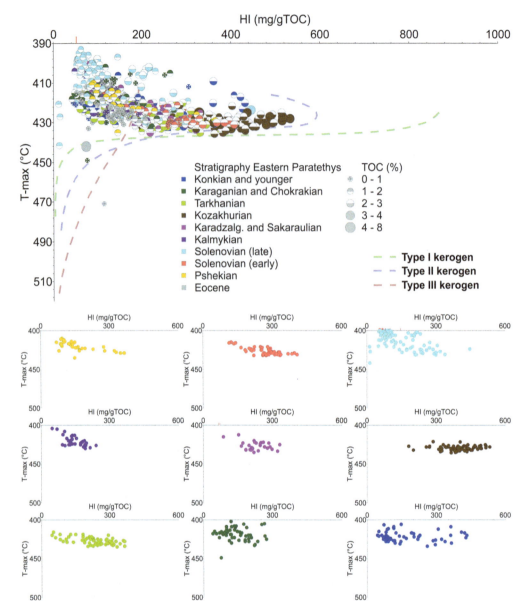

Fig. 10. (a) T_{max} v. HI plots for all samples. The symbol size reflects the TOC content. (b) T_{max} v. HI plots for samples from different time stages.

and towards the north include carbonate contents up to 50%. This difference reflects strong clastic input along the canyon axis. The upwards-increasing carbonate contents in Samotino More and Epsilon 1 may reflect the Late Kalmykian sea-level rise detected by Popov *et al.* (2010) (Fig. 10).

Kalmykian sediments contain varying amounts (0.5–2.2% TOC) of type III kerogen (average HI 150 mg HC/g TOC: Fig. 9). Despite strong clastic input, Kalmykian sediments in the canyon fill contain slightly more organic matter (1.4% TOC; HI 200 mg HC/g TOC) than rocks outside the canyon axis (1.1% TOC; HI 140 mg HC/g TOC). Negative correlations between TOC and carbonate contents (Figs 5–7) suggest that this is due to dilution by carbonate. The organic matter input is dominated by land plants, especially gymnosperms. Highest concentrations of land-plant-derived biomarkers are observed at the canyon axis. Pr/Ph ratios are low (average 0.8), indicating anoxic conditions.

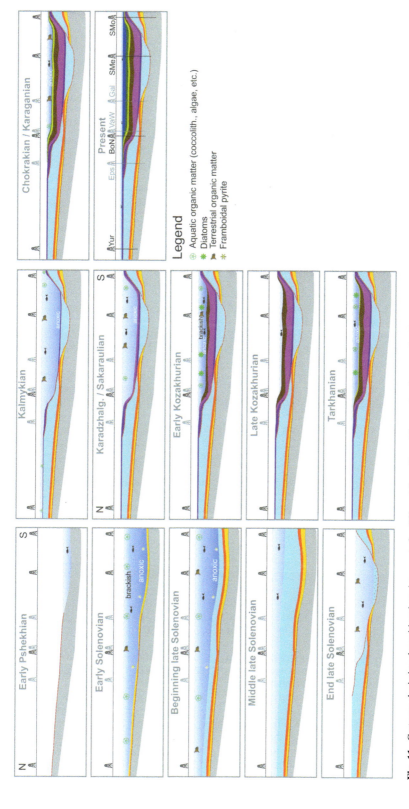

Fig. 11. Cartoons depicting depositional environments for different time slices. The sea-level curve after Popov *et al.* (2010) has been adapted in the Sakaraulian–Kozakhurian (dotted line) to account for a Kozakhurian transgression in the Izgrev and Ropotamo area. The colour scheme for the individual time intervals is provided in Figure 12.

Table 2. *TOC and HI values for different stratigraphic units*

Age	TOC (%)	TOC$_{mean}$ (%)	HI (mg HC/g TOC)	HI$_{mean}$ (mg HC/g TOC)	S2$_{mean}$ (mg HC/g rock)	No. of samples	Gen. Pot.	Gen. HC
Konkian and younger	0.4–15*	1.5*	40–430*	160*	2.4*	65		
Chokrakian/Karaganian	0.4–2.5	1.3	40–270	150	2.0	51	Fair	Gas
Tarkhanian	1.3–2.7	2.1	50–370	230	4.8	62	Good	Oil and gas
Kozakhurian	1.2–4.3	2.6	180–530	400	10.4	74	Very good	Oil
Karadzhalg./Sakaraul.	1.1–2.4	1.6	70–330	220	3.5	31	Fair to good	Gas and oil
Kalmykian	0.5–2.2	1.2	50–240	150	1.8	34	Fair	Gas
Late Solenovian	1.0–3.9	1.7	10–450	130	2.2	111	Fair to good	Gas
Early Solenovian	0.9–2.6	1.6	110–410	270	4.3	55	Good	Oil (and gas)
Pshekhian	0.4–2.3	1.5	40–370	160	2.4	33	Fair to good	Oil and gas

The classification of their generative potential (Gen. Pot.) and the type of generated hydrocarbons (Gen. HC) follows Peters (1986).
*Probably including contaminated samples.

Karadzhalgian–Sakaraulian. Popov *et al.* (2010) provided evidence for a gentle sea-level drop during Karadzhalgian–Sakaraulian time (Fig. 10). As a result of this drop, Karadzhalgian–Sakaraulian deposits are limited to the Kaliakra canyon. Rocks with relatively high organic matter contents and prevailing type III kerogen (1.6% TOC; HI 220 mg HC/g TOC: Fig. 9) have been deposited. Interbedded layers contain type II kerogen. Concentrations and ratios of di- and triterpenoids indicate strong input of terrestrial organic matter, dominated by gymnosperms. Absolute concentrations of terrestrial organic matter input are lower compared to the Kalmykian regiostage. Very low Pr/Ph ratios suggest an anoxic depositional environment.

Kozakhurian. The sea-level curve after Popov *et al.* (2010) (Fig. 10) shows a minor sea-level drop during the Early Kozakhurian followed by a major drop during Late Kozakhurian. Kozakhurian deposits transgressively overlie Lower Oligocene rocks in wells Izgrev and Ropotamo, indicating a sea-level highstand. Therefore, we assume a (? Sakaraulian to) Early Kozakhurian sea-level rise (see the modified trend in Fig. 10).

A major change in depositional environment at the base of the Kozakhurian regiostage is attested by a sudden increase in TOC/S ratios (average 7.0), which mainly results from a drop in sulphur content. The increase reflects a prominent drop in salinity, which can be related to the Kozakhurian isolation of the Eastern Paratethys (e.g. Popov *et al.* 2004a). Diatomites with very high biogenic opal contents (up to 87%, mean 44%) were deposited within the Kaliakra canyon and to its south. The thickness of the diatomites is significantly lower in the Izgrev–Ropotamo area compared to the Kaliakra canyon fill. Based on the spatial relationship with the shelf-break canyon, a model implying locally enhanced upwelling of subsurface water onto the continental shelf (e.g. Kaempf 2007) seems reasonable. Apart from opal, the Kozakhurian rocks are rich in organic matter classified as type II kerogen (Fig. 9) (2.6% TOC; HI 400 mg HC/g TOC). Maximum values are even higher (4.3% TOC; HI 530 mg HC/g TOC).

Maceral analysis reveals high percentages of alginite. The concentration of di- and triterpenoids are typically low, but high in Samotino (Melrose) 1 located at the canyon axis. This indicates a continuation of terrestrial organic matter input, mainly derived from gymnosperms, along the canyon axis. In wells outside the canyon axis, the proportion of angiosperms is higher. Low Pr/Ph ratios (mean 0.7) indicate anoxic conditions during deposition.

A major sea-level drop in Late Kozakhurian time formed an unconformity within the canyon fill and triggered severe mass-transport complexes within the Western Black Sea Basin, clearly visible in seismic data.

Tarkhanian. The erosional unconformity related to the Late Kozakhurian sea-level drop is defined as the Kozakhurian–Tarkhanian boundary. The biogenic opal content of these sediments locally exceeds 30% near the base of the Tarkhanian, but is typically below 20%. Although carbonate contents in the upper part of the unit can reach 21%, calcareous nannoplankton is missing.

TOC/S ratios (mean 2.8) vary strongly and reach maxima, typical for brackish-water conditions (>2.8), in Samotino (Melrose) 1. In Varna West, TOC/S ratios gradually decrease upwards from 2.8 to 1.8, reflecting a general increase in salinity.

TOC contents and HI values decrease upwards. Consequently, organic-matter-rich rocks with TOC contents exceeding 2.0% and type II kerogen (HI up to 370 mg HC/g TOC) prevail in the lower transgressive part, whereas rocks with TOC contents below

2.0% and type III kerogen (HI <200 mg HC/g TOC) dominate in its upper part.

High amounts of di- and triterpenoids reflect the input of land-plant material in the Samotino wells. In well Varna West, an upwards increase in land plants is inferred from biomarker data. Pr/Ph ratios (mean 0.8) are very low in samples from well Samotino (Melrose) 1. In most other wells, Pr/Ph variations indicate a trend towards less anoxic conditions.

Chokrakian–Karaganian. Chokrakian and Karaganian sediments are characterized by an upwards-increasing carbonate content (average 21%; max. 70%) and were deposited during a phase with high, unstable sea level (Figs 2 & 10). Very low TOC/S (mean 1.0) and Pr/Ph ratios (mean 0.5) suggest a return to fully marine, oxygen-depleted conditions. The sediments contain large amounts of organic matter classified as type III kerogen (1.3% TOC; HI 150 mg HC/g TOC: Fig. 9). Biomarker data are available from well Galata 1. The dominance of type III kerogen is supported by enhanced concentrations of land-plant-derived biomarkers. The ratio between di- and triterpenoids shows a dominance of gymnosperms and bioproductivity of siliceous organisms was limited.

Konkian and younger. Sediments deposited after the Karaganian regiostage are summarized in this unit. They show strongly varying carbonate contents displaying upwards-decreasing trends in northern wells (e.g. Bogdanov North) and upwards-increasing trends in the southern well Izgrev. Carbonate contents are relative low in Samotino (Melrose) 1, probably due to sedimentary input along the canyon axis. Biogenic opal contents are typically low to moderate. Due to possible contamination indicated by abnormal amounts of S1 hydrocarbons and specific biomarkers (e.g. methyl-catalenes) in several wells, the source potential and depositional environment were not evaluated.

A similar succession including sandy, clayey and carbonate rocks was investigated by Zdravkov *et al.* (2015) in an onshore well a few kilometres north of Yury Shimanov A. Within this well, Middle Miocene mudstones are rich in siliceous organisms and contain TOC contents of the order of 1–2%, locally up to 4%. The kerogen is mainly type II or type III.

Petroleum potential

TOC contents and the amount of S2 hydrocarbons obtained during RockEval pyrolysis are frequently used to classify the quality of potential source rocks. This paper follows the classification of Peters (1986) (Tables 3 & 4).

Table 3. *Parameters describing the generative potential of source rocks (after Peters 1986)*

Generative potential	TOC (wt%)	S2 (mg HC/g rock)
Poor	<0.5	<2.5
Fair	0.5–1.0	2.5–5.0
Good	1.0–2.0	5.0–10.0
Very good	>2.0	>10.0

Average TOC contents classify all Oligo-Miocene units as good or even very good (Kozakhurian, Tarkhanian) source rocks. In contrast, average S2 values are often low, indicating that many units contain poor or fair source rocks. In this paper, a combination of both parameters was used (e.g. good potential according to TOC and fair potential according to S2 is referred to as 'fair to good').

Oligocene. Based on average S2, TOC and HI values (Table 2), Pshekhian rocks are classified as 'fair to good' source rocks, which generate mainly gas and minor oil. A northwards increase in HI values indicates the presence of good oil-prone source rocks in the vicinity of well Yury Shimanov A. According to Py-GC results (Fig. 9), Pshekhian rocks from Yury Shimanov A will generate low wax paraffinic-naphthenic-aromatic oil with low sulphur contents.

The generative potential of Lower Solenovian rocks is classified as good and is higher than any other Oligocene unit. The best source-rock intervals contain 2.6% TOC and type II kerogen with a HI value of 410 mg HC/g TOC. Average HI values slightly below 300 indicate that the rocks will generate mainly oil and minor gas. The oil will have a mixed paraffinic-naphthenic-aromatic composition with low wax and moderately high sulphur contents (Fig. 9).

Upper Solenovian rocks contain large amounts of organic matter, but typically with low HI. Therefore, they host a fair to good generative potential to generate gas. Interbedded layers (e.g. in wells Bogdanov North and Varna West) contain a fair to good potential to produce oil.

Upper Oligocene (Kalmykian) rocks have been studied in wells Epsilon 1 and Samotino More 1.

Table 4. *Parameters describing the type of hydrocarbon generated (after Peters 1986)*

Type	HI (mg HC/g TOC)
Gas	<150
Gas and oil	150–300
Oil	>300

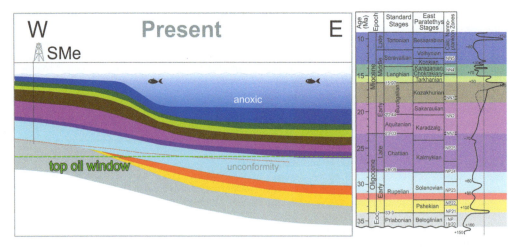

Fig. 12. Interpreted distribution of Oligocene and Miocene rocks from the shelf to deep water. The eroded section within the canyon is still present in the distal part of the basin (see also Fig. 3). The top of the oil window, as well as the Late Solenovian unconformity, are shown. Note that the Lower Oligocene source-rock intervals would be within the oil window.

Based on their average S2 and TOC values, their generative potential is classified as fair.

Miocene. Lower Miocene rocks with a Karadzhalgian–Sakaraulian age occur within the Kaliakra canyon. They host a fair to good potential to generate oil and gas. The overlying Kozakhurian rocks represent the best source rocks on the western Black Sea shelf and contain a very good generative potential. HI values up to 530 mg HC/g TOC show that the rocks are highly oil-prone. Depending on HI, a low wax paraffinic-naphthenic-aromatic oil or a high wax paraffinic oil, both with low sulphur contents, will be produced. TOC, HI and S2 values decrease within the overlying Middle Miocene (Tarkhanian, Chokrakian–Karaganian) units. Whereas Tarkhanian rocks are classified as good source rocks for oil and gas, Chokrakian–Karaganian are considered fair source rocks.

Maturity. Although all sampled Oligocene and Miocene source-rock samples are immature, biomarker data pinpoint a Tertiary source rock for the oil accumulations offshore Romania (Olaru-Florea *et al.* 2014). Furthermore, biomarker data indicate a Tertiary source rock for the Bulgarian Tyulenovo (onshore/offshore) oil field, according to Mayer *et al.* (2016).

A regional basin modelling study, comprising two different temperature scenarios, was carried out by Olaru-Florea *et al.* (2014) and clearly shows a mature source-rock kitchen for Lower Oligocene and Lower Miocene source rocks in the distal (deepwater) part of the Western Black Sea Basin. The Lower Oligocene source-rock interval is currently in the oil and gas window, and might be overmature in the deep basin (warm scenario). The Lower Miocene source-rock interval is currently in the oil and gas window, and might be overmature in the deepest part of the basin (warm scenario). Lateral hydrocarbon migration (>80 km) from this kitchen was proven by biomarker data.

Conclusions

The integration of seismic data, information from calcareous nannoplankton and geochemical data provide new insights into the development (Fig. 10) and hydrocarbon potential of Oligocene–Miocene ('Maikopian') rocks on the western Black Sea shelf.

Pshekian sediments overlie Eocene sediments with a major erosional unconformity. The amount of eroded section increases towards the north in the study area. Additionally, seismic data show a major west–east-trending shelf-break canyon offshore Bulgaria, termed Kaliakra canyon, which cuts into Lower Oligocene deposits (Pshekian, Lower and Upper Solenovian). Channel incision took place in middle Late Solenovian time. The canyon was filled with Upper Solenovian–Konkian (and younger) sediments. A major sea-level drop in the Late Kozakhurian triggered severe mass-transport complexes within the Western Black Sea Basin.

The studied Oligocene and Miocene succession is thermally immature on the shelf and contains potential source rocks in several horizons (Fig. 11), both outside and inside the Kaliakra canyon. Good and very good source rocks for oil occur in the Early Solenovian and Kozakhurian, respectively

(Table 2). Both intervals have been deposited during times of basin isolation with brackish-water conditions. Early Solenovian times are characterized by high productivity of calcareous nannoplankton. Bioproductivity during the Kozakhurian was dominated by diatoms, which is indicated by the high biogenic opal content (recognizable in well logs by low gamma radiation and low density).

Both of these source-rock intervals are within the oil and gas window in the deeper parts of the basin (Fig. 12). Long-distance migration from this highly productive kitchen is proven by published biomarker data (Olaru-Florea et al. 2014) and therefore highlights the potential of this petroleum system.

References

BANKS, C.J. & ROBINSON, A.G. 1997. Mesozoic strike-slip back-arc basins of the Western Black Sea region. In: ROBINSON, A.G. (ed.) Regional and Petroleum Geology of the Black Sea and Surrounding Region. American Association of Petroleum Geologists Memoirs, 68, 53–62.

BERNER, R.A. 1984. Sedimentary pyrite formation: an update. Geochimica et Cosmochimica Acta, 48, 605–615.

BONCEV, E. 1986. The Balkanides – Geotectonic Position and Development. Geologica Balcanica, Series Operum Singulorium, 1. Publication House of BAS, Sofia [in Bulgarian with English summary].

ČTYROKÝ, P. 1991. Mollusk finds in the Dynów Marlstones (Menilitic Formation) of the Ždánice unit. Geoscience Research Reports for 1991 [Zprávy o geologických výzkumech v roce 1991], Czech Ge logical Survey, Prague [in Czech].

DACHEV, C. & GEORGIEV, G. 1995. Rift tectonics problems in the Western Black Sea basin. In: IGCP Project No. 369: 'Comparative Evolution of Peri-Tethyan Rift Basins', 2nd Annual Meeting, Mamaya-Romania, Abstracts Volume.

DIDYK, B.M., SIMONEIT, B.R.T., BRASSELL, S.C. & EGLINTON, G. 1978. Organic geochemical indicators of paleoenvironmental conditions of sedimentation. Nature, 272, 216–222.

EGLINTON, T.I., SINNINGHE-DAMSTÉ, J.S., KOHNEN, M.E.L., DE LEEUW, J.W., LARTER, S.R. & PATIENCE, R.L. 1990. Analysis of maturity-related changes in the organic sulphur composition of kerogens by flash pyrolysis-gas chromatography. In: ORR, W.L. & WHITE, C.M. (eds) Geochemistry of Sulphur in Fossil Fuels. American Chemical Society, Symposium Series, 429, 529–565.

EMERY, M. & GEORGIEV, G. 1993. Tectonic evolution and hydrocarbon potential of the Southern Moesian platform and Balkan–Forebalkan regions of Northern Bulgaria. AAPG Bulletin, 77, 1620–1621 [Abstract].

ESPITALIÉ, J., LAPORTE, J.L., MADEC, M., MARQUIS, F., LEPLAT, P., PAULET, J. & BOUTEFEU, A. 1977. Méthode rapide de characterisation des roches mères de leur potential pétrolier et de leur degree d'evolution. Revue de l'Institut Francais du Pétrole, 32, 23–42.

FINETTI, I., BRICCHI, G., DEL BEN, A., PIPAN, M. & XUAN, Z. 1988. Geophysical study of the Black Sea area. Bullettino di Geofisica Teorica ed Applicata, 30, 197–324.

GEORGIEV, G. 2012. Geology and hydrocarbon systems in the Western Black Sea. Turkish Journal of Earth Sciences, 21, 723–754.

GORUR, N. 1988. Timing of opening of the Black Sea basin. Tectonophysics, 147, 247–262.

GRADSTEIN, F.M., OGG, J.G., SCMITZ, M.D. & OGG, G.M. 2012. The Geologic Time Scale. Elsevier, Amsterdam.

HORSFIELD, B. 1989. Practical criteria for classifying kerogens: some observations from pyrolysis-gas chromatography. Geochimica et Cosmochimica Acta, 53, 891–901.

HORSFIELD, B. 1997. The bulk composition of first-formed petroleum in source rocks. In: WELTE, D.H., HORSFIELD, B. & BACKER, D.R. (eds) Petroleum and Basin Evolution. Insights from Petroleum Geochemistry, Geology and Basin Modelling. Springer, Berlin, 335–402.

KAEMPF, J. 2007. On the magnitude of upwelling fluxes in shelf-break canyons. Continental Shelf Research, 27, 2211–2223.

MAYER, J., SACHSENHOFER, R.F., FLOODPAGE, J., UNGUREANU, C., KOZUHAROV, E. & TARI, G. 2016. Hydrocarbon generation and migration in the western Black Sea basin, insights from the Tyulenovo oil field (on- and offshore Bulgaria). Paper presented at the AAPG Petroleum Systems of Alpine–Mediterranean Fold Belts and Basins, 19–20 May 2016, Bucharest, Romania.

NAGYMAROSY, A. & VORONINA, A.A. 1993. Calcareous nannoplankton from the lower Maykopian Beds (Early Oligocene, Union of Independent States). Knihovnicka ZPN, 2, 189–223.

NIKISHIN, A.M., ZIEGLER, P.A. ET AL. 2001. Mesozoic and Cainozoic evolution of the Scythian Platform–Black Sea–Caucasus domain. In: ZIEGLER, P.A., CAVAZZA, W., ROBERTSON, A.H.F. & CRASQUIN-SOLEAU, S. (eds) Peri-Tethys Memoir 6: Peri-Tethyan Rift/Wrench Basins and Passive Margins. Memoires du Museum National d'Histoire Naturelle, 186, 295–346.

NIKISHIN, A.M., KOROTAEV, M., ERSHOV, A. & BRUNET, M. 2003. The Black Sea basin: tectonic history and Neogene–Quaternary rapid subsidence modelling. Sedimentary Geology, 156, 149–168.

NIKISHIN, A.M., OKAY, A., TÜYSÜZ, O., DEMIRER, A., WANNIER, M., AMELIN, N. & PETROV, E. 2015. The Black Sea basins structure and history: New model based on new deep penetration regional seismic data. Part 2: Tectonic history and paleogeography. Marine and Petroleum Geology, 59, 656–670.

OKAY, A.I., ŞENGOR, A.M.C. & GORUR, N. 1994. Kinematic history of the opening of the Black Sea and its effect on the surrounding regions. Geology, 22, 267–270.

OLARU-FLOREA, R., UNGUREANU, C. ET AL. 2014. Understanding of the petroleum system(s) of the western Black Sea: Insights from 3-D basin modelling. Search and Discovery Article 10686, AAPG International Conference & Exhibition, 14–17 September 2014, Istanbul, Turkey.

PAVICEVIC, M.K. & AMTHAUSER, G. 2000. Physikalisch-chemische Untersuchungsmethoden in den Geowissenschaften, Band 1. E. Schweizerbart'sche, Stuttgart.

PETERS, K.E. 1986. Guidelines for evaluating petroleum source rock using programmed pyrolysis. American

Association of Petroleum Geologists Bulletin, **70**, 318–329.

POPOV, S.V., AKHMETIEV, M.A. *ET AL.* 2001. Biogeography of the Northern Peri-Tethys from the late Eocene to the early Miocene, part. 1: late Eocene. *Paleontological Journal*, **35**(suppl. 1), 1–68.

POPOV, S.V., VORONINA, A.A. & GONTSCHAROVA, I.A. 1993. *Stratigraphy and Bivalves of the Oligocene–Lower Miocene of the Eastern Paratethys.* Publications of the Paleontological Institute, **256**. Russian Academy of Sciences, Moscow [in Russian].

POPOV, S.V., BUGROVA, E.M. *ET AL.* 2004*a*. Biogeography of the Northern Peri-Tethys from the Late Eocene to the Early Miocene. Part 3. Late Oligocene–Early Miocene. Marine basins. *Paleontological Journal*, **38**, 653–S716.

POPOV, S.V., RÖGL, F., ROZANOV, A.Y., STEININGER, F.F., SHCHERBA, I.G. & KOVAC, M. 2004*b*. *Lithological–Paleogeographic Maps of Paratethys: 10 Maps Late Eocene to Pliocene.* Courier Forschungsinstitut Senckenberg, Frankfurt, **250**.

POPOV, S.V., ANTIPOV, M.P., ZASTROZHNOV, A.S., KURINA, E.E. & PINCHUK, T.N. 2010. Sea-level fluctuations on the northern shelf of the Eastern Paratethys in the Oligocene–Neogene. *Stratigraphy and Geological Correlation*, **8**, 200–224.

POPOV, S.V. & STUDENCKA, B. 2015. Brackish-water Solenovian mollusks from the Lower Oligocene of the Polish Carpathians. *Paleontological Journal*, **49**, 342–355.

RADKE, M., WILLSCH, H. & WELTE, D.H. 1980. Preparative hydrocarbon group type determination by automated medium pressure liquid chromatography. *Analytical Chemistry*, **52**, 406–411.

ROBINSON, A., SPADINI, G., CLOETINGH, S. & RUDAT, J. 1995. Stratigraphic evolution of the Black Sea: inferences from basin modelling. *Marine and Petroleum Geology*, **12**, 821–835.

RÖGL, F. 1999. Mediterranean and Paratethys. Facts and hypotheses of an Oligocene to Miocene paleogeography (short overview). *Geologica Carpathica*, **50**, 339–349.

RUSU, A. 1999. Rupelian mollusk fauna of Solenovian type found in Eastern Carpathians (Romania). *Acta Palaeontologica Romaniae*, **2**, 449–452.

SACHSENHOFER, R.F., STUMMER, B., GEORGIEV, G., DELL-MOUR, R., BECHTEL, A., GRATZER, R. & CORIC, S. 2009. Depositional environment and hydrocarbon source potential of the Oligocene Ruslar Formation (Kamchia Depression; Western Black Sea). *Marine and Petroleum Geology*, **26**, 57–84.

SACHSENHOFER, R.F., POPOV, S.V. *ET AL.* 2017. Oligocene and Lower Miocene source rocks in the Paratethys: palaeogeographical and stratigraphic controls. *In*: SIMMONS, M.D., TARI, G.C. & OKAY, A.I. (eds) *Petroleum Geology of the Black Sea.* Geological Society, London, Special Publications, **464**, https://doi.org/10.1144/SP464.6

SAINT-GERMÈS, M. 1998. *Ètude sédimentologique et géochimique de la matière organique du bassin Maykopien (Oligocène–Miocène Inférieur) de la Crimée a l'Azerbaidjan.* Mémoires des Sciences de la Terre. PhD thesis, Académie de Paris Université Pierre et Marie Curie, Paris [in French].

SCHULZ, H.-M., BECHTEL, A., RAINER, T., SACHSENHOFER, R.F. & STRUCK, U. 2004. Paleoceanography of the western Central Paratethys during nannoplankton zone NP 23: the Dynow Marlstone in the Austrian Molasse Basin. *Geologica Carpathica*, **55**, 311–323.

SCHULZ, H.-M., BECHTEL, A. & SACHSENHOFER, R.F. 2005. The birth of the Paratethys during the early Oligocene: from Tethys to an ancient Black Sea analogue? *Global and Planetary Change*, **49**, 163–176.

SEIFERT, W.K. & MOLDOWAN, J.M. 1980. The effect of thermal stress on source rock quality as measured by hopane stereochemistry. *In*: DOUGLAS, A.G. & MAXWELL, J.R. (eds) *Advances in Organic Geochemistry 1979.* Pergamon Press, Oxford, 229–237.

TARI, G., DICEA, O., FAULKERSON, J., GEORGIEV, G., POPOV, S., STEFANESCU, M. & WEIR, G. 1997. Cimmerian and Alpine stratigraphy and structural evolution of the Moesian platform (Romania/Bulgaria). *In*: ROBINSON, A.G. (ed.) *Regional and Petroleum Geology of the Black Sea and Surrounding Regions.* American Association of Petroleum Geologists Memoirs, **68**, 63–90.

TAYLOR, G.H., TEICHMÜLLER, M., DAVIS, A., DIESSEL, C.F. K., LITTKE, R. & ROBERT, P. 1998. *Organic Petrology.* Gebrüder Borntraeger, Berlin.

TEN HAVEN, H.L., DE LEEUW, J.W., RULLKÖTTER, J. & SINNINGHE-DAMSTÉ, J. 1987. Restricted utility of the pristane/phytane ratio as a palaeoenvironmental indicator. *Nature*, **330**, 641–643.

TEXIDOR, P., GRIMALT, J.O., PUEYO, J.J. & RODRIGUEZ-VALERA, F. 1993. Isopranylglycerol diethers in non-alkaline evaporitic environments. *Geochimica et Cosmochimica Acta*, **57**, 4479–4489.

TUGOLESOV, D.A., GORSHKOV, A.S., MEYSNER, L.B., SOLO-VIOV, V.V. & KHAKHALEV, E.M. 1985. *Tectonics of Mesozoic Sediments in the Black Sea Basin.* Nedra, Moscow [in Russian].

VOLKMAN, J.K. & MAXWELL, J.R. 1986. Acyclic isoprenoids as biological markers. *In*: JOHNS, R.B. (ed.) *Biological Markers in the Sedimentary Record.* Elsevier, Amsterdam, 1–42.

VORONINA, A.A. & POPOV, S.V. 1984. Solenovian horizon from Eastern Paratethys. *Bulletin of the Academy of Sciences of the USSR Geologic Series*, **9**, 41–53 [in Russian].

ZACHOS, J.C., PAGANI, M., SLOAN, L., BILLUPS, K. & THOMAS, E. 2001. Trends, rhythms, and aberrations in global climate 65 Ma to present. *Science*, **292**, 686–693.

ZDRAVKOV, A., BECHTEL, A., CORIC, S. & SACHSENHOFER, R. F. 2015. Depositional environment, organic matter characterization and hydrocarbon potential of Middle Miocene sediments from northeastern Bulgaria (Varna-Balchik Depression). *Geologica Carpathica*, **66**, 409–426.

ZOLITSCHKA, B. 1998. *Paläoklimatische Bedeutung Laminierter Sedimente [Paleoclimatic implications of laminated sediments].* Relief Boden Paläoklima, **13**. Gebrüder Bornträger Verlag, Berlin [in German].

Source rock evaluation of Middle Eocene–Early Miocene mudstones from the NE margin of the Black Sea

STEPHEN J. VINCENT[1]* & MATTHEW N. D. KAYE[2]

[1]CASP, West Building, Madingley Rise, Madingley Road, Cambridge CB3 0UD, UK

[2]OceanGrove Geoscience Ltd, Unit 39, Howe Moss Avenue, Dyce, Aberdeen AB21 0GP, UK

*Correspondence: stephen.vincent@casp.cam.ac.uk

Abstract: This study comprises the source rock evaluation of 122 Late Middle Eocene–Early Miocene mudstones from the NE margin of the Black Sea. Samples are immature to early mature. The majority of samples have moderate to very good organic richness, poor to moderate source potential and a hydrogen-deficient to gas-prone source rock quality. However, a significant proportion of the samples have good to excellent organic richness and source potential, and an oil- and gas-prone quality derived from amorphous-rich kerogens. These samples would generate significant amounts of oil and associated gas where buried to peak maturity. They come from the lowermost (Rupelian) part of the Maykop Series and the late Bartonian–early Priabonian Kuma Suite or its stratigraphic equivalents. The Rupelian source-rock interval(s) in west Georgia is at least 60 m thick and potentially as much as 200 m thick. It has a source potential index (SPI) of 0.7–2.5 t HC m^{-2}. The thickness of the Kuma Suite-equivalent source rock interval south of the western Greater Caucasus is unconstrained. Maykop Series source rocks occur in the Black Sea Basin. Prospective Kuma Suite-equivalent samples on both the northern and southern margins of the Black Sea imply that similar sediments may also be present in the basin.

Supplementary material: Additional information on the geographical location and age determination of the samples discussed in this paper are available at https://doi.org/10.6084/m9.figshare.c.3841399

The Black Sea is one of the few remaining under-explored hydrocarbon basins in Europe. Exploration to date has been hampered by water depths typically in excess of 2 km. The basin is presumed to contain a world-class source rock, the Oligocene–Early Miocene Maykop Series. This unit forms the main source rock interval in producing fields in the adjacent South Caspian Basin, the north Caucasus foredeep and the Rioni Basin, Georgia (Inan *et al.* 1997; Katz *et al.* 2000; Bazhenova *et al.* 2002; Saint-Germes *et al.* 2002; Glumov *et al.* 2004; Isaksen *et al.* 2007). Basin modelling, together with evidence from the analysis of mud volcanoes and live oil seeps, indicate that the unit has achieved sufficient depth of burial and maturity level to generate hydrocarbons within the Black Sea region (Zabanbark & Konyukhov 1995; Geodekyan *et al.* 1996; Robinson *et al.* 1996; Glumov & Viginskiy 2000; Dimitrov 2002; Bohrmann *et al.* 2003; Kruglyakova *et al.* 2004; Andreev 2005; Afanasenkov *et al.* 2007; Wilson *et al.* 2007; Rainer *et al.* 2015). As a consequence, it is typically overpressured (Scott *et al.* 2009; Marín-Moreno *et al.* 2013).

Despite its economic importance, surprisingly little has been published in English language literature on the source rock potential of the Maykop Series, either within or along the margins of the Black Sea Basin. Prior to the works in this volume, the most comprehensive record was that of Sachsenhofer *et al.* (2009) who looked at Oligocene well material from the Bulgarian sector of the western Black Sea. Robinson *et al.* (1996) also carried out some screening of the Maykop Series, primarily from outcrops in the Greater Caucasus.

This study extends the work of Robinson *et al.* (1996) by evaluating the source rock potential of Maykop Series mudstones from along the southern flank of the western Greater Caucasus in Russia, and the margins of the Rioni Basin in west Georgia (Fig. 1). Mayer *et al.* (2017) extends the work of Sachsenhofer *et al.* (2009) by examining Oligo-Miocene material from the Bulgarian offshore, whilst Sachsenhofer *et al.* (2017*b*) summarizes our contribution and earlier Maykop Series studies.

Extensive work has been carried out on the stratigraphy (e.g. Akhmetiev *et al.* 1995), biostratigraphy (e.g. Zaporozhets 1999) and organic geochemistry (e.g. Saint-Germès 1998; Saint-Germes *et al.* 2000*b*, 2002; Bazhenova *et al.* 2002; Sachsenhofer *et al.* 2017*a*) of the Maykop Series at its type section

From: SIMMONS, M. D., TARI, G. C. & OKAY, A. I. (eds) 2018. *Petroleum Geology of the Black Sea.*
Geological Society, London, Special Publications, **464**, 329–363.
First published online September 25, 2017, https://doi.org/10.1144/SP464.7

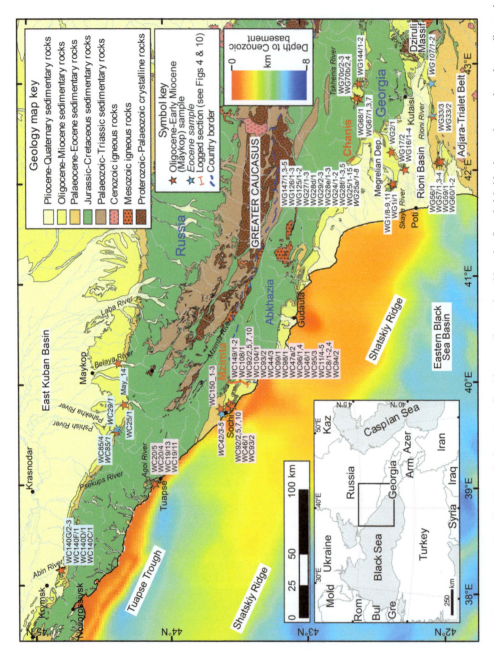

Fig. 1. Location map of the samples analysed in this study. Samples are listed in ascending stratigraphic order. Sample label backgrounds are coloured according to their geographical region: blue, Russia north; pink, Russia south; white, west Georgia. Samples from the Sochi region that have been incorporated into the Mzimta section are also listed there. The depth to Cenozoic basement of the Eastern Black Sea is from Tugolesov (1989).

along the Belaya River, south of the Russian city of Maykop, on the northern side of the western Greater Caucasus (Fig. 1). However, active uplift of the Caucasus during Maykop Series deposition (Kholodov & Nedumov 1996; Lozar & Polino 1997; Saintot et al. 2006; Vincent et al. 2007) means that these sediments were deposited in a separate sub-basin to those within the Black Sea and may have undergone a different geochemical evolution and burial/thermal history. Likewise, seismic data indicate that the Mid Black Sea High formed a bathymetric high during most or all of Maykop Series deposition (Finetti et al. 1988; Tugolesov 1989; Nikishin et al. 2015a, b), such that different sub-basins were developed in the western and eastern sectors of the Black Sea itself. As a consequence of this segmentation, this contribution forms the most comprehensive account of the source rock characteristics of the onshore equivalents of mudstones within the Eastern Black Sea.

A limited number of Middle–Late Eocene mudstones have also been analysed in this study in order to document a potential secondary source interval that is equivalent to the Kuma Suite on the northern side of the Greater Caucasus. Earlier work in English language literature on the Kuma Suite has been presented by Beniamovski et al. (2003), Distanova (2007) and Peshkov et al. (2016).

Geological background

The Kuma Suite and Maykop Series formed during a period of widespread palaeoenvironmental change. The Late Middle–early Late Eocene Kuma Suite can be traced from the northern side of the Crimean and Greater Caucasus mountains to the Aral Sea (Beniamovski et al. 2003). It was deposited following the closure of the northern Neotethys (Şengör & Yılmaz 1981; Okay & Şahintürk 1997; Kaymakcı et al. 2009; Nairn et al. 2013) and during a period of orogenic collapse and volcanism along the former Turkish–Iranian active margin (Vincent et al. 2005; Topuz et al. 2011; Aydınçakır & Şen 2013). Volcanic material may have contributed to increased organic productivity (Muzylöv 1996, cited in Beniamovski et al. 2003). On the southern side of the Russian western Greater Caucasus, age-equivalent strata to the Kuma Suite are known as the Navaginskaya Suite (Zakrevskaya et al. 2009, p. 271); in west Georgia they are unnamed.

The Maykop Series forms part of a wider Oligocene–Early Miocene organic-rich interval that extends from the Alps to the Caspian Sea (Vetö 1987; Sachsenhofer et al. 2009, 2017b). This formed during the inception of the Paratethys Sea as the initial Arabia–Eurasia collision resulted in restricted oceanic circulation, stratification and stagnant

bottom waters (Báldi 1984; Popov et al. 1993; Rögl 1999; Schulz et al. 2005; Allen & Armstrong 2008). The unit has assumed chronostratigraphic significance following its adoption as the basal Paratethyan stage (i.e. the Maykopian: Jones & Simmons 1996, 1997).

As a consequence of synchronous Caucasus uplift, the Maykop Series contains a large proportion of microfossils reworked from underlying units (Jones & Simmons 1996; Lozar & Polino 1997; Vincent et al. 2007, 2014). This has hampered accurate age determinations, and may explain why Robinson et al. (1996) considered the lowermost part of the series to be Middle and Late Eocene in age. Caucasian uplift created a local sediment source for potential reservoir-quality sandstones, particularly in the Tuapse Trough (Vincent et al. 2013, 2014), that because of their close association with organic-rich mudstones are likely to have an intraformational hydrocarbon source. An estimated sea-level fall of approximately 80–100 m at the base of the Oligocene section (Popov et al. 2010), along the northern shelf of eastern Paratethys, is similar to global (eustatic) values (Lear et al. 2004) and may have also contributed to this palaeo-oceanic reconfiguration (Popov et al. 1993).

Carbon storage associated with the Kuma Suite and the Maykop Series is estimated to be of the order of 6×10^{11} and 6×10^{13} t, respectively, and to have been partially responsible for periods of global cooling (Beniamovski et al. 2003; Allen & Armstrong 2008).

Materials and methods

Fourteen Late Middle–Late Eocene mudstone samples and 98 Oligocene–Early Miocene (Maykop Series) mudstone samples from the Russian and Georgian margins of the Eastern Black Sea were collected at outcrop for source rock prospectivity evaluation (Table 1; Fig. 1). Two Kuma Suite and eight Maykop Series mudstone samples were also collected from the northern side of the western Greater Caucasus in order to provide a comparison with our dataset and the results of earlier analyses of samples from this region (e.g. Saint-Germès 1998; Saint-Germes et al. 2002; Sachsenhofer et al. 2017a).

Late Middle Eocene–Early Miocene strata are poorly exposed along the flanks of the Caucasus. Samples were typically collected from isolated fresh river or road cuts or from local landslips. Samples were dug out of weathered outcrops. More continuous exposures were present along the Chanis River in NW west Georgia and in the Mzimta River valley, near Sochi, in southern Russia to allow the construction of logged sections through the whole

Table 1. Details of the TOC/Rock-Eval results from Eocene–Early Miocene mudstone samples from the NE margin of the Black Sea

Sample No.	Region	Section	Height (m) (Chanis or Mzimta)	Age	S_1 (kg t⁻¹)	S_2 (kg t⁻¹)	T_{max} (°C) (italics = unreliable)	OPI	AOC/ TOC (%)	TOC (%)	HI (mg S_2/ g TOC)
Russia north											
May_14		Belaya River		Oligocene	0.19	1.21	407	0.14	6.2	1.87	65
WC25/1		Pshekha river		Early Miocene	0.07	2.05	423	0.03	9.1	1.93	106
WC29/1		Pshekha river		Early Priabonian	0.58	17.94	413	0.03	34.9	4.40	408
WC85/4		Pshish River		Early Miocene	0.14	0.58	412	0.19	5.1	1.17	50
WC85/1		Pshish River		?Late Bartonian–Early Priabonian	0.96	22.27	405	0.04	21.0	9.16	243
WC140C/1		Abin River		Chattian	0.10	0.13	394	0.43	2.5	0.75	17
WC140D/1		Abin River		?Chattian	0.17	0.36	385	0.32	3.6	1.22	30
WC140F/1		Abin River		?Rupelian	0.33	6.60	417	0.05	28.9	1.99	332
WC140G/2		Abin River		Early Rupelian	0.11	0.82	419	0.12	8.6	0.90	91
WC140G/3		Abin River		Early Rupelian	0.15	1.87	419	0.07	9.5	1.77	106
Russia south											
WC8/1		Mzimta River	200.00	Early Rupelian	0.26	2.97	431	0.08	25.8	1.04	286
WC8/2		Mzimta River	201.00	Early Rupelian	1.63	24.91	432	0.06	44.0	5.01	497
WC8/4		Mzimta River	203.00	Early Rupelian	0.23	4.29	433	0.05	25.9	1.45	296
WC11/4		Mzimta River	250.00	mid Rupelian	0.23	1.30	435	0.15	11.9	1.07	121
WC11/5		Mzimta River	250.40	mid Rupelian	0.86	15.80	434	0.05	38.2	3.62	436
WC19/11	Tuapse			Rupelian	0.20	10.15	419	0.02	27.6	3.11	326
WC19/13	Tuapse			Rupelian	0.24	9.48	418	0.02	26.1	3.09	307
WC20/4	Tuapse			Rupelian	0.20	8.87	416	0.02	27.8	2.71	327
WC20/5	Tuapse			Rupelian	0.20	11.05	416	0.02	27.2	3.43	322
WC42/3	Sochi			Early Priabonian	0.01	0.17	410	0.06	18.7	0.08	213
WC42/4	Sochi			Early Priabonian	0.43	38.60	414	0.01	36.4	8.89	434
WC42/5	Sochi			Early Priabonian	1.11	64.71	415	0.02	43.8	12.47	519
WC44/3		Mzimta River	741.00	Late Rupelian	0.12	0.42	426	0.22	5.0	0.89	47
WC46/1		Mzimta River	400.00	Late Rupelian	0.01	0.15	428	0.06	3.7	0.36	42
WC47a/2		Mzimta River	554.00	Late Rupelian	0.02	0.04	426	0.33	10.0	0.05	80
WC92/2		Mzimta River	968.00	Late Rupelian–early Chattian	0.15	0.51	430	0.23	6.4	0.85	60
WC92/5		Mzimta River	982.00	Late Rupelian–early Chattian	0.18	1.88	432	0.09	8.0	2.15	87
WC92/7		Mzimta River	992.00	Late Rupelian–early Chattian	0.13	0.43	426	0.23	6.5	0.72	60
WC92_10		Mzimta River	1045.00	?Early Chattian	0.22	1.39	432	0.14	9.5	1.41	99
WC93/2		Mzimta River	780.00	Late Rupelian	0.31	2.38	432	0.12	10.3	2.17	110
WC94/2		Mzimta River	110.00	Early Rupelian	0.17	0.32	427	0.35	7.3	0.56	57
WC95/3		Mzimta River	310.00	Late Rupelian	0.05	0.24	431	0.17	6.3	0.38	63
WC96/1		Mzimta River	510.00	Late Rupelian	0.15	0.30	429	0.33	6.6	0.57	53
WC96/4		Mzimta River	539.00	Late Rupelian	0.02	0.24	424	0.08	3.1	0.69	35

Sample	Region	Locality	Depth	Age							
WC97/3		Mzimta River	581.00	Late Rupelian	0.08	0.57	424	0.12	5.6	0.96	59
WC98/1		Mzimta River	612.00	Late Rupelian	0.02	0.24	429	0.08	4.1	0.53	45
WC99/1		Mzimta River	673.00	Late Rupelian	0.02	0.27	422	0.07	3.9	0.62	44
WC104/1		Mzimta River	798.00	Late Rupelian	0.02	0.11	428	0.15	3.7	0.29	38
WC108/1		Mzimta River	1105.00	Aquitanian	0.07	0.36	422	0.16	4.9	0.73	49
WC149/1		Mzimta River	1520.00	Aquitanian	0.05	0.38	428	0.12	4.7	0.76	50
WC149/2		Mzimta River	1522.00	Aquitanian	0.05	0.32	427	0.14	4.5	0.69	46
WC150/1	Sochi			mid to Late Rupelian	0.04	0.19	424	0.17	3.5	0.55	35
WC150/2	Sochi			mid to Late Rupelian	0.04	0.22	422	0.15	4.2	0.52	42
WC150/3	Sochi			mid to Late Rupelian	0.09	0.25	418	0.26	4.8	0.59	42
West Georgia											
WG1/8	Central	Skaya River		Rupelian	0.63	18.71	409	0.03	21.8	7.37	254
WG1/9	Central	Skaya River		Rupelian	1.32	17.65	405	0.07	22.8	6.91	255
WG1/11	Central	Skaya River		?Oligocene	0.06	0.04	428	0.60	3.2	0.26	15
WG1i/1	Central	Skaya River		Oligocene	0.03	0.04	439	0.43	5.3	0.11	36
WG2/1	Central	Teykhuri River		?Early Miocene	0.12	1.71	427	0.07	6.5	2.32	74
WG16/1	Central	Tsivi River		?Oligocene–Early Miocene	0.20	2.16	416	0.08	7.2	2.72	79
WG16/2	Central	Tsivi River		?Oligocene–Early Miocene	0.32	2.80	419	0.10	11.5	2.26	124
WG16/3	Central	Tsivi River		?Oligocene–Early Miocene	0.18	0.85	396	0.17	3.8	2.25	38
WG16/4	Central	Tsivi River		?Oligocene–Early Miocene	0.54	2.53	396	0.18	8.2	3.09	82
WG17/2	Central	Tsivi River		?Oligocene–Early Miocene	0.28	2.38	417	0.11	8.5	2.61	91
WG24/1	Northwest	Chanis River	247.00	Late Rupelian	0.21	4.09	410	0.05	13.8	2.59	158
WG24/2	Northwest	Chanis River	248.00	Late Rupelian	0.15	2.47	422	0.06	7.4	2.94	84
WG25/1	Northwest	Chanis River	125.00	Early Rupelian	0.01	0.29	417	0.03	8.0	0.31	94
WG25/2	Northwest	Chanis River	130.00	Early Rupelian	0.23	5.38	423	0.04	19.9	2.34	230
WG25/3	Northwest	Chanis River	132.00	Early Rupelian	0.05	1.62	425	0.03	12.6	1.10	147
WG25/4	Northwest	Chanis River	134.00	Early Rupelian	0.06	1.50	425	0.04	12.5	1.04	144
WG25/5	Northwest	Chanis River	136.00	Early Rupelian	0.06	2.27	424	0.03	14.2	1.36	167
WG25/6	Northwest	Chanis River	138.00	Early Rupelian	0.04	1.48	427	0.03	10.4	1.21	122
WG25/7	Northwest	Chanis River	140.00	Early Rupelian	0.11	3.47	423	0.03	15.2	1.95	178
WG25/8	Northwest	Chanis River	142.00	Early Rupelian	0.06	2.21	424	0.03	13.3	1.42	156
WG25/9	Northwest	Chanis River	144.00	Early Rupelian	0.09	3.22	423	0.03	17.3	1.59	203
WG25/10	Northwest	Chanis River	146.00	Early Rupelian	0.09	3.11	424	0.03	17.2	1.54	202
WG25/11	Northwest	Chanis River	160.00	Early Rupelian	0.07	2.90	422	0.02	17.0	1.45	200
WG25/12	Northwest	Chanis River	162.00	Early Rupelian	0.10	3.56	420	0.03	22.3	1.36	262
WG25/13	Northwest	Chanis River	164.00	Early Rupelian	0.09	3.75	422	0.02	19.2	1.66	226
WG25/14	Northwest	Chanis River	166.00	Early Rupelian	0.08	3.80	422	0.02	20.1	1.60	238
WG25/15	Northwest	Chanis River	168.00	Early Rupelian	0.11	6.15	417	0.02	23.9	2.17	283
WG25a/1	Northwest	Chanis River	17.20	Late Bartonian–earliest Priabonian	0.08	12.66	418	0.01	28.7	3.69	343

(Continued)

Table 1. *Details of the TOC/Rock-Eval results from Eocene–Early Miocene mudstone samples from the NE margin of the Black Sea (Continued)*

Sample No.	Region	Section	Height (m) (Chanis or Mzimta)	Age	S_1 (kg t⁻¹)	S_2 (kg t⁻¹)	T_{max} (°C) (*italics* = unreliable)	OPI	AOC/ TOC (%)	TOC (%)	HI (mg S_2/ g TOC)
WG25a/2	Northwest	Chanis River	25.60	Late Bartonian–early Priabonian	0.00	0.59	426	0.00	6.6	0.74	80
WG25a/3	Northwest	Chanis River	33.60	Early Priabonian	0.00	0.02	417	0.00	1.7	0.10	20
WG25a/4	Northwest	Chanis River	50.30	Early Priabonian	0.00	0.09	420	0.00	3.6	0.21	43
WG25a/5	Northwest	Chanis River	67.10	Early Priabonian	0.00	0.10	419	0.00	3.3	0.25	40
WG25a/6	Northwest	Chanis River	83.90	Early Priabonian	0.00	0.03	418	0.00	2.3	0.11	27
WG25a/7	Northwest	Chanis River	100.60	Priabonian	0.00	0.09	419	0.00	3.4	0.22	41
WG25a/8	Northwest	Chanis River	117.40	Priabonian	0.00	0.09	422	0.00	3.7	0.20	45
WG27/1	Northwest	Chanis River	976.60	?Aquitanian	0.03	0.56	426	0.05	5.3	0.93	60
WG27/2	Northwest	Chanis River	982.40	?Aquitanian	0.04	0.44	425	0.08	4.9	0.81	54
WG27/3	Northwest	Chanis River	1011.50	?Aquitanian	0.03	0.41	432	0.07	3.7	0.99	41
WG28d/1	Northwest	Chanis River	744.50	Chattian	0.02	0.37	415	0.05	3.8	0.86	43
WG28e/1	Northwest	Chanis River	333.50	Late Rupelian	0.01	0.32	427	0.03	4.2	0.66	48
WG28e/2	Northwest	Chanis River	336.00	Late Rupelian	0.03	0.50	420	0.06	4.7	0.94	53
WG28e/3	Northwest	Chanis River	339.50	Late Rupelian	0.00	0.20	425	0.00	2.6	0.64	31
WG28f/1	Northwest	Chanis River	236.75	Late Rupelian	0.14	3.07	415	0.04	11.8	2.25	136
WG28f/2	Northwest	Chanis River	245.25	Late Rupelian	0.56	18.28	414	0.03	32.6	4.80	381
WG28f/3	Northwest	Chanis River	250.75	Late Rupelian	0.37	10.26	408	0.03	25.7	3.43	299
WG28f/5	Northwest	Chanis River	256.75	Late Rupelian	0.27	8.12	420	0.03	24.2	2.88	282
WG29/2	Northwest	Chanis River	390.00	?Early Chattian	0.03	0.55	426	0.05	6.5	0.74	74
WG29/3	Northwest	Chanis River	395.00	?Early Chattian	0.02	0.47	424	0.04	6.9	0.59	80
WG33/2	South	Kumuri River		Priabonian	0.01	0.05	423	0.17	6.2	0.08	63
WG33/3	South	Kumuri River		Rupelian	0.39	8.75	416	0.04	22.1	3.43	255
WG56/1	South			Oligocene	0.40	1.84	396	0.18	6.9	2.69	68
WG57/1	South			Oligocene	0.11	0.93	424	0.11	10.8	0.80	116
WG57/3	South			Oligocene	0.05	0.27	414	0.16	6.2	0.43	63
WG57/4	South			Oligocene	0.21	1.00	418	0.17	11.7	0.86	116
WG59/1	South			Oligocene	0.54	3.94	403	0.12	11.4	3.26	121
WG60/1	South			Oligocene–Early Miocene	0.20	0.22	416	0.48	1.7	2.09	11
WG60/2	South			Oligocene–Early Miocene	1.10	0.63	438	0.64	6.2	2.31	27
WG67/1	Northeast	Tskhenis River		Chattian	0.04	0.73	428	0.05	5.8	1.10	66
WG67/3	Northeast	Tskhenis River		Chattian	0.01	0.41	425	0.02	5.8	0.60	68
WG67/7	Northeast	Tskhenis River		Chattian	0.01	0.27	426	0.04	3.5	0.66	41
WG68/1	Northeast	Tskhenis River		Oligocene	0.05	0.11	472	0.31	3.1	0.43	26

Sample	Region	Location	Depth	Stratigraphy							
WG70b/2	Northeast	Tskhenis River		Chattian–Burdigalian	0.06	1.17	416	0.05	6.6	1.54	76
WG70b/4	Northeast	Tskhenis River		Aquitanian–Burdigalian	0.00	0.32	429	0.00	3.4	0.77	42
WG70c/2	Northeast	Tskhenis River		Aquitanian–Langhian	0.00	0.18	445	0.00	2.7	0.55	33
WG70c/3	Northeast	Tskhenis River		Aquitanian–Langhian	0.00	0.32	427	0.00	3.1	0.85	38
WG107/1	South			Priabonian	0.00	0.24	427	0.00	6.9	0.29	83
WG107/2	South			Priabonian	0.00	0.05	426	0.00	5.9	0.07	71
WG125/1	Northwest	Chanis River	1115.00	Aquitanian–Burdigalian	0.02	0.40	421	0.05	4.1	0.86	47
WG125/2	Northwest	Chanis River	1085.00	Aquitanian–Burdigalian	0.01	0.39	425	0.03	5.4	0.61	64
WG126/1	Northwest	Chanis River	1270.00	Aquitanian–Burdigalian	0.02	0.38	427	0.05	4.1	0.81	47
WG126/2	Northwest	Chanis River	1280.00	Aquitanian–Burdigalian	0.01	0.32	422	0.03	4.9	0.56	57
WG126/3	Northwest	Chanis River	1290.00	Aquitanian–Burdigalian	0.00	0.24	429	0.00	3.7	0.54	44
WG144/1	Northeast	Rioni tributary		Oligocene–Early Miocene	0.28	3.01	401	0.09	11.4	2.40	125
WG144/2	Northeast	Rioni tributary		Oligocene–Early Miocene	0.09	5.40	418	0.02	19.1	2.39	226
WG147/1	Northwest	Chanis River	1430.00	Aquitanian–Burdigalian	0.00	0.17	422	0.00	2.2	0.65	26
WG147/3	Northwest	Chanis River	1488.00	Aquitanian–Burdigalian	0.00	0.19	422	0.00	2.0	0.79	24
WG147/4	Northwest	Chanis River	1492.00	Aquitanian–Burdigalian	0.00	0.20	425	0.00	2.8	0.60	33
WG147/5	Northwest	Chanis River	1509.00	Aquitanian–Burdigalian	0.00	0.09	440	0.00	1.6	0.47	19

Samples are listed by geographical region and then in alphanumerical order; they are located in Figure 1. Additional details can be found in the Supplementary material.

of the Maykop Series at these localities. Sample ages are based on the determination of microfossils, nannofossils and palynomorphs, their stratigraphic position relative to other dated samples, and/or their mapped ages; details are provided in the Supplementary material.

All of the mudstone samples were cleaned, crushed and homogenized, and the 250–500 μm sieve fraction was selected for Rock-Eval pyrolysis analysis that was conducted using a Rock-Eval Oil Shows analyser. The following parameters were measured or calculated by the Rock-Eval analyser:

S_1 Liquid hydrocarbons (C_{7+}) evolved at 320°C, expressed in kg t^{-1} of rock

S_2 Hydrocarbons generated from kerogen cracking and evolved during temperature-programmed pyrolysis (from 320 to 600°C) of the sample kerogen, expressed in kg t^{-1} of rock. This peak also includes the thermal breakdown products of high molecular weight resins and asphaltenes not liberated during the S_1 analysis cycle

OPI   Oil production index, where OPI = $S_1/(S_1 + S_2)$

T_{max} The temperature in °C during kerogen pyrolysis at which the rate of hydrocarbon evolution (S_2 peak) reaches a maximum. This represents the thermal maturity of the sample

TOC   The total organic carbon content of the sample expressed as a percentage weight of the rock sample

ROC   Residual organic carbon content expressed as a percentage weight of the rock sample. This represents the dead or inert organic matter remaining in the sample after pyrolysis

AOC   Active organic carbon: that is, the carbon that is actively involved in the hydrocarbon generation process as the carbon in S_1 and S_2

HI   Hydrogen index given by the equation HI = $100 \times S_2/$TOC and expressed in mg S_2/g TOC. This represents the source rock quality of the sample (e.g. oil-prone, gas-prone)

T_{max} values regarded as anomalous are given in italics in Table 1. Tables 2–4 set out the schemes adopted for the purposes of describing and interpreting Rock-Eval data.

After reviewing the initial results (Table 1), 11 selected samples were submitted for further analysis to provide a more detailed source rock characterization, including:

- Kerogen typing and spore colour index (SCI) measurement by transmitted light and qualitative fluorescence microscopy used an Olympus BH2 microscope equipped with a UV/blue light fluorescence attachment and DP10 digital camera. The SCI technique is based on a 10-point colour scale of Munsell colour standards as the reference. The SCI scale is designed to give a straight-line correlation with vitrinite reflectance throughout the immature, mature and late mature zones.
- Pyrolysis-gas chromatography (PyGC) at 600°C for 10 s used a CDS 1000 flash pyrolyser coupled to a Shimadzu GC-14A gas chromatograph equipped with a 60 m × 0.32 mm CP-Sil-5-CB capillary column (0.25 μm phase thickness). Kerogens were thermally extracted at 320°C prior to analysis, in order to remove free hydrocarbons as far as possible.
- Solvent extraction used Soxtec extractor units and dichloromethane as the extraction solvent. Samples were extracted for 1 h and rinsed for 1 h followed by evaporation of the excess solvent under a stream of nitrogen prior to gravimetric analysis.
- Gas chromatography analysis (GC) of the whole extracts to obtain a detailed fingerprint of their mobile hydrocarbons, used a Shimadzu GC-14B gas chromatograph equipped with a 60 m × 0.32 mm CP-Sil-5-CB capillary column (0.25 μm phase thickness).
- Stable carbon isotope analysis of the solvent-extracted source rock kerogen and source rock hydrocarbons was conducted for correlation purposes. This used a Europa Scientific ANCA-GSL elemental analyser and a Europa Scientific GEO 20–20 isotope ratio mass spectrometer. The reference material used for analysis is a mineral oil standard traceable to NBS-22 (mineral oil), distributed by the IAEA. The results are reported with respect to PDB.
- Gas chromatography mass spectrometry (GC-MS) biomarker analysis of the de-asphaltened extracts used a ZABspec Ultima system coupled to a HP6890 gas chromatograph fitted with a fused silica DB-1 capillary column (30 m × 0.25 mm id, programmed at 70–320°C). Sample injections

Table 2. *The Rock-Eval interpretative schemes used in this study – source potential and TOC*

Source potential	Poor	Moderate	Good	Very good	Excellent
S_1 (kg t^{-1})	<0.5	0.5–1.0	1.0–2.0	2.0–3.0	>3.0
S_2 (kg t^{-1})	<2.0	2.0–5.0	5.0–10.0	10.0–20.0	>20.0
TOC (%)	<0.5	0.5–1.0	1.0–2.0	2.0–5.0	>5.0

Table 3. *The Rock-Eval interpretative schemes used in this study – thermal maturity*

Maturity	Immature	Early mature	Mid-mature	Late mature	Post-mature
T_{max} (°C)	<430	430–445	445–460	460–470	>470

were performed in splitless mode. The samples were analysed in multiple ion-detection mode (MID) at a scan cycle time of approximately 1.1 s. Results of these analyses are compiled in Table 5.

Source rock evaluation

Introduction

One hundred and twenty-two samples were screened by Rock-Eval pyrolysis analysis; their source rock properties are given in Table 1.

Samples show a broad range in TOC, with 44% falling within the good to very good range (1–5.0%) and 5% being excellent (>5.0%: Fig. 2). Maximum recorded TOC values for Eocene and Oligocene–Early Miocene samples were 12.5 and 7.4%, respectively. Thirty-five per cent of samples proved to have moderate to very good source potential (S_2 between 2 and 20 kg t^{-1}), with 3% having excellent source potential (S_2 >20 kg t^{-1}: Fig. 2). A maximum source potential of 64.7 and 24.9 kg t^{-1} was recorded for Eocene and Oligocene–Early Miocene samples, respectively.

Measured hydrocarbon index (HI) values indicate that 50% of samples have Type III gas-prone source quality (HI is mostly in the range 50–150 mg S_2/g TOC). However, 17% of the more prospective samples from across the region have a Type II oil- and gas-prone source quality (HI >250 mg S_2/g TOC) (Figs 2 & 3). Maximum HI values of 519 and 497 mg S_2/g TOC were recorded for the Eocene and Oligo-Miocene samples, respectively.

With respect to thermal maturity level, the most reliable T_{max} values provided a range of 396–438° C, which, in general, is indicative of immature to early mature organic matter (Fig. 3). The majority of samples provided a rather narrow T_{max} range

(416–428°C) indicative of immature organic matter (equivalent Ro <0.5%).

The following sections subdivide the samples by geographical position to evaluate their source rock properties in more detail, with a focus on the prospective samples in the dataset and their stratigraphic distribution.

West Georgia samples

Seventy-eight west Georgia samples were collected from various locations to the west of the Dziruli Massif. These include samples from along the Tskhenis and Rioni rivers in the NE of the region, from a logged section along the Chanis River in the NW, from outcrops at the southern margin of the Megrelian Depression in the central part of the region, and from the northern margin of the Adjara-Trialet Belt in the south (Fig. 1). Outcrops are typically mudstone-prone with deposition being interpreted to be due to low-density turbidite and hemipelagic settling. Mudstones are typically greenish black to greenish grey and may be both calcareous and non-calcareous. Abundant fish scales and jarosite staining (a weathering product of pyrite) characterize the Khadumian (*c.* Rupelian) part of the succession. Only in the Early Miocene part of the succession in northern outcrops do sandstones become more prevalent as shallower water, probably deltaic, facies were deposited.

Rock-Eval T_{max} values show a considerable overall range of 396–445°C (Fig. 3) with 80% being restricted to a rather narrow range (T_{max} 414–432°C) mostly indicative of the immature zone. The two most mature samples analysed (T_{max} 438 and 445° C) are early mature for oil generation.

The majority of the west Georgia samples have poor to very good organic richness (TOC mostly 0.1–3.7% extending up to 6.9–7.4%), poor to

Table 4. *The Rock-Eval interpretative schemes used in this study – source quality**

Source quality	Hydrogen-deficient	Gas-prone	Oil- and gas-prone	Oil-prone
HI (mg S_2/g TOC)	<50	50–250	250–600	>600
Source type	IV	III	II	I
Kerogen type	Inertinite	Vitrinite	Exinite	Liptinite
			Amorphous organic matter (AOM)	

*This also shows the broad correlation with the various kerogen classification schemes.

Table 5. *Results of the detailed geochemical evaluation carried out on selected prospective source rock mudstone samples of Eocene and Oligocene age from the NE margin of the Black Sea*

Sample No.		WC8/2	WC11/5	WC19/11	WC42/4	WC42/5	WC29/1	WC85/1	WG1/8	WG25a/1	WG25/15	WG28f/2
Area		Russia south					Russia north		West Georgia			
Region		Mzimta River		Tuapse	Sochi				Central	Northwest		
Section							Pshekha River	Pshish River	Skaya River		Chanis River	
Age		Early Rupelian	Mid Rupelian	Rupelian	Early Priabonian	Early Priabonian	Early Priabonian	Late Bartonian–Early Priabonian?	Rupelian	Late Bartonian–earliest Priabonian	Early Rupelian	Late Rupelian
Rock-Eval	S_1 (kg t^{-1})	1.63	0.86	0.20	0.43	1.11	0.58	0.96	0.63	0.08	0.11	0.56
	S_2 (kg t^{-1})	24.91	15.80	10.15	38.60	64.71	17.94	22.27	18.71	12.56	6.15	18.28
	T_{max} (°C)	432	434	419	414	415	413	405	409	418	417	414
	OPI	0.06	0.05	0.02	0.01	0.02	0.03	0.04	0.03	0.01	0.02	0.03
	TOC (%)	5.01	3.62	3.11	8.89	12.47	4.40	9.16	7.37	3.69	2.17	4.80
	HI (mg S_2/g TOC)	497	436	326	434	519	408	243	254	343	283	381
	SCI	–	5–6	2/2–3	–	–	1–2	–	2	2–3	3	2/2–3
Kerogen typing	Liptinite (%)	Tr	Tr	5	Tr	Tr	10	Tr	0	5	5	0
	Exinite (%)	Tr	Tr	5	Tr	Tr	Tr	Tr	5	0	30	5
	AOM (%)	100	100	70	100	100	90	100	90	85	65	90
	Vitrinite (%)	Tr	Tr	20	Tr	0	Tr	0	5	10	0	5
Py-GC*	Prist-1-ene/C_{17}	0.74	0.31	1.31	1.77	2.21	1.12	0.64	2.20	1.39	2.29	1.51

	1	2	3	4	5	6	7	8	9	10	11
Source rock HC yield and bulk composition											
Extract yield (mg/g rock)	2.2	–	0.5	–	–	1.1	1.9	2.0	1.0	1.1	2.1
Extract yield (mg/g TOC)	44.6	–	15.6	–	–	25.8	21.1	26.4	42.9	50.9	26.8
Saturates (%)	19.7	–	5.3	–	–	2.9	12.7	8.5	11.7	9.0	22.3
Aromatics (%)	32.6	–	5.9	–	–	12.6	12.0	7.1	35.2	11.5	18.5
Resins (%)	45.9	–	86.9	–	–	82.9	71.6	82.9	52.5	79.0	57.5
Asphaltenes (%)	1.8	–	1.9	–	–	1.6	3.7	1.5	0.6	0.5	1.7
HC/non-HC ratio	1.1	–	0.1	–	–	0.20	0.3	0.2	0.9	0.3	0.7
GC[†]											
Pristane/C_{17}	3.44	–	1.71	–	–	1.79	1.11	2.57	4.47	6.34	4.55
Phytane/C_{18}	1.13	–	2.05	–	–	3.32	2.17	7.01	3.02	10.71	6.52
Pristane/phytane	1.19	–	2.87	–	–	0.59	0.75	0.40	2.06	0.62	0.73
Biomarkers											
C_{30} $\beta\beta$ Hop (%)[‡]	–	–	0.63	–	–	0.41	–	0.64	0.55	0.37	0.3
C_{30} $\beta\alpha$ Hop (%)[§]	15.30	–	35.80	–	–	52.10	–	46.00	29.90	35.80	13.00
C_{32} Hop 22S (%)[¶]	57.40	–	–	–	–	–	–	–	–	–	–
C_{27} Ts/(Ts + Tm)	0.39	–	–	–	–	–	–	–	–	–	–
C_{29} Steranes 20S/(20S + 20R)	0.26	–	0.01	–	–	0.02	–	0	0.02	0.07	0.05
C_{29} Steranes $\beta\beta$/($\alpha\alpha + \beta\beta$)	0.23	–	–	–	–	0.17	–	–	–	0.24	0.15
$\delta^{13}C_{kerogen}$(‰)	-26.76	-25.95	-26.17	-27.07	-27.23	-26.19	-27.75	-25.18	-26.59	-25.26	-25.13
$\delta^{13}C_{whole\ extract}$ (‰)	-27.95	–	-29.04	–	–	-28.45	-29.30	-27.41	-28.85	-28.11	-28.04
$\Delta^{13}C$(‰)	1.19	–	2.87	–	–	2.26	1.55	2.23	2.26	2.85	2.91

*Pyrolosis-gas chromatography.
[†]Gas chromatography.
[‡]C_{30} $\beta\beta$ hopane/(C_{30} $\beta\beta$ + C_{30} hopane + C_{30} moretane).
[§]C_{30} $\beta\alpha$ moretane/(C_{30} moretane + C_{30} hopane).
[¶]C_{32} hopanes 22S/(22S + 22R). $\Delta^{13}C$ (‰) $= \delta^{13}C_{kerogen}$ (‰) $- \delta^{13}C_{whole\ extract}$ (‰)
Tr = trace.
Sample locations are shown in Figure 1. The analyses provide a more in-depth evaluation of the source rock kerogen and source rock hydrocarbon type.

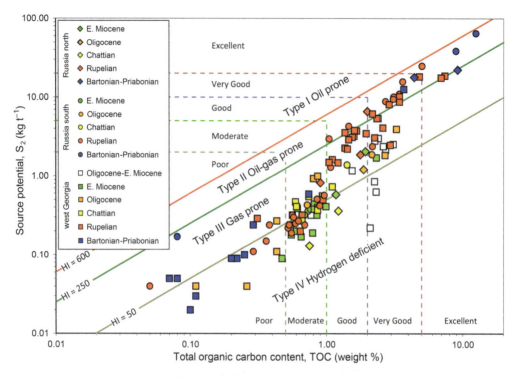

Fig. 2. Cross-plot of TOC v. S$_2$ for Eocene–Early Miocene mudstone samples from the NE margin of the Black Sea. Results show a wide variation in source quality from Type IV–III hydrogen-deficient/gas-prone to Type II oil- and gas-prone. Note that the best-quality source rocks occur in Bartonian–Priabonian and Rupelian strata.

moderate source potential extending to good to very good (S$_2$ mostly <0.1–4.1 kg t^{-1} extending up to 18.7 kg t^{-1}) and a Type IV–II hydrogen-deficient to oil- and gas-prone source quality (HI mostly 11–299 mg S$_2$/g TOC extending up to 381 mg S$_2$/g TOC) (Figs 2 & 3). Active organic carbon (AOC) accounts for 2–33% of the TOC, the remainder being dead/inert residual organic carbon (ROC).

A clear differentiation of organic richness and source potential by age is apparent in the west Georgia dataset (Fig. 2). Rupelian samples are the most prospective, with most Bartonian–Priabonian, Chattian, Oligocene and Early Miocene samples being less so. A group of undifferentiated Maykop Series samples are prospective and may, potentially, also be Rupelian in age. Further insight into variations in source rock character with age can be gained from the results from a logged section through late Bartonian–Langhian strata on the Chanis River (Fig. 4). Prospective source rock zones are identified at three levels near the base of the Chanis River section (denoted A, B and C on the accompanying log):

• Zone A is defined by one sample (WG25a/1) of very dark greyish brown calcareous mudstone of probable late Bartonian–earliest Priabonian age

at 17.2 m elevation. This is age-equivalent to the Kuma Suite on the northern side of the Greater Caucasus. The sample is organically rich (TOC 3.7%, AOC 29% of TOC) with very good source potential (S$_2$ 12.7 kg t^{-1}), and a Type II oil- and gas-prone source rock quality (HI 343 mg S$_2$/g TOC).

• Zone B, at an elevation of approximately 150 m, consists of a 38 m-thick interval of organically rich, dark greenish grey early Rupelian Maykop Series mudstones (TOC 1.0–2.5%, AOC 10–22% of TOC) with poor-good source potential (S$_2$ 1.5–6.2 kg t^{-1}) and a Type III–II gas-prone to oil- and gas-prone source rock quality (HI 122–283 mg S$_2$/g TOC). The results indicate that source rock prospectivity improves upwards through this interval.

• Zone C, at an elevation of approximately 250 m, consists of a 22 m-thick interval of organically very rich, dark grey to greenish grey Maykop Series mudstones (TOC 2–5%, AOC 7–33% of TOC) of probable late Rupelian age. The samples show a considerable variation in source potential from moderate to very good (S$_2$ 2.5–18.3 kg t^{-1}), and a Type III gas-prone to Type II oil- and gas-prone source rock quality (HI 84–381 mg S$_2$/g

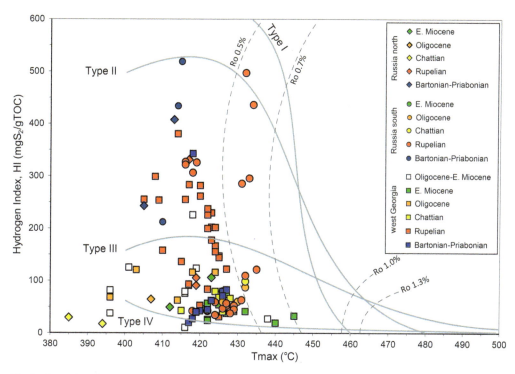

Fig. 3. Cross-plot of T_{max} v. HI for Eocene–Early Miocene mudstone samples from the NE margin of the Black Sea. Measured results show a wide variation in source quality from Type IV–III hydrogen-deficient/gas-prone to Type II oil- and gas-prone. Note that Type I kerogens, which have an abundance of strong C–C bonds, tend to show a rapid decrease in HI associated with a relatively narrow oil window with a high threshold maturity, whereas Type II and Type III kerogens, which have a more diverse chemical composition, show a more gradual decrease in HI with maturity associated with a broader hydrocarbon generation window. Note also that the relationship between Ro and T_{max} is dependent on the facies and heating rate (kinetics), and should be used here as a broad guide only. The cross-plot fields are after Cornford (1998) and are based mainly on data from the North Sea.

TOC) derived from amorphous-rich kerogen assemblages. Above this zone, mudstones are intercalated with coarse siliciclastic material and have poor source rock prospectivity.

The thickness of zone A is difficult to quantify. However, the organically rich mudstones within zones B and C have a cumulative thickness of 60 m and, taking into consideration the missing section between these zones, may be 125 m thick. Further missing section above zone C means that the total thickness of organic-rich mudstones may be as much as 200 m.

The most prospective samples encountered in the west Georgia dataset include the Bartonian–Priabonian (WG25a/1) and Rupelian samples (WG25/2, WG25/15, WG28f/2, WG28f/3 and WG28f/5) from the Chanis River section (Fig. 5) discussed above, as well as other Rupelian samples from the Skaya River (central region: WG1/8 and WG1/9) and the northern margin of the Adjara-Trialet Belt (southern region: WG33/3), plus an undifferentiated

Maykop Series sample from a tributary to the Rioni River (NE region: WG144/2) (Figs 1 & 5).

These samples have good to excellent organic richness (TOC 2.2–7.4%), good to very good source potential (S_2 5.4–18.7 kg t^{-1}), and a predominantly Type II oil- and gas-prone bulk source quality (HI 226–381 mg S_2/g TOC). The AOC comprises 19–33% of the TOC content.

The kerogen assemblages isolated from samples WG25a/1, WG25/15 and WG28f/2 from the Chanis River section, and from WG1/8 from the Skaya River, are dominated by amorphous organic matter (AOM: 65–90% of the assemblages: Table 5) characterized by a pale brown to denser medium brown structureless groundmass that exhibits a rather weak yellow through to brown fluorescence under blue light excitation (Fig. 6). The brighter yellow fluorescence is associated with microbially altered exinitic kerogen (Type II oil- and gas-prone: e.g. bisaccate pollen and algal material) as shown by the relict palynomorph structures when viewed under blue light, whilst the brown

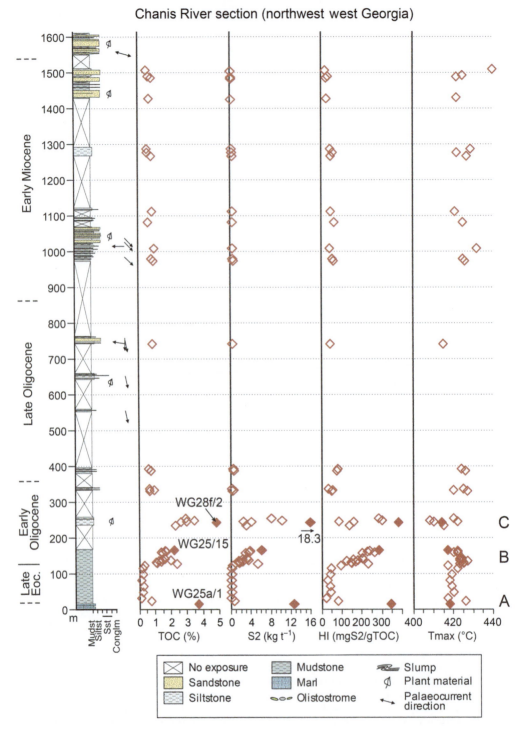

Fig. 4. Rock-Eval pyrolysis log for the Chanis River section, NW west Georgia. The stratigraphic log is adapted from Vincent *et al.* (2007). Samples selected for further analysis have a solid fill. The log position is located on Figure 1.

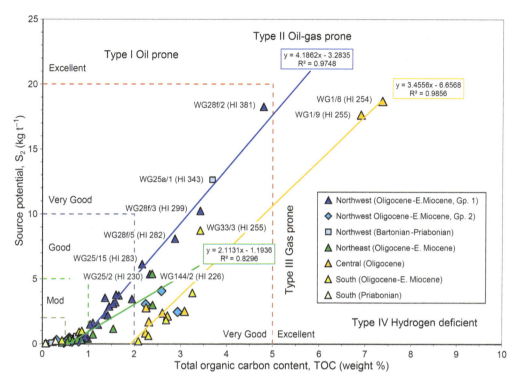

Fig. 5. Cross-plot of TOC v. S₂ for Eocene–Early Miocene mudstone samples from west Georgia.

fluorescent AOM is attributed to bacterially altered vitrinite and other humic debris (Type III gas-prone kerogen), evidence for which is given by relict structures of extensively altered vitrinite within the amorphous network.

Other macerals present in minor proportions include vitrinite, most of which is extensively sapropelized to an amorphous state, marine and non-marine algae (dinocysts, *Botryococcus* and *Pediastrum*), plus pollen and spores (fluorescent monolete spores and bisaccate pollen debris). The very pale colouration and strong greenish to light yellow palynomorph fluorescence characteristics (Fig. 7) reflects the immature nature of these samples. An additional notable feature is the presence of framboidal pyrite within the amorphous kerogen matrix; this is attributed to bacterial sulphate reduction during burial and early diagenesis.

The four west Georgia samples selected for kerogen typing were also submitted for kerogen pyrolysis-gas chromatography analysis to provide a detailed fingerprint of the kerogen pyrolysis products and, thus, investigate the gross hydrocarbon type likely to be generated at peak maturity. The chromatograms share a common Type II signature characterized by a strong C_1–C_5 gas fraction, a moderately pronounced alkene/alkane peak

distribution out to C_{30} and a moderately pronounced distribution of aromatic compounds (Fig. 8). This signature is typical of marine amorphous-rich kerogens with Type II oil- and gas-prone source rock quality, and indicates that a predominantly paraffinic (-naphthenic-aromatic) crude oil type and associated gas would be generated at peak maturity. The marked prist-1-ene peak is consistent with the thermally immature nature of these samples, whilst the prist-1-ene/C_{17} alkene ratio (Table 5) can be used as a broad indicator of maturity level for samples derived from similar source facies.

The three analysed Oligo-Miocene Maykop Series samples from west Georgia have a very similar kerogen carbon isotope composition ($\delta^{13}C_{kerogen}$ −25.1 to −25.3‰), whilst the Eocene sample (WG25a/1) proved to be isotopically lighter ($\delta^{13}C_{kerogen}$ −26.6‰) (Table 5; Fig. 9).

Further interpretation of the source rock quality can be made using the TOC v. S_2 cross-plot (Fig. 5) that takes into account the relatively inert kerogen fraction that is not involved in the hydrocarbon generation process and the effects of mineral matrix retention of hydrocarbons. The gradient of the linear trends observed through the various west Georgia datasets provides a better indication of the oil-prone source rock quality of the samples and

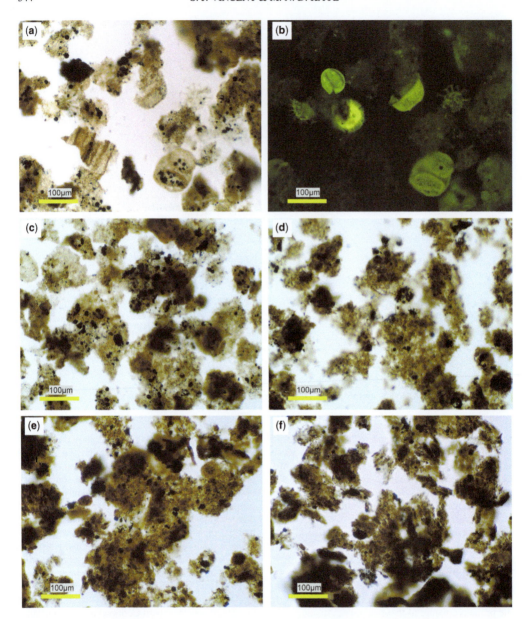

Fig. 6. Transmitted white light and fluorescence images of representative kerogen assemblages isolated from samples from the NE margin of the Black Sea. (**a**) & (**b**) Sample WG25/15 (early Rupelian); (**c**) sample WG28f/2 (late Rupelian); (**d**) sample WG25a/1 (late Bartonian–earliest Priabonian); (**e**) sample WC8/2 (early Rupelian); and (**f**) sample WC42/4 (early Priabonian). (a)–(d) are from the Chanis River section, NW west Georgia; (e) & (f) are from the southern Russian Caucasus, from the Mzimta River section and Sochi region, respectively. The kerogen assemblages are dominated by a medium brown groundmass of amorphous organic matter (AOM) with weak yellow to brown fluorescence. Other macerals include rare to minor amounts of fluorescent liptinite–exinite (mostly bisaccate pollen, plus marine algae: dinocysts and *Tasmanites*) and partially sapropelized vitrinite. Evidence of relict structures suggests that some of the AOM is derived from the bacterial alteration of these primary kerogen types. West Georgia sample WG25/15 shows a significant proportion of pollen and marine algal matter (strongly fluorescent palynomorphs: b), whilst the transmitted light image (a) shows their degraded nature (sapropelized) and partial transformation to amorphous kerogen. This image also shows a fragment of partially degraded vitrinite.

the hydrocarbon type likely to be generated at peak maturity (measured HI values merely provide in indication of the average quality of the kerogen assemblage as a whole). In this respect, the trend line constructed through the more prospective NW region samples gives a corrected hydrogen index value (HI' 419 mg S_2/g TOC) indicative of good Type II oil- and gas-prone source rock quality (Fig. 5). Similar trends constructed for the central and southern sample datasets suggest Type II oil- and gas-prone quality (HI' 346 mg S_2/g TOC) and predominantly Type III gas-prone quality (HI' 211 mg S_2/g TOC), respectively (Fig. 5).

In summary, the results indicate that significant oil and associated gas generation would be expected from several prospective Maykop Series horizons and one Eocene source rock horizon where these are buried into the main oil generation window (Ro 0.7–1.3%). However, negligible thermogenic hydrocarbon generation can be expected at the current low maturity level, as demonstrated by the mostly poor to moderate free oil yields (S_1 0.09–0.63 kg t^{-1}, max. 1.32 kg t^{-1}) and associated low OPI values (mostly <0.1). A few higher OPI values (0.1–0.2) recorded from the southern west Georgia samples could possibly be due to contamination and/or weathering.

Southern Russian western Greater Caucasus samples

Thirty-four samples were collected from the southern side of the Russian western Greater Caucasus from the Russian–Abkhazian border near Sochi westwards to the westernmost extent of Eocene and younger strata around Tuapse (Fig. 1). Many of the samples in the former region have been placed within a composite logged section constructed along the Mzimta River valley (Fig. 10). In comparison with outcrops in west Georgia, sediments here are typically more sandstone-prone (cf. Figs 4 & 10). Facies were typically deposited by high- and low-density turbidity currents and by hemipelagic settling. Olistostrome packages are developed near the base of the Oligocene intervals in both the Mzimta/Sochi (Fig. 10) and Tuapse regions (Vincent et al. 2007). Mudstones are typically grey or greenish grey, calcareous and silty. Palynology and nannofossil analysis, and mapped ages indicate that the majority of samples are early Rupelian in age.

Samples collected from the Mzimta, Sochi and Tuapse areas provided Rock-Eval T_{max} values of mostly 410–435°C (Fig. 3), indicative of the immature to early mature zones with respect to oil generation. Mzimta River samples have higher T_{max} values (T_{max} 422–435°C) that could be indicative of higher maturity or a change in source facies.

Several prospective source rock intervals were encountered on the southern side of the Russian

western Greater Caucasus. These include early Priabonian samples WC42/4 and WC42/5 from the Navaginskaya Suite in a road cut close to Sochi, Rupelian samples WC8/1, WC8/2, WC8/4 and WC11/5 from the lower part of the Mzimta River logged section (Fig. 10), and Rupelian samples WC19/11, WC19/13,WC20/4 and WC19/5 from the Agoi beach section, near Tuapse (Fig. 1). These intervals are detailed below.

Mzimta River area. Figure 10 provides a Rock-Eval data log of the 24 samples collected through the Maykop Series along the Mzimta River. Samples were largely collected from a structurally continuous transect within what is locally referred to as the Abkhazskaya tectonic zone (Lavrishchev et al. 2000, 2002). In this transect, an Early Oligocene potential source interval was not recognized. This is probably because this interval was poorly sampled and where it was, olistostromal deposits, which may not be typical of the unit as a whole, dominated the outcrops. However, five early–mid Rupelian mudstones were also sampled from a deformed tectonic sliver in the immediate footwall of the structurally overlying Akhtsu tectonic zone at the northern end of the Mzimta River section. These samples, from localities WC8 and WC11, include the most prospective Maykop Series samples collected from the Mzimta River section. They are approximately placed on the logged section based on their palynologically determined ages.

Samples WC8/2 and WC11/5 provided very good organic richness (TOC 3.6–5.0%), very good to excellent source potential (S_2 15.8–24.9 kg t^{-1}), and a Type II oil- and gas-prone source quality (HI 436–497 mg S_2/g TOC) derived from kerogen assemblages composed of a groundmass of AOM with yellow-brown/brown fluorescence and traces of liptinite–exinite (mostly marine algae) (Table 5; Fig. 6e). The AOC represents 38–44% of the TOC. Pyrolysis-gas chromatography of the kerogen isolated from these two samples provided Type II source rock signatures with a moderately well-defined distribution of alkene/alkane doublet peaks extending out to C_{30} plus moderately pronounced aromatic peaks (Fig. 11b). Significantly lower prist-1-ene/C_{17} alkene values, when compared with the rest of the dataset, appear consistent with the higher maturity level indicated by T_{max} values.

Samples WC8/1 and WC8/4 are considerably less prospective but have good organic richness (TOC 1.0–1.5%), moderate source potential (S_2 3.0–4.3 kg t^{-1}), and a poorer Type II oil- and gas-prone source quality (HI 286–296 mg S_2/g TOC). The AOC represents 26% of the TOC.

These four samples (WC11/5, WC8/1, WC8/2 and WC8/4) provide a well-defined linear trend on the TOC v. S_2 cross-plot (Fig. 12) from which a

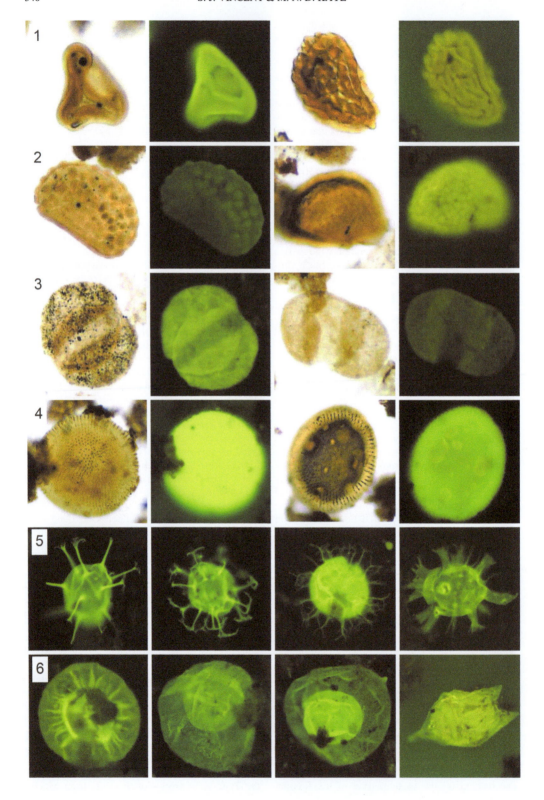

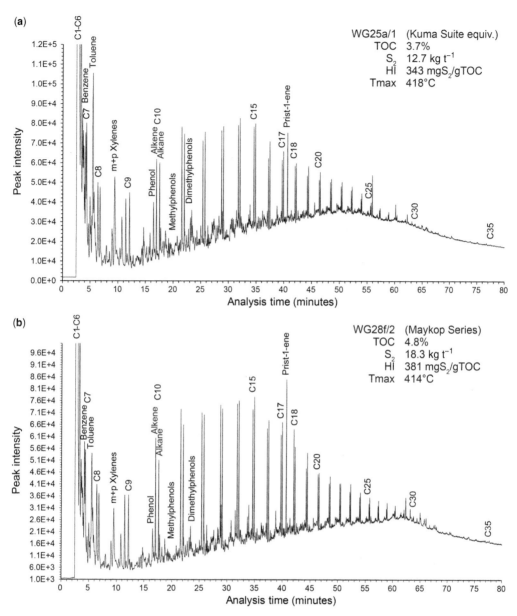

Fig. 8. Representative pyrolysis-gas chromatograms of prospective west Georgia source rock samples from the Chanis River section: (**a**) WG25a/1 (late Bartonian–earliest Priabonian) and (**b**) WG28f/2 (late Rupelian). These reveal a Type II oil- and gas-prone source rock signature characterized by a well-defined distribution of alkene/alkane doublet peaks out to C_{30} (labelled according to carbon number) with a moderately defined contribution of aromatic peaks. The diminishing height of the prist-1-ene peak relative to the neighbouring alkene/alkane peaks reflects increasing thermal maturation towards the onset of early maturity (i.e. Ro 0.5%).

Fig. 7. Transmitted white light and fluorescence images of various palynomorph types encountered in the kerogen assemblages of prospective source rock samples. Their light colour and strong greenish lemon yellow fluorescence are indicative of their thermally immature nature. (Row 1) Trilete spores (approximate diameter 70–80 µm); (row 2) monolete spores (~100–150 µm); (row 3) bisaccate pollen (~100–150 µm); (row 4) *Tasmanites* (~100–150 µm); (row 5) dinocysts (~70–80 µm); and (row 6) marine algal cysts (~150–175 µm).

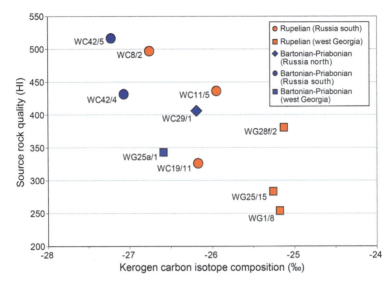

Fig. 9. Relationship between kerogen carbon isotope composition ($\delta^{13}C_{kerogen}$, ‰) and hydrogen index (HI) for Eocene and Oligocene mudstone samples from the NE margin of the Black Sea. Increasing source rock quality (HI) is broadly associated with lighter carbon isotope compositions (depleted ^{13}C) possibly due to an enhanced lipid content of the better-quality kerogen samples. The global increase in isotopic composition across the Eocene–Oligocene boundary could also contribute to the variation in isotope composition between the Eocene and Oligocene samples. Note that sample WC85/1 lies outside this broad trend, due to a possible erroneous isotope value, and is not plotted.

corrected HI value (HI′ 534 mg S_2/g TOC) indicates good Type II oil- and gas-prone source rock quality.

The remaining samples show a wide variation in organic richness (TOC <0.1–2.2%) but have mostly poor source potential (S_2 <0.1–1.9, max 2.4 kg t^{-1}) and a hydrogen-deficient to gas-prone source quality (HI 35–121 mg S_2/g TOC) with the AOC comprising 3–12% of the TOC. The majority of these poorer source rock samples form a linear trend on the TOC v. S_2 cross-plot giving a corrected HI value (HI′ 109 mg S_2/g TOC) indicative of Type III gas-prone source quality (Fig. 12).

Sochi area. Three mid–late Rupelian Maykop Series mudstones and three early Priabonian Navaginskaya Suite mudstones were analysed. The Maykop Series samples proved to have only moderate organic richness (TOC 0.5–0.6%) with poor source rock prospectivity and a Type IV hydrogen-deficient source rock quality. One of the early Priabonian samples is an organically lean non-source rock. The other two mudstones (WC42/4 and WC42/5) have excellent organic richness (TOC 8.9–12.5%), excellent source potential (S_2 39–65 kg t^{-1}), and a Type II oil- and gas-prone source rock quality (HI 434–519 mg S_2/g TOC). The kerogen assemblages of these two samples are dominated by a groundmass of medium brown AOM with a granular appearance and lemon yellow/brown fluorescence, and also contain traces of liptinite–exinite (bisaccate pollen

and marine algae: *Tasmanites* and dinocysts) and altered vitrinite (Table 5; Fig. 6f). The AOC comprises 36–44% of the TOC. The T_{max} values of approximately 415°C, together with the greenish yellow palynomorph fluorescence characteristics, indicate these samples are thermally immature. Pyrolysis-gas chromatography of the kerogen isolated from these two samples gave well-defined alkene/alkane peak distributions out to C_{30} with a reasonably marked contribution of aromatics and a well-defined prist-1-ene peak, characteristics that are consistent with a low maturity Type II oil- and gas-prone source quality (e.g. Fig. 11a).

Tuapse area. Four Rupelian Maykop Series mudstones (WC19/11, WC19/13, WC20/4 and WC20/5) were analysed from this region. They proved to be organically very rich (TOC 2.7–3.4%) with good to very good source potential (S_2 8.9–11.1 kg t^{-1}), and a Type II oil- and gas-prone source rock quality (HI 307–327 mg S_2/g TOC). The AOC comprises 26–28% of the TOC. The T_{max} values of 416–419°C combined with very rare SCI measurements of 2–3 accompanied by strong lemon yellow fluorescence characteristics (sample WC19/11) indicate that the organic matter in these samples is immature for thermogenic hydrocarbon generation (equivalent Ro 0.25–0.30%). Kerogen typing performed on sample WC19/11 indicates a predominance of moderately dense AOM (Table 5) with a

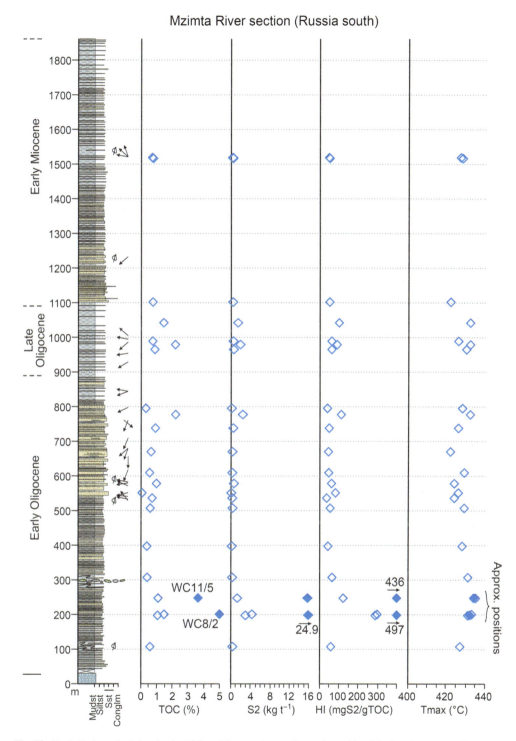

Fig. 10. Rock-Eval pyrolysis log for the Mzimta River section on the southern side of the Russian western Greater Caucasus. The samples selected for further analysis has a solid fill. The stratigraphic log, based on field observations and the work of Lavrishchev *et al.* (2000), is simplified from Vincent *et al.* (2007). The log position is located in Figure 1. The log key is shown on Figure 4.

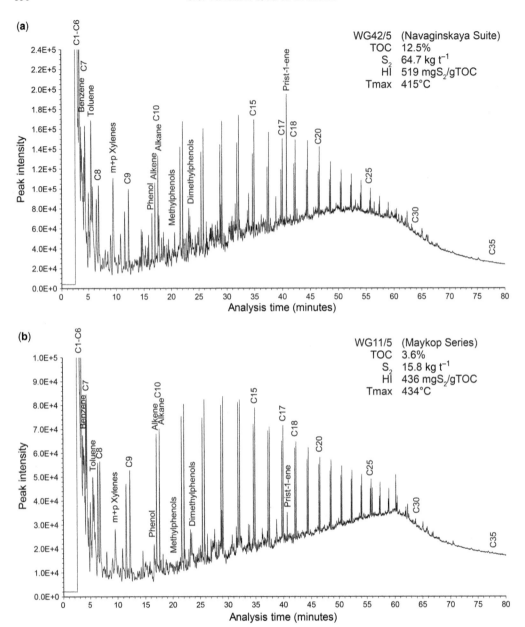

Fig. 11. Representative pyrolysis-gas chromatograms of the prospective southern Russian western Greater Caucasus samples: (**a**) WC42/5 (early Priabonian, Sochi) and (**b**) WC11/5 (mid Rupelian, Mzimta River). These resemble those from west Georgia in their Type II oil- and gas-prone source rock signature. The significantly smaller prist-1-ene peak in sample WC11/5 reflects the higher maturity level of this sample (T_{max} 434°C) compared to that of sample WC42/5 (T_{max} 415°C).

weak lemon yellow/brown fluorescence. Other macerals present in the kerogen assemblage include trace amounts of marine algal matter (*Tasmanites* and dinocysts) and land-derived palynomorphs (mainly fluorescent spores and bisaccate pollen grains).

Pyrolysis-gas chromatography of the kerogen isolated from sample WC19/11 provided a typical Type II oil- and gas-prone low maturity peak signature similar to those obtained for west Georgia and other western Greater Caucasus samples. Further

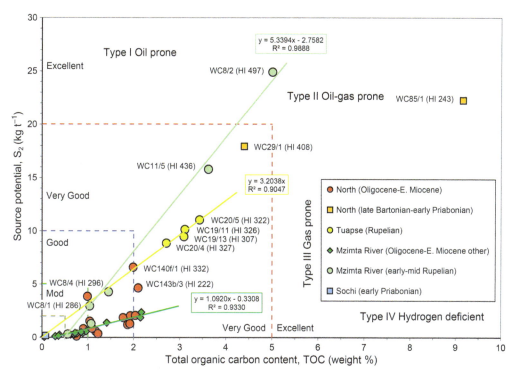

Fig. 12. Cross-plot of TOC v. S_2 cross-plot for Eocene–Early Miocene mudstone samples from the Russian western Greater Caucasus. Note, for the sake of clarity, that the two most prospective Eocene samples from the Sochi area are not shown.

interpretation of source rock quality using the TOC v. S_2 cross-plot reveals a well-defined linear trend through the Tuapse dataset, giving a corrected HI value (HI′) of 320 mg S_2/g TOC indicative of Type II oil- and gas-prone source rock quality (Fig. 12).

The kerogen assemblages isolated from prospective Maykop Series samples from the Tuapse and Mzimta River areas are isotopically slightly lighter ($\delta^{13}C_{kerogen}$ −26.8 to −26.0‰) than the west Georgia samples of the same age (Table 5; Fig. 9). Eocene samples from the Sochi region are even isotopically lighter ($\delta^{13}C_{kerogen}$ −27.2 to −27.1‰).

In summary, good oil and associated gas generation can be expected from the prospective Maykop Series and Eocene source rock horizons encountered in the Mzimta River, Sochi and Tuapse areas where these are buried into the main oil generation window.

Northern Russian western Greater Caucasus samples

Eight samples were collected from the northern side of the Russian western Greater Caucasus between the Abin River in the NW and the Belaya River in the SE (Fig. 1). Outcrops are mudstone-prone with

sandstones being extremely rare in all outcrops. Olistostrome packages are developed at the base of the Maykop Series succession on the Pshish River (locality WC85), as well as elsewhere (Vincent et al. 2007). The mudstones are typically very dark grey to grey and were deposited by hemipelagic settling in a basinal setting. Lower parts of the Maykop Series often contain fish scales and display jarosite staining.

The northern Russian western Greater Caucasus samples display very similar geochemical characteristics to the Eocene and Maykop Series samples from elsewhere in the region. All the samples analysed have moderate or better organic richness with two particularly rich Eocene samples (WC29/1 and WC85/1) having TOC values of 4.4 and 9.16%, respectively (Table 1; Fig. 2).

The relatively low T_{max} values (407–423°C), combined with the very strong greenish yellow and lemon yellow fluorescence characteristics of dinocysts and other marine algal palynomorphs (Fig. 7), indicate that all of the samples analysed are immature for thermogenic hydrocarbon generation. In general, these samples have a slightly lower maturity level and T_{max} distribution than the southern Russian western Caucasus and west Georgia samples (Fig. 3).

Within this sample set, the Rock-Eval data define two groups based on source rock quality. The three most prospective samples include two Eocene samples WC29/1 and WC85/1 from the Kuma Suite, the former sample containing a nannofossil assemblage indicative of a lower NP18, early Late Eocene age, and a single Maykop Series sample WC140f/1. The age control on the Maykop Series samples is less precise, with it probably being Early Oligocene in age. These three organically rich samples have good to excellent source potential (S_2 6.60–22.27 kg t^{-1}), and a predominantly Type II oil- and gas-prone source rock quality (HI 243–408 mg S_2/g TOC). Moderate oil and associated gas generation would be expected from these sampled horizons at peak maturity within the oil window (Ro 0.7–1.3%). The remainder of the samples have mostly poor source potential (S_2 0.36–2.05 kg t^{-1}) and a Type IV–III hydrogen-deficient to gas-prone quality (HI 17–102 mg S_2/g TOC).

Microscopy analysis of the Eocene samples WC29/1 and WC85/1 indicates kerogen assemblages dominated by AOM (Table 5) with a predominantly brown fluorescence, whilst pyrolysis-gas chromatography of the isolated kerogen provides evidence of a Type II source quality characterized by a moderately pronounced alkene/alkane distribution and well-defined aromatic peaks (e.g. toluene and xylenes). Sample WC85/1 shows a weaker peak distribution consistent with its relatively low HI value. The isolated bulk kerogen has a stable carbon isotope composition of $\delta^{13}C_{kerogen}$ −26.2‰ (WC29/1) and $\delta^{13}C_{kerogen}$ −27.8‰ (WC85/1) (Table 5; Fig. 9).

The shift towards isotopically lighter kerogens in the Eocene compared to Oligocene kerogens corresponds with the change to isotopically heavier TOC and a decrease in isotopic fractionation between organic carbon and carbonates as noted by Hayes et al. (1999) across the Eocene–Oligocene boundary. However, the trend towards isotopically lighter kerogens and better source quality/increasingly oil-prone character (higher HI value) (Fig. 9) can also be attributed to better preservation, in which bacterial action during burial and diagenesis effectively enhances the isotopically light lipid-rich content and reduces the isotopically heavier protein and carbohydrate fraction.

Hydrocarbon evaluation

The source rock extracts obtained from three Eocene samples (WC29/1, WC85/1 and WG25a/1) and five Maykop Series samples (WC8/2, WC19/11, WG1/8, WG25/15 and WG28f/2) were analysed by various techniques outlined in the 'Materials and methods' section and detailed below (see Table 5).

Extract yield and bulk chemical composition

The source rock samples selected for extraction are thermally immature (except for sample WC8/2, which is marginally early mature), organically rich mudstones (TOC 2.2–9.2%) with good to excellent source potential (S_2 6.0–25.0 kg t^{-1}), and a Type II oil- and gas-prone source rock quality (HI 243–497 mg S_2/ g TOC). The majority of samples provided extract yields in the range 1.0–2.2 mg TSE/g rock and 15.6–50.9 mg TSE/g TOC (where TSE is the total soluble extract), with the most mature sample WC8/2 (T_{max} 432°C, OPI 0.06) giving one of the highest extract yields as well as the highest saturates + aromatics/resins + asphaltene value (Fig. 13). This extract also has the most mature alkane signature (Fig. 14b). In contrast, the remaining samples (T_{max} <420°C, OPI 0.02–0.03) provided significantly lower free oil yields (S_1), variable solvent extract yields (mostly 1.1–2.0 mg/g of rock), and an extract bulk chemical composition dominated by resins (up to 86.9%) and polar aromatic compounds that tend to merge with the resins peak (e.g. Fig. 14a); these are all features typical of immature source rock extracts. Further maturation of these samples would result in progressive thermal degradation of the resin compounds to smaller and more mobile saturated and aromatic hydrocarbons.

Gas chromatogram characteristics

Gas chromatograms of the whole extracts recovered from samples WC19/11, WC29/1, WG1/8, WG25/15, WG28f/2 and WG25a/1 provide immature peak signatures (e.g. Fig. 14a) characterized by a weak C_{12}–C_{20+} n-alkane distribution, and well-defined pristane (Pr) and phytane (Ph) peaks relative to C_{17} and C_{18} (Fig. 15) with pristane/phytane values of <1.0 (except in the Chanis River Eocene extract WG25a/1). The C_{20+} n-alkane peak distribution is masked by additional peaks on the whole extract chromatograms, whilst the GC-MS traces (m/z 127 n-alkanes and isoprenoid compounds, m/z 211 alkanes: Fig. 16) reveal traces of high molecular weight alkanes out to C_{35} with a well-defined odd carbon number predominance in the C_{23}–C_{33} region (Fig. 14). These features are indicative of immature source rocks in which there has been negligible thermogenic hydrocarbon generation, and suggest a predominantly algal source facies with contributions of land-plant matter. The pristane/phytane values of <1.0 indicate an oxygen-deficient palaeoenvironment. The extract recovered from the Eocene sample WC85/1 from the northern part of

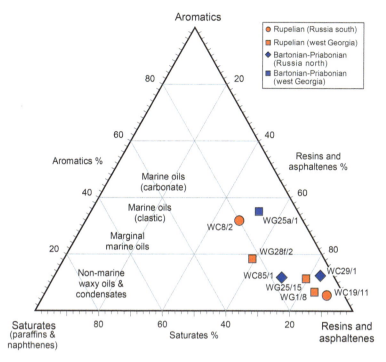

Fig. 13. Ternary plot showing the bulk chemical composition of source rock extracts from Eocene and Oligocene mudstone samples from the NE margin of the Black Sea. Note the oil types given on the ternary plot represent the compositions of mature oil types. Maturation of the source rock samples would result in thermal decomposition of the relatively immobile resin compounds and progressive enrichment of the more mobile saturated and aromatic hydrocarbon fractions.

the Russian western Greater Caucasus area differs in its stronger C_{12}–C_{20+} n-alkane distribution, with an even carbon number preference at C_{14} and C_{16}, but also appears to be derived from an algal-dominated source and an oxygen-deficient palaeoenvironment.

The whole extract chromatogram of the most mature extract (WC8/2, T_{max} 432°C: Fig. 14b) is characterized by a significantly better defined C_{12}–C_{30+} n-alkane distribution biased towards C_{14}–C_{19}, with a slight odd carbon number predominance at C_{23}, C_{25}, C_{27} and C_{29}, a pronounced pristane peak with Pr/Ph $\gg$1.0 (3.44) and a relatively well-defined distribution of aromatics peaks (naphthalenes and phenanthrenes). These characteristics are consistent with early thermogenic hydrocarbon generation (marginal–early mature for oil generation) and a mixed algal–terrigenous organic matter source input, which would be consistent with the generally greater input of coarse terrigenous material into the Mzimta region from the evolving Caucasus landmass than elsewhere during Early Oligocene time.

The isotopic composition of the whole extracts (−29.30 to −27.41‰) is broadly consistent with a mixed marine algal and land-plant organic matter

source, and lies within 1.2–2.9‰ of the source rock kerogens (−27.65 to −25.13‰) (Table 5).

Biomarker characteristics

The extracts, with the exception of WC8/2, provide immature biomarker signatures in which the terpanes show distinct contributions of hopenes and C_{29}–C_{31} $\beta\beta$ hopanes (Fig. 17a). During early diagenesis the hopenes are converted to hopanes, whilst the naturally occurring $\beta\beta$ hopanes are transformed to a mixture of $\beta\alpha$ (20R) moretanes and $\alpha\beta$ (20R) hopanes, such that the $\beta\beta$ hopanes are largely depleted at maturity levels below Ro 0.4% (Peters *et al.* 2005): that is, i.e. well before the onset of oil generation (Ro 0.5%). Tricyclic and tetracyclic terpanes are absent, whilst many of the pentacyclic terpanes (hopanes) normally observed in mature oils and extracts (e.g. WC8/2) are either present in low concentrations or partially obscured by other peaks.

The immature nature of these extracts is also reflected in the sterane fingerprints (m/z 217) by the predominance of regular steranes with a $\alpha\alpha\alpha$20R configuration (Fig. 17b). The relative proportions of

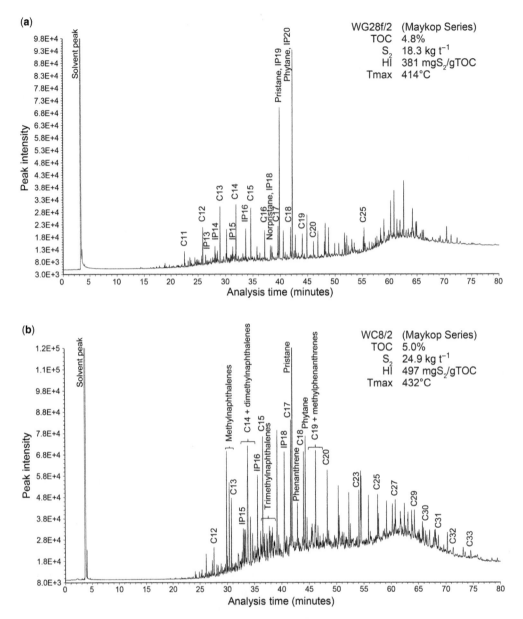

Fig. 14. Representative whole-extract chromatograms of source rock extracts recovered from the most prospective Rupelian Maykop Series field samples analysed: (**a**) WG28f/2 (Chanis River, NW west Georgia); and (**b**) WC8/2 (Mzimta River, southern Russia). The WG28f/2 trace reflects the majority of extracts derived from immature samples in its poorly defined n-alkane peak distribution focused on medium molecular weight components (C_{12}–C_{18} region) and well-defined pristane and phytane peaks (relative to C_{17} and C_{18}) with pristane/phytane <1.0. Peaks beyond C_{20} are very small and masked by other peaks on the whole-extract chromatograms. In contrast, the trace obtained from the marginally early mature sample WC8/2 reveals a stronger distribution of early thermogenic hydrocarbons and an alkane signature that suggests a predominantly marine algal source facies, whilst the relatively marked distribution of aromatics possibly reflects a contribution of terrestrial/humic organic matter.

C_{27}, C_{28} and C_{29} $\alpha\alpha\alpha$20R members show a broad distribution (Fig. 18) possibly attributed to differences in source facies, although the reliability

of this plot in determining source facies remains uncertain. The extracts provide well-defined distributions of methyldibenzothiophene peaks that reflect

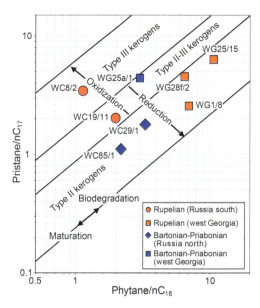

Fig. 15. Variation in Pr/C_{17} and Ph/C_{18} values of source rock extracts from Eocene and Oligocene mudstone samples from the NE margin of the Black Sea. The broad distribution shown by the majority of source rock extracts suggests a mixed facies, whilst the significantly higher Pr/Ph value of sample WC8/2 suggests greater terrestrial/humic organic matter input. The base diagram is redrawn from Shanmugam (1985).

the low maturity level. These extracts also yield well-defined distributions of methyl phenanthrene peaks and provide dibenzothiophene/phenanthrene values of 0.22–0.61. However, the extracts are too immature to provide meaningful maturity values (calculated vitrinite reflectance, Rc) from the phenanthrene and benzothiophene aromatics.

The extract recovered from the Mzimta River Maykop Series sample WC8/2 provides a biomarker signature that more closely resembles mature oils and extracts (Fig. 19). Several routine hopane and sterane maturity ratios (see Table 5) confirm the marginal early maturity obtained from measured source rock maturity data (T_{max}) and suggest an equivalent vitrinite reflectance of approximately Rc 0.55–0.60%. This is also supported by the methyldibenzothiophene ratio (4MDBT/1MDBT), which gives a calculated vitrinite reflectance of Rc 0.58% (where Rc = 0.51 + 0.073MDR: Radke & Welte 1983). With respect to source facies, the distributions of extended hopane peaks (m/z 191) and diasterane peaks (m/z 217) are consistent with a siliciclastic source, whilst the marked predominance of C_{29} $\alpha\alpha\alpha$ regular steranes within the C_{27}–C_{29} range suggests greater input of land-plant matter compared to the other samples analysed. This appears consistent

with the Pr/Ph ratio of $\gg 1.0$ and relatively high content of aromatic compounds (GC trace and bulk composition data).

Discussion

Late Middle–Late Eocene source potential

The latest Lutetian–early Priabonian (CP14–15a) Kuma Suite forms a potential source rock unit on the northern side of the Greater Caucasus (Beniamovski *et al.* 2003). The suite typically comprises dark grey calcareous mudstones or marls that range from approximately 20 to 80 m in thickness at outcrop and are up to 800 m in thickness in the west Kuban Basin (Beniamovski *et al.* 2003; Distanova 2007; Peshkov *et al.* 2016). Bentonites are present in the lower part of the suite. Distanova (2007) recorded that the Kuma Suite in the northern Caucasus region is thermally immature (T_{max} 415–427°C) with poor to excellent organic richness (TOC 0.2–10.3%, average 2.0%), poor to excellent genetic potential ($S_1 + S_2$ 0.4–30 kg t^{-1}), and a predominantly Type II oil- and gas-prone source rock quality (HI 20–612 mg S_2/g TOC, range Type IV–II). These are similar to the results recorded by this study. The two most prospective samples analysed in this study (samples WC29/1 and WC85/1) are from within the upper part of the Kuma Suite, on or close to the Pshish River. These samples are thermally immature with very good to excellent organic richness and source potential, and a Type II oil- and gas-prone source rock quality derived from kerogen assemblages dominated by AOM (Table 5). The chromatogram of the solvent-extracted hydrocarbons recovered from sample WC85/1 revealed an immature alkane distribution in the C_{12}–C_{17} range with a well-defined even carbon number preference at C_{14} and C_{16}, and a pristane/phytane value of 0.75. These characteristics are consistent with an immature marine algal-rich source facies. Detailed work is ongoing to better characterize the hydrocarbon potential of the Kuma Suite on the Belaya River section. This is consistent with or indicates even greater petroleum potential than the results cited above (M. Morton 2016 pers. comm.).

Only a limited number of age-equivalent mudstone samples have been analysed from the southern side of the Greater Caucasus. Of these, sample WG25a/1 from the base of the logged Chanis River section in NW west Georgia and samples WC42/4 and WC42/5 from the Navaginskaya Suite near Sochi, southern Russia, are the most prospective. These samples are thermally immature, have very good to excellent organic richness and source potential, and a Type II oil- and gas-prone source rock quality derived from kerogen

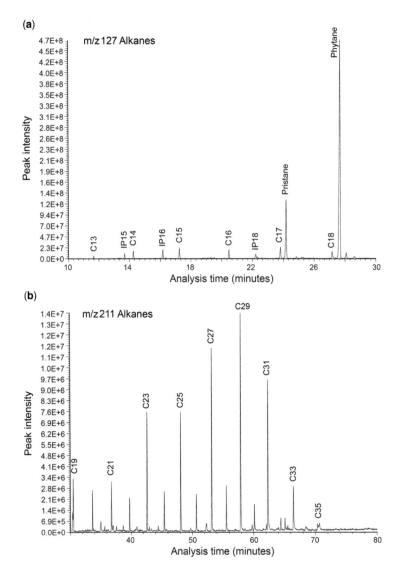

Fig. 16. Representative (**a**) m/z 127 and (**b**) m/z 211 biomarker signatures of Rupelian Maykop Series sample WG1/8 from central west Georgia (Skaya River) showing the distinctly immature n-alkane and isoprenoid compound distribution in the source rock extract. The pronounced odd carbon number predominance in the C_{23}–C_{33} n-alkane range could reflect the input of land-plant organic matter.

assemblages dominated by AOM (Table 5). Source rock hydrocarbons extracted from sample WG25a/1 reveal immature C_{12}–C_{20} alkane and biomarker signatures with a pristane/phytane value of 2.1 suggesting a slightly more oxygenated palaeoenvironment compared to the other source rock extracts analysed. The alkane distribution and almost equal proportions of C_{27}, C_{28} and C_{29} $\alpha\alpha\alpha$ regular steranes (Fig. 18) are consistent with a marine algal source facies, and compare closely with the results of earlier work by Distanova (2007) on carbonaceous samples

from the Navaginskaya Suite in the Sochi-Adler region in which a wide range in source rock prospectivity was recorded, including poor to excellent source potential (1.7–56.2 kg t^{-1}) and a Type III to Type I source rock quality (HI 160–790 mg S_2/g TOC).

Although pre-Maykop Series source potential is indicated by these analyses, further work is needed to establish the thickness of these productive intervals in the Eastern Black Sea region. Field observations of the Navaginskaya Suite suggest

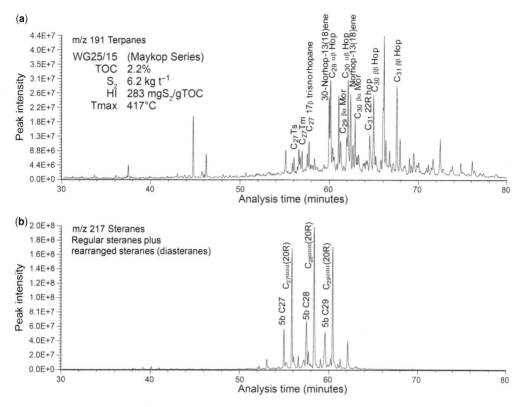

Fig. 17. Representative (**a**) m/z 191 (terpanes) and (**b**) m/z 217 (steranes) biomarker signatures of immature source rock extracts from early Rupelian Maykop Series sample WG25/15 from NW west Georgia (Chanis River). Note the prominent hopenes and $\beta\beta$ hopanes on the m/z 191 trace and the predominance of $\alpha\alpha\alpha$20R regular steranes on the m/z 217 trace suggest maturity levels of Ro <0.4% (Peters *et al.* 2005).

that the prospective intervals are only decimetre thick and that the bulk of the succession has a lesser potential (similar to sample WC42/3). In west Georgia, only the lowermost sample within the Chanis River section would appear to have encountered the Kuma Suite equivalent, such that its thickness here cannot be assessed.

Although Navaginskaya Suite samples (WC42/3–5) were deposited in the remnants of the Greater Caucasus Basin (see Vincent *et al.* 2016), the anoxic basin responsible for the Kuma Suite and its equivalents was much more extensive than this. As well as occurring on the shelfal margins of the Greater Caucasus Basin (i.e. the other Eocene samples analysed in this study), similar deposits extended between the Aral Sea in the east and at least Crimea in the west (Beniamovski *et al.* 2003). This would suggest that anoxia was not limited to restricted circulation sub-basins, but was developed across a wide area. Controls on increased bioproductivity are poorly constrained; Beniamovski *et al.* (2003) suggested increased nutrient input either due to enhanced continental runoff or volcanic ash input. Ar/Ar dating of

volcanic rocks in the Talysh and Adjara-Trialet basins of the Lesser Caucasus yield Bartonian (40.7–38.3 Ma: Vincent *et al.* 2005) and latest Lutetian–early Priabonian (41.5–37.3 Ma) ages, respectively, indicating that volcanism was contemporaneous with Kuma Suite deposition. Analyses from the Pontides indicate that middle Bartonian (P14) rocks in northern Turkey also form good to very good potential source rocks (TOC 1.8–3.8%, S_2 5.5–15.5 kg t^{-1} and HI 242–407 mg S_2/g TOC) (Aydemir *et al.* 2009). This would suggest that anoxia in the Kuma Basin also extended across the Eastern and Western Black Sea basins, thus increasing the source rock prospectivity of these basins.

Oligocene–Early Miocene source potential

Thirty-six Oligocene–Early Miocene Maykop Series mudstones from along the northern margin of the Eastern Black Sea have moderate to excellent organic richness and source potential (Fig. 2). These samples are (or potentially are) all Rupelian in age. The most prospective samples have organic richness >2%,

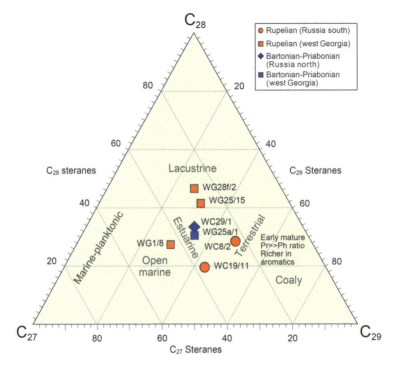

Fig. 18. Ternary plot of the C_{27}, C_{28} and C_{29} regular steranes ($\alpha\alpha\alpha$20R) recorded from source rock extracts from Eocene and Oligocene mudstone samples from the NE margin of the Black Sea. The sample set provides a poorly defined spread of data suggesting a broad range source facies and palaeoenvironment, while the significantly higher proportion of C_{29} steranes in sample WC8/2 could be consistent with greater terrestrial/humic organic matter input.

source potential >8 kg t^{-1} and corrected HI′ values of 346–534 mg S_2/g TOC (Figs 3, 5 & 12). They are typically thermally immature (T_{max} <430°C), although some samples in the Mzimta River section are early mature.

In NW west Georgia, the prospective interval(s) is at least 60 m thick and has the potential to be as much as 200 m thick. The source potential index (SPI) (Demaison & Huizinga 1994) of this interval (assuming an average density of 2.0 t m^{-3}) ranges somewhere between 0.7 and 2.5 t HC m^{-2}. In the much more sandstone-prone Maykop Series on the southern side of the Russian western Greater Caucasus, prospective samples of Rupelian age were only obtained from mudstone-rich intervals. This would suggest that siliciclastic dilution degrades source rock quality. It has not been possible to estimate the thickness of the prospective interval(s) in this region.

Robinson *et al.* (1996) summarized data from around the margins of the whole of the Eastern Black Sea. In the Greater Caucasus, they also identified the basal part of the Maykop Series as being the most prospective, with source potential measured as reaching 31 kg t^{-1}. These samples

were identified as being oil- and gas-prone (HI values up to 390 mg S_2/g TOC) and thermally immature (Ro 0.33–0.54%).

Both Saint-Germès and Sachsenhofer and their co-workers (Saint-Germès 1998; Saint-Germes *et al.* 2000*a*, 2002; Sachsenhofer *et al.* 2017*a*) have undertaken source rock analyses on the Maykop Series on the northern side of the Greater Caucasus. At the Maykop Series type-section on the Belaya River, these authors identified that the main source interval occurs in the late Rupelian immediately above a Paratethyan ('Solenovian') isolation event that is marked by an influx of freshwater and the deposition of the Polbian or Ostracoda beds. Palynological data would suggest that zone C in the Chanis River section (Fig. 4) occurs at the same stratigraphic level (i.e. within the Lower Morozkina Balka Formation). High organic matter preservation in the Belaya River section is thought to be due to high marine phytoplankton bioproductivity. The prospective interval here has TOC values of up to 3.8% and corrected HI values of up to 575 mg S_2/g TOC. It is, however, only about 5 m thick. Sachsenhofer *et al.* (2017*a*) calculated the SPI of the Maykop Series as a whole at its type section to be 2.0 t HC m^{-2}.

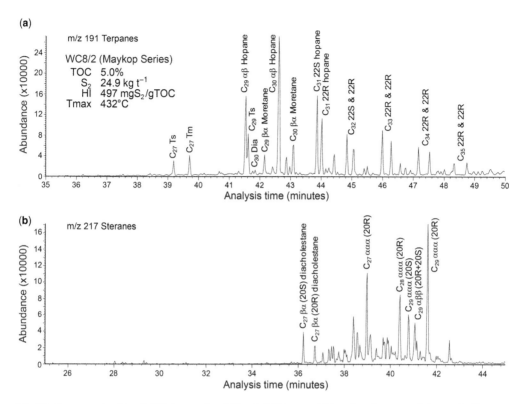

Fig. 19. Representative (**a**) m/z 191 (terpanes) and (**b**) m/z 217 (steranes) biomarker signatures of the marginally early mature source rock extract from early Rupelian Maykop Series sample WC8/2 from the southern Russian western Greater Caucasus (Mzimta River). These traces show a significantly more mature signature compared to that given in Figure 17.

Sachsenhofer *et al.* (2017*b*) summarizes the source rock potential of the Maykop Series (and its equivalents) from other parts of Paratethys. Around or within the Black Sea, SPI values are typically lower than on the Belaya River section. The only exception is the spatially restricted Kaliakra canyon, offshore Bulgaria (Mayer *et al.* 2017). Thus, the 0.7–2.5 t HC m^{-2} range estimate for the Chanis River section in NW west Georgia is promising given the relatively low, potentially terrestrially influenced, source quality of the Maykop Series here. The marine organic matter content, and, thus, the source potential and quality, would be expected to increase towards the SW away from the evolving Greater Caucasus. Some caution should be expressed, however, in extrapolating source rock properties from the margins of the western Greater Caucasus into the Eastern Black Sea because of the intervening Shatskiy Ridge. This developed as a bathymetric high during the Oligocene Epoch due to rotation and flexural loading of the northern Transcaucasus by the Greater Caucasus. It is possible that geochemical characteristics documented here may only

therefore be a guide to the source rock characteristics of the Maykop Series in the Tuapse Trough and its continuation south of Gudauta southeastwards to the Rioni Basin. Evidence for active petroleum systems in the Eastern Black Sea would, nevertheless, suggest either that the Maykop Series has as good or greater source potential than that documented in this study or that the Kuma Suite forms an important additional, previously unrecognized, source interval within the Eastern Black Sea Basin.

Conclusions

The systematic source rock evaluation of Late Middle Eocene–Early Miocene mudstones from the NE margin of the Black Sea has not been previously attempted. One hundred and twenty-two samples were screened by Rock-Eval pyrolysis (Table 1). Samples show a broad range of source rock characteristics, with most being thermally immature for oil generation (Fig. 3). The majority of samples have poor to moderate organic richness, poor source potential and a gas-prone (Type III) or hydrogen-deficient (Type IV) source quality (Fig. 2). However, 17 (14%) of the samples have very good to excellent organic richness, good to excellent source potential, and a Type II oil- and gas-prone quality. The majority of these samples (13) are Rupelian in age (Fig. 2), and crop out in the lowermost part of the Maykop Series in both west Georgia and Russia on the southern side of the Greater Caucasus. The remaining four samples are late Bartonian–early Priabonian in age: one is from north of the Greater Caucasus and occurs in the Kuma Suite; and three are from south of the Greater Caucasus and represent its stratigraphic equivalents. Maximum recorded TOC and source potential values for these late Bartonian–early Priabonian and Rupelian samples are 12.5 and 7.4%, and 64.7 and 24.9 kg t^{-1}, respectively. Interpreted hydrogen index trends clearly indicate good Type II oil- and gas-prone source quality within Rupelian strata in the Chanis River section in NW west Georgia (HI' 419 mg S$_2$/g TOC), and in the Mzimta River section (HI' 534 mg S$_2$/g TOC) and in the Tuapse region (HI' 320 mg S$_2$/g TOC), both in southern Russia (Figs 5 & 12). Kerogens are dominated by AOM.

Logged sections along the Chanis and Mzimta river sections enable greater insights to be gained into the stratigraphic distribution of source-prone intervals within the Rupelian part of the Maykop Series. In the Chanis River section in NW west Georgia, prospective source rock intervals occur in the earliest Rupelian (zone B) and in the late Rupelian (zone C) (Fig. 4). The latter is probably equivalent to the source-rock-prone interval related to the post-Solenovian isolation event at the Maykop Series type section on the Belaya River on the northern side of the Russian western Greater Caucasus. The prospective interval(s) on the Chanis River section is at least 60 m thick. Zone C is separated by intervals of non-exposure. If similar source rock properties occur throughout these non-exposed parts of the section, the prospective interval could be as much as 200 m thick. The source potential index (SPI) of the Maykop Series along the Chanis River ranges somewhere between 0.7 and 2.5 t HC m^{-2}.

In the much more sandstone-prone Maykop Series along the Mzimta River in Russia, prospective samples of early and mid Rupelian age were only obtained from allochthonous mudstone-rich intervals (Fig. 10). This would suggest that siliciclastic dilution degrades source rock quality. However, where mudstones are preserved, they have lighter δ^{13}C isotopic values than their Rupelian counterparts in west Georgia (Fig. 9). This possibly reflects a higher lipid content and better preservation of organic matter in these samples. It has not been possible to estimate the thickness of the prospective source rock interval(s) in this section.

Kuma Suite-equivalent prospective source rocks are present both on the southern side of the western Greater Caucasus, in Crimea and in the Pontides (as well as farther afield), raising the prospect that similar sediments were also deposited in the Black Sea. Maykop Series source rocks are known from the Black Sea. The source rock characteristics of mudstones deposited along the NE margin of the Black Sea provide insights into the potential of source rocks in the basin. However, the presence of the Shatskiy Ridge as a bathymetric high during the deposition of the Maykop Series means that there are uncertainties as to whether these insights apply only to the region to the NE of the ridge (i.e. the Tuapse Trough and Rioni Basin) or also to the Eastern Black Sea Basin itself.

Further work is required to fully document the source potential of the Early Oligocene portion of the Maykop Series and to properly characterize the Late Middle–Late Eocene Kuma Suite-equivalent on the southern side of the Greater Caucasus. This would be best achieved through the analysis of well samples (where available) from the Megrelian depression of NW Georgia. Alternatively, further outcrops of these intervals could be sought in NW west Georgia or in the autonomous region of Abkhazia, an area we were unable to gain access to.

The research would not have been possible without the field assistance of Mike Simmons, Simon Inger and the late Lara Voronova (all formerly CASP), Teimuraz Barabadze (Georgian Technical University, Georgia), Tanya Pinchuk (Krasnodarneftgas, Russia), and Vladimir Lavrishchev (Kavkazgeols'emka, Russia) or the funding of CASP's industrial sponsors. Biostratigraphic determinations were carried out by the late Bill Braham, Sam Gibbs (NOC, UK), Mike Bidgood (GSS International, UK), Paul Dodsworth and Donata Zucchi (both Millennia Ltd, UK), and Keith Richards (KrA Stratigraphic, UK). The majority of geochemical analyses were conducted by Julie Kaye and Matt Kaye (OceanGrove Geoscience Ltd, Aberdeen, UK), with GC-MS analyses being conducted by Nigel Goodwin/Malcolm Dee (ITS Ltd) and Gareth Harriman (GHGeochem Ltd), and carbon isotope analyses being conducted by Charles Belanger (Iso-Analytical Ltd) Their contributions are gratefully acknowledged. Reinhard Sachsenhofer and two anonymous reviewers, and editor Mike Simmons, are thanked for their constructive reviews and comments. This is Cambridge Earth Science contribution ESC 3754.

References

AFANASENKOV, A.P., NIKISHIN, A.M. & OBUKHOV, A.N. 2007. *Eastern Black Sea Basin: Geological Structure and Hydrocarbon Potential*. Science World, Moscow [in Russian].

AKHMETIEV, M.A., POPOV, S.V., KRHOVSKY, J., GONCHAROVA, I.A., ZAPOROZHETS, N.I., SYCHEVSKAYA, E.K. & RADIONOVA, E.P. 1995. *Palaeontology and Stratigraphy* of the Eocene-Miocene Sections of the Western Pre-Caucasia: Excursion Guidebook. IGCP Projects 326, 329 & 343. Russian Committee for IGCP, Moscow,

ALLEN, M.B. & ARMSTRONG, H.A. 2008. Arabia–Eurasia collision and the forcing of mid-Cenozoic global cooling. Palaeogeography, Palaeoclimatology, Palaeoecology, 265, 52–58, https://doi.org/10.1016/j.palaeo.2008.04.021

ANDREEV, V.M. 2005. Mud volcanoes and oil shows in the Tuapse Trough and Shatsky Ridge (Black Sea). Doklady Earth Sciences, 402, 516–519.

AYDEMIR, V., IZTAN, Y.H., SHIKHLINSKY, S., BATI, Z. & GÜRGEY, A. 2009. Middle Eocene aged source rock evidence from Black Sea, Turkey; BSP14-biozone. In: DALKILIÇ, M. (ed.) 2nd International Symposium on the Geology of the Black Sea Region, Abstracts. General Directorate of Mineral Research and Exploration (MTA), Ankara, 28–29.

AYDINÇAKIR, E. & ŞEN, C. 2013. Petrogenesis of the post-collisional volcanic rocks from the Borçka (Artvin) area: Implications for the evolution of the Eocene magmatism in the Eastern Pontides (NE Turkey). Lithos, 172–173, 98–117, https://doi.org/10.1016/j.lithos.2013.04.007

BÁLDI, T. 1984. The terminal Eocene and Early Oligocene events in Hungary and the separation of an anoxic, cold Paratethys. Eclogae Geologicae Helvetiae, 77, 1–27.

BAZHENOVA, O.K., FADEEVA, N.P., SAINT-GERMES, M.L., AREF'EV, O.V. & BAUDIN, F. 2002. Biomarkers of organic matter in Maykop rocks and oils of the Caucasian–Scythian region. Geochemistry International, 40, 899–913.

BENIAMOVSKI, V.N., ALEKSEEV, A.S., OVECHKINA, M.N. & OBERHÄNSLI, H. 2003. Middle to upper Eocene dysoxic–anoxic Kuma Formation (northeast Peri-Tethys): biostratigraphy and paleoenvironments. In: WING, S.L., GINGERICH, P.D., SCHMITZ, B. & THOMAS, E. (eds) Causes and Consequences of Globally Warm Climates in the Early Paleogene. Geological Society of America, Special Papers, 369, 95–112.

BOHRMANN, G., IVANOV, M. ET AL. 2003. Mud volcanoes and gas hydrates in the Black Sea: new data from Dvurechenskii and Odessa mud volcanoes. Geo-Marine Letters, 23, 239–249, https://doi.org/10.1007/s00367-003-0157-7

CORNFORD, C. 1998. Source rocks and hydrocarbons of the North Sea. In: GLENNIE, K.W. (ed.) Petroleum Geology of the North Sea, Basic Concepts and Recent Advances. Blackwell Science, Oxford, 376–462.

DEMAISON, G. & HUIZINGA, B.J. 1994. Genetic classification of petroleum systems using three factors: charge, migration and entrapment. In: MAGOON, L.B. & DOW, W.G. (eds) The Petroleum System, from Source to Trap. American Association of Petroleum Geologists, Memoirs, 60, 73–89.

DIMITROV, L.I. 2002. Mud volcanoes – the most important pathway for degrassing deeply buried sediments. Earth-Science Reviews, 59, 49–76, https://doi.org/10.1016/S0012-8252(02)00069-7

DISTANOVA, L.R. 2007. Conditions of oil source potential formation in the Eocene deposits in the basins of the Crimea and Caucasus regions. Moscow University Geology Bulletin, 62, 59–64.

FINETTI, I., BRICCHI, G., DEL BIN, A., PIPAN, M. & XUAN, Z. 1988. Geophysical study of the Black Sea. Bollettino di Geofisica Teorica ed Applicata, 30, 197–324.

GEODEKYAN, A.A., ZABANBARK, A. & KONYUKHOV, A.I. 1996. On oil and gas mother-rock potential of Maikop muds in the Black Sea depression. Doklady Akademii Nauk, 346, 222–225 [in Russian].

GLUMOV, I.F. & VIGINSKIY, V.A. 2000. Oil-gas prospects of Tuapse downwarp, Transcaucasus offshore (Black Sea). Petroleum Geology, 34, 67–73.

GLUMOV, I.F., MALOVITSKY, Y.P., NOVIKOVA, A.A. & SENIN, B.V. 2004. Regional geology and oil and gas content of the Caspian Sea. NEDRA, Moscow, 342 [in Russian].

HAYES, J.M., STRAUSS, H. & KAUFMAN, A.J. 1999. The abundance of 13C in marine organic matter and isotopic fractionation in the global biogeochemical cycle of carbon during the past 800 Ma. Chemical Geology, 161, 103–125, https://doi.org/10.1016/S0009-2541(99)00083-2

INAN, S., YALCIN, M.N., GULIEV, I.S., KULIEV, K. & FEIZULLAYEV, A.A. 1997. Deep petroleum occurrences in the lower Kura depression, south Caspian basin, Azerbaijan: an organic geochemical and basin modeling study. Marine and Petroleum Geology, 14, 731–762, https://doi.org/10.1016/s0264-8172(97)00058-5

ISAKSEN, G.H., ALIYEV, A., BARBOZA, S.A., PULS, D. & GULIYEV, I. 2007. Regional evaluation of source rock quality in Azerbaijan from the geochemistry of organic-rich rocks in mud-volcano ejecta. In: YILMAZ, P.O. & ISAKSEN, G.H. (eds) Oil and Gas of the Greater Caspian area. American Association of Petroleum Geologists, Studies in Geology, 55, 51–64.

JONES, R.W. & SIMMONS, M.D. 1996. A review of the stratigraphy of Eastern Paratethys (Oligocene–Holocene). Bulletin of the Natural History Museum London (Geology), 52, 25–49.

JONES, R.W. & SIMMONS, M.D. 1997. A review of the stratigraphy of Eastern Paratethys (Oligocene–Holocene), with particular emphasis on the Black Sea. In: ROBINSON, A.G. (ed.) Regional and Petroleum Geology of the Black Sea and Surrounding Region. American Association of Petroleum Geologists, Memoirs, 68, 39–52.

KATZ, B., RICHARDS, B., LONG, D. & LAWRENCE, W. 2000. A new look at the components of the petroleum system of the South Caspian Basin. Journal of Petroleum Science and Engineering, 28, 161–182, https://doi.org/10.1016/S0920-4105(00)00076-0

KAYMAKCI, N., OZCELIK, Y., WHITE, S.H. & VAN DIJK, P.M. 2009. Tectono-stratigraphy of the Cankiri Basin: Late Cretaceous to early Miocene evolution of the Neotethyan Suture Zone in Turkey. In: VAN HINSBERGEN, D.J.J., EDWARDS, M.A. & GOVERS, R. (eds) Collision and Collapse at the Africa–Arabia–Eurasia Subduction Zone. Geological Society, London, Special Publications, 311, 67–106, https://doi.org/10.1144/SP311.3

KHOLODOV, V.N. & NEDUMOV, R.I. 1996. Problems of the Caucasus paleoland existence during the Oligocene–Miocene time. Stratigraphy and Geological Correlation, 4, 181–190.

KRUGLYAKOVA, R.P., BYAKOV, Y.A., KRUGLYAKOVA, M.V., CHALENKO, L.A. & SHEVTSOVA, N.T. 2004. Natural oil and gas seeps on the Black Sea floor. Geo-Marine

Letters, **24**, 150–162, https://doi.org/10.1007/s00367-004-0171-4

LAVRISHCHEV, V.A., SEMENOV, V.M., ANDREEV, N.M. & GORSHKOV, A.S. 2000. *National Geological Map of the Russian Federation, Caucasus Series Sheet K-37-IV (Sochi)*. St Petersburg Cartographic Enterprise of VSEGEI, Moscow, 1:200 000 [in Russian].

LAVRISHCHEV, V.A., PRUTSKIY, N.I. & SEMENOV, V.M. 2002. *National Geological Map of the Russian Federation, Caucasus Series Sheet K-37-V (Krasnar Polyana)*. St Petersburg Cartographic Enterprise of VSEGEI, Moscow, 1:200 000 [in Russian].

LEAR, C.H., ROSENTHAL, Y., COXALL, H.K. & WILSON, P.A. 2004. Late Eocene to early Miocene ice sheet dynamics and the global carbon cycle. *Paleoceanography*, **19**, PA4015, https://doi.org/10.1029/2004PA001039

LOZAR, F. & POLINO, R. 1997. Early Cenozoic uprising of the Great Caucasus revealed by reworked calcareous nannofossils. EUG 9, 23–27 March 1997, Strasbourg. *Terra Nova*, **9**, (Abstract Suppl. 1), 141.

MARÍN-MORENO, H., MINSHULL, T.A. & EDWARDS, R.A. 2013. A disequilibrium compaction model constrained by seismic data and application to overpressure generation in the Eastern Black Sea Basin. *Basin Research*, **25**, 331–347, https://doi.org/10.1111/bre.12001

MAYER, J., RUPPRECHT, B.J. ET AL. 2017. Source potential and depositional environment of Oligocene and Miocene rocks offshore Bulgaria. *In*: SIMMONS, M.D., TARI, G.C. & OKAY, A.I. (eds) *Petroleum Geology of the Black Sea*. Geological Society, London, Special Publications, **464**, https://doi.org/10.1144/SP464.2

NAIRN, S.P., ROBERTSON, A.H.F., ÜNLÜGENÇ, U.C., TASLI, K. & İNAN, N. 2013. Tectonostratigraphic evolution of the Upper Cretaceous–Cenozoic central Anatolian basins: an integrated study of diachronous ocean basin closure and continental collision. *In*: ROBERTSON, A.H.F., PARLAK, O. & ÜNLÜGENÇ, U.C. (eds) *Geological Development of Anatolia and the Easternmost Mediterranean Region*. Geological Society, London, Special Publications, **372**, 343–384, https://doi.org/10.1144/SP372.9

NIKISHIN, A.M., OKAY, A.I., TÜYSÜZ, O., DEMIRER, A., AMELIN, N. & PETROV, E. 2015a. The Black Sea basins structure and history: new model based on new deep penetration regional seismic data. Part 1: Basins structure and fill. *Marine and Petroleum Geology*, **59**, 638–655, https://doi.org/10.1016/j.marpetgeo.2014.08.017

NIKISHIN, A.M., OKAY, A., TÜYSÜZ, O., DEMIRER, A., WANNIER, M., AMELIN, N. & PETROV, E. 2015b. The Black Sea basins structure and history: new model based on new deep penetration regional seismic data. Part 2: Tectonic history and paleogeography. *Marine and Petroleum Geology*, **59**, 656–670, https://doi.org/10.1016/j.marpetgeo.2014.08.018

OKAY, A.I. & ŞAHINTÜRK, O. 1997. Geology of the Eastern Pontides. *In*: ROBINSON, A.G. (ed.) *Regional and Petroleum Geology of the Black Sea and Surrounding Region*. American Association of Petroleum Geologists, Memoirs, **68**, 291–311.

PESHKOV, G.A., BARABANOV, N.N., BOLSHAKOVA, M.A., BORDUNOV, S.I., KOPAEVICH, L.F. & NIKISHIN, A.M. 2016. The oil and gas potential of the Kuma rocks of the Bakhchisarai region of Crimea. *Moscow University Geology Bulletin*, **71**, 262–268.

PETERS, K.E., WALTERS, C.C. & MOLDOWAN, J.M. 2005. *The Biomarker Guide, Biomarkers and Isotopes in Petroleum Exploration and Earth History, Volume 2*. Cambridge University Press, Cambridge.

POPOV, S.V., AKHMETIEV, M.A., ZAPOROZHETS, N.I., VORONINA, A.A. & STOLYAROV, A.S. 1993. Evolution of the Eastern Paratethys in the Late Eocene–Early Miocene. *Stratigraphy and Geological Correlation*, **1**, 572–600.

POPOV, S.V., ANTIPOV, M.P., ZASTROZHNOV, A.S., KURINA, E.E. & PINCHUK, T.N. 2010. Sea-level fluctuations on the northern shelf of the Eastern Paratethys in the Oligocene–Neogene. *Stratigraphy and Geological Correlation*, **18**, 200–224, https://doi.org/10.1134/S0869593810020073

RADKE, M. & WELTE, D.H. 1983. The methylphenanthrene index (MPI): a maturity parameter based on aromatic hydrocarbons. *In*: BJOROY, M. (ed.) *Advances in Organic Geochemistry*. Wiley, Chichester, 504–512.

RAINER, T., OLARU, R., KOSI, W., KREZSEK, C., UNGUREANU, C. & TARI, G. 2015. Thermal maturity and hydrocarbon generation potential of the western Black Sea. *In*: SIMMONS, M., DEMIRER, A., TARI, G., OKAY, A.I. & PANAIOTU, A. (eds) *Petroleum Geology of the Black Sea, Abstracts*. Geological Society, London, 29.

ROBINSON, A.G., RUDAT, J.H., BANKS, C.J. & WILES, R.L.F. 1996. Petroleum geology of the Black Sea. *Marine and Petroleum Geology*, **13**, 195–223, https://doi.org/10.1016/0264-8172(95)00042-9

RÖGL, F. 1999. Mediterranean and Paratethys. Facts and hypotheses of an Oligocene to Miocene paleogeography (short overview). *Geologica Carpathica*, **50**, 339–349.

SACHSENHOFER, R.F., STUMMER, B., GEORGIEV, G., DELLMOUR, R., BECHTELA, A., GRATZERA, R. & CORIĆ, S. 2009. Depositional environment and hydrocarbon source potential of the Oligocene Ruslar Formation (Kamchia Depression; Western Black Sea). *Marine and Petroleum Geology*, **26**, 57–84, https://doi.org/10.1016/j.marpetgeo.2007.08.004

SACHSENHOFER, R.F., POPOV, S.V. ET AL. 2017a. The type section of the Maikop Group (Oligocene–Lower Miocene) at the Belaya River (North Caucasus): Depositional environment and hydrocarbon potential. *AAPG Bulletin*, **101**, 289–319.

SACHSENHOFER, R.F., POPOV, S.V. ET AL. 2017b. Oligocene and Lower Miocene source rocks in the Paratethys: palaeogeographical and stratigraphic controls. *In*: SIMMONS, M.D., TARI, G.C. & OKAY, A.I. (eds) *Petroleum Geology of the Black Sea*. Geological Society, London, Special Publications, **464**, https://doi.org/10.1144/SP464.1

SAINT-GERMÈS, M. 1998. *Étude Sédimentologique et Géochimique de la Matière Organique du Bassin Maykopien (Oligocène-Miocène inférieur) de la Crimée a l'Azerbaidjan*. PhD thesis, Université Pierre et Marie Curie, Paris.

SAINT-GERMES, M.L., BAZHENOVA, O.K., BAUDIN, F., ZAPOROZHETS, N.I. & FADEEVA, N.P. 2000a. Organic matter in Oligocene Maikop Sequence of the North Caucasus. *Lithology and Mineral Resources*, **35**, 47–62, https://doi.org/10.1007/bf02788284

SAINT-GERMES, M., BOCHERENS, H., BAUDIN, F. & BAZHENOVA, O. 2000b. Evolution of the delta C-13 values of organic matter of the Maykop Series during Oligocene–Lower Miocene. *Bulletin de la Societe Geologique de France*, **171**, 13–21.

SAINT-GERMES, M., BAUDIN, F., BAZHENOVA, O., DERENNE, S., FADEEVA, N. & LARGEAU, C. 2002. Origin and preservation processes of amorphous organic matter in the Maykop Series (Oligocene–Lower Miocene) of Precaucasus and Azerbaijan. *Bulletin de la Société Géologique de France*, **173**, 423–436, https://doi.org/10.2113/173.5.423

SAINTOT, A., BRUNET, M.-F. *ET AL.* 2006. The Mesozoic–Cenozoic tectonic evolution of the Greater Caucasus. *In*: GEE, D. & STEPHENSON, R. (eds) *European Lithosphere Dynamics*. Geological Society, London, Memoirs, **32**, 277–289, https://doi.org/10.1144/GSL.MEM.2006.032.01.16

SCHULZ, H.-M., BECHTEL, A. & SACHSENHOFER, R.F. 2005. The birth of the Paratethys during the Early Oligocene: from Tethys to an ancient Black Sea analogue? *Global and Planetary Change*, **49**, 163–176.

SCOTT, C.L., SHILLINGTON, D.J., MINSHULL, T.A., EDWARDS, R.A., BROWN, P.J. & WHITE, N.J. 2009. Wide-angle seismic data reveal extensive overpressures in the Eastern Black Sea Basin. *Geophysical Journal International*, **178**, 1145–1163, https://doi.org/10.1111/j.1365-246X.2009.04215.x

ŞENGÖR, A.M.C. & YILMAZ, Y. 1981. Tethyan evolution of Turkey: a plate tectonic approach. *Tectonophysics*, **75**, 181–241.

SHANMUGAM, G. 1985. Significance of coniferous rain forests and related organic matter in generating commercial quantities of oil, Gippsland Basin, Australia. *AAPG Bulletin*, **69**, 1241–1254.

TOPUZ, G., OKAY, A.I. *ET AL.* 2011. Post-collisional adakite-like magmatism in the Ağvanis Massif and implications for the evolution of the Eocene magmatism in the Eastern Pontides (NE Turkey). *Lithos*, **125**, 131–150, https://doi.org/10.1016/j.lithos.2011.02.003

TUGOLESOV, D.A. 1989. *Album of Structural and Thickness Maps of Black Sea Basin Cenozoic Sediments; 1:1 500 000*. Main Administration of Geodesy and Cartography under the Council of Ministers of the USSR, Moscow [in Russian].

VETÖ, I. 1987. An oligocene sink for organic carbon: Upwelling in the paratethys? *Palaeogeography, Palaeoclimatology, Palaeoecology*, **60**, 143–153, https://doi.org/10.1016/0031-0182(87)90029-0

VINCENT, S.J., ALLEN, M.B., ISMAIL-ZADEH, A.D., FLECKER, R., FOLAND, K.A. & SIMMONS, M.D. 2005. Insights from the Talysh of Azerbaijan into the Paleogene evolution of the South Caspian region. *Geological Society of America Bulletin*, **117**, 1513–1533, https://doi.org/10.1130/B25690.1

VINCENT, S.J., MORTON, A.C., CARTER, A., GIBBS, S. & BARABADZE, T.G. 2007. Oligocene uplift of the Western Greater Caucasus; an effect of initial Arabia–Eurasia collision. *Terra Nova*, **19**, 160–166, https://doi.org/10.1111/j.1365-3121.2007.00731.x

VINCENT, S.J., MORTON, A.C., HYDEN, F. & FANNING, M. 2013. Insights from petrography, mineralogy and U–Pb zircon geochronology into the provenance and reservoir potential of Cenozoic siliciclastic depositional systems supplying the northern margin of the Eastern Black Sea. *Marine and Petroleum Geology*, **45**, 331–348, https://doi.org/10.1016/j.marpetgeo.2013.04.002

VINCENT, S.J., HYDEN, F. & BRAHAM, W. 2014. Along-strike variations in the composition of sandstones derived from the uplifting western Greater Caucasus – causes and implications for reservoir quality prediction in the Eastern Black Sea. *In*: SCOTT, R.A., SMYTH, H.R., MORTON, A.C. & RICHARDSON, N. (eds) *Sediment Provenance Studies in Hydrocarbon Exploration and Production*. Geological Society, London, Special Publications, **386**, 111–127, https://doi.org/10.1144/SP386.15

VINCENT, S.J., BRAHAM, W., LAVRISHCHEV, V.A., MAYNARD, J.R. & HARLAND, M. 2016. The formation and inversion of the western Greater Caucasus Basin and the uplift of the western Greater Caucasus: Implications for the wider Black Sea region. *Tectonics*, **35**, 2948–2962, https://doi.org/10.1002/2016TC004204

WILSON, R.J., MOUNTFORD, N., MAGUIRE, P. & HEDLEY, R. 2007. The impact of recent data on the interpretation of the geologic evolution and petroleum system of the Eastern Black Sea Basin, offshore Georgia. *In*: *AAPG and AAPG European Region Energy Conference and Exhibition, Official Programme and Abstract Volume, 18–21 November 2007, Athens, Greece*. American Association of Petroleum Geologists, Tulsa, OK, 142.

ZABANBARK, A. & KONYUKHOV, A.I. 1995. Submarine mud volcanoes and hydrocarbon potential of Maykop Formation in the Black Sea. *In*: *5th Zonenshain Conference on Plate Tectonics, Programme and Abstracts*, 22–25 November 1995, Moscow, RAS/GEOMAR, 124.

ZAKREVSKAYA, E., STUPIN, S. & BUGROVA, E. 2009. Biostratigraphy of larger foraminifera in the Eocene (upper Ypresian–lower Bartonian) sequences of the southern slope of the Western Caucasus (Russia, NE Black Sea). Correlation with regional and standard planktonic foraminiferal zones. *Geologica Acta*, **7**, 259–279.

ZAPOROZHETS, N.I. 1999. Palynostratigraphy and Dinocyst zonation of the Middle Eocene–Lower Miocene deposits at the Belaya River (Northern Caucasus). *Stratigraphy and Geological Correlation*, **7**, 161–178.

Messinian canyons in the Turkish western Black Sea

N. Ö. SIPAHIOĞLU* & Z. BATI

Turkish Petroleum, Söğütözü Mahallesi 2180. Cadde No. 10, 06530 Çankaya, Ankara, Turkey
Correspondence: osipahi@tp.gov.tr

Abstract: Several canyons are observed along the Turkish margin of the western Black Sea that are associated with a prominent unconformity and interpreted to be the manifestations of the sea-level fall during the Messinian salinity crisis in the Mediterranean. In this study, their morphology, geometry and fill characteristics, as well as downslope evolution, are compared and contrasted using four 3D seismic surveys and some 2D regional seismic lines.

Two types of canyon morphologies are observed in the study area: (1) shelf incising and (2) blind. Located in the western part of the study area and deeply incised into a wide shelf, the Karaburun Canyon extends roughly in a SW–NE direction. The fill of the canyon is almost absent on the shelf, where the canyon base is downlapped by a series of Pliocene clinoforms. A thin fill appears on the upper slope, which gets thicker towards the lower slope. The eastern part of the study area is dominated by a series of blind canyons (Boğaziçi canyons). They are typically confined to the continental rise, with their heads hardly reaching the lower slope. Their fill is entirely characterized by mass-transport complexes (MTCs).

It is concluded that during the Messinian lowstand, the sediments within the Karaburun Canyon bypassed the wide shelf and were funnelled down to the continental rise and abyssal plain through the slope, which was followed by progradation of the basin margin during the relative sea-level rise in the Pliocene. A minimal imprint by tectonics in that particular area might have helped establish more stable conditions for the development of a relatively mature sediment dispersal system extending from the hinterland down to the basin centre. In this area, the shelf-slope morphology was dominantly shaped by the depositional geometries of the sedimentary packages. Being fully confined to the continental rise, the Boğaziçi canyons are situated in an area where shelf-slope morphology is governed by the Late Cretaceous volcanic arc. Parallel with the coastline, these volcanic edifices have created fairly steep dips; thus, leading to the development of an unstable basin margin and favouring MTC deposition at least since the Early–Middle Miocene. The width and relief of the canyons display a decreasing trend from west to east, which may be attributed to their relative distance from a possible drainage system in the vicinity of the Bosporus that might have acted as the major sediment supplier during this period.

Being the major conduits for sediment transport from the continental shelf to continental rise, submarine canyons have long been the interest of many workers. Advances in seismic acquisition technologies, especially in the last 20 years, have allowed an enhanced imaging of the main architectural elements leading to a better understanding of the origin and evolution of such systems. Although the majority of studies in the last two decades have tended to focus on passive margins (Posamentier 2003; Posamentier & Kolla 2003; Lofi & Berne 2008; Hanquiez et al. 2010; Gong et al. 2011, 2014; Martínez et al. 2011; Nelson et al. 2011; Sutton & Mitchum 2011; He et al. 2013; Garcia et al. 2015; Suc et al. 2015; Tari et al. 2016), there has also been a considerable amount of work carried out in active margins (Roy Moulik & Prasad 2007; Covault & Graham 2008; Lu & Shipp 2011; Nelson et al. 2011; Gamberi et al. 2013; Sipahioglu et al. 2013; Sipahioğlu et al. 2013; Rossi et al. 2015). Submarine canyons form through a combination of erosion and deposition that occur as density flows pass over the submarine

slopes (e.g. Thomas & Bodin 2013). The fill of many submarine canyons are characterized by a very complex internal architecture due to the presence of variable lithologies and processes during cut-and-fill cycles (Mayall et al. 2006; Roy Moulik & Prasad 2007; Di Celma et al. 2010; Gamberi et al. 2011; Gong et al. 2011; Di Celma 2011; Figueiredo et al. 2013; He et al. 2013; Li et al. 2013; Nakajima et al. 2014; Tassy et al. 2014; Zhou et al. 2015; Allin et al. 2016). Various elements of canyon fills (such as channel-levee, debris flows, slumps, mass-transport complexes (MTCs), thalweg and basal deposits, lateral accretion packages, shale drapes, and relicts) have been identified and described using seismic and outcrop data (Posamentier & Kolla 2003; Roy Moulik & Prasad 2007; Gong et al. 2011; Figueiredo et al. 2013; He et al. 2013; Li et al. 2013; Nakajima et al. 2014; Tassy et al. 2014).

Although dense 2D and 3D seismic data show the presence of several canyon systems within the stratigraphic fill of the western Black Sea, only a few publications discuss them (Gillet et al. 2007; Sipahioglu

From: SIMMONS, M. D., TARI, G. C. & OKAY, A. I. (eds) 2018. *Petroleum Geology of the Black Sea.*
Geological Society, London, Special Publications, **464**, 365–387.
First published online September 25, 2017, https://doi.org/10.1144/SP464.12

& Çiftçi 2010; Suc *et al.* 2015; Tari *et al.* 2015, 2016; Kitchka *et al.* 2016). All of these studies address the canyons that are interpreted to be the manifestations of the sea-level fall during the so-called Messinian salinity crisis (MSC). Calibrating their high-resolution multi-channel seismic reflection data with the two existing Deep Sea Drilling Project (DSDP) wells 380A and 381 drilled in 1975, Gillet *et al.* (2007) concluded that the regional unconformity they observed in their seismic data represented the Messinian Unconformity (MU). They went on to say that the Black Sea underwent a severe sea-level fall during the Messinian, as suggested by Hsü & Giovanoli (1979). Sipahioglu & Çiftçi (2010) studied several canyons formed in response to a sea-level drop during the MSC in the Turkish sector of the western Black Sea and demonstrated the presence of MTCs within some of the canyons. They used age data from four exploration wells drilled by Turkish Petroleum (TP) in the Turkish sector of the western Black Sea (Karadeniz-1, İğneada-1, Limanköy-1 and Limanköy-2) in order to pinpoint the MU. Suc *et al.* (2015) investigated the possibility of a marine connection between the Sea of Marmara and the Black Sea during the MSC, and they verified the presence of subaerial erosion in the SW Black Sea during the peak of the MSC. Tari *et al.* (2015) revisited the core descriptions in the DSDP 380 well in relation to new seismic data in the basin. The so-called pebbly breccia interval penetrated by the well had been previously interpreted as the manifestation of a dramatic sea-level fall in the Mediterranean in the Messinian (Hsü & Giovanoli 1979). Using long-offset seismic data, Tari *et al.* (2015) demonstrated that the pebbly breccia unit was part of an MTC and, therefore, it did not indicate *in situ* shallow-water depositional setting. They concluded that the magnitude of the sea-level drop during the MSC was not as dramatic as the 1600 m previously suggested by Hsü & Giovanoli (1979). Kitchka *et al.* (2016) studied the Mid–Late Miocene MTCs in the Ukrainian Black Sea, concluding that erosion of the shelf-break areas and the formation of incised submarine valleys in response to sea-level falls in Mid–Late Miocene in the basin led to deposition of MTC packages. Based on 2D and 3D seismic and well data from the western Black Sea, Tari *et al.* (2016) concluded that the Black Sea never became desiccated; therefore, there was no evaporite deposition during the MSC. Here, using 2D and 3D seismic data, we studied two canyon systems formed during the MSC in the Turkish sector of the western Black Sea, and compared and contrasted the internal architecture of their fills and geometries based on their association with different basin-margin morphologies. Our aim is to show the effect of the basin-margin morphology on the fill types and architectures of two canyon systems,

which are only approximately 50 km apart and were initiated by the same dramatic sea-level fall (MSC).

Palaeogeographical setting

Located between Europe and Anatolia, the Black Sea is a back-arc basin formed as a result of the northwards subduction of the Tethyan Ocean beneath the Eurasian continent during Mid–Late Cretaceous to Paleocene (Boccaletti *et al.* 1974; Letouzey *et al.* 1977; Şengör & Yılmaz 1981; Zonenshain & Le Pichon 1986; Finetti *et al.* 1988; Görür 1988; Okay *et al.* 1994, 2001, 2013; Robinson *et al.* 1996; Spadini *et al.* 1996, 1997; Banks & Robinson 1997; Okay & Görür 2000; Nikishin *et al.* 2003, 2012, 2015*a*, *b*). During the Eocene, the closure of the northern branch of the Tethyan Ocean (the Izmir–Ankara Ocean) and a series of related collisions of continental masses led to the separation of the Mediterranean Sea from the Paratethys in the north (Gillet *et al.* 2007). Extending roughly in an east–west direction, the Paratethys covered a large area from the western Molasse Basin in Switzerland in the west to the Aral Sea in Central Asia in the east (Rögl 1999). The Black Sea, which became part of the Paratethys, has been characterized by deep-marine conditions from the Late Cretaceous to the present. Its southern margin was affected by Eocene thrusting and concomitant uplift. The Oligocene and younger stratigraphy of the Black Sea Basin is dominated by rapidly deposited turbidite systems (e.g. Nikishin *et al.* 2015*b*) with a variable temporal and spatial distribution (e.g. Sipahioğlu *et al.* 2013).

The study area is shown on the tectonic map of the Black Sea (Nikishin *et al.* 2015*a*) (Fig. 1). The western part of the study area is the Eocene–Oligocene Thrace Basin, whereas the eastern part of the study area is situated in a continental slope basin above the Late Cretaceous volcanic arc. These two different continental basin margins influenced the geometry of the canyon fills that will be discussed in this paper.

Study area

Extending from shelf to slope and continental rise in the Turkish western Black Sea, the study area covers approximately 6000 km^2, with water depths ranging from 80 to 2200 m (Fig. 2). The present-day shelf break lies at water depths of around 200 m. Compared to other margins of the Turkish Black Sea, this area is characterized by a relatively wider shelf and slope where the widths range between 15–45 and 40–50 km, respectively. The smoothness of the shelf and slope, as well as their width, increases from east to west. However, as the shelf widths

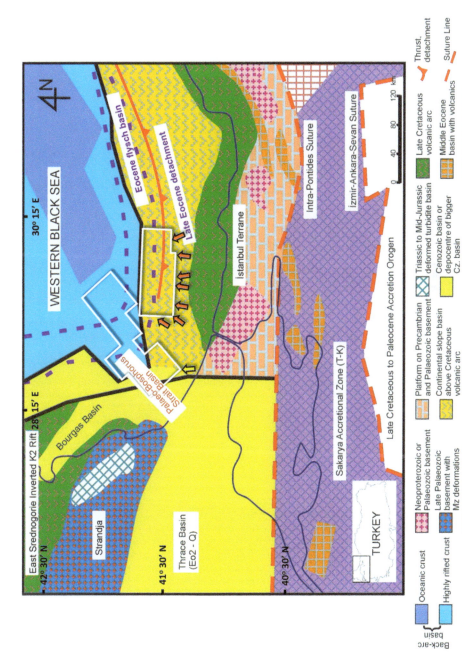

Fig. 1. Tectonic map of the Black Sea and outlines of the 3D seismic surveys used in this study (redrawn after Nikishin *et al.* 2015*a*). Yellow and red arrows show the sediment input directions in the Karaburun and Boğaziçi canyons, respectively.

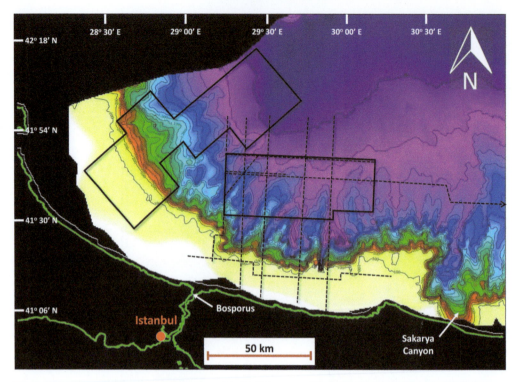

Fig. 2. Bathymetry map of the westernmost part of the Turkish western Black Sea. The dashed lines and black boxes show the 2D seismic lines and coverage areas of the 3D seismic surveys, respectively, used in this study.

range between 11 and 79 km in the Black Sea and the Mediterranean (Harris *et al.* 2014), the shelf in the study area can be considered as medium to narrow by the standards of these two landlocked basins. Similarly, the slope widths can also be considered as medium to narrow compared to the slope widths of 25.8–118 km in the Mediterranean and the Black Sea (Harris *et al.* 2014).

Results from several exploration wells drilled by Turkish Petroleum (TP) and its partners, and 2D and 3D seismic data in the Turkish sector of the western Black Sea, suggest that the present physiographical conditions in the area have not changed since the Early–Middle Miocene.

Data and methodology

Three-dimensional seismic data were primarily used in this study. Nineteen 2D seismic lines were also employed for regional correlations (Fig. 2). The data are petroleum exploration industry data and have been provided by TP. Three-dimensional data consist of four surveys that were acquired between 1995 and 2013. Three of these surveys are clustered in the western part of the study area. The data quality

is excellent with a 25 m inline and crossline spacing. All seismic interpretations were carried out using pre-stack time-migrated (PSTM) seismic cubes.

The base of the Messinian canyons was picked and interpreted in all 3D seismic surveys on the basis of heavily truncated reflectors (Fig. 3). It is a very prominent surface, indicating substantial incision into the strata below. The top of the canyons, where prominent, were also picked and interpreted in order to show not only thickness variation, but also to calculate interval amplitudes. Seismic reflection patterns of the canyon fills were dominantly used to compare and contrast different canyons. Standard amplitude extraction techniques were applied on PSTM seismic cubes to create amplitude maps that led to better observation of geometries, facies architectures, textural patterns and spatial distribution of the canyon fills.

The Messinian canyons observed in the study area are classified based on the classification suggested by Harris *et al.* (2014), who compiled the geomorphology of the world seas and subdivided submarine canyons into two categories: 'shelf incising' and 'blind'. As the term suggests, shelf-incising canyons typically incise the shelf above the shelf break where they display landwards-deflected isobaths. Blind

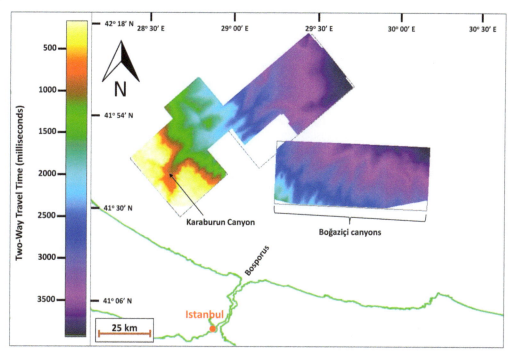

Fig. 3. Time–structure map of the Karaburun and Boğaziçi canyons as interpreted in 3D seismic surveys.

canyons, instead, are fully confined to the slope and, therefore, they do not show any incision landwards of the shelf break.

Submarine canyons

Several canyons developed along the southern margin of the western Black Sea during the MSC (Hsü & Giovanoli 1979; Can 1996; Gillet *et al.* 2007; Sipahioglu & Çiftçi 2010; Korucu *et al.* 2013; Suc *et al.* 2015; Tari *et al.* 2015, 2016). The association of these canyons with the MSC has been confirmed by several TP exploration wells (Karadeniz-1, İğneada-1, Limanköy-1 and Limanköy-2) and seismic data in the Turkish sector of the western Black Sea. The MU and its correlative conformity can be tracked easily on seismic data across the shelf, the slope, and even the continental rise and the abyssal plain (Can 1996; Gillet *et al.* 2007; Sipahioglu & Çiftçi 2010; Korucu *et al.* 2013; Suc *et al.* 2015). In the study area, these canyons are imaged in 3D seismic surveys (Fig. 3). The seismic surveys provide an opportunity to completely image one single canyon (Karaburun Canyon) from the shelf down to the continental rise. The other 3D seismic survey, Boğaziçi, is located to the east of the Karaburun Canyon, within which a series of canyons (Boğaziçi canyons) are partially observed. Underneath the present

slope, several other canyons of the same age exist between the Karaburun and Boğaziçi canyons; however, the lack of 3D seismic data in this area prevent a comprehensive and comparative study of these canyons.

The basin-margin morphology of the Karaburun and Boğaziçi canyons are and were different at present and during the Messinian. The Karaburun Canyon is situated along a more stable pre-Messinian basin margin characterized by a wide continental shelf (*c.* 45 km) and a gentle continental slope (0.5–3°), which accommodated the development of a prograding basin margin during the Pliocene. On the other hand, the Boğaziçi canyons are located along a basin margin where the continental shelf is quite narrow (*c.* 15 km) and the continental slope is fairly steep (7°–15°).

The fill of the canyons are divided into five types based on seismic facies and geometries:

- low-amplitude continuous (L-C);
- high-amplitude continuous (H-C);
- low-amplitude discontinuous (L-D);
- high amplitude discontinuous (H-D);
- chaotic (C).

The fill of the Karaburun Canyon is dominantly represented by L-C, H-C, L-D and H-D type seismic facies. The Boğaziçi canyons, on the other hand, are typically characterized by C type fills.

The Karaburun Canyon

The Karaburun Canyon is situated along the margin NW of the Bosporus characterized by a wide continental shelf (*c.* 45 km) and a gentle continental slope (0.5–3°), which accommodated a prograding basin margin during the Pliocene. It extends in a SW–NE direction and is the only shelf-incising canyon in the study area (Figs 3 & 4) (see Suc *et al.* 2015, figs 9 & 10). This unique incision initiates around 30 km landwards of the Messinian shelf break in the form of a few channels that evolve into a deep and wide canyon within about 10 km. Heavily truncating the underlying pre-Messinian strata, the canyon attains width and depths up to 9 km and 1000 m, respectively, at about the Messinian shelf edge (Fig. 5). The base of the canyon can be tracked as a truncation surface across the entire area and also on the shelf. Although the degree of incision of some canyons worldwide is so intense that they may extend landwards in the form of shelf valleys and directly connect with the terrestrial fluvial systems (Harris & Whiteway 2011), this is not observed in Karaburun Canyon, either because it is not present or it is below the resolution of the seismic data.

The Karaburun Canyon is subdivided into three transverse segments defined by different seismic facies (Fig. 4):

- upper segment;
- middle segment;
- lower segment.

Upper segment. The upper segment of the Karaburun Canyon truncates into the Messinian shelf where depositional dips of the underlying strata do not exceed 0.5°. In this segment, the canyon attains widths and reliefs up to 9 km and 1000 m, respectively (Figs 4 & 5). The canyon base and walls, as well as the entire Messinian surface, truncate the underlying reflectors at high angles (Figs 5 & 6). While parallel, sub-parallel, slightly concave L-C and H-C type seismic facies generally dominate its fill, L-D type seismic facies are also rarely observed. H-D type seismic facies, on the other hand, are almost absent. In dip sections, sigmoidal reflector geometries that show downlapping geometries onto the canyon base are very characteristic (Figs 6 & 7).

The upper segment is interpreted to define the part of the Karaburun Canyon where the intensity of erosion reached its peak. The extremely irregular character of the whole Messinian surface may be indicative of subaerial exposure of the shelf, including the floor of the upper canyon area, during the Messinian sea-level fall (Tari *et al.* 2016). A vast majority of the Karaburun Canyon in this segment is characterized by a very thin or complete absence

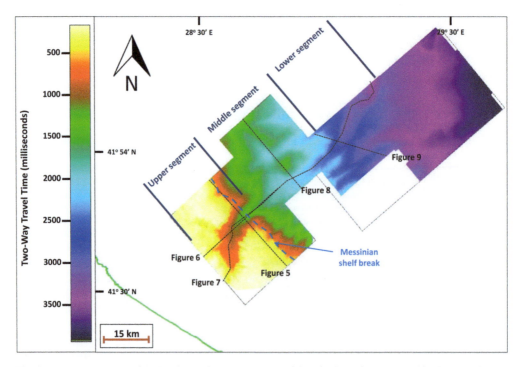

Fig. 4. Time–structure map of the Karaburun Canyon. Locations of the seismic sections presented in Figures 5–9 are also shown.

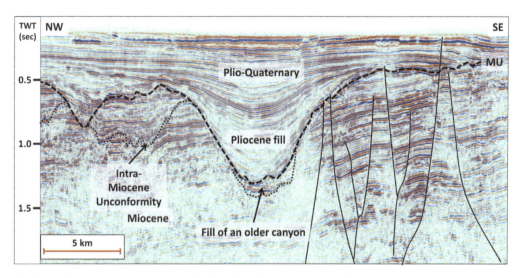

Fig. 5. A seismic section orientated perpendicular to the axis of the Karaburun Canyon in the upper segment (for the location, see Fig. 4). Note the presence of an older unconformity, named as the 'Intra-Miocene Unconformity', which was heavily truncated by the MU (cf. Tari *et al.* 2015, 2016). TWT, two-way travel time.

of a basal lag (below the seismic resolution) representing sediment bypass. It should be noted that in many parts of the upper segment, the canyon erodes into the fill of a pre-Messinian canyon; therefore, caution was taken not to interpret this older fill as part of the Karaburun Canyon (Fig. 5). The upper segment is directly filled by the clinoforms of the prograding margin in response to subsequent sea-level rise during the Pliocene (Fig. 6).

Middle segment. The middle segment describes the portion of the Karaburun Canyon located on the upper slope. Depositional dips within this zone are around 3°. The width of the canyon is 9 km

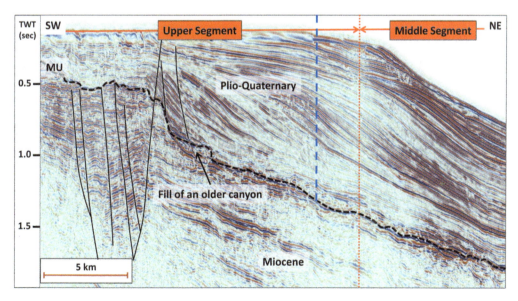

Fig. 6. A seismic section along the axis of the Karaburun Canyon through the upper and upper part of the middle segments (for the location, see Fig. 4). The blue dashed line shows the anticipated location of the shelf break during the MSC. TWT, two-way travel time.

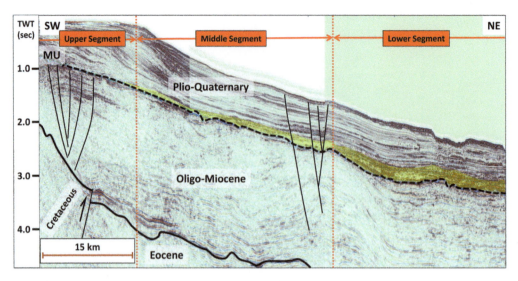

Fig. 7. An arbitrary seismic section along the axis of the Karaburun Canyon through the upper, middle and lower segments (for the location, see Fig. 4). Seismically visible canyon fill onlapping the canyon base is shown in yellow. TWT, two-way travel time.

maximum, with reliefs that do not exceed 500 m. The updip limit of the segment is the shelf break and the downdip limit is a point beyond which the dip of the canyon profile abruptly increases (Figs 4 & 7). Within this segment, the canyon attains a smoother outline, although it still exhibits intense truncation into the underlying reflectors (Fig. 8). In addition to the parallel to sub-parallel L-C and

H-C type seismic facies, L-D and H-D type seismic facies start to dominate the canyon fill in this middle segment.

A seismic line created along the canyon axis from the shelf to the lower slope show the presence of an older anticlinal structure between the middle and lower segment, which is truncated by the canyon (Fig. 7). Interpretation of the onlap patterns onto

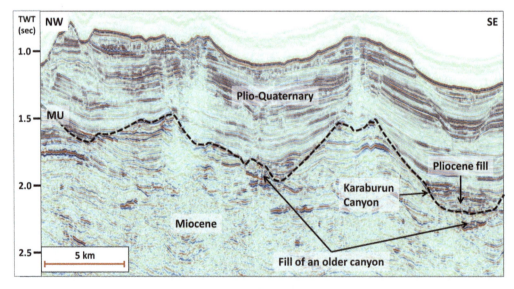

Fig. 8. A seismic section orientated perpendicular to the axis of the Karaburun Canyon in the middle segment (for the location, see Fig. 4). A well-developed fill starts to appear in the canyon at this segment. TWT, two-way travel time.

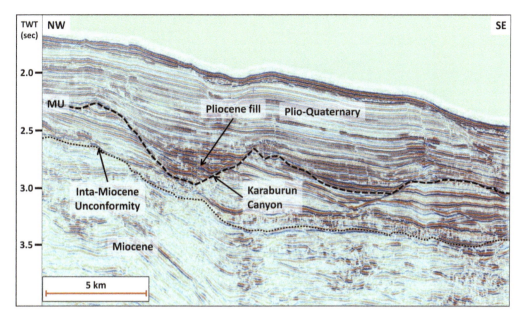

Fig. 9. A seismic section orientated perpendicular to the axis of the Karaburun Canyon in the lower segment (for the location, see Fig. 4). The canyon is still highly erosional with a thick fill. Note the presence of another Messinian canyon with a relatively lower relief and thinner fill to the SE. It is a blind canyon that has no connection with the Messinian shelf. TWT, two-way travel time.

the structure reveals that the structure might have already been emergent by the initiation of the Kara-burun Canyon. The anticline might have acted as a local high on the palaeo-seafloor when the canyon was active and modified the canyon profile by reducing the intensity of erosion, leading to smaller canyon axis dips. Fill deposits typically displaying L-D and H-D type seismic facies get more prominent within this segment. They are overlain by a series of clinoforms that represent the prograding basin margin during the Pliocene (Fig. 7). The erosional canyon base is still very prominent. There are a few more canyons orientated parallel to the Karaburun Canyon at this segment (Figs 4 & 8). All are blind canyons with no connection with the Messinian continental shelf.

Lower segment. The lower segment is the part of the Karaburun Canyon located on the lower slope of the Messinian continental margin (Figs 4 & 7), where depositional dips are around 1–1.5°. The maximum width and relief of the canyon at this segment is 8 km and 300 m, respectively. The updip limit of the segment is the point where depositional dips change from approximately 3° to 1–1.5°, which coincides with the anticlinal structure described above (Fig. 7). The downdip limit of the lower segment lies outside the 3D seismic coverage. The canyon fill is dominated by parallel to sub-parallel H-C and H-D type seismic facies. L-C and L-D type

seismic facies are also observed, but they are not as dominant as in upper and middle segments (Fig. 9).

Characterized by a thick fill, the lower segment constitutes the portion of the Karaburun Canyon where deposition took place (Fig. 9). The older anticlinal structure acted as a local high, causing an elevation difference between the middle and lower segments. Therefore, the energy of the turbidity current experienced a sudden decrease as it travelled from the middle to lower segment, which caused immediate deposition within the lower segment. The ongoing deposition, probably including the backfilling stage, might have caused the 'healing' of the slope, levelling the depositional dips of the middle and lower segment. This can clearly be observed by the onlapping canyon fill onto the downdip flank of the older anticline (Fig. 7). Although the intensity of erosion at this segment is less compared to the other two segments, the canyon still displays a considerable amount of truncation into underlying strata (Fig. 9). The fill gets thicker downdip as the depositional dips become gentler. Thickening of the fill is also accompanied by a series of more complex seismic geometries.

The Boğaziçi Canyons

A series of canyons extending in a roughly south–north direction dominate the toe-of-slope–continental rise NW of the Bosporus (Figs 3 & 10). Unlike

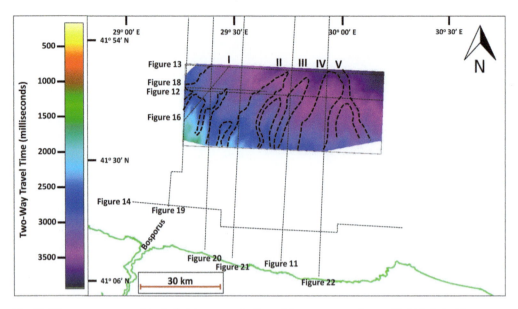

Fig. 10. Time–structure map of the Boğaziçi canyons. Outlines of the canyons are shown with thick dashed lines. The locations of the seismic sections presented in Figures 11–14, 16 & 18–22 are also shown.

the Karaburun Canyon region, the 3D seismic coverage here only covers the continental rise. However, several regional 2D seismic lines provide some insight into the extent of the canyons both downdip and updip. In contrast to the western part, this portion of the study area is characterized presently by a narrow shelf (c. 15 km) and a steep slope (c. 7–15°), which is underlain by the Upper Cretaceous volcanic rocks. The volcanic rocks, which form part of a

major volcanic arc, extend parallel with the coastline (Fig. 1) and have created a large escarpment between the shelf and the continental rise, as shown by the seismic geometry (Fig. 11).

A total of five Messinian canyons of various sizes, numbered I–V from west to east, are observed in the Boğaziçi seismic survey (Fig. 10). Their width and relief range between 8–12 km and 80–350 m, respectively (Fig. 12). The canyons coalesce

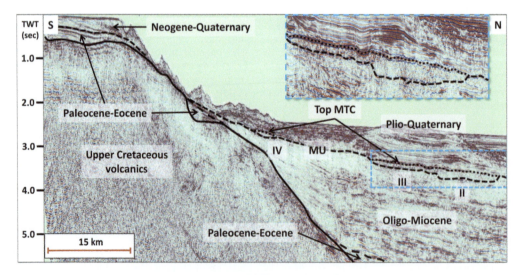

Fig. 11. A regional 2D seismic line extending through Boğaziçi Canyon IV, III and II (for the location, see Fig. 10). Canyons II and III are zoomed-in for a better view (upper right).

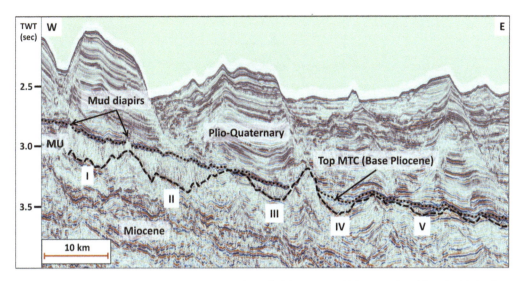

Fig. 12. A seismic section orientated perpendicular to the Boğaziçi canyons, which are numbered from I to V (for the location, see Fig. 10). The canyon reliefs display a decreasing trend from west to east. The fills of all the canyons are entirely dominated by chaotic–transparent reflectors, which are interpreted to be characterized by MTCs. The mud diapir observed in canyon I is interpreted to have resulted from the remobilization of MTC due to continuing deposition during the Plio-Quaternary. TWT, two-way travel time.

downdip, forming one large erosional body whose width reaches up to 50 km (Fig. 13). An east–west-orientated composite seismic section along the shelf to the south in the updip direction of the Boğaziçi canyons shows quite thin Cenozoic deposits over the Upper Cretaceous volcanics (Fig. 14). A clear erosional surface is observed within the Cenozoic section. This surface is interpreted to represent the MU, although there are no well data to confirm this interpretation. We are unable to observe any convincing erosional geometries along the MU that could prove that a connection between the Boğaziçi canyons and the shelf existed. Therefore, as we interpret that the Boğaziçi canyons were

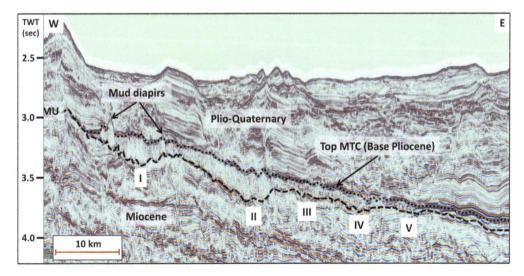

Fig. 13. A seismic section orientated perpendicular to the Boğaziçi canyons (I–V) at their most distal imageable portion by the 3D seismic data where the canyons merge to form one single MTC body (for the location, see Fig. 10). TWT, two-way travel time.

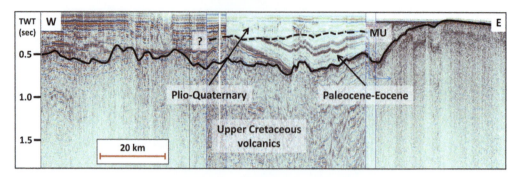

Fig. 14. An east–west-orientated composite seismic section along the shelf located to the south of the Boğaziçi canyons (for the location, see Fig. 10). A Late Cretaceous volcanic edifice has created a little accommodation space for deposition here at least since the Eocene. Despite the poor quality of the seismic line in the western portion, the MU is quite evident in the central portion. TWT, two-way travel time.

fully confined to the continental rise and abyssal plain with almost no seismically visible connectivity with the palaeo-slope, in that sense they are classified as blind canyons. The canyons initiate around the toe-of-slope and extend all the way down to the abyssal plain. They are characterized by smooth outlines in time–structure and time–thickness maps, with thicknesses ranging between 80 and 280 m (Figs 10 & 15). Distinctively transparent C type seismic facies dominate entirely the fill of the canyons, which are bounded at their top and base by H-C reflectors (Figs 12, 13 & 16).

The bases of the Boğaziçi canyons are marked by a H-C reflector truncating the underlying reflectors

and can be tracked as an erosional surface across the study area.

The C type fill patterns observed in all of the Boğaziçi canyons display quite a uniform and dominant distribution, although the dimensions of the fills decrease from canyon I to V (Figs 12, 13 & 16). The root mean square amplitude maps overlain with the structural contours of the basal surface reveal that the areas coinciding with the canyons are dominated by a very-low-amplitude interval (Fig. 17). The C-type reflection patterns and extremely low amplitudes of the Boğaziçi canyon fills indicate that they are MTCs.

The low amplitudes prevent a comprehensive interpretation of the syn- and post-depositional

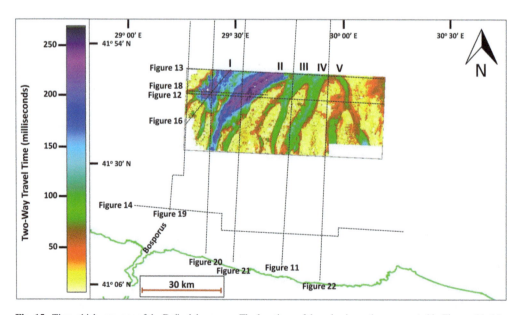

Fig. 15. Time–thickness map of the Boğaziçi canyons. The locations of the seismic sections presented in Figures 11–14, 16 & 18–22 are also shown.

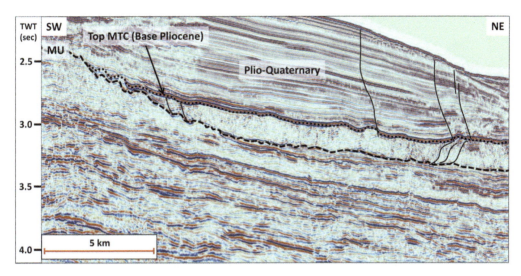

Fig. 16. A seismic section along the axis of the Boğaziçi Canyon I (for the location, see Figs 10 & 15). Some extensional and contractional faults associated with the canyon fill may be present, although the extremely chaotic and transparent nature of the fill prevent a clear interpretation. TWT, two-way travel time.

geometries associated with the Boğaziçi canyon fills. However, some extensional and contractional faults within the updip and downdip portions of the canyon fills can be inferred (Fig. 16). Although not clearly observed due to extremely low amplitudes of the fills, extensional faults are suspected in the headwall, or escarpment zone, which is the most proximal part of the MTC. Contractional faults, or toe-thrusts, on the other hand, may be associated with the terminal part of the MTC and are caused by structural shortening (e.g. Bøe *et al.* 2000; Posamentier & Martinsen 2011; Frey-Martínez *et al.* 2005; de la Vara *et al.*

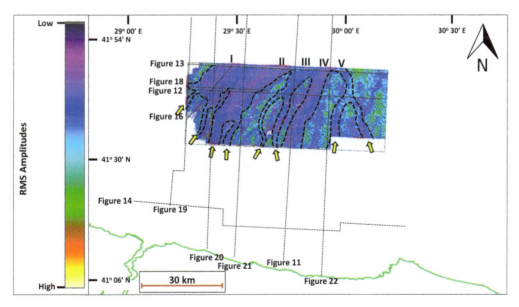

Fig. 17. RMS amplitude map of the sedimentary fill of the Boğaziçi canyons. Low amplitudes characterize the MTC-filled canyons. Outlines of the canyons are shown with dashed lines. Yellow arrows show anticipated sediment entry points. Locations of the seismic sections presented in Figures 11–14, 16 & 18–22 are also shown.

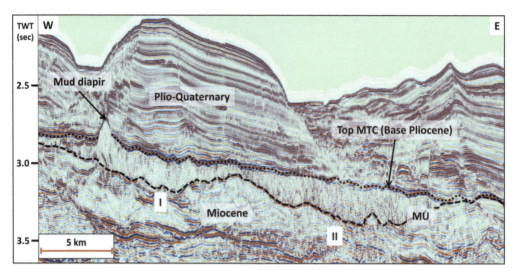

Fig. 18. A seismic section orientated perpendicular to the Boğaziçi canyons I and II (for the location, see Figs 10, 15 & 17). Note the presence of a well-developed mud diapir in Canyon I. TWT, two-way travel time.

2016). There is also clear evidence for post-depositional deformation of the MTCs, suggesting that some of the faults could be post-depositional. However, the MTCs have a tendency to get thicker downdip. We think that this thickening may not be entirely depositional, but related to toe-thrusting. Some rotational and deformation geometries are observed within MTC I (Figs 12, 13 & 18). They are interpreted as post-depositional and are attributed to the remobilization of the updip part of the MTC due to the weight of the overburden as a result of continuous deposition throughout the Plio-Pleistocene. Differential compaction over the MTCs appears to have affected the younger strata and shaped the present-day seabed.

The upper boundary of the MTCs is well defined and is characterized by a parallel to sub-parallel H-C reflector that caps all of the MTC fills of the Boğaziçi canyons, marking a synchronous termination of the MTC deposition in the area (Figs 12, 13, 16 & 18). This reflector is, in turn, overlain by parallel, H-C reflectors up to the present-day seabed that are interpreted as fine-grained deep-water sediments deposited during the Pliocene–Recent period. Several MTC intervals are also observed within this sedimentary package, but they are not coeval unlike the Boğaziçi canyons and are mostly isolated bodies within the stratigraphy.

The width and relief (and fill thickness) of the Boğaziçi canyons increase from east to west towards the Bosporus Strait (Figs 10, 12, 15 & 17). Similarly, the amount of shelf progradation at least since the Eocene, as well as the thickness of slope sediments overlying the Upper Cretaceous volcanics, show an increase in the same direction (Figs 11 & 19–22).

Discussion

Several canyons developed along the SW margin of the Turkish western Black Sea in response to the Messinian sea-level fall. These canyons were studied using four 3D seismic surveys and nineteen 2D seismic lines. Two major canyon systems (from west to east, Karaburun and Boğaziçi) are observed in the area (Fig. 3). The Karaburun Canyon is the only shelf-incising canyon covered by three 3D seismic surveys, which allowed the imaging of the entire canyon from shelf to lower slope (Figs 3 & 4). It is characterized by a well-developed erosional surface, which is, in turn, overlain by a fill whose thickness changes based on the physiographical position of the canyon on the Messinian continental margin. The Boğaziçi canyons, on the other hand, are imaged only partly by a 3D seismic survey and by a few 2D seismic lines (Figs 10, 15 & 17). Being confined to the continental rise, they can be classified as blind canyons with well-developed erosional bases and MTC-dominated fills.

The Karaburun Canyon extends in a SW–NE direction. The architectural elements observed within the canyon show variations based on the physiographical setting of the canyon segment. The upper segment, which is located on the Messinian shelf (Fig. 4), does not have a fill and is characterized by a highly erosional canyon base directly overlain by the clinoforms of the Pliocene prograding basin margin. The Pliocene sedimentary package is interpreted to have been deposited during the subsequent transgressive period when the basin margin displayed a rapid progradational trend (Figs 5 & 6). It is interpreted that this area was the site of sediment

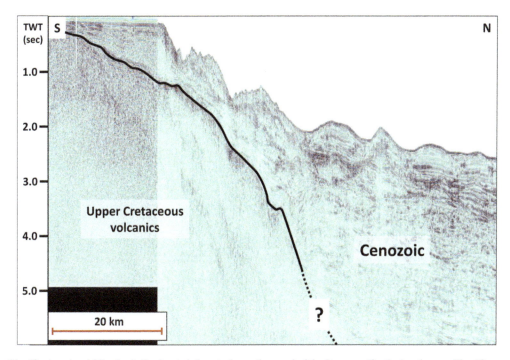

Fig. 19. A regional 2D seismic line located close to the northern end of the Bosporus (for the location, see Figs 10, 15 & 17). The poor quality of the seismic prevents the interpretation of the MU. Note that the shelf edge has prograded more throughout the Cenozoic at this location where the thickness of the Cenozoic section on the slope is greater. See Figures 11 & 20–22 for comparison. TWT, two-way travel time.

bypass until the Messinian shelf was flooded back and basin margin progradation took over in the Pliocene. The middle segment extends across the Messinian slope (Fig. 4) where the base of the canyon is still erosional but with a much smoother outline, on which a sedimentary fill starts to develop

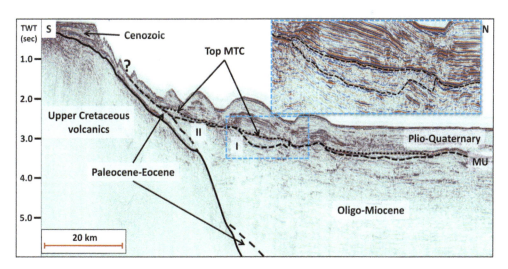

Fig. 20. A regional 2D seismic line extending through Boğaziçi canyons I and II (for the location, see Figs 10, 15 & 17). See Figures 11, 19, 21 & 22 for comparison. Canyons I is zoomed-in for a better view (upper right). TWT, two-way travel time.

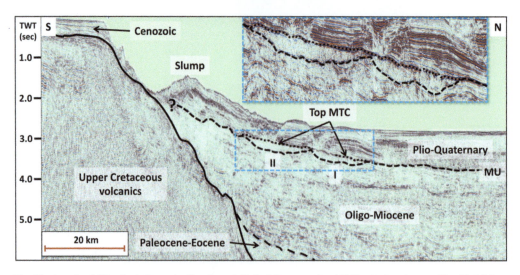

Fig. 21. A regional 2D seismic line extending through Boğaziçi canyons I and II (for the location, see Figs 10, 15 & 17). Note the possible large slump feature at the toe of the slope that almost exposed the Upper Cretaceous volcanics. See Figures 11, 19, 20 & 22 for comparison. Canyons I and II are zoomed-in for a better view (upper right). TWT, two-way travel time.

(Fig. 7). Canyon fills typically display higher-amplitude reflections with low continuity. Their upper boundary is generally uneven and, in turn, is overlain by parallel reflectors that are interpreted to represent thin-bedded turbidites deposited on the toe-of-slope/continental rise of the Pliocene basin margin. A pre-existing anticlinal structure at the downdip boundary of the middle segment is interpreted to have controlled depositional architecture within the segment by reducing the steepness of the gradient, thus leading to deposition of a fill (Fig. 7). The lower segment covers the Messinian toe-of-slope. At its updip boundary, the gradient increases abruptly downdip of the anticline, only for a short distance, and then becomes extremely smooth (Fig. 7). Although the erosional base is still quite prominent, this segment is typically characterized by a fill, which is depicted by

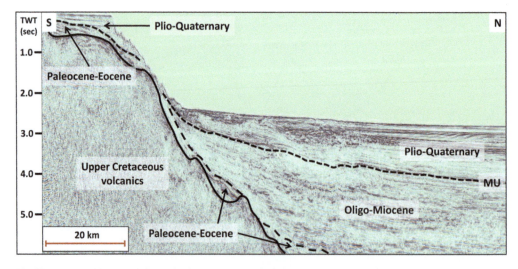

Fig. 22. A regional 2D seismic line extending close to the western margin of the Boğaziçi Canyon V (for the location, see Figs 10, 15 & 17). The fill of the canyon is not visible. See Figures 11 & 19–21 for comparison. TWT, two-way travel time.

high-amplitude, continuous/semi-continuous reflectors (Fig. 9). Having a wedge-shape geometry in dip sections, the canyon fill onlaps the pre-existing structural high and becomes thicker downdip. This area is interpreted to have been the active site of deposition during the Messinian lowstand where all the sediments that bypassed the Messinian shelf and the upper–middle portion of canyons were funnelled down through the slope and deposited in this area. An arbitrary seismic line created along the Karaburun Canyon axis from the shelf to the slope shows that the dips are relatively higher along the upper Messinian slope, which gradually turn lower downdip (Fig. 7). The transition from higher to lower dips marks the transition from erosion to fill deposition. The canyon does not contain any significant fill in the upper reaches, where the dips are higher. On the other hand, fill deposition is observed downdip from the point where the dips get gentler.

Shelf-incised canyons form by erosive turbidity flows sourced from fluvial and shelf sediment sources (Harris & Whiteway 2011). Despite being characterized by such erosive processes (e.g. Garcia et al. 2015), the overall evolution of a canyon is governed by the cycle of waxing–waning flow energy that promotes erosion and deposition (Gong et al. 2011; McHargue et al. 2011; Figueiredo et al. 2013; He et al. 2013; Macauley & Hubbard 2013), which is accompanied by downslope propagation due to erosive turbidity flows and/or upslope headwards retrogressive mass failures (Harris & Whiteway 2011). The evolution of the Karaburun Canyon can be subdivided into three broad periods: (1) erosion: (2) erosion–deposition; and (3) deposition. The erosion period corresponds to the time when the entire shelf, as well as the upper slope, were exposed (Tari et al. 2016). During this period, the shelf was heavily incised by bypassing sediments, which were funnelled down to the deeper part of the basin to be deposited. We believe that the lower slope never became subaerially exposed, although erosion, besides deposition, prevailed. The erosion–deposition period marks the late fall and early rise of the sea level when the backfilling of the canyon had already started even though the shelf was still being incised. During the deposition period, rapid rise of the sea level led to intense transgression of the coastline, drowning the entire canyon and allowing very thin lag deposition on the shelf.

River-associated shelf-incising canyons are more common in active margins and geographical areas where there is a relatively high rate of sediment supply into the basin (Harris & Whiteway 2011). Within the resolution of the 3D seismic data, the incision of the Karaburun Canyon into the shelf appears to have initiated around 30 km landwards of the shelf break. When tracked updip to the onshore (cf. Suc et al. 2015), this point coincides with a Miocene fluvial succession (the Ergene Formation of Siyako & Huvaz 2007, and the Middle–Upper Miocene units in the Akpınar and Kocayemiş localities of Suc et al. 2015), which could have acted as the sediment source (Fig. 23). However, more comprehensive work is needed in order to be able to prove the association of this sedimentary package with the Karaburun Canyon.

Covered by a 3D seismic survey, the Boğaziçi canyons include a total of five canyons that are roughly parallel to each other, extending in a south–north direction (Figs 10, 15 & 17). All of the canyons are confined to the continental rise and cannot be tracked updip towards the slope (Fig. 14). They are numbered as I–V from west to east, displaying a decreasing width and relief trend in that direction (Figs 10, 15 & 17). Despite only being located 50 km east of the Karaburun Canyon, the Boğaziçi canyons show different architectural geometries. Their basal and top surfaces are typically characterized by high-amplitude continuous reflections, and the canyon fills by very-low-amplitude to transparent and chaotic reflections, that are interpreted to represent MTCs (Figs 11–13, 16 & 18–21). Canyon bases are highly erosional. Although MTCs can also be the product of unconfined flow, and can comprise sheets and lobes (Posamentier & Martinsen 2011), we did not observe such geometries in our study area and interval. Variable-amplitude continuous reflections representing deep-water sediments of the Pliocene–Recent period dominate the overlying sedimentary package (Figs 11–13, 16 & 18–21). Some isolated MTC bodies are also observed within this interval.

Although several mechanisms have been proposed by several authors on the formation of MTCs, they cluster around the following: (1) sea-level fluctuations (Posamentier & Kolla 2003; Solheim et al. 2005; Algar et al. 2011; Diaz et al. 2011; King et al. 2011; Lu & Shipp 2011; Nelson et al. 2011; Sutton & Mitchum 2011; Martín-Merino et al. 2014); (2) slope topography (Moscardelli et al. 2006; Moscardelli & Wood 2007; Gamboa et al. 2010; Diaz et al. 2011; Gamberi et al. 2011; Lu & Shipp 2011; Mosher & Campbell 2011; Richardson et al. 2011; Zhu et al. 2011; Gong et al. 2014); (3) overpressure (Posamentier & Kolla 2003; Solheim et al. 2005; Richardson et al. 2011; Zhu et al. 2011); (4) seismicity (Moscardelli et al. 2006; Moscardelli & Wood 2007; Gamberi et al. 2011; Martínez et al. 2011; Mosher & Campbell 2011; Nelson et al. 2011; Richardson et al. 2011; Zhu et al. 2011; Gong et al. 2014; Zhang et al. 2016); (5) high sedimentation rates (Algar et al. 2011; Diaz et al. 2011; Mosher & Campbell 2011; Nelson et al. 2011; Sutton & Mitchum 2011; Zhu et al.

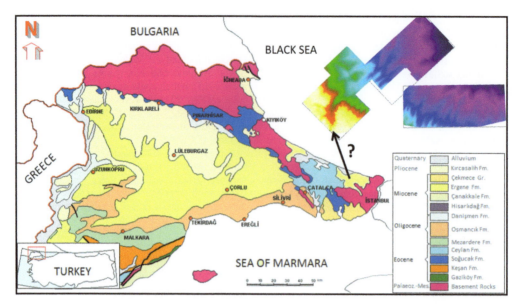

Fig. 23. Time map of the Karaburun and Boğaziçi canyons (see Fig. 3 for the colour scale) superimposed on the geology map of the Thrace Basin (NAFS: North Anatolian Fault System) (modified after Siyako 2006, according to Siyako & Huvaz 2007). The head of the Karaburun Canyon can be tracked updip to the onshore where it coincides with a Miocene fluvial succession (Ergene Formation).

2011; Gong *et al.* 2014); (6) oversteepening due to tectonic uplift (Algar *et al.* 2011; King *et al.* 2011; Sutton & Mitchum 2011; Martín-Merino *et al.* 2014); and (7) gas hydrate destabilization (Posamentier & Kolla 2003; Algar *et al.* 2011; Nelson *et al.* 2011; Richardson *et al.* 2011; Zhu *et al.* 2011). In fact, many of these factors are co-related and the occurrence of one is very likely to trigger another. In this paper, we have studied two coeval, intrabasinal canyon systems (Karaburun and Boğaziçi) that are only about 50 km apart, but are characterized by different types of sedimentary fills. Although the fill of the Karaburun Canyon is almost completely devoid of MTCs, the Boğaziçi canyons are entirely dominated by them. The Karaburun and Boğaziçi canyons are regarded to have formed in response to the MSC sea-level fall irrespective of its magnitude (1600 m, Hsü & Giovanoli 1979; 1300–1700 m, Munteanu *et al.* 2012; 500 m, Krezsek *et al.* 2016; Tari *et al.* 2016; several hundred metres, Gillet *et al.* 2007; Popov *et al.* 2010; 50–100 m, Krijgsman *et al.* 2010; tens of metres, Vasiliev *et al.* 2013). However, the prevalence of MTC deposition was mainly governed by other factors that were exclusive to Boğaziçi canyons: (1) steep continental slopes due to the presence of arc volcanics, which were further oversteepened during the Eocene compressional phase; and (2) seismicity of the area, both of which were triggered by the convergence of Anatolian and Arabian plates.

The evidence of the Eocene compressional deformation along the Turkish margin of the Black Sea is best observed in the offshore Akçakoca gas field, which is situated approximately 150 km east of the study area. Here, the intensely folded Eocene strata are unconformably overlain by Middle Miocene rocks. We constructed a regional composite 2D seismic section, constrained it by well data and extended this unconformity together with the MU to the Bosporus shelf through the Boğaziçi canyons (Fig. 24). However, it is extremely difficult to interpret these unconformities on the steep Bosporus slope where they converge and the MU appears to have truncated the Eocene unconformity as well. Therefore, we interpret that, unlike the Akçakoca area, the MU in the Bosporus area directly sits on the Paleocene–Eocene rocks, not only on the slope, but also on the shelf (Fig. 14). The high topography created by the uplifted Upper Cretaceous arc volcanics during the Eocene compression must have led to the development of limited accommodation space during the deposition of the post-Eocene strata on the shelf, triggering sediment by-pass and transport down the steep slope, which might have promoted MTC deposition on the continental rise and the abyssal plain. This has given rise to very thin sediment accumulation on the bypass-dominated slope since the Eocene (Fig. 24).

The width and relief (and fill thickness) of the Boğaziçi canyons show a decrease from west to

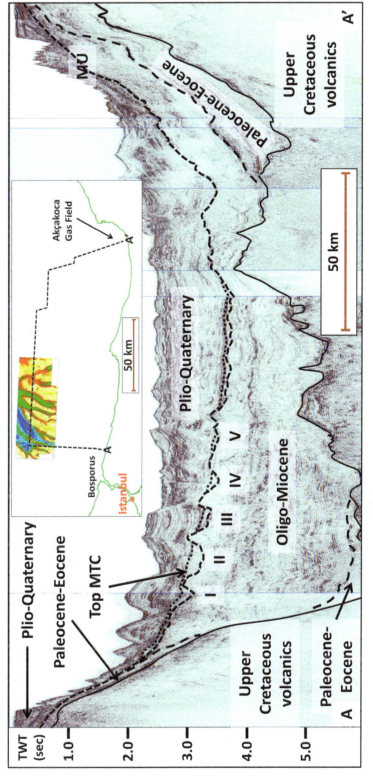

Fig. 24. A composite seismic section extending from the Bosporus margin through Boğaziçi canyons I–V to the Akçakoca margin. Abundant well data from the latter margin are used to pinpoint the ages and carry them to the study area, although both the great distance and the thinning of the Cenozoic package over the Bosporus margin cause uncertainties. The coloured part of the base map shows the time thicknesses of the Boğaziçi canyons. TWT, two-way travel time.

east (Figs 10, 12, 15 & 17). Several north–south-orientated regional 2D seismic lines passing through the shelf down to the abyssal plain clearly show that the amount of progradation of the shelf has been higher in the west than in the east at least since the Eocene (Figs 11 & 19–22): that is, the shelf to the west is wider than the shelf to the east. In parallel with this, the thickness of the slope sediments overlying the Upper Cretaceous volcanics is noticeably thicker in the west than in the east (cf. Figs 11 & 19–22). This is attributed to the presence of a relatively higher rate of sediment input from the SW of the Boğaziçi canyons. When correlated updip to the present-day coastline, this sediment source coincides with the general area of the Bosporus Strait. Although this observation might explain the possible role of a 'palaeo-Bosporus' drainage system in terms of sediment supply to the area, the lack of 3D seismic data on the shelf and the slope prevents a convincing conclusion.

Conclusions

In summary, two types of reflection patterns dominate the fills of the MSC-related canyons in the study area: (1) high–low amplitude, continuous–discontinuous (Karaburun Canyon); and (2) transparent–low-amplitude chaotic (Boğaziçi canyons). Although located not too distant from each other, these canyons were developed on different basin margins whose morphologies were defined by different tectonostratigraphic settings (Fig. 1). In the western part of the study area, where the Karaburun Canyon is located, the continental slope is primarily composed of sedimentary rocks. Seismic geometries suggest that the basin margin has prograded at least since the Early Miocene, leading to the development of a passive margin with a relatively wide shelf area (Figs 2–4, 6 & 7). In contrast, the eastern part of the study area, where the Boğaziçi canyons are located, is characterized by a very steep slope due to the Late Cretaceous volcanic arc. This arc has dominantly acted as a bypass surface with only a limited amount of sediment accumulation at least since the Eocene, and the sediments were transported from the shelf down into the deeper parts of the basin (Figs 11 & 19–22). For this reason, this area has been the site of sliding and slumping that must have promoted MTC deposition. The present-day seabed topography still indicates common erosion/remobilization of any sediment deposited intermittently on this oversteepened slope. Although there are indications that the palaeo-Bosporus Strait might have acted as a sediment fairway at least since the Late Miocene, we need better seismic coverage to substantiate this idea.

We are grateful to Turkish Petroleum and Shell for permitting the use of the data. We thank Sarp Karakaya for his assistance with seismic mapping, and Şakir Özsoy for drafting some of the figures. We are indebted to Jean-Pierre Suc and an anonymous reviewer for their comments and suggestions on the first draft, and to Gabor Tari and Aral I. Okay on the second draft. Their constructive and helpful suggestions have had invaluable contribution to this paper. Finally, we appreciate Mike Simmons and, once again, Gabor Tari for their persistent encouragement to write this paper.

References

ALGAR, S., MILTON, C., UPSHALL, H., ROESTENBURG, J. & CREVELLO, P. 2011. Mass-transport deposits of the deep-water northwestern Borneo margin – characterization from seismic-reflection, borehole, and core data with implications for hydrocarbon exploration and exploitation. *In*: SHIPP, R.C., WEIMER, P. & POSAMENTIER, H.W. (eds) *Mass-Transport Deposits in Deepwater Settings*. SEPM, Special Publications, **96**, 351–366.

ALLIN, J.R., HUNT, J.E., TALLING, P.J., CLARE, M.A., POPE, E. & MASSON, D.G. 2016. Different frequencies and triggers of canyon filling and flushing events in Nazaré Canyon, offshore Portugal. *Marine Geology*, **371**, 89–105.

BANKS, C.J. & ROBINSON, A.G. 1997. Mesozoic strike-slip back-arc basins of the Western Black Sea Region. *In*: ROBINSON, A.G. (ed.) *Regional and Petroleum Geology of the Black Sea and Surrounding Areas*. American Association of Petroleum Geologists, Memoirs, **68**, 53–62.

BOCCALETTI, M., GOCEV, P. & MANETTI, P. 1974. Mesozoic isopic zones in the Black Sea region. *Italian Journal of Geoscience*, **93**, 547–565.

BØE, R., HOVLAND, M., INSTANES, A., RISE, L. & VASSHUS, S. 2000. Submarine slide scars and mass movements in Karmsundet and Skudenesfjorden, southwestern Norway: morphology and evolution. *Marine Geology*, **167**, 147–165.

CAN, E. 1996. *Tectonic evolution of the southwestern Black Sea Margin, offshore Turkey*. MSc thesis, A&M University, Texas.

COVAULT, J.A. & GRAHAM, S.A. 2008. Turbidite architecture in proximal foreland basin system deep-water depocenters; insights from the Tertiary of western Europe. *Austrian Journal of Earth Sciences*, **101**, 36–51.

DE LA VARA, A., VAN BAAK, C.G., MARZOCCHI, A., GROTHE, A. & MEIJER, P.T. 2016. Quantitative analysis of Paratethys sea level change during the Messinian Salinity Crisis. *Marine Geology*, **379**, 39–51.

DI CELMA, C. 2011. Sedimentology, architecture, and depositional evolution of a coarse-grained submarine canyon fill from the Gelasian (early Pleistocene) of the Peri-Adriatic basin, Offida, central Italy. *Sedimentary Geology*, **238**, 233–253.

DI CELMA, C., CANTALAMESSA, G., DIDASKALOU, P. & LORI, P. 2010. Sedimentology, architecture, and sequence stratigraphy of coarse-grained, submarine canyon fills from the Pleistocene (Gelasian–Calabrian) of the Peri-Adriatic basin, central Italy. *Marine and Petroleum Geology*, **27**, 1340–1365.

DIAZ, J., WEIMER, P., BOUROULLEC, R. & DORN, G. 2011. 3-D seismic stratigraphic interpretation of Quaternary mass-transport deposits in the Mensa and Thunder Horse intraslope basins, Mississippi Canyon, northern deep Gulf of Mexico, U.S.A. *In*: SHIPP, R.C., WEIMER, P. & POSAMENTIER, H.W. (eds) *Mass-Transport Deposits in Deepwater Settings*. SEPM, Special Publications, **96**, 127–149.

FIGUEIREDO, J.J.P., HODGSON, D.M., FLINT, S.S. & KAVANAGH, J.P. 2013. Architecture of a channel complex formed and filled during long-term degradation and entrenchment on the upper submarine slope, Unit F, Fort Brown Fm., SW Karoo Basin, South Africa. *Marine and Petroleum Geology*, **41**, 104–116.

FINETTI, I., BRICCHI, G., DEL BEN, A., PIPAN, M. & XUAN, Z. 1988. Geophysical study of the Black Sea. *Bollettino di Geofisica Teorica e Applicata*, **30**, 197–324.

FREY-MARTÍNEZ, J., CARTWRIGHT, J. & HALL, B. 2005. 3D seismic interpretation of slump complexes: examples from the continental margin of Israel. *Basin Research*, **17**, 83–108.

GAMBERI, F., ROVERE, M. & MARANI, M. 2011. Mass-transport complex evolution in a tectonically active margin (Gioia basin, Southeastern Tyrrhenian Sea). *Marine Geology*, **279**, 98–110.

GAMBERI, F., ROVERE, M., DYKSTRA, M., KANE, I.A. & KNELLER, B. 2013. Integrating modern sea floor and outcrop data in the analysis of slope channel architecture and fill. *Marine and Petroleum Geology*, **41**, 83–103.

GAMBOA, D., ALVES, T., CARTWRIGHT, J. & TERRIHNA, P. 2010. MTD distribution on a 'passive' continental margin: the Espírito Santo Basin (SE Brazil) during the Palaeogene. *Marine and Petroleum Geology*, **27**, 1311–1324.

GARCIA, M., ERCILLA, G., ESTRADA, F., JANE, G., MENA, A., ALVES, T. & JUAN, C. 2015. Deep-water turbidite systems: a review of their elements, sedimentary processes and depositional models. Their characteristics on the Iberian margins. *Bolletin Geologico y Minero*, **126**, 189–218.

GILLET, H., LERICOLAIS, G. & RÉHAULT, J.-P. 2007. Messinian event in the Black Sea: Evidence of a Messinian erosional surface. *Marine Geology*, **244**, 142–165, https://doi.org/10.1016/j.margeo.2007.06.004

GONG, C., WANG, Y., ZHU, W., LI, W., XU, Q. & ZHANG, J. 2011. The Central Submarine Canyon in the Qiongdongnan Basin, northwestern South China Sea: architecture, sequence stratigraphy, and depositional processes. *Marine and Petroleum Geology*, **28**, 1690–1702.

GONG, C., WANG, Y., HODGSON, D.M., ZHU, W., LI, W., XU, Q. & LI, D. 2014. Origin and anatomy of two different types of mass-transport complexes: a 3D seismic case study from the northern South China Sea margin. *Marine and Petroleum Geology*, **54**, 198–215.

GÖRÜR, N. 1988. Timing of opening of the Black Sea basin. *Tectonophysics*, **147**, 247–262.

HANQUIEZ, V., MULDER, T., TOUCANNE, S., LECROART, P., BONNEL, C., MARCHES, E. & GONTHIER, E. 2010. The sandy channel-lobe depositional systems in the Gulf of Cadiz: gravity processes forced by contour current processes. *Sedimentary Geology*, **22**, 110–123.

HARRIS, P.T. & WHITEWAY, T. 2011. Global distribution of large submarine canyons: geomorphic differences between active and passive continental margins. *Marine Geology*, **285**, 69–86.

HARRIS, P.T., MACMILLAN-LAWLER, M., RUPP, J. & BAKER, E.K. 2014. Geomorphology of the oceans. *Marine Geology*, **352**, 4–24.

HE, Y.L., XIE, X.N., KNELLER, B.C., WANG, Z.F. & LI, X.S. 2013. Architecture and controlling factors of canyon fills on the shelf margin in the Qiongdongnan Basin, northern South China Sea. *Marine and Petroleum Geology*, **41**, 264–276.

HSÜ, K.J. & GIOVANOLI, F. 1979. Messinian event in the Black Sea. *Palaeogeography, Palaeoclimatology, Palaeoecology*, **29**, 75–93, https://doi.org/10.1016/0031-0182(79)90075-0

KING, P., ILG, B.R., ARNOT, M., BROWNE, G.H., STRACHAN, L.J., CRUNDWELL, M. & HELLE, K. 2011. Outcrop and seismic examples of mass-transport deposits from a Late Miocene deep-water succession, Taranaki Basin, New Zealand. *In*: SHIPP, R.C., WEIMER, P. & POSAMENTIER, H.W. (eds) *Mass-Transport Deposits in Deepwater Settings*. SEPM, Special Publications, **96**, 311–348.

KITCHKA, A.A., TYSHCHENKO, A.P. & LYSENKO, V.I. 2016. Mid-late Miocene sea level falls, gas hydrates decay, submarine sliding, and tsunamites in the Black Sea Basin. Paper presented at the *78th EAGE Conference & Exhibition 2016*, 30 May–2 June 2016, Vienna, Austria.

KORUCU, Ö., SIPAHIOĞLU, N.Ö., AKTEPE, S. & BENGÜ, E. 2013. Correlation and determination of Neogene sequences employing recent ultra-deep wells in western-central part of Turkish Black Sea. Paper presented at the *AAPG Europe Regional Conference*, Tbilisi, Georgia.

KREZSEK, C., SCHLEDER, Z., BEGA, Z., IONESCU, G. & TARI, G. 2016. The Messinian sea level fall in the western Black Sea: small or large? Insights from offshore Romania. *Petroleum Geoscience*, **22**, 392–399, https://doi.org/10.1144/petgeo2015-093

KRIJGSMAN, W., STOICA, M., VASILIEV, I. & POPOV, V. 2010. Rise and fall of the Paratethys Sea during the Messinian Salinity Crisis. *Earth and Planetary Science Letters*, **290**, 183–191, https://doi.org/10.1016/j.epsl.2009.12.020

LETOUZEY, J., BIJU-DUVAL, B., DORKEL, A., GONNARD, R., KRISCHEV, K., MONTADERT, L. & SUNGURLU, O. 1977. The Black Sea: a marginal basin: geophysical and geological data. *In*: BIJU-DUVAL, B. & MONTADERT, L. (eds) *Structural History of the Mediterranean Basins*. Editions Technip, Paris, 363–376.

LI, X., FAIRWEATHER, L. *ET AL.* 2013. Morphology, sedimentary features and evolution of a large palaeo submarine canyon in Qiongdongnan basin, Northern South China Sea. *Journal of Asian Earth Sciences*, **62**, 685–696.

LOFI, J. & BERNE, S. 2008. Evidences for Pre-Messinian submarine canyons incisions on the Gulf of Lions Miocene slope (Western Mediterranean). *Marine and Petroleum Geology*, **25**, 804–817.

LU, H. & SHIPP, C.R. 2011. Impact of a large mass-transport deposit on a field development in the upper slope of southwestern Sabah, Malaysia, offshore northwest Borneo. *In*: SHIPP, R.C., WEIMER, P. & POSAMENTIER, H.W. (eds) *Mass-Transport Deposits in Deepwater Settings*. SEPM, Special Publications, **96**, 199–218.

MACAULEY, R.V. & HUBBARD, S.M. 2013. Slope channel sedimentary processes and stratigraphic stacking, Cretaceous Tres Pasos Formation slope system, Chilean Patagonia. *Marine and Petroleum Geology*, **41**, 146–162.

MARTÍNEZ, J.F., BERTONI, C., GERARD, J. & MATIAS, H. 2011. Processes of submarine slope failure and fluid migration on the Ebro continental margin: Implications for offshore exploration and development. *In*: SHIPP, R.C., WEIMER, P. & POSAMENTIER, H.W. (eds) *Mass-Transport Deposits in Deepwater Settings*. SEPM, Special Publications, **96**, 181–198.

MARTÍN-MERINO, G., FERNÁNDEZ, L.P., COLMENERO, J.R. & BAHAMONDE, J.R. 2014. Mass-transport deposits in a Variscan wedge-top foreland basin (Pisuerga area, Cantabrian Zone, NW Spain). *Marine Geology*, **356**, 71–87.

MAYALL, M., JONES, E. & CASEY, M. 2006. Turbidite channel reservoirs – Key elements in facies prediction and effective development. *Marine and Petroleum Geology*, **23**, 821–841.

MCHARGUE, T., PYRCZ, M.J. *ET AL.* 2011. Architecture of turbidite channel systems on the continental slope: Patterns and predictions. *Marine and Petroleum Geology*, **28**, 728–743.

MOSCARDELLI, L. & WOOD, L. 2007. New classification system for mass transport complexes in offshore Trinidad. *Basin Research*, **20**, 73–98, https://doi.org/10.1111/j.1365-2117.2007.00340.x

MOSCARDELLI, L., WOOD, L. & MANN, P. 2006. Mass-transport complexes and associated processes in the offshore area of Trinidad and Venezuela. *AAPG Bulletin*, **9**, 1059–1088.

MOSHER, D.C. & CAMPBELL, C.D. 2011. The Barrington submarine mass-transport deposit, western Scotian slope, Canada. *In*: SHIPP, R.C., WEIMER, P. & POSAMENTIER, H.W. (eds) *Mass-Transport Deposits in Deepwater Settings*. SEPM, Special Publications, **96**, 151–159.

MUNTEANU, I., MATENCO, L., DINU, C. & CLOETINGH, S. 2012. Effects of large sea-level variations in connected basins: the Daciane Black Sea system of the Eastern Paratethys. *Basin Research*, **24**, 583e597, https://doi.org/10.1111/j.1365-2117.2012.00541.x

NAKAJIMA, T., KAKUWA, Y. *ET AL.* 2014. Formation of pockmarks and submarine canyons associated with dissociation of gas hydrates on the Joetsu Knoll, eastern margin of the Sea of Japan. *Journal of Asian Earth Sciences*, **90**, 228–242.

NELSON, C.H., ESCUTIA, C., DAMUTH, J.E. & TWICHELL Jr., D.C. 2011. Interplay of mass-transport and turbidite-system deposits in different active tectonic and passive continental margin settings; external and local controlling factors. *In*: SHIPP, R.C., WEIMER, P. & POSAMENTIER, H.W. (eds) *Mass-Transport Deposits in Deepwater Settings*. SEPM, Special Publications, **96**, 39–66.

NIKISHIN, A.M., KOROTAEV, M.V., ERSHOV, A.V. & BRUNET, M.-F. 2003. The Black Sea basin: tectonic history and Neogene–Quaternary rapid subsidence modeling. *Sedimentary Geology*, **156**, 149–168, https://doi.org/10.1016/S0037-0738(02)00286-5

NIKISHIN, A.M., ZIEGLER, P.A., BOLOTOV, S.N. & FOKIN, P.A. 2012. Late Palaeozoic to Cenozoic evolution of the Black Sea–Southern Eastern Europe region: a view from the Russian platform. *Turkish Journal of Earth Sciences*, **20**, 571–634, https://doi.org/10.3906/yer-1005-22

NIKISHIN, A.M., OKAY, A.I., TÜYSÜZ, O., DEMIRER, A., AMELIN, N. & PETROV, E. 2015a. The Black Sea basins structure and history: new model based on new deep penetration regional seismic data. Part 1: basins

structure and fill. *Marine and Petroleum Geology*, **59**, 38–655.

NIKISHIN, A.M., OKAY, A., TÜYSÜZ, O., DEMIRER, A., WANNIER, M., AMELIN, N. & PETROV, E. 2015b. The Black Sea basins structure and history: new model based on new deep penetration regional seismic data. Part 2: tectonic history and paleogeography. *Marine and Petroleum Geology*, **59**, 656–670.

OKAY, A.I. & GÖRÜR, N. 2000. Tectonic evolution models for the Black Sea. Paper presented at the *AAPG Regional Meeting*, 10–12 July 2000, Istanbul, Turkey.

OKAY, A.I., ŞENGÖR, A.M.C. & GÖRÜR, N. 1994. Kinematic history of the opening of the Black Sea and its effect on the surrounding regions. *Geology*, **22**, 267–270.

OKAY, A.I., TANSEL, İ. & TÜYSÜZ, O. 2001. Obduction, subduction and collision as reflected in the Upper Cretaceous–Lower Eocene sedimentary record of western Turkey. *Geological Magazine*, **138**, 117–142.

OKAY, A.I., SUNAL, G., SHERLOCK, S., ALTINER, D., TÜYSÜZ, O., KYLANDER-CLARK, A.R.C. & AYGÜL, M. 2013. Early Cretaceous sedimentation and orogeny on the active margin of Eurasia: southern Central Pontides, Turkey. *Tectonics*, **32**, 1247–1271.

POPOV, S.V., ANTIPOV, M.P., ZASTROZHNOV, A.S., KURINA, E.E. & PINCHUK, T.N. 2010. Sea- level fluctuations on the north shelf of the Eastern Paratethys in the Oligocene–Neogene. *Stratigraphy and Geological Correlation*, **18**, 200–224.

POSAMENTIER, H.W. 2003. Depositional elements associated with a basin floor channel–levee system: case study from the Gulf of Mexico. *Marine and Petroleum Geology*, **20**, 677–690.

POSAMENTIER, H.W. & KOLLA, V. 2003. Seismic geomorphology and stratigraphy of depositional elements in deep water settings. *Journal of Sedimentary Research*, **73**, 367–388.

POSAMENTIER, H.W. & MARTINSEN, O.J. 2011. The character and genesis of submarine mass-transport deposits: insights from outcrop and 3D seismic data. *In*: SHIPP, R.C., WEIMER, P. & POSAMENTIER, H.W. (eds) *Mass-Transport Deposits in Deepwater Settings*. SEPM, Special Publications, **96**, 7–38.

RICHARDSON, S.E.J., DAVIES, R.J., ALLEN, M.B. & GRANT, S.F. 2011. Structure and evolution of mass transport deposits in the South Caspian Basin, Azerbaijan. *Basin Research*, **23**, 702–719.

ROBINSON, A.G., RUDAT, J.H., BANKS, C.J. & WILES, R.L.F. 1996. Petroleum geology of the Black Sea. *Marine and Petroleum Geology*, **13**, 195–223.

RÖGL, F. 1999. Mediterranean and Paratethys. Facts and hypotheses of an Oligocene to Miocene paleogeography (short overview). *Geologica Carpathica*, **50**, 339–349.

ROSSI, M., MINERVINI, M., GHIELMI, M. & ROGLEDI, S. 2015. Messinian and Pliocene erosional surfaces in the Po Plain-Adriatic Basin: insights from allostratigraphy and sequence stratigraphy in assessing play concepts related to accommodation and gateway turnarounds in tectonically active margins. *Marine and Petroleum Geology*, **66**, 192–216.

ROY MOULIK, S.K. & PRASAD, G.K. 2007. Seismic expression of the canyon fills facies and its geological significance – A case study from Ariyalur–Pondicherry subbasin, Cauvery basin. Search and Discovery Article

#10125 presented at the *AAPG Annual Convention*, 1–4 April 2007, Long Beach, California.

ŞENGÖR, A.M.C. & YILMAZ, Y. 1981. Tethyan evolution of Turkey: a plate tectonic approach. *Tectonophysics*, **75**, 181–241.

SIPAHIOGLU, N.Ö. & ÇIFTÇI, S.Y. 2010. Seismic stratigraphic analysis of Late Oligocene–Recent deltaic–turbiditic systems, Kırklareli and Bogazici 3D seismic survey areas, Western Black Sea. Paper presented at the *AAPG European Region Annual Conference: Exploration in the Black Sea and Caspian Regions*, 17–19 October 2010, Kiev.

SIPAHIOGLU, N.O., KARAHANOGLU, N. & ALTINER, D. 2013. Analysis of Plio-Quaternary deep marine systems and their evolution in a compressional tectonic regime, Eastern Black Sea Basin. *Marine and Petroleum Geology*, **43**, 187–207.

SIPAHIOĞLU, Ö., KORUCU, Ö., AKTEPE, S. & BENGÜ, E. 2013. Westerly-sourced Late Oligocene–Middle Miocene axial sediment dispersal system in Turkish Western Black Sea: myth or reality? Paper presented at the *19th International Petroleum and Natural Gas Congress and Exhibition of Turkey*, 15–17 May 2013, Ankara, Turkey.

SIYAKO, M. 2006. Tertiary rock units of the Thrace Basin. *In*: ALTINER, D., BATI, Z., TUNAY, G., ŞENEL, M. & EKMEKÇI, E. (eds) *Lithostratigraphic Units of the Thrace Region*. General Directorate of Mineral Research and Exploration, Ankara, 43–83 [in Turkish].

SIYAKO, M. & HUVAZ, O. 2007. Eocene stratigraphic evolution of the Thrace Basin, Turkey. *Sedimentary Geology*, **198**, 75–91.

SOLHEIM, A., BERG, K., FORSBERG, C.F. & BRYN, P. 2005. The Storegga Slide complex: repetitive large scale sliding with similar cause and development. *Marine and Petroleum Geology*, **22**, 97–107.

SPADINI, G., ROBINSON, A. & CLOETINGH, S. 1996. Western v. Eastern Black Sea tectonic evolution: pre-rift lithospheric controls on basin formation. *Tectonophysics*, **266**, 139–154.

SPADINI, G., ROBINSON, A.G. & CLOETINGH, S. 1997. Thermo-mechanical modelling of Black Sea basin formation, subsidence and sedimentation. *In*: ROBINSON, A.G. (ed.) *Regional and Petroleum Geology of the Black Sea and Surrounding Areas*. American Association of Petroleum Geologists, Memoirs, **68**, 19–38.

SUC, J.-P., GILLET, H. *ET AL.* 2015. The region of the Strandja Sill (North Turkey) and the Messinian events. *Marine and Petroleum Geology*, **66**, 149–164.

SUTTON, J.P. & MITCHUM, R.M. 2011. Upper Quaternary seafloor mass-transport deposits at the base of slope, offshore Niger Delta, deepwater Nigeria. *In*: SHIPP, R.C., WEIMER, P. & POSAMENTIER, H.W. (eds) *Mass-Transport Deposits in Deepwater Settings*. SEPM, Special Publications, **96**, 85–110.

TARI, G., FALLAH, M., KOSI, W., FLOODPAGE, J., BAUR, J., BATI, Z. & SIPAHIOĞLU, N.Ö. 2015. Is the impact of the Messinian Salinity Crisis in the Black Sea comparable to that of the Mediterranean? *Marine and Petroleum Geology*, **66**, 135–148.

TARI, G., FALLAH, M. *ET AL.* 2016. Why are there no Messinian evaporates in the Black Sea? *Petroleum Geoscience*, **22**, 381–391, https://doi.org/10.1144/petgeo2016-003

TASSY, A., FOURNIER, F. *ET AL.* 2014. Discovery of Messinian canyons and new seismic stratigraphic model, offshore Provence (SE France): implications for the hydrographic network reconstruction. *Marine and Petroleum Geology*, **57**, 25–50.

THOMAS, M.F.H. & BODIN, S. 2013. Architecture and evolution of the Finale channel system, the Numidian Flysch Formation of Sicily; insights from a hierarchical approach. *Marine and Petroleum Geology*, **41**, 163–185.

VASILIEV, I., REICHART, G.-J. & KRIJGSMAN, W. 2013. Impact of the Messinian Salinity Crisis on Black Sea hydrology – Insights from hydrogen isotopes analysis on biomarkers. *Earth and Planetary Science Letters*, **362**, 272–282, https://doi.org/10.1016/j.epsl.2012.11.038

ZHANG, C., WEI, W., ZHANG, S., WU, C. & FU, X. 2016. Architecture of lacustrine mass-transport complexes in the Mesozoic Songliao Basin, China. *Marine and Petroleum Geology*, **78**, 826–835, https://doi.org/10.1016/j.marpetgeo.2016.07.001

ZHOU, W., WANG, Y. *ET AL.* 2015. Architecture, evolution history and controlling factors of the Baiyun submarine canyon system from the middle Miocene to Quaternary in the Pearl River Mouth Basin, northern South China Sea. *Marine and Petroleum Geology*, **67**, 389–407.

ZHU, M., GRAHAM, S. & MCHARGUE, T. 2011. Characterization of mass-transport deposits on a Pliocene siliciclastic continental slope, northwestern south China Sea. *In*: SHIPP, R.C., WEIMER, P. & POSAMENTIER, H.W. (eds) *Mass-transport Deposits in Deepwater Settings*. SEPM, Special Publications, **96**, 111–125.

ZONENSHAIN, L.P. & LE PICHON, X. 1986. Deep basins of the Black Sea and Caspian Sea as remnants of Mesozoic back-arc basins. *Tectonophysics*, **123**, 181–211.

Holocene source rock deposition in the Black Sea, insights from a dropcore study offshore Bulgaria

M. FALLAH[1]*, J. MAYER[1], G. TARI[1] & J. BAUR[2]

[1]*OMV Exploration and Production GmbH, Trabrennstraße 6–8, A-1020 Vienna, Austria*

[2]*Total, Place Jean Millier, La Defense 6, Paris 92078, France*

*Correspondence: Mohammad.fallah@omv.com

Abstract: One of the main issues in source rock evaluation has always been the availability of thermally immature samples, which would represent the same source rock quality and facies as the mature source rock within the deeper parts of the basin. Forty dropcore sample locations from shallow depths beneath the present-day seafloor were selected and analysed for mineral composition and bulk geochemical parameters. The water depths of the samples range from shelfal to bathyal environments. The quartz content of the samples clearly decreases with increasing distance from sedimentary input sources (e.g. river deltas), whereas clay content increases towards the distal areas. Mass movements (e.g. slides and debris flows) along the present-day shelf are recognizable on the bathymetry, as well as in the mineral content. Bulk geochemical parameters show that currently only poor to fair gas-prone source rocks are deposited within the study area. This lack of source rock quality, as well as organic content, is attributed to the fine-grained sedimentary input from the Danube river. These fine-grained sediments decrease the organic productivity due to dulling (decrease in the thickness of the photic zone) of the water column, and dilute the currently deposited source rock with low TOC sediments. These effects decrease with distance from the Danube delta, as indicated by published data from outside the study area. Additionally mass movements along the present-day shelf rework possible source rocks. The results of this study clearly show that anoxic conditions alone are not sufficient for source rock deposition. Distance from major sedimentary input and basin geometry are of major importance, and should be considered in basin modelling.

The Black Sea is one of the most important analogues for recent source rock deposition, due to a high level of nutrients and high organic productivity, as well as good conditions for the preservation of organic matter (e.g. water stratification and anoxia). Recent deposition of sediments with a high organic content is described by several authors (e.g. Arthur & Sageman 2004). This paper concentrates on 40 dropcore samples taken within the Han Asparuh block, offshore Bulgaria in 2014. The samples were acquired from 195 to 1981 m water depth, therefore ranging from shelfal to bathyal environments. Published material from previous sampling campaigns, undertaken between 1969 and 2004, were used to bring the results into a regional context (Arthur & Sageman 2004). Bulk geochemical and mineralogical results were interpreted in accordance with high-resolution bathymetry to identify relevant geological processes for source rock deposition.

Geological overview

The study area is located close to the junction of the Moesian Platform and the Western Black Sea Basin (Fig. 1). The Moesian Platform is formed by the foreland of the Carpathians in the north and the Balkan thrust-fold belt in the south. Its sedimentary cover comprises three main structural sequences: (1) Palaeozoic, (2) Permo-Triassic and (3) Jurassic–Cenozoic, separated by Hercynian and Cimmerian unconformities (Tari *et al.* 1997).

The east–west-trending Balkan thrust-fold belt represents a segment of the Alpine Orogen (Boncev 1986). It consists of a stack of dominantly north-verging thrust sheets that developed during multiphase collisional events along a long-lived convergent continental margin. Compression culminated towards the end of the Early Cretaceous (first phase) and in the early Middle Eocene (second phase: Emery & Georgiev 1993). Loading of the Moesian Platform by the Balkan thrust sheets resulted in the formation of the Kamchia Depression filled with Middle Eocene–Quaternary deposits (Robinson *et al.* 1995; Georgiev 2012). The Balkan Orogen is made up of a southern uplifted overthrust zone (Balkan or Stara Plania) and a northern subsided Fore-Balkan thrust-fold zone (Georgiev 2012). The Fore-Balkan zone subsides beneath the Cenozoic fill of the Kamchia Depression. Offshore, the East Balkan Orogen first turns towards the SE, then shifts considerably to the south (Georgiev 2012).

From: Simmons, M. D., Tari, G. C. & Okay, A. I. (eds) 2018. *Petroleum Geology of the Black Sea.*
Geological Society, London, Special Publications, **464**, 389–401.
First published online September 15, 2017, https://doi.org/10.1144/SP464.11

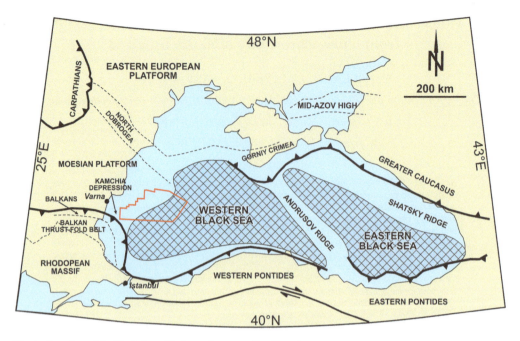

Fig. 1. Overview of the Black Sea with the study area outlined in red.

The Black Sea is considered by many authors to be a Late Cretaceous–Palaeogene back-arc extensional basin that developed north of the Pontide magmatic arc, which formed by northwards subduction of the Neo-Tethys Ocean, initiated in the Albian (Tugolesov *et al.* 1985; Finetti *et al.* 1988; Gorur 1988; Okay *et al.* 1994; Dachev & Georgiev 1995; Robinson *et al.* 1995; Banks & Robinson 1997; Nikishin *et al.* 2001, 2003). According to Nikishin *et al.* (2015), the main regional rifting event is Late Barremian–Albian in age and oceanic crust was formed from Cenomanian to Mid-Santonian time.

During Cenozoic time, depositional environments in the western Black Sea area have been strongly influenced by separations of the Paratethys Sea from the Mediterranean Sea. Basin separation occurred several times during the Cenozoic (Zachos *et al.* 2001; Popov *et al.* 2010). These separations favoured the development of endemic faunas, water stratification and anoxia within the water column (Popov *et al.* 2004a, b; Schulz *et al.* 2005).

The last basin separations ended at 9.4 ka. This event is indicated by a change in mollusk assemblages and faunal colonization (Major *et al.* 2006). The following sea-level highstand induced the end of the activity of the Danube turbidite system (Lericolais *et al.* 2013; Friedrich *et al.* 2002). Stratification of the Black Sea water occurred prior to this last separation (14.5 ka BP (before present) according to Bahr *et al.* 2008).

Anoxia and high productivity developed in the Holocene Black Sea around 7.6 ka, leading to deposition of a sapropel with up to 20 wt% organic carbon. The development of eutrophic conditions coincided with rising sea level and overflow of saline waters from the Mediterranean Sea (Arthur & Sageman 2004). Trapping of river-derived nutrients in the Black Sea behind the shallow sill to the Mediterranean and the high freshwater flux to the Black Sea Basin during climatic warming are additional causes of eutrophication associated with the global mid-Holocene transgression. Organic carbon contents increase towards the basin centre (Fig. 2) because of lower clastic dilution and focusing of organic carbon transport from the margins to the centre (Arthur & Sageman 2004).

Materials and methodology

Sixty dropcores were acquired by the Fugro Group (GeoConsulting & OceanSismica) with a standard piston gravity corer (Kullenberg trigger and 6 m core barrel). The dropcores were photographed, described and sampled on site. In total, 340 m of sediment core were recovered. Samples from 40 dropcore locations were chosen to be used in this study in accordance with their position to provide an even spread throughout the whole study area. The locations of the sampled dropcores are shown in

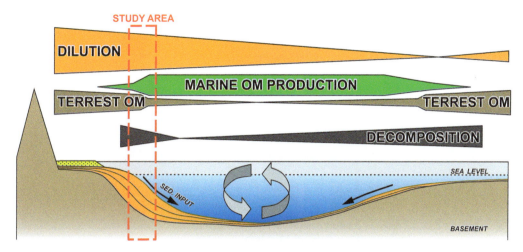

Fig. 2. Diagrammatic cross-section of an epicontinental basin bounded by an orogen showing the spatial distribution of factors involved in the production and preservation of organic carbon in sediments during sea-level-rise events (Arthur & Sageman 2004).

Figure 3. Three sample subsets – (A) deepest part, (B) middle part and (C) shallowest part – were taken from each dropcore. To sample sediments with the same or similar ages, the shallowest samples of each dropcore was chosen to be analysed.

All samples have been analysed in duplicate for total sulphur (S), total carbon (TC) and total organic carbon (after removal of carbonate minerals: TOC) using an 'Eltra' instrument. Assuming calcite to be the only carbonate mineral present, the calcite equivalent percentage was calculated (Calcite$_{eq}$ = (TC − TOC) × 8.333). Pyrolysis measurements were carried out in duplicate using a 'Rock-Eval 6' instrument. The S1 and S2 peaks (mg HC/g rock) were used to calculate the hydrogen index (HI = S2 × 100/TOC [mg HC/g TOC]) and the production index (PI = S1/(S1 + S2): Espitalié *et al.* 1977). T_{max} was measured as a maturity indicator. The source rock quality assignment in this paper follows Peters (1986).

To determine the bulk mineralogical composition, a Bruker AXS D8 Advance X-ray diffraction (XRD) spectrometer was used. Organic material was removed using hydrogen peroxide (H_2O_2). Different mineral phases were identified with the DIFRAC.EVA V3 software. Quantification of minerals is based on peak heights within the spectrum based on the method of Schulz (1964) and OMV internal standards. The contents of quartz, potassium-feldspar, plagioclase, calcite, dolomite, ankerite, siderite, anhydrite, pyrite, total clay and mica in the sediments were obtained.

To determine the ages of the sediments, a detailed literature study about sedimentation rates was carried out. According to the stratigraphy of Ross & Degens

(1974), late Quaternary (<25 ka) Black Sea sediments are subdivided into two Holocene marine units. 2720 ± 160 and 7540 ± 130 years BP were the ages assigned to the Unit I–II and Unit II–III boundaries, respectively. Arthur & Dean (1998) distinguished between three different units. However, instead of using the highest sapropel layer as the boundary between units 1 and 2 (Ross & Degens 1974), they used the lowest occurrence of white coccolith-rich laminae. Ages of 1.63 and 7.8 ka for the Unit 1–2 and Unit 2–3 boundaries were assigned, respectively. Generally, age assignments are hampered by different results due to different methods and corrections (e.g. reservoir effect) used by different authors. Due to the differences in age dating and sedimentary input, no age sedimentation rates and sedimentary ages were calculated for individual samples. The samples used within this study were taken from similar depths, but are most likely not completely age equivalent. However, based on published age dating, the maximum age differences in-between the sampled sediments is in the range of thousands of years, which, if used as an analogue for previously deposited source rocks (e.g. Mesozoic source rocks), is below the resolution of any age-dating method (Jones 1990). Therefore, the samples are treated as pseudo-age equivalent within this study.

Results

The sediments in the cores sampled in the Han Asparuh study area were fairly uniform, which is remarkable for an area covering over 14 000 km^2 and ranging in depths from <100 m to over

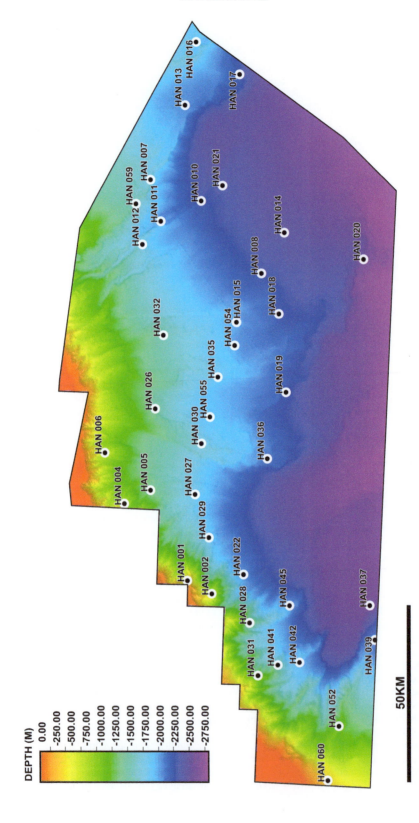

Fig. 3. Bathymetry map and location of samples within the study area.

2000 m. Seafloor sediments were very loosely compacted clay, with very high moisture content. The uppermost section included many thin white layers of coccoliths and, perhaps, other shelly debris. In some cores, there were 0.5–4 cm-thick olive grey clay layers of what was interpreted to be sapropel. In most Han Asparuh cores, there was a 1.7 m-thick sapropel deposit that consisted of two different layers. The upper sapropel layer was lighter, softer, and had several thin white layers near the top that appear to contain grits and shell debris. The lower sapropel layer was dark olive grey to olive black to chocolate brown, elastic, and often split along the sediment layers, probably due to gas expansion.

Six facies groups were classified to correlate these facies laterally across the area of interest. Pictures of selected dropcores and schematic drawings of observed structures are shown in Figure 4. A core log was prepared for each core which had a compilation of core photographs, classified facies, core description, depth of core, and date and time of acquiring:

- Facies I – characterized by grey to light olive coloured, laminated, weak to stiff clays. Darker siltier layers and occasional black organic streaks may occur.
- Facies II – made up of dark greenish to grey/blackish, firm to stiff clay. These sediments show a banded characteristic.
- Facies III – comprising dark yellowish brown, disturbed, friable to weak, sapropel grey clay bands with white shelly streaks occuring within this facies. A very strong odour of H_2S was observed.
- Facies IV – characterized by grey, structurally disintegrated, 'mushy' uniform clay. Greenish bands and occasional black streaks, as well as degassing blisters and gas fractures, occurred within this facies. A H_2S smell was observed in some samples.
- Facies V – made up of black to dark greenish grey, weak and sticky clay. Expansion gaps and evidence of dissociated gas, as well as a slight H_2S smell, were observed in samples from this facies.
- Facies VI – comprises weak to soft clay with silt and little fine sand.

The TOC contents were generally poor to fair; three samples showed good TOC contents and one sample had a very good TOC content. A very slight decrease of TOC v. water depth can be observed. Sulphur contents show a wider range than TOC contents and no obvious trend. They range from below 0.1% up to 2.2%. The hydrogen index was generally low (<100 mg HC/g TOC). These low values indicate a gas-prone source rock. Only two samples show a hydrogen index above 150 mg HC/g TOC. These two samples would be considered to be oil- and gas-prone source rocks. A minor increase in HI with depth could be recognized; however, this trend could also reflect increasing measurement uncertainties due to the decrease in TOC. All samples are immature according to their T_{max} values (<430°C).

Selected XRD data are plotted in a map-based view in Figure 5 and TOC values in Figure 6. A negative relationship between quartz content and TOC is recognizable. Clay contents and TOC contents show no clear relationship. Generally, the clay content increases with water depth and the quartz content decreases. Due to the similar TOC and HI values of all the samples, detailed correlations were not possible. The sulphur contents correspond well with the amount of pyrite in individual samples. An increase in pyrite, recognizable by XRD, leads to an increase in the sulphur content, measured using an ELTRA analyser, indicating that most of the sulphur in the sediment is bound in pyrite minerals. The carbonate content is generally low (calcite equivalent percentage, as well as calcite measured by XRD).

Interpretation

Geochemical data clearly indicate that only poor to fair source rocks are currently being deposited in the area of interest. The low source rock quality is most likely due to dilution of the sediments by relatively high sedimentary input from the Danube. The Danube cone reaches into the northernmost part of the study area and is clearly visible on the bathymetry (Fig. 3). A thickening of the uppermost sedimentary layer in the dropcores towards the north has been observed. Most probably, the fine-grained sediments of the Danube affect the present-day source rock generation in two ways.

The high present-day input of sediments with a low organic content dilutes the organic-rich sediments, which are currently produced in the Western Black Sea Basin. High organic contents in young shallow sediments are described by Arthur & Sageman (2004) outside of the study area of this paper. Present-day anoxic conditions (negative Eh values) within the water column in large areas of the Western Black Sea Basin were described by Kempe et al. (1988). Secondly, the incoming fine-grained sediments decrease the thickness of the photic zone by dulling the water column. The decreased photic zone results in a retardation of organic productivity within the basin. These mechanisms could also explain the low carbonate content, which most likely is also due to the dulling of the water column.

The results of this study clearly show that an anoxic basin with currently active biological productivity does not necessarily mean that a high-quality source rock will be deposited. The general assumption that a source rock will always improve towards the distal parts of a basin could only be observed in published data. Over the large study area (>14 000 km²)

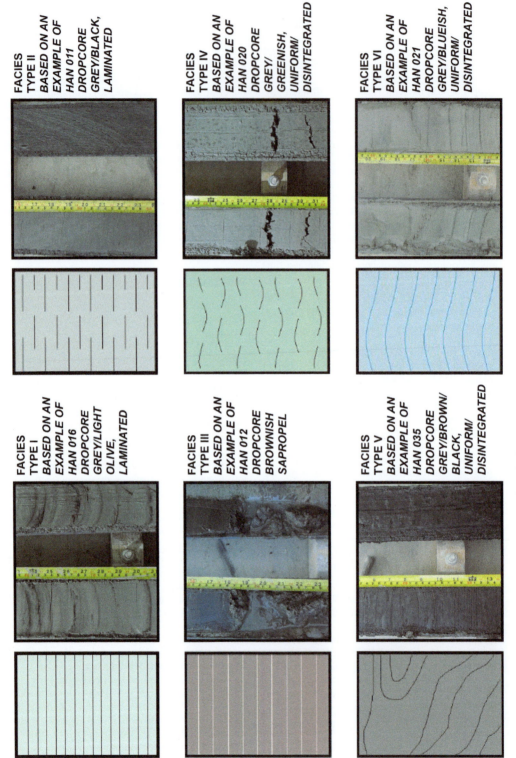

Fig. 4. Description and examples of encountered sedimentary facies in dropcores.

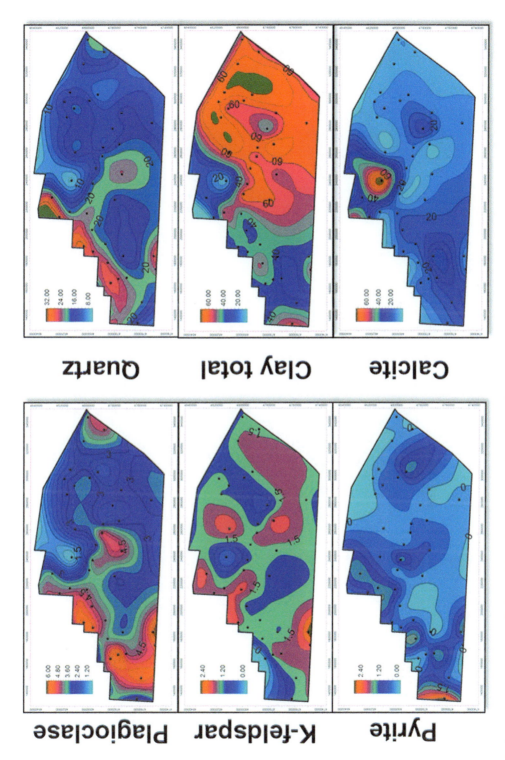

Fig. 5. Gridded maps of selected XRD data. Blue colours are used for low values and red colours for high values.

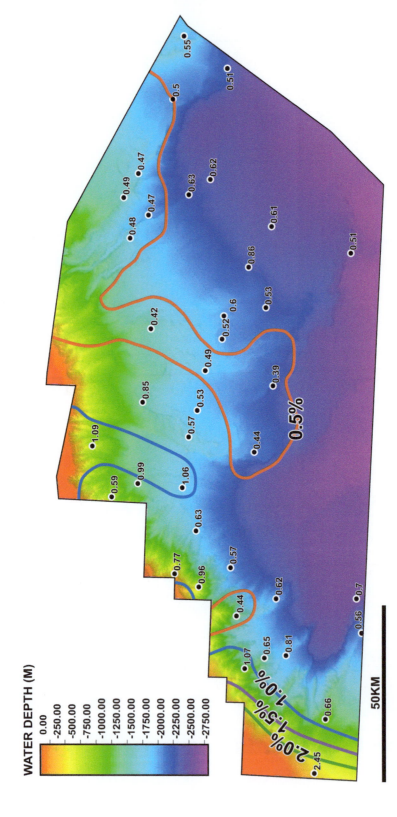

Fig. 6. Bathymetry map with TOC values of samples. The TOC content was contoured by hand in accordance with measured values and implications from bathymetry data.

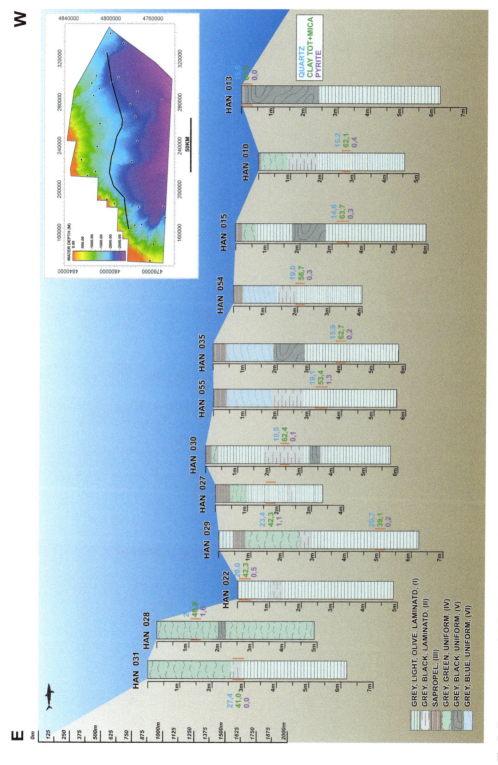

Fig. 7. East–west cross-section (location shown on the map) with sedimentary facies, depth of samples and selected data.

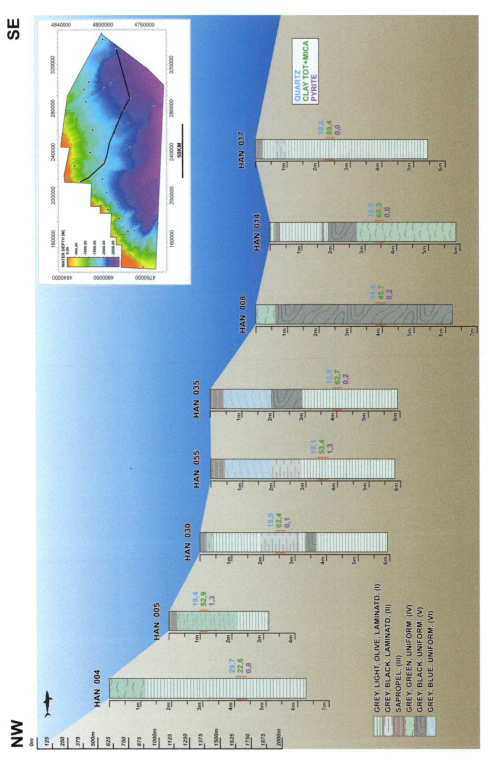

Fig. 8. NW–SE cross-section (location shown on the map) with sedimentary facies, depth of samples and selected data.

and comprising dropcores from shelfal to bathyal environments, no significant variation in organic content or quality could be observed.

The highest measures of organic content in sample HAN_060 is most likely due to terrigenous organic matter input from nearby rivers. A hand-contoured map is shown in Figure 6 and clearly highlights the impact of mass movements along the shelf. These mass slides and channels are also visible on the bathymetry shown in the background. Resedimentation of thinner organic-rich intervals and subsequent mixing with surrounding sediments could also be responsible for an additional deterioration of the source rock quality of the sampled sediments.

Published data points from other parts of the Western Black Sea Basin (e.g. Arthur & Sageman 2004) show that the poor source rock generation encountered in our area is limited to the surroundings of the Danube cone. The more distal settings have not been affected by the sedimentary input of the Danube.

To interpret the lateral variability of the mineralogical content of the sampled sediments, as well as the aerial extent of the assigned facies, seven cross-sections were created. Sample locations, as well as quartz, clay and pyrite content, were marked in the corresponding sample depths.

An east–west profile (Fig. 7) depicts 12 dropcores and their assigned facies. The overall distance of this profile is roughly 150 km, and water depths range from 872 to 1750 m. All dropcores, except HAN_028, are dominated by facies I. A sapropel-rich layer was encountered in all dropcores, except HAN_031, HAN_028 and HAN_022. The absence of the sapropel-rich layer could be due to the shallower water depths and the closer proximity to sedimentary input. Overall, the sapropel-rich layer thins towards the east. Sediments with indications of slumping were encountered in dropcores HAN_035, HAN_015 and HAN_013. The encountered facies V is interpreted as being due to movements along the slope.

A roughly NW–SE-trending profile (Fig. 8) comprises eight dropcores and their assigned facies. The overall distance of this profile is roughly 150 km, and water depths range from 567 m down to 1898 m. As already observed in the east–west profile, most dropcores are dominated by facies I. Sapropel-rich intervals occur in nearly all deeper samples and thin towards the SE. HAN_008 shows significantly different facies to all surrounding dropcores. This change in facies is attributed to sedimentary reworking (e.g. mass slides) along the shelf. This is indicated by a predominance of facies V compared to other dropcores. Thinner intervals of facies V were encountered in dropcores HAN_030, HAN_035 and HAN_014.

Conclusions

Forty dropcore samples from shallow depths beneath the present-day seafloor were selected and analysed for mineral composition and bulk geochemical parameters. The water depths of the samples range from shelfal to bathyal environments. The quartz content of the samples clearly decreases with increasing distance from sedimentary input sources. The sedimentary sources comprise minor contributions from smaller Bulgarian rivers and a dominant impact of the present-day Danube delta. The cone of the Danube delta is recognizable within the northern part of the study area. Clay content increases towards the distal areas. Mass movements are recognizable on the bathymetry maps used, and consist of slides and debris flows. These events which occur along the present-day shelf are also recognizable in the mineral content.

Bulk geochemical parameters show that currently only poor to fair gas-prone source rocks are deposited within the study area. However, published data (e.g. Arthur & Sageman 2004) describe the deposition of sediments with a high source rock potential outside of the study area. The lack of source rock quality, as well as organic content, within sediments in the study area is attributed to the relatively greater depositional load of fine-grained sedimentary input from the Danube river. These fine-grained sediments decrease the organic productivity due to dulling of the water column, and dilute the currently deposited source rock. These effects decrease with distance from the Danube delta, as indicated by published data from outside the study area. In addition, mass movements along the present-day shelf rework possible source rocks and dilute them with overlying and underlying sediments. The results of this study clearly show that anoxic conditions alone are not sufficient for source rock deposition. Proximity to the major sedimentary input, resulting in dilution of organic-rich sediments as well as decreased organic productivity, is of major importance and should be considered in basin modelling.

This research benefitted from discussions with many colleagues working in the Han Asparuh Joint Venture such as Thomas Eder, Walter Kosi (OMV), Flavia Allache, Francois Courteix (TOTAL) and Hugo Matias, Ioan Munteanu, Ines Perez Baroja (Repsol).

We also thank Professors Bernhard Grasemann and Jörn Peckmann (University of Vienna) for their technical support with the methodology of the analyses.

References

ARTHUR, M.A. & DEAN, W.E. 1998. Organic-matter production and preservation and evolution of anoxia in the Holocene Black Sea. *Paleoceanography*, **13**, 395–411.

ARTHUR, M.A. & SAGEMAN, B.B. 2004. Sea-level control on source rock development: perspectives from the Holocene Black Sea, the mid-Cretaceous western interior basin of North America and the late Devonian Appalachian basin. *In*: HARRIS, N.B. (ed.) *The Deposition of Organic Carbon-rich Sediments: Models, Mechanisms and Consequences*. SEPM, Special Publications, **82**, 35–59.

BAHR, A., LAMY, F., ARZ, H., MAJOR, C., KWIECIEN, O. & WEFER, G. 2008. Abrupt changes of temperature and water chemistry in the late Pleistocene and early Holocene Black Sea. *Geochemistry, Geophysics, Geosystems*, **9**, Q01004.

BANKS, C.J. & ROBINSON, A.G. 1997. Mesozoic strike-slip back-arc basins of the Western Black Sea region. *In*: ROBINSON, A.G. (ed.) *Regional and Petroleum Geology of the Black Sea and Surrounding Region*. American Association of Petroleum Geologists, Memoirs, **68**, 53–62.

BONCEV, E. 1986. *The Balkanides – Geotectonic Position and Development*. Geologica Balcanica: Series Operum Singulorium, **1**. Bulgarian Academy of Sciences, Sofia [in Bulgarian with English summary].

DACHEV, C. & GEORGIEV, G. 1995. Rift tectonics problems in the Western Black Sea basin. *In*: *Field Guidebook, Central and North Dobrogea, Romania, Comparative Evolution of Peri-Tethyan Rift Basins*. IGCP Project No. 369. Geological Institute of Romania, Bucharest.

EMERY, M. & GEORGIEV, G. 1993. Tectonic evolution and hydrocarbon potential of the Southern Moesian platform and Balkan–Forebalkan regions of Northern Bulgaria. *AAPG Bulletin*, **77**, 1620–1621 [Abstract].

ESPITALIÉ, J., LAPORTE, J.L., MADEC, M., MARQUIS, F., LEPLAT, P., PAULET, J. & BOUTEFEU, A. 1977. Méthode rapide de characterisation des roches mères de leur potential pétrolier et de leur degree d'evolution. *Revue de l'Institut Francais du Pétrole*, **32**, 23–42.

FINETTI, I., BRICCHI, G., DEL BEN, A., PIPAN, M. & XUAN, Z. 1988. Geophysical study of the Black Sea area. *Bullettino di Geofisica Teorica ed Applicata*, **30**, 197–324.

FRIEDRICH, J., DINKEL, C. *ET AL.* 2002. Benthic nutrient cycling and diagenetic pathways in the North-Western Black Sea. *Estuarine, Coastal and Shelf Science*, **54**, 369–383.

GEORGIEV, G. 2012. Geology and Hydrocarbon Systems in the Western Black Sea. *Turkish Journal of Earth Sciences*, **21**, 723–754.

GORUR, N. 1988. Timing of opening of the Black Sea basin. *Tectonophysics*, **147**, 247–262.

JONES, G.A. 1990. AMS radiocarbon dating of sediments and waters from the Black Sea. *Eos, Transactions of the American Geophysical Union*, **71**, 152.

KEMPE, S., LIEBZEIT, G., DIERCKS, A., NICHOLSON, J., GAGNON, A., WOODWARD, B. & REALANDER, M. 1988. *Water Column Analysis in Temporal and Spatial Variability in Sedimentation in the Black Sea*, Cruise Report.

LERICOLAIS, G., BOURGETB, J., POPESCUC, I., JERMANNAUDD, P., MULDERE, T., JORRYA, S. & PANIN, N. 2013. Late Quaternary deep-sea sedimentation in the western Black Sea: new insights from recent coring and seismic data in the deep basin. *Global and Planetary Change*, **103**, 232–247.

MAJOR, C.O., GOLDSTEIN, S.L., RYAN, W.B.F., LERICOLAIS, G., PIOTROWSKI, A.M. & HAJDAS, I. 2006. The co-evolution of Black Sea level and composition through the last deglaciation and its paleoclimatic significance. *Quaternary Science Reviews*, **25**, 2031–2047.

NIKISHIN, A.M., ZIEGLER, P.A. *ET AL.* 2001. Mesozoic and Cainozoic evolution of the Scythian Platform–Black Sea–Caucasus domain. *In*: ZIEGLER, P.A., CAVAZZA, W., ROBERTSON, A.H.F. & CRASQUIN-SOLEAU, S. (eds) *Peri-Tethys Memoir 6: Peri-Tethyan Rift/Wrench Basins and Passive Margins*. Memoires du Museum National d'Histoire Naturelle, **186**, 295–346.

NIKISHIN, A., KOROTAEV, M., ERSHOV, A. & BRUNET, M. 2003. The Black Sea basin: tectonic history and Neogene–Quaternary rapid subsidence modelling. *Sedimentary Geology*, **156**, 149–168.

NIKISHIN, A.M., OKAY, A., TÜYSÜZ, O., DEMIRER, A., WANNIER, M., AMELIN, N. & PETROV, E. 2015. The Black Sea basins structure and history: new model based on new deep penetration regional seismic data. Part 2: tectonic history and paleogeography. *Marine and Petroleum Geology*, **59**, 656–670.

OKAY, A.I., ŞENGOR, A.M.C. & GORUR, N. 1994. Kinematic history of the opening of the Black Sea and its effect on the surrounding regions. *Geology*, **22**, 267–270.

PETERS, K.E. 1986. Guidelines for evaluating petroleum source rock using programmed pyrolysis. *AAPG Bulletin*, **70**, 318–329.

POPOV, S.V., BUGROVA, E.M. *ET AL.* 2004*a*. Biogeography of the northern Peri-Tethys from the Late Eocene to the Early Miocene. Part 3. Late Oligocene–Early Miocene. Marine basins. *Paleontological Journal*, **38**, S653–S716.

POPOV, S.V., RÖGL, F., ROZANOV, A.Y., STEININGER, F.F., SHCHERBA, I.G. & KOVAC, M. 2004*b*. *Lithological–Paleogeographic Maps of Paratethys: 10 Maps Late Eocene to Pliocene*. Courier Forschungsinstitut Senckenberg, Frankfurt, **250**.

POPOV, S.V., ANTIPOV, M.P., ZASTROZHNOV, A.S., KURINA, E.E. & PINCHUK, T.N. 2010. Sea-level fluctuations on the Northern Shelf of the Eastern Paratethys in the Oligocene–Neogene. *Stratigraphy and Geological Correlation*, **8**, 200–224.

ROBINSON, A., SPADINI, G., CLOETINGH, S. & RUDAT, J. 1995. Stratigraphic evolution of the Black Sea: inferences from basin modelling. *Marine and Petroleum Geology*, **12**, 821–835.

ROSS, D.A. & DEGENS, E.T. 1974. Recent sediments of the Black Sea. *In*: DEGENS, E.T. & ROSS, D.A. (eds) *The Black Sea – Geology, Chemistry, and Biology*. American Association of Petroleum Geologists, Memoirs, **20**, 183–199.

SCHULZ, L.G. 1964. Quantitative interpretation of mineralogical composition from X-ray and chemical data for the Pierre Shale. *USGS Professional Paper, 391-C*, 31.

SCHULZ, H.-M., BECHTEL, A. & SACHSENHOFER, R.F. 2005. The birth of the Paratethys during the early Oligocene: from Tethys to an ancient Black Sea analogue? *Global and Planetary Change*, **49**, 163–176.

TARI, G., DICEA, O., FAULKERSON, J., GEORGIEV, G., POPOV, S., STEFANESCU, M. & WEIR, G. 1997. Cimmerian and Alpine stratigraphy and structural evolution of the Moesian platform (Romania/Bulgaria). *In*: ROBINSON, A.G.

(ed.) *Regional and Petroleum Geology of the Black Sea and Surrounding Regions.* American Association of Petroleum Geologists, Memoirs, **68**, 63–90.

TUGOLESOV, D.A., GORSHKOV, A.S., MEYSNER, L.B., SOLO-VIOV, V.V. & KHAKHALEV, E.M. 1985. *Tectonics of*

Mesozoic Sediments in the Black Sea Basin. Nedra, Moscow [in Russian].

ZACHOS, J.C., PAGANI, M., SLOAN, L., BILLUPS, K. & THOMAS, E. 2001. Trends, rhythms, and aberrations in global climate 65 Ma to present. *Science*, **292**, 686–693.

Stratigraphy, structure and petroleum exploration play types of the Rioni Basin, Georgia

G. TARI[1]*, D. VAKHANIA[2], G. TATISHVILI[3], V. MIKELADZE[3], K. GOGRITCHIANI[3],
S. VACHARADZE[3], J. MAYER[1], C. SHEYA[1], W. SIEDL[1],
J. J. M. BANON[4] & J. L. TRIGO SANCHEZ[4]

[1]*OMV, Trabrennstrasse 6-8, 1220 Vienna, Austria*

[2]*Consultant, M. Aleksidze Street 1/9, Tbilisi 0193, Georgia*

[3]*Agency for Mineral Resources, Ministry of Energy, Sanapiro, 2 Zviad Gamsakhurdia,
T'bilisi 0105, Georgia*

[4]*REPSOL, Mendez Alvaro Street 44, Madrid, Spain*

*Correspondence: Gabor.Tari@omv.com

Abstract: The Rioni Basin is an underexplored petroliferous basin located at the Georgian margin of the Black Sea flanked by two folded belts (the Greater Caucasus and the Achara–Trialet Belt). Whereas the stratigraphy of the northern onshore Rioni Basin has elements which are common with that of the offshore Shatsky Ridge, the southern onshore Rioni Basin segment is both stratigraphically and structurally akin to the offshore Gurian folded belt in the eastern Black Sea. In the northern basin segment, the existing oil fields (East and West Chaladidi) and an undeveloped oil discovery (Okumi) are related to either post-salt or pre-salt antiformal traps in detachment folds or in poorly understood stratigraphic pinchouts beneath a regional Upper Jurassic evaporite sequence. In the southern Rioni Basin, the oil in existing fields has either anticlinal four-way closures (Supsa) or a subthrust trap (Shromisubani) related to the leading edge of the north-vergent Achara–Trialet folded belt. Despite the long history of petroleum exploration in the Rioni Basin, these proven plays are not fully understood and systematically explored using modern technology. The existence of an Upper Jurassic regional evaporite seal highlights the possibility of pre-salt plays in the northern part of the basin.

The geology of western Georgia at the eastern end of the Black Sea (Fig. 1) is dominated by a foreland basin complex extending from east to west and flanked to north and south by compressional fold belts of the Greater Caucasus and the Achara–Trialet Belt (e.g. Adamia *et al.* 1992, 2015; Fig. 2). Moreover, the north-vergent Achara–Trialet thrust–fold belt is superimposed on an earlier extensional basin that probably began to rift during the Cretaceous (Derman 2011; Nikishin *et al.* 2015*a*, *b*) or in the Late Paleocene (Robinson *et al.* 1997), and was compressively deformed mainly during the Miocene (Banks *et al.* 1997). This folded belt can be followed to the Georgian offshore area as the Gurian folded belt (or Gurian Trough of others, e.g. Nikishin *et al.* 2015*a*, *b*).

The present paper has two goals. Firstly, to describe the stratigraphy and the structure of the Rioni Basin as it relates to the Eastern Black Sea. Secondly, this work offers an overview of the relatively well-established petroleum geology of the onshore Rioni Basin with its small (i.e. 1–2 mbbo reserves) producing oil fields (Supsa, Shromisubani and Chaladidi). The discussion of the existing fields and play types is supplemented by vintage seismic reflection and well log data.

As well as the existing oil fields in the onshore Rioni Basin, the presence of an unexplored active petroleum system in the nearby Georgian offshore is already proven by active and well-documented seeps on the seafloor (e.g. Reitz *et al.* 2011; Evtushenko & Ivanov 2013; Dembicki 2014). Furthermore, numerous oil shows have been encountered in two deep-water wells drilled in neighbouring parts of offshore Turkey (e.g. HPX-1 and Sürmene-1 wells, Tari & Simmons 2018).

Despite the presence of at least one effective petroleum system, there are obvious risks associated with the existing plays in the Rioni Basin. However, we outline some new, speculative plays for future exploration efforts in this underexplored basin.

From: SIMMONS, M. D., TARI, G. C. & OKAY, A. I. (eds) 2018. *Petroleum Geology of the Black Sea.*
Geological Society, London, Special Publications, **464**, 403–438.
First published online May 4, 2018, https://doi.org/10.1144/SP464.14

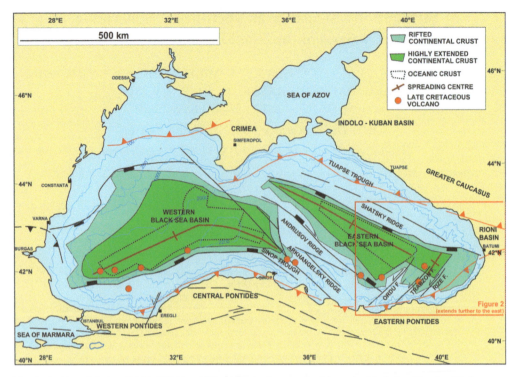

Fig. 1. Position of the broader Rioni Basin of Georgia in the framework of the entire Black Sea. The simplified tectonic map of the Black Sea was adapted from Nikishin *et al.* (2015a). Red rectangle shows the location of Figure 2.

Regional geodynamic framework of the Rioni Basin in the context of the Eastern Black Sea

The regional setting of western Georgia is typically described in the context of entire Georgia located between the Greater and Lesser Caucasus thrust–fold belts (e.g. Nemčok *et al.* 2013; Adamia *et al.* 2015). In this work the emphasis is placed on the relations of the stratigraphy and structural framework between the Eastern Black Sea basin and the Rioni flexural basin.

The Western and Eastern Black Sea basins are separated by two major en échelon basement ridges, the Archangelsky and Andrusov structural highs, collectively called the Mid-Black Sea High (Fig. 1). The sediment cover above these ridges has been described by extensive academic and industry seismic reflection data (e.g. Belousov *et al.* 1988; Robinson *et al.* 1996; Rangin *et al.* 2002; Nikishin *et al.* 2015a, b), crustal-scale wide-angle seismic data (e.g. Edwards *et al.* 2009; Shillington *et al.* 2017), regional potential field data (e.g. Starostenko *et al.* 2004), seafloor dredging (MacGregor *et al.* 1993)

and one deep-water well over the Andrusov High (Aydemir & Demirer 2013).

On the other side of the Eastern Black Sea Basin the very prominent NW–SE-trending Shatsky basement high defines the triangular shape of the basin on its NE margin (Fig. 1). The Shatsky Ridge extends about 600 km across the entire basin, with a large escarpment on its SW flank which appears to be fault-bounded. As noted by many (e.g. Banks *et al.* 1997), the Shatsky Ridge extends into onshore Georgia, projecting below the Rioni Basin (Fig. 2). Along the coast line of the Georgian Black Sea, two prominent structural culminations are located on the overall crest of the Shatsky Ridge: the Gudauta and Ochamchira highs (Fig. 2). The Gudauta High was interpreted by Nikishin *et al.* (2017) as the result of Neogene compressional reactivation of a pre-existing normal fault, the Ordu–Pitsunda Fault (Fig. 1), which acted as a transform fault zone segmenting the basin during its Cretaceous opening. Based on a compilation of existing publications and our own regional seismic mapping, we present here a more specific depiction of this basin-scale fault projected on a regional top Cretaceous surface (Fig. 2). The map-view trace of the Ordu–Pitsunda

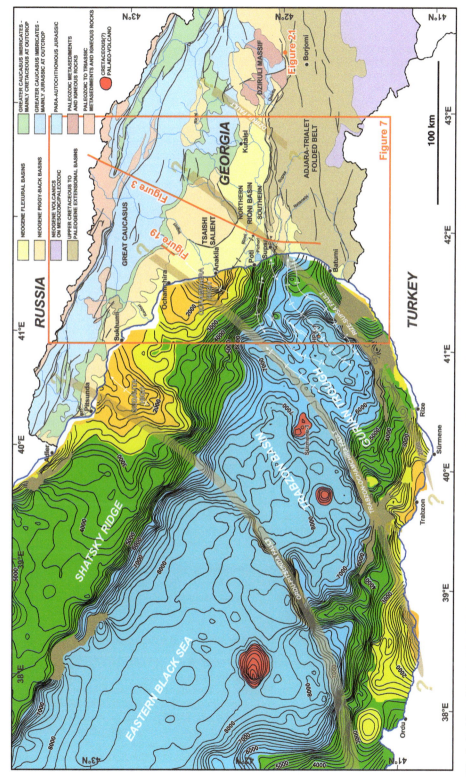

Fig. 2. Position of the Rioni Basin in the regional framework of the Eastern Black Sea, western Georgia, the Greater Caucasus and the Achara–Trialet thrust–fold belt. The contours in the Black Sea, with 400 m increments, show the depth structure of an inferred top Cretaceous seismic horizon compiled from different sources (Tugolesov *et al.* 1985; Finetti *et al.* 1988; Meisner & Tugolesov 2003; Shillington *et al.* 2008) and our own regional mapping. The simplified onshore geology of Georgia has been adapted from Banks *et al.* (1997). The red lines show the location of Figures 3 and 18, whereas the red rectangle refers to the location of Figure 7.

Fault has a small-circle geometry, consistent with a basin opening rotation pole located close to Crimea. A rotation pole for the opening of the Eastern Black Sea Basin, located just east of Crimea, has already been suggested by Okay *et al.* (1994).

Our systematic interpretation of various offshore industry seismic reflection datasets also better defined the Trabzon Fault (Fig. 1) which we have correlated to the NE, to the Georgian coastline near Ochamchira. We therefore define this regional fault in the Eastern Black Sea as the Trabzon–Ochamchira Fault (Fig. 2). Based on their map-view characteristics and association with large transform fault segments, the Gudauta and Ochamchira highs appear to be structurally analogous and interpreted as the result of possibly transpressive reactivation during the Neogene–Recent mountain building in the Greater Caucasus.

We have also correlated the Rize Fault (Fig. 1) using regional offshore seismic datasets to the Georgian coast, making a landfall near Supsa and extending beyond to the NE beneath the Rioni Basin (Fig. 2). This transform/transfer fault is interpreted to be responsible for the location of the SE edge of the Tsaitsi salient.

Given its size and geometry, we consider the Dziruli High (Fig. 2) to be structurally analogous to the Gudauta and Ochamchira highs; we therefore tentatively interpret a major fault delimiting it on its NW margin (Fig. 2). Due to the lack of sufficient subsurface data, we only speculate that this fault played the same kinematic role in the opening of the Eastern Black Sea as the Ordu–Pitsura, Trabzon-Ochamchira and Rize-Supsa faults (Fig. 2). All these major transform/transfer faults might have been reactivated during the Neogene, creating positively inverted sub-regional structural highs on their SE flanks. Note that the exceptionally straight SW edge of the Shatsky Ridge, extending across most of the Eastern Black Sea (Figs 1, 2), becomes offset to the NE along the above-described major faults in the offshore and onshore Georgian segment of the Eastern Black Sea (Fig. 2). We therefore interpret these progressive left-stepping offsets along the crest of the Shatsky Ridge from NW to SE, as it projects beneath the Rioni Basin (Fig. 2), as one of the structural signatures of the hard collision between the Greater and Lesser Caucasus, commencing at *c.* 5 Ma (Avdeev & Niemi 2011; Forte *et al.* 2014; Rolland 2017). Present-day and historic seismicity in the Rioni Basin also appears to confirm the reactivation of these NE–SW-trending faults (cf. Tsereteli *et al.* 2016).

As to the nature of the triangle-shaped deep basin located between the Andrusov High and the Shatsky Ridge (Fig. 1), different interpretations have been published. Given the absence of clear magnetic anomalies and wells reaching anywhere close to the deepest part of this basin, all the models proposed so far remain speculative. The alternative interpretations range from suggesting: (a) a combination of oceanic crust and extended continental crust (e.g. Edwards *et al.* 2009; Shillington *et al.* 2009) or (b) mostly oceanic crust beneath the central part of the Eastern Black Sea (Fig. 1, e.g. Nikishin *et al.* 2015a, b). Furthermore, the geometry of the observed large normal faults in the basin suggests a relatively small crustal extension, on the order of only $\beta = 1.19$, which cannot explain the inferred opening of an oceanic domain (Meredith & Egan 2002).

An even more controversial issue is the age of rifting in the Eastern Black Sea, as the suggested time intervals for its opening range from the Jurassic to the Eocene. A Jurassic age was suggested by Golmshtok *et al.* (1992) and Zonenshain & Le Pichon (1986), whereas Okay *et al.* (1994, 2013), Tüysüz (1999), Hippolyte *et al.* (2010) and Nikishin *et al.* (2003, 2015a, b) preferred a Cretaceous opening for this part of the basin. Moreover, Paleocene– Early Eocene and Eocene opening ages were suggested by Robinson *et al.* (1995a, b), Spadini *et al.* (1996), Banks *et al.* (1997), Kazmin *et al.* (2000) and Vincent *et al.* (2005). Based on the datasets available to us to conduct this regional study of the Rioni Basin at the margin of the SE Black Sea, we could not differentiate between these alternative models in favour of one in particular.

The Late Cretaceous, predominantly extrusive volcanism along the Pontide magmatic arc was all submarine (e.g. Okay *et al.* 2017). There are large offshore volcanic centres along the entire Turkish Black Sea seen on modern deep-water seismic reflection data (Nikishin *et al.* 2015a, b; Tari & Simmons 2018). At least three of these large Paleocene (Robinson *et al.* 1995a) or Campanian (Nikishin *et al.* 2015a, b) palaeo-volcanoes are situated within the SE Black Sea (Fig. 2). As all of these large volcanoes have a cone-shaped geometry without any sign of erosion, they were emplaced in a deep-water environment on the palaeo-basin floor. In our opinion, their formation post-dated the opening of the Eastern Black Sea basin.

The geological subdivision of the Rioni Basin

The Rioni Basin of western Georgia is subdivided areally into a northern v. a southern basin segment, the dividing line more or less being the Rioni River itself (Fig. 2). The northern basin segment, underlain by the Shatsky Ridge, is much wider than the southern segment. Both sub-basins of the Rioni Basin are flexural basins due to the Late Cenozoic loading of the Great Caucasus in the north and the Achara–Trialet thrust belt in the south (Fig. 3). The

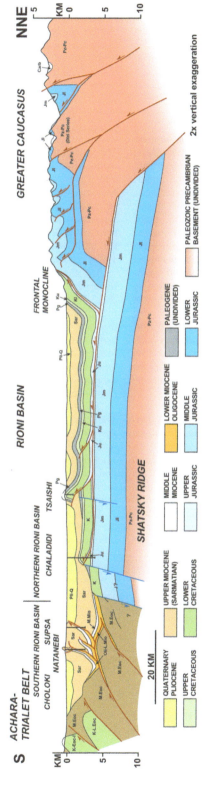

Fig. 3. Regional transect across the onshore Rioni Basin in western Georgia (Banks *et al.* 1997). For location see Figure 2. Note the major differences between the northern Rioni Basin, with the onshore continuation of the Mesozoic Shatsky Ridge underneath it, and the southern Rioni Basin, which has a dominantly Cenozoic stratigraphic framework. Vertical exaggeration is two-fold.

folded belts bounding the Rioni Basin define the triangular Rioni Basin between the Black Sea coast in the west and the outcrops of the Dziruli basement high to the east (Fig. 2). The Dziruli Massif, as the continuation of the Shatsky Ridge, has a sedimentary cover of Upper Jurassic–Lower Cretaceous rocks resting unconformably on Middle Jurassic volcanics and a Hercynian metasedimentary and igneous basement (e.g. Adamia *et al.* 2017).

The several-kilometre-thick Eocene turbidite sequence, which progressively onlaps the northern flank of the Shatsky Ridge beneath the Rioni Basin Dziruli Massif, corresponds to the early underfilled 'flysch' stage of a compressional tectonic regime (e.g. Banks *et al.* 1997). The regional shortening in the Greater Caucasus region is related to final closure of the Neotethys Ocean. Propagation of the thrust faulting and folding in both the Greater Caucasus and the Achara–Trialet folded belts continued into the Miocene and, as a result, the two segments of the Rioni Basin became overfilled foreland basins with 'molasse' sediments. The Neogene–Quaternary sedimentary fill of the Rioni Basin is therefore mostly made up of non-marine or restricted-marine to shallow-water sediments. The Dziruli Massif became sub-aerially exposed during the Early Miocene, separating the Rioni Basin to the west from the Kartli and Kura basins to the east towards the Caspian Sea (Adamia *et al.* 1992, 2010; Robinson *et al.* 1997).

As the result of ongoing shortening in the Rioni Basin (Tibaldi *et al.* 2017a, b), the SW-vergent frontal thrusts of the Great Caucasus presently plunge into the Eastern Black Sea in the Russian sector around Sochi, whereas the N-vergent Achara–Trialet folded belt, extending through southern Georgia, projects beneath the Georgian Black Sea at Supsa (Fig. 2). The offshore and onshore continuation of the Achara–Trialet belt essentially becomes the Eastern Pontides in NE Turkey (Robinson *et al.* 1995a, b; Banks *et al.* 1997; Okay & Sahintürk 1997; Yilmaz *et al.* 2000).

Stratigraphy of the Rioni Basin

The dividing line between the unequal northern and southern basin segments is approximately the E–W-trending Rioni River itself (Figs 2, 3). The larger northern part of the basin is dominated by thin-skinned thrusts and anticlines as the SW-vergent Great Caucasus propagates into its foreland and is underlain by the onshore continuation of the Mesozoic Shatsky Ridge (Banks *et al.* 1997). In contrast, the smaller southern Rioni Basin has a dominantly Cenozoic stratigraphic framework; it is defined as the area between the leading edge of the N-vergent Achara–Trialet thrust belt and the crest of the Shatsky Ridge (Fig. 2).

The entire Rioni Basin is situated in the area of the Eastern Paratethys with limited links to the world's ocean systems throughout most of its post-Eocene development (e.g. Popov *et al.* 2010). Instead of the global stage names, local stratigraphic stage names (e.g. Jones & Simmons 1997) are therefore being used in Georgia for the Upper Cenozoic sequence (Fig. 4).

North Rioni Basin

The stratigraphic elements of the northern part of the Rioni Basin have been summarized in two separate figures, one for the Cenozoic and one for the Mesozoic strata (Figs 4, 5). The compilation was performed using various existing lithostratigraphic summaries (e.g. Morariu & Noual 2009; Adamia *et al.* 2010, 2015) based on outcrops situated along the basin margin and the numerous (more than 100) exploration wells drilled in the basin.

The Cenozoic stratigraphy of the North Rioni Basin is dominated by siliciclastic sequences (Fig. 4). The dominance of coarser clastics since the Middle Miocene is the result of the filling of the accommodation space in a foredeep basin between two folded belts. The deposition occurred in an overfilled basin, with the dominance of molasse-type clastics with coastal, fluvial and alluvial lithofacies. Some of the shallow-marine sandstones are proven reservoirs, such as Maeotian and Sarmatian sandstones in the Supsa and Shromisubani fields. The more shale-dominated sequences in between are proven seal units. As potential source rocks within the Miocene, the lower part of the Middle Miocene (the Tarkmanian shales) may be considered, but there is not yet any detailed information as to their geochemical characteristics (e.g. TOC, HI index, etc.).

Compared to the Upper and Middle Miocene the Lower Miocene succession has more shale units in it, corresponding to deposition on the palaeo-shelf of the Black Sea not yet influenced by proximal sedimentary influx from the future Great Caucasus folded belt. In particular, the Oligocene–Lower Miocene Maykop sequence is dominated by shales, with some siltstone and sandstone intercalations. A NW–SE-trending depositional centre of the Maykop Group is interpreted to occur on the palaeo-shelf of the Black Sea, east of the Shatsky Ridge. The thickness of the Maykop sequence is typically no more than just a few hundred metres in the onshore Rioni Basin, as opposed to the much thicker Maykop (up to a few kilometres thickness) located in the deep-water basin west of the Shatsky Ridge. The Maykop sequence is considered as one of the main source rocks in the broader area (e.g. Mayer *et al.* in press; Sachsenhofer *et al.* 2017a, b; in press). Note that Oligocene Khadumi

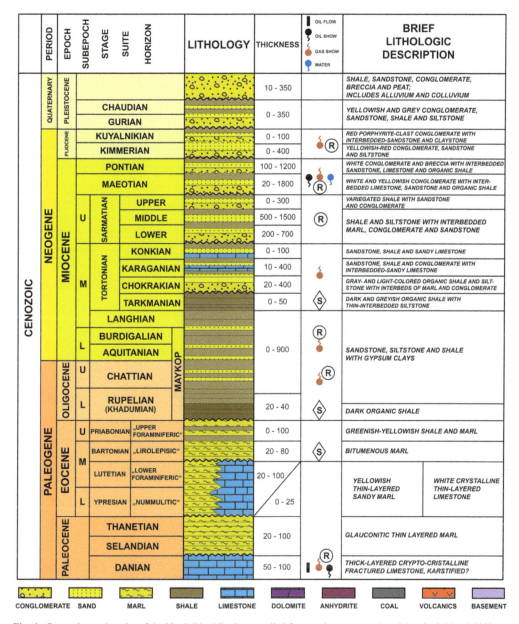

Fig. 4. Cenozoic stratigraphy of the North Rioni Basin, compiled from various sources (e.g. Morariu & Noual 2009; Adamia *et al.* 2010, 2011*a*, *b*, 2015). For detailed explanation see text. The Mesozoic part of the stratigraphy of the North Rioni Basin is shown in Figure 5.

local stage is used frequently as the local equivalent of the Lower Rupelian (Fig. 6).

The euxinic Maykop sequence has an unconformable contact with the underlying marl-dominated Eocene sequence (Figs 4, 6). The Upper Eocene marls and shales are interpreted to be deposited on the upper slope and/or the distal parts of the palaeo-shelf. Of particular importance is the

Bartonian 'Lirolepisic' sequence with its bituminous marls which appear to correspond to the proven Kuma source-rock interval well known from the NE Black Sea and the broader Caucasus area (e.g. Beniamovski *et al.* 2003; Distanova & Bazhenova 2007; Gavrilov *et al.* 2017; Pupp *et al.* in press). This 20–80 m sequence, deposited mostly in anoxic conditions, is considered to be a secondary

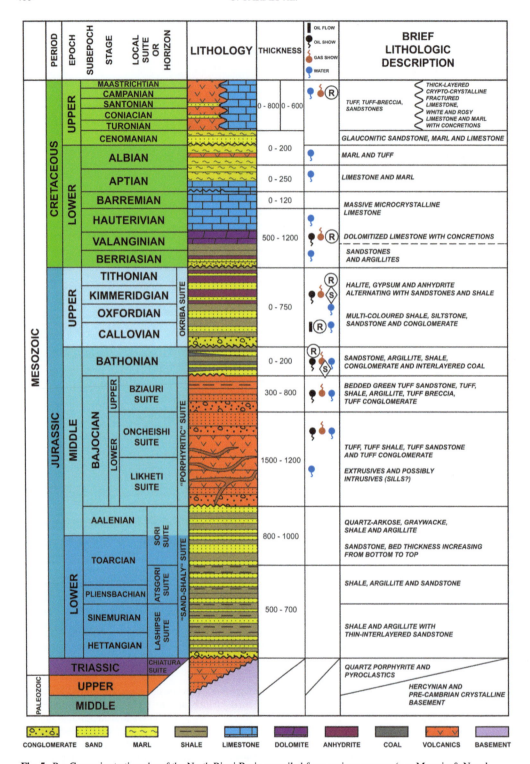

Fig. 5. Pre-Cenozoic stratigraphy of the North Rioni Basin compiled from various sources (e.g. Morariu & Noual 2009; Adamia *et al.* 2010, 2011*a*, *b*, 2015). For detailed explanation see text.

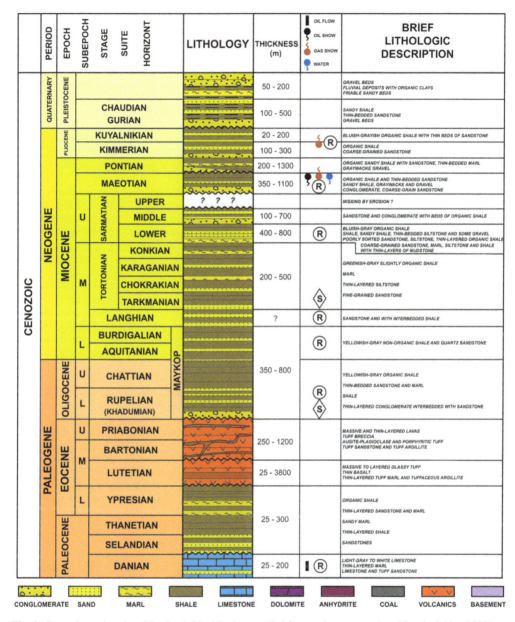

Fig. 6. Cenozoic stratigraphy of the South Rioni Basin, compiled from various sources (e.g. Morariu & Noual 2009; Adamia *et al.* 2010, 2011*a*, *b*, 2015). Compare it with that of the North Rioni Basin shown in Figure 4. For detailed explanation see text.

source rock, contributing to the generation of hydrocarbons in the southern Rioni Basin (Mayer *et al.* in press).

The underlying Middle and Lower Eocene sequence is present either in a marly facies or in a neritic carbonate sequence with *Nummulites*. Beneath an unconformity, the upper part of the Paleocene sequence developed in a glauconitic marl facies

deposited in deeper water, possibly on the upper slope and/or the distal parts of the palaeo-shelf. These marls may act as the top seal sequence for the underlying thick Cretaceous–Paleocene (Turonian–Danian) thick-layered, fractured (and possibly karstified?) limestone strata which is the proven reservoir in the Chalididi Field. These neritic limestone beds (or effectively chalks, Robinson *et al.* 1997) may interfinger laterally with Senonian volcanics (tuffs and tuff breccias).

The Upper Aptian, Albian and Cenomanian sequence is dominated by siliciclastics. The unconformity between the Cenomanian glauconitic sandstones and the Albian marls is one potential candidate for the break-up unconformity for the Eastern Black Sea if the model of Cretaceous opening by Nikishin *et al.* (2003) is accepted. The Valanginian–Lower Aptian neritic carbonate platform limestones and dolomites are potential reservoirs in the Rioni Basin, as many wells (e.g. Ochamchira-1, -2 and -5) tested high water inflows in this interval.

The basal Cretaceous succession includes siliciclastics deposited in fluvial-coastal environments. Beneath a prominent unconformity, the Tithonian and Kimmeridgian of the Upper Jurassic has a mixture of siliciclastics such as shales, siltstones, sandstones and conglomerates. Importantly, there is also a Tithonian evaporite succession with interlayered halite, gypsum and anhydrite. The evaporites form the top seal in the case of the Okumi oil discovery for the underlying sandstones. Upper Jurassic evaporites are about *c.* 250 m thick in wells near Ochamchira, but they pinch out to the east in the Rioni Basin (Robinson *et al.* 1997). Using analogies from the western Neotethys, the Kimmeridgian shales are assumed to have source-rock potential but there are no detailed data to support this interpretation at present.

The underlying Oxfordian and Callovian sequence, as part of the Okriba Suite, has lots of potential sandstone reservoirs in it with intercalated shale units.

The unconformity-bounded, maximum 200 m thick Bathonian sequence has a distinct coal sequence with intercalated conglomerates, sandstones and shales in its upper part. The unconformity surface on top of the Bathonian is either interpreted as the footprint of the 'Meso-Cimmerian' Orogeny (e.g. Adamia *et al.* 1992) or, alternatively, the break-up unconformity for a synrift period on the NE-facing passive margin of a large Middle Jurassic basin system (e.g. Afanasenkov *et al.* 2007) that became the Great Caucasus during the Late Cenozoic.

The underlying 'Porphyritic Suite' is a very thick sequence (up to 2000 m locally) of basalts with some calc-alkaline chemistry suggesting a back-arc setting related to the northwards subduction of the Palaeotethys oceanic plate (e.g. McCann *et al.* 2010). This sequence may be regarded as the 'economic' basement for hydrocarbon exploration in the basin, as many wells were terminated in the top part of the volcanics. Consequently, there is no information about the thickness variations and lateral extent of the Middle Jurassic volcanics in the North Rioni Basin.

The lowermost Middle–Lower Jurassic (Aalenian–Hettangian) sequence, the 'Sand-Shaly' Suite, is a siliciclastic sequence estimated to have a maximum thickness of *c.* 1300–1700 m based on outcrops located at the eastern basin margin. The dominantly continental deposition during this period took place in depocentres corresponding to an embryonic synrift extensional system (e.g. McCann *et al.* 2010).

The Triassic succession is separated from the Jurassic succession by the 'Eo-Cimmerian' unconformity and is characterized by volcanoclastics. The pre-Mesozoic basement of the Rioni Basin outcrops in the Dziruli Massif (Fig. 2). It has Hercynian granitic metamorphics in its core, overlain by Devonian–Carboniferous phyllites (Robinson *et al.* 1997; Adamia *et al.* 2015). Most of the massif is intruded by granitoids and covered with tuffs and lavas of Middle Jurassic age (Adamia *et al.* 2015). This indicates that the Dziruli Massif was also part of the Middle Jurassic magmatic province seen throughout the broader Caucasus region (McCann *et al.* 2010) and the Eastern Pontides (Okay & Nikishin 2015).

In conclusion, the Cenozoic–Mesozoic sedimentary column appears to be relatively thin in the northern part of the Rioni Basin, along the frontal folds of the Greater Caucasus thrust–fold belt near the Black Sea coast. The stratigraphy of the North Rioni Basin is interpreted to be very similar to that of the Shatsky Ridge area offshore (Robinson *et al.* 1996). The latter therefore seems to have been essentially the NE-facing, SW passive margin of the Mesozoic Greater Caucasus Basin, accumulating inner- to outer-shelf sediments during the post-rift subsidence as deep-marine turbidites accumulated along the basin axis to the north and east, exposed today in the Greater Caucasus (e.g. Meisner *et al.* 2011; Nikishin *et al.* 2017).

South Rioni Basin

As opposed to the northern part of the onshore Rioni Basin (Figs 4, 5), the stratigraphic elements of the southern part of the Rioni Basin have only been summarized in one figure (Fig. 6). The compilation covers only the Cenozoic sequence as the underlying Mesozoic strata are poorly constrained by exploration wells, and this basin segment largely lacks outcrops older than the Eocene volcanic sequence (Fig. 2).

Along the Achara–Trialet thrust–fold belt along the southern margin of the Rioni Basin, Eocene volcanics and volcanoclastics predominate at outcrop (Figs 2, 3). However, the fill of the Rioni foreland basin is Late Miocene–Quaternary in age, all units thickening southwards towards the Achara–Trialet folds and onlapping the onshore continuation of the Shatsky Ridge towards the north (Fig. 3).

Similar to the stratigraphy depicted in Figure 4, the Middle–Late Miocene to Quaternary strata is a molasse basin sequence with an overall coarsening-upwards trend, ending with Quaternary continental, alluvial and fluvial siliciclastics in an overfilled fore-deep basin (Fig. 6).

In the Upper Miocene sequence, some of the shallow-marine sandstones are proven reservoirs such as the sandstones in the Maeotian and Sarmatian of the Supsa and Shromisubani fields. The more shale-dominated sequences in between are proven seal units. Interestingly, the upper part of the Sarmatian might be completely missing (Fig. 6), probably due to a pronounced shortening period along the Achara–Trialet folded belt.

The Middle Miocene–Oligocene sequence is very similar to that of the northern basin segment.

In the southern basin, the euxinic Maykop Group sequence has an unconformable, peneplaned contact with the underlying thick volcanic Eocene sequence. The various volcanics and volcanoclastics are the continuation of the extensive arc volcanics known from the Eastern Pontides of Turkey (e.g. Robinson *et al.* 1995a; Okay & Sahintürk 1997; Yilmaz *et al.* 2000).

The Eocene volcanics could be as thick as a few kilometres and may therefore have the same role in the southern Rioni Basin as the Jurassic volcanics in the northern part of the basin, that is, providing the 'economic basement' for hydrocarbon exploration efforts. Indeed, most of the wells reached their total depth in the volcanoclastic sequence.

The few wells which penetrated the entire volcanic sequence encountered a sequence similar to that of the northern Rioni Basin, including a clastic Lower Eocene–Upper Paleocene succession dominated by marls and shales. The underlying Danian–Senonian carbonate platform limestones are identical to those drilled in the northern basin segment.

Structure of the Rioni Basin

Similarly to the stratigraphic description of the basin, the structural styles are also subdivided areally into a northern v. southern Rioni Basin segment, the dividing line being approximately the Rioni River itself (Fig. 7).

North Rioni Basin

Some 50 km SW of the main Greater Caucasus deformation front, there is a highly arcuate salient of anticlinal structures with prominent surface expression (Fig. 7). Structurally all these surface anticlines are ramp anticlines with the thrusts detaching at the level of the Upper Jurassic evaporites (Banks *et al.* 1997), forming a detachment folded belt (Figs 3, 8).

Most of these anticlines have been explored in detail with several wells and a grid of seismic lines (e.g. Vakhania 2008a, b). The most prominent surface anticlines are the Tsaishi and Senaki anticlines (Fig. 9). While the NW segment of the Tsaishi anticline is dominated by a SW-vergent thrust fault, its continuation to the SE displays a curve in the anticlinal axis which suggests the reversal of the vergence to the NE. The digital elevation map (DEM) also displays hogbacks made up of Lower Miocene sandstones that are steeper on the NE flank (Fig. 9).

The elongated Senaki anticline is a complex feature with left-stepping en échelon anticlinal and synclinal structural segments well documented by the surface geology map and the DEM (Fig. 9). The map-view character of these compressional elements is interpreted as the result of left-lateral transpression along the SE margin of the Tsaishi salient (Fig. 7).

The subsurface geology of the area shown in Figure 9 is constrained by about 20 exploration wells and 200 km vintage 2D reflection seismic data. As the seismic reflection sections were only available as hard copies of the original sections, acquired and processed in 1988, they were scanned and georeferenced in a seismic database and interpreted on a workstation. The integration of the numerous wells with the moderate- to poor-quality legacy seismic data allowed the systematic mapping of the major stratigraphic units, including the critical Upper Jurassic evaporites. Based on this work, the thin-skinned character of the thrust–fold belt (Banks *et al.* 1997) has been confirmed.

To illustrate the seismic expression of the detachment folded belt of the Tsaishi salient (Fig. 7), the Khobi and Kvaloni anticlines were chosen as they are both subsurface features (Fig. 8). The Khobi anticline, detached on the Upper Jurassic evaporites, is on trend with the Senaki anticline and has a southerly vergence (Fig. 10). The leading edge of the Senaki thrust can be seen on the northern end of the seismic profile, but is somewhat distorted due to the curved map-view nature of the profile (Fig. 8). The easterly end of the Kvaloni detachment anticline shows a northerly vergence on the southern end of the section. The evaporites appear to overlie a sequence below with extensional faults. Despite the significant velocity pull-up, a structural high can be mapped beneath the anticline with a large normal fault on its northern flank. We interpret the undrilled deeper sequence offset by the large normal fault and some others as the synrift Dogger succession of the Rioni Basin, including the volcanics (Fig. 5).

Another seismic example, some 5 km to the west, shows the Kvaloni anticline with a N-vergent thrust at its core (Fig. 11). The presence of NE-vergent anticlines among the detachment folds in the area is also taken as indirect evidence for the underlying Upper Jurassic evaporite sequence acting as the ultimate

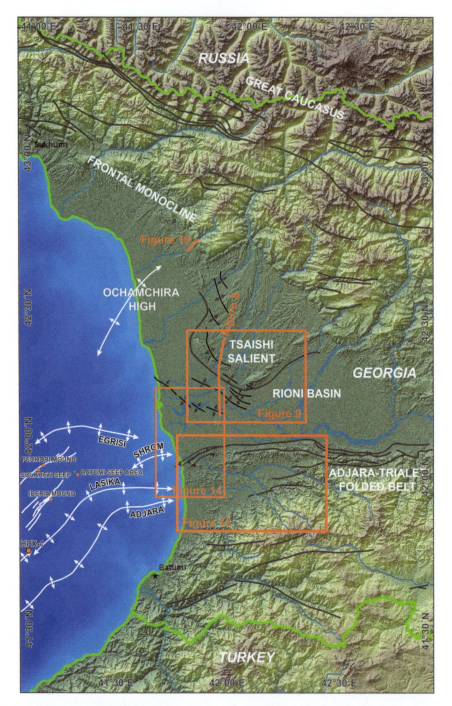

Fig. 7. Digital elevation model (DEM) of the Rioni Basin area using a 30 m grid. The Tsaishi and Senaki detachment anticlines have a very clear surface expression in the northern part of the basin, delineating a structural salient where detachment folds developed, in contrast to the basement-involved thrusting along-strike of the Greater Caucasus producing a frontal monocline along the northern edge of the basin (Fig. 3). Similarly, the N-vergent folded belt along the southern perimeter of the basin also has prominent surface anticlines. The general E–W trend of the Achara–Trialet turns into dominantly SW–NE-trending anticlines in the nearby offshore area. Red lines and rectangles show the location of subsequent figures.

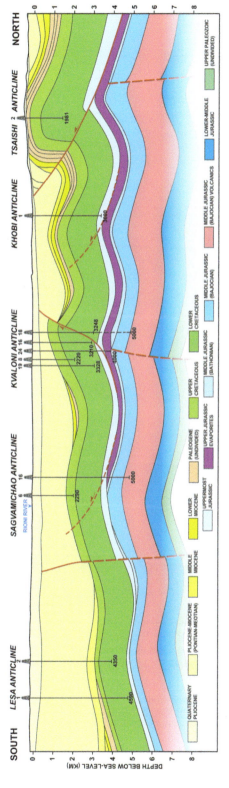

Fig. 8. Sub-regional structural transect across the Kvaloni and Tsaishi detachment anticlines compiled by Davit Vakhania in 2004. The cross-section is constrained by many exploration wells, for a location see Figures 7 and 9. The structure of the strata beneath the Upper Jurassic evaporites highlighted in magenta remains enigmatic due to the general lack of deep (>5 km) wells and poor seismic imaging at that depth.

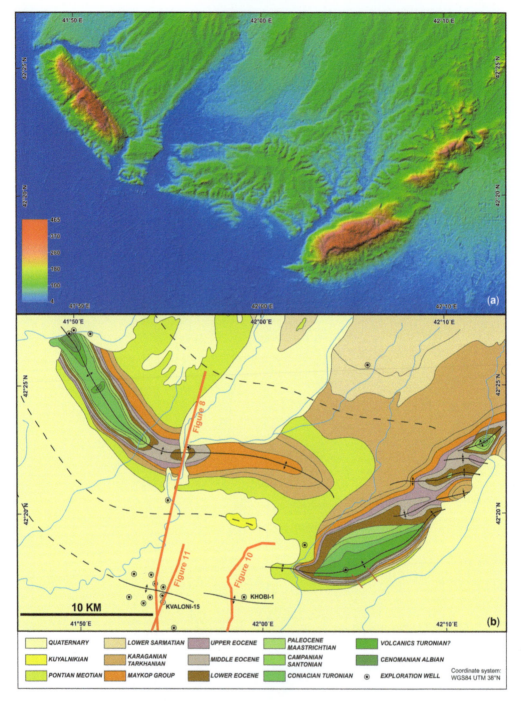

Fig. 9. (a) Digital elevation model (DEM) of the Tsaishi and Senaki anticlines in the northern Rioni Basin area (for location see Fig. 7). (b) Surface geology of the Tsaishi and Senaki anticlines based on 1:100 000 scale geologic maps compiled by Davit Vakhania. The traces of the sub-regional transect of Figure 8 and two seismic illustrations of Figures 10 and 11 are indicated by red lines.

décollement surface (Fig. 8). Unlike Adamia *et al.* (2010) and Tibaldi *et al.* (2017*a, b*), we do not see enough seismic and/or well evidence for the imbrication of the Middle Jurassic sequence beneath the Upper Jurassic evaporite detachment.

Instead, based on our seismic interpretation of the vintage 2D seismic data in the area, we interpret a regional NE–SW structural high which is responsible for the trend of the Senaki anticline. This high is correlated with the Rize–Supsa transform fault

(Fig. 2), with a relatively large (800–1200 m) normal offset on its NW flank.

The anticlines depicted on the seismic profiles (Figs 10, 11) display an early phase of growth by the onlaps within the Eocene sequence on their back-limb. However, most of the structural growth occurred during the latest Miocene–Pliocene. The Quaternary strata is not imaged by the vintage seismic data; however, based on surface geomophological and microtectonic observations, Tibaldi *et al.*

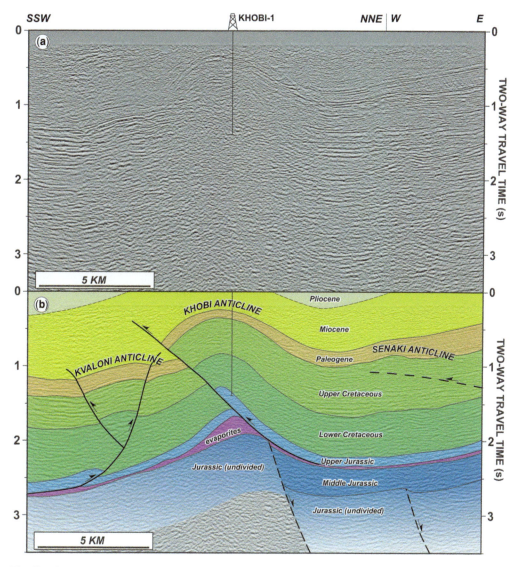

Fig. 10. Vintage 2D seismic data across the Khobi anticline (for location see Fig. 9), (**a**) uninterpeted and (**b**) interpreted. The S-vergence of this anticline is obvious and the underlying detachment is interpreted to be the Upper Jurassic evaporite sequence shown in magenta. Note the velocity pull-up beneath the anticline and the tentative interpretation of a large normal fault-bounded structural high.

(2017*a*, *b*) documented the neotectonic activity associated with the surface anticlines. The neotectonic activity is also supported by the analysis of historic and present-day earthquakes (Tsereteli *et al.* 2016; Adamia *et al.* 2017).

South Rioni Basin

A simplified geological map of the southwestern part of the Rioni Basin and the frontal folds of the Achara–Trialet belt (Fig. 12) suggests a different structural style (Banks *et al.* 1997). The overall structure of this area is best illustrated by a regional geological cross-section (Fig. 3). The leading edge anticline at Supsa exposed Sarmatian strata on the surface along its crest, thrust over a thick Late Miocene–Quaternary succession (Fig. 13). Whereas the quality of the few available legacy seismic profiles is relatively poor, the structural transect is well constrained by the large number of wells drilled in

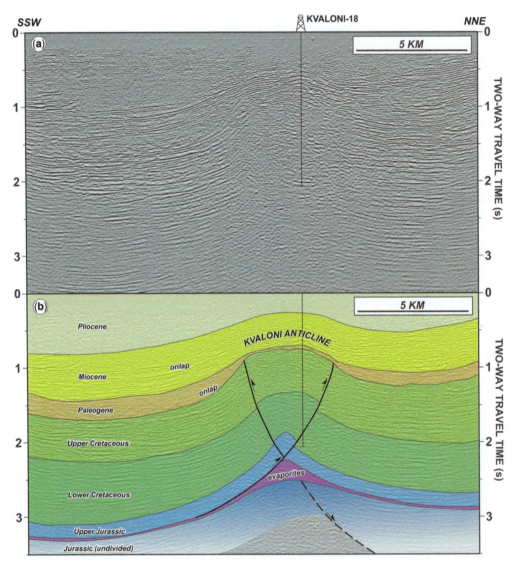

Fig. 11. Seismic expression of the Kvaloni anticline seen on vintage 2D seismic data (for location see Fig. 9), (**a**) uninterpeted and (**b**) interpreted. The N-vergence of this anticline is obvious. The frequently found opposing vergence of individual anticlines in the North Rioni Basin (Fig. 8) is considered in this work as indirect evidence of the presence of a regional detachment situated within the Upper Jurassic evaporites (Fig. 5).

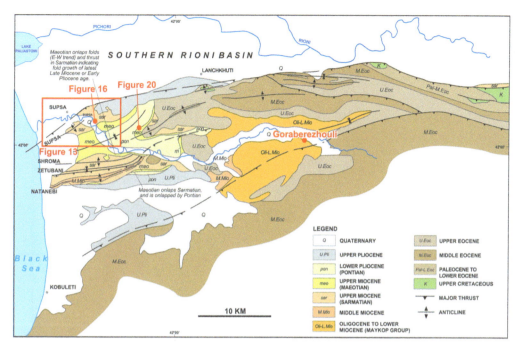

Fig. 12. Structural elements of the South Rioni Basin and the frontal part of the Achara–Trialet thrust–fold belt adapted from Banks *et al.* (1997). Structurally, this area is very different from the northern Rioni Basin as the detachment levels run within lower Cenozoic strata, most probably within the Maykop succession. This onshore folded belt also serves as a good exploration analogue for the offshore Gurian thrust–fold belt located to the west (Figs 2, 7).

the area. The Oligocene–Lower Miocene Maykop Group crops out repeatedly in the frontal fold belt and, due to its shale-prone dominant lithology, it provides a regional décollement level for the N-vergent anticlinal structures. Some other anticlines such as the Natanebi anticline (Fig. 12) were interpreted by Banks *et al.* (1997) as pop-up structures bound by both N- and S-vergent thrusts. Basically the same structural style was documented by Alania *et al.* (2017) along the frontal part of the Adjara–Trialet zone further to the east in the Kura Basin.

Exploration history and existing oil fields of the Rioni Basin

The first three known wells in the Rioni Basin were drilled by Anglo Belgium Oil on the Supsa prospect in the frontal folds of the Achara–Trialet thrust belt (Figs 7, 14) to depths of 220–280 m in 1886 (Benton 1997). Oil flows of up to 6 bo/d were recorded but, as further drilling was unsuccessful, operations were discontinued. The state company Gruzneft began more systematic exploratory drilling in the Rioni Basin in 1930. In 1939 Supsa was 'rediscovered' with a test of 22 bo/d of 26° API (American

Petroleum Institute) crude from Sarmatian (Upper Miocene) sandstones. After World War II, follow-up drilling targeted anticlines identified by seismic data. The initial reservoir objectives were Miocene sandstones, but progressively deeper objectives were targeted. Gruzneft struck oil at Chaladidi in 1969, and recorded flows of up to 365 bo/d of 29° API crude from Upper Cretaceous–Paleocene carbonates. As all other wildcats penetrating Cretaceous sequences were dry or water wet, Cenozoic objectives continued to be tested. The Shromisubani discovery was drilled in 1974, when the discovery well tested up to 730 bo/d from Maeotian (Pliocene) clastics. In 1975, the Ochamchira-1 well encountered a sub-commercial oil find in a Jurassic (Bathonian) reservoir. During the following almost two decades, no more discoveries were made in the onshore Rioni Basin despite various exploration programs. However, when Gruzneft drilled Okumi-1 some 30 km east of Ochamchira in 1991 (Fig. 2), 126 bo/d rate was achieved from an Upper Jurassic clastic sequence.

After the break-up of the Soviet Union the Georgian British Oil Company (GBOC), a joint venture between JKX Oil & Gas and Saknavtobi, was granted exploration and production licenses

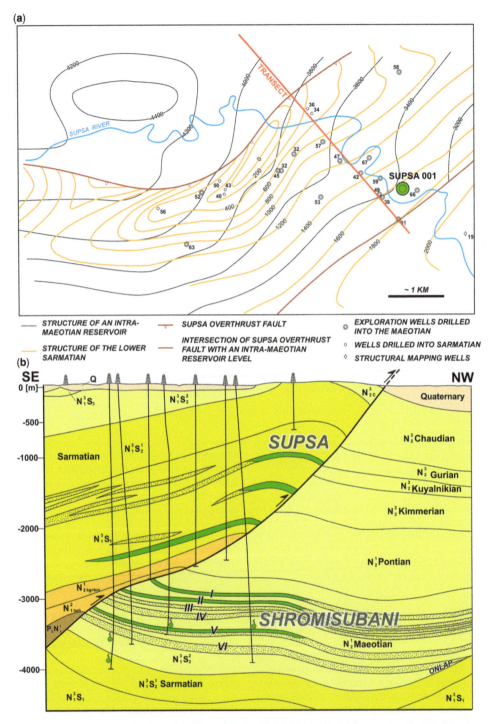

Fig. 13. (**a**) The Supsa anticline with its four-way closure is well constrained by the numerous wells drilled into it. The trap of the Shromisubani Field underlying the Supsa anticline is defined by the depth contours of an intra-Maeotian reservoir level terminating against the thrust fault plane. (**b**) The cross-section shows the relative position of the Supsa Field with its Sarmatian reservoir sequence and the underlying Shromisubani Field with its Maeotian reservoirs. The major frontal thrust may detach southwards into the Maykop sequence (cf. Fig. 3).

Fig. 14. Location of the Supsa, Shromisubani and Chaladidi oil fields in the Rioni Basin. The Okumi Field, not shown on this map, is located some 60 km to the NNE from the Chaladidi Field in Abkhazia (see Figs 2, 7).

covering much of the Rioni Basin in 1993. In 1995 GBOC started appraisal drilling in the Shromisubani Field, which from 1974 to 1993 produced a cumulative 400 000 barrels of oil from eight Miocene reservoir units.

The oil fields in the onshore Rioni Basin are discussed below in order of their discovery: Supsa (1939), Chaladidi (1969), Okumi (1991) and Shromisubani (1974). All these fields are clustered within 10–30 km from the Black Sea coast (Fig. 14). While all these fields are generally small (i.e. 2–4 MMbbl recoverable), they highlight the presence of at least

one proven and effective petroleum system with an oil-generating hydrocarbon kitchen.

Supsa Field

The Supsa area was already explored in the 1880s (Benton 1997) by some shallow wells and oil was found. The first wells were drilled on an obvious surface anticline (Fig. 14), which represents the N-vergent leading edge of the Achara–Trialet thrust–fold belt. The Miocene clastics in the Supsa Field contain about 29 MMbbl oil in place

(Robinson *et al.* 1997) in a number of stacked Sarmatian clastic units (Fig. 15). The actual trap of the Supsa Field, with its Sarmatian sediments, is a 14–18 km^2 sized fault-bend fold (Fig. 13), one among a few others along the southern margin of the Rioni Basin, east of Supsa (Fig. 12).

Outcrop conditions are very poor in this area and the depositional setting of the reservoir units is therefore poorly known. One outcrop along the Supsa River exposes the Upper Sarmatian section (Fig. 12) beneath a major unconformity with the overlying Maeotian sequence (Fig. 16a). The very fine-grained

silty sandstone has metre-scale synsedimentary recumbent folds in it (Fig. 16b), which may be interpreted as the manifestation of the growth of the structure during the Sarmatian. A thin-section from a spot sample taken at this outcrop (Fig. 16c) highlights the subhorizontal parallel alignment of elongate grains indicating lamination/orientation induced by sediment flow within a fine-grained, argillaceous-rich volcanic litharenite. The quartz content is relatively low in this sandstone (*c.* 9%) due to the overwhelming dominance of volcanic and schistoidal (metamorphic) rock fragments. Most

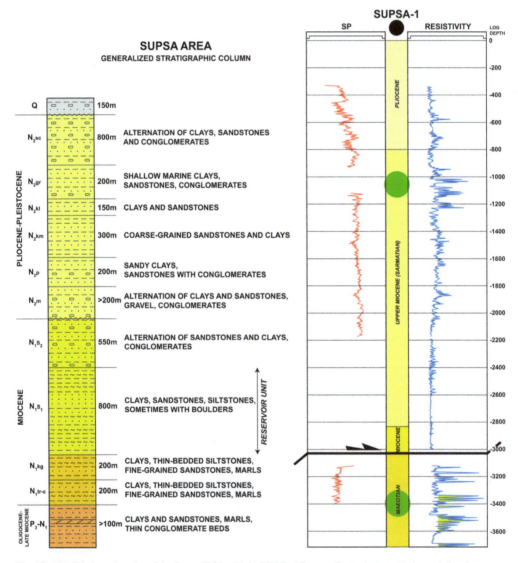

Fig. 15. Simplified stratigraphy of the Supsa Field, with its Middle Miocene (Sarmatian) reservoirs and also the underlying Shromisubani Field, with its slightly younger Upper Miocene (Maeotian) reservoirs. The type log displays the character of the producing sandstones embedded in an overall silty and shaly Miocene and Pliocene sequence.

Fig. 16. Upper Sarmatian sandstone cropping out close to Shromishubani at 42° 1′ 57.53″ N, 41° 51′ 13.37″ E (for location see Fig. 12). This outcrop is located on the backlimb of the Supsa anticline. (**a**) The Upper Sarmatian section is separated by a major unconformity from the overlying Maeotian sequence. (**b**) The very fine-grained silty Sarmatian sandstone has metre-scale synsedimentary recumbent folds in it which may be interpreted as the manifestation of the growth of the structure during deposition. (**c**) A thin-section from a spot sample taken at this outcrop highlights the subhorizontal parallel alignment of elongate grains, indicating lamination/orientation induced by sediment flow within a fine-grained, argillaceous-rich volcanic litharenite. Due to the friable nature of the outcrop sample, the macroporosities counted in excess of 10% are most likely artefacts of the expansion of the sandstone fabric during thin-section preparation.

high-birefringent clasts represent micaceous and schist-like metamorphic specimens, suggesting sediment delivery from a mixed volcanic and metamorphic terrain. The abundance of Eocene volcanogenic material indicates that the main provenance of the sediment was the Achara–Trialet volcanogenic region to the south of the basin (cf. the Hopa-1 well results, Tari & Simmons 2018). Due to the friable nature of the outcrop sample, the macroporosities counted in excess of 10% are most likely artefacts of the expansion of the sandstone fabric during thin-section preparation.

The ranges of effective porosity and permeability in the Supsa Field Sarmatian reservoirs are 5–22% and 5–40 mD, respectively. The gas-oil-ratio is 242 scf/bbl. The typical production flow rates from these moderate to poor reservoirs are in the range of 20–30 bo/day.

The API of the Supsa oils has a range of 22–29° due to partial biodegradation as a result of relatively shallow present-day depth (i.e. 215–460 m, Fig. 13a). The net pay in the reservoir is 4–30 m. The oil itself is likely sourced from both the Maykop sequence (Dembicki 2014) and the Middle Eocene Kuma sequences (Mayer *et al.* in press). Although Robinson *et al.* (1997) suggested an Upper Eocene source rock generating the hydrocarbons in the onshore Rioni fields, this could be the result of the insufficient biostratigraphic control on the age of the samples they worked with at that point of time (cf. Sachsenhofer *et al.* 2017*a*; Vincent & Kaye 2017).

Chaladidi Field

As opposed to the Miocene siliciclastic sequence of the Supsa Field, the Chaladidi Field is reservoired in an Upper Cretaceous–Lower Paleocene carbonate sequence (Fig. 17) in the northern Rioni Basin (Fig. 14). Similarly to Supsa, the two Chaladidi oil field segments (the smaller West and the larger East Chaladidi) are located in two adjacent en échelon four-way closures on compressional ramp anticlines. These anticlines are very close to the southwestern edge of the Tsaishi salient (Figs 2, 3), caused by the propagation of the Great Caucasus deformational front along an Upper Jurassic evaporite décollement level (Banks *et al.* 1997). The column height in the oil field is about 200 m. As the lithology of the reservoir is chalk with a low matrix porosity, it is the fracture porosity which makes this rock a reservoir. The ranges of effective porosity and permeability in the Chaladidi field reservoirs are 5–12% and 1–400 mD, respectively. The typical production flow rates from these moderate-quality reservoirs are in the range of 200–350 bo/day.

The seal for the Chaladidi oil accumulation is provided by overlying Eocene marls and shales. The source for the 26–28° API oil produced from 1800 to 2200 m depth is assumed to be the Maykop sequence. The production from the East

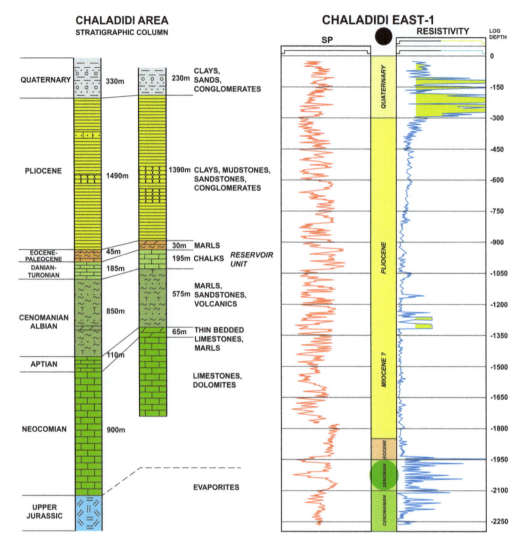

Fig. 17. Simplified stratigraphy of the Chaladidi Field, with its Danian–Turonian carbonate reservoirs. The type log displays the character of the producing thin-bedded limestones (or chalks, according to Robinson *et al.* 1997) embedded in an overall marly sequence.

Chaladidi Field is currently suspended. Interestingly, many of the other wells drilled on structurally analogous detachment anticlines in the area of the Tsaishi salient (Fig. 14) were unsuccessful in finding hydrocarbons.

Okumi oil discovery

The Okumi (or Oqumi) oil discovery was a fortunate side-product of coal exploration efforts in the north-ernmost part of the Rioni Basin (Fig. 2) along the frontal monocline of the Great Caucasus (Robinson *et al.* 1997). The Okumi oil was encountered in Upper Jurassic shallow-marine sands at 1500–1600 m depth beneath the top seal Tithonian–Kimmeridgian evaporite sequence (Fig. 18). The range of effective porosity in the Okumi reservoirs is about 6–14%. There is no data on the permeability of the reservoir sands. The typical initial flow rates from these moderate-quality reservoirs were in the range of 60–100 bo/day.

The Okumi oil is very light (API 42.9°). The pristane/phytane ratio (1.9) is moderate and suggests a normal marine source rock, whereas the diasterane content is high, indicating a significant clastic component (Robinson *et al.* 1996). In addition, the

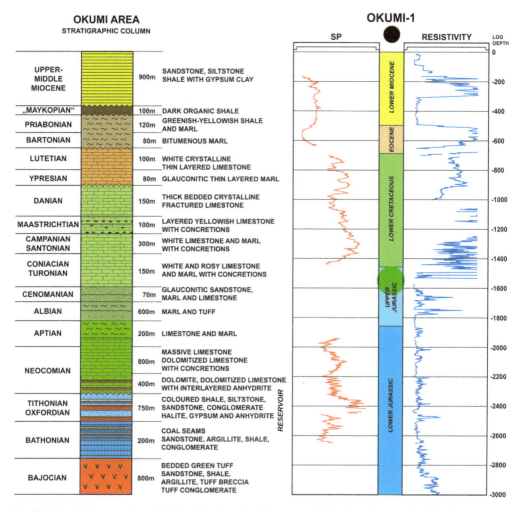

Fig. 18. Simplified stratigraphy of the Okumi (or Oqumi) oil discovery, with its Upper Jurassic (Oxfordian/ Callovian?) reservoirs beneath the Tithonian to Kimmeridgian evaporites serving as the top seal for this poorly understood accumulation. The type log displays the character of the reservoirs beneath an evaporitic seal sequence (see Fig. 4). The tested interval produced 126 bo/d from an Upper Jurassic clastic sequence between 1517 and 1540 m.

Okumi oil contains no oleanane, an angiosperm biomarker; sourcing from the Maykop or Kuma sequences can therefore be excluded. A likely source-rock candidate is the organic-rich Kimmeridgian shale sequence and/or the Bathonian coals and coaly shales (Fig. 5); however, no detailed geochemical work has been conducted to correlate the oil with these potential source rocks. Regardless of the uncertainty about the pre-Tithonian source rock, the Okumi oil highlights the presence of a 'pre-salt' petroleum system which is independent of the Maykop and Kuma systems.

The exact trapping geometry along the margin of the northern Rioni Basin is not understood, as there are no seismic reflection data in the area and only a few wells were drilled (Fig. 19). One possible trapping mechanism might be the updip pinch-out of the reservoir sands beneath the evaporites (Fig. 19). This oil discovery has not been developed to date.

Shromisubani Field

Beneath the major thrust responsible for the Supsa anticline, the Shromisubani Field/Tskaltsminda Field is a subthrust accumulation with an area of about 6 km^2 (Fig. 13a, b). The oil is reservoired in Pliocene (Maeotian) clastics (Fig. 6) with an updip seal provided by the thrust plane. Based on the wireline log data character from the Miocene (Maeotian)

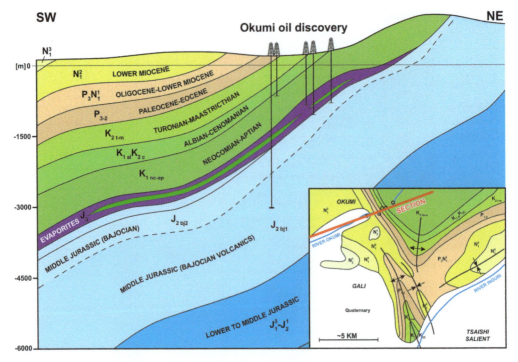

Fig. 19. Cross-section through the Okumi oil discovery (for location see Figs 2, 7) redrawn after Gudushauri & Abaiadze (2015). This oil find (highlighted schematically in green) was found along the frontal monocline of the Greater Caucasus (Fig. 2); the nature of the trap therefore remains enigmatic. One possible trapping mechanism may be the updip pinchout of the reservoir sands within or beneath the Upper Jurassic evaporite seal.

reservoir sands in Shromisubani-101 well (Fig. 15), these units are interpreted to be deposited in fluvial to shallow-marine environments (Robinson *et al.* 1997).

Despite the significantly deeper depth than the overlying Supsa Field (around 3280 m) the API of the Shromisubani oil ranges over 20–35°; this might be due to partial biodegradation. The gas-oil-ratio is 242 scf/bbl. The oil itself is not dominantly sourced from the Maykop Group (Dembicki 2014) but rather from the Kuma sequence (Mayer *et al.* in press). Similarly to the Sarmatian reservoirs of the overlying Supsa Field, the quality of the Pliocene (Maeotian) sandstones appears to limit the productivity in this field. The net pay in the reservoirs sands is between 9 and 50 m. The ranges of effective porosity and permeability in the Shromisubani field reservoirs are 14–20% and 7–60 mD, respectively. The typical initial flow rates from these poor- to moderate-quality reservoirs are in the range of 150–700 bo/day.

An outcrop spot sample taken along a tributary of the Supsa River (Fig. 12) was analysed to address the reservoir quality issue of the Maeotian sequence (Fig. 20a). The sample was taken from a fine- to very-fine-grained white-coloured Maeotian sandstone intercalated between shalier beds (Fig. 20b). The thin-section image of the highly macroporous (*c.* 17%) volcanic clast-rich litharenite (Fig. 20c) shows the lack of any detrital clay and cementation preserving a well-connected pore network. The 'dirty' appearance of the grains is the consequence of the immature composition of the sandstone and the abundance of highly unstable volcanic detritus. In particular the volcanic glass detritus is prone to alteration processes such as chloritization and/or dissolution. Most grains possess a reddish-brown grain-coating phase, inferred to be a partially Fe-oxidized clay coating. Highly birefringent monocrystalline grains represent abundant pyroxene content in this sandstone (*c.* 14%), consistent with the inferred volcanic provenance area. The quality of such a sandstone reservoir at the burial depth of 3300 m would limit the viable production of an oil pool, in line with the observations made in the Shromisubani Field itself.

Other potential reservoir units in the Rioni Basin

An undeveloped gas find at Goraberezhouli in the Trialet–Achara folded belt (Fig. 12) has an

Fig. 20. Maeotian sandstone cropping out in the Achara–Trialet thrust–fold belt at 42° 1′ 30.00″ N, 41° 54′ 9.00″ E (for location see Fig. 12). (**a**) This Maeotian outcrop is located on the frontal limb of an anticline SE of Supsa. (**b**) The sample was taken from a fine- to very-fine-grained white-coloured Maeotian sandstone intercalated between shalier beds. (**c**) The thin-section image of the highly macroporous (*c.* 17%) volcanic-clast-rich litharenite shows the lack of any detrital clay and cementation preserving a well-connected pore network. The 'dirty' appearance of the grains is the consequence of the immature composition of the sandstone and the abundance of highly unstable volcanic detritus.

Oligocene reservoir interval at the depth of 700–900 m. The effective porosity of the sandstone reservoir within the Maykop is about 17% with a permeability of 2–250 mD. In order to address the reservoir quality of similar intra-Maykop sandstone reservoirs within the Rioni Basin, a spot sample was taken from an Oligocene outcrop south of the

Fig. 21. Oligocene sandstone cropping out south of Dziruli at 42° 1′ 14.08″ N, 43° 33′ 56.23″ E, north of Chaschuri (for location see Fig. 2). (**a**) The well-bedded Oligocene strata cropping out at the northern edge of the Achara–Trialet thrust–fold belt. (**b**) The outcrop is dominated by thick sandstone beds with some shales in between. (**c**) The thin-section of the sample taken shows a fine-grained arenite with abundant vesicular or dissolution macroporosity. The rock fragments are mostly of volcanic origin, highlighting the large volcanic contribution in this rock.

Dziruli Massif (Fig. 21). The well-bedded sandstone (Fig. 21a, b) is a fine-grained arenite that shows abundant vesicular or dissolution macroporosity (Fig. 21c). Rounded, bubble-shaped macropores (stained blue) within the matrix may represent former glass vesicles that have been dissolved. The rock fragments are mostly of volcanic origin or represent monocrystalline feldspar types (mostly potassium-feldspar). Note also the angular rock fragment shapes, possibly former glass shards, further highlighting the large volcanic contribution in these deposits. Some of the macropores take the shape of glass shards (Y-shaped and curved forms, suggesting dissolution). The lithic fragments are embedded within a diagenetically altered chloritic/smectitic matrix. These findings, based on a single outcrop sample, are in keeping with those of Vincent *et al.* (2013; 2014) and demonstrate the impact of reworked Eocene volcanics on the reservoir quality of Oligocene–Miocene sandstones in the broader Eastern Black Sea region.

Another potential reservoir unit in the northern Rioni Basin is the Neocomian neritic carbonate succession with Valanginian dolomites and Hauterivian–Barremian limestones (Fig. 5). A total of 100–500 b/day of water was produced from Neocomian dolomites in the Ochamchira wells of the northern Rioni Basin near the Black Sea coast (Fig. 2). Although there were no oil finds in these carbonates in the entire basin, the high water flow rates imply a potentially moderate quality reservoir. To have an independent view of this reservoir, a spot sample was taken from Lower Cretaceous (Barremian) limestones in an outcrop at the southern edge of the Dziruli Massif, near the village of Chumateleti (for location see Fig. 7). The spot sample was taken just a few metres above the unconformity between the basement composed of Upper Paleozoic granites and the overlying carbonates (Fig. 22a) from well-bedded limestones (Fig. 22b). Based on the thin-section expression of the sample it represents a non-reservoir, algal-rich dolobindstone, which displays no visible macroporosity and at best supports an ineffective and minor intercrystalline microporosity within the microsparitic matrix (Fig. 22c). We therefore believe that the Lower Cretaceous carbonates need to be either fractured and/or karstified in order to have reservoir potential in the Rioni Basin.

There is an analogue for this reservoir, that is, the Tyulenovo oil field at the Bulgarian Black Sea coast is producing from a karstified Valanginian carbonate reservoir (e.g. Georgiev 2012). Note also that the main reservoir target in the as-yet undrilled Gudauta High (Fig. 2) is the Kimmeridgian–Barremian neritic carbonate sequence (Nikishin *et al.* 2017), as many large karst caves exist in the Neocomian of nearby Russia and Georgia.

Proven and possible source rocks in the Rioni Basin

While the onshore oil fields and undeveloped oil discoveries are smallish (i.e. on the order of a few million barrels recoverable) they are important in the sense that they prove the petroliferous character of the onshore and offshore Rioni Basin. Table 1 is an attempt to summarize all the proven and possible

Fig. 22. Neocomian limestones in outcrop at the southern edge of the Dziruli Massif near the village of Chumateleti at 42° 2′ 14.39″ N, 43° 31′ 6.00″ E (for location see Fig. 7). (**a**) The well-bedded limestones overlie the weathered basement composed of Upper Paleozoic granites. (**b**) Close-up of the typical outcrop expression of these Barremian limestones. (**c**) The thin-section expression of this Barremian limestone displays no primary porosity. Overview photo of the dolomitized bindstone that comprises common fragments of microbial carbonate clasts (i.e. green algae) sitting within a recrystallized microsparitic matrix.

source rocks in the basin. The Oligocene–Lower Miocene Maykop group is the main source rock for the Chaladidi Field (e.g. Robinson *et al.* 1996; Sachsenhofer *et al.* 2017*b*). Based on the detailed geochemical analysis of the Shromisubani oil, it derived dominantly from the Eocene Kuma formation (Mayer *et al.* in press).

Table 1. *Source rock characteristics in the Rioni Basin*

Formation / Group	Location, age	TOC Min	TOC Average	TOC Max	HI Min	HI Average	HI Max	Source
Maikop Group	Rioni Basin, Pshekian		2.7	5.2		280	460	Pupp et al. (in press)
	Rioni Basin, Solenovian and Lower Oligocene		2			140		Pupp et al. (in press)
	Channis River Section, Rupelian	0.31	1.8	4.8	30	180	380	Vincent & Kaye (2017)
	Channis River Section, Chattian and younger	0.54	0.74	0.99	20	50	80	Vincent & Kaye (2017)
	Georgia (undiff.), Oligocene–Lower Miocene	0.45*		2.4*	10*		640*	Gudushauri & Abaiadze (2015)
Kuma Formation	Martvili section, Upper Eocene	1.1	3.3	7.2	250	390	600	Pupp et al. (in press)
	Khobi section, Upper Eocene	0.5	2.9	5.4	300	380	600	Pupp et al. (in press)
	Channis River Section, Bartonian–Priabonian	0.1	0.7	3.7	20	80	340	Vincent & Kaye (2017)
	Georgia (undiff.), Upper Eocene	0.37		2.95	60		250	Gudushauri & Abaiadze (2015)
Jurassic	Middle–Upper Jurassic (Bathonian?) coals and coaly shales	35.25		46.93			250	This study
Jurassic	Lower Jurassic (Toarcian?)	0.47†		1.3†	240			Gudushauri & Abaiadze (2015)

*Abnormal T_{max} (<400°C) as well as PI (up to 2.9) values indicate contamination of samples.
†Post-mature according to authors.

There are no data on the source-rock characteristics of the Upper Jurassic (Kimmeridgian) shales known in the northern Rioni Basin (Fig. 5). However, our own geochemical data, based on two coaly Bathonian(?) shale samples, suggest that these are excellent oil- and gas-prone source rocks (Table 1). We correlate these coaly shales with those described in the Eastern Pontides, south of Trabzon (Mann et al. 1998). Although the age dating of the Turkish samples collected near at Salansa is imprecise, the fact that they are stratigraphically above the Dogger volcanics makes them analogous to their Georgian counterparts (Fig. 5). Further to the west in the Central Pontides of Turkey, Derman & Iztan (1997) described Bajocian–Bathonian coals and shales of the Himmetpaşa Formation as potential gas-prone source rocks.

The Lower Jurassic (Toarcian?) sequence of the northern Rioni Basin margin has black shales (Fig. 5), but very little is known geochemically about these to date (e.g. Gudushauri & Abaiadze 2015).

The importance of these Jurassic source rocks is that at least one of them has to be responsible for the charging of the Okumi oil find (Robinson et al. 1996). We therefore infer in this work the presence of an effective pre-Tithonian petroleum system in order to formulate pre-salt petroleum play types.

Proven and speculative play types of the onshore Rioni Basin

The main play types of the northern Rioni Basin, proven and speculative, have been summarized graphically and also in text format in Figure 23 & Table 2, respectively. The conceptual play type cartoon covers an area from the Black Sea shelf to the onshore parts of the northern Rioni Basin in order to capture the sub-regional context of the play types.

The play concepts described below are decidedly broader and more 'creative' than the actual proven plays documented by the fields described above. This approach allows for more exploration targets to be identified by: (a) either more in-depth technical work on the existing geological and geophysical datasets; or (b) acquiring new, modern data, especially seismic reflection data, in order to progress with exploration in this part of the Rioni Basin.

The most important play type proven in the Chaladidi Field (Fig. 17) is the oil play reservoired in Turonian–Danian platform carbonates and trapped in the four-way closures of detachment anticlines (#1 in Fig. 23). Separated from the reservoir of this play type by Albian–Cenomanian shales and marls, there is a deeper potential target with Aptian–Neocomian carbonate platform reservoirs trapped in the crestal areas of the detachment anticlines (#2 in Fig. 23).

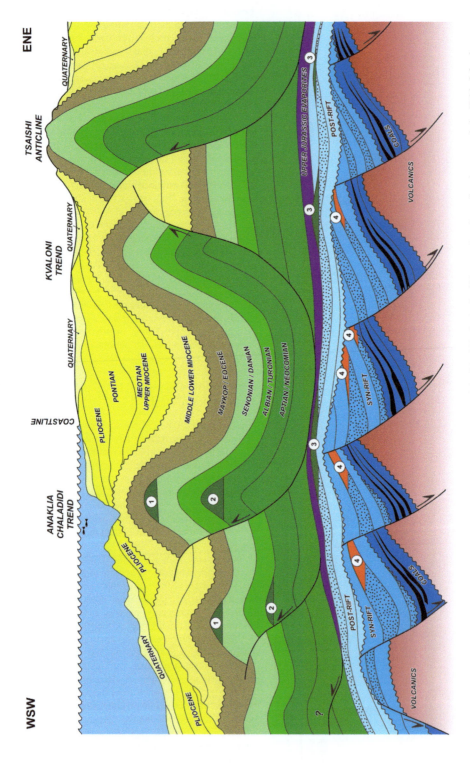

Fig. 23. Cartoon summary of proven and speculative play types in the northern part of the Rioni Basin, both offshore and onshore. See the text and also Table 2 for the explanation of the various play types. The lack of hydrocarbons in the already-drilled Tsaishi and Kvaloni anticlines may be explained either by breaching due to their neotectonic character (Tibaldi *et al.* 2017*a*, *b*) and/or by the tortuous and long migration from the inferred offshore hydrocarbon kitchen. The anticlinal trend of Chaladidi (onshore) and Anaklia (offshore) closer to the coastline may not have these limitations.

Table 2. *Petroleum plays in the Rioni Basin*

Play name	Source	Reservoir	Seal	Critical risk	Field, well or analogue
Late Cretaceous–Early Paleocene shallow-marine carbonates in detachment folds	Oligocene–Early Miocene Maykop Formation deep-marine organic-rich mudstones (oil and gas)	Late Cretaceous–Early Paleocene shallow-marine carbonates	Eocene deep-marine mudstones, deep-marine mudstones	Migration pathway, reservoir-quality (poroperm)	Chaladidi Field
Early Cretaceous shallow-marine carbonates in detachment folds	Oligocene–Early Miocene Maykop Formation deep-marine organic-rich mudstones (oil and gas)	Early Cretaceous shallow-marine fractured carbonates	Albian–Cenomanian deep-marine mudstones	Migration pathway, reservoir-quality (poroperm)	Tyulenovo Field, Bulgaria, western Black Sea
Late Jurassic pre-salt clastics in four-way closures	Early–Middle Jurassic coals, Late Jurassic deep-marine organic-rich mudstones	Late Jurassic sandstones	Kimmeridgian–Tithonian evaporites	Trap presence and charge	Okumi Field
Middle Jurassic synrift fault blocks	Early–Middle Jurassic coals, Late Jurassic deep-marine organic-rich mudstones	Middle Jurassic sandstones	Late Jurassic shales	Trap presence and charge	Ochamchira-1 well, subcommercial oil find

A proven, although poorly understood, play is that represented by the Okumi (or Oqumi) wells (Fig. 2), where oil is reservoired in Upper Jurassic sandstones (Fig. 19). This undeveloped oil find is sealed by overlying Uppermost Jurassic evaporites and trapped in an assumed updip reservoir pinchout. Further to the west beneath the Tsaishi salient we assume subtle four-way closures, as the result of differental compaction and/or minor inversion, to provide the traps of this play (#3 in Fig. 23). This play underlines the possibility of 'pre-salt' plays in the northern segment of the Rioni Basin where the uppermost Jurassic evaporites are present. Also, pre-salt plays may include the broader offshore areas in the Georgian and Russian sectors of the Black Sea (see the Gudauta prospect of Nikishin *et al.* 2017). As mentioned earlier, the distinct geochemical signature of the Okumi oil (Robinson *et al.* 1996) suggests a different charge for the pre-salt play compared to the 'post-salt' play types. Petroleum charge from the Maykop or the Kuma sequences to the Upper Jurassic strata, across the evaporite seal, is a very unlikely scenario (Fig. 23).

The deeper pre-salt gas or oil play within the synrift Middle–Upper Jurassic sandstones (#4 in Fig. 23) is also speculative. This play type assumes an almost unexplored pre-salt petroleum system (cf. Okumi) and it is envisioned to rely most probably on charge either from the Kimmeridgian shales or from the underlying Bathonian coal sequence (Fig. 5). Our spot samples taken from Bathonian coaly shales suggest excellent oil- and gas-generating potential for these rocks (Table 1).

We consider a very useful structural analogue for the detached anticlines of the northern Rioni Basin (Fig. 8) on the other side of the Caucasus, in the Terek–Caspian foredeep basin in Russia. In particular, in the case of the very productive Sunzha Anticline, Sobornov (1994) recognized the involvement and interplay of two detachment levels, including Upper Jurassic evaporites and Oligocene Maykop shales.

The common element between the anticlines in the northern Rioni Basin of Georgia and those in the Terek–Caspian foredeep basin of Russia is the presence of the same major detachment levels (Figs 3, 8) creating a variety of potential traps. The assumed structural analogy could therefore help not only with the exploration efforts in the underexplored onshore Rioni Basin, but also in the undrilled offshore Georgian segment of the Black Sea.

Active hydrocarbon systems in the Georgian Black Sea

Several sites of hydrocarbon seepage were documented to the west of the Rioni Basin (Fig. 7) in offshore Georgia (Pape *et al.* 2010; Reitz *et al.* 2011; Evtushenko & Ivanov 2013; Dembicki 2014). The Batumi, Pechori, Iberia and Colkheti seeps are all located in the offshore continuation of the Achara–Trialet thrust–fold belt (Fig. 7). The seep sites are located within an 20 km² area (Batumi seep area), showing some differences with respect to element concentrations and oxygen, hydrogen, strontium and chlorine isotope signatures in pore waters, as well as impregnation of sediments with petroleum and hydrocarbon potential. However, strontium isotope ratios indicate that the fluids associated with the seeps all originate from the Upper Oligocene

(Maykop) strata (Reitz *et al.* 2011). The presence of oleanane, an angiosperm biomarker, also suggests that the hydrocarbon source rocks belong to the Maykop Group.

Isotope data prove the presence of a working thermogenic petroleum system in the Georgian offshore region (Reitz *et al.* 2011). Note that Derman & Iztan (1997), Derman (2011, 2014), Sipahioglu *et al.* (2013) and Afanasenkov *et al.* (2007) also reported many similar offshore and onshore seeps and mud volcanoes from the adjacent Turkish and Russian sectors of the Eastern Black Sea, respectively.

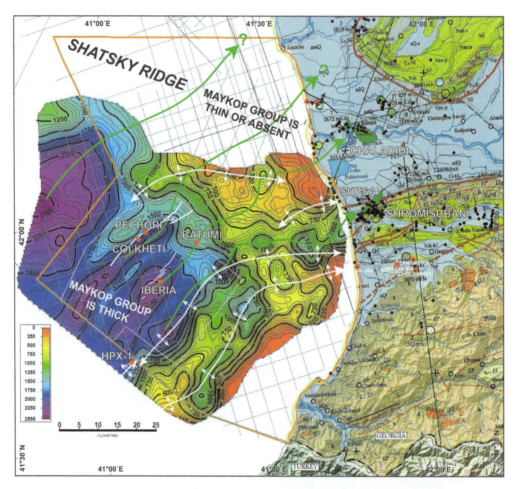

Fig. 24. Simplified isopach map of the Maykop Group in the Georgian part of the Black Sea based on the interpretation of a regional 2D seismic reflection dataset. The oil fields in the onshore Rioni Basin (e.g. Chaladidi) are interpreted to be largely sourced from the deep-water segment of the Georgian offshore where the Maykop Group is buried deeply enough to generate oil. The oil fields in the southern part of the Rioni Basin (e.g. Supsa and Shromisubani) may also have additional hydrocarbon charge from the Adjara–Trialet folded belt to the south, from the Eocene Kuma source-rock sequence (Mayer *et al.* in press). The geologic map of western Georgia was adapted from Gudjabidze & Gamkrelidze (2003). Green lines show conceptual lateral and updip migration routes from the deep-water oil-generating kitchen.

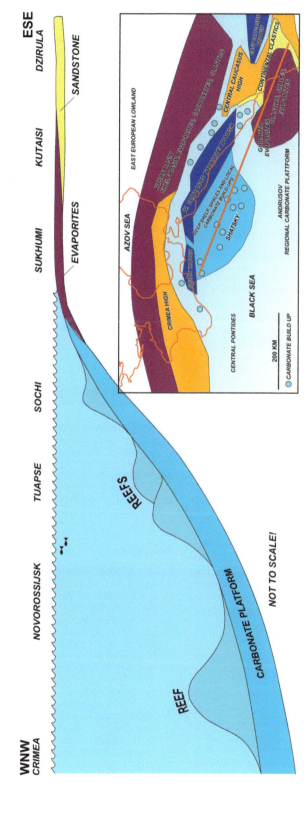

Fig. 25. Cartoonish transect along the crest of the NW–SE-trending Shatsky Ridge, emphasizing the along-strike variation in the dominant Upper Jurassic lithology. While large Upper Jurassic reefs exist in the Crimean and Russian segment of the Eastern Black Sea region (Afanasenkov *et al.* 2005, 2007; Guo *et al.* 2011; Meisner *et al.* 2009, 2011; Nikishin *et al.* 2017), these are replaced towards the SE by siliciclastics, including evaporites (e.g. Robinson *et al.* 1995*a*), in the Georgian offshore and in the Rioni Basin (Fig. 5).

Our regional basin modelling efforts in the broader Rioni Basin clearly indicate that the source of the oil in the onshore Chaladidi Field has to be in the nearby offshore; the average burial depth of the Maykop (i.e. 1–2 km) onshore is not sufficient to reach the oil generation window. Further, the Maykop sequence is very thin (100–200 m) or even absent over the crestal area of the Shatsky Ridge, both offshore and offshore (Figs 10, 11), as opposed to the deeper part of the Eastern Black Sea where it is an order of magnitude thicker (2000–2500 m, Fig. 24). The inferred charge scenario from the offshore area to the western part of the onshore Rioni Basin assumes lateral and updip migration on the scale of 50–100 km.

Jurassic reef play along the Shatsky Ridge in offshore and onshore Georgia?

There are no Upper Jurassic reef lithologies documented so far in the northern Rioni Basin (Fig. 5), which could be considered as the exploration analogues for those described in the Russian segment of the Shatsky Ridge (e.g. Afanasenkov *et al.* 2005, 2007; Nikishin *et al.* 2017; Tari & Simmons 2018). Given the presence of large Upper Jurassic reefs described in the Crimea (e.g. Nikishin *et al.* 2017) and in the Russian western Caucasus (Guo *et al.* 2011), this suggests significant along-strike palaeoenvironmental changes along the Shatsky Ridge (Fig. 25).

We captured this along-strike variation in a cartoon (Fig. 25) which reflects the consequences of the Late Jurassic palaeogeography in the Eastern Black Sea based on Robinson *et al.* (1996), Meisner *et al.* (2011) and Nikishin *et al.* (2017). The large-scale lithological changes from slope to shelf and to coastal environments means that there is a zero-edge for the distribution of Upper Cretaceous reefs to the SE in the offshore, somewhere between the latitudes of Sochi and Sukhumi (Fig. 25). As the Upper Jurassic reefs in the Russian segment of the Black Sea are replaced by age-equivalent evaporitic basins in the Rioni Basin of Georgia (Fig. 25), the reef play (e.g. Nikishin *et al.* 2017) or the 'Maria play' (Tari & Simmons 2018) is replaced by the pre-salt plays as described in this work (Fig. 23).

Conclusions and exploration outlook

The existing oil fields in the onshore Rioni Basin are generally small, but they underline the presence of at least two active and effective petroleum systems with a dominantly oil charge. The existence of an active petroleum system in the offshore area adjacent to the Rioni Basin of Georgia is also proven by active and well-documented seeps on the seafloor.

Numerous oil shows were encountered in the deep-water wells drilled in the neighbouring parts of offshore Turkey (HPX-1 and Sürmene-1).

The overview of the stratigraphy and the structure of the northern and southern segments of the Rioni Basin offer a systematic description of the proven and potential reservoirs and source rocks in this basin. Moreover, a summary of the play types in the northern Rioni Basin provides an outlook for future exploration efforts. Some of the proven and speculative plays defined in this work could also, for the first time, be followed into the undrilled Georgian segment of the Eastern Black Sea.

While the established plays have a fairly well-understood exploration risk profile, the new conceptual plays beneath an Upper Jurassic evaporite sequence would require the acquisition of new 2D and 3D seismic reflection and various other geological and geophysical datasets prior to drilling. The integration of the new datasets with existing datasets and a careful and detailed evaluation of exploration failures and successes in the Rioni Basin are needed as the next steps in this underexplored basin at the eastern margin of the Black Sea.

Special thanks are due to Mike Simmons and Andrew Davies for their valuable reviews of the first version of this paper. Andrew Robinson and Mike Simmons kindly provided additional feedback on the second, extended version of the manuscript. We appreciate technical contributions by Xu Qianhui, Piotr Dzido and Tam Lovett during this study. Conversations with Michael Nibladze about the history of petroleum exploration in Georgia were very helpful. Reinhard Sachsenhofer kindly offered his insights into the various source rocks around the Black Sea. Wolfgang Hujer and Anja Kobstädt are thanked for the analysis of the thin-sections. Many of the figures in this paper were kindly drafted by Peter Pernegr, and Leslie Jessen provided the DEM images used in this paper.

References

ADAMIA, S., AKHVLEDIANI, K.T., KILASONIA, V.M., NAIRN, A.E.M., PAPAVA, D. & PATTON, D.K. 1992. Geology of the republic of Georgia: a review. *International Geology Review*, **34**, 447–476.

ADAMIA, S., ALANIA, V., CHABUKIANI, A., CHICHUA, G., ENUKIDZE, O. & SADRADZE, N. 2010. Evolution of the Late Cenozoic basins of Georgia (SW Caucasus): a review. *In*: SOSSON, M., KAYMAKCI, N., STEPHENSON, R.A., BERGERAT, F. & STAROSTENKO, V. (eds) *Sedimentary Basin Tectonics from the Black Sea and Caucasus to the Arabian Platform*. Geological Society of London, Special Publications, **340**, 239–259.

ADAMIA, S., ZAKARIADZE, G., CHKHOTUA, T., SADRADZE, N., TSERETELI, N., CHABUKIANI, A. & GVENTSADZE, A. 2011*a*. Geology of the Caucasus: a review. *Turkish Journal of Earth Sciences*, **20**, 489–544.

ADAMIA, S., ALANIA, V., CHABUKIANI, A., KUTELIA, Z. & SADRADZE, N. 2011*b*. Great Caucasus (Cavcasioni): a long-lived north-Tethyan back-arc basin. *Turkish Journal of Earth Sciences*, **20**, 611–628.

ADAMIA, S.A., CHKHOTUA, T.T. *ET AL.* 2015. Tectonic setting of Georgia–eastern Black Sea: a review. *In*: SOSSON, M., STEPHENSON, R.A. & ADAMIA, S.A. (eds) *Tectonic Evolution of the Eastern Black Sea and Caucasus*. Geological Society of London, Special Publications, **428**, 11–40.

ADAMIA, S., ALANIA, V., TSERETELI, N., VARAZANASHVILI, O., SADRADZE, N., LURSMANASHVILI, N. & GVENTSADZE, A. 2017. Postcollisional tectonics and seismicity of Georgia. *In*: SORKHABI, R. (ed.) *Tectonic Evolution, Collision, and Seismicity of Southwest Asia: In Honor of Manuel Berberian's Forty-Five Years of Research Contributions*. Geological Society of America, Special Paper, **525**, 1–38.

AFANASENKOV, A.P., NIKISHIN, A.M. & OBUKHOV, A.N. 2005. The system of Late Jurassic carbonate buildups in the northern Shatsky swell (Black Sea). *Doklady Earth Sciences*, **403**, 696–699.

AFANASENKOV, A.P., NIKISHIN, A.M. & OBUKHOV, A.N. 2007. *Geology of the Eastern Black Sea*. Scientific World, Moscow [in Russian with English summary].

ALANIA, V.M., CHABUKIANI, A.O., CHAGELISHVILI, R.L., ENUKIDZE, O.V., GOGRICHIANI, K.O., RAZMADZE, A.N. & TSERETELI, N.S. 2017. Growth structures, piggy-back basins and growth strata of the Georgian part of the Kura foreland fold–thrust belt: implications for Late Alpine kinematic evolution. *In*: SOSSON, M., STEPHENSON, R.A. & ADAMIA, S.A. (eds) *Tectonic Evolution of the Eastern Black Sea and Caucasus*. Geological Society of London, Special Publications, **428**, 171–185, https://doi.org/10.1144/SP428.5

AVDEEV, B. & NIEMI, N.A. 2011. Rapid Pliocene exhumation of the Central Greater Caucasus constrained by low-temperature thermochronometry. *Tectonics*, **30**.

AYDEMIR, V. & DEMIRER, A. 2013. Upper Cretaceous and Paleocene Shallow Water Carbonates along the Pontide Belt. *19th International Petroleum and Natural Gas Congress and Exhibition of Turkey*, Ankara, 284–290, abstract book.

BANKS, C.J., ROBINSON, A.G. & WILLIAMS, M.P. 1997. Structure and regional tectonics of the Achara-Trialet fold belt and the adjacent Rioni and Kartli foreland basins, Republic of Georgia. *In*: *Regional and Petroleum Geology of the Black Sea and Surrounding Region*. AAPG, Memoirs, Tulsa, Oklahoma, **68**, 331–345.

BELOUSOV, V.V., VOLVOVSKY, B.S. *ET AL.* 1988. Structure and evolution of the Earth's crust and upper mantle of the Black Sea. *Bollettino di Geofisica Teorica e Applicata*, **30**, 109–196.

BENIAMOVSKI, V.N., ALEKSEEV, A.S., OVECHKINA, M.N. & OBERHÄNSLI, H. 2003. Middle to upper Eocene dysoxic--anoxic Kuma Formation (northeast Peri-Tethys): biostratigraphy and paleoenvironments. *In*: WING, S.L., GINGERICH, P.D., SCHMITZ, B. & THOMAS, E. (eds) *Causes and Consequences of Globally Warm Climates in the Early Paleogene*. Geological Society of America Special Paper, **369**, 95–112.

BENTON, J. 1997. Exploration history of the Black Sea province. *In*: ROBINSON, A.G. (ed.) *Regional and Petroleum Geology of the Black Sea and Surrounding Region*. AAPG, Memoirs, Tulsa, Oklahoma, **68**, 7–18.

DEMBICKI, H. JR. 2014. Confirming the presence of a working petroleum system in the Eastern Black Sea Basin, offshore Georgia using SAR imaging, sea surface slick sampling, and geophysical seafloor characterization. AAPG Search and Discovery, paper #10610.

DERMAN, A.S. 2011. Importance of knowing paleogeography in order to understand the opening of the Black Sea. AAPG Search and Discovery, paper #90109.

DERMAN, A.S. 2014. Petroleum Systems of Turkish Basins. *In*: MARLOW, L., KENDALL, C.C.G. & YOSE, L.A. (eds) *Petroleum Systems of the Tethyan Region*. AAPG Memoirs, **106**, 469–504.

DERMAN, A.S. & IZTAN, Y.H. 1997. Results of geochemical analysis of seeps and potential source rocks from Northern Turkey and the Turkish Black Sea. *In*: ROBINSON, A.G. (ed.) *Regional and Petroleum Geology of the Black Sea and Surrounding Region*. AAPG, Memoirs, Tulsa, Oklahoma, **68**, 313–330.

DISTANOVA, L. & BAZHENOVA, O.K. 2007. Conditions of source rock potential formation in Eocene deposits of Caucasian-Scythian Region, extended abstract. *EAGE 69th Conference & Exhibition*, 11–14 June 2007, London, UK.

EDWARDS, R.A., SCOTT, C.L., SHILLINGTON, D.J., MINSHULL, T.A., BROWN, P.J. & WHITE, N.J. 2009. Wide-angle seismic data reveal sedimentary and crustal structure of the Eastern Black Sea. *The Leading Edge*, **28**, 1056–1065.

EVTUSHENKO, N.V. & IVANOV, A.Y. 2013. Oil seeps in the southeastern Black Sea studied using satellite synthetic aperture radar images, Izvestiya. *Atmospheric and Oceanic Physics*, **49**, 913–918.

FINETTI, I., BRICCHI, G., DEL BEN, A., PIPAN, M. & XUAN, Z. 1988. Geophysical study of the Black Sea area. *Bollettino di Geofisica Teorica e Applicata*, **30**, 197–324.

FORTE, A.M., COWGILL, E. & WHIPPLE, K.X. 2014. Transition from a singly vergent to doubly vergent wedge in a young orogen: the Greater Caucasus. *Tectonics*, **33**, 2077–2101.

GAVRILOV, Y.O., SHCHEPETOVA, E.V., SHCHERBININA, E.A., GOLOVANOVA, O.V., NEDUMOV, R.I. & POKROVSKY, B.G. 2017. Sedimentary environments and geochemistry of Upper Eocene and Lower Oligocene rocks in the northeastern Caucasus. *Lithology and Mineral Resources*, **52**, 447–466.

GEORGIEV, G. 2012. Geology and hydrocarbon systems in the Western Black Sea. *Turkish Journal of Earth Sciences*, **21**, 723–754.

GOLMSHTOK, A.Y., ZONENSHAIN, L.P., TEREKHOV, A.A. & SHAINUROV, R.V. 1992. Age, thermal evolution and history of the Black Sea Basin based on heat flow and multichannel reflection data. *Tectonophysics*, **210**, 273–293.

GUDJABIDZE, G.E. & GAMKRELIDZE, I.P. 2003. Geological Map of Georgia, scale 1:500 000, Tbilisi, Georgia.

GUDUSHAURI, S. & ABAIADZE, A. 2015. Estimation of geological conditions of discovery of hydrocarbon fields in the Georgian offshore. *Istanbul Conference, Turkey*.

GUO, L., VINCENT, S.J. & LAVRISHCHEV, V. 2011. Upper Jurassic reefs from the Russian western Caucasus, implications for the eastern Black Sea. *Turkish Journal of Earth Sciences*, **20**, 629–653.

HIPPOLYTE, J.C., MÜLLER, C., KAYMAKCI, N. & SANGU, E. 2010. Dating of the Black Sea Basin: new nannoplankton ages from its inverted margin in the Central Pontides (Turkey). *In*: SOSSON, M., KAYMAKCI, N.,

STEPHENSON, R.A., BERGERAT, F. & STAROSTENKO, V. (eds) *Sedimentary Basin Tectonics from the Black Sea and Caucasus to the Arabian Platform*. Geological Society of London, Special Publications, **340**, 113–136, https://doi.org/10.1144/SP340.7

JONES, R.W. & SIMMONS, M.D. 1997. A review of the stratigraphy of Eastern Paratethys (Oligocene–Holocene), with particular emphasis on the Black Sea. *In*: ROBINSON, A.G. (ed.) *Regional and Petroleum Geology of the Black Sea and Surrounding Region*. AAPG, Memoirs, Tulsa, Oklahoma, **68**, 39–52.

KAZMIN, V.G., SCHREIDER, A.A. & BULYCHEV, A.A. 2000. Early stages of evolution of the Black Sea. *In*: BOZKURT, E., WINCHESTER, J.A. & PIPER, J.D.A. (eds) *Tectonics and Magmatism in Turkey and the Surrounding Area*. Geological Society, London, Special Publications, **173**, 235–249.

MACGREGOR, D.S., RUDAT, J.H. & IGNATOV, A.M. 1993. Unconventional exploration techniques in high cost deepwater basin: a case study from the Black Sea. *SEG Technical Program Expanded Abstracts*. Society of Exploration Geophysicists, 1351.

MANN, U., KORKMAZ, S., BOREHAM, C.J., HERTLE, M., RADKE, M. & WILKES, H. 1998. Regional geology, depositional environment and maturity of organic matter of Early to Middle Jurassic coals, coaly shales, shales and claystones from the Eastern Pontides, NE Turkey. *International Journal of Coal Geology*, **37**, 257–286.

MAYER, J., SACHSENHOFER, R.F., UNGUREANU, C. & TARI, G. In press. Petroleum charge and migration in the Black Sea: insights from oil and source rock geochemistry. *Journal of Petroleum Geology*, **41**.

MCCANN, T., CHALOT-PRAT, F. & SAINTOT, A. 2010. The Early Mesozoic evolution of the Western Greater Caucasus (Russia): Triassic–Jurassic sedimentary and magmatic history. *In*: SOSSON, M., KAYMAKCI, N., STEPHENSON, R.A., BERGERAT, F. & STAROSTENKO, V. (eds) *Sedimentary Basin Tectonics from the Black Sea and Caucasus to the Arabian Platform*. Geological Society of London, Special Publications, **340**, 181–238, https://doi.org/10.1144/SP340.10

MEISNER, L.B. & TUGOLESOV, D.A. 2003. Key reflecting horizons in sedimentary fill seismic records of the Black Sea basin (correlation and stratigraphic position). *Stratigraphy and Geological Correlation*, **11**, 606–619.

MEISNER, A., KRYLOV, O. & NEMČOK, M. 2009. Development and structural architecture of the Eastern Black Sea. *The Leading Edge*, **28**, 1046–1055.

MEISNER, A., SHEYA, C. & NEMCOK, N. 2011. Ancient depositional environments of the Eastern Black Sea. *AAPG Search and Discovery*, article #50388.

MEREDITH, D.J. & EGAN, S.S. 2002. The geological and geodynamic evolution of the eastern Black Sea basin: insights from 2-D and 3-D tectonic modelling. *Tectonophysics*, **350**, 157–179.

MORARIU, D. & NOUAL, V. 2009. Cretaceous play – new exploration potential in the Eastern Georgia. Neftegasovaa geologia. Teoria i practika 4, unpaginated.

NEMČOK, M., GLONTI, B., YUKLER, A. & MARTON, B. 2013. Development history of the foreland plate trapped between two converging orogens; Kura Valley, Georgia, case study. *In*: NEMČOK, M., MORA, A. & COSGROVE, J.W. (eds) *Thick-Skin-Dominated Orogens: From Initial Inversion to Full Accretion*. Geological Society,

London, Special Publications, **377**, 159–188, https://doi.org/10.1144/SP377.9

NIKISHIN, A.M., KOROTAEV, M.V., ERSHOV, A.V. & BRUNET, M.F. 2003. The Black Sea basin: tectonic history and Neogene–Quaternary rapid subsidence modelling. *Sedimentary Geology*, **156**, 149–168.

NIKISHIN, A.M., OKAY, A.I., TÜYSÜZ, O., DEMIRER, A., AMELIN, N. & PETROV, E. 2015a. The Black Sea basins structure and history: new model based on new deep penetration regional seismic data. Part 1: basins structure and fill. *Marine and Petroleum Geology*, **59**, 638–655.

NIKISHIN, A.M., OKAY, A., TÜYSÜZ, O., DEMIRER, A., WANNIER, M., AMELIN, N. & PETROV, E. 2015b. The Black Sea basins structure and history: new model based on new deep penetration regional seismic data. Part 2: tectonic history and paleogeography. *Marine and Petroleum Geology*, **59**, 656–670.

NIKISHIN, A.M., WANNIER, M. *ET AL.* 2017. Mesozoic to recent geological history of southern Crimea and the Eastern Black Sea region. *In*: SOSSON, M., STEPHENSON, R.A. & ADAMIA, S.A. (eds) *Tectonic Evolution of the Eastern Black Sea and Caucasus*. Geological Society of London, Special Publications, **428**, 241–264, https://doi.org/10.1144/SP428.1

OKAY, A.I. & NIKISHIN, A.M. 2015. Tectonic evolution of the southern margin of Laurasia in the Black Sea region. *International Geology Review*, **57**, 1051–1076.

OKAY, A.I. & SAHINTÜRK, Ö. 1997. Geology of the Eastern Pontides. *In*: ROBINSON, A.G. (ed.) *Regional and Petroleum Geology of the Black Sea and Surrounding Region*. AAPG, Memoirs, Tulsa, Oklahoma, **68**, 291–312.

OKAY, A.I., ŞENGÖR, A.C. & GÖRÜR, N. 1994. Kinematic history of the opening of the Black Sea and its effect on the surrounding regions. *Geology*, **22**, 267–270.

OKAY, A.I., SUNAL, G., SHERLOCK, S., ALTINER, D., TÜYSÜZ, O., KYLANDER-CLARK, A.R. & AYGÜL, M. 2013. Early Cretaceous sedimentation and orogeny on the active margin of Eurasia: Southern Central Pontides, Turkey. *Tectonics*, **32**, 1247–1271.

OKAY, A.I., ALTINER, D., SUNAL, G., AYGÜL, M., AKDOĞAN, R., ALTINER, S. & SIMMONS, M. 2017. Geological evolution of the Central Pontides. *In*: SIMMONS, M.D., TARI, G.C. & OKAY, A.I. (eds) *Petroleum Geology of the Black Sea*. Geological Society, London, Special Publications, **464**. First published online September 15, 2017, https://doi.org/10.1144/SP464.3

PAPE, T., BAHR, A. *ET AL.* 2010. Molecular and isotopic partitioning of low-molecular-weight hydrocarbons during migration and gas hydrate precipitation in deposits of a high-flux seepage site. *Chemical Geology*, **269**, 350–363.

POPOV, S.V., ANTIPOV, M.P., ZASTROZHNOV, A.S., KURINA, E.E. & PINCHUK, T.N. 2010. Sea-level fluctuations on the north shelf of the Eastern Paratethys in the Oligocene–Neogene. *Stratigraphy and Geological Correlation*, **18**, 200–224.

PUPP, M., BECHTEL, A., ĆORIĆ, S., GRATZER, R., RUSTAMOV, J. & SACHSENHOFER, R.F. In press. Eocene and Oligomiocene source rocks in the Rioni and Kura basins of Georgia: depositional environment and petroleum potential. *Journal of Petroleum Geology*, **41**.

RANGIN, C., BADER, A.G., PASCAL, G., ECEVITOGLU, B. & GÖRÜR, N. 2002. Deep structure of the Mid Black

Sea High (offshore Turkey) imaged by multi-channel seismic survey (BLACKSIS cruise). *Marine Geology*, **182**, 265–278.

REITZ, A., PAPE, T. *ET AL*. 2011. Sources of fluids and gases expelled at cold seeps offshore Georgia, eastern Black Sea. *Geochimica et Cosmochimica Acta*, **75**, 3250–3268.

ROBINSON, A.G., BANKS, C.J., RUTHERFORD, M.M. & HIRST, J.P.P. 1995*a*. Stratigraphic and structural development of the eastern Pontides, Turkey. *Journal of the Geological Society of London*, **152**, 861–872.

ROBINSON, A., SPADINI, G., CLOETINGH, S., RUDAT, J., CLOETINGH, S., DURAND, B. & PUIGDEFABREGAS, C. 1995*b*. Stratigraphic evolution of the Black Sea: inferences from basin modelling. *Marine and Petroleum Geology*, **12**, 821–835.

ROBINSON, A.G., RUDAT, J.H., BANKS, C.J. & WILES, R.L.F. 1996. Petroleum geology of the Black Sea. *Marine and Petroleum Geology*, **13**, 195–223.

ROBINSON, A.G., GRIFFTH, E.T., GARDINER, A.R. & HOME, A.K. 1997. Petroleum geology of the Georgian fold and thrust belts and foreland basins. *In*: ROBINSON, A.G. (ed.) *Regional and Petroleum Geology of the Black Sea and Surrounding Region*. AAPG, Memoirs, Tulsa, Oklahoma, **68**, 347–367.

ROLLAND, Y. 2017. Caucasus collisional history: review of data from East Anatolia to West Iran. *Gondwana Research*, **49**, 130–146.

SACHSENHOFER, R.F., POPOV, S.V. *ET AL*. 2017*a*. The type section of the Maikop Group (Oligocene–lower Miocene) at the Belaya River (North Caucasus): depositional environment and hydrocarbon potential. *AAPG Bulletin*, **101**, 289–319.

SACHSENHOFER, R.F., POPOV, S.V. *ET AL*. 2017*b*. Oligocene and Lower Miocene source rocks in the Paratethys: Palaeogeographic and stratigraphic controls. *In*: SIMMONS, M.D., TARI, G.C. & OKAY, A.I. (eds) *Petroleum Geology of the Black Sea*. Geological Society, London, Special Publications, **464**. First published online September 7, 2017 https://doi.org/10.1144/SP464.1

SACHSENHOFER, R.F., POPOV, S.V. *ET AL*. In press. 'Paratethyan' petroleum source rocks: an overview. *Journal of Petroleum Geology*, **41**.

SHILLINGTON, D.J., WHITE, N., MINSHULL, T.A., EDWARDS, G.R.H., JONES, S., EDWARDS, R.A. & SCOTT, C.L. 2008. Cenozoic evolution of the eastern Black Sea: a test of depth-dependent stretching models. *Earth and Planetary Science Letters*, **265**, 360–378.

SHILLINGTON, D.J., SCOTT, C.L., MINSHULL, T.A., EDWARDS, R.A., BROWN, P.J. & WHITE, N. 2009. Abrupt transition from magma-starved to magma-rich rifting in the eastern Black Sea. *Geology*, **37**, 7–10.

SHILLINGTON, D.J., MINSHULL, T.A., EDWARDS, R.A. & WHITE, N. 2017. Crustal structure of the Mid Black Sea High from wide-angle seismic data. *In*: SIMMONS, M.D., TARI, G.C. & OKAY, A.I. (eds) *Petroleum Geology of the Black Sea*. Geological Society, London, Special Publications, **464**. First published online October 4, 2017, https://doi.org/10.1144/SP464.6

SIPAHIOGLU, N.O., KARAHANOGLU, N. & ALTINER, D. 2013. Analysis of Plio-Quaternary deep marine systems and their evolution in a compressional tectonic regime, Eastern Black Sea Basin. *Marine and Petroleum Geology*, **43**, 187–207.

SOBORNOV, K.O. 1994. Structure and petroleum potential of the Dagestan thrust belt, northeastern Caucasus, Russia. *Bulletin of Canadian Petroleum Geology*, **42**, 352–364.

SPADINI, G., ROBINSON, A. & CLOETINGH, S. 1996. Western versus Eastern Black Sea tectonic evolution: pre-rift lithospheric controls on basin formation. *Tectonophysics*, **266**, 139–154.

STAROSTENKO, V., BURYANOV, V. *ET AL*. 2004. Topography of the crust-mantle boundary beneath the Black Sea Basin. *Tectonophysics*, **381**, 211–233.

TARI, G.C. & SIMMONS, M.D. 2018. History of deepwater exploration in the Black Sea and an overview of deepwater petroleum play types. *In*: SIMMONS, M.D., TARI, G.C. & OKAY, A.I. (eds) *Petroleum Geology of the Black Sea*. Geological Society, London, Special Publications, **464**. First published online May 4, 2018, https://doi.org/10.1144/SP464.16

TIBALDI, A., ALANIA, V., BONALI, F.L., ENUKIDZE, O., TSERETELI, N., KVAVADZE, N. & VARAZANASHVILI, O. 2017*a*. Active inversion tectonics, simple shear folding and back-thrusting at Rioni Basin, Georgia. *Journal of Structural Geology*, **96**, 35–53.

TIBALDI, A., RUSSO, E., BONALI, F.L., ALANIA, V., CHABUKIANI, A., ENUKIDZE, O. & TSERETELI, N. 2017*b*. 3-D anatomy of an active fault-propagation fold: a multidisciplinary case study from Tsaishi, western Caucasus (Georgia). *Tectonophysics*, **717**, 253–269.

TSERETELI, N., TIBALDI, A., ALANIA, V., GVENTSADSE, A., ENUKIDZE, O., VARAZANASHVILI, O. & MÜLLER, B.I.R. 2016. Active tectonics of central-western Caucasus, Georgia. *Tectonophysics*, **691**, 328–344.

TUGOLESOV, D.A., GORSHKOV, A.S. *ET AL*. 1985. *Tectonics of the Mesozoic Sediments of the Black Sea Basin*. Nedra, Moscow [in Russian].

VAKHANIA, D. 2008*a*. General overview of gas and oil saturation of sedimentary cover of the West (Kolkhida) zone of the sinking Transcaucasian intermountain region against the background of oil and gas generation problems. *Georgian Oil and Gas*, **22**, 87–107.

VAKHANIA, D. 2008*b*. Geodynamic evolution of oil and gas deposit generation within the Western (Kolkhida) zone of the sinking Transcaucasian intermountain region. *Georgian Oil and Gas*, **22**, 108–130 [in Russian with English summary].

TÜYSÜZ, O. 1999. Geology of the Cretaceous sedimentary basins of the Western Pontides. *Geological Journal*, **34**, 75–93.

VINCENT, S.J. & KAYE, M.N.D. 2017. Source rock evaluation of Middle Eocene–Early Miocene mudstones from the NE margin of the Black Sea. *In*: SIMMONS, M.D., TARI, G.C. & OKAY, A.I. (eds) *Petroleum Geology of the Black Sea*. Geological Society, London, Special Publications, **464**. First published online September 25, 2017, https://doi.org/10.1144/SP464.7

VINCENT, S.J., ALLEN, M.B., ISMAIL-ZADEH, A.D., FLECKER, R., FOLAND, K.A. & SIMMONS, M.D. 2005. Insights from the Talysh of Azerbaijan into the Paleogene evolution of the South Caspian region. *Geological Society of America Bulletin*, **117**, 1513–1533.

VINCENT, S.J., MORTON, A.C., HYDEN, F. & FANNING, M. 2013. Insights from petrography, mineralogy and U–Pb zircon geochronology into the provenance and reservoir potential of Cenozoic siliciclastic depositional systems supplying the northern margin of the

Eastern Black Sea. *Marine and Petroleum Geology*, **45**, 331–348.

VINCENT, S.J., HYDEN, F. & BRAHAM, W. 2014. Along-strike variations in the composition of sandstones derived from the uplifting western Greater Caucasus: causes and implications for reservoir quality prediction in the Eastern Black Sea. *In*: SCOTT, R.A., SMYTH, H.R., MORTON, A.C. & RICHARDSON, N. (eds) *Sediment Provenance Studies in Hydrocarbon Exploration and Production*. Geological Society, London, Special Publications, **386**, 111–127, https://doi.org/10.1144/SP386.15

YILMAZ, A., ADAMIA, S., CHABUKIANI, A., CHKHOTUA, T., ERDOGAN, K., TUZCU, K. & KARABIYIKOGLU, M. 2000. Structural correlation of the Southern Transcaucasus (Georgia)-Eastern Pontides (Turkey). *In*: BOZKURT, E., WINCHESTER, J.A. & PIPER, J.D.A. (eds) *Tectonics and Magmatism in Turkey and the Surrounding Area*. Geological Society London, Special Publications, **173**, 171–182.

ZONENSHAIN, L.P. & LE PICHON, X. 1986. Deep basins of the Black Sea and Caspian Sea as remnants of Mesozoic back-arc basins. *Tectonophysics*, **13**, 181–211.

History of deepwater exploration in the Black Sea and an overview of deepwater petroleum play types

G. C. TARI[1]* & M. D. SIMMONS[2]

[1]*OMV Exploration & Production GmbH, Trabrennstrasse 6-8, A-1020 Vienna, Austria*

[2]*Halliburton, 97 Jubilee Avenue, Milton Park, Abingdon, OX14 4RW, UK*

*Correspondence: Gabor.Tari@omv.com

Abstract: Deepwater hydrocarbon exploration drilling only began in the Black Sea less than 20 years ago, primarily because of the economical/technological challenges associated with mobilizing suitable rigs through the Bosporus. However, to date (end 2017), *c.* 20 deepwater wells have now been drilled, targeting a large variety of plays in this underexplored basin. The deepwater wells drilled to date are categorized by their main play objectives, within either the sag/post-rift or syn-rift basin fill of the Black Sea. The sag/post-rift play types have proven to be more successful, finding either biogenic gas in Miocene to Pliocene reservoirs associated with the Paleo-Danube and Paleo-Dnieper/Dniester or oil in Oligocene deepwater siliciclastic systems. Syn-rift or early post-rift plays, in contrast, assumed mostly shallow water carbonate reservoir targets. Only one well targeted pre-rift stratigraphy. Most of the exploration failures to date are directly related to the lack of reservoir at the targeted stratigraphic levels. However, the recent discoveries have underlined the presence of at least two active and effective petroleum systems that cover large parts of the deepwater Black Sea Basin.

The petroleum exploration history of the Black Sea and its surroundings before the mid-1990s has been summarized by Benton (1997). Because no exploration wells had been drilled deeper than 100 m water depth in the entire basin at that point, his overview covered wells drilled on the continental shelves or in the nearby onshore areas. Some of these important oil and gas discoveries include Tyulenovo (1951) in Bulgaria, Golitsyna (1975) in Ukraine, Akcakoca (1976) in Turkey, Galata (1976) in Bulgaria, and Lebada East (1981) in Romania (Fig. 1).

An understanding of the petroleum geology of the Black Sea was summarized within the seminal paper of Robinson *et al.* (1996) and by the subsequent book on the broader Black Sea region (Robinson 1997). Afanasenkov *et al.* (2007) provided a more recent review of aspects of the petroleum geology of the Eastern Black Sea region, Georgiev (2012) provided a review for parts of the Western Black Sea region, and Tari *et al.* (2009, 2011) summarized key exploration concepts for the deepwater Black Sea in general.

Since 1997, many more wells have been drilled on the shelf around the Black Sea, with additional discoveries and/or important geological results. The most significant of these include Subbotina-1 (2005) in Ukraine; Samotino More Melrose-1 (2007) in Bulgaria; Doina-1 (1995) and Ana-1 (2007) in Romania; Izgrev-1, Obzor-1, and Ropotamo-1 (2007–2008) in Bulgaria; Istranca-1 (2012) in Turkey and Marina-1 (2014) in Romania. As will be shown, the geological results of these wells have been instrumental in encourgaing exploration for similar play concepts in the deepwater, or for associated new plays down systems-tract.

Three scientific research wells were drilled in the deepwater Black Sea as early as 1975 (Ross & Neprochnov 1978). These Deep Sea Drilling Project (DSDP 379, 380 and 381) wells were drilled during Leg 42 (Ross *et al.* 1978) in the southwestern and central part of the Black Sea (Fig. 1), at water depths of 2165, 1728 and 2107 m, respectively.

Because of the technical and economic challenges of mobilizing suitable drilling units into the Black Sea under the bridges across the Bosporus in Istanbul, deepwater exploration drilling in the Black Sea did not begin until 1999. Since then, *c.* 20 deepwater wells have been drilled targeting a variety of plays in this underexplored region, one of the last major frontier basins in the world (Fig. 1).

The Black Sea is clasically divided into two separate basins, the Western and Eastern Black Sea Basins (WBSB and EBSB), with the divide formed by the Andrusov and Akhangelsky Ridges and the Tetyaev High (collectively, the Mid-Black Sea High) that trend approximately north–south in the central part of the Black Sea (Fig. 1). The tectonic histories of these two basins are perceived by most as being broadly similar, although the details continue to form the subject of lively debate in the literature (e.g. Nikishin *et al.* 2015*a, b*; Okay & Nikishin 2015; Tari 2015; Simmons *et al.* 2018).

From: SIMMONS, M. D., TARI, G. C. & OKAY, A. I. (eds) 2018. *Petroleum Geology of the Black Sea.*
Geological Society, London, Special Publications, **464**, 439–475.
First published online May 4, 2018, https://doi.org/10.1144/SP464.16

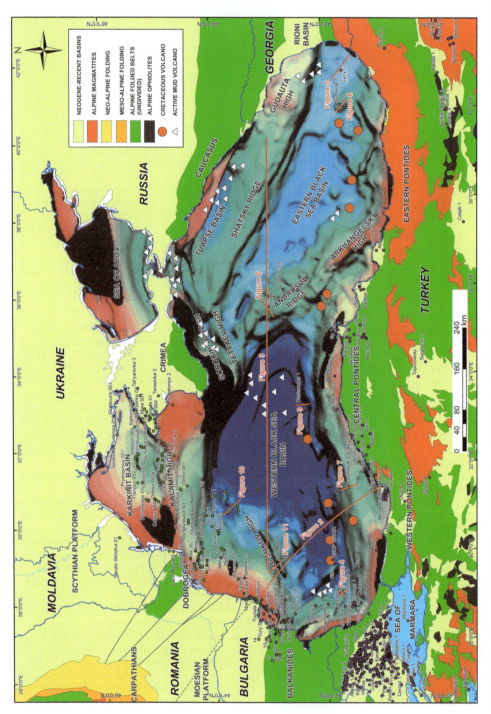

Fig. 1. Simplified structural map of the Black Sea with the locations of the deepwater wells discussed in this chapter (after Routh *et al.* 2017; Munteanu *et al.* 2017). Within the Black Sea itself, the depth to break-up unconformity is shown, adapted from Robinson (1997). White triangles represent offshore mud volcanoes compiled by Dimitrov (2002*a*, *b*), Krastel *et al.* (2003), Starostenko *et al.* (2010) and Dembicki (2014). Red dots represent Cretaceous palaeovolcanoes as presented by Nikishin *et al.* (2015*a*). The approximate location of the regional geoseismic and chronostratigraphic transects, shown in Figures 2 and 3, are shown by red lines. The locations of the play-type cartoons of Figures 4–15 are highlighted.

Most deepwater drilling has occurred in the WBSB, whereas only three wells have been drilled in the EBSB to date (Fig. 1). In terms of deepwater drilling by country, the Romanian sector of the Black Sea has undergone the most activity (10 wells), with the Turkish sector being a close second (eight wells). Only two truly deepwater (i.e. deeper than 500 m water depth) wells have been drilled in the Bulgarian sector to date. In contrast, the Ukrainian, Russian and Georgian segments have had no deepwater exploration wells so far; consequently, they remain practically unexplored. However, one well (Maria-1) is being drilled by ENI and Rosneft in the Russian sector over the Shatsky Ridge (Fig. 1) as of December 2017 (Bloomberg 2017).

In this contribution, first the regional structure and tectonostratigraphy of the deepwater Black Sea are briefly reviewed, followed by a description of the deepwater wells drilled to date. This brief account of the deepwater wells is presented in a chronological order to emphasize the historical character of the exploration efforts. All wells are classified into play type categories, based on the primary pre-drill exploration target in any given well. Because of data confidentiality, the emphasis in this paper is placed on the generic description of the play concepts, rather than the technical details of any given well. Therefore, these play types are illustrated by play-type cartoons in a consistent graphic style.

Finally, a summary of all of the play concepts for the deepwater Black Sea is provided. Two decades after the landmark publication of Robinson (1997), a new systematic compilation of the tested and speculative deepwater petroleum plays in the Black Sea provides a useful summary and starting point for future exploration efforts.

Brief summary of Black Sea geodynamic history

The Black Sea, in terms of crustal structure, consists of two rift basins: the WBSB and EBSB (Fig. 1), separated by the Andrusov Ridge (or Mid-Black Sea Ridge/High of some authors). These basins differ in terms of their sizes, structures and sedimentary thicknesses (e.g. Finetti *et al.* 1988; Robinson 1997; Meisner and Tugolesov 2003; Nikishin *et al.* 2015*a*, *b*).

The current consensus in the published literature is that the larger WBSB is underlain by oceanic crust and contains a sedimentary cover up to 14–18 km thick (Belousov *et al.* 1988; Görür 1988; Okay *et al.* 1994; Graham *et al.* 2013; Nikishin *et al.* 2015*a*, *b*; Schleder *et al.* 2015; Tari *et al.* 2015*a*; Fig. 2). In contrast, the same aged (e.g. Nikishin *et al.* 2015*b*) or supposedly younger (e.g. Robinson *et al.* 1996) EBSB has thinned continental

or 'incipient' oceanic crust and contains up to 12 km thickness of sediments (Shillington *et al.* 2017). Note that direct geophysical evidence for the presence of normal oceanic crust in these basins remains elusive despite the large number of papers addressing this issue (e.g. Finetti *et al.* 1988; Starostenko *et al.* 2004, 2015; Shillington *et al.* 2008; Nikishin *et al.* 2015*a*, *b*). Regardless, the two basins coalesced during the early Cenozoic, forming the present-day basin.

The greatest challenge for understanding the geological history of the Black Sea is the absence of well penetrations in its very thick basin fill. A regional geoseismic transect across the WBSB illustrates this problem. Only one well (Polshkov-1) may have reached the syn-rift to pre-rift succession over the Polskhov High (Fig. 2). However, even at this location, the Mesozoic strata that might have been encountered (Tari *et al.* 2009) provide an incomplete stratigraphic record of what lies in the basin centre.

Therefore, regarding the opening history of the Black Sea, very different models were published during the last three decades. Because the age of rifting is crucially important for the prediction for the maturation of source rocks and the development of reservoir sequences in the deepwater areas (Simmons *et al.* 2018), the existing models are briefly summarized in the following paragraphs.

Tugolesov *et al.* (1985) proposed a Paleocene–Eocene age for the formation of both the WBSB and EBSB. In their model, these basins were separated by the Andrusov Ridge until the Pliocene to Quaternary when the two sub-basins became one.

Görür (1988) proposed that the Black Sea Basin began opening as a back-arc basin by the rifting of a young continental margin magmatic arc during the Aptian–Cenomanian period. Banks & Robinson (1997) suggested an Early Cretaceous opening for the WBSB, followed by the rifting in the EBSB during the Paleocene/Eocene.

Nikishin *et al.* (2003) placed the age of initial rifting in the Albian for both basins, but dated the main opening phase as Cenomanian/Coniacian. In their model, the Campanian to Middle Eocene interval represents the post-rift phase, interrupted by renewed subsidence as a result of regional compression across the entire basin. These authors interpreted the syn-compressional downbending in the basin as the reason for the coalescence of the WBSB and EBSB during the Pliocene.

It is now understood that rifting in the WBSB began in the Barremian (e.g. Munteanu *et al.* 2011), with the deposition of syn-rift clastics and carbonates identified in the Pontide Mountains (Hippolyte *et al.* 2010, 2016; Okay *et al.* 2017), Crimea (Nikishin *et al.* 2017) and offshore Romania (Ionescu 2002; Boote 2017). The process involved the Istanbul Terrane (modern day western Pontides) splitting away from Moesia as a consequence of

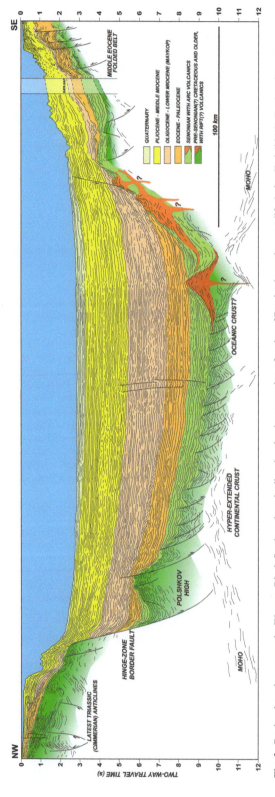

Fig. 2. Regional section across the Western Black Sea based on the line drawing interpretation of various 2D seismic sections modified after Tari (2015). For location, see Figure 1. The undrilled basin centre may contain a sedimentary fill of up to 14 km thick. Note the presence of palaeovolcanoes in the deepwater parts of the Turkish margin. Tari (2015) speculated that these volcanic sequences, defined only by seismic data and no well control at all, may correspond to the two Upper Cretaceous volcanic units known from the Pontides (Fig. 3).

subduction of the Neo-Tethyan Ocean to the south (Okay *et al.* 1994, 2001; Banks & Robinson 1997). The exact timing and kinematics of this basin opening is not fully resolved (Tari 2015), and subduction-related volcanics did not occur until the Turonian (Tüysüz *et al.* 2012; Keskin & Tüysüz 2017; Tüysüz 2017). The inferred seafloor spreading is assumed to have begun during the Coniacian–Santonian, again without any direct geophysical evidence to date. Based on analogues from the Pontides, deepwater sedimentation is assumed to be established in the WBSB by the Coniacian with an associated island arc contributing volcaniclastic material (Görür 1988, 1997; Görür & Tüysüz 1997; Georgiev *et al.* 2001; Hippolyte *et al.* 2010; Okay & Nikishin 2015; Nikishin *et al.* 2015*a, b*; Okay *et al.* 2013, 2017). Some assume that ocean spreading ceased by the Early Cenozoic, and the rate of sedimentation increased, with the incoming sediments ultimately sourced from erosion of the surrounding orogens (e.g. Maynard *et al.* 2012).

The main phase of the opening of the EBSB has been variously interpreted as coeval with the Western Black Sea (Okay *et al.* 1994; Nikishin *et al.* 2003, 2015*a, b*; Stephenson & Schellart 2010), as late Campanian to Danian (Vincent *et al.* 2016), Paleocene–Early Eocene (Robinson *et al.* 1995*a, b*, 1996; Spadini *et al.* 1996; Shillington *et al.* 2008) or Eocene (Kazmin *et al.* 2000). It is notable that the rift-related successions of the western and central Pontides (e.g. Çağlayan Formation) have no equivalents in the eastern Pontides (Okay & Şahintürk 1997; Hippolyte *et al.* 2015).

Given that the Western Greater Caucasus Basin to the north of the Shatsky Ridge opened in the Early Jurassic as a result of Neotethyan subduction (Saintot *et al.* 2006), there may have also been a proto Eastern Black Sea south of the Shatsky Ridge during the Jurassic, or at least an initial phase of rifting (Zonenshain & Le Pichon 1986).

The Black Sea is traditionally thought to be a marginal or back-arc basin with active rifting beginning in the middle Cretaceous (Letouzey *et al.* 1977; Kazmin *et al.* 1986; Zonenshain & Le Pichon 1986; Finetti *et al.* 1988; Golmshtok *et al.* 1992; Nikishin *et al.* 2015*a*). Attempts have been made to explain the formation of the Black Sea basins in terms of geodynamic models of modern back-arc basin formation, in which extension is driven by slab roll-back (e.g. Stephenson & Schellart 2010). However, debate in the literature exists as to not only the geodynamic reason for the basin opening, but also its precise timing and kinematics (e.g. Tari 2015). As to the magmatic arc associated with the southern margin of the basin, however, there are still some open-ended questions regarding its relation to the opening of the Black Sea basins (e.g. Keskin & Tüysüz 2017).

Deepwater stratigraphy of the Black Sea basins

As a preamble to the history of deepwater drilling and the overview of petroleum play types, a summary chronostratigraphic chart was compiled to illustrate the inferred lithostratigraphy in both the WBSB and the EBSB (Fig. 3). This simplified summary is explained in the following paragraphs, following a bottom-to-top stratigraphic order. Because only a few wells have penetrated the entire Cenozoic sequence in the deepwater Black Sea (Fig. 1), this infererred lithostratigraphic framework is largely based on the extrapolation of the known stratigraphic relations from the basin margins projected into a west–east transect across the entire basin system (Fig. 3).

The Jurassic stratigraphy on the western basin margin is mostly constrained by the wells drilled on the Bulgarian and Romanian shelf of the Black Sea (Moroşanu 1996, 2012; Harbury & Cohen 1997; Ionescu *et al.* 2002; Dinu *et al.* 2005; Georgiev 2012; Boote 2017; Krezsek *et al.* 2017). Jurassic outcrops around the Black Sea were also studied in detail, such as in the Pontides (e.g. Okay *et al.* 2017), Crimea (e.g. Nikishin *et al.* 2017) and the Caucasus (e.g. Guo *et al.* 2011).

On the western end of the regional chronostratigraphic transect (Fig. 3), the Middle Jurassic strata are generally characterized by a siliciclastic sequence overlying the regional Cimmerian unconformity (Georgiev & Atanasov 1993; Tari 1997). Because normal faults controlled the deposition of a succession of continental clastics, with coals transitioning to shallow water carbonates, this sedimentary package can be interpreted as a syn-rift sequence. The top syn-rift unconformity appears to be located between the Oxfordian and Kimmeridgian. Harbury & Cohen (1997) reported Tithonian normal faulting, but not at a regional scale. The overlying Kimmeridgian to Valanginian sequence has a carbonate platform character with some unconformities resulting from subaerial exposure.

The Jurassic strata in the middle part of the regional chronostratigraphic transect is largely based on the geology exposed in Crimea (Fig. 3). The 'Taurian series' is defined as a thick sequence of terrigenous flysch and flyschoid deposits of Late Triassic and Early Jurassic age, which locally also comprise volcanic units, large limestone blocks as olistoliths and frequently coarse clastic interbeds, such as conglomerates (e.g. Nikishin *et al.* 2017). Recent work showed that a large part of what has been mapped as Taurian (or Tavrik) is something different and actually much younger (Sheremet *et al.* 2017). This particular lithofacies succession can be correlated across the Eastern Black Sea with that

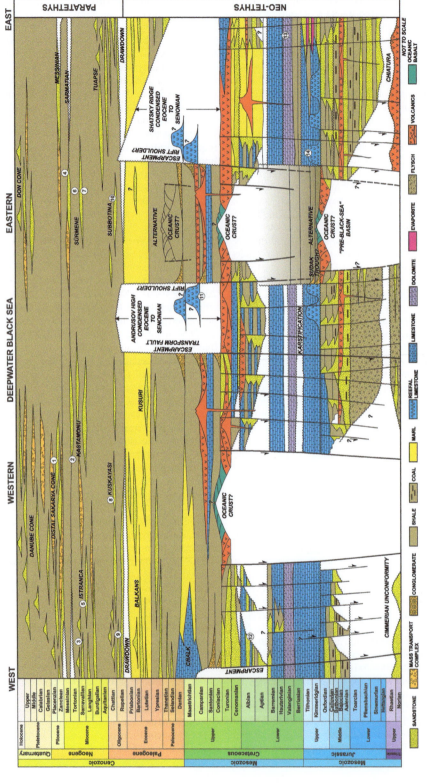

Fig. 3. Simplified chronostratigraphic scheme of the deepwater Black Sea along a regional west–east transect. For an approximate location, see Fig. 1, but note that the transect was compiled not to horizontal scale. Geological timescale is adapted from Cohen *et al.* (2016). For a detailed explanation, see text.

outcropping in the Western Great Caucasus around Krasnaya Polyana north of Sochi (Gabdullin *et al.* 2014). Above a regional unconformity, the Middle Jurassic sequence appears to be a syn-rift sequence formed from continental clastics, but with a thick and regionally well-developed volcanic sequence. The products of the regional Bajocian volcanism can be found around the Eastern Black Sea region, such as in Crimea (Meijers *et al.* 2010; Nikishin *et al.* 2017), the Great Caucasus (McCann *et al.* 2010) and the Central and Eastern Pontides (Çimen *et al.* 2016, 2017; Okay *et al.* 2014, 2017). Recent work by Akdoğan *et al.* (2018) using detrital zircon agfe dating also showed that the volcanic activity in the Eastern Pontides was longer than described before spanning the Sinemurian to Oxfordian period. The geochemistry of these Jurassic volcanics suggests an arc-related origin.

Above the volcanic sequence, on the margin of the Eastern Black Sea, a regional Bathonian unconformity is observed in both Russia and Georgia (Saintot *et al.* 2006; Afanasenkov *et al.* 2007; Adamia *et al.* 2017; Tari *et al.* 2018, respectively). This unconformity has been interpreted as the signature of the end of active rifting in the Great Caucasus Basin to the east (e.g. Saintot *et al.* 2006).

The age of the formation of the EBSB is debated (see Simmons *et al.* 2018; Tari *et al.* 2018 for a summary). Despite the dominant current views that the Eastern Black Sea opened up as an oceanic basin during the Late Cretaceous (e.g. Nikishin *et al.* 2015*a*, *b*) or younger (e.g. Robinson *et al.* 1995*a*, *b*, 1996), our stratigraphic compilation provides an alternative interpretation, assuming a precursor basin (Fig. 3). The presence of a Jurassic 'pre-Black Sea oceanic basin' *sensu* Zonenshain & Le Pichon (1986) would help to alleliviate some of the apparent kinematic problems; for example, the small crustal extension (on the order of only $\beta = 1.19$) is deduced from the geometry of the observed large Cretaceous normal faults bounding the EBSB (Meredith & Egan 2002).

A fairly distinct feature of the Late Jurassic stratigraphy of the northeastern Black Sea is the presence of large Kimmeridgian to Tithonian reefs (Fig. 3). These reefs are described from onshore in Crimea and the Russian Great Caucasus (e.g. Guo *et al.* 2011; Nikishin *et al.* 2017; Sheremet *et al.* 2017) and from the northwestern part of the Shatsky Ridge in the Russian segment of the EBSB (Afanasenkov *et al.* 2005, 2007). If there was a Middle Jurassic predecessor to the Eastern Black Sea rift basin, then these reefs could be interpreted as preferentially developed on the rift shoulders. In addition, a pre-Oxfordian rifting may explain the regional unconformity present in the Rioni Basin of onshore Georgia (e.g. Tari *et al.* 2018) as a break-up unconformity.

Tithonian to Berriasian platform carbonates seem to be ubiquitous around the Black Sea, including the Bulgarian (Georgiev 2012), Romanian (Krezsek *et al.* 2017), Ukrainian (Gozhik *et al.* 2008), Russian (Afanasenkov *et al.* 2007) and Turkish (Okay *et al.* 2017) segments (Fig. 3). The Black Sea region was probably the site of a widespread carbonate platform on the northern margin of the Tethys, corresponding to a post-rift phase with only relatively minor bathymetric variations (Simmons *et al.* 2018; Vincent *et al.* 2018). A notable exception is the late Jurassic deepwater Sudak Trough with its shale, turbidite and mass transport complex basin fill outcropping in Crimea (Nikishin *et al.* 2017). This exception was already interpreted by Saintot *et al.* (2006) as a possible along-strike analogue for the Late Jurassic oceanic basin that may be present beneath the Eastern Black Sea (Fig. 3).

Uplift and erosion occurred in the Valanginian and Hauterivian, and the regional Upper Jurassic to Lower Cretaceous carbonate platform was subaerially exposed and karstified in multiple episodes (Simmons *et al.* 2018; Vincent *et al.* 2018). Another rifting phase began in the Barremian, but not regionally. Evidence for the onset of localized extension is documented on the western and northern margin of the Black Sea in Bulgaria (Harbury & Cohen 1997; Georgiev 2012), Romania (Moroşanu 1996, 2012; Boote 2017; Krezsek *et al.* 2017), on the Odessa shelf (Gozhik *et al.* 2006, 2008) and in Crimea (Nikishin *et al.* 2017).

It was, however, during the Aptian when regional wide-rift style extension (*sensu* Buck 1991) began around the WBSB (Tari 2015). This widely distributed extensional period, spanning the entire Aptian–Albian period (Fig. 3), is well documented in the Romanian segment of the Black Sea, both onshore and offshore, by mixed clastic and carbonate deposition in Romania (Ionescu *et al.* 2002; Moroşanu 2012; Boote 2017; Krezsek *et al.* 2017). The initial extension in the NW Black Sea was SW–NE and may have been influenced by the pre-existing structural fabric in that area (Schleder *et al.* 2015; Krezsek *et al.* 2017). Boote (2017) suggested a west-facing basin margin during Aptian times using datasets in the Romanian offshore. If correct, this early rifting episode may not have had the same geodynamic framework as the subsequent rifting episodes in the Black Sea Basin proper located generally to the east from his study area.

In the southern part of the Central Pontides, a thick succession of turbidites (more than 2 km) were deposited in a forearc setting (Okay *et al.* 2013, 2017; Akdoğan *et al.* 2017), whereas along the present day Black Sea coast, shallow marine and continental deposition occurred (Yilmaz & Altiner 2007; Masse *et al.* 2009). Large syn-rift fault blocks (e.g. Kozlu High) of assumed Early

Cretaceous age were mapped in the offshore Zongul-dak area (Menlikli *et al.* 2009). Lithostratigraphic evidence from Crimea and from the Eastern Black Sea coastal area in Russia and Georgia also indicates Early Cretaceous rifting of broad areas (Nikishin *et al.* 2015*a*, *b*; Adamia *et al.* 2010, 2017).

We have tentatively indicated the presence of late Albian andesitic volcanism in the central and eastern part of the chronostratigraphic overview (Fig. 3). This is primarily based on surface (Nikishin *et al.* 2013) and subsurface evidence (Gozhik *et al.* 2006) from Crimea and the nearby Odessa shelf. These Albian trending east–west/WSW–ENE palaeovolcanoes seem to be related to the Karkinit Trough, which was rifting during Albian (Robinson 1997). Interestingly, Nikishin *et al.* (2013, 2015*a*, *b*) also provided seismic examples of possible Albian volcanoes over the Shatsky Ridge, but these remain undrilled and are therefore speculative (Fig. 3).

We have designated the Aptian–Albian period as 'syn-rift 1' (after Tari 2015). The top Albian regional unconformity is a basinwide feature (Fig. 3) which marks the end of the wide-rift style of extension in the region. For example, numerous exploration wells drilled in the East Lebada Field offshore Romania documented the break-up uncon-formity on top of slightly deformed and eroded Albian strata (Ionescu *et al.* 2002). The overlying Cenomanian to Maastricthian strata with its deeper water depositional facies displays a post-rift charac-ter with no signs of ongoing rifting. Therefore the age of the break-up unconformity is well constrained by the biostratigraphically well-dated, nearly com-plete Albian sequence from the unconformably overlying Cenomanian (Avram *et al.* 1988, 1996; Krezsek *et al.* 2017).

The first, wide-rift style phase of rifting in the WBSB during the Aptian–Albian can be explained by asymmetric rifting at the southern margin of the European plate without invoking back-arc extension (Tari 2015). Based on extensive industry seismic datasets in and around the basin, the Ukrainian, Romanian and Bulgarian margins are described as parts of a lower plate continental margin, whereas the conjugate Turkish margin is best understood in terms of an upper plate continental margin (Tari *et al.* 2015*a*).

The stratigraphic record of the next phase of rift-ing on the conjugate margins of the WBSB is mark-edly different (Tari 2015). On the Turkish side, in the Pontides, a significant part of the Upper Cretaceous sequence appears to be missing, either by erosion or by non-deposition. The typical absence of the Cenomanian strata could be attributed to either uplift and erosion on a rift-shoulder (Hippolyte *et al.* 2010) or to uplift and erosion resulting from collision to the south of the Central Pontides during Turonian–Con-iacian times (Okay *et al.* 2006).

In contrast, Cenomanian clastics are widespread and tend to overstep the previous Aptian–Albian depocentres along the NW margin of the Black Sea in offshore Bulgaria, Romania, and the Ukraine (e.g. Moroşanu 1996, 2012; Georgiev *et al.* 2001; Gozhik *et al.* 2006, 2008; Georgiev 2012; Krezsek *et al.* 2017). The stratigraphic record of the Black Sea rifting on this margin is much more complete, based on the detailed stratigraphic schemes estab-lished onshore and on the nearby shelf areas (e.g. Robinson *et al.* 1996; Dinu *et al.* 2005; Gozhik *et al.* 2006, 2008; Bergerat *et al.* 2010; Khriachtch-evskaia *et al.* 2010). In our compilation of the inferred stratigraphy in the deepwater parts of the Black Sea, the near-complete presence of a Cenoma-nian–Turonian sequence is assumed to correspond to a new rifting period (Fig. 3). This 'syn-rift 2' exten-sional period in the WBSB was interpreted by Tari (2015) as one in a narrow-rift style *sensu* Buck (1991) where the extensional strain, formerly distrib-uted regionally in numerous troughs, was replaced by rifting in a single main graben system. It was this NE–SW trending rift basin system, correspond-ing to NW–SE-directed extension, that is assumed to reach the stage of initial oceanic spreading by Conia-cian times (Fig. 3).

Upper Cretaceous volcanics in the Black Sea region are largely confined to the southern Turkish margin of the basin, both offshore and onshore (Fig. 1). The first volcanics that correspond to the 'syn-rift 2' stage appeared during the Turonian (Der-eköy Formation). The mostly basaltic volcanics out-cropping in the Pontides are thought to have been deposited within extensional grabens (Tüysüz *et al.* 2012), contemporaneously with the rifting in the axis of the future WBSB. The geochemical character of these volcanics suggests an arc-related origin (e.g. Keskin & Tüysüz 2017). Therefore, this is the phase in the formation of the WBS that could be considered as the onset of a back-arc basin formation *sensu stricto* (Tari 2015). These volcanics, and the distinct younger Campanian volcanics, can be speculatively interpreted on offshore seismic data (Fig. 2).

Importantly, Turonian effusive volcanics were drilled on the southern edge of the Odessa shelf, in the Ilichevsk-2 well (Fig. 1), between 1172 and 1714 m, overlain by Santonian pelagic limestones (Gozhik *et al.* 2006; Yakushyn & Ishchenko 2014). The location of these Turonian basalts is *c.* 400 km to the north from the outcrops of the age-equivalent Dereköy Formation in the Central Pontides. The very large present-day separation of the same volcanic units implies a largely post-Turonian opening of the WBSB.

The presence of oceanic crust was interpreted beneath both the WBSB and EBSB using regional long-offset, deep-penetration reflection seismic data acquired a few years ago (Graham *et al.* 2013;

Nikishin *et al.* 2015*a, b*). Whereas the seismic signature of the inferred oceanic crust at the acoustic basement level has some elements of the expected geometry, there are no signs of a spreading centre that could be mapped along the axis of either basin segment. In addition, as was noted by many before, there are no clear magnetic delineations detected anywhere in the Black Sea to support regional oceanic spreading. Furthermore, there are no clear seaward-dipping reflectors or obvious exhumed mantle visible on the superb reflection seismic data which illuminated for the first time the entire basin fill (Graham *et al.* 2013; Nikishin *et al.* 2015*a, b*).

Based on these observations, the presence of only a relatively narrow, incipient oceanic crust is assumed in both the WBSB and EBSB corresponding to a brief spreading period (only about 5–6 Ma) during the Coniacian and early Santonian (Fig. 3). In the case of the EBSB, this also assumes that this basin segment of the Black Sea opened during the Cretaceous synchronous with the WBSB, rather than in the Jurassic or in the Paleogene that are alternatives (Fig. 3). Regardless, the oceanic crust, if present, in both basins is overlain by a pelagic post-rift shale sequence.

The younger, volumetrically more important and widespread Upper Cretaceous volcanic unit in the Pontides is called the Cambu Formation (e.g. Okay *et al.* 2017). According to biostratigraphic data from overlying and underlying sediments, it was deposited as the product of arc magmatism during the post-rift evolution of the Black Sea throughout the Campanian (Tüysüz 1999; Tüysüz *et al.* 2012) or during the Santonian (Hippolyte *et al.* 2010; Fig. 3). Until recently, very little has been published regarding the offshore extent of the Cretaceous volcanic sequences of the Pontides. Nikishin *et al.* (2015*a*) suggested a Campanian to Santonian age for large palaeovolcanoes (at least 12 of them) observed on seismic data in the southern part of the Black Sea, tracking the Turkish coastline by an average 100 km distance (Figs 1 & 2). These large volcanoes, forming the peri-Pontides volcanic belt (Nikishin *et al.* 2015*a*), show no signs of subaerial erosion and therefore had to be formed in a submarine setting in water depth of at least 2 km. This observation suggests that deepwater Black Sea basins were already open by the time these (Campanian to Santonian?) volcanoes were emplaced.

Post-Coniacian, the deepwater environment during the entire post-rift sedimentation of both the WBSB and the EBSB resulted in a thick (more than 10 km) basin fill, especially in the WBSB (Fig. 1). These strata are dominated by pelagic shales and only locally by marls and sandstones (Fig. 3). The two major marly intervals in the deepwater Black Sea span the Campanian to Danian and Lutetian to Priabonian intervals (Fig. 3) when carbonate

platforms prevailed on the shallow-water margins of the Black Sea (e.g. Tambrea *et al.* 2002; Gozhik *et al.* 2006; Kopaevich *et al.* 2008; Menlikli *et al.* 2009; Less *et al.* 2011; Georgiev 2012; Aydemir & Demirer 2013; Adamia *et al.* 2017). However, there were entry points to the deepwater basin supplying reservoir-quality clastic turbidites. These include Middle Eocene sands on the Bulgarian shelf documented in several wells (Sinclair *et al.* 1997; Georgiev 2012) and similar-aged slope fans on the Turkish shelf as proven gas reservoirs in the Akcakoca and Ayazli discoveries (Robinson *et al.* 1996). The Middle Eocene Kusuri Formation has been well studied at outcrop in the Central Pontides (Janbu *et al.* 2007).

The Coniacian to Lower Oligocene basin fill onlaps the prominent basement highs inherited from the earlier syn-rift periods of the Black Sea, such as the Andrusov High and Shatsky Ridge (Fig. 3). Because these structures had a major bathymetric expression on the palaeoseafloor, their flanks formed escarpments, shedding locally coarse talus sediments. At the same time, condensed sedimentation occurred on top of these structures (Fig. 3).

Note that regarding the EBSB segment of our regional chronostratigraphic transect, we have noted the alternative models of Robinson *et al.* (1995*a, b*), Banks *et al.* (1997), Kazmin *et al.* (2000) and Vincent *et al.* (2005), suggesting Paleocene to Early Eocene and Eocene opening ages (Fig. 3). If correct, then the shallow water Campanian to Maastrichtian platform carbonates (Akveren Formation) described in the Pontides (e.g. Yilmaz *et al.* 1997) could also be present over the Andrusov High and the Shatsky Ridge as a pre-rift succession (Fig. 3). We argue here that the deepwater Sinop-1 well drilled on the Andrusov High in 2010 (see later) disproved this notion.

From the latest Eocene onward, the progressive convergence of Arabia towards Europe and the successive collision of terranes (e.g. the Anatolia–Tauride block) with the southern margin of Eurasia closed the Neotethys Ocean and uplifted the Caucasus, Pontide and Alborz mountain ranges (Cavazza *et al.* 2012; Espurt *et al.* 2014). This closure isolated the Black Sea and Caspian Sea basins from the closing Tethys to the south and created the Paratethys realm in Early Oligocene times (Báldi 1980; Popov *et al.* 1993; Jones & Simmons 1997; Rögl 1999; Schulz *et al.* 2005; Popov *et al.* 2010).

Tari *et al.* (2013) observed a regional erosional unconformity between the Eocene and Oligocene strata along the margins of the Black Sea that can also be followed into the deepwater basin (Fig. 3). Based on basin-scale seismic intepretation, they proposed a very large base-level drop (more than 2000 m!) in the Black Sea Basin at the end of the Eocene. The 'Messinian-style' dessication in the basin may have been partially triggered by the docking of the Balkans folded belt onto the Moesian Platform, closing the marine connection between the Tethys in the west and what became the Paratethys in the east.

The Oligocene to Recent deepwater sedimentation in the Black Sea is dominated by pelagic shales (Fig. 3). Owing to the lack of major river systems around the Black Sea during the entire Cenozoic until the inception of the Danube c. 4 Ma ago (De Leeuw et al. 2018; Olariu et al. 2018), only a fraction of the Cenozoic deepwater Black Sea strata contains reservoir-quality sandstones (Maynard et al. 2012). However, occasional and short-lived sedimentary entry points to the Black Sea existed, providing deepwater sand systems that are critical for exploration prospectivity (Fig. 3).

Within the Oligocene to Lower Miocene Maykop Suite succession, at least four deepwater channel–levee/basin floor systems have been described so far. Tari et al. (2009) provided the first images of an Oligocene slope channel–levee system in the Bulgarian sector using 3D seismic data. Rees et al. (2017) provided a discussion of the source-to-sink system associated with this. Menlikli et al. (2009), Sipahioğlu et al. (2013a), and Aktepe et al. (2013) showed a Late Oligocene terminal fan complex (Kuşkayası) in the Turkish sector, also using 3D seismic data.

Multiple Oligocene slope sand reservoirs were penetrated by the Ukrainian Subbotina oil discovery on the shelf of the Kerch Peninsula (Stovba et al. 2009; Gozhik et al. 2010). In the Russian sector, Afanasenkov et al. (2007) and Mityukov et al. (2011) described intra-Maykop channel–levee and slope fan systems using 3D seismic data in the Tuapse Trough.

In a regional correlation of the Miocene sequence of the Turkish offshore, Korucu et al. (2013) described a prograding deltaic system that was encountered in the Karadeniz-1 and Istranca-1 wells on the western side of the Turkish shelf. These form the reservoir in the Istranca-1 gas discovery (Fig. 1). Other deepwater Miocene sand systems were documented in the Kastamonou-1 and Sürmene-1 wells (Korucu et al. 2013). Bega & Ionescu (2009) described multiple Miocene and Pliocene deepwater sand systems before the discovery of the Domino gas accumulation (Fig. 3). The 3D seismic imaging of the reservoir architecture of these slope sands is further complicated by the very complex Pliocene to Quaternary overburden (Routh et al. 2017).

During the Late Miocene, two major sea-level drops (Popov et al. 2010) produced the Tortonian (Sarmatian) and the Messinian (Pontian) unconformities of regional extent across the Black Sea. The Sarmatian sea-level drop (Afanasenkov et al. 2007; Meisner et al. 2009, 2011) appears to be more prevalent on the basin margins and may have produced deepwater fan systems (Fig. 3). During the Miocene, many mass transport complexes (MTCs) formed in all segments of the Black Sea (Tari et al. 2009, 2015b, 2016; Sipahioğlu et al. 2013a; Krezsek et al. 2016; Kitchka et al. 2016a, b; Sipahioğlu & Bati 2017).

Whereas the Messinian sea-level drop has been envisioned by many (e.g. Hsü & Giovanoli 1979) to be on the order of 1600 m, it was recently shown to be of a significantly less spectacular amplitude (Tari et al. 2015a, b), on the order of 500–600 m (Krezsek et al. 2016). Regardless, it probably produced deepwater sands systems in the basin (e.g. Konerding et al. 2010; Boote 2017), such as in the Sakarya Cone (Tari et al. 2016), which may have been the Messinian Bosporus for the Black Sea (Pfannenstiel 1944).

The Pliocene to Quaternary history of the WBSB is dominated by the Danube, reaching the basin c. 4 Ma ago (De Leeuw et al. 2018). The Danube, along with the Dniester and Dnieper rivers, built the very wide shelf in the NW Black Sea (Fig. 1), and the corresponding slope channel–levee systems extend far out onto the basin floor (e.g. Wong et al. 1994; Popescu et al. 2001; Lericolais et al. 2013).

Finally, the Black Sea remains a largely deepwater basin today (maximum water depth 2245 m) c. 75 Ma after its inception (if it indeed opened during the Late Cretaceous) with mostly pelagic shale sedimentation on its basin floor (Fig. 3).

Deepwater wells and play types tested to date (as of December 2017)

The following description of deepwater wells drilled in the Black Sea is presented in a chronological order with a play-type cartoon describing the primary elements of the play(s) envisioned in the pre-drill stage. It is important to emphasize that only pre-drill ideas are captured in the cartoons; in some of the wells, these ideas were revised significantly after the drilling results. In addition, there is considerable overlap between some of the play types drilled, for example, in the cases of Sürmene-1 v. Sile-1. In these cases, only one generic play-type cartoon is shown.

Limanköy-1 and -2 (1999)

The Limanköy-1 and -2 wells (Fig. 4), drilled in Turkish waters in 1999 (Fig. 1), were the first deepwater exploration wells in the entire Black Sea. Whereas the outcome of the exploration programme did not meet the economic expectations, the complex drilling project was an operational success.

In the framework of the Western Black Sea joint project between Turkiye Petrolleri (TPAO) and

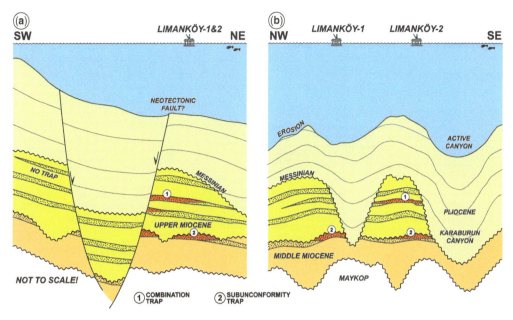

Fig. 4. Play concept cartoon of the Limanköy-1 and -2 wells, drilled in Turkish waters in 1999, that were the first deepwater exploration wells in the entire Black Sea. For location, see Figure 1. For a detailed explanation, see text.

ARCO Turkey Inc., 1127 km of 2D seismic was acquired in 1996. The interpretation of this new data identified the Kiyiköy, Limanköy, Karaburun and Eregli leads. The Kiyiköy–Limanköy area was considered the most important exploration target, and a 613 km² 3D seismic survey was shot, processed and interpreted in 1997. Because of the high mobilization costs of a drilling rig in the Black Sea, a decision was made to drill two back-to-back wells. The SEDCO 700 rig was mobilized from West Africa to Istanbul and crossed the Bosporus on 20 June 1999.

Although TPAO was the operator of this joint venture, ARCO (later BP, in 1999) acted as the drilling superintendent for the two wells to be drilled. The drilling of the Limanköy-1 and -2 wells, located in water depths of 820 and 720 m, respectively, occurred between the beginning of July to the end of September 1999.

The primary Miocene targets in the Limanköy wells (Fig. 4) had uncalibrated class III amplitude v. offset (AVO) anomalies based on the pre-drill analysis of the 3D seismic data. The trap for these prospects was provided by a combination of three-way fault-bound closures in the footwall of large normal faults dipping landward (Fig. 4a) and a large erosional unconformity at the base of the Pliocene (Fig. 4b; the Messinian unconformity). This we term the *Limanköy play*. Seismic illustrations of the well locations have been presented by Menlikli

et al. (2009), Korucu *et al.* (2013), Tari *et al.* (2015*a*, *b*) and Sipahioğlu & Bati (2017).

The drilling results showed that the relatively shallow, high-reflectivity events with the AVO anomalies were indeed gas charged, but the predicted reservoirs were diatomites and diatomitic shales, with no effective permeability (Menlikli *et al.* 2009; Korucu *et al.* 2013).

Deeper, secondary reservoir targets were Lower Miocene semi-consolidated deepwater sands. Gas samples of thermogenic and mixed thermogenic–biogenic origin recovered by a modular formation dynamics tester from these sands were thermally immature at their present-day burial depths and had a dominantly methane composition. The importance of these two wells was that they demonstrated the presence of a deepwater mixed thermo- and biogenic gas-prone petroleum system beneath the present-day continental slope, for the first time in the WBSB.

HPX-1 (2005)

After a five-year break, the next deepwater well was drilled by the joint venture of BP, TPAO and Chevron in the Eastern Black Sea in Turkey, just a few kilometres from the maritime border with Georgia (Fig. 1). The HPX-1 well (Fig. 5) operated by BP was drilled in significantly deeper water (1529 m) than the Limanköy wells. The Global Santa Fe Explorer drillship spudded the well in July and the

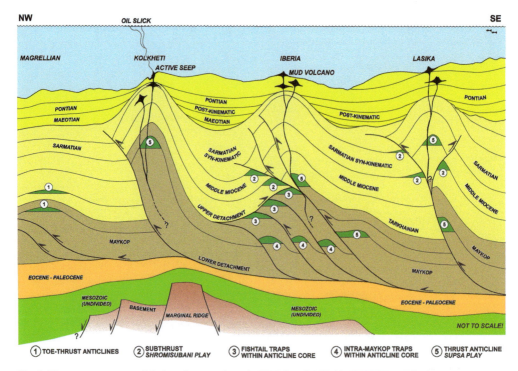

Fig. 5. Play concept cartoon of the broader area where the HPX-1 well drilled in Turkish waters in 2005. For location, see Figure 1.

total depth (TD) at 4700 m was reached by December 2005.

The high-profile HPX-1 well drilled the Hopa prospect mapped on 3D seismic data on the crest of a NW-vergent anticline in the easternmost part of the Turkish Black Sea (Menlikli *et al.* 2009). Because several Miocene to Pliocene anticlines exist in the border zone between Turkey and Georgia, forming the offshore continuation of the Achara–Trialet (Gurian) folded belt (e.g. Robinson *et al.* 1996; Adamia *et al.* 2017; Tari *et al.* 2018), the well was regarded by the petroleum industry as a potential play-opener with several follow-up prospects in the area. The drilling targets in the HPX-1 well were Upper Miocene deepwater sand units, all in four-way closures mapped on 3D seismic data (Menlikli *et al.* 2009). This we term the *Supsa play* based on the analogue of the working play in the Georgian onshore (Benton 1997; Tari *et al.* 2018). The presence of an oil-prone petroleum system was assumed, based on the well-known offshore oil seeps in nearby Turkey (e.g. Derman & Iztan 1997) and in Georgia (e.g. Pape *et al.* 2010; Reitz *et al.* 2011; Evtushenko & Ivanov 2013).

The post-drill evaluation of the well identified reservoir quality as the key issue in this part of the Black Sea. The provenance area of the Miocene deepwater sands was most likely from the Lesser Caucasus, which contains multiple volcanic sequences that would have been exposed and eroded at the time of deposition. Therefore, the provenance of the clastic reservoirs in the southeastern corner of the Eastern Black Sea is a major risk for reservoir quality (e.g. Vincent *et al.* 2013, 2014). Some thermogenic fluids were sampled from sidewall cores in the well (Menlikli *et al.* 2009) and proved the presence of an effective petroleum system.

The plays in the Gurian thrust-fold belt are summarized by a subregional cartoon transect drawn decidedly not to scale to illustrate the various play types in the SE Black Sea (Fig. 5). At the deepwater leading edge of the Gurian thrust-fold belt, subtle toe-thrust anticlines are considered with mostly intra-Maykop Suite reservoirs trapped in large, simple, but low-amplitude four-way closures (play 1). Inboard from the toe-thrust system, large detachment anticlines provide various trap types beneath the Upper to Middle Miocene sandstone reservoirs tested by the HPX-1 well, in four-way closures. Deeper untested traps include subthrust traps (play 2), proven by the onshore Shromisubani Field (see Tari *et al.* 2018; the *Shromisubani play*). Intra-Maykop, Lower Miocene deepwater sandstones may be trapped in thrust imbricates against an upper detachment in a zig-zag (or fishtail) structure (play 3). Deeper intra-Maykop (i.e. Oligocene) deepwater sandstones may be trapped in thrust imbricates in the core of the Gurian anticlines (play 4). Finally, the hanging-wall traps of the large detachment thrust faults provide relatively simple traps with intra-Maykop deepwater sandstone reservoirs (play 5). This particular play type includes all of the 'simple' structural targets, i.e. three-way closures with fault control. Given the reservoir quality risk, the reservoirs in the primary drilling targets must be older (but not necessarily deeper) than those encountered in the HPX-1 well (Fig. 5).

Sinop-1 (2010)

The failure of the HPX-1 well to find producable hydrocarbons put on hold any follow-up deepwater exploration projects for the next five years. In 2010, a structural high on the Andrusov Ridge seperating the Western and Eastern Black Sea (Fig. 1) became the new potential play opener target in the basin. The prominent NW–SE structural trend of syn-rift highs in the centre of the basin was long known (e.g. Tugolesov *et al.* 1985; Finetti *et al.* 1988; Robinson *et al.* 1996; Rangin *et al.* 2002) as a potential exploration trend. Therefore, a joint venture between Petrobras, TPAO and ExxonMobil spudded the Sinop-1 well in February 2010 in a water depth of 2182 m, practically at the basin floor of the Black Sea (Fig. 6). The TD of the well at 5531 m was reached by end of June.

The Sinop-1 well had two primary reservoir targets (Aydemir & Demirer 2013): (a) post-rift isolated platform or reefal carbonates (Late Cretaceous to Paleocene Akveren Formation); and (b) syn-rift sandstones (Aptian to Albian Velibey Formation) with pelagic shales units located between these targets (Fig. 6). The equivalent of the syn-rift play on the shelf is the proven *Lebada play* in the Romanian sector (e.g. Robinson *et al.* 1996; Ionescu *et al.* 2002; Boote 2017; Krezsek *et al.* 2017).

For the Late Cretaceous to Paleocene carbonates, which we designate here as the *Sinop play*, a lateral charge was assumed from the regionally proven source rocks within the Maykop Suite onlapping the Andrusov High on both of its flanks (Fig. 3). The large gas chimneys evident on the 2D seismic data over the crest of the Andrusov High (e.g. Robinson *et al.* 1996) were interpreted at the pre-drill stage as indirect evidence for an active petroleum system.

The Sinop-1 well failed as an exploration play opener because it did not encounter viable reservoirs in the section drilled (Aydemir & Demirer 2013). Although Late Creceous shallow-water rudist-bearing limestone units exist in the Pontides (e.g. Menlikli *et al.* 2009; Özer *et al.* 2009; Okay *et al.*

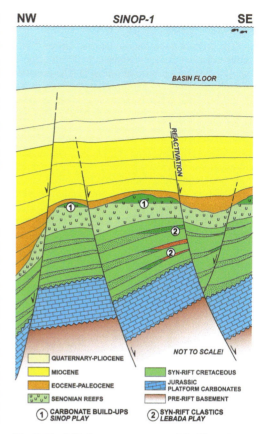

Fig. 6. Pre-drill play concept cartoon of the Sinop-1 well drilled on the Turkish side of the Andrusov Ridge. For location, see Figure 1.

2017), their presence was not confirmed in the basin centre, even at the drilling location over the crest of the prominent palaeohigh mapped on 3D seismic data (Fig. 6). The predicted deeper Cretaceous syn-rift siliciclastics were also not penetrated; instead, a thick sequence of Cretaceous volcanics (Gürsökü Formation) was found between 5126 and 5531 m (Aydemir & Demirer 2013). Consequently, the Sinop-1 well failed to meet its exploration objectives, primarily because of reservoir presence issues. Given the drilling results, the partnership of Exxon-Mobil, TPAO and Petrobras decided not to drill the follow-up prospect along the Andrusov High.

Yassihöyük-1 (2010)

The Yassihöyük-1 well was drilled immediately after the Sinop-1 well by the joint venture of TPAO and Chevron in the Turkish sector of Western Black Sea (Fig. 1) using the same deepwater semi-submersible rig (Leiv Eriksson) as was used for Sinop-1. The well was spudded in August 2010 at a water depth of 2020 m, and the TD was reached at 5155 m by early November.

The primary target was an isolated Upper Cretaceous to Lower Eocene carbonate build-up mapped on 3D reflection seismic data (Fig. 7). The presence of isolated carbonate platforms of this age in the deepwater Turkish Black Sea was generally assumed before drilling Sinop-1 and Yassihöyük-1. This pre-drill interpretation was supported by various typical geometries indicated on seismic data, such as back-reef pinnacles, lowstand build-ups, pronounced asymmetry with raised rims, back-stepping, aggradation and prograding clinoforms (Menlikli *et al.* 2009). These more than 800 m thick carbonate platforms were assumed to grow over footwall blocks of syn-rift faults. The predicted reservoir facies was based upon the occurrence of similar facies at outcrops in the Pontides (e.g. Menlikli *et al.* 2009; Özer *et al.* 2009; Okay *et al.* 2017). The pre-drill interpretation, based on 3D seismic data, also suggested that the top of the carbonate platform was subaerially exposed and karstified. Therefore, good reservoir properties were expected. Early Cretaceous syn-rift sandstones, as in the proven *Lebada play*

(e.g. Robinson *et al.* 1996) were regarded as a potential secondary reservoir target (Fig. 7).

The primary pre-drill risk was assumed to be petroleum charge because the predicted carbonate reservoirs required a lateral charge, assuming an effective Maykop Suite source rock interval (Menlikli *et al.* 2009). Vertical charge through the main bounding faults from potential Cretaceous syn-rift source rocks was also considered (Fig. 7). The top seal was expected to be the Middle Eocene marl sequence draping the carbonates.

Similar to the Sinop-1 well, the Yassihöyük-1 well drilled an unexpected sequence of Cretaceous volcanics (Aydemir & Demirer 2013) beneath just a few tens of metres of the Upper Cretaceous limestones of the Akvaren Formation. The carbonates at this location were developed in a shallow-water, but non-reservoir facies. The well reached TD within Upper Cretaceous volcanics and did not reach the syn-rift succession (Fig. 7). The post-drill evaluation of the well highlighted the lack of Upper Cretaceous carbonate reservoirs as the primary reason for failure.

The well did not test the deeper Cretaceous syn-rift siliciclastic and/or carbonate plays (Fig. 7) that, based on outcrop evidence in the Central Pontides (e.g. Okay *et al.* 2017) and by analogue to offshore Romania (Ionescu 2002; Boote 2017; Krezsek *et al.* 2017), have potential (Simmons *et al.* 2018). This play remains untested. The presence of an effective petroleum charge system at the well location remains elusive, although elevated gas readings were found within the volcanic sequence.

The failure to distinguish carbonate build-ups/reefs from palaeovolcanoes using modern 3D seismic data prompted the post-drill analysis by Posamentier *et al.* (2014) to define objective seismic interpretation criteria to distinguish between these features.

Sürmene-1 (2010/11)

The Sürmene-1 deepwater well was drilled in the Eastern Black Sea (Fig. 1) by the Leiv Eiriksson rig, immediately after Yassihöyük-1. This exploration drilling project was operated solely by TPAO. The well was spudded in 1800 m water depth in mid-November of 2010 and reached 4830 m with oil shows at *c.* 4800 m (Atalay *et al.* 2012).

Because the well had experienced some drilling problems (subsea wellhead damage), it was temporarily abandoned with a re-entry planned for Q4 2011 to drill another 800 m. After procuring a wellhead connector and a wellhead adapter, eight months later, in September 2011, it was tied back and drilled to the depth of 5650 m by the Deepwater Champion drillship. The well was abandoned at TD because of a

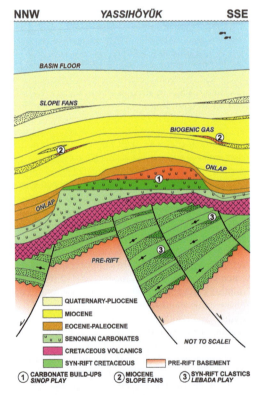

Fig. 7. Pre-drill play concept cartoon of the Yassihöyük-1 well drilled in the Turkish sector of the Western Black Sea. For location, see Figure 1.

very low drilling margin (less than 0.5 ppg, Atalay *et al.* 2012).

The somewhat unusual primary play concept of the Sürmene-1 well was a first in Black Sea exploration history because it was based on multiple four-way closures mapped on 3D seismic data above a large Cretaceous palaeovolcano (Fig. 8). Because many Cretaceous palaeovolcanoes exist along the Turkish margin (Fig. 1), based on the regional compilation by Nikishin *et al.* (2015*a*, *b*), many similar prospects could be defined in the southern Black Sea; therefore we name this play generically as the *Sürmene play*.

Because of the size of the four-way closures above a cone-shaped palaeovolcano, the trap formation cannot entirely be a result of simple differential compaction within the very thick post-rift basin fill. Other potential trap-forming structural elements include early Cenozoic extensional detachment faults on the basinward flank of the volcano with subsequent positive inversion (Fig. 8) during the Late Cenozoic. In any case, the reservoir elements in this play at the Sürmene-1 location were assumed to be Middle to Upper Miocene channelized sheet sands derived from the north, from the palaeo-Caucasus (Öztaş 2010). Because of a northerly provenance (e.g. Vincent *et al.* 2013, 2014), the reservoir quality at the

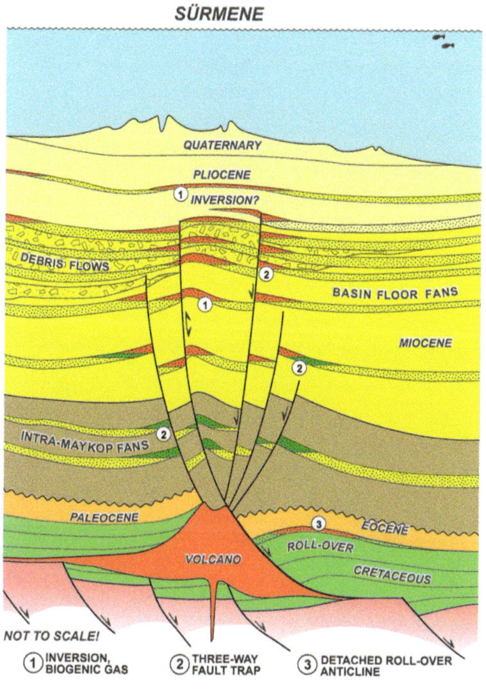

Fig. 8. Pre-drill play concept cartoon of the Sürmene-1 well drilled in the Turkish sector of the Eastern Black Sea. For location, see Figure 1.

Sürmene-1 well was expected to be much better than those of the sandstones penetrated in the HPX-1 well (Fig. 4) that included a substantial reworked volcaniclastic content. The presence of an active petroleum system was assumed, based on the oil seeps in the broader area (e.g. Derman & Iztan 1997), including the oil shows found in the HPX-1 well.

Despite the multiple oil shows encountered, the Sürmene-1 well was not successful in opening up a Miocene turbidite play fairway in the Eastern Black Sea. The most probable reason for the lack of an economic discovery is suspected to be reservoir quality, perhaps in a way similar to the case of the HPX-1 well (e.g. Tari *et al.* 2018).

Kastamonu-1 (2011)

The Kastamonu-1 well was drilled by the joint venture of ExxonMobil and TPAO in central offshore Turkey, WBSB (Fig. 1). The Deepwater Champion, a sixth generation drillship, was mobilized for the first time to the Black Sea for this project. The well was spudded in 2200 m of water, on the basin floor, in April 2011. After a very complex drilling operation, involving overpressure in the penetrated Cenozoic sequence, the well was completed by September. The Kastamonu-1 well is considered to be the first deepwater technical discovery in the Black Sea because it encountered and tested thermogenic gas from Pliocene to Miocene sandstones.

The trap for the *Kastamonu play* (Fig. 9) involves large and elongated shale-cored anticlinal closures that trend WSW to ENE in a fairly large area in the central Western Black Sea between the Central Pontides and Crimea (Fig. 1). Because many lookalike anticlines with large four-way closures (on the order of tens of square kilometres) were mapped on 2D seismic reflection data in the broader area, the Kastamonou well was considered to be a potential play opener with many follow-up prospects (Menlikli *et al.* 2009).

The anticlinal apex of the Kastamonu prospect had prominent crestal normal faults (Fig. 9), some of them being very young, perhaps neotectonic in character, slightly offsetting even the seafloor. The reservoir in this play was assumed to be Pliocene–Miocene deepwater sandstones intercalated in the dominantly shaly strata. The presence of active charge was evident, based on the obvious direct hydrocarbon indicators observed pre-drill on 2D seismic data, such as frequent flags in the footwall of the normal faults and flat spots with seismic phase changes at their ends (Menlikli *et al.* 2009). The brightness of the reflectors suggested the presence of gas in the mapped traps in the pre-drill evaluation. Pelagic shale sequences between the proposed reservoir intervals were assumed to be multiple seal sequences.

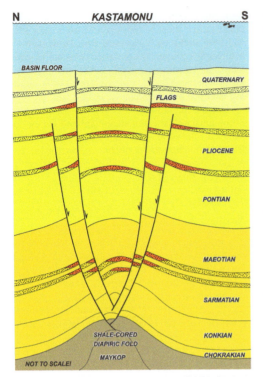

Fig. 9. Pre-drill play concept cartoon of the Kastamonu-1 well drilled in the Turkish sector of the Western Black Sea. For location, see Figure 1.

The Kastamonu-1 well found multiple gas-charged sandstones, but with low gas saturation. The traps may have been partially breached by the late crestal faulting above the anticline (Fig. 9). Ongoing faulting, possibly connected to shale tectonics beneath, could be inferred by the presence of an active mud volcano field (Xing & Spiess 2015) further to the north of the Kastamonu area (Fig. 1). The overpressured Oligocene to Lower Miocene Maykop sequence is interpreted to provide the source for the mud volcanism.

The existence of other anticlines in the adjacent area where the anticlinal closures have not been compromised by neotectonics remains to be seen. The Kastamonu-1 well was drilled based only on 2D seismic data. Future exploration projects targeting other anticlines in the large regional trap fairway between the Central Pontides and Crimea (Fig. 1) will require 3D seismic data to determine the possible extent of trap-breaching in the Kastamonu play (Fig. 9).

Domino-1 (2012)

Operated by ExxonMobil on behalf of the Petrom-ExxonMobil joint venture, the Domino-1 well was the first deepwater exploration well drilled

offshore Romania (Fig. 1). The well was drilled by the same Deepwater Champion drillship that drilled the Sürmene-1 re-entry and Kastamonu-1 wells the year before in deepwater offshore Turkey. The Domino-1 well was spudded in the Neptun Deep exploration block, 170 km offshore in deepwater (930 m), in January 2012.

The *Domino play* concept (Fig. 10) was described by Bega & Ionescu (2009), although the play itself and the name of the flagship Domino prospect was defined during a joint venture project between Petrom and TotalFinaElf using a regional 2D seismic dataset in the early 2000s. The closure for the traps results from a late Neogene inversional event above a basement high. The reservoirs of the Domino play are numerous Miocene to Pliocene deepwater clastic systems deposited on the palaeo-slope (Bega & Ionescu 2009). The source for the biogenic gas in the area is believed to be within Miocene shales as in the nearby small gas fields at the shelf edge (e.g. Duley & Fogg 2009; Pottorf et al. 2010; Olaru et al. in press).

The Domino-1 well was the first deepwater exploration well in the offshore Romanian segment of the western Black Sea (Fig. 1) and it provided the first deepwater economic discovery in the entire Black Sea Basin by encountering 70.7 m of net gas pay. The follow-up wells, operated by ExxonMobil

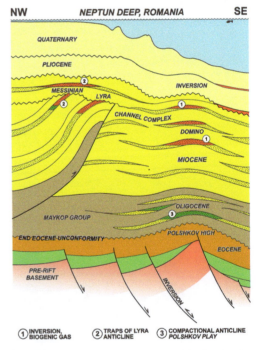

Fig. 10. Pre-drill play concept cartoon of the Domino-1 well drilled in the Romanian sector of the Western Black Sea. For location, see Figure 1.

during the second exploration drilling campaign in the Neptun Deep block, were completed in January 2016. This campaign also included the exploration well Domino-2, which was completed in October 2014, and additional deepwater wells, such as Pelican South, Dolphin, Flamingo, Domino-4 and Califar (Routh *et al.* 2017). The majority of the seven-well campaign encountered gas, and the joint venture of ExxonMobil and Petrom is now continuing its analysis of reservoir data to determine the potential of the block.

The significance of the biogenic gas play in the broader Black Sea was already known as a result of the presence of small producing fields on the Bulgarian (Galata, Kaliakra and Kavarna) and Turkish (Akcakoca and Ayazli) shelf. However, the Domino-1 well was a play opener for biogenic gas exploration in the deepwater Black Sea, and this play is not limited to the Romanian sector.

Şile-1 (2015)

Driven by the subcommercial Istranca gas discovery on the Turkish shelf, indicating a thermogenic kitchen in the southwestern Black Sea, the Şile-1 deepwater well was drilled downdip (Fig. 1) by the joint venture of Shell and Turkish Petroleum using the Noble Globetrotter II drillship. The well was spudded in January 2015 in 2100 m water depth and completed in July. In essence, Şile-1 targeted the previously described *Sürmene play*, i.e. multiple complex four-way closures developed above another large Cretaceous palaeovolcano (cf. Fig. 7) named *Ivanov* by Nikishin *et al.* (2015a, b).

In general, the offshore Cretaceous volcanics may correspond to two distinct reflection seismic sequences (Fig. 2), interpreted as the offshore equivalents of the Turonian Dereköy and the Campanian Cambu Formations (Tari 2015) known from the onshore part of the Pontides (e.g. Keskin & Tüysüz 2017). On a regional scale, the volcanics are part of a Late Cretaceous magmatic arc extending more than 1000 km from the Apuseni Mountains of Romania, through Serbia and Bulgaria to the Black Sea (Gallhofer *et al.* 2015). This arc represents the westernmost segment in the Alpine–Himalayan orogenic system related to the northward subduction of the Neotethys.

The numerous Cretaceous palaeovolcanoes along the Turkish margin of the Black Sea have significantly affected hydrocarbon exploration efforts in the basin because the Cenozoic sequence above them has already been tested in the Sürmene-1 and Şile-1 wells (Fig. 8). Many of these very large cone-shaped palaeovolcanoes have large detachment faults on their flanks compensating for the pronounced differential compaction above them in the locally very thick (6–8 km) post-Late Cretaceous basin fill. These extensional faults, propagating very high into the Oligocene and Miocene sedimentary strata above the volcanoes, can have three-way closure traps in their footwall blocks (Fig. 7). This somewhat unusual speculative trap style remains untested in the deepwater area of the Turkish Black Sea. An even deeper trap type, unlikely to be tested any time soon, is provided by roll-over structures on the volcano flank detachment faults (Fig. 7).

At the time of writing, the results from the Şile-1 well have not been made public.

Polskhov-1 (2016)

The Polskhov-1 well, located *c.* 128 km offshore Bulgaria (Fig. 1) and operated by TOTAL on behalf of joint venture partners OMV and Repsol, was initially expected to spud in 2014, but this was delayed to May 2016. The well was drilled by the Noble Globetrotter II drillship to a target depth of *c.* 5615 m in a water depth of 1911 m, and required four months to complete. In its third-quarter 2016 report, TOTAL announced that the well made an oil discovery, opening up a new play fairway. A second well, Rubin-1, was being drilled in the same general area in late 2017.

Tari *et al.* (2009) published the exploration play concept for the Polshkov High. At least six different deepwater play types (Fig. 11) have been identified near the Polshkov High by these authors. One of these, the Varna Fan play (play 1), was defined by OMV, based on 2D seismic reflection data, in the deepwater exploration of NE Bulgaria in the early 2000s. The trap for this play is provided by the unconformity surface of the Messinian event, which, similar to the Mediterranean Basin, corresponds to a period of subaerial exposure and erosion in the Black Sea Basin (e.g. Gillet *et al.* 2007). Based on a 2D seismic dataset, the clastic source for this potential reservoir was assumed to be derived from the Kamchia Trough to the SW. However, the subsequent 3D survey showed that the Varna Fan was sourced from the shelf, from a NW direction (Tari *et al.* 2009). This finding significantly increased the reservoir quality risk associated with this play.

Further out in the deepwater, the traps of plays 2 and 3 are associated with the compactional anticline above the prominent syn-rift Polshkov High. The four-way closure within the Cenozoic succession is mappable from approximately the top Oligocene downward and becomes significant (i.e. with a vertical closure of more than 100 m) within the lower part of the Paleogene sequence. We designate the name *Polshkov play* to any low-amplitude, four-way closures associated with differential compaction above basement highs in the Black Sea.

The reservoir sequence in play 2 was assumed to be sourced from the Kamchia Trough to the west

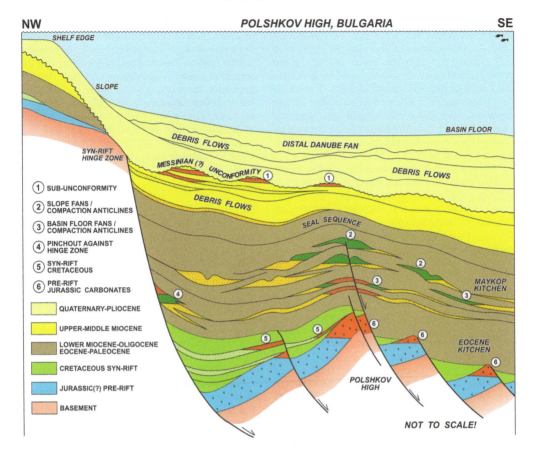

Fig. 11. Pre-drill play concept cartoon of the Polshkov-1 well drilled in the Bulgarian sector of the Western Black Sea. For location, see Figure 1. Adapted from Tari *et al.* (2009).

(Rees *et al.* 2017), forming slope fans with channel–levee units within the Maykop Suite. These slope fans transition to basin-floor fans in the underlying Eocene and Paleocene (play 3) as distal turbidites of the Kamchia foredeep basin in the SW.

It was difficult to determine the exact position of the Western Black Sea break-up unconformity inboard from the Polshkov High on the 2D seismic grid. However, post-rift Cretaceous seismic reflectors onlap against the overall syn-rift high, regardless of their tectonostratigraphic position. The basin floor fans onlapping the northwestern flank of the Polshkov High define play 5. The source rocks for all of these play types were assumed to be the shales of the Maykop Suite, the deepwater equivalent of the Ruslar Formation described from the wells drilled on the Bulgarian shelf (Sachsenhofer *et al.* 2009) or deeper Eocene shales.

The deepest target is the Polshkov High itself, which represents a large, back-rotated syn-rift fault block. In play 6, the target is the Mesozoic pre-rift succession, in relation to the late Cretaceous opening

of the WBSB (Fig. 3), preserved along the SW–NE elongated apex of the structural high. The trap is defined by several fault blocks in which the assumed Jurassic carbonate platform sequence may have a facies (e.g. neritic carbonates with oolites) suitable for gas production. The target of this play is older than the syn-rift ones targeted by the industry over the mid-Black Sea High (e.g. Sinop-1 over the Andrusov Ridge) offshore Turkey or over the Tetyaev High offshore Crimea.

At the time of writing, the results from the Polshkov-1 and the recently drilled Rubin-1 wells have not been made public.

Lira-1X (2015)

The Romanian deepwater blocks Trident and Est Rapsodia were awarded to a joint venture led by Lukoil in 2011 (Picarelli 2017). After a disappointing dry hole (Helena-1X) burdened with several drilling issues on the Est Rapsodia block in 2014/15 (Kasumov 2017), it was relinquished in 2016.

In contrast, after the 2780 m-deep first well on the Trident block in 2015, Daria-1X, the second well, drilled in 700 m of water depth in 2015, resulted in a gas discovery, very near the border of the Neptun Deep block (Fig. 1). This well, Lira-1X, was drilled by the semi-submersible drilling rig TransOcean Development Driller II between August and October 2015. The well was drilled to a TD of 2700 m and found 46 m of gas pay in Miocene to Pliocene reservoirs. The discovery, made by the joint venture of Lukoil, PanAtlantic Petroleum and Romgaz, is reported to be c. 1 Tcf in scale (Picarelli 2017).

The Lira-1X well was drilled on trend with the Califar-1 gas discovery on the Neptun Deep block (Fig. 1). The area of the structural closure for the gas find was reported to be c. 39 km^2. The anticlinal structure providing the trap for the Lira discovery is associated with a gravity-driven system detached within the Maykop succession (Fig. 10), as illustrated by Bega & Ionescu (2009), Schleder et al. (2016) and Munteanu et al. (2011, 2017). This is a version of the Supsa play known from the EBSB.

Overview of tested and speculative deepwater exploration plays in the Black Sea

To provide an inventory and a starting point for play-based exploration efforts in the deepwater Black Sea, we compiled the most important plays in a play table (Table 1). Many of the plays have been previously described herein as part of the historic account of the deepwater wells drilled to date. These include the *Limanköy, Supsa, Shromisubani, Sinop, Lebada, Kastamonu, Sürmene, Domino* and *Polshkov* plays.

However, other speculative plays are included in this article to be tested in the future; these plays, briefly described in the following subsections, include the *Maria, Gudauta, Subbotina, Kuskayasi* and *MTC* plays.

Table 1 includes a listing of 14 deepwater plays in a stratigraphic order, based on the age of the main reservoir target.

General deepwater play elements

A common element among these plays is the assumed effective source rocks that are specific to the Paratethys region. The Paratethys region experienced restricted and episodic connection to the open ocean, primarily by means of the Dnieper–Donets Basin and Pripyat Strait to the North Sea Basin or the Birladsky Strait to the peri-Carpathian Basin (Popov et al. 2004). During times of high global sea-level, this configuration led to the development of a stratified water column and the deposition and preservation of the organic matter (Popov & Stolyarov 1996; Schulz et al. 2005; Sachsenhofer et al. 2009,

2017a, b; Mayer et al. 2017). Therefore, the Oligo-cene to Lower Miocene Maykop Suite is one of the most important source rocks common to almost all of the plays (Table 1). The other source rock sequences, whose importance has just recently been fully appreciated, is the Eocene Kuma Formation (e.g. Beniamovski et al. 2003; Distanova & Bazhenova 2007; Peshkov et al. 2016; Sachsenhofer et al. 2017b; Vincent & Kaye 2017; Mayer et al. in press; Pupp et al. in press).

With regard to the Cenozoic reservoirs of plays 1–10 (Table 1), source-to-sink relationships in the Black Sea within the broader Paratethys region play an important role. As the Paratethys domain developed, so did the hinterland surrounding the WBSB (Jones & Simmons 1997). During the Oligo-cene, the Scythian and Moesian Platforms were exposed to the north and NW, and the Balkanides were a topographic feature to the west. The Pontides were a growing mountain range to the south, as evidenced by a region-wide peneplain (Yilmaz et al. 1997), and several other small highs, such as the Strandja Massif in the SW, were also topographic features. With regard to the WBSB, the shelf and seafloor were dominated by fault structures inherited from the opening of the basin in the Cretaceous (Banks & Robinson 1997; Khriachtchevskaia et al. 2009; Hippolyte et al. 2010; Konerding et al. 2010). Intra-basin features, such as the Kozlu Ridge, Polshkov High, Histria, Kamchia and Burgas basins, all heavily influenced sediment distribution patterns during deposition of the Maykop (Fig. 3). Similarly, in the EBSB, the basin configuration determined by the Shatsky Ridge, Tetyaev High, Sorokin and Tuapse Troughs largely influenced sediment dispersal (Afanasenkov et al. 2007; Khlebnikova et al. 2014; Meisner et al. 2009, 2011; Mityukov et al. 2011; Nikishin et al. 2017). The plays with Mesozoic reservoirs (plays 11 through 14, Table 1), depend on the syn-rift and pre-rift basin configurations in the Black Sea region (Fig. 3).

For much of its history, sedimentation in the deepwater Black Sea was dominated by pelagic shales (Fig. 3). Consequently, there are numerous intra-formational seals for plays 1–12. However, the seals for pre-rift carbonate reservoirs of plays 13 and 14 may have some top seal risks (Table 1). Most of traps in the deepwater Black Sea Basin are structural in character, with the exception of the combination trap of the *Limanköy play* and the stratigraphic traps of the *Kuskayasi* and *Maria plays* (Table 1).

Shatsky Ridge and the Maria play

One of the untested plays that may provide significant reserves in the deepwater is a belt of Jurassic reefs scattered along the crest of the Shatsky Ridge

Table 1. *Play table summary of the most significant tested and speculative plays in the deepwater Black Sea. Most of the plays listed here have a corresponding play-type cartoon. For a detailed explanation, see text*

Play no.	Play name	Source	Reservoir	Seal	Critical Risk	Field, well or analogue
1	Pleistocene–Miocene turbidite sandstones in mass-transport complex-related traps (**MTC play**)	Oligocene–Miocene Maykop Formation organic-rich deep-marine mudstones, biogenic gas	Pleistocene to Miocene turbidite sandstones	Pleistocene Miocene deep-marine mudstones and MTC units	Updip and lateral seal, migration pathway, reservoir-quality (poroperm)	None to date
2	Miocene deep marine sandstones over shale diapiric walls (**Kastamonu play**)	Middle–Late Eocene Kuma Formation deep marine organic-rich shales (oil), Oligocene–Early Miocene Maykop Formation deep marine organic-rich shales (oil), biogenic gas	Miocene–Pliocene deep marine sandstones	Miocene–Pliocene intraformational deep marine shales	Reservoir-quality (poroperm), migration pathway, trap integrity	Kastamonu-1 uncommercial gas discovery
3	Pliocene–Miocene turbidite sandstones in inverted traps (**Domino play**)	Oligocene–Miocene Maykop Formation organic-rich deep-marine mudstones, biogenic gas	Pliocene–Miocene turbidite sandstones	Pliocene–Miocene deep-marine mudstones	Updip and lateral seal, migration pathway, reservoir-quality (poroperm)	Domino and Cobalcescu gas discoveries, Romanian Black Sea
4	Pliocene–Miocene turbidite sandstones in subthrust trap (**Shromisubani play**)	Oligocene–Miocene Maykop Formation organic-rich deep-marine mudstones	Pliocene Miocene deepwater turbidite sandstones	Pliocene deep-lacustrine and deep-marine mudstones	Updip fault closure, Migration pathway, reservoir – quality (poroperm)	Shromisubani oil field, onshore Georgia
5	Pliocene–Miocene turbidite sandstones in combination traps (**Limanköy play**)	Oligocene–Miocene Maykop Formation organic-rich deep-marine mudstones, biogenic gas	Pliocene–Miocene deepwater turbidite sandstones	Pliocene deep-lacustrine and deep-marine mudstones	Updip fault closure, migration pathway, reservoir-quality (poroperm)	Limanköy, non-commercial gas discovery, Turkey; Ana, Doina gas finds in Romania
6	Miocene sandstones in anticlines (**Supsa play**)	Middle–Late Eocene Kuma Formation deep marine organic-rich shales (oil), Oligocene–Early Miocene Maykop Formation deep marine organic-rich shales (oil)	Miocene sandstones	Oligocene–Early Miocene (Maykop) intraformational deep marine shales	Migration pathway, trap integrity, reservoir quality (poroperm)	Supsa oil field, onshore Georgia, HPX-1 oil shows, offshore Turkey; Lira gas discovery, offshore Romania
7	Miocene deep marine sandstones over Cretaceous palaeovolcanoes (**Sürmene play**)	Middle–Late Eocene Kuma Formation deep marine organic-rich shales (oil), Oligocene–Early Miocene Maykop Formation deep marine organic-rich shales (oil), biogenic gas	Miocene–Pliocene deep marine sandstones	Miocene–Pliocene intraformational deep marine shales	Reservoir-quality (poroperm), migration pathway, trap integrity	Sürmene-1 and Sile-1 wells with oil shows, offshore Turkey

#	Play	Source	Reservoir	Seal	Key risks	Discoveries
8	Lower Miocene to Oligocene (Maykop) deep marine sandstones in stratigraphic traps (**Kuskayasi play**)	Middle–Late Eocene Kuma Formation deep marine organic-rich shales (oil), Oligocene–Early Miocene Maykop Formation deep marine organic-rich shales (oil), biogenic gas	Oligocene–Early Miocene (Maykop) deep marine sandstones	Oligocene–Early Miocene (Maykop) intraformational deep marine shales	Trap presence, migration pathway, reservoir quality (poroperm)	None to date
9	Lower Miocene to Oligocene (Maykop) deep marine sandstones in compactional anticlines (**Polshkov play**)	Middle–Late Eocene Kuma Formation deep marine organic-rich shales (oil), Oligocene–Early Miocene Maykop Formation deep marine organic-rich shales (oil), biogenic gas	Oligocene–Early Miocene (Maykop) deep marine sandstones	Oligocene–Early Miocene (Maykop) intraformational deep marine shales	Trap presence, migration pathway, reservoir quality (poroperm)	Polshkov-1 oil discovery, Bulgarian Black Sea
10	Eocene–Early Miocene deep marine sandstones in anticlines (**Subbotina play**)	Middle–Late Eocene Kuma Formation deep marine organic-rich shales (oil), Oligocene–Early Miocene Maykop Formation deep marine organic-rich shales (oil), biogenic gas	Eocene–Early Miocene deep marine sandstones	Eocene–Early Miocene intraformational deep marine shales	Migration pathway, trap integrity	Subbotina oil discovery offshore Kerch Peninsula, Samotino. More condensate discovery, offshore Bulgaria
11	Late Cretaceous shallow-marine platformal carbonates in footwall blocks (**Sinop play**)	Early–Middle Jurassic Etropole Formation organic-rich shallow-marine mudstones and marlstones	Late Cretaceous shallow-marine platformal carbonates	Late Cretaceous intraformational shallow-marine carbonates and mudstones, Paleocene–Eocene deep-marine mudstones	Reservoir-quality (poroperm), lateral seal – fault plane, migration pathway. Distinguishing v. volcanics on seismic	None to date
12	Early Cretaceous shallow-marine and fluviatile sandstones in footwall blocks (**Lebada play**)	Early–Middle Jurassic Etropole Formation organic-rich shallow-marine mudstones and marlstones	Early Cretaceous shallow-marine and fluviatile sandstones	Early Cretaceous shallow-marine mudstones	Lateral seal – fault plane, migration pathway, reservoir quality (poroperm)	Lebada Field, Romanian Black Sea
13	Valangian shallow-marine platformal carbonates in inverted fault block traps (**Gudauta play**)	Early–Middle Jurassic Etropole Formation organic-rich shallow-marine mudstones and marlstones	Valangian shallow-marine platformal carbonates	Valangian intraformational shallow-marine carbonates and mudstones	Reservoir quality (poroperm), top seal presence, migration pathway	Tjulenovo Field, onshore Bulgaria
14	Middle–Late Jurassic carbonate build-ups (**Maria play**)	Early Jurassic coals, Late Jurassic deep-marine organic-rich mudstones, Oligocene–Early Miocene Maykop Formation deep-marine organic-rich mudstones (oil and gas)	Middle–Late Jurassic carbonate build-ups	Aptian–Albian deep-marine mudstones, Late Cretaceous deep-marine carbonates, Oligocene–Early Miocene Maykop Formation deep-marine mudstones	Migration pathway, reservoir presence (volcanics?) and quality (poroperm), top seal presence	Being tested offshore Russia at the time of writing

in the EBSB (e.g. Afanasenkov *et al.* 2005, 2007; Meisner *et al.* 2009). Numerous possible isolated reef complexes have been identified by Russian 2D seismic surveys (Fig. 12). No offshore well data exists about the type or depositional environment of these reefs, although it is generally assumed that they were deposited in deeper-water settings (Afanasenkov *et al.* 2005; Guo *et al.* 2011). If so, siliceous sponges and microbialites probably formed major contributors to reef growth (Guo *et al.* 2011). Alternatively, if the reefs developed in shallower water, higher diversity coral-dominated reef lithologies may be present, possibly with better reservoir qualities. Such shallow water reefs may have also been periodically exposed by relative sea-level falls, leading to porosity and permeability development associated with karstification and meteoric diagenesis. In particular, exposure may have occurred associated with Valanaginian–Hauterivian uplift, ending widespread carbonate deposition (Fig. 3).

The regional play-type cartoon (Fig. 12) summarizing the main deepwater play types in the Russian sector of the Eastern Black Sea depicts this play as the primary target in that part of the basin (play 1). Given the larger number and fairly large size of these reefs on 2D and 3D seismic data (Afanasenkov

et al. 2005, 2007) mapped, this play type provides repeatability and therefore potentially very large resources. Oil charge is assumed to be from either the Maykop Suite or the slightly older Kuma sequence connected via faults to the reefs (Fig. 12). Reservoir parameters are difficult to predict for this carbonate reservoir, and the producability would strongly depend on the lithofacies. The seal for most of these reefs is provided by pelagic Upper Cretaceous shales of the early post-rift sequence (assuming a Cretaceous opening of the EBSB, Fig. 3).

Perhaps one of the most convincing examples of these Jurassic reefs is the Maria prospect, shown by Afanasenkov *et al.* (2005, 2007); therefore we refer to this untested play as the *Maria play*. The critical risk associated with this play is similar to that of the Yassihöyük prospect (Fig. 7), i.e. whether carbonate build-ups are really imaged on the seismic data or they are palaeovolcanoes (cf. Posamentier *et al.* 2014).

Another play type (play 4) in the area corresponds to the syn-rift Cretaceous sequence that may include inverted structural traps formed during the Eocene when regional-scale compression affected the entire Black Sea (e.g. Robinson 1997; Stovba *et al.* 2009). Many of the syn-rift normal

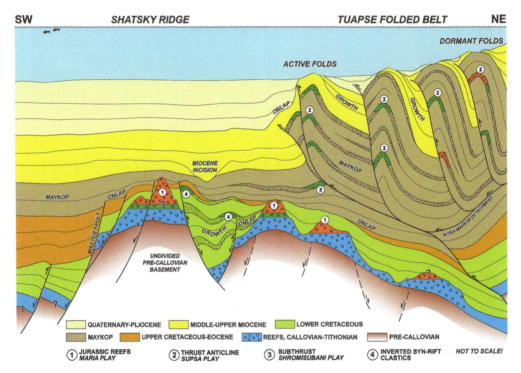

Fig. 12. Cartoon summary of the untested play types along the Shatsky Ridge in the Russian sector of the Eastern Black Sea, including the *Maria play*, compiled after Afanasenkov *et al.* (2005, 2007) and Nikishin *et al.* (2017). For an approximate location, see Figure 16.

faults were inverted during this time, as especially well documented along the basin margins (Yilmaz et al. 1997; Hippolyte et al. 2016; Okay et al. 2006, 2017). This untested play type is interpreted to be analogous to the *Lebada play* on the Romanian shelf (e.g. Ionescu et al. 2002).

Inversion of marginal Ridges and the Gudauta play

The Gudauta High in the Abkhazian segment of the EBSB (Fig. 1) has been known for many years (e.g. Tugolesov et al. 1985). As an exploration target located mostly on the shelf, it has already been highlighted by Afanasenkov et al. (2007) and Nikishin et al. (2017). The large four-way structural closure (on the order of 400–500 km^2) at the level of the top Neocomian carbonate platform results from Neogene inversion on the northeastern end of the pre-existing Ordu–Pitsunda regional transform fault (Nikishin et al. 2017). Because similar large transform faults exist, segmenting the southeastern part of the EBSB region (Tari et al. 2018), the deepwater segments of the corresponding marginal ridges (e.g. Trabzon Ridge and Ochamchira High, in Turkey and Georgia, respectively) may provide future exploration targets.

Thrust-fold belts around the Black Sea and the Subbotina play

Several thrust-fold belt segments exist in the Black Sea, including the Balkans (e.g. Stuart et al. 2011) in the Bulgarian offshore, the Sudak belt offshore Crimea (e.g. Sheremet et al. 2016), the Great Caucasus in the Russian offshore (e.g. Nikishin et al. 2017), the Adjara–Trialet zone (or Gurian 'Trough') in the Georgian offshore (e.g. Banks et al. 1997; Robinson et al. 1997) and the offshore Akcakoca segment of the Pontides in the Turkish offshore (Robinson et al. 1996). Among these thrust-fold belts, in their deepwater segments, only the Gurian folded belt was tested by the HPX-1 well in Turkey. The reasons for the failure of that well to become a play opener were previously discussed herein. Because the other thrust fold belts remain untested and have a different growth history, and because the potential resevoir intervals had different provenance, as compared with the HPX-1 well, they are described individually in the following paragraphs.

The SE-vergent folded belt offshore southeastern Crimea and the Kerch Peninsula (informally called the Sudak folded belt by Tari (2010)) trends ENE to WSW; it extends onto the present-day shelf (Finetti et al. 1988; Sheremet et al. 2016), bounding the Sorokin foredeep basin from the north (Fig. 1). The Subbotina oil discovery (Stovba et al. 2009; Gozhik et al. 2010; Kitchka et al. 2012) was made in 2005 on one of the anticlines of the Sudak folded belt situated on the shelf (Fig. 1); it is considered as the shallow-water example of the *Subbotina play* concept (Fig. 13, play 1). The geometry of individual thrust imbricates shows a south-vergent, thin-skinned folded belt involving mostly Cenozoic strata, but probably also includes some of the Upper Cretaceous sequence (Tari 2010). The timing of the onset of the folded belt is difficult to estimate, but it is interpreted to coincide with the Eocene, then Oligocene, initiation of uplift of the Crimean Highlands (Panek et al. 2009; Nikishin et al. 2017). The folded belt shows very recent activity in the western part of the area, where the geometry of the seafloor is strongly influenced by the growing anticlines. However, in the eastern part, the offshore Kerch peninsula, Miocene to Quaternary sediments of the Palaeo-Don Fan appear to seal the dormant leading edge of the folded belt (Fig. 13).

Many elongated anticlines (play 1, Fig. 13) within the folded belt were mapped on vintage 2D seismic data with typical map-view dimensions of 4 × 12 km. The average structural closure is c. 20–40 km^2 (Stovba et al. 2009; Gozhik et al. 2010).

The reservoir sequences in the Sudak anticlines are expected to be primarily Maykop age turbidites, in some cases truncated at the base of the Pliocene (play 2, Fig. 13). The basin floor sheet sand reservoir facies also should extend into the deeper part of the folded belt strata. An additional play type associated with the Sudak folded belt is the localized clastic fan systems at the leading edge of the folded belt terminating against the thrust front (play 3, Fig. 13).

The overall Sudak folded belt shows a large sedimentary wedge where imbrications resulted in the overthickening of the sedimentary cover above the potential source rock intervals during the Miocene interval (Fig. 13). The extra overburden triggered an active petroleum system, as evidenced by numerous mud volcanoes in the Sorokin Trough (Fig. 1) on the present-day seafloor (Krastel et al. 2003).

The *Subbotina play* can be followed to the Tuapse Basin of the EBSB (Fig. 1). The Tuapse Trough to the SE of the Shatsky Ridge is largely occupied by a thin-skinned folded belt (Fig. 12) that corresponds to the leading edge of the Great Caucasus propagating into the Eastern Black Sea (Fig. 1). The bulk of the pre-kinematic strata in the SE-vergent anticlines consists of the Maykop Suite (e.g. Nikishin et al. 2010, 2017). Although no offshore wells have been drilled so far to provide direct calibration, the syn-kinematic sequence appears to be Middle Miocene to Pliocene (Meisner et al. 2009; Mityukov et al. 2011). The anticlinal crests closer to the leading edge of the folded belt have a bathymetric expression on the modern seafloor (Fig. 12), and are therefore neotectonic in nature (Almendinger et al. 2011).

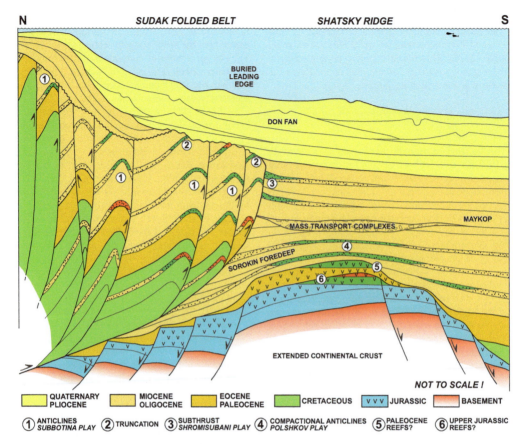

Fig. 13. The untested deepwater play types in the area south of Crimea and the Kerch Peninsula, adapted from Tari (2010). For an approximate location, see Figure 16.

Several play types are associated with the Tuapse folded belt (Fig. 12). All of these assume Maykop deepwater sand reservoirs receiving charge from either the Maykop succession itself or from the Kuma sequence beneath. The simplest traps are located within the anticlinal crests with multiple intra-Maykop turbidite targets (play 2 in Fig. 12), whereas more challenging and deeper traps may be targeted in subthrust settings (play 3 in Fig. 12). Given the large number of anticlinal traps in the Tuapse Trough, a deepwater play opener discovery would have a major effect in the so-far undrilled Russian sector. The critical risk for many of the anticlinal targets may be trap timing, given their young age and possible ongoing deformation (Almendinger *et al.* 2011).

The Middle Eocene thrust-fold belt of the Pontides extends from the onshore to the offshore (Yilmaz *et al.* 1997; Sunal & Tüysüz 2002; Şen 2013; Espurt *et al.* 2014; Hippolyte *et al.* 2016; Okay *et al.* 2017). The offshore anticlines of this folded belt in the Eregli embayment were targeted by several wells on the shelf and led to the Akcakoca and Ayazli biogenic gas discoveries (Robinson *et al.* 1996; Derman 2014), which became producing fields. The main reservoir is the Middle Eocene Kusuri Formation (Fig. 3). The extent of these deepwater sands is well-documented onshore (Janbu *et al.* 2007), but beyond the shelf edge in the present deepwater basin, it is not fully understood.

The large east–west-trending Balkans thrust fold belt onshore Bulgaria (e.g. Vangelov *et al.* 2013) continues into the offshore with a gradual rotation of the structural grain to NW–SE, (e.g. Doglioni *et al.* 1996; Sinclair *et al.* 1997). The existing wells drilled on the Bulgarian shelf in this trend have provided encouragement for exploration with an uneconomic condensate discovery in Eocene deepwater sands on one of the Balkan anticlines (i.e. Samotino More: Benton 1997; Georgiev 2012). The offshore segment of the Balkans folded belt shows multiple stages of growth, not only during the Paleocene and Eocene, but also during the Miocene (Stuart *et al.* 2011). Shell recently acquired a large, *c.* 4500 km^2, 3D seismic reflection survey in their Khan Kubrit (formerly Silistar) block that covers most of the Bulgarian part of the folded belt; the company plans to drill a deepwater well in 2019.

Paleocene–Eocene reefs in the deepwater?

Given the very few deepwater wells drilled in the Black Sea to date, ongoing exploration efforts frequently use reservoir analogues from the well-known coastal and shelf areas of the Black Sea. In particular, the well-documented cases of Eocene neritic limestone units with potentially good reservoir characteristics along the coast of the Black Sea, for example, at Karaburun, NW Turkey (e.g. Less *et al.* 2011), or at Belogorsk, Crimea (e.g. Kopaevich *et al.* 2008; Nikishin *et al.* 2017), highlight the possibility of targeting synchronous and similar rocks in offshore exploration. The age of the shallow water carbonate units ranges from Middle to Late Eocene, and they have locally good reservoir characteristics (e.g. porosity up to 25%).

Some documented cases of offshore Eocene carbonate reservoirs exist on the present-day shelf of the Black Sea, for example, in the Romanian sector (Tambrea *et al.* 2002). However, even within the palaeoshelf, a clear map-view separation typically exists between potential reservoir facies (mixed carbonate and siliciclastic facies with Nummulites) and non-reservoir marls. In the Bulgarian part of the Black Sea, the potential Eocene reservoir rock facies also transitions outboard to non-reservoir units with increasingly higher marl content (Harbury & Cohen 1997).

A question then arises of whether similar shallow-water Eocene carbonate reservoirs can be expected in the present-day deepwater of the Black Sea (e.g. Fig. 13). The Tetyaev High could include a sequence of Paleocene to Eocene (?) carbonates (play 5, Fig. 13), similar to those cropping out in Crimea (Kopaevich *et al.* 2008). However, this play assumes that the EBSB opened in the Eocene; otherwise, neritic carbonates cannot be considered as reservoirs in a deepwater setting. If the EBSB opened during the Mesozoic, then only the *Maria play* of Upper Jurassic reefs (play 6, Fig. 13) would be valid on the Tetyaev High.

At least one deepwater well (Sinop-1) tested the Eocene sequence in a palaeodeepwater setting, drilling on the Andrusov High (Fig. 1). The Eocene strata turned out to be a deepwater, pelagic non-reservoir marl sequence deposited on the palaeobasin floor (Aydemir & Demirer 2013). If the Black Sea opened up during the Mesozoic, then the Eocene sequence is part of the already deepwater post-rift basin fill (Fig. 3). Therefore, the absence of shallow-water, reservoir facies Eocene carbonate units (either reefal build-ups or nummulitic banks) should not be surprising in a drilling location anywhere basinward of the Eocene shelf break. The Eocene neritic carbonate play over the prominent intra-basin structural highs of the Black Sea Basin, such as the Polshkov, Andrusov, Arkhangelsky, Tetyaev (Fig. 13) and Shatsky Highs, should be considered very risky, at best.

Compactional anticlines of the Polshkov play over structural highs

Intra-Maykop deepwater basin floor fans could form low-amplitude, four-way closures above the prominent syn-rift structural highs in the Black Sea. In the case of the Tetyaev High (Fig. 13, play 4), an analogue is the *Polshkov play* as play 2 over the Polshkov High (Fig. 12). The deeper part of the Cenozoic post-rift sedimentary fill over the Tetyaev High has a subtle closure within just a few hundred metres of the apex of the syn-rift structure (Fig. 13). The area of the intra-Cenozoic closure is *c.* 100 km^2 with a vertical amplitude of only *c.* 50–100 m as the result of differential compaction over the Tetyaev structures (Tari 2010). The generic *Polshkov play* may be considered above other intra-basin highs of the Black Sea Basin, such as the Andrusov, Arkhangelsky and Shatsky highs; however, the presence and quality of Oligocene to Miocene reservoirs over these highs becomes an important risk element.

Kuskayasi stratigraphic trap play

With the deepwater Black Sea Basin becoming an emerging focus for exploration, stratigraphic traps merit consideration alongside the structural and combination traps previously discussed. The first of these was introduced by Menlikli *et al.* (2009) as the Kuskayasi exploration target in the Turkish sector, north of Zonguldak. Sipahioğlu *et al.* (2013a) described this Upper Oligocene play as an amalgamated set of basin floor fans at the eastern end of an axial, east–west-trending intra-Maykop deepwater system outboard of the hinge zone (Fig. 14). Based on a 3D seismic reflection survey, the multiple terminal fan sand packages pinch out abruptly downdip and updip within Maykop shales. Because there is no structural closure on any of the fan segments, the play-type cartoon (Fig. 14) depicts a stratigraphic trap with a critically important updip lateral seal. Lateral and vertical hydrocarbon charge is envisioned to occur from the deepwater Maykop shales located basinward and beneath. 1D synthetic seismic modelling of selected parts of the Kuskayasi prospect showed that the brightness of the amalgamated fan system seen on the 3D seismic data is associated with hydrocarbon (gas) presence, rather than lithology (Aktepe *et al.* 2013). This interpretation is also supported by the unusually slow seismic

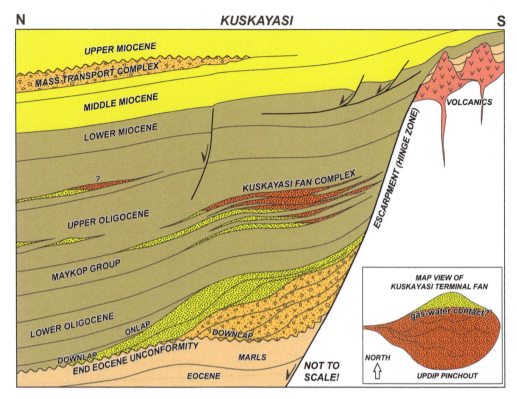

Fig. 14. The untested Kuskayasi stratigraphic play, adapted from Menlikli *et al.* (2009), Sipahioğlu *et al.* (2013*a*) and Aktepe *et al.* (2013). For an approximate location, see Figure 16.

velocities observed within the target fan complex and by the conformance of amplitudes to a structural level.

Larger packages of Lower Maykop coarse silici-clastics form a large wedge against the hinge zone, possibly related to the end-Eocene drawdown episode in the Black Sea (Tari *et al.* 2013). This clastic wedge cannot be considered as an exploration target because the sands and conglomerates are in contact with the large normal fault responsible for the hinge zone (Fig. 14).

Mass-transport complex plays

The presence of debris flow units or MTCs in the basin fill of the Black Sea has only been recognized during the last decade (e.g. Tari *et al.* 2009; Dondurur *et al.* 2013; Sipahioğlu *et al.* 2013*b*). In the DSDP 380A well, drilled in the WBSB in 1975 (Fig. 1), a Miocene 'pebbly breccia' was described containing largely angular clasts, up to cobble size; it was interpreted as fluvial sediments corresponding to a major sea-level drop and subaerial exposure (Hsü & Giovanoli 1979). With the benefit of modern 2D seismic reflection data across the well location, it

is clear now that the well penetrated the updip segments of at least two large MTCs (Tari *et al.* 2015*a*, *b*, 2016). The geometry of the MTCs defined by the seismic data is consistent with gravity slides on the palaeoslope. Sipahioğlu & Bati (2017) provided additional 3D reflection seismic evidence for the existence of MTCs in the Turkish sector of the basin. Kitchka *et al.* (2016*a*, *b*) described roughly age-equivalent Pliocene to Miocene examples of MTCs from the conjugate Ukrainian margin of the Black Sea (Fig. 3).

The recognition of large mass transport complexes in this basin has very important consequences for ongoing exploration efforts (Fig. 15). Whereas the lithological composition of MTCs typically translates to a non-reservoir facies, they could act both as lateral and top seals for underlying hydrocarbon traps (e.g. Gong *et al.* 2014). Based on our seismic mapping in the WBS, we see two potential play types associated with MTCs (Fig. 15). At the leading edge of major MTCs, a singular thrust fault can be frequently seen ramping up on pre-existing strata. Therefore, play 1 is a subthrust play corresponding to this frontal thrust of an MTC. Other possibilities are the low-amplitude four-way closures developed

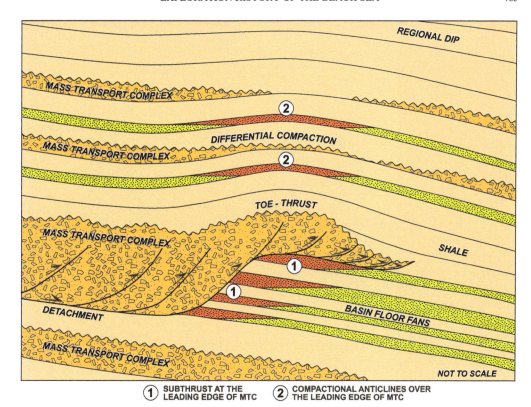

Fig. 15. Speculative MTC-related plays. Because these untested plays could be present in many parts of the deepwater Black Sea, no particular location for this cartoon is indicated on Figure 16.

above the toe-thrust anticlines as a result of differential compaction (play 2, Fig. 15). None of these plays have been targeted so far in the Black Sea.

Biogenic gas and gas hydrates

The Miocene to Pliocene biogenic gas fairway has been delimited primarily by numerous gas flares in the water column, shallow gas accumulations, and numerous bottom-simulating reflectors in Bulgaria, Romania and Ukraine (e.g. Naudts *et al.* 2009; Starostenko *et al.* 2010). Many gas discoveries on the Romanian shelf have a 99% methane composition, attesting to a biogenic origin (Bega & Ionescu 2009; Duley & Fogg 2009). Other shallow gas- and methane-venting features were documented in the Russian, Georgian and Turkish segments of the Black Sea by Mazzini *et al.* (2008), Dembicki (2014) and Ergün *et al.* (2002), respectively.

Although shallow gas and gas hydrates are currently considered to be a drilling hazard for deepwater exploration drilling in the Black Sea, a technological breakthrough of methane production from gas hydrate reservoirs would translate to very large resources in several segments of the basin.

The most prospective zone seems to be the Danube Cone in the NW Black Sea, where gas hydrate reservoir targets can be found between 60 and 100 m below the seafloor in water depths of *c.* 1500 m (Zander *et al.* 2017*a*, *b*).

A summary of exploration play concepts in the deepwater Black Sea

We have summarized the most important 14 deepwater play types in a play table (Table 1), focusing on the petroleum system elements and the critical risk factors. We also tabulated the existing fields, undeveloped discoveries or any analogues for the particular play types. A decidedly simplified map view of the possible extent of these 14 plays across the Black Sea is also provided (Fig. 16a, b).

To date, at least two proven petroleum plays exist in the deepwater Black Sea, i.e. the *Domino* and the *Polshkov plays* (Figs 10 & 11). There are underexplored play types, such as the *Sürmene* and *Kastamonu plays* (Figs 8 & 9), where additional drilling may prove the validity of the play after the first technical, but uneconomic, discoveries. There are also

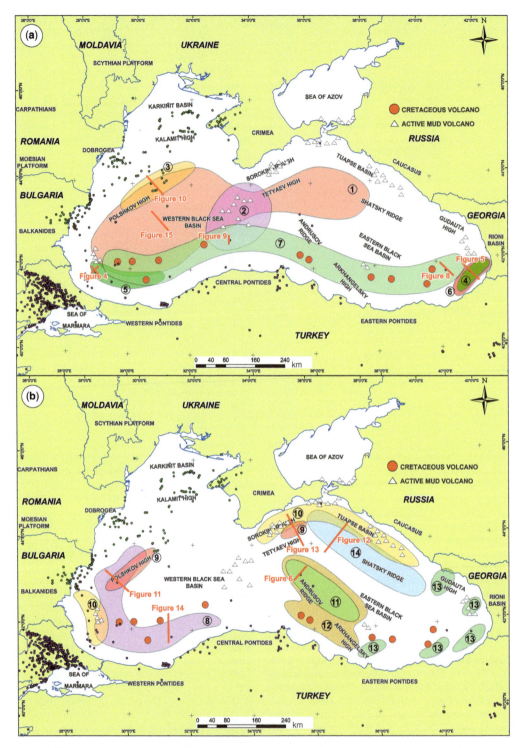

Fig. 16. Simplified map view of the possible extent of the tested and speculative deepwater plays described. The numbers correspond to the plays listed in Table 1.

deepwater plays, such as the *Sinop play* (Fig. 6), where the predicted carbonate reservoirs may be present in other segments of the basin. Furthermore, several untested, speculative deepwater plays exist in the Black Sea (*MTC, Kuskayasi, Gudauta and Maria plays*) that will require exploration drilling and testing to either prove or discard.

The successful *Domino play*, with its biogenic gas, will be developed in the Romanian sector. The Black Sea appears to be at the same exploration stage as the Eastern Mediterranean; that is, after finding biogenic gas accumulations, the thermogenic petroleum system(s) must be proven by a new set of exploration wells in the deepwater.

The prevailing industry perception of an overall lack of reservoir influx into the basin during the post-rift phase, except for the palaeo-Danube (e.g. Maynard *et al.* 2012), must be re-evaluated because some of the recent wells yielded either oil discoveries (*Polshkov play*) or oil shows (*Sürmene play*) in Miocene to Oligocene deepwater sand systems in certain segments of the basin where relatively short-lived sedimentary entry points existed on the basin margin (Fig. 3). Whether or not these oil discoveries will be appraised and developed remains to be seen. Alongside a better understanding of sediment flux into the basin, the provenance and hence quality of this sediment remain a critical concern (Simmons *et al.* 2018).

In general, most of the deepwater exploration failures in the Black Sea to date are directly related to the lack of reservoir at the targeted stratigraphic levels. A few more deepwater wells will be drilled in 2018/19, hopefully, with oil and/or thermogenic gas discoveries to further emphasize the emerging nature of the Black Sea. The way forward is clearly the careful evaluation of failures and successes in other parts of the Black Sea to plan future hydrocarbon exploration wells in this large, underexplored frontier basin system.

Summary and conclusions

Deepwater hydrocarbon exploration drilling began in the Black Sea less than 20 years ago, primarily because of the economical/technological challenges associated with mobilizing suitable drilling rigs through the Bosporus. To date, however, *c.* 20 deepwater wells have now been drilled, targeting a large variety of plays in this very large underexplored basin complex.

The numerous deepwater Black Sea play types can be subdivided into pre-, syn- and post-rift/sag plays. The sag/post-rift play types have proven to be more successful, finding either biogenic gas in Miocene to Pliocene reservoirs associated with the Palaeo-Danube and Palaeo-Dniper/Dnestr, or oil in

Oligocene deepwater siliciclastic systems. Syn-rift or early post-rift plays, in contrast, assumed mostly shallow water carbonate reservoir targets. Just one well has targeted the pre-rift stratigraphy.

The largest targets are syn-rift fault blocks, such as the Andrusov and Tetyaev highs in the centre of the Black Sea. Although their internal stratigraphy is still very poorly constrained (i.e. proportion of pre-rift and syn-rift v. pre-rift basement), this translates to not only reservoir presence risk, but also to reservoir quality risk. The trap sizes are very large. In addition, the assumed lateral charge from the Miocene–Oligocene Maykop Formation and from the Middle Eocene Kuma Formation makes these structures attractive.

The overall structure of the Shatsky Ridge in the Russian sector of the basin is not as clear because it includes elements of a large Jurassic carbonate platform on top. The Polshkov High is unique in the sense that it represents a large rotated syn-rift fault block along the lower plate edge of the Western Black Sea in Bulgaria. On the conjugate upper plate margin, very large inverted syn-rift structures, such as the Kozlu Anticline, are recognized in the Turkish sector. On top of some of these syn-rift highs, various carbonate geometries were interpreted on seismic data, which turned out to be associated with Cretaceous volcanism.

In the post-rift and sag basin fill, several intra-Cenozoic reservoirs are being targeted in the compactional anticlines above the large syn-rift highs. Another play associated with deepwater sands of the Maykop sequence is the deepwater extension of the Subbotina discovery offshore Crimea. The Subbotina structure is just one compressional anticline among many others situated in a dominantly Miocene, south-vergent folded belt offshore Kerch Peninsula. Similar thrust-fold belts are also known in the Russian, Georgian, Turkish and Bulgarian sectors of the Black Sea. However, reservoir presence and quality will remain a definite risk for the Cenozoic reservoir intervals in certain segments of the Black Sea as the function of intermittent sedimentary entry points and the corresponding provenance areas.

This work has benefited from conversations with many colleagues working on the geology of the Black Sea, some of whom are listed here alphabetically: Zühtü Bati, Jan Baur, Zamir Bega, Steve Bottomley, David Boote, Francois Courteix, Tom Cousins, Ali Demirer, Sami Derman, Thomas Eder, Kia Fallah, Jonathan Floodpage, Georgi Georgiev, Gary Ingram, Gabriel Ionescu, Sasha Kitchka, Emil Kozhuharov, Csaba Krezsek, Oxana Khriachtchevskaia, Andrew Lavender, Damian Leslie, Hugo Matias, Cem Menlikli, Ioan Munteanu, Anatoly Nikishin, Aral Okay, Radu Olaru, Igor Popadyuk, Thomas Rainer, Emily Rees, Dave Roberts, Andrew Robinson, Reinhard Sachsenhofer, Zsolt Schleder, Özgür Sipahioğlu, Cameron Sheya, Gerald Stern, Sergy Stovba, Giorgi Tatishvili, Cristina Ungureanu,

Davit Vakhania, Dian Vangelov and Steve Vincent. This list is clearly not a comprehensive one, given the much larger number of geoscientists with whom we have interacted regarding the Black Sea during the last two decades. All of the plays described in this paper reflect our own interpretative ideas, admittedly with some personal bias included; therefore, they remain solely the responsibity of the authors. Randell Stephenson and Özgür Sipahioğlu provided helpful and constructive reviews of the draft manuscript. Peter Pernegr is thanked for drafting most of the figures.

References

ADAMIA, S., ALANIA, V., CHABUKIANI, A., CHICHUA, G., ENUKIDZE, O. & SADRADZE, N. 2010. Evolution of the Late Cenozoic basins of Georgia (SW Caucasus): a review. *In*: SOSSON, M., KAYMAKCI, N., STEPHENSON, R., BERGERAT, F. & STAROSTENKO, V. (eds) *Sedimentary basin tectonics from the Black Sea and Caucasus to the Arabian Platform*. Geological Society of London, Special Publications, **340**, 239–259.

ADAMIA, S.A., CHKHOTUA, T.G. ET AL. 2017. Tectonic setting of Georgia–eastern Black Sea: a review. *In*: SOSSON, M., STEPHENSON, R.A. & ADAMIA, S.A. (eds) *Tectonic Evolution of the Eastern Black Sea and Caucasus*. Geological Society, London, Special Publications, **428**, 11–40, https://doi.org/10.1144/SP428.6

AFANASENKOV, A.P., NIKISHIN, A.M. & OBUKHOV, A.N. 2005. The system of Late Jurassic carbonate buildups in the northern Shatsky swell (Black Sea). *Doklady Earth Sciences*, **403**, 696–699.

AFANASENKOV, A.P., NIKISHIN, A.M. & OBUKHOV, A.N. 2007. *Geology of the Eastern Black Sea*. Scientific World, Moscow. [In Russian, with English summary].

AKDOĞAN, R., OKAY, A.I., SUNAL, G., TARI, G., MEINHOLD, G. & KYLANDER-CLARK, A.R. 2017. Provenance of a large Lower Cretaceous turbidite submarine fan complex on the active Laurasian margin: Central Pontides, northern Turkey. *Journal of Asian Earth Sciences*, **134**, 309–329.

AKDOĞAN, R., OKAY, A.I. & DUNKL, I. 2018. Triassic–Jurassic arc magmatism in the Pontides as revealed by the U-Pb detrital zircon ages in the Jurassic sandstones of northeastern Turkey. *Turkish Journal of Earth Sciences*, **27**, 87–109.

AKTEPE, S., KORUCU, Ö., SIPAHIOĞLU, Ö. & BENGÜ, E. 2013. How much can we learn from 1D seismic forward model? An Example of Turkish Western Black Sea. *In*: *19th International Petroleum and Natural Gas Congress and Exhibition of Turkey*, Ankara, Abstracts.

ALMENDINGER, O.A., MITYUKOV, A.V., MYASOEDOV, N.K. & NIKISHIN, A.M. 2011. Modern erosion and sedimentation processes in the deep-water part of the Tuapse Trough based on the data of 3D seismic survey. *Doklady Earth Sciences*, **439**, 76–78.

ATALAY, R., PAMIR, S.K. & ÖZCAN, B. 2012. Deepwater well drilled by two MODU in Black Sea, Turkey. *Proceedings of the Twenty-second International Offshore and Polar Engineering Conference, Rhodes, Greece, June 17–22, 2012*, International Society of Offshore and Polar Engineering, 1107–1112.

AVRAM, E., DRĂGĂNESCU, A., SZASZ, L. & NEAGU, Th. 1988. Stratigraphy of the outcropping Cretaceous deposits in the Southern Dobrogea (SE Romania). *Mémoires, Institut de Géologie et de Géophysique*, **33**, 5–43.

AVRAM, E., COSTEA, I., DRAGASTAN, O., MUTIU, R., NEAGU, Th., ŞINDILAR, V. & VINOGRADOV, C. 1996. Distribution of the Middle-Upper Jurassic and Cretaceous facies in the Romanian eastern part of the Moesian Platform. *Revue Roumaine de Géologie*, **39–40**, 3–33.

AYDEMIR, V. & DEMIRER, A. 2013. Upper Cretaceous and Paleocene Shallow Water Carbonates along the Pontide Belt. *19th International Petroleum and Natural Gas Congress and Exhibition of Turkey*, Ankara, abstract book, 284–290.

BÁLDI, T. 1980. The early history of the Paratethys. *Bulletin of the Hungarian Geological Society*, **110**, 456–472.

BANKS, C.J. & ROBINSON, A.G. 1997. Mesozoic strike-slip back-arc basins of the Western Black Sea region. *In*: ROBINSON, A.G. (ed.) *Regional and Petroleum Geology of the Black Sea and Surrounding Region*. AAPG, Memoirs, Tulsa, **68**, 53–62.

BANKS, C.J., ROBINSON, A.G. & WILLIAMS, M.P. 1997. Structure and regional tectonics of the Achara–Trialet fold belt and the adjacent Rioni and Kartli foreland basins, Republic of Georgia. *In*: ROBINSON, A.G. (ed.) *Regional and Petroleum Geology of the Black Sea and Surrounding Region*. AAPG, Memoirs, Tulsa, **68**, 331–345.

BEGA, Z. & IONESCU, G. 2009. Neogene structural styles of the NW Black Sea region, offshore Romania. *The Leading Edge*, **28**, 1082–1089.

BELOUSOV, V.V., VOLVOVSKY, B.S. ET AL. 1988. Structure and evolution of the Earth's crust and upper mantle of the Black Sea. *Bollettino Geofisica Teorica ed Applicata*, **30**, 109–196.

BENIAMOVSKI, V.N., ALEKSEEV, A.S., OVECHKINA, M.N. & OBERHÄNSLI, H. 2003. Middle to upper Eocene dysoxic-anoxic Kuma Formation (northeast Peri-Tethys): biostratigraphy and paleoenvironments. *In*: WING, S.L., GINGERICH, P.D., SCHMITZ, B. & THOMAS, E. (eds) *Causes and Consequences of Globally Warm Climates in the Early Paleogene*. Geological Society of America, Special Papers, Boulder, Colorado, **369**, 95–112.

BENTON, J. 1997. Exploration history of the Black Sea Province. *In*: ROBINSON, A.G. (ed.) *Regional and Petroleum Geology of the Black Sea and Surrounding Region*. AAPG, Memoirs, Tulsa, **68**, 7–18.

BERGERAT, F., VANGELOV, D. & DIMOV, D. 2010. Brittle deformation, palaeostress field reconstruction and tectonic evolution of the Eastern Balkanides (Bulgaria) during Mesozoic and Cenozoic times. *In*: SOSSON, M., KAYMAKCI, N., STEPHENSON, R.A., BERGERAT, F. & STAROSTENKO, V. (eds) *Sedimentary Basin Tectonics from the Black Sea and Caucasus to the Arabian Platform*. Geological Society, London, Special Publications, **340**, 77–111, https://doi.org/10.1144/SP340.6

BLOOMBERG 2017. blog: https://www.bloomberg.com/news/articles/2017-12-19/sanctions-proof-oil-rig-thwarts-u-s-policy-from-cuba-to-russia

BOOTE, D. 2017. The geological history of the Istria 'Depression', Romanian Black Sea shelf: tectonic controls on second-/third-order sequence architecture. *In*: SIMMONS, M.D., TARI, G.C. & OKAY, A.I. (eds) *Petroleum Geology of the Black Sea*. Geological Society, London, Special Publications, **464**. First published online October 9, 2017, https://doi.org/10.1144/SP464.8

BUCK, W.R. 1991. Modes of continental lithospheric extension. *Journal of Geophysical Research: Solid Earth*, **96**, 20 161–20 178.

CAVAZZA, W., FEDERICI, I., OKAY, A.I. & ZATTIN, M. 2012. Apatite fission-track thermochronology of the Western Pontides (NW Turkey). *Geological Magazine*, **149**, 133–140.

ÇIMEN, O., GÖNCÜOĞLU, M.C. & SAYIT, K. 2016. Geochemistry of the metavolcanic rocks from the Çangaldağ Complex in the Central Pontides: implications for the Middle Jurassic arc-back-arc system in the Neotethyan Intra-Pontide Ocean. *Turkish Journal of Earth Sciences*, **25**, 491–512.

ÇIMEN, O., GÖNCÜOĞLU, M.C., SIMONETTI, A. & SAYIT, K. 2017. Whole rock geochemistry, Zircon U–Pb and Hf isotope systematics of the Çangaldağ Pluton: evidences for Middle Jurassic Continental Arc Magmatism in the Central Pontides, Turkey. *Lithos*, **290**, 136–155.

COHEN, K.M., FINNEY, S.C., GIBBARD, P.L. & FAN, J.-X. 2016. The ICS International Chronostratigraphic Chart. *Episodes*, **36**, 199–204.

DE LEEUW, A., MORTON, A., VAN BAAK, C.G. & VINCENT, S.J. 2018. Timing of arrival of the Danube to the Black Sea: provenance of sediments from DSDP Site 380/380A. *Terra Nova*, **30**, 114–124, https://doi.org/10.1111/ter.12314

DEMBICKI, H., JR., 2014. Confirming the presence of a working petroleum system in the Eastern Black Sea Basin, offshore Georgia using SAR imaging, sea surface slick sampling, and geophysical seafloor characterization. *AAPG Search and Discovery*, paper 10610.

DERMAN, A.S. 2014. *Petroleum Systems of Turkish Basins*. AAPG, Tulsa, OK, Memoirs, **106**, 469–504.

DERMAN, A.S. & IZTAN, Y.H. 1997. Results of geochemical analysis of seeps and potential source rocks from Northern Turkey and the Turkish Black Sea. *In*: ROBINSON, A.G. (ed.) *Regional and Petroleum Geology of the Black Sea and Surrounding Region*. AAPG, Memoirs, Tulsa, **68**, 313–330.

DIMITROV, L.I. 2002*a*. Mud volcanoes – the most important pathway for degassing deeply buried sediments. *Earth Science Reviews*, **59**, 49–76.

DIMITROV, L. 2002*b*. Contribution to atmospheric methane by natural seepages on the Bulgarian continental shelf. *Continental Shelf Research*, **22**, 2429–2442.

DINU, C., WONG, H.K., TAMBREA, D. & MATENCO, L. 2005. Stratigraphic and structural characteristics of the Romanian Black Sea shelf. *Tectonophysics*, **410**, 417–435.

DISTANOVA, L. & BAZHENOVA, O.K. 2007. Conditions of Source Rock Potential Formation in Eocene Deposits of Caucasian-Scythian Region, extended abstract. *In*: *EAGE 69th Conference and Exhibition*, 11–14 June 2007, London.

DOGLIONI, C., BUSATTA, C., BOLIS, G., MARIANINI, L. & ZANELLA, M. 1996. Structural evolution of the eastern Balkans (Bulgaria). *Marine and Petroleum Geology*, **13**, 225–251.

DONDURUR, D., KÜÇÜK, H.M. & ÇIFÇI, G. 2013. Quaternary mass wasting on the western Black Sea margin, offshore of Amasra. *Global and Planetary Change*, **103**, 248–260.

DULEY, P. & FOGG, A. 2009. Old dogs and new tricks; unlocking the hydrocarbon potential of the Romanian Black Sea: Ana and Doina gas fields and the role of inversion in derisking. *The Leading Edge*, **28**, 1090–1096.

ERGÜN, M., DONDURUR, D. & ÇIFÇI, G. 2002. Acoustic evidence for shallow gas accumulations in the sediments of the Eastern Black Sea. *Terra Nova*, **14**, 313–320.

ESPURT, N., HIPPOLYTE, J.-C., KAYMAKCI, N. & SANGU, E. 2014. Lithospheric structural control on inversion of the southern margin of the Black Sea Basin, Central Pontides, Turkey. *Lithosphere*, **6**, 26–34.

EVTUSHENKO, N.V. & IVANOV, A.Y. 2013. Oil seeps in the Southeastern Black Sea studied using satellite synthetic aperture radar images, izvestiya. *Atmospheric and Oceanic Physics*, **49**, 913–918.

FINETTI, I., BRICCHI, G., DEL BEN, A. & PIPAN, M. 1988. Geophysical study of the Black Sea area. *Bollettino di Geofisica Teoretica e Applicata*, **30**, 197–324.

GABDULLIN, R.R., SAMARIN, E.N., IVANOV, A.V., BADULINA, N.V. & AFONIN, M.A. 2014. Lithogeochemical characterization of depositional environments in the Crimean–Caucasian trough during the Early Jurassic-Aalenian (Kacha Uplift and Krasnaya Polyana). *Moscow University Geology Bulletin*, **69**, 410–423.

GALLHOFER, D., QUADT, A.V., PEYTCHEVA, I., SCHMID, S.M. & HEINRICH, C.A. 2015. Tectonic, magmatic, and metallogenic evolution of the Late Cretaceous arc in the Carpathian-Balkan orogen. *Tectonics*, **34**, 1813–1836.

GEORGIEV, G. 2012. Geology and hydrocarbon systems in the Western Black Sea. *Turkish Journal of Earth Sciences*, **21**, 723–754.

GEORGIEV, G. & ATANASOV, A. 1993. The importance of the Triassic-Jurassic unconformity to the hydrocarbon potential of Bulgaria. *First Break*, **11**, 489–497.

GEORGIEV, G., DABOVSKI, C. & STANISHEVA-VASSILEVA, G. 2001. East Srednogorie-Balkan rift zone. *Mémoires du Muséum national d'histoire naturelle*, **186**, 259–293.

GILLET, H., LERICOLAIS, G. & RÉHAULT, J.P. 2007. Messinian event in the Black Sea: evidence of a Messinian erosional surface. *Marine Geology*, **244**, 142–165.

GOLMSHTOK, A.Y., ZONENSHAIN, L.P., TEREKHOV, A.A. & SHAINUROV, R.V. 1992. Age, thermal evolution and history of the Black Sea Basin based on heat flow and multichannel reflection data. *Tectonophysics*, **210**, 273–293.

GONG, C., WANG, Y., HODGSON, D.M., ZHU, W., LI, W., XU, Q. & LI, D. 2014. Origin and anatomy of two different types of mass–transport complexes: a 3D seismic case study from the northern South China Sea margin. *Marine and Petroleum Geology*, **54**, 198–215.

GÖRÜR, N. 1988. Timing of opening of the Black Sea basin. *Tectonophysics*, **147**, 247–262.

GÖRÜR, N. 1997. Cretaceous syn- to postrift sedimentation on the southern continental margin of the western Black Sea basin. *In*: ROBINSON, A.G. (ed.) *Regional and Petroleum Geology of the Black Sea and Surrounding Region*. AAPG, Memoirs, Tulsa, **68**, 227–240.

GÖRÜR, N. & TÜYSÜZ, O. 1997. Petroleum geology of the southern continental margin of the Black Sea. *In*: ROBINSON, A.G. (ed.) *Regional and Petroleum Geology of the Black Sea and Surrounding Region*. AAPG, Memoirs, **68**, 241–254.

GOZHIK, P.F., MASLUN, N.V., PLOTNIKOVA, L.F., IVANIK, M.M. & YAKUSCHIN, L.M. 2006. *The Stratigraphy of Meso-Cenozoic Deposits of the Black Sea Northwestern Shelf*. Geological Institute, Kiev, Ukraine [in Ukrainian with English abstract].

GOZHIK, P.F., MASLUN, N.V., IVANIK, M.M., PLOTNIKOVA, L.F. & YAKUSHIN, L.N. 2008. Stratigraphic model of the Mesozoic and Cenozoic of the western Black Sea basin. *Geology and Minerals of the World's Oceans*, **1**, 55–69.

GOZHIK, P.F., MASLUN, N.V., BOICISKHY, Z.I., IVANIK, M.M. & KHLUSHINA, G.V. 2010. Regional Cenozoic stratigraphy of the Pre-Kerch shelf and the Eastern Black Sea Basin. *Geological Journal of the National Academy of Sciences of Ukraine*, **1**, 7–41 [in Ukrainian with English abstract].

GRAHAM, R., KAYMAKCI, N. & HORN, B. 2013. The Black Sea: something different? *GeoExpro*, **10**, 57–62.

GUO, L., VINCENT, S.J. & LAVRISHCHEV, V. 2011. Upper Jurassic reefs from the Russian western Caucasus, implications for the eastern Black Sea. *Turkish Journal of Earth Sciences*, **20**, 629–653.

HARBURY, N. & COHEN, M. 1997. Sedimentary history of the Late Jurassic-Paleogene of Northeast Bulgaria and the Bulgarian Black Sea. *In*: ROBINSON, A.G. (ed.) *Regional and Petroleum Geology of the Black Sea and Surrounding Region*. AAPG, Memoirs, Tulsa, **68**, 129–168.

HIPPOLYTE, J.C., MÜLLER, C., KAYMAKCI, V. & SANGU, E. 2010. Dating of the Black Sea Basin: new nannoplankton ages from its inverted margin in the Central Pontides (Turkey). *In*: SOSSON, M., KAYMAKCI, N., STEPHENSON, R.A., BERGERAT, F. & STAROSTENKO, V. (eds) *Sedimentary Basin Tectonics from the Black Sea and Caucasus to the Arabian Platform*. Geological Society, London, Special Publications, **340**, 113–136, https://doi.org/10.1144/SP340.7

HIPPOLYTE, J.C., MÜLLER, C., SANGU, E. & KAYMAKCI, N. 2015. Stratigraphic comparisons along the Pontides (Turkey) based on new nannoplankton age determinations in the Eastern Pontides: geodynamic implications. *In*: SOSSON, M., STEPHENSON, R.A. & ADAMIA, S.A. (eds) *Tectonic Evolution of the Eastern Black Sea and Caucasus*. Geological Society, London, Special Publications, **428**. First published online October 27, 2015, https://doi.org/10.1144/SP428.9

HIPPOLYTE, J.C., ESPURT, N., KAYMAKCI, N., SANGU, E. & MÜLLER, C. 2016. Cross-sectional anatomy and geodynamic evolution of the Central Pontide orogenic belt (northern Turkey). *International Journal of Earth Sciences*, **105**, 81–106.

HSÜ, K.J. & GIOVANOLI, F. 1979. Messinian event in the Black Sea. *Palaeogeography, Palaeoclimatology, Palaeoecology*, **29**, 75–93.

IONESCU, G. 2002. Facies architecture and sequence stratigraphy of the Black Sea offshore Romania. *In*: DINU, C. & MOCANU, V. (eds) *Geology and Tectonics of the Romanian Black Sea Shelf and Its Hydrocarbon Potential*. Bucharest Geoscience Forum, Special Volumes, **2**, 43–51.

IONESCU, G., SISMAN, M. & CATARAIANI, R. 2002. Source and reservoir rocks and trapping mechanism on the romanian Black Sea shelf. *In*: DINU, C. & MOCANU, V. (eds) *Geology and Tectonics of the Romanian Black Sea Shelf and Its Hydrocarbon Potential*. Bucharest Geoscience Forum, Special Volumes, **2**, 67–83.

JANBU, N.E., NEMEC, W., KIRMAN, E. & ÖZAKSOY, V. 2007. Facies anatomy of a sand-rich channelized turbiditic system: the Eocene Kusuri Formation in the Sinop Basin, north-central Turkey. *In*: NICHOLS, G., WILLIAMS, E. & PAOLA, C. (eds) *Sedimentary Processes Environments and Basins*. International Association of Sedimentologists, Special Publications, **38**, 457–517.

JONES, R.W. & SIMMONS, M.D. 1997. A review of the stratigraphy of Eastern Paratethys (Oligocene–Holocene), with particular emphasis on the Black Sea. *In*: ROBINSON, A.G. (ed.) *Regional and Petroleum Geology of the Black Sea and Surrounding Region*. AAPG, Memoirs, Tulsa, **68**, 39–52.

KASUMOV, T. 2017. Lukoil improves drilling performance in Black Sea with integrated real-time technology and virtual teams. Presentation given at the *SIS Global Forum*, 13–15 September, Paris, https://http://www.software.slb.com/sis-global-forum-2017?tab=Technical Presentations (last accessed 15 January 2018).

KAZMIN, V.G., SBORTSHIKOV, I.M., RICOU, L.E., ZONENSHAIN, L.P., BOULIN, J. & KNIPPER, A.L. 1986. Volcanic belts as markers of the Mesozoic-Cenozoic active margin of Eurasia. *Tectonophysics*, **123**, 123–152.

KAZMIN, V.G., SCHREIDER, A.A. & BULYCHEV, A.A. 2000. Early stages of evolution of the Black Sea. *In*: BOZKURT, E., WINCHESTER, J.A. & PIPER, J.D.A. (eds) *Tectonics and Magmatism in Turkey and the Surrounding Area*. Geological Society, London, Special Publications, **173**, 235–249, https://doi.org/10.1144/GSL.SP.2000.173.01.12

KESKIN, M. & TÜYSÜZ, O. 2017. Stratigraphy, petrogenesis and geodynamic setting of Late Cretaceous volcanism on the SW margin of the Black Sea, Turkey. *In*: SIMMONS, M.D., TARI, G.C. & OKAY, A.I. (eds) *Petroleum Geology of the Black Sea*. Geological Society, London, Special Publications, **464**. First published online September 28, 2017, https://doi.org/10.1144/SP464.5

KHLEBNIKOVA, O.A., NIKISHIN, A.M., MITYUKOV, A.V., RUBTSOVA, E.V., FOKIN, P.A., KOPAEVICH, L.F. & ZAPOROZHETS, N.I. 2014. The composition of the sandstone from the Oligocene turbidite of the Tuapse marginal trough. *Moscow University Geology Bulletin*, **69**, 399–409.

KHRIACHTCHEVSKAIA, O., STOVBA, S. & POPADYUK, I. 2009. Hydrocarbon prospects of the Ukrainian Western Black Sea. *The Leading Edge*, **28**, 1024–1029.

KHRIACHTCHEVSKAIA, O., STOVBA, S. & STEPHENSON, R. 2010. Cretaceous–Neogene tectonic evolution of the northern margin of the Black Sea from seismic reflection data and tectonic subsidence analysis. *In*: SOSSON, M., KAYMAKCI, N., STEPHENSON, R.A., BERGERAT, F. & STAROSTENKO, V. (eds) *Sedimentary Basin Tectonics from the Black Sea and Caucasus to the Arabian Platform*. Geological Society, London, Special Publications, **340**, 137–157, https://doi.org/10.1144/SP340.8

KITCHKA, A., KHARCHENKO, M.V., VAKARCHUK, S.G. & DOVZHOK, T.E. 2012. Sedimentary architecture and reservoir rocks origin of the Maykop Formation challenge exploration in the Black Sea. *In*: *74th EAGE Conference and Exhibition Incorporating EUROPEC 2012*.

KITCHKA, O.A., TYSHCHENKO, A.P., LYSENKO, V.I., BEZKHYZHKO, O.M. & ISHCHENKO, I.I. 2016a. Neogene submarine rock sliding and development of mass transport deposits in the Ukrainian sector of Black sea basin. *In*: *15th EAGE International Conference on Geoinformatics – Theoretical and Applied Aspects*, Kiev, Ukraine.

KITCHKA, A.A., TYSHCHENKO, A.P. & LYSENKO, V.I. 2016*b*. Mid–Late Miocene sea level falls, gas hydrates decay, submarine sliding, and tsunamites in the Black Sea Basin. *In*: *78th EAGE Conference and Exhibition 2016*.

KONERDING, C., DINU, C. & WONG, H.K. 2010. Seismic sequence stratigraphy, structure and subsidence history of the Romanian Black Sea shelf. *In*: SOSSON, M., KAYMAKCI, N., STEPHENSON, R.A., BERGERAT, F. & STAROSTENKO, V. (eds) *Sedimentary Basin Tectonics from the Black Sea and Caucasus to the Arabian Platform*. Geological Society, London, Special Publications, **340**, 159–180, https://doi.org/10.1144/SP340.9

KOPAEVICH, L.F., ALISOVA, E.A., NIKISHIN, A.M. & YAKOVISHINA, E.V. 2008. The Eocene Nummulitic Bank in the Crimea. *Moscow University Geology Bulletin*, **63**, 195–198.

KORUCU, Ö., SIPAHIOĞLU, N.Ö., AKTEPE, S. & BENGÜ, E. 2013. Correlation and determination of Neogene sequences employing recent ultra-deep wells in Western-Central part of Turkish Black Sea. *In*: *Abstract Volume, AAPG Europe Regional Conference*, Tbilisi, 12–19.

KRASTEL, S., SPIESS, V., IVANOV, M., WEISNREBE, W., BOHRMANN, G., SHASHSKIN, P. & HEIDERSDORF, F. 2003. Acoustic investigations of mud volcanoes in the Sorokin Trough, Black Sea. *Geo-Marine Letters*, **23**, 230–238.

KREZSEK, C., SCHLEDER, Z., BEGA, Z., IONESCU, G. & TARI, G. 2016. The Messinian sea-level fall in the western Black Sea: small or large? Insights from offshore Romania. *Petroleum Geoscience*, **22**, 392–399, https://doi.org/10.1144/petgeo2015-093

KREZSEK, C., BERCEA, R.I., TARI, G. & IONESCU, G. 2017. Cretaceous sedimentation along the Romanian margin of the Black Sea: inferences from onshore to offshore correlations. *In*: SIMMONS, M.D., TARI, G.C. & OKAY, A.I. (eds) *Petroleum Geology of the Black Sea*. Geological Society, London, Special Publications, **464**. First published online September 7, 2017, https://doi.org/10.1144/SP464.10

LERICOLAIS, G., BOURGET, J., POPESCU, I., JERMANNAUD, P., MULDER, T., JORRY, S. & PANIN, N. 2013. Late Quaternary deep-sea sedimentation in the western Black Sea: new insights from recent coring and seismic data in the deep basin. *Global and Planetary Change*, **103**, 232–247.

LESS, G., ÖZCAN, E. & OKAY, A.I. 2011. Stratigraphy and larger Foraminifera of the Middle Eocene to Lower Oligocene shallow-marine units in the northern and eastern parts of the Thrace Basin, NW Turkey. *Turkish Journal of Geosciences*, **20**, 793–845.

LETOUZEY, J., BIJU-DUVAL, B., DORKEL, A., GONNARD, R., KRISTCHEV, K., MONTADERT, L. & SUNGURLU, O. 1977. The Black Sea: a marginal basin; geophysical and geological data. *In*: *International Symposium on the Structural History of the Mediterranean Basins*, Technip, Paris, 363–376.

MASSE, J.-P., TÜYSÜZ, O., FENERCI-MASSE, M., ÖZER, S. & SARI, B. 2009. Stratigraphic organisation, spatial distribution, palaeoenvironmental reconstruction, and demise of Lower Cretaceous (Barremian – Lower Aptian) carbonate platforms of the Western Pontides (Black Sea region, Turkey). *Cretaceous Research*, **30**, 1170–1180.

MAYER, J., RUPPRECHT, B.J. ET AL. 2017. Source potential and depositional environment of Oligocene and Miocene rocks offshore Bulgaria. *In*: SIMMONS, M.D., TARI, G.C. & OKAY, A.I. (eds) *Petroleum Geology of the Black Sea*. Geological Society, London, Special Publications, **464**. First published online September 25, 2017, https://doi.org/10.1144/SP464

MAYER, J., SACHSENHOFER, R.F., UNGUREANU, C. & TARI, G. In press. Petroleum charge and migration in the Black Sea: insights from oil and source rock geochemistry. *Journal of Petroleum Geology*, **41**.

MAYNARD, J.R., ARDIC, C. & MCALLISTER, N. 2012. Source to sink assessment of Oligocene to Pleistocene sediment supply in the Black Sea. *GCSSEPM Conference Houston Transactions*, **32**, 664–700.

MAZZINI, A., IVANOV, M.K., NERMOEN, A., BAHR, A., BOHRMANN, G., SVENSEN, H. & PLANKE, S. 2008. Complex plumbing systems in the near subsurface: geometries of authigenic carbonates from Dolgovskoy Mound (Black Sea) constrained by analogue experiments. *Marine and Petroleum Geology*, **25**, 457–472.

MCCANN, T., CHALOT-PRAT, F. & SAINTOT, A. 2010. The Early Mesozoic evolution of the Western Greater Caucasus (Russia): Triassic–Jurassic sedimentary and magmatic history. *In*: SOSSON, M., KAYMAKCI, N., STEPHENSON, R.A., BERGERAT, F. & STAROSTENKO, V. (eds) *Sedimentary Basin Tectonics from the Black Sea and Caucasus to the Arabian Platform*. Geological Society, London, Special Publications, **340**, 181–238, https://doi.org/10.1144/SP340.10

MEIJERS, M.J.M., VROUWE, B. ET AL. 2010. Jurassic arc volcanism on Crimea (Ukraine): implications for the paleo-subduction zone configuration of the Black Sea region. *Lithos*, **119**, 412–426.

MEISNER, A., KRYLOV, O. & NEMČOK, M. 2009. Development and structural architecture of the Eastern Black Sea. *The Leading Edge*, **28**, 1046–1055.

MEISNER, A., SHEYA, C. & NEMCOK, N. 2011. Ancient depositional environments of the Eastern Black Sea. *AAPG Search and Discovery Article*, 50388.

MEISNER, L.B. & TUGOLESOV, D.A. 2003. Key reflecting horizons in sedimentary fill seismic records of the Black Sea basin (correlation and stratigraphic position). *Stratigraphy and Geological Correlation*, **11**, 606–619.

MENLIKLI, C., DEMIRER, A., SIPAHIOĞLU, Ö., KÖRPE, L. & AYDEMIR, V. 2009. Exploration plays in the Turkish Black Sea. *The Leading Edge*, **28**, 1066–1075.

MEREDITH, D.J. & EGAN, S.S. 2002. The geological and geodynamic evolution of the eastern Black Sea basin: insights from 2-D and 3-D tectonic modelling. *Tectonophysics*, **350**, 157–179.

MITYUKOV, A.V., ALMENDINGER, O.A., MYASOEDOV, N.K., NIKISHIN, A.M. & GAIDUK, V.V. 2011. The sedimentation model of the Tuapse Trough (Black Sea). *Doklady Earth Sciences*, **440**, 1245–1248.

MOROŞANU, I. 1996. Tectonic setting of the Romanian offshore area at the pre-Albian level. *In*: WESSELY, G. & LIEBL, W. (eds) *Oil and Gas in Alpidic Thrustbelts and Basins of Central and Eastern Europe*. EAGE, Special Publications, London, **5**, 315–323.

MOROŞANU, I. 2012. The hydrocarbon potential of the Romanian Black Sea continental plateau. *Romanian Journal of Earth Sciences*, **86**, 91–109.

MUNTEANU, I., MATENCO, L., DINU, C. & CLOETINGH, S.A.P.L. 2011. Kinematics of back-arc inversion of the Western Black Sea Basin. *Tectonics*, **30**, TC5004.

MUNTEANU, I., DIVIACCO, P., SAULI, C., DINU, C., BURCĂ, M., PANIN, N. & BRANCATELLI, G. 2017. New insights into the Black Sea Basin, in the light of the reprocessing of vintage regional seismic data. *In*: FINKL, C. & MAKOWSKI, C. (eds) *Diversity in Coastal Marine Sciences*. Coastal Research Library, **23**, Springer, Cham, Switzerland, 91–114.

NAUDTS, L., DE BATIST, M., GREINERT, J. & ARTEMOV, Y. 2009. Geo-and hydro-acoustic manifestations of shallow gas and gas seeps in the Dnepr paleodelta, northwestern Black Sea. *The Leading Edge*, **28**, 1030–1040.

NIKISHIN, A.M., KOROTAEV, M.V., ERSHOV, A.V. & BRUNET, M.F. 2003. The Black Sea basin: tectonic history and Neogene–Quaternary rapid subsidence modelling. *Sedimentary Geology*, **156**, 149–168.

NIKISHIN, A.M., ERSHOV, A.V. & NIKISHIN, V.A. 2010. Geological history of the western Caucasus and adjacent foredeeps based on analysis of the regional balanced section. *Doklady Earth Sciences*, **430**, 155–157.

NIKISHIN, A.M., KHOTYLEV, A.O., BYCHKOV, A.Y., KOPAEVICH, L.F., PETROV, E.I. & YAPASKURT, V.O. 2013. Cretaceous volcanic belts and the evolution of the Black Sea Basin. *Moscow University Geology Bulletin*, **68**, 141–154.

NIKISHIN, A.M., OKAY, A.I., TÜYSÜZ, O., DEMIRER, A., AMELIN, N. & PETROV, E. 2015*a*. The Black Sea basins structure and history: new model based on new deep penetration regional seismic data. Part 1: Basins structure and fill. *Marine and Petroleum Geology*, **59**, 638–655.

NIKISHIN, A.M., OKAY, A., TÜYSÜZ, O., DEMIRER, A., WANNIER, M., AMELIN, N. & PETROV, E. 2015*b*. The Black Sea basins structure and history: New model based on new deep penetration regional seismic data. Part 2: Tectonic history and paleogeography. *Marine and Petroleum Geology*, **59**, 656–670.

NIKISHIN, A.M., WANNIER, M. *ET AL.* 2017. Mesozoic to Recent geological history of southern Crimea and the Eastern Black Sea region. *In*: SOSSON, M., STEPHENSON, R.A. & ADAMIA, S.A. (eds) *Tectonic Evolution of the Eastern Black Sea and Caucasus*. Geological Society, London, Special Publications, **428**, 241–264, https://doi.org/10.1144/SP428.1

OKAY, A.I. & NIKISHIN, A.M. 2015. Tectonic evolution of the southern margin of Laurasia in the Black Sea region. *International Geology Review*, **57**, 1051–1076.

OKAY, A.I. & ŞAHINTÜRK, Ö. 1997. Geology of the Eastern Pontides. *In*: ROBINSON, A.G. (ed.) *Regional and Petroleum Geology of the Black Sea and Surrounding Region*. AAPG, Memoir, Tulsa, **68**, 291–312.

OKAY, A.I., ŞENGÖR, A.C. & GÖRÜR, N. 1994. Kinematic history of the opening of the Black Sea and its effect on the surrounding regions. *Geology*, **22**, 267–270.

OKAY, A.I., TANSEL, I. & TÜYSÜZ, O. 2001. Obduction, subduction and collision as reflected in the Upper

Cretaceous–Lower Eocene sedimentary record of western Turkey. *Geological Magazine*, **138**, 117–142.

OKAY, A.I., TUYSUZ, O., SATIR, M., OZKAN-ALTINER, S., ALTINER, D., SHERLOCK, S. & EREN, R.H. 2006. Cretaceous and Triassic subduction-accretion, high-pressure–low-temperature metamorphism, and continental growth in the Central Pontides, Turkey. *Geological Society of America Bulletin*, **118**, 1247–1269.

OKAY, A.I., SUNAL, G., SHERLOCK, S., ALTINER, D., TÜYSÜZ, O., KYLANDER-CLARK, A.R. & AYGÜL, M. 2013. Early Cretaceous sedimentation and orogeny on the active margin of Eurasia: Southern Central Pontides, Turkey. *Tectonics*, **32**, 1247–1271.

OKAY, A.I., SUNAL, G., TÜYSÜZ, O., SHERLOCK, S., KESKIN, M. & KYLANDER-CLARK, A.R.C. 2014. Low-pressure–high temperature metamorphism during extension in a Jurassic magmatic arc, Central Pontides, Turkey. *Journal of Metamorphic Geology*, **32**, 49–69.

OKAY, A.I., ALTINER, D., SUNAL, G., AYGÜL, M., AKDOĞAN, R., ALTINER, S. & SIMMONS, M. 2017. Geological evolution of the Central Pontides. *In*: SIMMONS, M.D., TARI, G.C. & OKAY, A.I. (eds) *Petroleum Geology of the Black Sea*. Geological Society, London, Special Publications, **464**. First published online September 15, 2018, https://doi.org/10.1144/SP464.3

OLARIU, C., KREZSEK, CS. & JIPA, D. 2018. The Danube River inception: evidence for a 4 Ma continental-scale river born from segmented ParaTethys basins. *Terra Nova*, **30**, 63–71.

OLARU, R., KRÉZSEK, CS., RAINER, T.M., UNGUREANU, C., TURI, V., IONESCU, G. & TARI, G. In press. 3D basin modelling of the Oligocene–Miocene Maykop source rocks in the Western Black Sea with focus on offshore Romania. *Journal of Petroleum Geology*, **41**.

ÖZER, S., MERIÇ, E., GÖRMÜŞ, M. & KANBUR, S. 2009. Biogeographic distribution of rudists and benthic foraminifera: an approach to Campanian-Maastrichtian paleobiogeography of Turkey. *Geobios*, **42**, 623–638.

ÖZTAŞ, Y. 2010. TPAO and activities in Thrace and Black Sea Basins. Presentation given at the *16th Balkan and Black Sea Petroleum Association Conference*, 15–16 April 2010, Vienna.

PANEK, T., DANIŠÍK, M., HRADECKÝ, J. & FRISCH, W. 2009. Morpho-tectonic evolution of the Crimean mountains (Ukraine) as constrained by apatite fission track data. *Terra Nova*, **21**, 271–278.

PAPE, T., BAHR, A. *ET AL.* 2010. Molecular and isotopic partitioning of low-molecular-weight hydrocarbons during migration and gas hydrate precipitation in deposits of a high-flux seepage site. *Chemical Geology*, **269**, 350–363.

PESHKOV, G.A., BARABANOV, N.N., BOLSHAKOVA, M.A., BORDUNOV, S.I., KOPAEVICH, L.F. & NIKISHIN, A.M. 2016. The oil and gas potential of the Kuma rocks of the Bakhchisarai region of Crimea. *Moscow University Geology Bulletin*, **71**, 262–268.

PFANNENSTIEL, M. 1944. Die diluvialen Entwicklungsstadien und die Urgeschichte von Dardanellen, Marmarameer und Bosporus. *Geologische Rundschau*, **34**, 342–434.

PICARELLI, A. 2017. E&P activity in the Romanian and Bulgarian waters of the Black Sea. Petroleum Exploration Society of Great Britain Newsletter, September, 38–41. https://issuu.com/pesgb64/docs/september_2017

POPESCU, I., LERICOLAIS, G., PANIN, N., WONG, H.K. & DROZ, L. 2001. Late Quaternary channel avulsions on the Danube deep-sea fan. *Marine Geology*, **179**, 25–37.

POPOV, S.V. & STOLYAROV, A.S. 1996. Paleogeography and anoxic environments of the Oligocene-Early Miocene eastern Paratethys. *Israeli Journal of Earth Science*, **45**, 161–167.

POPOV, S.V., AKHMET'EV, M.A. & ZAPOROZHETS, N.I. 1993. The Late Eocene–Early Miocene Evolution of the Eastern Paratethys. *Stratigrafiya, Geologicheskaya Korrelyatsiya*, **1**, 10–39.

POPOV, S.V., RÖGL, F., ROZANOV, A.Y., STEININGER, F.F., SHCHERBA, I.G. & KOVAC, M. 2004. *Lithological-Paleogeographic maps of Paratethys. 10 Maps Late Eocene to Pliocene*. Courier Forschungsinstitut Senckenberg, Frankfurt.

POPOV, S.V., ANTIPOV, M.P., ZASTROZHNOV, A.S., KURINA, E.E. & PINCHUK, T.N. 2010. Sea-level fluctuations on the north shelf of the Eastern Paratethys in the Oligocene–Neogene. *Stratigraphy and Geological Correlation*, **18**, 200–224.

POSAMENTIER, H., AYDEMIR, V. ET AL. 2014. Volcanic deposits in the Black Sea – seismic recognition criteria for differentiating volcanics from carbonates. *AAPG Search and Discovery*, 90189, Presented at the AAPG Annual Convention and Exhibition, 6–9 April 2014, Houston, TX.

POTTORF, R.J., WENGER, L.M., MANKIEWICZ, P.J., JAFAROV, E.J. & BEGA, Z. 2010. Oil and source rock correlations within the Delfin Nord 1 well, offshore Romania. *AAPG Search and Discovery*, 90109, Presented at the European Region Annual Conference, Kiev, Ukraine, 17–19 October 2010.

PUPP, M., BECHTEL, A., ĆORIĆ, S., GRATZER, R., RUSTAMOV, J. & SACHSENHOFER, R.F. In press. Eocene and Oligo-Miocene source rocks in the Rioni and Kura basins of Georgia: depositional environment and petroleum potential. *Journal of Petroleum Geology*, **41**.

RANGIN, C., BADER, A.G., PASCAL, G., ECEVITOĞLU, B. & GÖRÜR, N. 2002. Deep structure of the Mid Black Sea High (offshore Turkey) imaged by multi-channel seismic survey (BLACKSIS cruise) 1, 2. *Marine Geology*, **182**, 265–278.

REES, E.V.L., SIMMONS, M.D. & WILSON, J.W.P. 2017. Deep-water plays in the western Black Sea: insights into sediment supply within the Maykop depositional system. *In*: SIMMONS, M.D., TARI, G.C. & OKAY, A.I. (eds) *Petroleum Geology of the Black Sea*. Geological Society, London, Special Publications, **464**. First published online September 15, 2017, https://doi.org/10.1144/SP464.13

REITZ, A., PAPE, T. ET AL. 2011. Sources of fluids and gases expelled at cold seeps offshore Georgia, eastern Black Sea. *Geochimica et Cosmochimica Acta*, **75**, 3250–3268.

ROBINSON, A.G. 1997. *Regional and petroleum geology of the Black Sea and surrounding region*. AAPG Memoir, Tulsa, **68**, 367.

ROBINSON, A.G., BANKS, C.J., RUTHERFORD, M.M. & HIRST, J.P.P. 1995a. Stratigraphic and structural development of the eastern Pontides, Turkey. *Journal of the Geological Society of London*, **152**, 861–872.

ROBINSON, A., SPADINI, G., CLOETINGH, S., RUDAT, J., CLOETINGH, S., DURAND, B. & PUIGDEFABREGAS, C. 1995b. Stratigraphic evolution of the Black Sea; inferences from basin modelling. *Marine and Petroleum Geology*, **12**, 821–835.

ROBINSON, A.G., RUDAT, J.H., BANKS, C.J. & WILES, R.L.F. 1996. Petroleum geology of the Black Sea. *Marine and Petroleum Geology*, **13**, 195–223.

ROBINSON, A.G., GRIFFTH, E.T., GARDINER, A.R. & HOME, A.K. 1997. Petroleum geology of the Georgian fold and thrust belts and foreland basins. *In*: ROBINSON, A.G. (ed.) *Regional and Petroleum Geology of the Black Sea and Surrounding Region*. AAPG, Memoirs, Tulsa, **68**, 347–367.

RÖGL, F. 1999. Mediterranean and Paratethys. Facts and hypotheses of an Oligocene to Miocene paleogeography (short overview). *Geologica Carpathica*, **50**, 339–349.

ROSS, D.A. & NEPROCHNOV, Y.P. 1978. *Initial reports of the Deep Sea Drilling Project*, 42(2). U.S. Government Printing Office, Washington D.C.

ROSS, D.A., YURI, P.P., ET AL. 1978. Site 380. *In*: ROSS, D.A. & NEPROCHNOV, Y.P. (eds) *Initial reports of the Deep Sea Drilling Project*, 42(2). U.S. Government Printing Office, Washington D.C., 119–291.

ROUTH, P., NEELAMANI, R. ET AL. 2017. Impact of high-resolution FWI in the Western Black Sea: revealing overburden and reservoir complexity. *The Leading Edge*, **36**, 60–66.

SACHSENHOFER, R.F., STUMMER, B., GEORGIEV, G., DELLMOUR, R., BECHTEL, A., GRATZER, R. & ĆORIĆ, S. 2009. Depositional environment and hydrocarbon source potential of the Oligocene Ruslar Formation (Kamchia Depression; western Black Sea). *Marine and Petroleum Geology*, **26**, 57–84.

SACHSENHOFER, R.F., POPOV, S.V. ET AL. 2017a. The type section of the Maikop Group (Oligocene–lower Miocene) at the Belaya River (North Caucasus): depositional environment and hydrocarbon potential. *AAPG Bulletin*, **101**, 289–319.

SACHSENHOFER, R.F., POPOV, S.V. ET AL. 2017b. Oligocene and Lower Miocene source rocks in the Paratethys: Palaeogeographic and stratigraphic controls. *In*: SIMMONS, M.D., TARI, G.C. & OKAY, A.I. (eds) *Petroleum Geology of the Black Sea*. Geological Society, London, Special Publications, **464**. First published online September 7, 2017, https://doi.org/10.1144/SP464.1

SAINTOT, A., BRUNET, M.F. ET AL. 2006. The Mesozoic-Cenozoic tectonic evolution of the Greater Caucasus. *In*: GEE, D.G. & STEPHENSON, R.A. (eds) *European Lithosphere Dynamics*. Geological Society, London, Memoirs, **32**, 277–289, https://doi.org/10.1144/GSL.MEM.2006.032.01.16

SCHLEDER, Z., KREZSEK, C., TURI, V., TARI, G., KOSI, W. & FALLAH, M. 2015. Regional structure of the Western Black Sea Basin: constraints from cross-section balancing. *In*: *Petroleum Systems in Rift Basins, the 34th Annual GCSEPM Foundation Perkins–Rosen Research Conference*, 13–16.

SCHLEDER, Z., KREZSEK, C., LAPADAT, A., BEGA, Z., IONESCU, G. & TARI, G. 2016. Structural style in a Messinian (intra-Pontian) gravity-driven deformation system, western Black Sea, offshore Romania. *Petroleum Geoscience*, **22**, 400–410, https://doi.org/10.1144/petgeo2015-094

SCHULZ, H.M., BECHTEL, A. & SACHSENHOFER, R.F. 2005. The birth of the Paratethys during the Early Oligocene: from Tethys to an ancient Black Sea analogue? *Global and Planetary Change*, **49**, 163–176.

ŞEN, Ş. 2013. New evidences for the formation of and for petroleum exploration in the fold-thrust zones of the central Black Sea Basin of Turkey. *AAPG Bulletin*, **97**, 465–485.

SHEREMET, Y., SOSSON, M., RATZOV, G., SYDORENKO, G., VOITSITSKIY, Z., YEGOROVA, T. & MUROVSKAYA, A. 2016. An offshore–onland transect across the northeastern Black Sea basin (Crimean margin): evidence of Paleocene to Pliocene two-stage compression. *Tectonophysics*, **688**, 84–100.

SHEREMET, Y., SOSSON, M., MULLER, C., GINTOV, O., MUROVSKAYA, A. & YEGOROVA, T. 2017. Key problems of stratigraphy in the Eastern Crimea Peninsula: some insights from new dating and structural data. *In*: SOSSON, M., STEPHENSON, R.A. & ADAMIA, S.A. (eds) *Tectonic Evolution of the Eastern Black Sea and Caucasus*. Geological Society, London, Special Publications, **428**, 265–306, https://doi.org/10.1144/SP428.14

SHILLINGTON, D.J., WHITE, N., MINSHULL, T.A., EDWARDS, G.R.H., JONES, S., EDWARDS, R.A. & SCOTT, C.L. 2008. Cenozoic evolution of the eastern Black Sea: a test of depth-dependent stretching models. *Earth and Planetary Science Letters*, **265**, 360–378.

SHILLINGTON, D.J., MINSHULL, T.A., EDWARDS, R.A. & WHITE, N. 2017. Crustal structure of the Mid Black Sea High from wide-angle seismic data. *In*: SIMMONS, M.D., TARI, G.C. & OKAY, A.I. (eds) *Petroleum Geology of the Black Sea*. Geological Society, London, Special Publications, **464**. First published online October 4, 2017, https://doi.org/10.1144/SP464.4

SIMMONS, M.D., TARI, G.C. & OKAY, A.I. 2018. The petroleum geology of the Black Sea: introduction. *In*: SIMMONS, M.D., TARI, G.C. & OKAY, A.I. (eds) *Petroleum Geology of the Black Sea*. Geological Society, London, Special Publications, **464**. First published online May 4, 2018, https://doi.org/10.1144/SP464.15

SINCLAIR, H.D., GEORGIEV, S.J.G., BYRNE, P. & MOUNTNEY, N.P. 1997. The Balkan thrust wedge and foreland basin of Eastern Bulgaria: structural and stratigraphic development. *In*: ROBINSON, A.G. (ed.) *Regional and Petroleum Geology of the Black Sea and Surrounding Region*. AAPG, Memoirs, Tulsa, **68**, 91–114.

SIPAHIOĞLU, N.Ö. & BATI, Z. 2017. Messinian canyons in the Turkish western Black Sea. *In*: SIMMONS, M.D., TARI, G.C. & OKAY, A.I. (eds) *Petroleum Geology of the Black Sea*. Geological Society, London, Special Publications, **464**. First published online September 25, 2017, https://doi.org/10.1144/SP464.12

SIPAHIOĞLU, N.O., KORUCU, Ö., AKTEPE, S. & BENGÜ, E. 2013*a*. Westerly-sourced Late Oligocene–Middle Miocene axial sediment dispersal system in Turkish Western Black Sea: myth or reality? Paper presented at the *19th International Petroleum and Natural Gas Congress and Exhibition of Turkey*, 15–17 May 2013, Ankara, Turkey.

SIPAHIOĞLU, N.O., KARAHANOGLU, N. & ALTINER, D. 2013*b*. Analysis of Plio-Quaternary deep marine systems and their evolution in a compressional tectonic regime, Eastern Black Sea Basin. *Marine and Petroleum Geology*, **43**, 187–207.

SPADINI, G., ROBINSON, A. & CLOETINGH, S. 1996. Western versus Eastern Black Sea tectonic evolution: pre-rift lithospheric controls on basin formation. *Tectonophysics*, **266**, 139–154.

STAROSTENKO, V., BURYANOV, V. *ET AL*. 2004. Topography of the crust–mantle boundary beneath the Black Sea Basin. *Tectonophysics*, **381**, 211–233.

STAROSTENKO, V.I., RUSAKOV, O.M., SHNYUKOV, E.F., KOBOLEV, V.P. & KUTAS, R.I. 2010. Methane in the northern Black Sea: characterization of its geomorphological and geological environments. *In*: SOSSON, M., KAYMAKCI, N., STEPHENSON, R., BERGERAT, F. & STAROSTENKO, V. (eds) *Sedimentary Basin Tectonics from the Black Sea and Caucasus to the Arabian Platform*. Geological Society, London, Special Publications, **340**, 57–75, https://doi.org/10.1144/SP340.5

STAROSTENKO, V.I., RUSAKOV, O.M. *ET AL*. 2015. Heterogeneous structure of the lithosphere in the Black Sea from a multidisciplinary analysis of geophysical fields. *Geofizicheskiy Zhurnal*, **37**, 3–28.

STEPHENSON, R. & SCHELLART, W.P. 2010. The Black Sea back-arc basin: insights to its origin from geodynamic models of modern analogues. *In*: SOSSON, M., KAYMAKCI, N., STEPHENSON, R., BERGERAT, F. & STAROSTENKO, V. (eds) *Sedimentary Basin Tectonics from the Black Sea and Caucasus to the Arabian Platform*. Geological Society, London, Special Publications, **340**, 11–21, https://doi.org/10.1144/SP340.2

STOVBA, S., KHRIACHTCHEVSKAIA, O. & POPADYUK, I. 2009. Hydrocarbon bearing areas in the eastern part of the Ukrainian Black Sea. *The Leading Edge*, **28**, 1042–1045.

STUART, C.J., NEMCOK, M., VANGELOV, D., HIGGINS, E.R., WELKER, C. & MEAUX, D.P. 2011. Structural and depositional evolution of the East Balkan thrust belt, Bulgaria. *AAPG Bulletin*, **95**, 649–673.

SUNAL, G. & TÜYSÜZ, O. 2002. Palaeostress analysis of Tertiary post-collisional structures in the Western Pontides, northern Turkey. *Geological Magazine*, **139**, 343–359.

TAMBREA, D., RAILEANU, A. & BOROSI, V. 2002. Seismic facies and depositional framework of Eocene deposits. *In*: DINU, C. & MOCANU, V. (eds) *Geology and Tectonics of the Romanian Black Sea Shelf and its Hydrocarbon Potential*. Bucharest Geoscience Forum, Special Volumes, **2**, 85–100.

TARI, G. 2010. Exploration country focus: Ukraine. *American Association of Petroleum Geologists European Region Newsletter*, **5**, 4–6.

TARI, G. 2015. Is the Black Sea really a back-arc basin? *In*: *Transactions of the GCSEPM Foundation Perkins–Rosen 34th Annual Research Conference 'Petroleum Systems in Rift Basins'*, 510–520.

TARI, G., DICEA, O., FAULKERSON, J., GEORGIEV, G., POPOV, S., STEFANESCU, M. & WEIR, G. 1997. Cimmerian and Alpine stratigraphy and structural evolution of the Moesian Platform (Romania/Bulgaria). *In*: ROBINSON, A.G. (ed.) *Regional and Petroleum Geology of the Black Sea and Surrounding Region*. AAPG, Memoirs, Tulsa, **68**, 63–90.

TARI, G., DAVIES, J., DELLMOUR, R., LARRATT, E., NOVOTNY, B. & KOZHUHAROV, E. 2009. Play types and hydrocarbon potential of the deepwater Black Sea, NE Bulgaria. *The Leading Edge*, **28**, 1076–1081.

TARI, G., MENLIKLI, C. & DERMAN, S. 2011. Deepwater play types of the Black Sea: a brief overview. *AAPG Search and Discovery Article*, 10310.

TARI, G., KOSI, W., FALLAH, M., SIEDL, W., BEGA, Z., KREZSEK, C. & KOZHUHAROV, E. 2013. The end Eocene drawdown in the Black Sea: the deepwater record of the birth of the Paratethys. *In*: *American Association of Petroleum Geologists European Regional Conference & Exhibition*, Tbilisi, Abstract Book.

TARI, G., FALLAH, M., KOSI, W., SCHLEDER, Z., TURI, V. & KREZSEK, C. 2015a. Regional rift structure of the Western Black Sea Basin: map-view kinematics. *In*: *Petroleum Systems in Rift Basins: 34th Annual GCSEPM Foundation Perkins–Rosen Research Conference*, 372–395.

TARI, G., FALLAH, M., KOSI, W., FLOODPAGE, J., BAUR, J., BATI, Z. & SIPAHIOĞLU, N.O. 2015b. Is the impact of the Messinian Salinity Crisis in the Black Sea comparable to that of the Mediterranean? *Marine and Petroleum Geology*, 66, 135–148.

TARI, G., FALLAH, M. ET AL. 2016. Why are there no Messinian evaporites in the Black Sea? *Petroleum Geoscience*, 22, 381–391, https://doi.org/10.1144/petgeo2016-003

TARI, G.C., VAKHANIA, D. ET AL. 2018. Stratigraphy, structure and petroleum exploration play types of the Rioni Basin, Georgia. *In*: SIMMONS, M.D., TARI, G.C. & OKAY, A.I. (eds) *Petroleum Geology of the Black Sea*. Geological Society, London, Special Publications, 464. First published online May 4, 2018, https://doi.org/10.1144/SP464.14

TUGOLESOV, D.A., GORSHKOV, A.S. ET AL. 1985. *Tectonics of the Mesozoic Sediments of the Black Sea Basin*. Nedra, Moscow [in Russian].

TÜYSÜZ, O. 1999. Geology of the Cretaceous sedimentary basins of the Western Pontides. *Geological Journal*, 34, 75–93.

TÜYSÜZ, O. 2017. Cretaceous geological evolution of the Pontides. *In*: SIMMONS, M.D., TARI, G.C. & OKAY, A.I. (eds) *Petroleum Geology of the Black Sea*. Geological Society, London, Special Publications, 464. First published online September 8, 2017, https://doi.org/10.1144/SP464.9

TÜYSÜZ, O., YILMAZ, I.Ö., SVABENICKA, L. & KIRICI, S. 2012. The Unaz Formation: a key unit in the Western Black Sea region, N Turkey. *Turkish Journal of Earth Sciences*, 21, 1009–1028.

VANGELOV, D., GERDJIKOV, Y., KOUNOV, A. & LAZAROVA, A. 2013. The Balkan Fold-Thrust Belt: an overview of the main features. *Geologica Balcanica*, 42, 29–47.

VINCENT, S.J. & KAYE, M.N.D. 2017. Source rock evaluation of Middle Eocene–Early Miocene mudstones from the NE margin of the Black Sea. *In*: SIMMONS, M.D., TARI, G.C. & OKAY, A.I. (eds) *Petroleum Geology of the Black Sea*. Geological Society, London, Special Publications, 464. First published September 25, 2017, https://doi.org/10.1144/SP464.7

VINCENT, S.J., ALLEN, M.B., ISMAIL-ZADEH, A.D., FLECKER, R., FOLAND, K.A. & SIMMONS, M.D. 2005. Insights from the Talysh of Azerbaijan into the Paleogene evolution of the South Caspian region. *Geological Society of America Bulletin*, 117, 1513–1533.

VINCENT, S.J., MORTON, A.C., HYDEN, F. & FANNING, M. 2013. Insights from petrography, mineralogy and U–Pb zircon geochronology into the provenance and reservoir potential of Cenozoic siliciclastic depositional systems supplying the northern margin of the Eastern Black Sea. *Marine and Petroleum Geology*, 45, 331–348.

VINCENT, S.J., HYDEN, F. & BRAHAM, W. 2014. Along-strike variations in the composition of sandstones derived from the uplifting western Greater Caucasus: causes and implications for reservoir quality prediction in the Eastern Black Sea. *In*: SCOTT, R.A., SMYTH, H.R., MORTON, A.C. & RICHARDSON, N. (eds) *Sediment Provenance Studies in Hydrocarbon Exploration and Production*. Geological Society, London, Special Publications, 386, 111–127, https://doi.org/10.1144/SP386.15

VINCENT, S.J., BRAHAM, W., LAVRISHCHEV, V.A., MAYNARD, J.R. & HARLAND, M. 2016. The formation and inversion of the western Greater Caucasus Basin and the uplift of the western Greater Caucasus: Implications for the wider Black Sea region. *Tectonics*, 35, 2948–2962.

VINCENT, S.J., GUO, L., FLECKER, R., BOUDAGHER-FADEL, M.K., ELLAM, R.M. & KANDEMIR, R. 2018. Age constraints on intra-formational unconformities in Upper Jurassic–Lower Cretaceous carbonates in northeast Turkey; geodynamic and hydrocarbon Implications. *Marine and Petroleum Geology*, 91, 639–657.

WONG, H.K., PANIN, N., DINU, C., GEORGESCU, P. & RAHN, C. 1994. Morphology and post-Chaudian (Late Pleistocene) evolution of the submarine Danube fan complex. *Terra Nova*, 6, 502–511.

XING, J. & SPIESS, V. 2015. Shallow gas transport and reservoirs in the vicinity of deeply rooted mud volcanoes in the central Black Sea. *Marine Geology*, 369, 67–78.

YAKUSHYN, L.M. & ISHCHENKO, I.I. 2014. Stratigraphic scheme of the Cretaceous sediments of the Ukrainian part of the Black and Azov Seas as a basis for further exploration work for oil and gas. *The Oil and Gas Industry of Ukraine*, 4, 35–41.

YILMAZ, I.Ö. & ALTINER, D. 2007. Cyclostratigraphy and sequence boundaries of inner platform mixed carbonate–siliciclastic successions (Barremian–Aptian) (Zonguldak, NW Turkey). *Journal of Asian Earth Sciences*, 30, 253–270.

YILMAZ, Y., TÜYSÜZ, O., YIGITBAS, E., GENC, S.C. & SENGÖR, A.M.C. 1997. Geology and tectonic evolution of the pontides. *In*: ROBINSON, A.G. (ed.) *Regional and Petroleum Geology of the Black Sea and Surrounding Region*. AAPG, Memoirs, Tulsa, 68, 183–226.

ZANDER, T., HAECKEL, M. ET AL. 2017a. On the origin of multiple BSRs in the Danube deep-sea fan, Black Sea. *Earth and Planetary Science Letters*, 462, 15–25.

ZANDER, T., CHOI, J.C., VANNESTE, M., BERNDT, C., DANNOWSKI, A., CARLTON, B. & BIALAS, J. 2017b. Potential impacts of gas hydrate exploitation on slope stability in the Danube deep-sea fan, Black Sea. *Marine and Petroleum Geology*, https://doi.org/10.1016/j.marpetgeo.2017.08.010

ZONENSHAIN, L.P. & LE PICHON, X. 1986. Deep basins of the Black Sea and Caspian Sea as remnants of Mesozoic back-arc basins. *Tectonophysics*, 123, 181–212.

Index

Page numbers in *italics* refer to Figures. Page numbers in **bold** refer to Tables.